MODERN OPTICS

ROBERT D. GUENTHER
DUKE UNIVERSITY

JOHN WILEY & SONS

NEW YORK CHICHESTER BRISBANE TORONTO SINGAPORE

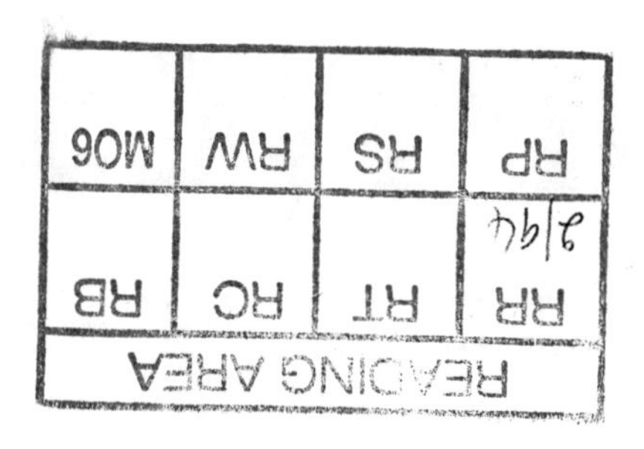

Library of Congress Cataloging in Publication Data:

Guenther, Robert D.
Modern optics / Robert D. Guenther.
p. cm.
Includes bibliographical references.

1. Optics. I. Title.
QC355.2.G84 1990
535–dc20 89–37809
CIP

Printed in the United States of America

10 9 8 7 6 5 4 3 2

Modern Optics

For Sharon, Valerie and Brett

Preface

This textbook is designed for use in a standard physics course on optics. The book is the result of a one-semester elective course that has been taught to juniors, seniors, and first-year graduate students in physics and engineering at Duke University for several years. Students who take this course should have completed an introductory physics course and math courses through differential equations. Electricity and magnetism can be taken concurrently.

Modern Optics differs from the classical approach of most textbooks on this subject in that its treatment of optics includes some material that is not found in more conventional textbooks. These topics include nonlinear optics, guided waves, Gaussian beams, and light modulators. Moreover, a selection of optional material is provided for the instructor so that the course content can reflect the interest of the instructor and the students. Basic derivations are included to make the book appealing to physics departments, and design concepts are included to make the book appealing to engineering departments. Because of the material covered here, the electrical engineering and biomedical engineering departments at Duke have made the corresponding optics course a prerequisite for some of their advanced courses in optical communications and medical imaging.

Before the 1960s, the only contact that the average person had with optics was a camera lens or eyeglasses. Geometric optics was quite adequate for the design of these systems, and it was natural to emphasize this aspect of optics in a curriculum. The approach used introduced the students to the theory and to examples of the application of the theory, accomplished by a description of a large variety of optical instruments. The reason for this approach was that lens design is found to be quite tedious, and the optimization of a lens design is more easily described than accomplished.

Today the student is exposed to many more optical systems. Everyone encounters supermarket scanners, copying machines, compact disk players, holograms, and discussions of fiber optic communications. In the research environment, lasers, optical modulators, fiber optic interconnects, and nonlinear optics have become important tools. Upon graduation, many students will be called on to participate in the use or the development of these modern optical systems. An elementary discussion of geometrical optics and a review of classical optical instruments will not adequately prepare the student for these demands.

This book was written to provide both a fundamental study of the principles of optics and an exposure to actual optics engineering problems and solutions. To include new material has meant that some of the topics covered in classical texts had to be removed. A large portion of the conventional treatment of geometrical optics was deleted, along with a discussion of classical optical systems. In their place were a geometrical optics discussion of fiber optics and a discussion of holography. Rather than describe a host of optical systems, a few optical systems, such as the Fabry–Perot interferometer, are examined using a variety of theories. This book emphasizes diffraction and the use of Fourier theory to describe the operation of an optical system.

To allow the development of a one-year course in modern optics, a number of topics have been added or expanded. A discussion of electrooptic and magnetooptic effects is used to introduce optical modulators, and a discussion of nonlinear optics is constructed around second harmonic generation. Because of the importance of birefringence in optical modulators and nonlinear optics, an expanded discussion of optical anisotropy has been included. This is a departure from most texts that ignore anisotropy because of the need to use tensors. In modern optics, anisotropy is an important design tool, and its treatment allows a discussion of the design of optical modulators and phase matching in nonlinear materials.

The first two chapters review wave theory and electromagnetic theory. Except for the section on polarization in Chapter 2, these chapters could be used as reading assignments for well-prepared students. Chapter 3 discusses reflection and refraction and utilizes the boundary conditions of Maxwell's equations to obtain the fraction of light reflected and refracted at a surface.

Chapter 4 discusses interference of waves and describes several instruments that are used to measure interference. Two of the interferometers, Young's two-slit experiment and the Fabry–Perot interferometer, should receive emphasis in discussions of this chapter because of the role they play in later discussions. An appendix to this chapter provides a brief introduction to some of the design techniques that are used to produce multilayer interference filters. All of the appendices in the book are included to fill in the gaps in students' knowledge and to provide some flexibility for the instructor. The appendices may, therefore, be ignored or used as the subject matter for special assignments.

The treatment of geometrical optics, presented in Chapter 5, is not traditional. It was through the reduction of traditional subject matter that space was obtained to introduce more modern topics. A brief introduction to the matrix formalism used in lens design is presented, and its use is demonstrated by analyzing a confocal Fabry–Perot resonator. Geometrical optics and the concept of interference are used to analyze the propagation of light in a fiber. This introduction to fiber optics is then extended through the use of the Lagrangian formulation to propagation in a graded-index optical fiber. The first part of the chapter demonstrates the formal connection between geometric optics and wave theory. Most students would rather not cover this material and, therefore, it is usually omitted. The connection between the matrix equations and the more familiar lens equations is established in

Appendix 5-A. Because of their importance in the Graduate Records Exam, aberrations are treated in Appendix 5-B.

The Fourier theory in Chapter 6 is presented as a review of and refresher on the subject. It is an important element in the discussion of the concept of coherence in Chapter 8, and Fraunhofer diffraction in Chapter 10. The discussions of optical signal processing, Appendix 10-B, and imaging, Appendix 10-C, draw heavily on Fourier theory.

The discussion of dispersion given in Chapter 7 could be delayed and combined with the other chapters on material interactions (Chapters 13 and 15). It is included here to justify the discussion of coherence in Chapter 8. The discussion of dispersion in materials had as its objective the development by the student of a unifying view of the interaction of light and matter.

The development of coherence theory in Chapter 8 is built around applications of the theory to spectroscopy and astronomy. It is a very difficult subject, but building the theory around the methods used to measure coherence should make the subject more intelligible.

Both the Fresnel and Gaussian wave formalism of diffraction are introduced in Chapter 9. The Gaussian wave formalism is used to analyze a Fabry–Perot cavity and thin lens. This chapter can be skipped, and the material introducing the Fresnel–Huygens integral can be covered in a single lecture.

The Fresnel formalism is expanded and discussed in Chapters 10 and 11. Fraunhofer diffraction is treated from a linear-system viewpoint in Chapter 10, and applications of the theory to signal processing and imaging are presented in Appendices 10-B and 10-C. These two appendices are the most important in the book. Fresnel Diffraction is introduced in Chapter 11, where it is used to interpret Fermat's principle and analyze zone plates and pinhole cameras. In Chapter 12, Fresnel theory is used to discuss the operation of a hologram. Chapter 12 also includes a simple, quasigeometric theory that is used to highlight the fundamental properties of a hologram.

Chapter 13 uses the introduction of polarizers and retarders as a basis for the development of the theory of the propagation of light in anisotropic materials. The treatment of anisotropic materials is expanded over the conventional presentation to allow an easy transition into the discussion of light modulators in Chapter 14. The many geometrical constructions used in the discussion of anisotropy are confusing to everyone. To try to make the material understandable, only one construction is used in Chapter 13. To provide the student with reference material to aid in reading other books and papers, the other constructions are discussed in the appendices.

The discussion of modulators in Chapter 14 provides an application-based introduction to electro- and magnetooptic interactions. The design of an electrooptic modulator provides the student with an example of the use of tensors. The material interactions presented in Chapters 14 and 15 require the use of tensors, a subject normally avoided in an undergraduate curriculum. Tensor notation has been used in this book because it is key in the understanding of many optical devices. Some familiarity with tensors removes much of the "magic" associated with the design of modulators and the application of phase matching discussed in Chapter 15.

The subject of nonlinear optics in Chapter 15 is developed by using examples based on frequency doubling. Only a few brief comments are made about third-order nonlinearities. The additional discussion of third-order processes is best presented by using a quantum mechanical viewpoint.

It was thought that this would be best done in a separate course. The material presented in this chapter would prepare the student to immediately undertake a course in nonlinear optics.

Enough material has been included for a one-year course in optics. Chapters 2 to 4, 6, 9 (excluding Gaussian waves), and 10 contain the core material and could be used in a one-quarter course. By adding Chapter 7 and 8 along with Appendices 10-B and 10-C, a one-semester course can be created. The instructor can alter the subjects discussed from year to year by adding topics such as Appendix 4-A, the guided wave discussion of Chapter 5, or the discussion of holography in Chapter 12 in place of Appendices 10-B and 10-C. A less demanding one-semester course can be created by ignoring Chapter 8 and by substituting Chapter 5, Appendix 4-A, or possibly Chapter 12. In anticipation of developing skill and knowledge, the subject matter and problems increase in difficulty as the student moves through the book.

A number of people provided help in the preparation of this book. Those who provided photos or drawings are identified in the figure captions. Their generosity is most appreciated. Dr. Frank DeLucia provided the initial motivation for writing the book. Many ideas and concepts are the result of breakfast discussions with A. VanderLugt. The book would never have gone past the note stage without the equation writer, MacΣqn, written by Dennis Venable. Thomas Stone provided ideas, photographic skills, and encouragement during the preparation of most of the photos in this book. His enthusiasm kept me working.

A very special thanks must go to Nicholas George. He loaned me equipment and lab space to prepare many of the photos. His encouragement prevented me from shelving the project, and his technical discussions provided me with an improved understanding of optics.

Robert D. Guenther

Contents

CHAPTER 1 WAVE THEORY 1

Traveling Waves 2
Wave Equation 4
Transmission of Energy 6
Transmission of Momentum 8
Three Dimensions 9
Attenuation of Waves 13
Summary 13
Problems 14
Appendix 1-A Harmonic Motion 17
Appendix 1-B Complex Numbers 22

CHAPTER 2 ELECTROMAGNETIC THEORY 25

Maxwell's Equations 26
Energy Density and Flow 33
Momentum 36
Polarization 38
Stokes Parameters 46
Jones Vector 49
Propagation in a Conducting Medium 49
Summary 53
Problems 54
Appendix 2-A Vectors 56
Appendix 2-B Electromagnetic Units 59

CHAPTER 3 REFLECTION AND REFRACTION 61

Reflection and Transmission at a Discontinuity 62
Laws of Reflection and Refraction 65

Fresnel's Formulae 67
Reflected and Transmitted Energy 73
Normal Incidence 75
Polarization by Reflection 77
Total Reflection 78
Reflection from a Conductor 83
Summary 84
Problems 85

CHAPTER 4 INTERFERENCE 87

Addition of Waves 88
Interference 91
Young's Interference 94
Dielectric Layer 97
Michelson Interferometer 102
Interference by Multiple Reflection 106
Fabry–Perot Interferometer 111
Summary 115
Problems 117

Appendix 4-A Multilayer Dielectric Coatings 120

CHAPTER 5 GEOMETRICAL OPTICS 129

Eikonal Equation 131
Fermat's Principle 133
Applications of Fermat's Principle 136
Lens Design and Matrix Algebra 138
Geometric Optics of Resonators 144
Guided Waves 148
Parametric Characterization of a Light Guide 163
Lagrangian Formulation of Optics 166
Propagation in a Graded Index Optical Fiber 169
Summary 176
Problems 178

Appendix 5-A ABCD Matrix 182
Appendix 5-B Aberrations 191

Wavefront Aberration Coefficients 195
Transverse Ray Coefficients 197
Spherical Aberrations 198
Coma 203
Astigmatism 206
Field Curvature 208
Distortion 208
Aberration Reduction 209

CHAPTER 6 FOURIER ANALYSIS 213

Fourier Series 214
Periodic Square Wave 218

The Fourier Integral 221
Rectangular Pulse 224
Pulse Modulation-Wave Trains 226
Dirac Delta Function 229
Correlation 235
Convolution Integrals 237
Linear System Theory 239
Fourier Transforms in Two Dimensions 241
Summary 243
Problems 245
Appendix 6-A Fourier Transform Properties 248
Convolution Properties 249

CHAPTER 7 DISPERSION 251

Stiff Strings 253
Group Velocity 255
Dispersion of Guided Waves 260
Material Dispersion 262
Signal Velocity 279
Summary 282
Problems 284
Appendix 7-A Chromatic Aberrations 287

CHAPTER 8 COHERENCE 291

Photoelectric Mixing 293
Interference Spectroscopy 295
Fourier Transform Spectroscopy 297
Fringe Contract and Coherence 301
Temporal Coherence Time 301
Autocorrelation Function 303
Spatial Coherence 305
A Line Source 308
Spatial Coherence Length 311
Stellar Interferometer 312
Intensity Interferometry 313
Summary 317
Problems 319

CHAPTER 9 DIFFRACTION AND GAUSSIAN BEAMS 323

Huygens' Principle 326
Fresnel Formulation 329
The Obliquity Factor 331
Gaussian Beams 336
The *ABCD* Law 343
Summary 349
Problems 350

Appendix 9-A Fresnel-Kirchoff Diffraction 352
Appendix 9-B Rayleigh-Sommerfeld Formula 357
Appendix 9-C Green's Theorem 359

CHAPTER 10 FRAUNHOFER DIFFRACTION 361

Fraunhofer Diffraction 363
Fourier Transforms via a Lens 366
Plane Wave Representation 369
Diffraction by a Rectangular Aperture 370
Diffracton from a Circular Aperture 371
Array Theorem 375
N Rectangular Slits 377
Summary 386
Problems 388
Appendix 10-A Fraunhofer Diffraction by a Lens 391
Appendix 10-B Abbe's Theory of Imaging 397
Abbe's Theory of Imaging 397
Amplitude Spatial Filtering 408
Apodization 411
Phase Filtering 413
Phase and Amplitude Filter 415
Appendix 10-C Incoherent Imaging 420
Coherent Imaging 428

CHAPTER 11 FRESNEL DIFFRACTION 433

Fresnel Approximation 434
Rectangular Apertures 438
Fresnel Zones 444
Circular Aperture 449
Opaque Screen 453
Zone Plate 454
Fermat's Principle 458
Summary 459
Problems 461
Appendix 11-A Babinet's Principles 463
Appendix 11-B 467

CHAPTER 12 HOLOGRAPHY 469

Holographic Recording 473
Off-Axis Holography 480
Spatial Spectrum of Off-Axis Holograms 483
Classification of Holograms 485
Diffraction Efficiency 489
Holography and Zone Plates 491
Resolution Requirements 494
Imaging Properties of Off-Axis Holograms 496
Fresnel Hologram 499

Fourier Transform Hologram 501
Coherency Requirements 503
Summary 508
Problems 509
Appendix 12-A Phase Holograms 512
Appendix 12-B Vanderlugt Filter 515

CHAPTER 13 ANISOTROPY 521

Dichroic Polarizers 523
Reflection Polarizers 526
Polarization by Birefringence 529
Optical Indicatrix 534
Fresnel's Equation 537
Retarder 541
Mueller Calculus 543
Jones Calculus 544
Optical Activity 545
Summary 552
Problems 554
Appendix 13-A Tensors 556
Appendix 13-B Poynting Vector in Anisotropic Dielectric 560
Ray Ellipsoid 561
Appendix 13-C Normal Surfaces 563
Refraction in Crystals 565
Ray Surfaces 567

CHAPTER 14 OPTICAL MODULATION 569

Electrooptic Effect 571
Electrooptic Indicatrix 573
Kerr Effect 576
Amplitude Modulation 579
Modulator Design 587
Magnetooptic Effect 590
Photoelastic Effect 596
Acoustooptics 601
Summary 613
Problems 614
Appendix 14-A 616
Appendix 14-B Phenomenological Acoustooptic Theory 620
Appendix 14-C Acoustic Figure of Merit 630

CHAPTER 15 NONLINEAR OPTICS 633

Nonlinear Polarization 636
Nonlinear Optical Coefficient 638
Symmetry Properties 639
Wave Propagation in a Nonlinear Medium 641
Conservation of Energy 644

Conservation of Momentum 645
Second Harmonic Generation 649
Methods of Phase Matching 651
Phase Conjugation 659
Summary 665
Problems 668
Appendix 15-A Generalized Linear Theory 670
Nonlinear Equation of Motion 671
Perturbation Technique 672
Second Order Nonlinearity 674
Appendix 15-B Miller's Rule 677
Appendix 15-C Nonlinear Polarization in the 32-Point Group 679

REFERENCES 681

INDEX 685

Wave Theory

The theory of wave motion is an important mathematical model in many areas of physics. A large number of seemingly unrelated phenomena can be explained using the solution of the wave equation, the basic equation of wave theory. The wave theory is a fundamental part of modern quantum theory and the solutions of the wave equation are used to explain a number of classical phenomena. Familiarity with the wave theory developed in the study of light will aid in the understanding of such diverse physical processes as water waves, vibrating drums and strings, traffic dynamics, and seismic waves.

Mathematically, the basis of wave theory is a second-order, partial differential equation called the wave equation. In this chapter, a vibrating string will be used as an illustration to aid in visualizing the various aspects of the wave theory. Initially, a traveling wave on a string will be used to find the functional form of a one-dimensional wave and to derive the wave equation. Following a discussion of the energy and momentum associated with the traveling wave, the one-dimensional model associated with the string illustration will be expanded to three dimensions. The displacement of the wave discussed in this chapter is assumed to be a scalar function and the theory is called a *scalar wave theory*. In the next chapter, the vector wave theory will be discussed.

Christian Huygens (1629–1695) developed the wave theory of light in 1678. **Isaac Newton (1642–1727)** proposed a counter theory based on a particle view of light. Newton's scientific stature resulted in only a few scientists during the 18th century, for example **Leonard Euler (1707–1783)** and **Benjamin Franklin (1706–1790)**, accepting the wave theory and rejecting the particle theory of Newton. In 1801 **Thomas Young (1773–1829)** and in 1814 **Augustin Jean Fresnel (1788–1827)** utilized experiments to demonstrate interference and diffraction of light and presented a theoretical explanation of the experiments through the use of the wave theory. Fresnel was able to explain rectilinear propagation using the wave theory, thereby removing Newton's main objection to the wave theory. The acceptance of

Fresnel's theory was very slow, and the final rejection of Newton's theory did not come until the measurement of the speed of light in water and air by **Jean Bernard Léon Foucault (1819–1868)**. The velocity measurements were a key element in the rejection of Newton's theory because the particle theory required the speed of light in a medium to exceed the speed of light in a vacuum in order to explain refraction. The measurements by Foucault showed the propagation velocity in a vacuum to exceed the velocity in water.

TRAVELING WAVES

Before the equation of motion of a wave is discussed, a mathematical expression for a wave will be obtained. We will assume that a disturbance propagates without change along a string and that each point on the string undergoes simple harmonic motion (see Appendix 1-A for a brief review of harmonic motion). This assumption will allow us to obtain a simple mathematical expression for a wave that will be used to define the parameters characterizing a wave.

A guitar string is plucked creating a pulse that travels to the right and left along the x axis at a constant speed c. In Figure 1-1, the pulse traveling toward the right is shown. The pulse's amplitude is defined as $y \equiv f(x, t)$ and equals y_1 at position x_1 and time t_1. This amplitude travels a distance $c(t_2 - t_1)$ to the right of x_1 and is described mathematically by

$$y \equiv f(x, t)$$

Assume the pulse does not change in amplitude as it propagates

$$f(x_1, t_1) = f(x_2, t_2)$$

where $x_2 = x_1 + c(t_2 - t_1)$. If the function has the form

$$y = f(ct - x) \tag{1-1}$$

then the requirement that the pulse does not change is satisfied because

$$f(x_1, t_1) = f(ct_1 - x_1)$$

$$f(x_2, t_2) = f(ct_2 - x_2) = f[ct_2 - x_1 - c(t_2 - t_1)]$$

$$= f(ct_1 - x_1)$$

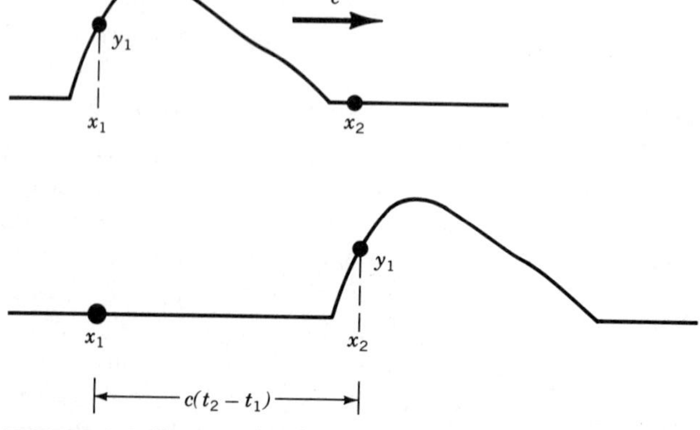

FIGURE 1-1. Propagation of a pulse on a guitar string. The amplitude does not change as the pulse propagates along the string.

Using the same reasoning, we can show that an unchanging pulse traveling to the left, along the x axis, with speed c is described by

$$y = g(x + ct)$$

The expression $y = f(ct - x)$ is a shorthand notation to denote a function that contains x and t only in the combination $(ct - x)$, i.e., the function can contain combinations of the form $2(ct - x)$, $(t \pm x/c)$, $(x - ct)$, $(ct - x)^2$, $\sin(ct - x)$, etc., but not expressions such as $(2ct - x)$ or $(ct^2 - x^2)$.

To the assumption of an unchanging propagating disturbance is now added the requirement that each point on the guitar string oscillate transversely, i.e., perpendicular to the direction of propagation, with simple harmonic motion. The string in Figure 1-1 lies along the x axis and the harmonic motion will be in the y direction. The point on the string at the origin ($x = 0$) undergoes simple harmonic motion with amplitude Y and frequency ω (the angular frequency $\omega = 2\pi\nu$ will be used throughout this book; the linear frequency ν is defined in Appendix 1-A). The equation describing the motion of the origin is

$$y = Y \cos \omega t$$

The origin acts as a source of a continuous train of pulses (a wave train) moving to the right.

A function of $(ct - x)$ that will reduce to harmonic motion at $x = 0$ is

$$y = f(ct - x) = Y \cos\left[\frac{\omega}{c}(ct - x)\right]$$

This is called a *harmonic wave*.

A number of different notations are used for a harmonic wave; the one used in this book involves a constant

$$k = \frac{\omega}{c} \tag{1-2}$$

called the *propagation constant* or the *wave number* and is written

$$y = Y \cos (\omega t - kx) \tag{1-3}$$

The values of x for which the phase $(\omega t - kx)$ changes by 2π is the *spatial period* and is called the *wavelength* λ. Let $x_2 = x_1 + \lambda$, so that

$$\omega t - kx_2 = \omega t - kx_1 - k\lambda = \omega t - kx_1 - 2\pi$$

thus

$$k = \frac{2\pi}{\lambda} \tag{1-4}$$

since $k = \omega/c = 2\pi\nu/c$, we also have the relationship $c = \nu\lambda$.

To determine the speed of the wave in space, a point on the wave is selected and the time it takes to go some distance is measured. This is equivalent to asking how fast a given value of phase propagates in space. Assume that in the time $\Delta t = (t_2 - t_1)$, the disturbance y_1 travels a distance $\Delta x = (x_2 - x_1)$, as is shown in Figure 1-1. Since the disturbance at the two points is the same, i.e., y_1, then the phases must be equal

$$\omega t - kx = \omega(t + \Delta t) - k(x + \Delta x)$$

$$\frac{\Delta x}{\Delta t} = \frac{\omega}{k}$$

In the limit as $\Delta t \to 0$, we obtain the *phase velocity*

$$c \equiv \frac{dx}{dt} = \frac{\omega}{k}$$

The adjective "phase" is used because this velocity describes the motion of a preselected phase of the wave. Another method that can be used to obtain the propagation speed associated with a wave is to define the phase velocity using the result from partial differential calculus

$$\left(\frac{\partial x}{\partial t}\right)_y = -\frac{\left(\frac{\partial y}{\partial t}\right)_x}{\left(\frac{\partial y}{\partial x}\right)_t} = \frac{\omega}{k}$$

This equation may be verified by applying it to **(1-3)**.

WAVE EQUATION

To generate the differential equation of motion of a wave propagating along a string, we must look at a small section of the string as a pulse passes by. We are going to assume that we only have small amplitude pulses so that the tension in the string is not changed appreciably as the pulse passes by. As a consequence of this assumption, we have $\partial y/\partial x << 1$; therefore, the deflected string shown in Figure 1-2 makes an angle θ with the horizontal such that $\cos\theta \approx 1$ and $\sin\theta \approx \tan\theta = \partial y/\partial x$ (we use partial derivatives because the deflection is a function of both time and position; in this derivation, we hold time constant). With these approximations, the components of tension at position x in Figure 1-2 are

$$T_x = T\cos\theta \approx T$$

$$T_y = T\sin\theta \approx T\left(\frac{\partial y}{\partial x}\right)$$

At position $x + \Delta x$ in Figure 1-2, the slope is also small since $\Delta\theta$ is small $[\cos(\theta + \Delta\theta) \approx 1]$, resulting in

$$T_x + \Delta T_x \approx T$$

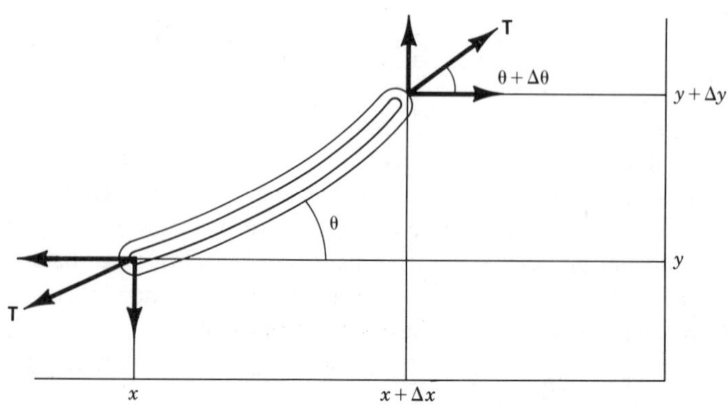

FIGURE 1-2. The string is deflected as the pulse passes by. The tension T is decomposed into components in the x and y directions.

Our assumptions give us a horizontal force that is very nearly zero. The net vertical force is obtained by subtracting the y components of tension at position x and $x + \Delta x$

$$T_y + \Delta T_y - T_y = \Delta T_y$$

If Δx is small, then the change in tension ΔT_y is

$$\Delta T_y = \frac{\partial T_y}{\partial x}\Delta x$$

$$\frac{\partial T_y}{\partial x} = \frac{\partial}{\partial x}\left(T\frac{\partial y}{\partial x}\right) = T\frac{\partial^2 y}{\partial x^2}$$

$$\Delta T_y = T\frac{\partial^2 y}{\partial x^2}\Delta x$$

The mass of this segment of the string is $\Delta m = \mu\Delta x$, where μ is the mass/unit length. The vertical acceleration is $a_y = \partial^2 y/\partial t^2$ so

$$T\frac{\partial^2 y}{\partial x^2}\Delta x = \Delta T_y = \Delta m a_y = \mu\Delta x\frac{\partial^2 y}{\partial t^2}$$

$$\frac{\partial^2 y}{\partial t^2} = \frac{T}{\mu}\frac{\partial^2 y}{\partial x^2} \tag{1-5}$$

This differential equation is, in the language of partial differential equations, a hyperbolic equation. It is also called the wave equation for a nondispersive medium. The term "nondispersive" means that the pulse does not change shape as it propagates. If the wave equation were *dispersive*, then waves of different frequencies would travel at different velocities. We will have more to say about dispersive waves in Chapter 7.

The physical significance of the ratio T/μ can be understood through a dimensional analysis.

$$T \text{ is a force} \Rightarrow \text{kg}\cdot\text{m/sec}^2$$

$$\mu \text{ is a mass/unit length} \Rightarrow \text{kg/m}$$

$$T/\mu \Rightarrow (\text{kg}\cdot\text{m/sec}^2)/(\text{kg/m}) = \text{m}^2/\text{sec}^2 = (\text{velocity})^2$$

Thus by a dimensional analysis, we can identify the ratio

$$\sqrt{\frac{T}{\mu}}$$

as the velocity of the wave c along the string. We may rewrite the wave equation as

$$\frac{\partial^2 y}{\partial x^2} = \frac{1}{c^2}\frac{\partial^2 y}{\partial t^2} \tag{1-6}$$

This differential equation describes a number of physical situations: the vibrating string we have just analyzed, sound waves, a vibrating drum head, elastic waves in solids such as seismic waves in the earth, electric signals in a cable, and electromagnetic waves—the subject of this book.

We said earlier that the motion of a vibrating string was described by equations of the form **(1-1)**

$$y_1 = f(ct - x) \qquad \text{or} \qquad y_2 = g(ct + x)$$

To show that y_1 is a solution, we substitute y_1 into the wave equation [we will denote derivatives with respect to the argument of the function f, i.e., $(ct - x)$, by a prime]

$$\frac{\partial f}{\partial t} = \frac{\partial f}{\partial(ct-x)}\frac{\partial(ct-x)}{\partial t} = c \cdot f', \qquad \frac{\partial f}{\partial x} = \frac{\partial f}{\partial(ct-x)}\frac{\partial(ct-x)}{\partial x} = (-1)f'$$

$$\frac{\partial^2 f}{\partial t^2} = c^2 f'', \qquad \frac{\partial^2 f}{\partial x^2} = f''$$

$$\frac{\partial^2 f}{\partial x^2} - \frac{1}{c^2}\frac{\partial^2 f}{\partial t^2} = f'' - \frac{1}{c^2}(c^2 f'') = 0$$

The same procedure is used to prove $g(ct + x)$ is a solution.

Another important property of the wave equation is the *superposition principle*. The principle states that the sum of two solutions is also a solution. Proof of the superposition principle is the objective of Problem 1-10.

TRANSMISSION OF ENERGY

Each element of a string, with a harmonic wave propagating along it, moves up and down in the y direction, undergoing simple harmonic motion. This can be seen by selecting a coordinate position to observe, say x_1, and replacing kx_1 in **(1-3)** by the constant δ, i.e., $y = Y\cos(\omega t - \delta)$. Equation **(1-3)** now assumes the same form as the equation for a harmonic oscillator **(1A-5)**. The elements of the string do not move in the direction of wave motion (the x direction). Even though the elements are not translated along the direction of propagation, energy is transmitted. An aid to understanding how this occurs can be obtained by imagining yourself at the end of a long line of people waiting to purchase tickets. To communicate with a friend at the front of the line, you could pass a note from hand to hand until it reaches your friend at the front of the line. No one need move in the direction of your friend and yet the note arrives.

Energy is transmitted by a wave in the same way as the note. To discover the wave characteristics that determine the energy transmitted by a wave, consider the string shown in Figure 1-3. Point a on the string has work done on it by the point to its left, and it does work on the point to its right. The work done on point a is

$$dW = F \cdot dy = (T \sin\theta)(v\,dt)$$

The work done is equal to the change in energy, allowing us to write the instantaneous power transmitted as

$$P = \frac{dE}{dt} = Tv \sin\theta$$

From **(1-3)**, we can write

$$\frac{dE}{dt} = YT \sin\theta[-\omega \sin(\omega t - kx)]$$

We continue to assume small amplitude waves so that the tension at point a in Figure 1-3 is parallel to the slope of the wave at a

$$\sin\theta \approx \tan\theta \approx -\tan\phi = -\frac{dy}{dx} = -kY\sin(\omega t - kx)$$

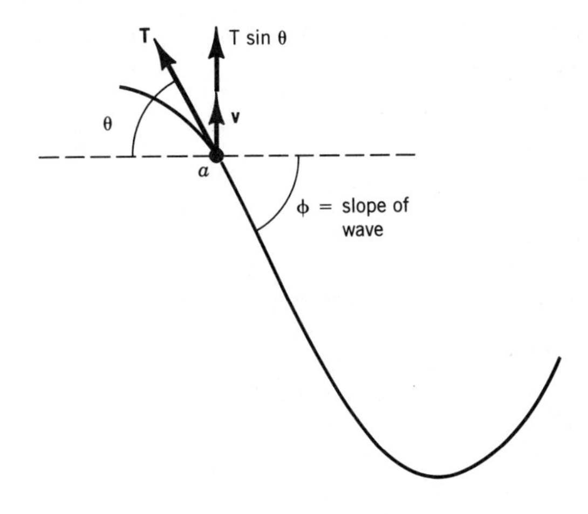

FIGURE 1-3. Forces acting on point a of the string.

$$\frac{dE}{dt} = T[-kY \sin(\omega t - kx)][-\omega Y \sin(\omega t - kx)]$$

$$\frac{dE}{dt} = Tk\omega Y^2 \sin^2(\omega t - kx)$$

The average power transmitted in one period $(2\pi/\omega)$ is

$$P_{\text{avg}} = \langle E \rangle = \frac{\omega}{2\pi}\int_0^{2\pi/\omega} P \, dt = \frac{Tk\omega^2 Y^2}{2\pi}\int_0^{2\pi/\omega} \sin^2(\omega t - kx) \, dt$$

$$P_{\text{avg}} = \frac{Tk\omega Y^2}{2} = \frac{T\omega^2 Y^2}{2c} \tag{1-7}$$

For a given string, the tension T and the phase velocity c are constants and the average power transmitted is proportional to the square of the frequency and the square of the amplitude. In Figure 1-4, we plot the wave and its instantaneous power, demonstrating that the power travels along with the wave.

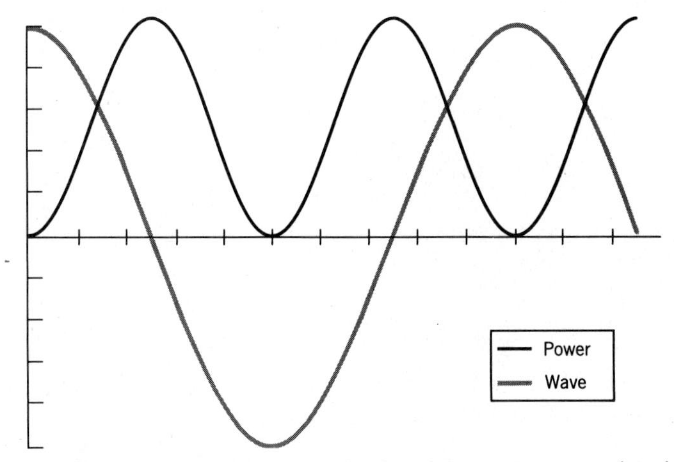

FIGURE 1-4. A traveling wave and its instantaneous power are plotted with their amplitudes normalized to 1.

TRANSMISSION OF MOMENTUM

Classically, we can associate with a wave the transmission of both momentum and energy. **Larmor** developed an indirect proof of the existence of momentum in a classical wave that we will restate.

Assume that a wave

$$Y_i \cos(\omega_i t - k_i x)$$

is incident on a totally reflecting surface moving toward the wave source at a velocity v, where $v << c$. The reflected wave is identical to the incident wave but travels in the opposite direction; thus, the reflected wave is

$$Y_i \cos(\omega_i t - k_i x) = Y_r \cos(\omega_r t + k_r x) \tag{1-8}$$

Since **(1-8)** must hold for all times, we may equate the phases of **(1-8)** on the mirror surface at a time t_1. At time t_1, the position of the mirror surface is $x = -vt_1$, allowing us to write the phases as

$$k_i\left[\left(\frac{\omega_i}{k_i}\right) + v\right] = k_r\left[\left(\frac{\omega_r}{k_r}\right) - v\right]$$

We know that

$$\frac{\omega_i}{k_i} = \frac{\omega_r}{k_r} = c$$

so we can rewrite the equality, by multiplying both sides by c,

$$\omega_i(c + v) = \omega_r(c - v)$$

The average power (or equivalently, the energy per unit time) transmitted by a wave $\langle E \rangle$ was shown in **(1-7)** to be proportional to ω^2. The ratio of the reflected energy $\langle E_r \rangle$ to the incident energy $\langle E_i \rangle$ is then

$$\frac{\langle E_r \rangle}{\langle E_i \rangle} \propto \frac{\omega_r^2}{\omega_i^2} = \frac{(c + v)^2}{(c - v)^2} = \frac{\left(1 + \frac{v}{c}\right)^2}{\left(1 - \frac{v}{c}\right)^2}$$

$$\langle E_r \rangle - \langle E_i \rangle = \left(\frac{\langle E_r \rangle}{\langle E_i \rangle} - 1\right)\langle E_i \rangle = \left[\frac{\left(1 + \frac{v}{c}\right)^2}{\left(1 - \frac{v}{c}\right)^2} - 1\right]\langle E_i \rangle$$

For $v/c << 1$, a normal situation

$$\langle E_r \rangle - \langle E_i \rangle \approx \frac{2v}{c}\langle E_i \rangle$$

Therefore, the reflected wave has an excess of energy that must come from work done, by the moving mirror, on the wave

$$\frac{dW}{dt} = F\frac{ds}{dt} = Fv = \left(\frac{2\langle E_i \rangle}{c}\right)v$$

From this relationship, we see that

$$F = \frac{2\langle E_i \rangle}{c} = \frac{dp}{dt} \tag{1-9}$$

Here, we have the desired result that a momentum p is associated with the incident wave. Note that the force is independent of the motion of the reflecting surface that was used to derive this result. Since the velocity of the mirror surface does not appear in **(1-9)**, the result can be applied to stationary surfaces. All of our calculations are per unit area, which means that the force due to the incident wave **(1-9)** can be thought of as a pressure. The momentum in the wave can therefore be observed by measuring the pressure exerted on a surface that reflects the wave.

THREE DIMENSIONS

Although the wave traveling along a string can be described with only one dimension, we will need a three-dimensional model in our discussion of light. The one-dimensional model **(1-6)** generalizes to a three-dimensional wave equation

$$\frac{\partial^2 f(\mathbf{r}, t)}{\partial x^2} + \frac{\partial^2 f(\mathbf{r}, t)}{\partial y^2} + \frac{\partial^2 f(\mathbf{r}, t)}{\partial z^2} = \frac{1}{c^2}\frac{\partial^2 f(\mathbf{r}, t)}{\partial t^2} \qquad (1\text{-}10)$$

or in vector notation (see Appendix 2-A)

$$\nabla^2 f(\mathbf{r}, t) = \frac{1}{c^2}\frac{\partial^2 f(\mathbf{r}, t)}{\partial t^2} \qquad (1\text{-}11)$$

A generalized harmonic wave solution of **(1-11)** is

$$f(\mathbf{r}, t) = E(\mathbf{r}) \cos[\omega t - \mathcal{S}(\mathbf{r})]$$

At a fixed time, the surfaces for which $\mathcal{S}(\mathbf{r}) =$ constant are called *wavefronts*. If $E(\mathbf{r})$, the amplitude of the wave, is a constant over the wavefront, the wave is said to be *homogeneous*; if not, then the wave is *inhomogeneous*. The average power defined by **(1-7)** is defined for the three-dimensional case as the average power transmitted per unit cross-sectional area. This quantity is called the *intensity* of the wave.

Another notation we will find useful in the study of waves is the complex notation introduced in Appendix 1-A to describe the motion of a simple harmonic oscillator (a brief review of complex numbers is found in Appendix 1-B). The generalized solution of the wave equation can be expressed in complex notation as

$$f(\mathbf{r}, t) = \mathcal{E}(\mathbf{r})e^{i\omega t}$$

where

$$\mathcal{E}(\mathbf{r}) = E(\mathbf{r}) \exp\{-i[\mathcal{S}(\mathbf{r}) + \delta]\}$$

and δ is an arbitrary phase angle. If we substitute $f(\mathbf{r}, t)$ into the wave equation, we obtain

$$\frac{\partial^2 f(\mathbf{r}, t)}{\partial t^2} = -\omega^2 \mathcal{E}(\mathbf{r})e^{i\omega t}$$

$$\nabla^2 f(\mathbf{r}, t) = e^{i\omega t}\nabla^2 \mathcal{E}(\mathbf{r})$$

If we use the relationship $\omega/c = k$, the wave equation becomes

$$(\nabla^2 + k^2)\, \mathcal{E}(\mathbf{r}) = 0 \qquad (1\text{-}12)$$

If we are interested in the spatial properties of the wave but not the temporal, we need only seek solutions of this equation, which is called the *Helmholtz equation*.

The determination of the phase velocity of the three-dimensional wave is more complicated than it was for the one-dimensional case because we must monitor the motion of a surface. As we show below, it is possible to derive the phase velocity of an arbitrary wavefront. In this book, however, we will restrict our consideration to the two simple wavefronts: a plane and a sphere. The three-dimensional wave with a plane phase front is called a *plane wave* and the wave with a spherical phase front is called a *spherical wave*.

To calculate the phase velocity of an arbitrary wavefront, we must follow a point of constant phase as the wave moves through space and time. If the phase at $(\mathbf{r}, t)$ is equal to the phase at $(\mathbf{r} + \Delta\mathbf{r}, t + \Delta t)$, then

$$\omega t - \mathcal{S}(\mathbf{r}) = \omega t + \omega\Delta t - \mathcal{S}(\mathbf{r} + \Delta\mathbf{r})$$

$$\omega\Delta t - \left[\frac{\mathcal{S}(\mathbf{r} + \Delta\mathbf{r}) - \mathcal{S}(\mathbf{r})}{\Delta\mathbf{r}}\right]\bullet\Delta\mathbf{r} = 0$$

and in the limit as $\Delta t, \Delta\mathbf{r} \to 0$, the term in brackets becomes the gradient of $\mathcal{S}(\mathbf{r})$

$$\omega dt - \nabla\mathcal{S}(\mathbf{r})\bullet d\mathbf{r} = 0 \tag{1-13}$$

If we define a unit vector $\hat{\mathbf{n}}$ in the direction of $d\mathbf{r}$, $d\mathbf{r} = \hat{\mathbf{n}}ds$, where ds is the distance between surface $\mathcal{S}(\mathbf{r})$ and $S(\mathbf{r} + d\mathbf{r})$, then the velocity of the surface is

$$\frac{ds}{dt} = \frac{\omega}{\hat{\mathbf{n}}\bullet\nabla\mathcal{S}(\mathbf{r})}$$

Figure 1-5 shows these parameters for a plane surface. This derivative has its smallest value when $\hat{\mathbf{n}}$ is normal to the wavefront

$$\hat{\mathbf{n}} = \frac{\nabla\mathcal{S}(\mathbf{r})}{|\nabla\mathcal{S}(\mathbf{r})|}$$

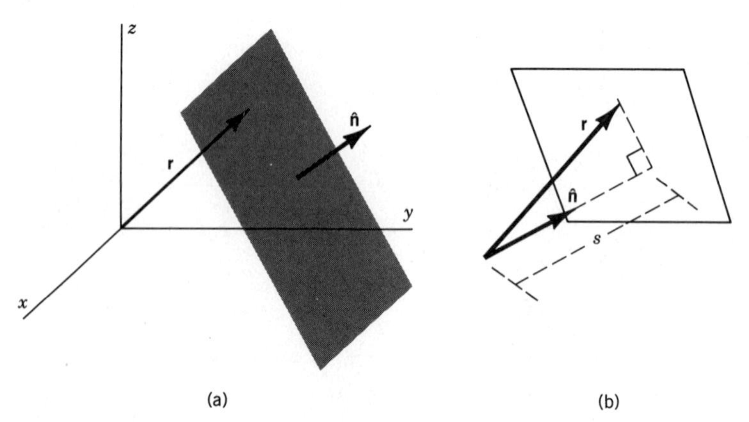

FIGURE 1-5. (a) A plane wave. Its normal is the unit vector $\hat{\mathbf{n}}$ that points in the direction of propagation. A surface of constant phase is the shaded plane passing through the point defined by the vector $\mathbf{r}$. (b) The projection of $\mathbf{r}$ on the plane's normal defines the distance s from the origin.

We then have an equation for the phase velocity of the wavefront c

$$\frac{ds}{dt} = c = \frac{\omega}{|\nabla \mathcal{S}(\mathbf{r})|} \tag{1-14}$$

Note that the phase velocity is not a vector. We must exercise care in assigning any physical significance to this velocity.

Mathematically, a plane is described by the equation

$$\mathbf{r}\cdot\hat{\mathbf{n}} = s$$

where $\mathbf{r}$ is a position vector of a point in the plane, $\hat{\mathbf{n}}$ is a unit vector normal to the plane, and s is a constant that is equal to the distance from the origin to the plane, as shown in Figure 1-5.

For a wavefront to be a plane, we must have $\mathcal{S}(\mathbf{r}) = k(\hat{\mathbf{n}}\cdot\mathbf{r})$. For convenience, we define the wavevector $\mathbf{k} = k\hat{\mathbf{n}}$ and we write our plane wave solution of the wave equation as

$$f(\omega t - \mathbf{k}\cdot\mathbf{r}) + g(\omega t + \mathbf{k}\cdot\mathbf{r})$$

The harmonic, plane wave solution is

$$f(\mathbf{r}, t) = E(\mathbf{r})\cos(\omega t - \mathbf{k}\cdot\mathbf{r}) \tag{1-15}$$

In complex notation, the harmonic plane wave is written as

$$f(\mathbf{r}, t) = E(\mathbf{r})e^{i(\omega t - \mathbf{k}\cdot\mathbf{r})} \tag{1-16}$$

We will find this harmonic plane wave important because, as we will see later, any three-dimensional wave can be written as a combination of plane waves of different amplitudes, directions, and frequencies.

The second three-dimensional wave we will find useful is one with spherical symmetry

$$f(\mathbf{r}, t) = f(r, \theta, \phi, t) = f(r, t)$$

where

$$r = \sqrt{x^2 + y^2 + z^2}$$

For example, a wave from a point source located at the origin would produce a wave with a wavefront that is a sphere. In this case, the wavefront is given by $\mathcal{S}(\mathbf{r}) = \mathcal{S}(r) = kr =$ constant (the equation for a sphere). The wave equation can be obtained by converting from rectangular to spherical coordinates. We only have to obtain the r component because f is not a function of θ and ϕ if it has spherical symmetry.

$$\frac{\partial f}{\partial x} = \frac{\partial f}{\partial r}\frac{\partial r}{\partial x} = \frac{x}{r}\frac{\partial f}{\partial r}$$

$$\frac{\partial^2 f}{\partial x^2} = \frac{\partial}{\partial r}\left(\frac{x}{r}\frac{\partial f}{\partial r}\right)\frac{\partial r}{\partial x} = \frac{x}{r}\left[\frac{x}{r}\frac{\partial^2 f}{\partial r^2} + \frac{\partial f}{\partial r}\frac{\partial}{\partial r}\left(\frac{x}{r}\right)\right]$$

$$= \frac{x^2}{r^2}\frac{\partial^2 f}{\partial r^2} + \left(\frac{1}{r} - \frac{x^2}{r^3}\right)\frac{\partial f}{\partial r} \tag{1-17}$$

We do the same for the derivatives with respect to y and z and add the results to **(1-17)**.

$$\frac{\partial^2 f}{\partial x^2} + \frac{\partial^2 f}{\partial y^2} + \frac{\partial^2 f}{\partial z^2} = \frac{1}{r^2}\frac{\partial^2 f}{\partial r^2}(x^2 + y^2 + z^2) + \frac{3}{r}\frac{\partial f}{\partial r} - \frac{x^2 + y^2 + z^2}{r^3}\frac{\partial f}{\partial r}$$

$$= \frac{\partial^2 f}{\partial r^2} + \frac{2}{r}\frac{\partial f}{\partial r}$$

The wave equation for spherically symmetric solutions becomes

$$\frac{\partial^2 f}{\partial r^2} + \frac{2}{r}\frac{\partial f}{\partial r} = \frac{1}{c^2}\frac{\partial^2 f}{\partial t^2}$$

or equivalently

$$\frac{\partial^2}{\partial r^2}(rf) = \frac{1}{c^2}\frac{\partial^2}{\partial t^2}(rf)$$

This is the one-dimensional wave equation with a general solution

$$rf(r, t) = f(ct - r) + g(ct + r)$$

The harmonic spherical wave is

$$f(r, t) = \frac{A}{r}\cos(\omega t - kr) \tag{1-18}$$

It is easy to find physical examples of the spherical wave. One realization that everyone has seen is the water wave formed on the surface of a still pool by a drop of water striking the surface. As the wave moves out from its source, it forms circles of ever-increasing radii. This water wave can be thought of as a representation of the intersection of the spherical wavefront and a plane that cuts through the wave and contains the source. If we assume that there are no losses, then the total power in the wave is a constant P_0 given by the product of the power per unit area P and the area of the spherical wave $4\pi r^2$

$$P_0 = P(4\pi r^2)$$

This leads to the conclusion that the power per unit area is inversely proportional to the radius r of the spherical wave's phase front

$$P = \frac{P_0}{4\pi r^2}$$

Since r is also the distance that the wave has traveled from the source, we see that the power per unit area in a spherical wave is inversely proportional to the square of the distance the wave has propagated. The power per unit area of a spherical wave can also be obtained by using **(1-18)** in **(1-7)** and is found to be given by

$$P \propto \frac{A^2}{r^2}$$

Conservation of energy and the wave model both yield the inverse square law behavior.

ATTENUATION OF WAVES

Physical systems cause losses that would decrease a wave's energy as it propagates. We can take the effect of a damping force into account by adding a loss term to the wave equation. We will use the same functional form for the loss term as is used in Appendix 1-A, in which losses in a harmonic oscillator are considered, **(1A-9)**.

$$\frac{\partial^2 f}{\partial x^2} - \gamma \frac{\partial f}{\partial t} - \frac{1}{c^2}\frac{\partial^2 f}{\partial t^2} = 0 \tag{1-19}$$

where γ is the resistance per unit length. Is this equation still a wave equation? We can find out by testing if the one-dimensional representation of a plane wave **(1-16)**

$$f(x, t) = Ae^{i(\omega t - kx + \phi)} \tag{1-20}$$

is a solution of **(1-19)**.

If we differentiate **(1-20)**, we get

$$\frac{\partial^n f}{\partial x^n} = (-ik)^n Ae^{i(\omega t - kx + \phi)} = (-ik)^n f(x, t) \tag{1-21}$$

$$\frac{\partial^n f}{\partial t^n} = (i\omega)^n Ae^{i(\omega t - kx + \phi)} = (i\omega)^n f(x, t) \tag{1-22}$$

Here, the usefulness of complex notation becomes apparent. The untidy equations obtained when differentiating sine and cosine functions are avoided. Substituting the above results into **(1-19)** yields

$$k^2 = \left(\frac{\omega}{c}\right)^2 - i\omega\gamma$$

This type of equation is called a *dispersion equation*. We see that for **(1-19)** to have a plane wave solution of the form **(1-20)**, k must be complex. If we write the complex k as

$$k = \kappa_1 - i\kappa_2$$

and substitute this complex k into **(1-20)**, we obtain

$$f(x, t) = Ae^{-\kappa_2 x} e^{i(\omega t - \kappa_1 x + \phi)}$$

A solution of **(1-19)** is a harmonic wave whose amplitude is attenuated as it propagates in the positive x direction. κ_2 is an attenuation constant that equals the distance the wave will propagate before its amplitude falls to $1/e$ of the value it had at $x = 0$.

SUMMARY

We have developed a one-dimensional model of a wave whose properties are described by the wave equation

$$\frac{\partial^2 f(x, t)}{\partial x^2} = \frac{1}{c^2}\frac{\partial^2 f(x, t)}{\partial t^2}$$

We extended this equation to three dimensions

$$\nabla^2 f(\mathbf{r}, t) = \frac{1}{c^2}\frac{\partial^2 f(\mathbf{r}, t)}{\partial t^2}$$

A special case of the wave equation called the Helmholtz equation

$$(\nabla^2 + k^2)\,\mathcal{E}(\mathbf{r}) = 0$$

was introduced for those situations when only the spatial properties of the wave are to be discussed.

The most important solution to the wave equation for the discussions in this book is the harmonic plane wave

$$f(\mathbf{r}, t) = E(\mathbf{r}) \cos(\omega t - \mathbf{k} \cdot \mathbf{r})$$

or in complex notation

$$f(\mathbf{r}, t) = E(\mathbf{r}) e^{i(\omega t - \mathbf{k} \cdot \mathbf{r})}$$

The solution to the wave equation is modified if the propagation medium has some loss mechanism. The plane harmonic wave propagating in the x direction then has an amplitude that is attenuated as it propagates

$$f(x, t) = A e^{-\kappa_2 x} e^{i(\omega t - \kappa_1 x + \phi)}$$

where

$$\underset{\sim}{k} = \kappa_1 - i\kappa_2$$

is the complex wave vector for the wave propagating in the loss medium.

The average power per unit area (equivalently the energy per unit time per unit area) was defined as the intensity of the wave and shown to be proportional to the amplitude squared. It was also demonstrated that a momentum could be associated with a wave.

PROBLEMS

1-1. Assume a sine wave with an amplitude of 10 cm and a wavelength of 200 cm moving with a velocity of 100 cm/sec:

(a) What is the frequency of both ν and ω?

(b) What is the value of k?

(c) What is the wave equation?

(d) We assume the left end is at the origin and moving down at $t = 0$. What is the equation of motion at the left end?

(e) What is the equation of motion of the point 150 cm to the right of the origin?

1-2. The wavelength of light ranges from 390 to 780 nm and its velocity (in a vacuum) is about 3×10^8 m/sec. What is the corresponding frequency range?

1-3. Given the complex number

$$z_1 = \frac{1}{2i}(1 - 4i)$$

what is its real part and its modulus? Calculate

$$z_1 z_1^*$$

What is the imaginary part of

$$z_2 = \frac{e^{i\omega t} - e^{-i\omega t}}{2} \quad ?$$

1-4. If

$$f(x, t) = A e^{i(kx - \omega t)}$$

what are the expressions for $\mathcal{Re}\{|f|^2\}$ and $[\mathcal{Re}\{f\}]^2$?

1-5. Show that

$$s(x, t) = Ae^{-(2x+3t)^2}$$

is a solution of the wave equation. What is the velocity? In what direction is the wave moving?

1-6. The thickness of a human hair is about 4×10^{-2} mm. Compare its dimension to that of a light wave.

1-7. Find the direction of travel of the following two waves:

$$s(x, t) = A \sin(kx - \omega t), \qquad s(x, t) = A \cos(\omega t - kx)$$

1-8. If we write the wave function in complex notation

$$s(x, t) = Ae^{i\phi}$$

show that s is unchanged when its phase increases or decreases by 2π.

1-9. Given the wave

$$\varphi(x, t) = 10 \cos 2\pi \left[\frac{x}{2 \times 10^{-7}} - (1.5 \times 10^{15})\, t \right]$$

determine the speed, wavelength, and frequency in mks units.

1-10. If s_1 and s_2 are solutions of the wave equation

$$\frac{\partial^2 s}{\partial x^2} = \frac{1}{v^2} \frac{\partial^2 s}{\partial t^2}$$

prove the superposition principle, i.e., $a s_1 + b s_2$ is also a solution, where a and b are both constants.

1-11. A simple harmonic oscillator has a mass m of 0.01 kg and a force constant s of 36 N/m. At $t = 0$, the mass is displaced 50 mm to the right of its equilibrium position and is moving to the right at a speed of 1.7 m/s. Calculate (a) the frequency, (b) the period, (c) the amplitude, and (d) the phase constant.

1-12. The HCl molecule to first order acts like a simple harmonic oscillator. The hydrogen, which does all the vibrating, has a mass of 1.67×10^{-27} kg and the force constant is 840 N/m. What is the vibration frequency?

1-13. Assume the motion of an oscillator is described by

$$y = 2 \cos 6\pi t + \sin 5\pi t$$

If the mass of the oscillator is 10 g, what are the maximum and the minimum kinetic energies?

1-14. When playing a guitar, harmonics of the open string are generated by placing a finger lightly on the string above the 12th fret (this is at the halfway point between the supports of the guitar string, bridge, and saddle). What other positions, in terms of string length, would be useful for harmonic generation?

1-15. A mass of 10 g is attached to a spring and causes it to stretch 1 cm downward. If the attached mass is moved downward another 1 cm and released, what will be its period and maximum velocity?

1-16. A spring has a spring constant of $k = 2.5$ N/m and a damping force constant of $b = 0.1$ kg/sec. A mass of 25 g is attached to the spring and the spring released at $t = 0$ from a position at $x = -5$ cm. Find (a) the frequency and (b) the epoch angle. (c) Write an expression for the displacement as a function of time of the oscillator.

1-17. Find the times when the maximum displacements occur and the values of the maximum displacement for the first three excursions of the oscillator of Problem 1-16.

1-18. A particle of mass m that is constrained to move along the x axis experiences a force toward the origin equal to mkx. Upon release, the mass undergoes simple harmonic motion about the origin with a period of 6 sec. At the time $t = 2$ sec, the mass reaches the origin. At time $t = 3$ sec, the particle is found to have a velocity of 5 cm/sec. (a) What is the force constant? (b) What is the maximum distance from the origin reached by the particle?

1-19. A string 3 m long and weighing 300 g is held in a tension of 10 N. How long does it take a wave to travel the string's length?

1-20. A string is held under a tension by passing the string over a pulley and attaching a 2 kgm weight. An 8 m long segment of the string that weighs 600 gm has a wave of amplitude 10 cm and wavelength 3m traveling along its length. (a) What is the velocity of the wave? (b) What is the maximum transverse velocity of a point on the wave?

1-21. A string with $\mu = 0.1$ kg/m and $T = 90$ n has a wave of amplitude 30 cm and frequency 1 Hz propagating along it. How much energy is transmitted by the wave?

Appendix 1-A

HARMONIC MOTION

Physical systems can, under proper conditions, vibrate, i.e., move to and fro about some fixed point, repeating the motion in a regular fashion. We will calculate the equations of motion of a body vibrating with simple harmonic motion. The mechanical model we will use to explore this type of vibration will be of importance to us in our discussion of light and how it propagates through matter. We assume we have a mass moving between $\pm A$ about an equilibrium point O as shown in Figure 1A-1. The force on the body is an elastic restoring force, i.e., linearly dependent on the displacement coordinate

$$F = -sx = ma = m\frac{d^2x}{dt^2}$$

The constant of proportionality s is often called the spring constant. The equation of motion for this system with a linear restoring force is

$$m\frac{d^2x}{dt^2} + sx = 0 \tag{1A-1}$$

It would appear that such a simple, second-order, differential equation would have limited applicability, but such is not the case.

The force acting on a mass can be written in terms of the potential energy function

$$F = -\frac{dV}{dx}$$

(We use the potential function because it is a scalar and thus easier to manipulate than a vector.) We see that our force $F = -sx$ is due to a potential energy function $V(x)$ that is proportional to x^2 as illustrated in Figure 1A-1. If we have a more complicated potential function, we can expand the potential function, about the equilibrium point, in a Taylor series

$$V = V_0 + \frac{1}{2}\left(\frac{d^2V}{dx^2}\right)x^2 + \frac{1}{6}\left(\frac{d^3V}{dx^3}\right)x^3 + \cdots$$

There is no term involving x because $V(x)$ is a minimum at the equilibrium position. We see our model is applicable when the Taylor expansion beyond the first term is not needed, i.e., when the displacements x of the oscillations about the equilibrium position are small.

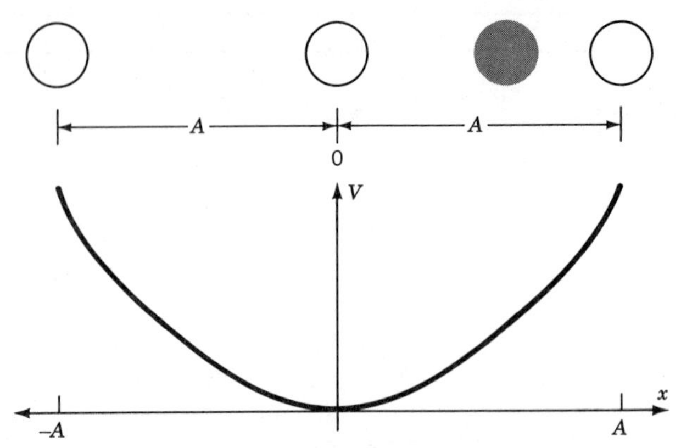

FIGURE 1A-1. A mass m is moving between points A and $-A$ around the equilibrium point O. The potential energy V for a restoring force linearly dependent on the displacement from equilibrium x is shown.

One way to solve the equation of motion is to convert it to a first order differential equation by substitution of

$$\frac{d^2x}{dt^2} = \frac{d}{dt}\left(\frac{dx}{dt}\right) = \frac{dv}{dt} = \frac{dv}{dx}\frac{dx}{dt} = v\frac{dv}{dx}$$

resulting in the first-order equation

$$mv\frac{dv}{dx} + sx = 0$$

The solution can be obtained by a simple integration

$$\frac{mv^2}{2} + \frac{sx^2}{2} = E$$

Here, the first term on the left is kinetic energy and the second term potential energy. The constant of integration is E, the total energy. When we have zero kinetic energy

$$x_{max} = A = \sqrt{\frac{2E}{s}}$$

the velocity of the mass is given by

$$v = \frac{dx}{dt} = \sqrt{\frac{s}{m}}\sqrt{A^2 - x^2}$$

After integration of the velocity equation, we obtain

$$\sin^{-1}\left(\frac{x}{A}\right) = \sqrt{\frac{s}{m}}t + \delta$$

where δ is the constant of integration. The quantity

$$\sqrt{\frac{s}{m}}t + \delta$$

is called the phase of the oscillation; as you can see, at $t = 0$, the phase is δ and for this reason, δ is called the *epoch* or *initial phase angle*. If we assume that at $t = 0$ we have $x = x_0$, then

$$\sin \delta = \frac{x_0}{A}$$

In our discussions, we will assume that $x_0 = A$ so that $\delta = \pi/2$ and

$$x = A \cos\left(\sqrt{\frac{s}{m}}t\right) \tag{1A-2}$$

The *period* T is the time required to complete one oscillation. The value of x at time t and $(t + T)$ must be equal; thus, the two phases are

$$\left[\sqrt{\frac{s}{m}}(t + T) + \delta\right] = \left[\sqrt{\frac{s}{m}}t + \delta\right] + 2\pi$$

$$T = 2\pi\sqrt{\frac{m}{s}} \tag{1A-3}$$

The *frequency* of oscillation, that is, the number of times x has the same value in a unit of time is

$$\nu = \frac{1}{T} = \frac{1}{2\pi}\sqrt{\frac{s}{m}} \tag{1A-4}$$

Throughout this book, we will use the angular frequency $\omega = 2\pi\nu$ to reduce the number of 2π's that must be written. The equations describing the position, velocity, and acceleration of a particle undergoing harmonic motion can be written as

$$x = A \cos(\omega_0 t + \delta) \tag{1A-5}$$

$$v = -\omega_0 A \sin(\omega_0 t + \delta) \tag{1A-6}$$

$$a = -\omega_0^2 A \cos(\omega_0 t + \delta) \tag{1A-7}$$

where an arbitrary epoch angle has been added. These functions are plotted in Figure 1A-2 when $\delta = 0$.

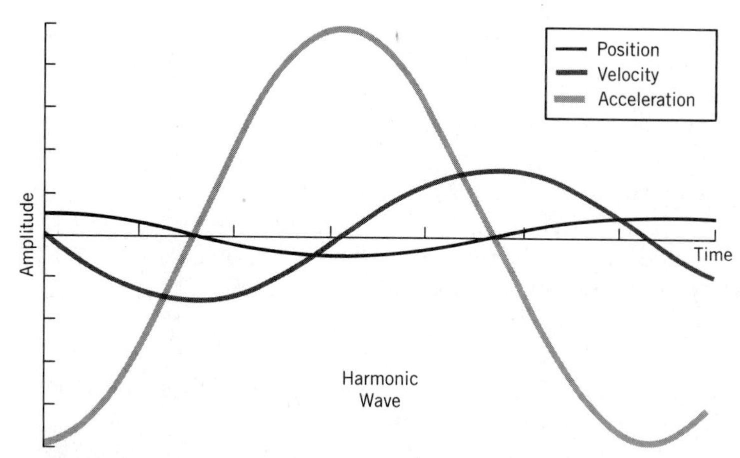

FIGURE 1A-2. Position, velocity, and acceleration from **(1A-5)**, **(1A-6)**, and **(1A-7)** for a frequency of 2 Hz.

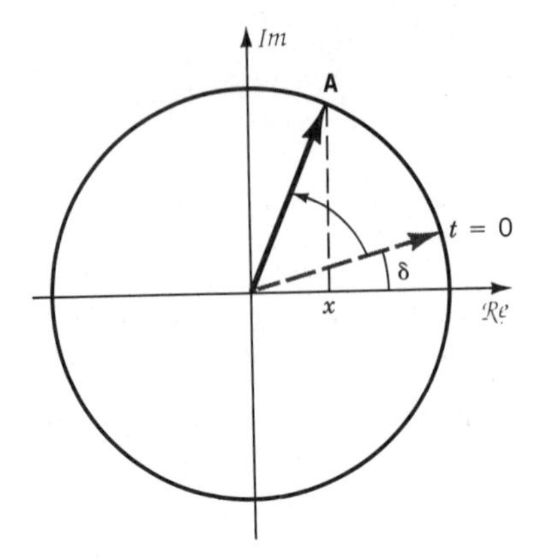

FIGURE 1A-3. This Argand diagram shows the displacement in complex notation. The real part of the complex displacement is the projection on the real axis. The point at x undergoes simple harmonic motion as the vector **A** rotates about the origin at an angular velocity, $\omega_0 t$.

We can use complex notation to write the equation of simple harmonic motion (see Appendix 1-B for a brief review of complex notation).

$$x = Ae^{i(\omega_0 t + \delta)}$$

with equivalent expressions for the velocity and acceleration. Complex notation simplifies the mathematical manipulations required and allows us to use a graphics representation that aids in understanding harmonic motion; see Figure 1A-3.

Damped Oscillator

No oscillator will run forever; there will be losses that dissipate the energy. A first guess as to the form of the loss term would be a constant, independent of the motion. However, a constant loss term yields an equation of motion of the same form as the undamped oscillator case. Another reasonable loss term is a damping force that is proportional to the velocity

$$F_d = -b\frac{dy}{dt} \tag{1A-8}$$

where b is a positive constant of proportionality with units of kg/sec called the *damping constant*. The new equation of motion is then

$$m\frac{d^2y}{dt^2} = -sy - b\frac{dy}{dt}$$

Dividing through by the mass gives

$$\frac{d^2y}{dt^2} + \gamma\frac{dy}{dt} + \omega_0^2 y = 0 \tag{1A-9}$$

where we have defined a new constant $\gamma = b/m$ with units of sec^{-1}, called the *damping factor*. If we assume a solution of the form

$$y = Ce^{pt}$$

and substitute this solution into the linear homogeneous differential equation **(1-9)**, we find p must satisfy the quadratic equation

$$p^2 + \gamma p + \omega_0^2 = 0$$

Thus,

$$p = -\frac{\gamma}{2} \pm \sqrt{\frac{\gamma^2}{4} - \omega_0^2}$$

We can divide the solution into three cases depending on the relative values of γ and ω_0. The solutions when $\gamma < 2\omega_0$ are the only solutions of interest to us; the other two cases, $\gamma \geq 2\omega_0$, do not lead to oscillations but rather aperiodic motion. When $\gamma < \omega_0$, the solution becomes

$$y = Ce^{-\gamma(t/2)}e^{i\omega_d t} \tag{1A-10}$$

where

$$\omega_d = \sqrt{\omega_0^2 - \frac{\gamma^2}{4}}$$

Whenever $\gamma << \omega_0$, we use ω_0 in place of ω_d without significant error. The amplitude of the oscillations described by **(1A-10)** decay exponentially (if we had assumed the constant loss term, the decay would have been linear). The amplitude is reduced to $1/e$ of its original value in a time given by $2/\gamma$.

A useful characterization of this damped oscillator can be made by using the quality factor

$$Q = \frac{\omega_0}{\gamma} \tag{1A-11}$$

The energy in the oscillator falls to $1/e$ of its initial value in about $Q/2\pi$ vibrations.

Appendix 1-B

COMPLEX NUMBERS

Since we will use complex numbers extensively in our discussion of light, we should review a few important properties of complex numbers. Complex numbers are ordered pairs of real numbers (x, y) that can be written as

$$z = x + iy \qquad (i = \sqrt{-1})$$

An integral part of the definition of complex numbers are the rules of calculating the sum and product of two of these ordered pairs

$$z_1 + z_2 = (x_1 + x_2) + i\,(y_1 + y_2) \tag{1B-1}$$

$$z_1 z_2 = (x_1 x_2) - (y_1 y_2) + i\,(x_1 y_2 + x_2 y_1) \tag{1B-2}$$

In analogy with the representation of real numbers as points on a straight line, a geometric representation of complex numbers can be created by associating the ordered pair with points in the xy plane of a rectangular Cartesian coordinate system. The xy plane is called the complex plane and the ordered pair can be thought of as a "vector" from the origin to the point (x, y). One must be cautious in thinking of complex numbers as vectors because multiplication is not defined the same for normal vectors and complex numbers.

Extending the geometrical representation to polar format yields a second notation

$$z = r\,(\cos\phi + i\sin\phi) \tag{1B-3}$$

where

$$r = \sqrt{x^2 + y^2} \qquad \text{and} \qquad \tan\phi = \frac{y}{x}$$

The geometrical representation is illustrated in Figure 1B-1.

The Euler formula

$$e^{i\phi} = \cos\phi + i\sin\phi \tag{1B-4}$$

provides an exponential formulation of the polar format that we will use extensively in our discussion of optics. In the exponential formulation, the complex number is

$$z = re^{i\phi} \tag{1B-5}$$

It is useful to note that multiplying by i is geometrically equivalent to rotating through 90°, and multiplying by $i^2 = -1$ is equivalent to rotating through 180°, moving a point from the positive to the negative real axis. Mathematically, we can express these observations as

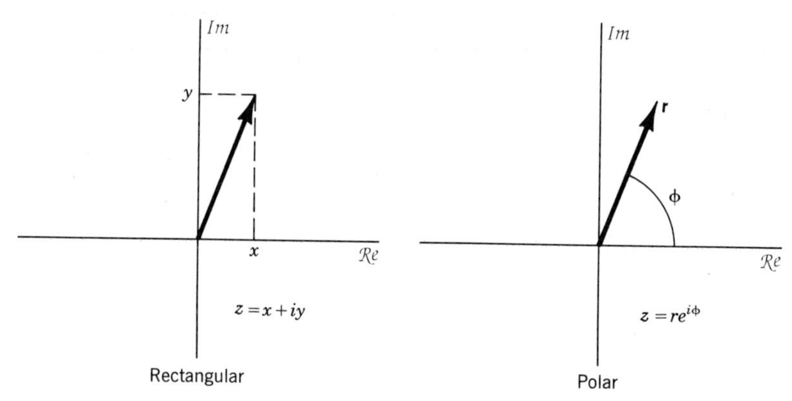

FIGURE 1B-1. Geometrical representation of complex numbers shown in rectangular and polar format.

$$e^{i2\pi} = 1, \qquad e^{i\pi} = e^{-i\pi} = -1, \qquad e^{\pm i(\pi/2)} = \pm i$$

The real and imaginary parts of the complex number are

$$\mathcal{Re}\{z\} = x = r\cos\phi = \frac{z + z^*}{2} \tag{1B-6}$$

$$\mathcal{Im}\{z\} = y = r\sin\phi = \frac{z - z^*}{2i} \tag{1B-7}$$

where

$$z^* = re^{-i\phi}$$

The modulus (absolute value) of the complex number is r and is given by

$$r = |z| = \sqrt{zz^*} = \sqrt{\left[\mathcal{Re}\{z\}\right]^2 + \left[\mathcal{Im}\{z\}\right]^2} = \sqrt{x^2 + y^2} \tag{1B-8}$$

The modulus can be interpreted geometrically as the length of the vector, that is, the distance from the origin to the point (x, y); it allows us to order complex numbers, for instance,

$$z_1 > z_2 \quad \text{if} \quad |z_1| > |z_2|$$

The use of complex notation simplifies the mathematical manipulation we must perform and the number of trigonometric identities we must recall. Remember that the complex notation is only a mathematical convenience. We use only one of the two components of the complex number when representing an experimentally observed parameter and, by convention, we use the real part of the complex number.

When linear mathematical operations are to be made, we may use complex notation throughout the problem as long as we remember to convert the answer to the real component when representing the physical property. Nonlinear operations such as squaring require that we use the real component throughout the problem (see Problem 1-4). To retain the mathematical simplification of complex notation, the notation shown in **(1B-6)** is used for these problems.

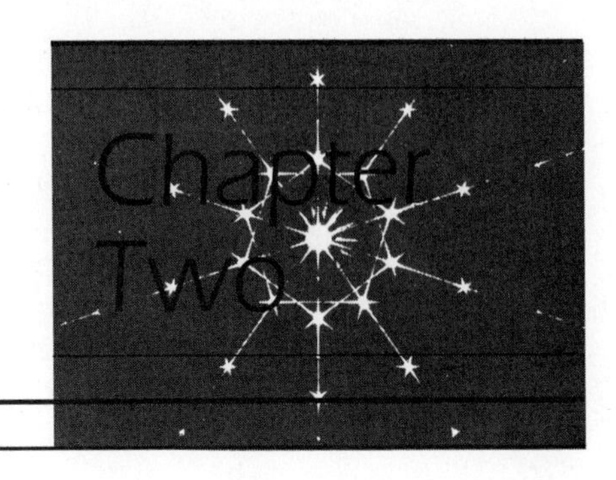

Electromagnetic Theory

The scalar wave theory that we discussed in Chapter 1 was applied to the study of light before the development of the theory of electromagnetism. At that time, it was assumed that the light waves were longitudinal in analogy with sound waves, i.e., the wave's displacements were in the direction of propagation. A further assumption, that light propagated through some type of medium, was made because the scientists of that time approached all problems from a mechanistic point of view. The scalar theory was successful in explaining diffraction (see Chapter 9), but problems arose in interpretation of the effects of polarization in interference experiments (discussed in Chapter 4). Young was able to resolve the difficulties by suggesting that the waves could be transverse as are the waves on a vibrating string. Using this idea, Fresnel developed a mechanistic description of light that could explain the amount of reflected and transmitted light from the interface between two media (see Chapter 3).

Independent of this activity, the theory of electromagnetism was under development. **Michael Faraday (1791–1867)** observed in 1845 that a magnetic field would rotate the plane of polarization of light waves passing through the magnetized region. This observation led Faraday to associate light with electromagnetic radiation, but he was unable to quantify this association. Faraday attempted to develop electromagnetic theory by treating the field as lines pointing in the direction of the force that the field would exert on a test charge. The lines were given a mechanical interpretation with a tension along each line and a pressure normal to the line. **James Clerk Maxwell (1831–1879)** furnished a mathematical framework for Faraday's model in a paper read in 1864 and published a year later.[1] In this paper, Maxwell identified light as "an electromagnetic disturbance in the form of waves propagated through the electromagnetic field according to electromagnetic laws" and demonstrated that the propagation velocity of light was given by the electromagnetic properties of the material.

Maxwell was not the first to recognize the connection between the electromagnetic properties of materials and the speed of light. **Kirchhoff** recognized in 1857 that the speed of light could be obtained from electromagnetic properties. **Riemann** in 1858 assumed that electromagnetic forces propagated at a finite velocity and derived a propagation velocity given by the electromagnetic properties of the medium. However, it was Maxwell who demonstrated that the electric and magnetic fields are waves that travel at the speed of light. It was not until 1887 that an experimental observation of electromagnetic waves other than light was obtained by **Heinrich Rudolf Hertz (1857–1894)**.

The classical electromagnetic theory is successful in explaining all of the experimental observations to be discussed in this book. There are, however, experiments that cannot be explained by classical wave theory, especially those conducted at short wavelengths or very low light levels. Quantum electrodynamics is capable of predicting the outcome of all optical experiments; its shortcoming is that it does not explain why or how. An excellent elementary introduction to quantum electrodynamics has been written by Richard Feynman.[2]

In this chapter, we will borrow, from electromagnetic theory, Maxwell's equations and Poynting's theorem to derive properties of light waves. Details of the origins of these fundamental electromagnetic relationships are not needed for our study of light, but can be obtained by consulting any electricity and magnetism text.[3]

The basic properties that will be derived are (1) the wave nature of light, (2) the fact that light is a transverse wave, (3) the velocity of light in terms of fundamental electromagnetic properties of materials, (4) the relative magnitude of the electric and magnetic fields and relationships between the two fields, and finally, (5) the momentum and energy associated with a light wave.

The concept of polarized light and a geometrical construction used to visualize its behavior will be introduced. Both a vector and matrix notation used to describe polarization will be presented in this chapter, but details on the manipulation of light's polarization will not be discussed until Chapter 13. The chapter will conclude with a discussion of the propagation of light in a conducting medium.

MAXWELL'S EQUATIONS

The basis of electromagnetic theory is Maxwell's equations. They allow the derivation of the properties of light. In our study of optics, we will treat these equations as axioms but provide the reader with a source that can be consulted if information on the origin of the equations is desired. In rationalized mks units, Maxwell's equations are the following.

Gauss's (Coulomb's) Law for the Electric Field *Coulomb's law* provides a means for calculating the force between two charges (see Chapter 2 of Wangsness[3])

$$\mathbf{F} = \frac{q_0}{4\pi\epsilon_0}\int \frac{dq}{r^2}\hat{\mathbf{n}}$$

where dq is the charge on an infinitesimal surface and $\hat{\mathbf{n}}$ is a unit vector connecting charges q_0 and dq. The electric field

$$\mathbf{E} = \frac{\mathbf{F}}{q_0}$$

is obtained using Coulomb's law (see Chapter 3, Wangsness). We view this field, as Michael Faraday did, as lines of flux, called lines of force, originating on positive charges and terminating on negative charges. Gauss's law states that the quantity of charge contained within a closed surface is equal to the number of flux lines passing outward through the surface (Chapter 4, Wangsness). This view of the electric field leads to

$$\nabla \cdot \mathbf{D} = \rho \tag{2-1}$$

where ρ is the charge density and $\mathbf{D}$ is the electric displacement (Chapter 10, Section 5 of Wangsness). The use of the displacement allows the equation to be applied to any material.

Gauss's Law for the Magnetic Field Charges at rest led to **(2-1)**. Charges in motion, that is, a current $\mathbf{i}$ or a current density $\mathbf{J}$, create a magnetic field $\mathbf{B}$ (Chapter 14, Wangsness). As we did for the electric field, we treat the magnetic field as flux lines, called lines of induction, and we assume that the current density is a constant so that $\nabla \cdot \mathbf{J} = 0$. This leads to (Chapter 16, Wangsness)

$$\nabla \cdot \mathbf{B} = 0 \tag{2-2}$$

The zero results from the fact that the magnetic equivalent of a single charge has never been observed.

Faraday's Law The previous two equations are associated with electric and magnetic fields that are constant with respect to time. The next equation, an experimentally derived equation, deals with a magnetic field that is time-varying or equivalently a conductor moving through a static magnetic field. In terms of the concept of flux, it states that an electric field around a circuit is associated with a change in the magnetic flux contained within the circuit.

$$\nabla \times \mathbf{E} + \frac{\partial \mathbf{B}}{\partial t} = 0 \tag{2-3}$$

Ampère's Law (Law of Biot and Savart) An electric charge in motion creates a magnetic field around its path. The law of Biot–Savart allows us to calculate the magnetic field at a point located a distance $\mathbf{R}$ from a conductor carrying a current density $\mathbf{J}$. Ampère's law is the inverse relationship used to calculate the current in a conductor due to the magnetic field contained in a loop about the conductor. Neither relationship is adequate when the current is a function of time. Maxwell's major contribution to physics was to observe that the addition of a *displacement current* to Ampère's law allowed fluctuating currents to be explained. The relationship became (see Chapter 21 of Wangsness)

$$\nabla \times \mathbf{H} = \mathbf{J} + \frac{\partial \mathbf{D}}{\partial t} \tag{2-4}$$

As discussed in Appendix 2-B, the constants in Maxwell's equations depend on the units used. Many optics books use cgs units that result in a form for Maxwell's equations shown in Appendix 2-B.

The dynamic response of atoms and molecules in the propagation medium is taken into account through what are called the *constitutive relations*.

$$\mathbf{D} = f(\mathbf{E})$$

$$\mathbf{J} = g(\mathbf{E})$$

$$\mathbf{B} = h(\mathbf{H})$$

Here, we will assume that the functional relations are independent of space and time and we will write the constitutive relations as

$$\mathbf{D} = \epsilon\mathbf{E}, \qquad \epsilon = \text{dielectric constant}$$

$$\mathbf{J} = \sigma\mathbf{E}, \qquad \sigma = \text{conductivity (Ohms law)}$$

$$\mathbf{B} = \mu\mathbf{H}, \qquad \mu = \text{permeability}$$

where the constants ϵ, σ, and μ contain the description of the material. Later, we will explore the effects resulting from the constitutive relations having a temporal or spatial dependence.

Often, **D** and **B** are defined as

$$\mathbf{D} = \epsilon_0\mathbf{E} + \mathbf{P} \tag{2-5}$$

$$\mathbf{B} = \mu_0\mathbf{H} + \mathbf{M}$$

where **P** is the polarization and **M** the magnetization. This formulation emphasizes that the internal field of a material is due not only to the applied field, but also a field created by the atoms and molecules that make up the material. We will find **(2-5)** useful in Chapters 7 and 15. We will not use the relationship involving **M** in this book.

By manipulating Maxwell's equations, we can obtain a number of the properties of light such as its wave nature, the fact that it is a transverse wave, and the relationship between the **E** and **B** fields. We will make a number of simplifying assumptions about the medium in which light is propagating to allow a quick derivation of the properties of light. Later, we will see what happens if we modify these assumptions.

We assume that the light is propagating in a medium we will call *free space* that has the following properties:

1. Uniform: ϵ and μ have the same value at all points.
2. Isotropic: ϵ and μ do not depend on the direction of propagation.
3. Nonconducting: $\sigma = 0$, thus $\mathbf{J} = 0$.
4. Free from charge: $\rho = 0$.
5. Nondispersive: ϵ and μ are not functions of the frequency, i.e., they have no time dependence.

Our definition departs somewhat from other definitions of free space in that we include in the definition not only the vacuum, where $\epsilon = \epsilon_0$ and $\mu = \mu_0$, but also dielectrics, where $\sigma = 0$ but the other electromagnetic constants can have arbitrary values.

If we use the above assumptions, Maxwell's equations and the constitutive relations simplify to

$$\nabla \cdot \mathbf{E} = 0 \tag{2-6a}$$

$$\nabla \cdot \mathbf{B} = 0 \tag{2-6b}$$

$$\nabla \times \mathbf{E} = -\frac{\partial \mathbf{B}}{\partial t} \tag{2-6c}$$

$$\nabla \times \mathbf{H} = \frac{\partial \mathbf{D}}{\partial t} \tag{2-6d}$$

$$\mathbf{B} = \mu \mathbf{H} \tag{2-6e}$$

$$\epsilon \mathbf{E} = \mathbf{D} \tag{2-6f}$$

These simplified equations can now be used to derive some of the basic properties of a light wave.

Wave Equation

To find how the electromagnetic wave described by **(2-6)** propagates in free space, Maxwell's equations must be rearranged to display explicitly the time and coordinate dependence. Using **(2-6e)** and **(2-6f)** we can rewrite **(2-6d)** as

$$\frac{1}{\mu} \nabla \times \mathbf{B} = \epsilon \frac{\partial \mathbf{E}}{\partial t}$$

The curl of **(2-6c)** is taken and the magnetic field dependence eliminated by using the rewritten **(2-6d)**

$$\nabla \times (\nabla \times \mathbf{E}) = \nabla \times \left(-\frac{\partial \mathbf{B}}{\partial t} \right) = -\frac{\partial}{\partial t}(\nabla \times \mathbf{B}) = -\frac{\partial}{\partial t}\left(\epsilon \mu \frac{\partial \mathbf{E}}{\partial t} \right)$$

The assumption that ϵ and μ are independent of time allows the equation to be rewritten

$$\nabla \times (\nabla \times \mathbf{E}) = -\epsilon \mu \frac{\partial^2 \mathbf{E}}{\partial t^2}$$

Using the vector identity **(2A-12)**, we can write

$$\nabla(\nabla \cdot \mathbf{E}) - \nabla^2 \mathbf{E} = -\epsilon \mu \frac{\partial^2 \mathbf{E}}{\partial t^2}$$

Because free space is free of charge $\nabla \cdot \mathbf{E} = 0$, giving us

$$\nabla^2 \mathbf{E} = \mu \epsilon \frac{\partial^2 \mathbf{E}}{\partial t^2} \tag{2-7}$$

We can use the same procedure to obtain

$$\nabla^2 \mathbf{B} = \mu \epsilon \frac{\partial^2 \mathbf{B}}{\partial t^2} \tag{2-8}$$

These equations are wave equations, with the wave's velocity given by

$$v = \frac{1}{\sqrt{\mu \epsilon}} \tag{2-9}$$

The connection of the velocity of light with the electric and magnetic properties of a material was one of the most important results of Maxwell's theory. In a vacuum,

$$\mu_0\epsilon_0 = (4\pi \times 10^{-7})(8.8542 \times 10^{-12})$$

$$= 1.113 \times 10^{-17}\ \frac{\text{sec}^2}{\text{m}^2}$$

$$\frac{1}{\sqrt{\mu_0\epsilon_0}} = 2.998 \times 10^8\ \frac{\text{m}}{\text{sec}} = c \tag{2-10}$$

In a material, the velocity of light is less than c. We can characterize a material by defining the *index of refraction*, the ratio of the speed of light in a vacuum to its speed in a medium.

$$n = \frac{c}{v} = \sqrt{\frac{\epsilon\mu}{\epsilon_0\mu_0}} \tag{2-11}$$

TABLE 2.1 Representative Magnetic Permeability

Material	μ/μ_0	Class
Silver	0.99998	Diamagnetic
Copper	0.99999	Diamagnetic
Water	0.99999	Diamagnetic
Air	1.00000036	Paramagnetic
Aluminum	1.000021	Paramagnetic
Iron	5000	Ferromagnetic
Nickel	600	Ferromagnetic

The data in Table 2.1 demonstrate that if magnetic materials are not considered, then $\mu/\mu_0 \approx 1$ so that

$$n = \sqrt{\frac{\epsilon}{\epsilon_0}}$$

The data displayed in Table 2.2 demonstrate that, at least for some materials, the theory agrees with experimental results. The materials whose indices are listed in Table 2.2 have been specially selected to demonstrate good agreement; we will see in Chapter 7 that the assumption that ϵ, μ, and σ are independent of the frequency results in a theory that neglects the response time of the system to the electromagnetic signal.

Transverse Waves

Hooke postulated, in the 17th century, that light waves might be transverse but his idea was forgotten. Young and Fresnel made the same claim in the 19th century and accompanied their postulation with a theoretical description of light based on transverse waves. Forty years later, Maxwell proved that light must be a transverse wave. We can demonstrate the transverse nature of light by substitution of the plane wave solution of the wave equation into Gauss's law

$$\nabla \cdot \mathbf{E} = \frac{\partial E_x}{\partial x} + \frac{\partial E_y}{\partial y} + \frac{\partial E_z}{\partial z} = 0$$

TABLE 2.2 Selected Index of Refraction

Material	n (yellow light)	$(\epsilon/\epsilon_0)^{1/2}$ (static)
Air	1.000294	1.000295
CO_2	1.000449	1.000473
C_6H_6 (benzene)	1.482	1.489
He	1.000036	1.000034
H_2	1.000131	1.000132

To complete the demonstration, consider the divergence of the electric component of the plane wave. We will examine only the x coordinate of the divergence in detail

$$\frac{\partial E_x}{\partial x} = \frac{\partial}{\partial x}\left[E_{0x}e^{\,i(\omega t - \mathbf{k}\cdot\mathbf{r} + \phi)}\right] = iE_{0x}e^{\,i(\omega t - \mathbf{k}\cdot\mathbf{r} + \phi)}\frac{\partial}{\partial x}(\omega t - k_x x - k_y y - k_z z + \phi)$$

$$\frac{\partial E_x}{\partial x} = -ik_x E_x$$

We easily obtain similar results for E_y and E_z, allowing the divergence of $\mathbf{E}$ to be rewritten as a dot product of $\mathbf{k}$ and $\mathbf{E}$. Gauss's law for the electric field states that the divergence of $\mathbf{E}$ is zero, which for a plane wave can be written

$$\nabla\cdot\mathbf{E} = -i\mathbf{k}\cdot\mathbf{E} = 0 \tag{2-12}$$

If the dot product of two vectors $\mathbf{E}$ and $\mathbf{k}$ is zero, then the vectors $\mathbf{E}$ and $\mathbf{k}$ must be perpendicular [see **(2A-1)**]. In the same manner, substituting the plane wave into $\nabla\cdot\mathbf{B} = 0$ yields $\mathbf{k}\cdot\mathbf{B} = 0$. Therefore, Maxwell's equations require light to be a transverse wave, i.e., the vector displacements $\mathbf{E}$ and $\mathbf{B}$ are perpendicular to the direction of propagation $\mathbf{k}$.

Interdependence of E and B

The electric and magnetic fields are not independent as we can see by continuing our examination of the plane wave solutions of Maxwell's equations. First, let us calculate several derivatives of the plane wave. We will need

$$\frac{\partial \mathbf{B}}{\partial t} = \frac{\partial}{\partial t}\mathbf{B}_0 e^{\,i(\omega t - \mathbf{k}\cdot\mathbf{r} + \phi)} = i\,\mathbf{B}\frac{\partial}{\partial t}(\omega t - \mathbf{k}\cdot\mathbf{r} + \phi)$$

$$\frac{\partial \mathbf{B}}{\partial t} = i\omega\mathbf{B} \tag{2-13}$$

and in a similar manner

$$\frac{\partial \mathbf{E}}{\partial t} = i\omega\mathbf{E} \tag{2-14}$$

A simple expression for the curl of $\mathbf{E}$, $\nabla\times\mathbf{E}$, can be obtained when we use the derivatives just calculated. The expression for the curl of $\mathbf{E}$ is given by **(2A-7)** and is rewritten here

$$\nabla\times\mathbf{E} = \left(\frac{\partial E_z}{\partial y} - \frac{\partial E_y}{\partial z}\right)\hat{\mathbf{i}} + \left(\frac{\partial E_x}{\partial z} - \frac{\partial E_z}{\partial x}\right)\hat{\mathbf{j}} + \left(\frac{\partial E_y}{\partial x} - \frac{\partial E_x}{\partial y}\right)\hat{\mathbf{k}}$$

The terms making up the x component of the curl are

$$\frac{\partial E_z}{\partial y} = E_{0z}\frac{\partial}{\partial y}e^{\,i(\omega t - \mathbf{k}\cdot\mathbf{r} + \phi)} = -i\,k_y E_z$$

$$\frac{\partial E_y}{\partial z} = -i\,k_z E_y$$

By the evaluation of each component, we find that the curl of $\mathbf{E}$ for a plane wave is

$$\nabla\times\mathbf{E} = -i\,\mathbf{k}\times\mathbf{E} \tag{2-15}$$

A similar derivation leads to the curl of **B** for a plane wave

$$\nabla\times\mathbf{B} = -i\,\mathbf{k}\times\mathbf{B} \tag{2-16}$$

With these vector operations on a plane wave defined, we can evaluate **(2-6c)** for a plane wave. The left side of

$$\nabla\times\mathbf{E} = -\frac{\partial\mathbf{B}}{\partial t}$$

is replaced with **(2-15)** and the right side by **(2-13)**, resulting in an equation connecting the electric and magnetic field

$$-i\mathbf{k}\times\mathbf{E} = -i\omega\mathbf{B}$$

Using the relationship between ω and **k** given by **(1-2)** and the relationship for the wave velocity in terms of the electromagnetic properties of the material **(2-9)**, we can write

$$\frac{\sqrt{\mu\epsilon}}{k}\mathbf{k}\times\mathbf{E} = \mathbf{B} \tag{2-17}$$

A second relationship between the magnetic and electric fields can be generated by using the same procedure to rewrite

$$\nabla\times\mathbf{B} = \mu\epsilon\frac{\partial\mathbf{E}}{\partial t}$$

for a plane wave as

$$-i\,\mathbf{k}\times\mathbf{B} = i\epsilon\mu\omega\mathbf{E}$$

$$\frac{1}{k\sqrt{\mu\epsilon}}\mathbf{k}\times\mathbf{B} = -\mathbf{E} \tag{2-18}$$

From the definition of the cross product given in Appendix 2-A **(2A-2)**, we see that the electric and magnetic fields are perpendicular to each other, in phase, and form a right-handed coordinate system with the propagation direction **k**; see Figure 2-1.

If we are only interested in the magnitude of the two fields, we can use **(2-11)** to write

$$n|\mathbf{E}| = c|\mathbf{B}| \tag{2-19}$$

In a vacuum, $n = 1$ so that for a vacuum

$$|\mathbf{E}| = c|\mathbf{B}| \tag{2-20}$$

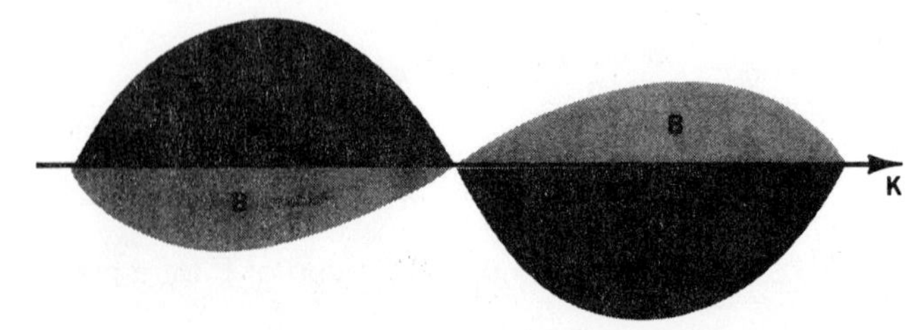

FIGURE 2-1. Graphical representation of an electromagnetic plane wave. Note **E** and **B** are perpendicular to each other, individually perpendicular to the propagation vector **k**, in phase and form a right-handed coordinate system as required by **(2-17)** and **(2-18)**.

For our plane wave, the ratio of the field magnitudes is

$$\frac{|\mathbf{E}|}{|\mathbf{H}|} = \sqrt{\frac{\mu}{\epsilon}}$$

This ratio has the units of ohms (Ω) ($\mu \rightarrow ml/Q^2$, $\epsilon \rightarrow Q^2t^2/ml^3$, and $\Omega \rightarrow ml^2/Q^2t$) and is called the impedance of the medium. In a vacuum,

$$Z_0 = \sqrt{\frac{\mu_0}{\epsilon_0}} = 377\ \Omega$$

When the ratio is a real quantity, as it is here, then **E** and **H** are in phase.

ENERGY DENSITY AND FLOW

We saw in our discussion of waves propagating along strings that the power transmitted by a wave is proportional to the square of the amplitude of the wave. Any text on electromagnetic theory (see Chapter 21 of Wangsness, for example) demonstrates that the energy density (in J/m^3) associated with an electromagnetic wave is given by

$$U = \frac{(\mathbf{D}\cdot\mathbf{E} + \mathbf{B}\cdot\mathbf{H})}{2} \tag{2-21}$$

We can simplify **(2-21)** by using the simple constitutive relations $\mathbf{D} = \epsilon\mathbf{E}$ and $\mathbf{B} = \mu\mathbf{H}$, if they apply to the propagation medium

$$U = \frac{1}{2}\left(\epsilon E^2 + \frac{B^2}{\mu}\right) = \frac{1}{2}\left(\epsilon + \frac{1}{\mu c^2}\right)E^2$$

In a vacuum, further simplification is possible

$$U = \epsilon_0 E^2 = \frac{B^2}{\mu_0}$$

John Henry Poynting (1852–1914) demonstrated that the presence of both an electric and magnetic field at the same point in space results in a flow of the field energy. This fact is called the Poynting theorem and the *Poynting vector* completely describes the flow

$$\mathbf{S} = \mathbf{E}\times\mathbf{H} \tag{2-22}$$

The units of the Poynting vector are J/(m^2·sec). We will use a plane wave to determine some of the properties of this vector. Since **S** will involve terms quadratic in **E**, it will be necessary to use the real form of **E** (see Problem 1-4).

$$\mathbf{H} = \frac{\mathbf{B}}{\mu} = \frac{\sqrt{\mu\epsilon}}{\mu k}\mathbf{k}\times\mathbf{E}$$

where

$$\mathbf{E} = \mathbf{E}_0 \cos(\omega t - \mathbf{k}\cdot\mathbf{r} + \phi)$$

$$\mathbf{S} = \frac{\sqrt{\mu\epsilon}}{\mu k}\mathbf{E}_0 \times (\mathbf{k}\times\mathbf{E}_0) \cos^2(\omega t - \mathbf{k}\cdot\mathbf{r} + \phi)$$

$$= \frac{n}{\mu c}|\mathbf{E}_0|^2\frac{\mathbf{k}}{k} \cos^2(\omega t - \mathbf{k}\cdot\mathbf{r} + \phi) \tag{2-23}$$

Note that the energy is flowing in the direction of propagation (denoted by the unit vector $\mathbf{k}/k$).

We normally do not detect $\mathbf{S}$ at the very high frequencies associated with light ($\approx 10^{15}$ Hz) but rather detect a temporal average of $\mathbf{S}$ with the average taken over a time T determined by the response time of the detector used. We must obtain the time average of $\mathbf{S}$ to relate theory to actual measurements. The time average of $\mathbf{S}$ is called the *flux density* and has units of W/m^2. We will call this quantity the *intensity* of the light wave

$$I = \left|\langle\mathbf{S}\rangle\right| = \frac{1}{T}\int_{t_0}^{t_0+T} \mathrm{A}\cos^2(\omega t - \mathbf{k}\cdot\mathbf{r} + \phi)\, dt \tag{2-24}$$

where we have defined

$$\mathrm{A} = \frac{n}{\mu c}\left|\mathbf{E}_0\right|^2\frac{\mathbf{k}}{k}$$

to simplify the notation.

The units used for the flux density are a confusing mess in optics. One area of optics is interested in measuring the physical effects of light and the measurement of energy is called *radiometry*. In radiometry, the flux density is called the *irradiance* with units of W/m^2. Another area of optics is interested in the psychophysical effects of light and the measurement of energy is called *photometry*. For this group, the flux density is called *illuminance* with units of lumen/m^2 or lux. Each of these two group has its own set of units to measure the energy flow of a field that is not well defined in frequency or phase. Much of the research in modern optics belongs to a third area of optics that is associated with the use of a light source that has both a well-defined frequency and phase—the laser. For this area of optics, common usage defines the flux density as the intensity. In this book, all of the waves discussed are uniquely defined in terms of the electric field and the electromagnetic properties of the material in which the wave is propagating. To emphasize that the results of our theory are only immediately applicable to a light source with a well-defined frequency and phase, we will use the term intensity for the magnitude of the Poynting vector.

We will assume that $\mathbf{k}$ is independent of time over the period T

$$\langle\mathbf{S}\rangle = \frac{\mathrm{A}}{\omega T}\int_{t_0\omega}^{(t_0+T)\omega} \cos^2(\omega t - \mathbf{k}\cdot\mathbf{r} + \phi)\; d\,(\omega t)$$

Using the trig identity

$$\cos^2\theta = \frac{1}{2}(1 + \cos 2\theta)$$

and evaluating the integral result in the expression

$$\langle\mathbf{S}\rangle = \frac{\mathrm{A}}{2} + \frac{\mathrm{A}}{4\omega T}\left[\sin 2(\omega t_0 + \omega T - \mathbf{k}\cdot\mathbf{r} + \phi) - \sin 2(\omega t_0 - \mathbf{k}\cdot\mathbf{r} + \phi)\right] \tag{2-25}$$

The largest value the term in brackets can assume is 2. The period T is the response time of the detector to the light wave. Normally, it is much longer than the period of light oscillations so that $\omega T >> 1$ and we can neglect the second term of **(2-25)**. As an example, suppose our detection system has a 1 GHz bandwidth yielding a response time of $T = 10^{-9}$ sec (the reciprocal of the bandwidth). Green light has a frequency of $\nu = 6 \times 10^{14}$ Hz or $\omega \approx 4 \times 10^{15}$. With these values, $\omega T = 4 \times 10^5$ and the neglected term would be no larger than 10^{-6} of the first term. Therefore, in optics the assumption that $\omega T >> 1$ is reasonable and allows the average Poynting vector to be written as

$$\langle \mathbf{S} \rangle = \frac{\mathbf{A}}{2} = \frac{n}{2\mu c}\left|\mathbf{E}_0\right|^2 \frac{\mathbf{k}}{k} \tag{2-26}$$

Just as we saw in our discussion of the vibrating string, the energy per unit time per unit area depends on the square of the amplitude of the wave.

The calculation was made with plane waves of **E** and **H** that are in phase. We will later see that materials, where the conductivity $\sigma \neq 0$, will yield a complex impedance because **E** and **H** are no longer in phase. If the two waves are 90° out of phase, then the integral in **(2-24)** will contain $\sin x \cos x$ as its integrand, resulting in $\langle \mathbf{S} \rangle = 0$. Therefore, no energy is transmitted.

The energy crossing a unit area A in time Δt is contained in a volume $A(v\Delta t)$ (in a vacuum $v = c$) as shown in Figure 2-2. To find the magnitude of this energy, we must multiply this volume by the average energy density $\langle U \rangle$. Thus we expect the energy flow to be given by

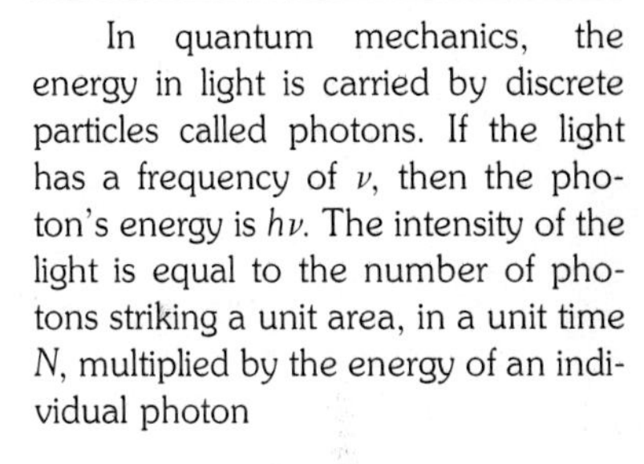

In quantum mechanics, the energy in light is carried by discrete particles called photons. If the light has a frequency of ν, then the photon's energy is $h\nu$. The intensity of the light is equal to the number of photons striking a unit area, in a unit time N, multiplied by the energy of an individual photon

$$I = Nh\nu$$

$$\left|\langle \mathbf{S} \rangle\right| = \frac{\text{energy}}{A\Delta t} \propto \frac{Av\Delta t \langle U \rangle}{A\Delta t} = v\langle U \rangle$$

We may use the definitions of the wave velocity

$$v = \frac{1}{\sqrt{\mu\epsilon}}$$

and index of refraction $n = c/v$ to rewrite **(2-26)**

$$\left|\langle \mathbf{S} \rangle\right| = \frac{\epsilon v E_0^2}{2} = v\langle U \rangle \tag{2-27}$$

giving the expected result that the energy is flowing through space at the speed of light in the medium. The relationship defined by **(2-27)**,

$$(\text{energy flow}) = (\text{wave velocity}) \cdot (\text{energy density})$$

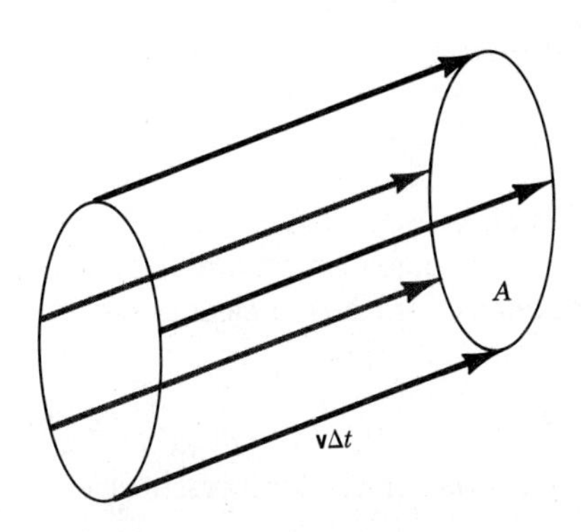

FIGURE 2-2. The energy or momentum of a wave crossing a unit area A in a time Δt.

is a general property of waves and could have been derived using the string model of Chapter 1.

At the Earth's surface, the flux density of full sunlight is 1.34×10^3 J/(m^2·sec). It is not completely correct, but we will associate this flux with the time average of the Poynting vector; the electric field associated with the sunlight is then $E_0 = 10^3$ V/m.

Our discussion of the time average of the Poynting vector provides an opportunity to discover one of the advantages of the use of complex notation. To obtain the time average of the product of two waves A and B, where

$$A = \mathcal{Re}\{\mathcal{A}\} = \mathcal{Re}\{A_0 e^{i(\omega t + \phi_1)}\}$$

$$B = \mathcal{Re}\{\mathcal{B}\} = \mathcal{Re}\{B_0 e^{i(\omega t + \phi_2)}\}$$

use **(1B-6)** to write the average over one period as

$$\langle AB \rangle = \frac{1}{T}\int_0^T \left(\frac{\mathcal{A} + \mathcal{A}^*}{2}\right)\left(\frac{\mathcal{B} + \mathcal{B}^*}{2}\right) dt$$

$$(\mathcal{A} + \mathcal{A}^*)(\mathcal{B} + \mathcal{B}^*) = \mathcal{A}\mathcal{B} + \mathcal{A}^*\mathcal{B}^* + \mathcal{A}\mathcal{B}^* + \mathcal{A}^*\mathcal{B}$$

where

$$\mathcal{A}\mathcal{B} = A_0 B_0 e^{i(2\omega t + \phi_1 + \phi_2)}$$

and

$$\mathcal{A}^*\mathcal{B}^* = A_0 B_0 e^{-i(2\omega t + \phi_1 + \phi_2)}$$

The time averages of the latter two terms are zero and we are left with

$$\langle AB \rangle = \frac{1}{T}\int_0^T \frac{\mathcal{A}\mathcal{B}^* + \mathcal{A}^*\mathcal{B}}{4} dt$$

From **(1B-6)**, we may rewrite this as

$$\langle AB \rangle = \frac{1}{2}\mathcal{Re}\left\{\mathcal{A}\mathcal{B}^*\right\} \tag{2-28}$$

The reader may find this quite general relation easier to use than performing an integration such as **(2-24)**.

MOMENTUM

The origin of momentum, associated with an electromagnetic wave, is easier to understand than is the source of momentum associated with an abstract wave **(1-9)**. The electric field of the electromagnetic wave acts on a charged particle in the material with a force

$$\mathbf{F}_E = q\mathbf{E} \tag{2-29}$$

This force accelerates the charged particle to a velocity v in a direction transverse to the direction of light propagation and parallel to the electric field. The moving charges interact with the magnetic field of the electromagnetic wave with a force, parallel to the propagation vector, of

$$\mathbf{F}_H = q(\mathbf{v} \times \mathbf{B}) \tag{2-30}$$

The combined action of these two forces creates a radiation pressure.

Although the actual derivation will not be carried out here, it is possible, by the comparison of the sum of these two forces $\mathbf{F}_E + \mathbf{F}_H$ (called the *Lorentz force*) with results that can be derived from Maxwell's theory,[3] to postulate a momentum density associated with the electromagnetic wave, given by

$$\mathbf{g} = \frac{\mathbf{S}}{c^2} \tag{2-31}$$

A dimensional analysis can be used to verify that **(2-31)** is a momentum density; the units of $\mathbf{g}$ are

$$\frac{\text{J/(m}^2\cdot\text{sec)}}{(\text{m/sec})^2} = \frac{\text{kgm}\cdot\text{m/sec}}{\text{m}^3}$$

The pressure on a surface of area A is defined as

$$P = \frac{\mathbf{F}\bullet\hat{\mathbf{n}}}{A} = \frac{\dfrac{\Delta\mathbf{p}}{\Delta t}\bullet\hat{\mathbf{n}}}{A}$$

We assume the light is totally absorbed, i.e., the momentum change is equal to the total momentum contained in the light wave, $\Delta\mathbf{p} = \mathbf{p}$. The total momentum in the light wave is given by the momentum density $\mathbf{g}$, multiplied by a unit volume $V, \mathbf{p} = \mathbf{g}V$.

$$P = \frac{\dfrac{V\mathbf{g}}{\Delta t}\bullet\hat{\mathbf{n}}}{A}$$

We will choose a volume $c\Delta t$ long with a cross-sectional area of A (see Figure 2-2), enabling the pressure to be expressed as

$$P = \frac{\left(\dfrac{\mathbf{g}\bullet\hat{\mathbf{n}}}{\Delta t}\right) c\cdot\Delta t\cdot A}{A} = \frac{\mathbf{S}\bullet\hat{\mathbf{n}}}{c} = \frac{I}{c} \tag{2-32}$$

At the Earth's surface and normal to it, sunlight has a flux density of 1.34×10^3 J/(m^2·sec). We will again make the incorrect assumption that the flux density of sunlight is equal to the Poynting vector, allowing the use of **(2-32)** to estimate the pressure of sunlight to be

$$P = 4.46\times10^{-6}\frac{\text{N}}{\text{m}^2}$$

As a point of reference, atmospheric pressure is about 10^5 N/m^2.

Substituting **(2-27)** into **(2-32)**, we discover that the radiation pressure is equal to the energy density of the incident radiation

$$P = \langle U\rangle$$

Combining **(2-27)** with **(2-31)** suggests that it is proper to associate momentum with the ratio of wave energy to velocity. This is consistent with relativistic principles. In the theory of relativity, the energy is given by

$$U = mc^2$$

which implies a mass of U/c^2 and a momentum of U/c. The idea is also consistent with quantum theory, where $U = h\nu$ so that

$$p = \frac{h}{\lambda} = \frac{h\nu}{c} = \frac{U}{c}$$

POLARIZATION

The displacement of a transverse wave is a vector quantity. We must therefore specify not only the frequency, phase, and direction of the wave but also the magnitude and direction of the displacement. The direction of the displacement vector is called the *direction of polarization* and the plane containing the direction of polarization and the propagation vector is called the *plane of polarization*. This quantity has the same name as the field quantity introduced in **(2-5)**. Because the two terms describe completely different physical phenomena, there should be no danger of confusion.

From our study of Maxwell's equations, we know that **E** and **H**, for a plane wave in free space, are mutually perpendicular and lie in a plane normal to the direction of propagation **k**. We also know that, given one of the two vectors, we can use **(2-17)** to obtain the other. Convention requires that we use the electric vector to label the direction of the electromagnetic wave's polarization. The selection of the electric field is not completely arbitrary. From **(2-29)** and **(2-30)**, we can write the ratio of the forces on a moving charge in an electromagnetic field due to the electric and magnetic fields as

$$\frac{F_E}{F_H} = \frac{eE}{evB}$$

We can replace B, using **(2-19)** to obtain

$$\frac{F_E}{F_H} = \frac{c}{nv} \tag{2-33}$$

where v is the velocity of the moving charge. Assume that a charged particle is traveling in air at the speed of sound so that $v = 335$ m/sec; then the force due to the electric field of a light wave on that particle would be 8.9×10^5 times larger than the force due to the magnetic field. The size of these numbers demonstrates that except in relativistic situations, when $v \approx c$, the interaction of the electromagnetic wave with matter will be dominated by the electric field.

A conventional vector notation is used to describe the polarization of a light wave; however, to visualize the behavior of the electric field vector as light propagates, a geometrical construction is useful. The geometrical construction, called a Lissajous' figure, describes the path followed by the tip of the electric field vector.

Polarization Ellipse

Assume that a plane wave is propagating in the z direction and the electric field, determining the direction of polarization, is oriented in the x, y plane. In complex notation, the plane wave is given by

$$\mathrm{E} = \mathbf{E}_0 e^{i(\omega t - \mathbf{k}\cdot\mathbf{r} + \phi)} = \mathbf{E}_0 e^{i(\omega t - kz + \phi)}$$

This wave can be written in terms of the x and y components of $\mathbf{E}_0$

$$\mathrm{E} = E_{0x} e^{i(\omega t - kz + \phi_1)}\,\hat{\mathbf{i}} + E_{0y} e^{i(\omega t - kz + \phi_2)}\,\hat{\mathbf{j}} \tag{2-34}$$

(We will use only the real part of **E** for manipulation to prevent errors.) We divide each component of the electric field by its maximum value so that the problem is reduced to one of the following two sinusoidally varying unit vectors:

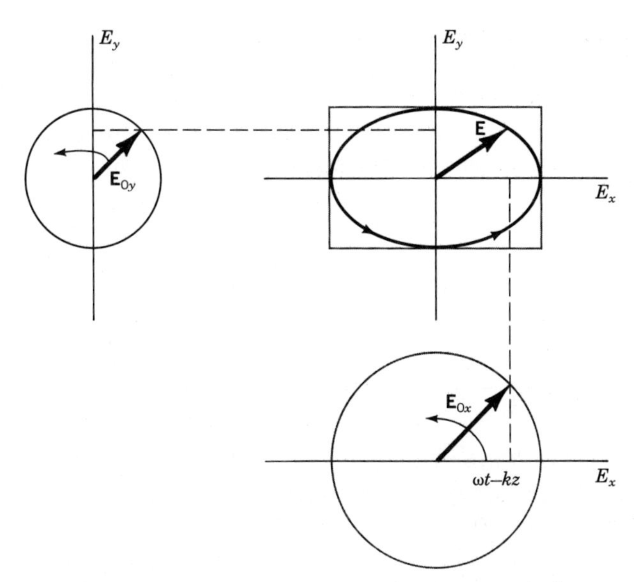

FIGURE 2-3. Geometrical construction showing how the Lissajous' figures are constructed from harmonic motion along the x and y coordinate axes. The harmonic motion along each coordinated axis is created by projecting a vector rotating around a circle onto the axis according to the technique discussed in Appendix 1B-1.

$$\frac{E_x}{E_{0x}} = \cos(\omega t - kz + \phi_1) = \cos(\omega t - kz)\cos\phi_1 - \sin(\omega t - kz)\sin\phi_1$$

$$\frac{E_y}{E_{0y}} = \cos(\omega t - kz)\cos\phi_2 - \sin(\omega t - kz)\sin\phi_2$$

When these unit vectors are added together, the result will be a set of figures called *Lissajous' figures*. The geometrical construction shown in Figure 2-3 can be used to visualize the generation of the Lissajous' figure. The harmonic motion along the x axis is found by projecting a vector rotating around a circle of diameter E_{0x} onto the x axis. The harmonic motion along the y axis is generated the same way using a circle of diameter E_{0y}. The resulting x and y components are added to obtain **E**. In Figure 2-3, the two harmonic oscillators both have the same frequency $(\omega t - kz)$, but differ in phase by

$$\delta = \phi_2 - \phi_1 = -\frac{\pi}{2}$$

The tip of the electric field **E** in Figure 2-3 traces out an ellipse, with its axes aligned with the coordinate axes. To determine the direction of the rotation of the vector, assume that $\phi_1 = 0, \phi_2 = -\pi/2$, and $z = 0$ so that

$$\frac{E_x}{E_{0x}} = \cos\omega t \qquad \frac{E_y}{E_{0y}} = \sin\omega t$$

$$\mathbf{E} = \left(\frac{E_x}{E_{0x}}\right)\hat{\mathbf{i}} + \left(\frac{E_y}{E_{0y}}\right)\hat{\mathbf{j}}$$

The normalized vector **E** can easily be evaluated at a number of values of ωt to discover the direction of rotation. Table 2.3 shows the value of the vector

TABLE 2.3 Rotating E-Field Vector

ωt	$\mathbf{E}$
0	$\hat{\mathbf{i}}$
$\frac{\pi}{4}$	$\frac{1}{\sqrt{2}}(\hat{\mathbf{i}}+\hat{\mathbf{j}})$
$\frac{\pi}{2}$	$\hat{\mathbf{j}}$
$\frac{3\pi}{4}$	$\frac{1}{\sqrt{2}}(-\hat{\mathbf{i}}+\hat{\mathbf{j}})$
π	$-\hat{\mathbf{i}}$

as ωt increases. The rotation of the vector $\mathbf{E}$ in Figure 2-3 is seen to be in a counterclockwise direction, moving from the positive x direction, to the y direction, and finally to the negative x direction.

To obtain the equation for the Lissajous' figure, we eliminate the dependence of the unit vectors on $(\omega t - kz)$. First, multiply the equations by $\sin \phi_2$ and $\sin \phi_1$, respectively, and then subtract the resulting equations. Second, multiply the two equations by $\cos \phi_2$ and $\cos \phi_1$, respectively, and then subtract the new equations. These two operations yield the following pair of equations:

$$\frac{E_x}{E_{0x}} \sin \phi_2 - \frac{E_y}{E_{0y}} \sin \phi_1 = \cos(\omega t - kz)(\cos \phi_1 \sin \phi_2 - \sin \phi_1 \cos \phi_2)$$

$$\frac{E_x}{E_{0x}} \cos \phi_2 - \frac{E_y}{E_{0y}} \cos \phi_1 = \sin(\omega t - kz)(\cos \phi_1 \sin \phi_2 - \sin \phi_1 \cos \phi_2)$$

The term in parens can be simplified using the trig identity

$$\sin \delta = \sin(\phi_2 - \phi_1) = \cos \phi_1 \sin \phi_2 - \sin \phi_1 \cos \phi_2$$

After replacing the term in parens by $\sin \delta$, the two equations are squared and added, yielding the equation for the Lissajous' figure

$$\left(\frac{E_x}{E_{0x}}\right)^2 + \left(\frac{E_y}{E_{0y}}\right)^2 - \left(\frac{2E_xE_y}{E_{0x}E_{0y}}\right) \cos \delta = \sin^2 \delta \tag{2-35}$$

The trig identity

$$\cos \delta = \cos(\phi_2 - \phi_1) = \cos \phi_1 \cos \phi_2 + \sin \phi_1 \sin \phi_2$$

was also used to further simplify **(2-35)**.

Equation **(2-35)** has the same form as the equation of a conic

$$Ax^2 + Bxy + Cy^2 + Dx + Ey + F = 0$$

Geometry defines the conic as an ellipse because from **(2-35)**,

$$B^2 - 4AC = \frac{4}{E_{0x}^2 E_{0y}^2}(\cos^2 \delta - 1) < 0$$

This ellipse is called the *polarization ellipse*. The orientation of the ellipse with respect to the x axis is

$$\tan 2\theta = \frac{B}{A - C} = \frac{2E_{0x}E_{0y} \cos \delta}{E_{0x}^2 - E_{0y}^2} \tag{2-36}$$

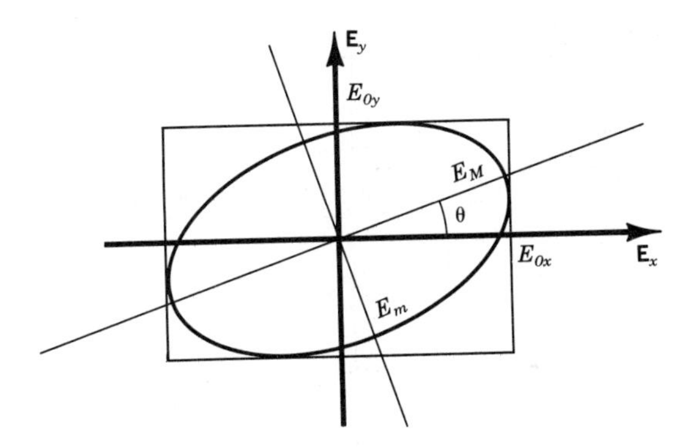

FIGURE 2-4. General form of the ellipse described by **(2-35)**.

If $A = C$ and $B \neq 0$, then $\theta = 45°$. When $\delta = \pm\pi/2$, then $\theta = 0°$ as shown in Figure 2-3.

The tip of the resultant electric field vector obtained from **(2-34)** traces out the polarization ellipse in the plane normal to **k**, as predicted by **(2-35)**. A generalized polarization ellipse is shown in Figure 2-4. The x and y coordinates of the electric field are bounded by $\pm E_{0x}$ and $\pm E_{0y}$. The rectangle in Figure 2-4 illustrates those limits. The component of the electric field along the major axis of the ellipse is

$$E_M = E_x \cos\theta + E_y \sin\theta$$

and along the minor axis of the ellipse is

$$E_m = -E_x \sin\theta + E_y \cos\theta$$

where θ is obtained from **(2-36)**. The ratio of the length of the minor to the major axis of the ellipse is equal to the ellipticity φ, i.e., the amount of deviation of the ellipse from a circle

$$\tan\varphi = \pm\left(\frac{E_m}{E_M}\right) = \frac{E_{0x}\sin\phi_1\sin\theta - E_{0y}\sin\phi_2\cos\theta}{E_{0x}\cos\phi_1\cos\theta + E_{0y}\cos\phi_2\sin\theta} \qquad (2\text{-}37)$$

To find the time dependence of the vector **E**, rewrite **(2-34)** in complex form

$$\mathbf{E} = e^{i(\omega t - kz)}(\hat{\mathbf{i}}E_{0x}e^{i\phi_1} + \hat{\mathbf{j}}E_{0y}e^{i\phi_2}) \qquad (2\text{-}38)$$

This equation shows explicitly that the electric vector moves about the ellipse in a sinusoidal motion.

By specifying the parameters that characterize the polarization ellipse (θ and φ), we completely characterize a wave's polarization. A review of two special cases will aid in understanding the polarization ellipse.

Linear Polarization

First consider when $\delta = 0$ or π; then **(2-35)** becomes

$$\left(\frac{E_x}{E_{0x}}\right)^2 + \left(\frac{E_y}{E_{0y}}\right)^2 \pm \left(\frac{2E_xE_y}{E_{0x}E_{0y}}\right) = 0$$

The ellipse collapses into a straight line with slope E_{0y}/E_{0x}. The equation of the straight line is

$$\frac{E_x}{E_{0x}} = \pm\frac{E_y}{E_{0y}}$$

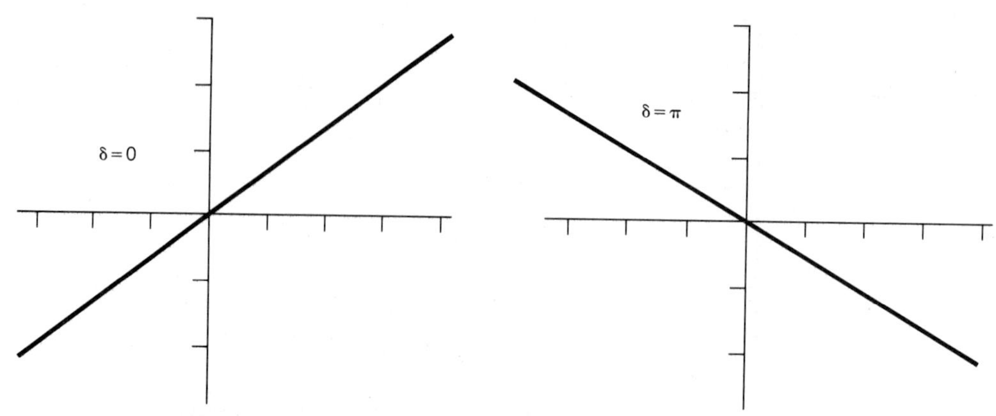

FIGURE 2-5. Lissajous' figures for phase differences between the y and x components of oscillation of 0 and π.

Figure 2-5 displays the straight-line Lissajous' figures for the two phase differences. The θ parameter of the ellipse is the slope of the straight line

$$\tan\,\theta = \frac{E_{0y}}{E_{0x}}$$

resulting in the value of **(2-36)** being given by

$$\tan\,2\theta = \frac{2\,\tan\,\theta}{1 - \tan^2\,\theta} = \frac{2E_{0x}E_{0y}}{E_{0x}^2 - E_{0y}^2}$$

The φ parameter is given by **(2-37)** as $\tan\varphi = 0$.

The time dependence of the **E** vector shown in Figure 2-5 is given by **(2-38)**. The real component is

$$\mathbf{E} = (E_{0x}\hat{\mathbf{i}} \pm E_{0y}\hat{\mathbf{j}})\cos(\omega t - kz)$$

At a fixed point in space, the x and y components oscillate in phase (or 180° out of phase) according to the equation

$$\mathbf{E} = (E_{0x}\hat{\mathbf{i}} \pm E_{0y}\hat{\mathbf{j}})\cos(\omega t - \phi)$$

The electric vector undergoes simple harmonic motion along the line defined by E_{0x} and E_{0y}. At a fixed time, the electric field varies sinusoidally along the propagation path (the z axis) according to the equation

$$\mathbf{E} = (E_{0x}\hat{\mathbf{i}} \pm E_{0y}\hat{\mathbf{j}})\cos(\phi - kz)$$

This light is said to be *linearly polarized.*

Circular Polarization

The second case occurs when $E_{0x} = E_{0y} = E_0$ and $\delta = \pm\pi/2$. From **(2-35)**,

$$\left(\frac{E_x}{E_0}\right)^2 + \left(\frac{E_y}{E_0}\right)^2 = 1$$

The ellipse becomes a circle as shown in Figure 2-6. For this polarization, $\tan\,2\theta$ is indeterminate and $\tan\,\varphi = 1$.

From **(2-38)**, the temporal behavior is given by

$$\mathbf{E} = E_0[\cos(\omega t - kz)\,\hat{\mathbf{i}} \pm \sin(\omega t - kz)\,\hat{\mathbf{j}}\,]$$

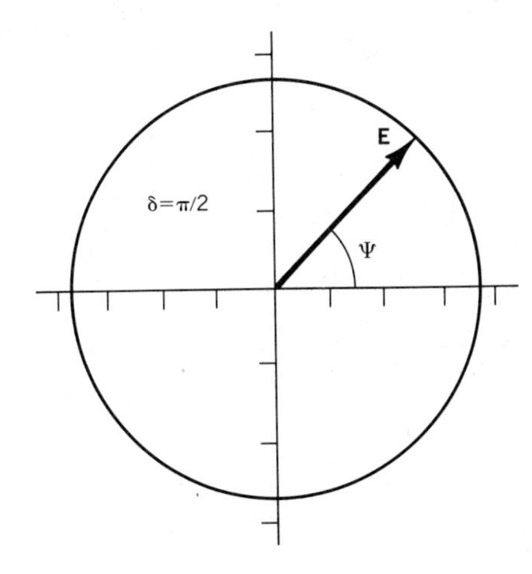

FIGURE 2-6. Lissajous' figures for the case when the phase difference between the y and x components of oscillation differ by $\pm(\pi/2)$ and the amplitudes of the two components are equal. The tip of the electric field vector shown moves along the circle.

The time dependence of the angle Ψ that the **E** field makes with the x axis in Figure 2-6 can be obtained by finding the tangent of the angle Ψ.

$$\tan \Psi = \frac{E_y}{E_x} = \pm \frac{\sin(\omega t - kz)}{\cos(\omega t - kz)} = \pm \tan(\omega t - kz)$$

The interpretation of this result is that at a fixed point in space, the **E** vector rotates in a clockwise direction if $\delta = \pi/2$ and a counterclockwise direction if $\delta = -\pi/2$.

In particle physics, the light would be said to have a negative helicity if it rotated in a clockwise direction. If we look at the source, the electric vector seems to follow the threads of a left-handed screw, agreeing with the nomenclature that left-handed quantities are negative. However, in optics the light that rotates clockwise as we view it traveling toward us from the source is said to be *right-circularly polarized*. The counterclockwise rotating light is *left-circularly polarized*.

The association of right-circularly polarized light with "right handedness" in optics came about by looking at the path of the electric vector in space at a fixed time; then, $\tan \Psi = \tan(\phi - kz)$. See Figure 2-7. As shown in Figure 2-7, right-circular polarized light at a fixed time seems to spiral in a counterclockwise fashion along the z direction, following the threads of a right-handed screw.

This motion can be generalized to include elliptical polarized light when $E_{0x} \neq E_{0y}$. Figure 2-3 schematically displays the generation of the Lissajous' figure for the case of $\delta = \pi/2$, but with unequal values of E_{0x} and E_{0y}. Figure 2-8 shows two calculated Lissajous' figures. If the electric vector moves around the ellipse in a clockwise direction, as we face the source, then the phase difference and ellipticity are

$$0 \leq \delta \leq \pi \quad \text{and} \quad 0 < \varphi < \frac{\pi}{4}$$

and the polarization is right-handed. If the motion of the electric vector is moving in a counterclockwise direction, then the phase difference and ellipticity are

$$-\pi \leq \delta \leq 0 \quad \text{and} \quad -\frac{\pi}{4} < \varphi < 0$$

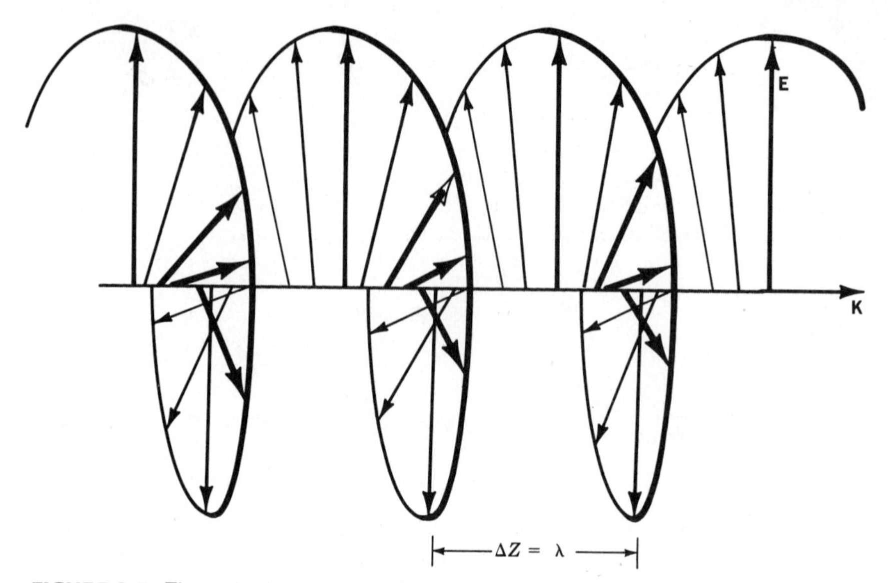

FIGURE 2-7. The path of the electric vector of right-circular polarized light at a fixed time.

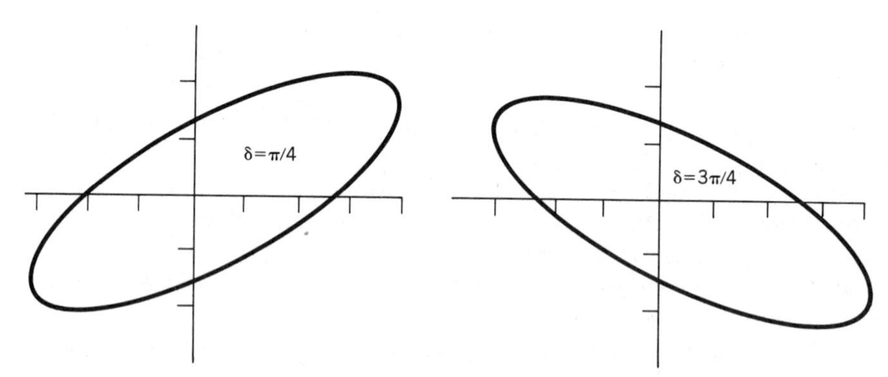

FIGURE 2-8. Lissajous' figures for elliptical polarized light. They were calculated with $E_{0x} = 0.75$ and $E_{0y} = 0.25$.

The orientation of either ellipse with respect to the x axis will be given by **(2-36)** and will depend upon the relative magnitudes of E_{0x} and E_{0y}.

The procedure used to decompose an arbitrary polarization into polarizations parallel to two axes of a Cartesian coordinate system is a technique used extensively in vector algebra to simplify mathematical calculations. According to the mathematical formalism associated with this technique, the polarization is described in terms of a set of basis vectors $\mathbf{e}_i$. An arbitrary polarization would be expressed as

$$\mathbf{E} = \sum_{i=1}^{2} a_i \mathbf{e}_i \qquad (2\text{-}39)$$

The set of basis vectors $\mathbf{e}_i$ is orthonormal, i.e.,

$$\mathbf{e}_i\mathbf{e}_j^* = \delta_{ij} = \begin{cases} 1, & i = j \\ 0, & i \neq j \end{cases}$$

where we have assumed that the basis vectors could be complex. We mention this mathematical formalism because an identical formalism is encountered in elementary particle physics in which it is used to describe spin.[4]

In a Cartesian coordinate system, the $\mathbf{e}_i$'s are the unit vectors $\hat{\mathbf{i}}$, $\hat{\mathbf{j}}$, $\hat{\mathbf{k}}$. The summation in **(2-39)** extends over only two terms because the electromagnetic wave is transverse, confining **E** to a plane normal to the direction of propagation (according to the coordinate convention we have selected, the **E** field is in the x, y plane).

The polarization could also be described in terms of a right-circularly polarized component

$$\mathbf{E}_{\mathcal{R}} = E_{0\mathcal{R}}\left[\hat{\mathbf{i}} \cos(\omega t - kz) - \hat{\mathbf{j}} \sin(\omega t - kz)\right]$$

and a left-circularly polarized component

$$\mathbf{E}_{\mathcal{L}} = E_{0\mathcal{L}}\left[\hat{\mathbf{i}} \cos(\omega t - kz) + \hat{\mathbf{j}} \sin(\omega t - kz)\right]$$

An arbitrary elliptical polarization would then be written as

$$\begin{aligned} \mathbf{E} &= \mathbf{E}_{\mathcal{R}} + \mathbf{E}_{\mathcal{L}} \\ &= \hat{\mathbf{i}}\,(\mathbf{E}_{0\mathcal{R}} + \mathbf{E}_{0\mathcal{L}}) \cos(\omega t - kz) - \hat{\mathbf{j}}\,(\mathbf{E}_{0\mathcal{R}} - \mathbf{E}_{0\mathcal{L}}) \sin(\omega t - kz) \end{aligned} \qquad (2\text{-}40)$$

The geometrical construction that demonstrates the expression of an arbitrary elliptical polarized light wave in terms of right and left circularly polarized waves is shown in Figure 2-9. The use of circular polarized waves as the basis set for describing polarization is discussed by Klein.[5]

In the formalism associated with **(2-39)**, the expansion coefficients a_i can be used to form a 2–2 matrix, which in statistical mechanics is called the *density matrix* and in optics the *coherency matrix*.[6] The elements of the matrix are formed by the rule

$$\rho_{ij} = \mathbf{a}_i\mathbf{a}_j^*$$

We will not develop the theory of polarization using the coherency matrix, but simply use the coherency matrix to justify the need for four independent measurements to characterize polarization. There is no unique set of measurements required by theory but normally measurements made are of the *Stokes parameters*, which are directly related to the polarization ellipse of Figure 2-4. (We will see in a few moments that only three of the four measurements are independent. This will be in agreement with the definition of the coherency matrix where $\rho_{ij} = \rho_{ji}^*$, i.e., the matrix is Hermitian.)

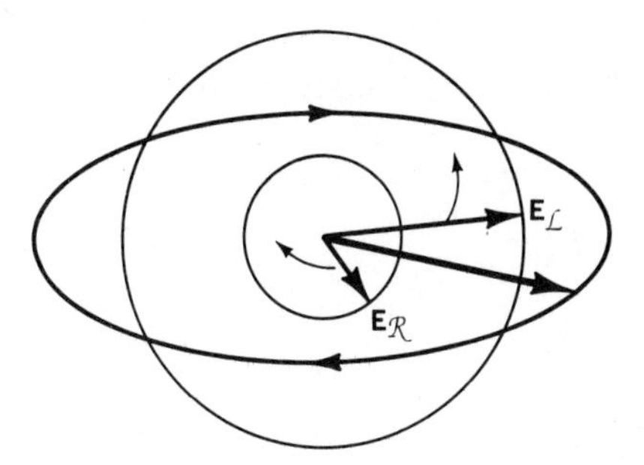

FIGURE 2-9. Construction of elliptical polarized light from two circularly polarized waves.

STOKES PARAMETERS

The Stokes parameters[7] of a light wave are measurable quantities, defined as

$s_0 \rightarrow$ Total flux density.

$s_1 \rightarrow$ Difference between flux density transmitted by a linear polarizer oriented parallel to the x axis and one oriented parallel to the y axis. The x and y axes are usually selected to be parallel to the horizontal and vertical directions in the laboratory.

$s_2 \rightarrow$ Difference between flux density transmitted by a linear polarizer oriented at 45° to the x axis and one oriented at 135°.

$s_3 \rightarrow$ Difference between flux density transmitted by a right-circular polarizer and a left-circular polarizer.

The physical instruments that can be used to measure the Stokes parameters will be discussed in Chapter 13.

If the Stokes parameters are to characterize the polarization of a wave, they must be related to the parameters of the polarization ellipse. It is therefore important to establish that the Stokes parameters are variables of the polarization ellipse **(2-35)**.

In its current form, **(2-35)** contains no measurable quantities and thus must be modified if it is to be associated with the Stokes parameters. In the discussion of the Poynting vector, it was pointed out that the time average of the Poynting vector is the quantity observed when measurements are made of light waves. We must, therefore, find the time average of **(2-35)** if we wish to relate its parameters to observable quantities. To simplify the discussion, assume that the amplitudes of the orthogonally polarized waves, E_{0x} and E_{0y} and their relative phase δ are constants. We will also use the shorthand notation for a time average introduced in **(2-24)**

$$\langle E_x^2 \rangle = \frac{1}{T}\int_{t_0}^{t_0+T} E_{0x}^2\left[\cos(\omega t - kz)\cos\phi_1 - \sin(\omega t - kz)\sin\phi_1\right]^2 dt$$

The time average of **(2-35)** can now be written

$$\frac{\langle E_x^2 \rangle}{E_{0x}^2} + \frac{\langle E_y^2 \rangle}{E_{0y}^2} - 2\frac{\langle E_x E_y \rangle}{E_{0x}E_{0y}}\cos\delta = \sin^2\delta \tag{2-41}$$

Multiplying both sides of **(2-41)** by $(2E_{0x}E_{0y})^2$ removes the terms in the denominators of **(2-41)**

$$4E_{0y}^2\langle E_x^2 \rangle + 4E_{0x}^2\langle E_y^2 \rangle - 8E_{0x}E_{0y}\langle E_x E_y \rangle\cos\delta = (2E_{0x}E_{0y}\sin\delta)^2$$

The same argument that was used to simplify **(2-25)** can be used to obtain the time averages for the first two terms

$$\langle E_x^2 \rangle = \frac{E_{0x}^2}{2}, \qquad \langle E_y^2 \rangle = \frac{E_{0y}^2}{2}$$

The calculation of the time average in the third term

$$\langle E_x E_y \rangle = \frac{1}{2}E_{0x}E_{0y}\cos\delta \tag{2-42}$$

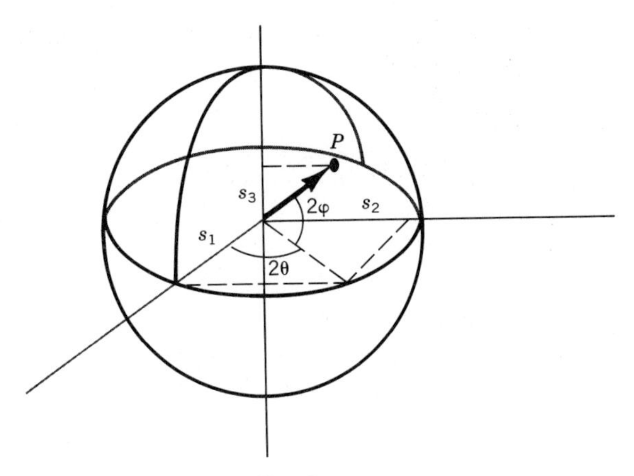

FIGURE 2-10. Poincaré's sphere.

Before it was discovered that the Stokes parameters could be treated as elements of a column matrix, a geometric construction was used to determine the effect of an anisotropic medium on polarized light. The parameters s_1, s_2, s_3 are viewed as the Cartesian coordinates of a point on a sphere of radius s_0. This sphere is called the *Poincaré sphere*[8] and is shown in Figure 2-10.

On the sphere, right-hand polarized light is represented by points on the upper half-surface. Linear polarization is represented by points on the equator. Circular polarization is represented by the poles. With the development of the matrix view of polarization, the usefulness of the Poincaré's sphere has decreased and it is now, for many people, only of historical interest.

is left as a problem, Problem 2-12. With these time averages, **(2-41)** can be written as

$$4E_{0x}^2 E_{0y}^2 - (2E_{0x}E_{0y} \cos\,\delta)^2 = (2E_{0x}E_{0y} \sin\,\delta)^2$$

If $E_{0x}^2 + E_{0y}^2$ is added to both sides of this equation, it can be rewritten

$$(E_{0x}^2 + E_{0y}^2)^2 - (E_{0x}^2 - E_{0y}^2)^2 - (2E_{0x}E_{0y} \cos\,\delta)^2 = (2E_{0x}E_{0y} \sin\,\delta)^2 \quad (2\text{-}43)$$

Each term in this equation can be identified with a Stokes parameter.

In our derivation, we required that the amplitudes and relative phase of the two orthogonally polarized waves be a constant, but we can relax this requirement and instead define the Stokes parameters as temporal averages. With this modification, the terms of **(2-43)** become

$$\begin{aligned} s_0 &= \langle E_{0x}^2 \rangle + \langle E_{0y}^2 \rangle, & s_1 &= \langle E_{0x}^2 \rangle - \langle E_{0y}^2 \rangle \\ s_2 &= \langle 2E_{0x}E_{0y} \cos\,\delta \rangle, & s_3 &= \langle 2E_{0x}E_{0y} \sin\,\delta \rangle \end{aligned} \quad (2\text{-}44)$$

Equation **(2-43)** can now be written as

$$s_0^2 - s_1^2 - s_2^2 = s_3^2 \quad (2\text{-}45)$$

For a polarized wave, only three of the Stokes parameters are independent. This agrees with the requirement placed on elements of the Hermitian coherency matrix introduced above.

With this demonstration of the connection between the Stokes parameters and the polarization ellipse, the Stokes parameters can be written in terms of the parameters of the polarization ellipse in Figure 2-4.

$$\begin{aligned} s_1 &= s_0 \cos\,2\varphi \cos\,2\theta \\ s_2 &= s_0 \cos\,2\varphi \sin\,2\theta \\ s_3 &= s_0 \sin\,2\varphi \end{aligned} \quad (2\text{-}46)$$

It is this close relationship between the Stokes parameters and the polarization ellipse that makes the Stokes parameters a useful characterization of polarization.

The Stokes parameters can be used to describe the *degree of polarization* defined as

$$V = \frac{1}{s_0}\sqrt{s_1^2 + s_2^2 + s_3^2} \tag{2-47}$$

[The equality of **(2-45)** applies to completely polarized light when $V = 1$.] The degree of polarization can be used to characterize any light source that is physically realizable. If the time averages used in the definition of the Stokes parameters s_2 and s_3 are zero,

$$\langle E_{0x}^2 \rangle = \langle E_{0y}^2 \rangle \qquad \text{and} \qquad s_0 = 2\langle E_{0x}^2 \rangle$$

then the light wave is said to be unpolarized and $V = 0$.

H. Mueller[7] pointed out that the Stokes parameters can be thought of as elements of a column matrix or a 4-vector; see Table 2.4.

$$\begin{pmatrix} s_0 \\ s_1 \\ s_2 \\ s_3 \end{pmatrix}$$

TABLE 2.4 Jones and Stokes Vectors

Polarization	Jones Vector	Stokes Vector
Horizontal Polarization	$\begin{bmatrix} 1 \\ 0 \end{bmatrix}$	$\begin{bmatrix} 1 \\ 1 \\ 0 \\ 0 \end{bmatrix}$
Vertical Polarization	$\begin{bmatrix} 0 \\ 1 \end{bmatrix}$	$\begin{bmatrix} 1 \\ -1 \\ 0 \\ 0 \end{bmatrix}$
+45° Polarization	$\frac{1}{\sqrt{2}}\begin{bmatrix} 1 \\ 1 \end{bmatrix}$	$\begin{bmatrix} 1 \\ 0 \\ 1 \\ 0 \end{bmatrix}$
−45° Polarization	$\frac{1}{\sqrt{2}}\begin{bmatrix} 1 \\ -1 \end{bmatrix}$	$\begin{bmatrix} 1 \\ 0 \\ -1 \\ 0 \end{bmatrix}$
Right-Circular Polarization	$\frac{1}{\sqrt{2}}\begin{bmatrix} 1 \\ i \end{bmatrix}$	$\begin{bmatrix} 1 \\ 0 \\ 0 \\ 1 \end{bmatrix}$
Left-Circular Polarization	$\frac{1}{\sqrt{2}}\begin{bmatrix} 1 \\ -i \end{bmatrix}$	$\begin{bmatrix} 1 \\ 0 \\ 0 \\ -1 \end{bmatrix}$

This view will allow us to follow a polarized wave through a series of optical devices through the use of matrix algebra as we will see later.

JONES VECTOR

There is one other representation of polarized light, complementary to the Stokes parameters, developed by **R. Clark Jones** in 1941 and called the *Jones vector.* It is superior to the Stokes vector in that it handles light of a known phase and amplitude with a reduced number of parameters. It is inferior to the Stokes vector in that, unlike the Stokes representation, that is experimentally determined, the Jones representation cannot handle unpolarized or partially polarized light. The Jones vector is a theoretical construct that can only describe light with a well-defined phase and frequency. The density matrix formalism can be used to correct the shortcomings of the Jones vector, but then the simplicity of the Jones representation is lost.

If we assume that the coordinate system is such that the electromagnetic wave is propagating along the z axis, it was shown earlier that any polarization could be decomposed into two orthogonal $\mathbf{E}$ vectors, say for this discussion, parallel to the x and y directions. The Jones vector is defined as a two-row, column matrix consisting of the complex components in the x and y direction

$$\mathbf{E} = \begin{bmatrix} E_{0x} \exp\{i(\omega t - \mathbf{k}\cdot\mathbf{r} + \phi_1)\} \\ E_{0y} \exp\{i(\omega t - \mathbf{k}\cdot\mathbf{r} + \phi_2)\} \end{bmatrix} \tag{2-48}$$

If absolute phase is not an issue, then we may normalize the vector by dividing by that number (real or complex) that simplifies the components but keeps the sum of the square of the components equal to 1. For example,

$$\mathbf{E} = \frac{E_{0x}}{\sqrt{E_{0x}^2 + E_{0y}^2}} \exp[i(\omega t - \mathbf{k}\cdot\mathbf{r} + \phi_1)] \begin{bmatrix} 1 \\ \dfrac{E_{0x}}{E_{0y}} e^{i\delta} \end{bmatrix}$$

The normalized vector would be the terms contained within the bracket, each divided by $1/\sqrt{2}$ if $E_{0x} = E_{0y}$. The general form of the Jones vector is

$$\mathbf{E} = \begin{bmatrix} \mathbf{A} \\ \mathbf{B} \end{bmatrix}, \qquad \mathbf{E}^* = \begin{bmatrix} \mathbf{A}^* \ \mathbf{B}^* \end{bmatrix}$$

Some examples of Jones vectors (on the left) and Stokes vectors (on the right) are shown in Table 2.4.

PROPAGATION IN A CONDUCTING MEDIUM

In Chapter 1, we discussed the propagation of a wave with attenuation. In our discussion of the propagation of light, however, we have ensured that we would experience no loss by assuming $\sigma = 0$. We now relax that assumption and allow $\sigma \neq 0$. Maxwell's equations become

$$\nabla\cdot\mathbf{D} = 0, \qquad \nabla\cdot\mathbf{B} = 0$$

$$\nabla\times\mathbf{H} = \mathbf{J} + \frac{\partial\mathbf{D}}{\partial t}, \qquad \nabla\times\mathbf{E} = -\frac{\partial\mathbf{B}}{\partial t}$$

We continue to neglect dynamic or resonant effects so that we may use the simple constitutive relations

$$\mathbf{J} = \sigma\mathbf{E}, \qquad \mathbf{D} = \epsilon\mathbf{E}, \qquad \mathbf{B} = \mu\mathbf{H}$$

where ϵ, μ, and σ are independent of time. Maxwell's equations in a medium with dissipation can be rewritten using these constitutive relations as

$$\nabla \cdot \mathbf{E} = 0, \qquad \nabla \cdot \mathbf{H} = 0$$
$$\nabla \times \mathbf{H} = \sigma \mathbf{E} + \epsilon \frac{\partial \mathbf{E}}{\partial t}, \qquad \nabla \times \mathbf{E} = -\mu \frac{\partial \mathbf{H}}{\partial t} \tag{2-49}$$

We now apply the same procedure used to derive the wave equation for free space

$$\nabla \times (\nabla \times \mathbf{E}) = \nabla \times \left(-\mu \frac{\partial \mathbf{H}}{\partial t} \right) = -\mu \frac{\partial}{\partial t} (\nabla \times \mathbf{H})$$

$$-\mu \frac{\partial}{\partial t} (\nabla \times \mathbf{H}) = -\mu \frac{\partial}{\partial t} \left(\sigma \mathbf{E} + \epsilon \frac{\partial \mathbf{E}}{\partial t} \right)$$

$$\nabla \times (\nabla \times \mathbf{E}) = \nabla (\nabla \cdot \mathbf{E}) - \nabla^2 \mathbf{E}$$

yielding the wave equation in a conducting medium

$$\nabla^2 \mathbf{E} = \mu \sigma \frac{\partial \mathbf{E}}{\partial t} + \mu \epsilon \frac{\partial^2 \mathbf{E}}{\partial t^2} \tag{2-50}$$

This wave equation is of the same form as **(1-19)**. We can derive a similar equation for the magnetic field

$$\nabla^2 \mathbf{B} = \mu \sigma \frac{\partial \mathbf{B}}{\partial t} + \mu \epsilon \frac{\partial^2 \mathbf{B}}{\partial t^2} \tag{2-51}$$

Equations **(2-50)** and **(2-51)** are called the *telegraph equations*. They are wave equations derived to explain the propagation of pulses on telegraph lines.

We see that the wave equation **(2-50)** contains a damping term $\partial \mathbf{E}/\partial t$ when we allow $\sigma \neq 0$. By comparing **(2-50)** to **(1-19)**, we can state that the solutions of **(2-50)** will be electromagnetic waves that will experience attenuation proportional to $\mu\sigma$ as it propagates. Using **(1-21)** and **(1-22)**, we may rewrite **(2-49)**, for plane wave solutions, as

$$\nabla \times \mathbf{E} = -\mu \frac{\partial \mathbf{H}}{\partial t} \quad \rightarrow \quad \nabla \times \mathbf{E} = -i\omega\mu \mathbf{H}$$
$$\nabla \times \mathbf{H} = \sigma \mathbf{E} + \epsilon \frac{\partial \mathbf{E}}{\partial t} \quad \rightarrow \quad \nabla \times \mathbf{H} = i\omega \left(\epsilon - \frac{i\sigma}{\omega} \right) \mathbf{E} \tag{2-52}$$

We rewrite **(2-50)** in terms of these expressions for the curl of **E** and **H**

$$\nabla^2 \mathbf{E} + \omega^2 \mu \left(\epsilon - \frac{i\sigma}{\omega} \right) \mathbf{E} = 0 \tag{2-53}$$

This has the form of the Helmholtz equation **(1-12)** if we replace k^2 by the complex function

$$k^2 = \omega^2 \mu \left(\epsilon - i \frac{\sigma}{\omega} \right) \tag{2-54}$$

We use the identity

$$k = \frac{n\omega}{c} = \omega \sqrt{\mu\epsilon}$$

to demonstrate that the equations for conducting media are identical to those derived for nonconducting media if the dielectric constant ϵ is replaced by a complex dielectric constant

$$\tilde{\epsilon} = \epsilon - i\left(\frac{\sigma}{\omega}\right) \tag{2-55}$$

This equation suggests that σ may contain a frequency dependence (in fact, in the cgs system, the units of σ are sec^{-1}; for copper in cgs units, $\sigma = 5.14 \times 10^{17}/\text{sec}$). In solid-state physics, one finds that the mobility of the electrons creates a frequency dependence that shows up in σ.

Since we have replaced k by the complex quantity,

$$\mathcal{k} = \omega\sqrt{\mu\left(\epsilon - i\frac{\sigma}{\omega}\right)}$$

we must replace the index of refraction by a complex index. In the literature, this is accomplished in two ways

$$\mathcal{N} = n(1 - i\kappa) \tag{2-56}$$

$$\mathcal{N} = n_1 - in_2$$

We will use the notation displayed in **(2-56)**.

To find out how the plane wave propagates in this conductive medium, we simply replace the propagation constant k by

$$\mathcal{k} = \mathcal{N}\frac{\omega}{c} = \left(\frac{\omega n}{c}\right)(1 - i\kappa)$$

as we did in Chapter 2. κ is called the *extinction coefficient* and $n\kappa$ the *absorption coefficient.*

If we assume $\mathbf{k}$ is parallel to the z axis, then the plane wave is

$$\mathbf{E} = \mathbf{E}_0 e^{i\omega t} \exp\{-i\omega(n/c)(1 - i\kappa)z\}$$

$$\mathbf{E} = \mathbf{E}_0 \exp\{-(\omega n\kappa/c)z\} \exp\{i\omega(t - nz/c)\} \tag{2-57}$$

$$\mathcal{Re}\{\mathbf{E}\} = \mathbf{E}_0 \exp\{-(\omega n\kappa/c)z\} \cos(\omega t - kz) \tag{2-58}$$

The wave described by **(2-58)** is a plane wave, attenuated by the exponent

$$\exp\{-(\omega/c)n\kappa z\} \tag{2-59}$$

Figure 2-11 displays the exponential decay of a light wave propagating in an absorbing medium.

In Figure 2-11, a layer of xylene floats on water containing the dye Rhodamine 6G in solution. The Rhodamine strongly absorbs a beam of blue light from a HeCd laser (442 nm). As can be seen in Figure 2-11, the blue light is rapidly attenuated once it enters the water. Some of the energy absorbed by the Rhodamine is reemitted at longer wavelengths. The reemitted light travels in all directions, as it has no memory of the direction traveled by the blue light. For this reason, the beam of light in the water appears diffuse and, as can be seen in the color insert, is orange in color.

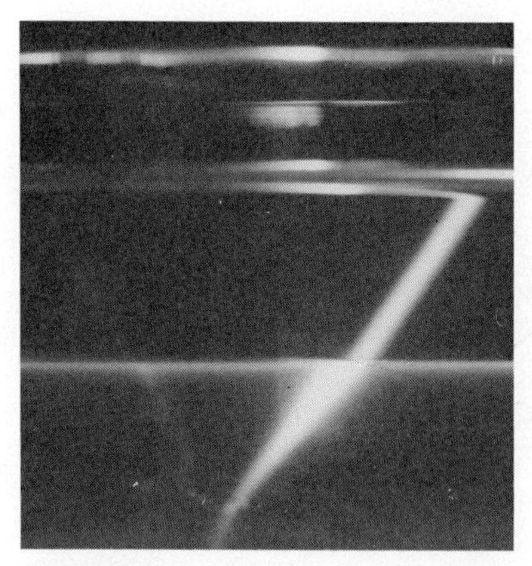

FIGURE 2-11. Blue laser light is shown propagating in xylene (above) and water (below). The water contains the dye Rhodamine 6G in solution. The red Rhodamine dye absorbs the blue light and the beam rapidly decays to zero. Some of the energy absorbed by the dye is reemitted in the yellow to red region of the spectrum. This reemitted light caused the diffuse appearance of the light as it propagates in the water. A color image of this figure can be seen in the insert.

To evaluate the absorption coefficient $n\kappa$ in terms of electromagnetic properties of the medium, we will derive a relationship between $n\kappa$ and σ. We rewrite **(2-56)** as

$$\mathcal{N}^2 = n^2(1 - \kappa^2 - 2i\kappa) = \frac{c^2}{\omega^2}\mathcal{k}^2$$

Equation **(2-54)** can be used to express $\mathcal{N}^2$ in terms of the constants of the material

$$\mathcal{N}^2 = c^2\mu\left(\epsilon - i\frac{\sigma}{\omega}\right) \tag{2-60}$$

Equating real and imaginary terms, we obtain

$$n^2(1-\kappa^2) = c^2\mu\epsilon, \qquad 2n^2\kappa = c^2\frac{\mu\sigma}{\omega}$$

We can use these two relationships to find

$$n^2 = \frac{c^2}{2}\left[\sqrt{\mu^2\epsilon^2 + \left(\frac{\mu\sigma}{\omega}\right)^2} + \mu\epsilon\right] \tag{2-61}$$

$$n^2\kappa^2 = \frac{c^2}{2}\left[\sqrt{\mu^2\epsilon^2 + \left(\frac{\mu\sigma}{\omega}\right)^2} - \mu\epsilon\right] \tag{2-62}$$

Note that when $\sigma = 0$, $\kappa = 0$, and we obtain the free space result **(2-11)**

$$n^2 = \frac{\mu\epsilon}{\mu_0\epsilon_0}$$

An estimate of the magnitude of the quantities under the radicals in **(2-61)** and **(2-62)** can be obtained by using values for copper, where in the mks units $\sigma = 5.8 \times 10^7$ mhos/m and $n = 0.62$ at $\lambda = 589.3$ nm. (The index of refraction is less than 1, which implies that the phase velocity is greater than the speed of light. This apparent contradiction of a fundamental postulate of the theory of relativity will be discussed during the study of dispersion in Chapter 7.) The two terms under the radical are

$$\frac{\mu\sigma}{\omega} = \frac{[(4\pi\times10^{-7})(5.8\times10^{7})(5.893\times10^{-7})]}{(2\pi)(3\times10^{8})} = 2.3\times10^{-14}\,\frac{\text{sec}^2}{\text{m}^2}$$

$$\mu\epsilon = \mu_0\epsilon_0 n^2 = (4\pi\times10^{-7})(8.8542\times10^{-12})(0.62)^2 = 4.3\times10^{-18}\,\frac{\text{sec}^2}{\text{m}^2}$$

By comparing the relative magnitude of these two terms, we are justified in assuming that $\sigma/\omega >> \epsilon$ and can make the approximation

$$n^2\kappa^2 \approx \frac{c^2\mu\sigma}{2\omega}$$

$$n\kappa = c\sqrt{\frac{\mu\sigma}{2\omega}} \tag{2-63}$$

We use **(2-63)** to find the depth at which an electromagnetic wave is attenuated to $1/e$ of its original energy when propagating into a conductor. At that depth, denoted by d, the exponent in **(2-59)** will equal 1; thus,

$$\frac{\omega}{c}n\kappa d = \frac{2\pi}{\lambda_0}n\kappa d = 1$$

$$d = \frac{\lambda_0}{2\pi n\kappa} \approx \frac{\lambda_0}{2\pi c}\sqrt{\frac{2\omega}{\mu\sigma}}$$

$$d = \sqrt{\frac{2}{\mu\sigma\omega}} \tag{2-64}$$

TABLE 2.5 Skin Depth for Copper

λ_0	d
10^{-7} m	6.2×10^{-10} m
10^{-5}	6.2×10^{-9} m
10^{-1}	6.2×10^{-7} m
10^{3}	6.2×10^{-5} m

The depth d is called the *skin depth*. The skin depth for copper at a number of wavelengths is shown in Table 2.5.

SUMMARY

In this chapter, Maxwell's equations were used to obtain the wave equation for free space

$$\nabla^2 \mathbf{E} = \mu\epsilon \frac{\partial^2 \mathbf{E}}{\partial t^2}$$

and a conductive medium

$$\nabla^2 \mathbf{E} = \mu\sigma \frac{\partial \mathbf{E}}{\partial t} + \mu\epsilon \frac{\partial^2 \mathbf{E}}{\partial t^2}$$

Although the form of the two equations appears quite different, we demonstrated that plane wave solutions existed for both equations when the dielectric constant of the conductive medium was replaced by a complex constant

$$\tilde{\epsilon} = \epsilon - i\left(\frac{\sigma}{\omega}\right)$$

This replacement means that the optical properties of a conductive material are described by a complex index of refraction

$$\mathcal{N} = n(1 - i\kappa)$$

The plane wave propagating in the conductive material has an amplitude attenuated by the exponent

$$e^{-(\omega/c)n\kappa z}$$

By manipulation of Maxwell's equations, we were able to show that the propagation velocity of a light wave is governed by the electrical properties of the medium. The index of refraction was used to indicate the propagation velocity in the medium, relative to the propagation velocity in a vacuum

$$n = \frac{c}{v} = \sqrt{\frac{\epsilon\mu}{\epsilon_0\mu_0}}$$

Also, we were able to demonstrate that the light wave must be a transverse wave and that the magnitude of the electric and magnetic fields is related by

$$n|\mathbf{E}| = c|\mathbf{B}|$$

By comparing the forces experienced by a charged particle in an electromagnetic field, we found that we could describe the polarization of an electromagnetic wave by the electric field vector. We developed the formalism necessary for discussion of polarization, but delayed a discussion of the manipulation of a light wave's polarization until Chapter 13.

PROBLEMS

2-1. Light is traveling in glass ($n = 1.5$). If the amplitude of the electric field of the light is 100 V/m (volts/meter), what is the amplitude of the magnetic field? What is the magnitude of the Poynting vector?

2-2. A 60 W monochromatic point source is radiating equally in all directions in a vacuum. What is the electric field amplitude 2 m from the source?

2-3. The flux density at the Earth's surface due to sunlight is $I = 1.34 \times 10^3$ J/(m^2·sec). Calculate the electric and magnetic fields at the Earth's surface by assuming that the average Poynting vector is equal to this flux density.

2-4. What is the flux density of light needed to keep a glass sphere of 10^{-8} g and 2×10^{-5} m in diameter floating in midair?

2-5. An 85 kg astronaut has only a flashlight to propel him in space. If the flashlight emits 1 W of light in a parallel beam for 1 hr, how fast will the astronaut be going at the end of the hour, assuming he started at rest?

2-6. What is the polarization of the following waves?

$$\mathbf{E} = E_0\left[\hat{\mathbf{i}}\cos(\omega t - kz) + \hat{\mathbf{j}}\cos\left(\omega t - kz + \frac{5\pi}{4}\right)\right]$$

$$\mathbf{E} = E_0\left[\hat{\mathbf{i}}\cos(\omega t + kz) + \hat{\mathbf{j}}\cos\left(\omega t + kz - \frac{\pi}{4}\right)\right]$$

$$\mathbf{E} = E_0\left[\hat{\mathbf{i}}\cos(\omega t - kz) - \hat{\mathbf{j}}\cos\left(\omega t - kz + \frac{\pi}{6}\right)\right]$$

2-7. Show that the addition of two elliptical polarized waves propagating along the z axis results in another elliptical polarized wave.

2-8. Write an expression in mks units for a plane electromagnetic wave, with a wavelength of 500 nm and an intensity of 53.2 W/m^2, propagating in the z direction. Assume that the wave is linearly polarized at an angle of 45° to the x axis.

2-9. Using conventional vector notation, prove that a right- and a left-circularly polarized wave can combine to yield a linearly polarized wave. Carry out the same demonstration using the Jones vector notation. What requirement must be placed on the two circularly polarized waves? Sketch the geometrical construction that demonstrates the combination of circularly polarized waves to generate a linearly polarized wave.

2-10. Write the equation for a plane wave propagating in the positive z direction that has right–elliptical polarization with the major axis of the ellipse parallel to the x axis. Use both the conventional vector and the Jones vector notation.

2-11. Describe the polarization of a wave with the Jones vector

$$\begin{pmatrix} -i \\ 2 \end{pmatrix}$$

Write the Jones vector that is orthogonal to this vector and describe its polarization.

2-12. Prove that **(2-42)** is correct using the expressions

$$E_x = E_{0x}\left[\cos(\omega t - kz)\cos\phi_1 - \sin(\omega t - kz)\sin\phi_1\right]$$

$$E_y = E_{0y}\left[\cos(\omega t - kz)\cos\phi_2 - \sin(\omega t - kz)\sin\phi_2\right]$$

for the two orthogonally polarized electric fields.

2-13. Demonstrate using the Jones vector notation that right- and left-circularly polarized light waves are orthogonal.

2-14. Find the skin-depth for sea water with a resistivity of $\rho = 0.20\ \Omega/\text{m}$ for $\nu = 30$ KHz and 30 MHz. What frequency should we use to communicate with a submarine that will not be deeper than 100 m?

2-15. At what frequency would the approximation used to obtain **(2-26)** produce a 10% error?

2-16. If a 1 kw laser beam is focused to a spot with an area of $10^{-9}\ \text{m}^2$, what is the amplitude of the electric field at the focus?

2-17. The human eye is sensitive to light of wavelengths from approximately 600 nm (red) to 400 nm (blue). (a) Calculate the frequency of both wavelengths. (b) Find the energy of the photons associated with the red and blue wavelength limits.

2-18. If green light, 500 nm, could be frequency-modulated to 0.1% of the light wave's frequency, calculate the number of 6MHz bandwidth TV channels that could be carried by the modulation.

2-19. Given the following Stokes vector:

$$\begin{pmatrix} 1 \\ 0 \\ -\frac{3}{5} \\ \frac{4}{5} \end{pmatrix}$$

(a) Calculate the degree of polarization, (b) determine the orthogonal vector, and (c) draw the polarization ellipse.

2-20. How thin must a sheet of iron be if it is one skin-depth thick? How many atoms thick is such a sheet?

Appendix 2-A

VECTORS

We will review a few properties of vectors that will be of use in our discussion of light. A vector is a quantity with both magnitude and direction; it can be defined in terms of unit vectors along the three orthogonal axes of a Cartesian coordinate system

$$\mathbf{E} = \hat{\mathbf{i}}E_x + \hat{\mathbf{j}}E_y + \hat{\mathbf{k}}E_z$$

To add two vectors, we add like components

$$\mathbf{E}_1 + \mathbf{E}_2 = (E_{1x} + E_{2x})\,\hat{\mathbf{i}} + (E_{1y} + E_{2y})\,\hat{\mathbf{j}} + (E_{1z} + E_{2z})\,\hat{\mathbf{k}}$$

There are two ways to multiply vectors.

Scalar (Dot) Product

$$\mathbf{E} \cdot \mathbf{H} = EH \cos\theta = E_xH_x + E_yH_y + E_zH_z \tag{2A-1}$$

where θ is the angle between **E** and **H**. This product is a scalar quantity that gives the projection of one vector onto the second vector. If **E** and **H** are perpendicular, then the dot product is zero.

Vector (Cross) Product

$$\mathbf{E} \times \mathbf{H} = (EH \sin\theta)\,\hat{\mathbf{n}}$$

where $\hat{\mathbf{n}}$ is a unit vector normal to the plane formed by **E** and **H**. The cross product is a vector with a magnitude equal to the area of the parallelogram formed by **E** and **H**; it is zero if the two vectors are parallel. The calculation of the components of this new vector is as follows:

$$\mathbf{E} \times \mathbf{H} = \begin{vmatrix} \hat{\mathbf{i}} & \hat{\mathbf{j}} & \hat{\mathbf{k}} \\ E_x & E_y & E_z \\ H_x & H_y & H_z \end{vmatrix}$$

$$\mathbf{E} \times \mathbf{H} = (E_yH_z - E_zH_y)\,\hat{\mathbf{i}} - (E_xH_z - E_zH_x)\,\hat{\mathbf{j}} + (E_xH_y - E_yH_x)\,\hat{\mathbf{k}} \tag{2A-2}$$

We will find use for the vector triple-product relationship

$$\mathbf{A} \times \mathbf{B} \times \mathbf{C} = (\mathbf{A} \cdot \mathbf{C})\,\mathbf{B} - (\mathbf{A} \cdot \mathbf{B})\,\mathbf{C} \tag{2A-3}$$

Derivatives

A vector operator called the *del operator* is defined in Cartesian notation as

$$\nabla = \hat{\mathbf{i}}\frac{\partial}{\partial x} + \hat{\mathbf{j}}\frac{\partial}{\partial y} + \hat{\mathbf{k}}\frac{\partial}{\partial z} \tag{2A-4}$$

We can treat the operator as a vector and calculate three products of use in optics.

Gradient of a Scalar

$$\nabla V = \hat{\mathbf{i}}\frac{\partial V}{\partial x} + \hat{\mathbf{j}}\frac{\partial V}{\partial y} + \hat{\mathbf{k}}\frac{\partial V}{\partial z} \tag{2A-5}$$

The gradient is a vector giving the magnitude and direction of the fastest rate of change of the scalar quantity V.

Divergence of a Vector

$$\nabla \boldsymbol{\cdot} \mathbf{E} = \frac{\partial E_x}{\partial x} + \frac{\partial E_y}{\partial y} + \frac{\partial E_z}{\partial z} \tag{2A-6}$$

The divergence gives the amount of flux flowing toward (negative) or away from (positive) a point. If the divergence is zero, there are no sources or sinks in the volume.

Curl of a Vector

$$\nabla \times \mathbf{E} = \hat{\mathbf{i}}\left(\frac{\partial E_z}{\partial y} - \frac{\partial E_y}{\partial z}\right) + \hat{\mathbf{j}}\left(\frac{\partial E_x}{\partial z} - \frac{\partial E_z}{\partial x}\right) + \hat{\mathbf{k}}\left(\frac{\partial E_y}{\partial x} - \frac{\partial E_x}{\partial y}\right) \tag{2A-7}$$

A physical interpretation of this operation can be made easily if the vector is a velocity; then when the curl of the velocity is nonzero, rotation is also occurring.

If we calculate the divergence of a gradient of a scalar, we obtain the *Laplacian*

$$\nabla \cdot \nabla V = \nabla^2 V = \frac{\partial^2 V}{\partial x^2} + \frac{\partial^2 V}{\partial y^2} + \frac{\partial^2 V}{\partial z^2} \tag{2A-8}$$

for a scalar function and for a vector quantity

$$\nabla^2 \mathbf{E} = \hat{\mathbf{i}}\nabla^2 E_x + \hat{\mathbf{j}}\nabla^2 E_y + \hat{\mathbf{k}}\nabla^2 E_z \tag{2A-9}$$

Several other identities involving the del operator will come in handy in our study of optics

$$\nabla \times \nabla V = 0 \tag{2A-10}$$

$$\nabla \boldsymbol{\cdot} \nabla \times \mathbf{E} = 0 \tag{2A-11}$$

$$\nabla \times \nabla \times \mathbf{E} = \nabla(\nabla \boldsymbol{\cdot} \mathbf{E}) - \nabla^2 \mathbf{E} \tag{2A-12}$$

$$\nabla \boldsymbol{\cdot} (\mathbf{A} \times \mathbf{B}) = \mathbf{A} \boldsymbol{\cdot} (\nabla \times \mathbf{B}) + \mathbf{B} \boldsymbol{\cdot} (\nabla \times \mathbf{A}) \tag{2A-13}$$

We have displayed all of the above relations in a Cartesian coordinate system; similar expressions can be derived using spherical coordinates. We list them in Table 2A.1 for the reader's convenience.

TABLE 2A.1 Vector Operations in Spherical Coordinates

Operator	r comp	θ comp	ϕ comp
∇V	$\dfrac{\partial V}{\partial r}$	$\dfrac{1}{r}\left(\dfrac{\partial V}{\partial \theta}\right)$	$\dfrac{1}{r\sin\theta}\left(\dfrac{\partial V}{\partial \phi}\right)$
$\nabla\cdot\mathbf{E}$	$\dfrac{1}{r}\left(\dfrac{\partial rE_r}{\partial r}\right)$	$\dfrac{1}{r\sin\theta}\left(\dfrac{\partial\left[E_\theta\sin\theta\right]}{\partial\theta}\right)$	$\dfrac{1}{r\sin\theta}\dfrac{\partial E_\phi}{\partial\phi}$
$\nabla^2 V$	$\dfrac{1}{r^2}\dfrac{\partial}{\partial r}\left(r^2\dfrac{\partial V}{\partial r}\right)$	$\dfrac{1}{r^2\sin\theta}\dfrac{\partial}{\partial\theta}\left(\sin\theta\dfrac{\partial V}{\partial\theta}\right)$	$\dfrac{1}{(r\sin\theta)^2}\dfrac{\partial^2 V}{\partial\phi^2}$
$(\nabla\times\mathbf{E})_r$	$\dfrac{1}{r\sin\theta}\left[\dfrac{\partial}{\partial\theta}(E_\phi\sin\theta)-\dfrac{\partial E_\theta}{\partial\phi}\right]$		
$(\nabla\times\mathbf{E})_\theta$		$\dfrac{1}{r}\left[\dfrac{1}{\sin\theta}\dfrac{\partial E_r}{\partial\phi}-\dfrac{\partial}{\partial r}(rE_\phi)\right]$	
$(\nabla\times\mathbf{E})_\phi$			$\dfrac{1}{r}\left[\dfrac{\partial}{\partial r}(rE_\theta)-\dfrac{\partial E_r}{\partial\theta}\right]$

Appendix 2-B

ELECTROMAGNETIC UNITS

Of all the topics in physics, the one that introduces the most confusion is the subject of electromagnetic units. In this book, we have used mks units, but the reader will find most older optics books use cgs units.

In the rationalized mks units, ϵ and μ in a vacuum have the following values:

$$\epsilon_0 = 8.8542 \times 10^{-12} \frac{C^2}{Nm^2} (F/m)$$

$$\epsilon_0 \approx \left(\frac{1}{36\pi}\right) \times 10^{-9}$$

$$\mu_0 = 4\pi \times 10^{-7} \frac{Ns^2}{C^2} (H/m)$$

In cgs units, $\epsilon_0 = \mu_0 = 1$.

Maxwell's equations in cgs units are written

$$\nabla \cdot \mathbf{D} = 4\pi\rho$$

$$\nabla \cdot \mathbf{B} = 0$$

$$\nabla \times \mathbf{H} = \frac{4\pi}{c}\mathbf{J} + \frac{1}{c}\frac{\partial \mathbf{D}}{\partial t}$$

$$\nabla \times \mathbf{E} + \frac{1}{c}\frac{\partial \mathbf{B}}{\partial t} = 0$$

A good discussion of the subject of units can be found in Jackson.[3] Here, we provide Table 2B.1 to aid the reader in converting from one system to the other.

TABLE 2B.1 Electromagnetic Units

cgs	*Units*	*mks*	*Units*
c	$\frac{\text{cm}}{\text{sec}}$	$\frac{1}{\sqrt{\mu_0\epsilon_0}}$	$\frac{\text{m}}{\text{sec}}$
D	$12\pi \times 10^{-5}\left(\frac{\text{statC}}{\text{cm}^2}\right)$	$\sqrt{4\pi\epsilon_0}\mathbf{D}$	$\frac{\text{C}}{\text{m}^2}$
B	10^4 G	$\sqrt{\frac{4\pi}{\mu_0}}\mathbf{B}$	$\frac{\text{weber}}{\text{m}^2}$

TABLE 2B.1 (continued)

cgs	*Units*	*mks*	*Units*
$\mathbf{H}$	$4\pi \times 10^{-3}$ O	$\sqrt{4\pi\mu_0}\mathbf{H}$	$\frac{\text{A-turn}}{\text{m}}$
$\mathbf{E}$	$\frac{1}{3} \times 10^{-4} \frac{\text{statC}}{\text{cm}}$	$\sqrt{4\pi\epsilon_0}\mathbf{E}$	$\frac{\text{V}}{\text{m}}$
$\mathbf{J}$	$3 \times 10^{5} \frac{\text{statA}}{\text{cm}^2}$	$\frac{\mathbf{J}}{\sqrt{4\pi\epsilon_0}}$	$\frac{\text{A}}{\text{m}^2}$
σ	$\frac{9 \times 10^{9}}{\text{sec}}$	$\frac{\sigma}{4\pi\epsilon_0}$	$\frac{\text{mho}}{\text{m}}$
ρ	$3 \times 10^{3} \frac{\text{statC}}{\text{cm}^3}$	$\frac{\rho}{\sqrt{4\pi\epsilon_0}}$	$\frac{\text{C}}{\text{m}^3}$
ϵ	ϵ_r	$\frac{\epsilon}{\epsilon_0}$	
μ	μ_r	$\frac{\mu}{\mu_0}$	

Reflection and Refraction

In Chapter 2, we treated the propagation of light in a uniform medium using Maxwell's equations. In this chapter, we wish to explore what happens to the propagation of a light wave when the electrical properties of the medium change in a discontinuous way. We will find that the wave will experience reflection at the boundary between two media with different electromagnetic properties. The light transmitted across the boundary will undergo a change in propagation direction. The direction change is called *refraction*. We will make the first of several derivations of the laws of refraction and reflection, here relying only on the wave properties of light to obtain the laws. To obtain the amplitude of the light waves reflected and refracted, we will use boundary conditions developed in classical electromagnetic theory for Maxwell's equations.

Once equations for the reflected and transmitted amplitudes are obtained, we will consider light incident normal to the boundary to simplify the equations for the amplitudes of reflected and transmitted waves. With this simplification, it will become obvious that the fractions of the wave reflected and transmitted at the boundary between two media depend on the relative propagation velocities of the wave in the two media.

We will find that there is an angle, called Brewster's angle, for which light reflected from a boundary will be linearly polarized. There is also a set of conditions for which all light incident on a boundary will be reflected. A few of the properties of this reflected wave, called a totally reflected wave, will be discussed.

The use of reflection existed before written history, as is evidenced by the discovery of a mirror from the period around 1900 B.C. Some of the earliest written comments about reflection can be found in *Exodus 38:*8 and *Job 37:*18. **Euclid** in about 300 B.C. discussed the focus of a spherical mirror in *Catoptrics*. **Cleomedes** (50 A.D.) discussed the refraction of light at an air-water interface. He described an experiment whereby a coin at the bottom of a bowl and hidden by the bowl's sides could be made visible by pouring water into the bowl.

Claudius Ptolemy of Alexandria (139 A.D.) made tables of the angles of incidence and refraction. His work is one of the few examples of experiment during that time. The concept of a sine of an angle was not yet developed so his tables were only approximately correct. One of the most interesting individuals active in optics during the Middle Ages was **Alhazen** (Abû Alî al Hasan ibn al Hasan ibn Al Haitam) **(965–1038)** who developed optics during the golden age of the Arabic empire. He added to the law of reflection developed by Ptolemy the fact that the incident angle and the reflected angle lie in the same plane, called the plane of incidence. He also corrected Ptolemy's tables of incident and refracted angles. Alhazen failed, however, to discover the law of refraction. (Alhazen was a successful optical scientist but some of his civil engineering projects got him into trouble with his caliph, and Alhazen had to feign insanity and hide to escape the wrath of the caliph.)

Vitello (in 1270) repeated the experiments of Ptolemy but also failed to discover the law of refraction. **Johannes Kepler (1571–1630)** gave a broad outline of the correct theory of the telescope and discussed total reflection without knowledge of the law of refraction. He used an empirical expression $\theta_i = n\theta_r$, where $n = 3/2$.

The law of refraction was discovered, evidently through experimentation, by **Willebrord Snell van Royen (1591–1626)**, a professor of mechanics at Leyden. He never published but Huygens and Isaak Voss claimed to have examined Snell's manuscript. **Descartes** in 1637 deduced the law of refraction theoretically and expressed it in its present form. It is interesting that one of his assumptions was wrong. He often confidently deduced theory without allowing himself to be disturbed by any possible discrepancy between his final conclusions and the actual facts. Later, **Pierre de Fermat (1601–1665)** deduced the law of refraction from the assumption that light travels from a point in one medium to a point in another medium in the least time.

REFLECTION AND TRANSMISSION AT A DISCONTINUITY

To develop the law of reflection and refraction, we will first consider the one-dimensional problem of a vibrating string in which the string has a discontinuous change in its mass/unit length at the origin of an x, y coordinate system; see Figure 3-1. The string is stretched along the x direction and a wave propagating along the string in the positive x direction creates a displacement in the y direction. An example of such a string might be an "A" string of a guitar with part of its winding removed. From the discussion

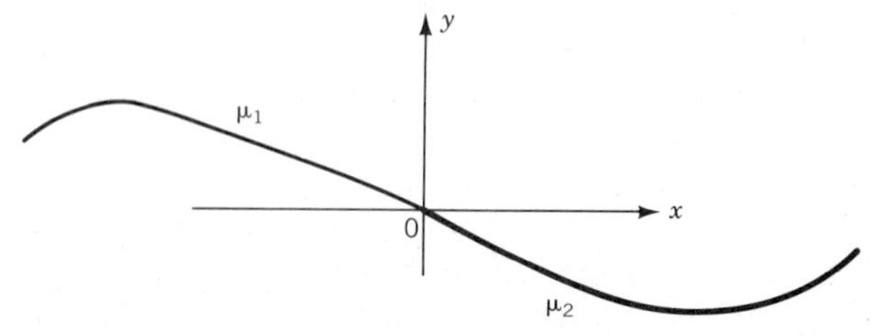

FIGURE 3-1. String with a nonuniform mass/unit length. For $x < 0$, the mass/unit length is μ_1, whereas for $x > 0$, the mass/unit length is μ_2. The tension is the same throughout the string.

leading to **(1-6)**, it is easy to see that a string with a nonuniform mass/unit length will have a nonuniform wave propagation velocity. What happens to the wave motion on this string with a nonuniform propagation velocity? Assume that the tension is the same throughout the string and that the mass/unit length is μ_1 for $x < 0$ and μ_2 for $x > 0$, as shown in Figure 3-1. We have two boundary conditions that must be satisfied at $x = 0$.

1. The displacement at $x = 0$ must be the same for the two strings. If this were not the case, the string would be broken.
2. The slopes of the two strings must be the same at $x = 0$. We assume, as we did in Chapter 1, that the string undergoes small displacements so $\sin\theta \approx \tan\theta = dy/dx$. The vertical force at any point is therefore $T_y = T dy/dx$. If the slopes were not equal, there would be a finite vertical force acting on an infinitesimal mass and the acceleration at $x = 0$ would be infinite.

Initially, we will assume that there are two waves present. One wave is incident from the left and propagating toward $x = 0$, traveling from left to right in Figure 3-1. This incident wave is given by

$$y_i = Y_i \cos(\omega_i t - k_i x) \qquad x < 0$$

The speed of the wave is

$$c_1 = \sqrt{\frac{T}{\mu_1}}$$

At the origin $x = 0$, the wave and its first derivative are

$$y_i\Big|_{x=0} = Y_i \cos\omega_i t \qquad \frac{dy_i}{dx}\Big|_{x=0} = k_i Y_i \sin\omega_i t$$

The second wave, called the transmitted wave, is one that has propagated past $x = 0$, also traveling from left to right in Figure 3-1. The transmitted wave is represented by

$$y_t = Y_t \cos(\omega_t t - k_t x) \qquad x > 0$$

with a propagation velocity of

$$c_2 = \sqrt{\frac{T}{\mu_2}}$$

At $x = 0$, the wave and its first derivative are

$$y_t\Big|_{x=0} = Y_t \cos\omega_t t, \qquad \frac{dy_t}{dx}\Big|_{x=0} = k_t Y_t \sin\omega_t t$$

We will attempt to satisfy the boundary conditions with these two waves, but will discover that a third wave will be needed if the boundary conditions are to be satisfied.

The first boundary condition requires that the waves be equal at $x = 0$ and is met if $Y_i = Y_t$. Since the boundary condition must hold for all time, the frequencies of the waves propagating at the two different velocities must be the same, $\omega_i = \omega_t$.

The second boundary condition requires that the first derivatives of the two waves must be equal at $x = 0$ and we find again that $\omega_i = \omega_t$, since the equality must hold for all time. The second boundary condition also requires that

$$k_i Y_i = k_t Y_t$$

but because $Y_i = Y_t$ from the first boundary condition, we must have $k_i = k_t$. The definition **(1-2)** allows the replacement of the propagation constants by $\omega_i/c_1 = \omega_t/c_2$. Because the two frequencies are required to be equal by the boundary conditions, we are led to a contradiction of the initial assumption that the propagation velocities in the two dissimilar string segments are unequal, $c_1 \neq c_2$.

To satisfy the boundary conditions, a reflected wave must be introduced, traveling from right to left and originating at $x = 0$. The reflected wave is defined as

$$y_r = Y_r \cos(\omega_r t + k_r x) \qquad x < 0$$

Its propagation velocity is c_1. The evaluation of the wave and its first derivative at the origin yields

$$y_r\big|_{x=0} = Y_r \cos \omega_r t \qquad \left.\frac{dy_r}{dx}\right|_{x=0} = -k_r Y_r \sin \omega_r t$$

With the new wave, the boundary conditions are written

$$y_i + y_r - y_t\big|_{x=0} = 0, \qquad \left.\frac{dy_i}{dt} + \frac{dy_r}{dt} - \frac{dy_t}{dt}\right|_{x=0} = 0$$

These two equations are satisfied if

$$\omega_i = \omega_r = \omega_t \tag{3-1}$$

$$Y_i + Y_r = Y_t \tag{3-2}$$

$$k_i Y_i - k_r Y_r = k_t Y_t \tag{3-3}$$

Solving **(3-2)** and **(3-3)** simultaneously yields

$$Y_t = \frac{k_i + k_r}{k_t + k_r} Y_i = \frac{\omega\left(\dfrac{1}{c_1} + \dfrac{1}{c_1}\right)}{\omega\left(\dfrac{1}{c_1} + \dfrac{1}{c_2}\right)} Y_i = \frac{2\dfrac{\omega}{c_1}}{\left(\dfrac{\omega}{c_1 c_2}\right)(c_1 + c_2)} Y_i$$

The transmission coefficient at the junction of the two dissimilar strings is defined as $\tau = Y_t/Y_i$ and is given by

$$\tau = \frac{2c_2}{c_1 + c_2} \tag{3-4}$$

In the same way, a reflection coefficient can also be defined at the junction where the mass/unit length of the string changes

$$\rho = \frac{Y_r}{Y_i} = \frac{c_2 - c_1}{c_1 + c_2} \tag{3-5}$$

If $c_1 = c_2$, then $Y_r = 0$, there is no reflected wave, and the incident and transmitted waves have equal amplitudes, $Y_i = Y_t$. The wave propagates past the point $x = 0$ without change. Note that the transmission and reflection coefficients are only a function of the wave velocities. This same dependence on propagation velocity will be found for light waves.

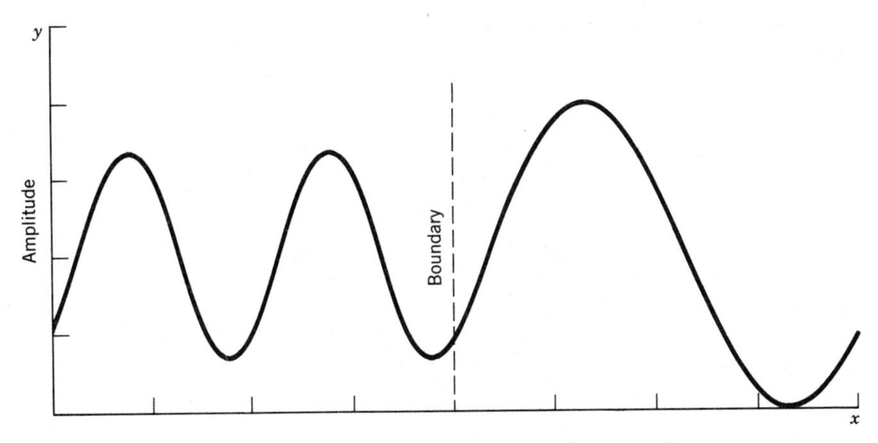

FIGURE 3-2. Two strings of unequal mass/unit length joined at $x=0$. The wave velocity of the string to the left is 10 m/sec and the right is 20 m/sec. The wave is incident onto the junction from the left and has an amplitude of 3 cm and a wavelength of 1 m. The plot shows the resultant waves: on the right of $x=0$ the transmitted wave and on the left of $x=0$ the sum of the incident and reflected wave.

The frequency across the discontinuity is required, by the boundary conditions, to be a constant but the velocity changes. This means that the wavelength must also change as is shown in Figure 3-2.

There are two waves to the left of the origin. From the principle of superposition, the displacement of the string at any time is a sum of these two waves: the incident and the reflected wave. The addition of the incident and the reflected wave leads to a standing wave if the wave is totally reflected at the discontinuity.

Total reflection will occur if the string on the right of Figure 3-1 is replaced by a rigid support. The rigid support is equivalent to a string with an infinite mass/unit length, $\mu_2 = \infty$. Since T, the tension in the string, is finite, the propagation velocity is zero, $c_2 = 0$. The transmitted amplitude is then $Y_t = 0$ and $Y_i = -Y_r = Y$. The wave is totally reflected, i.e., $\rho = -1$ and $\tau = 0$. From the superposition principle, the incident and reflected waves add together to produce the resultant wave, a standing wave

$$\begin{aligned} y &= y_i + y_r \\ &= Y\left[\cos(\omega t - kx) - \cos(\omega t + kx)\right] \\ &= 2Y \sin \omega t \sin kx \end{aligned} \tag{3-6}$$

(The addition of waves will be discussed in more detail in Chapter 4.)

LAWS OF REFLECTION AND REFRACTION

The three-dimensional problem of a wave propagating across a boundary, where a discontinuous change of propagation velocity occurs, can be studied by using the coordinate system shown in Figure 3-3. The result will be an extension, to three dimensions, of the one dimensional laws of reflection and refraction we have just derived. We will perform the derivation by only requiring the phase of the wave to vary smoothly across the boundary.

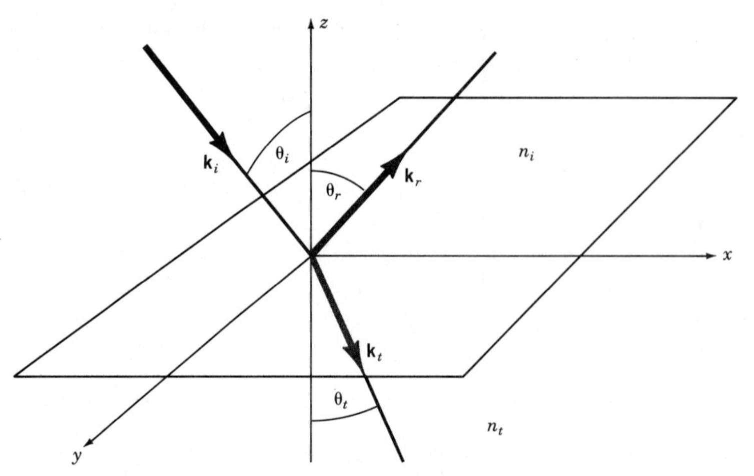

FIGURE 3-3. Coordinate diagram for light reflection and transmission across a boundary. The index of refraction in the upper half-plane is n_i and in the lower half-plane is n_t.

The wave velocity in the medium in the upper half-plane of Figure 3-3 is v_i. For a light wave, this propagation velocity can be indicated by characterizing the medium with the index of refraction $n_i = c/v_i$ **(2-11)**.

The incident wave is assumed to be a plane wave, defined in **(1-15)** to be of the form $F(\omega t - \mathbf{k} \cdot \mathbf{r})$. To learn something about the geometry of the reflective and refractive waves, we require that the phases of the three wave functions—incident, reflected, and transmitted—be the same on the boundary between the two half-planes (at $z = 0$),

$$\omega_i t - \mathbf{k}_i \cdot \mathbf{r} \big|_{z=0} = \omega_i t - \mathbf{k}_r \cdot \mathbf{r} \big|_{z=0} = \omega_t t - \mathbf{k}_t \cdot \mathbf{r} \big|_{z=0} \tag{3-7}$$

Since the equality of **(3-7)** must hold for all time, the frequency of the wave across the boundary does not change $\omega_i = \omega_t = \omega$. At $z = 0$, we have

$$\mathbf{r} \rightarrow (x, y, 0), \qquad \mathbf{k}_i \rightarrow \left(\frac{n_i \omega}{c}\right)(k_x^i, k_y^i, k_z^i)$$

$$\mathbf{k}_r \rightarrow \left(\frac{n_i \omega}{c}\right)(k_x^r, k_y^r, k_z^r), \qquad \mathbf{k}_t \rightarrow \left(\frac{n_t \omega}{c}\right)(k_x^t, k_y^t, k_z^t)$$

where the (k_x^i, k_y^i, k_z^i) and (k_x^t, k_y^t, k_z^t) are the direction cosines of the incident and transmitted wave, i.e., for the incident wave in a rectangular coordinate system.

$$\mathbf{k}_i = \left(\frac{n_i \omega}{c}\right)(k_x^i \,\hat{\mathbf{i}} + k_y^i \,\hat{\mathbf{j}} + k_z^i \,\hat{\mathbf{k}})$$

The spatial part of the phase must meet the equality specified by **(3-7)**, independent of the temporal part; therefore, since the boundary has been positioned at $z = 0$,

$$\frac{n_i \omega}{c}(k_x^i x + k_y^i y) = \frac{n_i \omega}{c}(k_x^r x + k_y^r y) = \frac{n_t \omega}{c}(k_x^t x + k_y^t y)$$

This relationship must hold independently for all values of x and y

$$n_i k_x^i = n_i k_x^r = n_t k_x^t, \qquad n_i k_y^i = n_i k_y^r = n_t k_y^t \tag{3-8}$$

The relationships in **(3-8)** require that both the transmitted and reflected waves lie in the same plane as the incident wave. We call the plane the *plane of incidence* and define it as the plane containing the incident wave vector $\mathbf{k}_i$ and the normal to the boundary, at the point where $\mathbf{k}_i$ intersects the boundary.

In this discussion, the boundary is taken to be the $z = 0$, x, y plane; the normal to the plane is the unit vector $\hat{\mathbf{k}}$ parallel to the z axis. (Please note that $\hat{\mathbf{k}}$ is the unit vector along the z direction and $\mathbf{k}$ is the wave vector. The caret above the unit vector and the context of the discussion should prevent confusion.) The plane of incidence is arranged to lie in the x, z plane, as shown in Figure 3-3. The direction cosines associated with the propagation vectors are thus

$$k_x^i = \sin\,\theta_i, \qquad k_y^i = 0, \qquad k_z^i = \cos\,\theta_i$$

$$k_x^r = \sin\,\theta_r, \qquad k_y^r = 0, \qquad k_z^r = \cos\,\theta_r$$

$$k_x^t = \sin\,\theta_t, \qquad k_y^t = 0, \qquad k_z^t = \cos\,\theta_t$$

Equation **(3-8)** can now be written

$$n_i\,\sin\,\theta_i = n_i\,\sin\,\theta_r = n_t\,\sin\,\theta_t$$

From this expression, we have in the first medium

$$\sin\,\theta_i = \sin\,\theta_r \tag{3-9}$$

If the coordinate system is selected so that $\mathbf{k}_i$ is propagating in the positive direction and $\cos\,\theta_i \geq 0$, then it is apparent from Figure 3-3 that $\mathbf{k}_r$ is in the negative direction and $\cos\,\theta_r \leq 0$, resulting in $\theta_r = \pi - \theta_i$. The statement that the reflected wave is in the same plane as the incident wave and **(3-9)** together form the *law of reflection*.

If we return to **(3-8)**, the second relationship is

$$n_i\,\sin\,\theta_i = n_t\,\sin\,\theta_t \tag{3-10}$$

which is the *law of refraction* or *Snell's law*.

By requiring phase continuity across the boundary, we have obtained the laws of reflection and refraction. These relations hold for any solution of the wave equation and are not dependent on the electromagnetic properties of light waves. Later, we will use other approaches to obtain these laws.

FRESNEL'S FORMULAE

The geometry of reflected and transmitted waves has been obtained using only the wave character of light; however, nothing about the amplitudes of the waves has been determined. We must use Maxwell's equations and the boundary conditions associated with these equations to learn about the amplitudes of the reflected and transmitted waves. The geometry to be used in this discussion is shown in Figure 3-4. Two media are separated by an interface, the x, y plane at $z = 0$, whose normal $\hat{\mathbf{n}} = \hat{\mathbf{k}}$ is the unit vector along the z direction. The incident wave is labeled with an i, the reflected wave by an r, and the transmitted wave by a t. The incident wave's propagation vector $\mathbf{k}_i$, which we assume to lie in the x, z plane, and the normal to the interface establish the plane of incidence.

The electric field vectors for each of the three waves have been decomposed into two components: one in the plane of incidence, labeled P, and

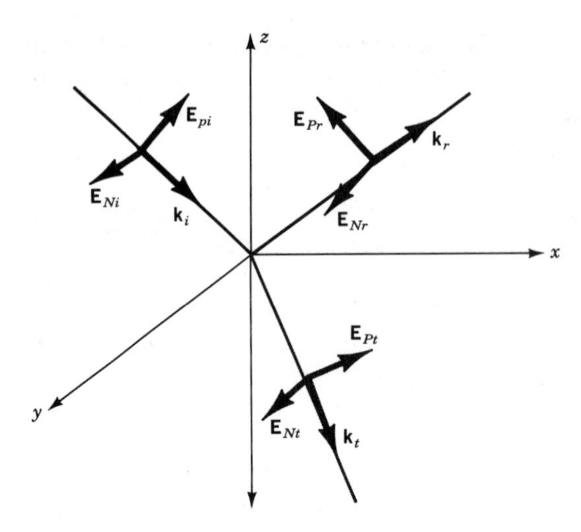

FIGURE 3-4. Orientation of the electric field and wave vectors in the coordinate system we selected for the discussion of reflection and refraction. The plane of incidence is the x, z plane.

one normal to the plane and parallel to the unit vector $\hat{\mathbf{j}}$ along the y axis, labeled N. This is an extension of the technique, discussed in Chapter 2, of using orthogonal vectors to describe the polarization of a light wave. (According to custom, the two polarizations are labeled π for parallel to the plane of incidence and σ for perpendicular to the plane of incidence. The Greek letter σ denotes perpendicular because s is the first letter of the German word *senkrecht*. We will use N and P in this book in place of the Greek letters.) The upper half-plane has a velocity of propagation v_i and an index of n_i and the lower half-plane has a velocity of propagation v_t and an index of n_t.

The actual vectors to be used are as follows:

Incident Wave

$$\mathbf{k}_i = k_i\,(\hat{\mathbf{i}} \sin\theta_i - \hat{\mathbf{k}} \cos\theta_i) \tag{3-11}$$

$$\mathbf{E}_i = E_{Pi}\,(\hat{\mathbf{i}} \cos\theta_i + \hat{\mathbf{k}} \sin\theta_i) + E_{Ni}\,\hat{\mathbf{j}} \tag{3-12}$$

Reflected Wave

$$\mathbf{k}_r = k_i\,(\hat{\mathbf{i}} \sin\theta_i + \hat{\mathbf{k}} \cos\theta_i) \tag{3-13}$$

$$\mathbf{E}_r = E_{Pr}\,(-\hat{\mathbf{i}} \cos\theta_i + \hat{\mathbf{k}} \sin\theta_i) + E_{Nr}\,\hat{\mathbf{j}} \tag{3-14}$$

Transmitted Wave

$$\mathbf{k}_t = k_t\,(\hat{\mathbf{i}} \sin\theta_t - \hat{\mathbf{k}} \cos\theta_t) \tag{3-15}$$

$$\mathbf{E}_t = E_{Pt}\,(\hat{\mathbf{i}} \cos\theta_t + \hat{\mathbf{k}} \sin\theta_t) + E_{Nt}\,\hat{\mathbf{j}} \tag{3-16}$$

The boundary conditions associated with Maxwell's equations are the following:

1. From $\nabla \cdot \mathbf{D} = \rho$, the normal components of $\mathbf{D}$ must be continuous if there are no surface charges. We use $\mathbf{D} = \epsilon\mathbf{E}$ to write this boundary condition

$$\left[\epsilon_i\,(\mathbf{E}_i + \mathbf{E}_r) - \epsilon_t \mathbf{E}_t\right] \cdot \hat{\mathbf{n}} = 0 \tag{3-17}$$

Evaluating the dot product yields

$$\epsilon_i \sin\theta_i\,(E_{Pi} + E_{Pr}) = \epsilon_t \sin\theta_t E_{Pt} \tag{3-18}$$

2. From the Maxwell's equation containing $\nabla \times \mathbf{E}$, we see that the tangential component of $\mathbf{E}$ is continuous. This boundary condition is written

$$(\mathbf{E}_i + \mathbf{E}_r + \mathbf{E}_t)\times\hat{\mathbf{n}} = 0 \tag{3-19}$$

Each one of the vector products is of the form

$$\mathbf{E}\times\hat{\mathbf{n}} = E_y\,\hat{\mathbf{i}} - E_x\,\hat{\mathbf{j}}$$

Evaluating the cross product yields

$$(E_{Ni} + E_{Nr} - E_{Nt})\,\hat{\mathbf{i}} - (E_{Pi}\cos\,\theta_i - E_{Pr}\cos\,\theta_i - E_{Pt}\cos\,\theta_t)\,\hat{\mathbf{j}} = 0$$

Both vector components must be equal to zero, thus,

$$E_{Ni} + E_{Nr} = E_{Nt} \tag{3-20}$$

$$(E_{Pi} - E_{Pr})\cos\,\theta_i = E_{Pt}\cos\,\theta_t \tag{3-21}$$

3. From $\nabla\cdot\mathbf{B} = 0$, the normal component of **B** must be continuous. We use **(2-17)** to rewrite the normal component of **B** in terms of **E**

$$\mathbf{B}\cdot\hat{\mathbf{n}} = \left(\frac{\sqrt{\mu\epsilon}}{k}\right)\,\mathbf{k}\times\mathbf{E}\cdot\hat{\mathbf{n}}$$

The boundary condition is then written

$$\left[\frac{\sqrt{\mu_i\epsilon_i}}{k_i}(\mathbf{k}_i\times\mathbf{E}_i + \mathbf{k}_r\times\mathbf{E}_r) - \frac{\sqrt{\mu_t\epsilon_t}}{k_t}(\mathbf{k}_t\times\mathbf{E}_t)\right]\cdot\hat{\mathbf{n}} = 0 \tag{3-22}$$

Each of the vectors will be of the form

$$(\mathbf{k}\times\mathbf{E})\cdot\hat{\mathbf{n}} = \left[(E_yk_z - E_zk_y)\,\hat{\mathbf{i}} + (E_zk_x - E_xk_z)\,\hat{\mathbf{j}} + (E_xk_y - E_yk_x)\,\hat{\mathbf{k}}\right]\cdot\hat{\mathbf{k}} = -E_yk_x$$

$$\sqrt{\mu_i\epsilon_i}\,(E_{Ni} + E_{Nr})\sin\,\theta_i = \sqrt{\mu_t\epsilon_t}\,E_{Nt}\sin\,\theta_t \tag{3-23}$$

We can simplify this relationship by rewriting Snell's law **(3-10)** as

$$\frac{\sin\,\theta_t}{\sin\,\theta_i} = \sqrt{\frac{\mu_i\epsilon_i}{\mu_t\epsilon_t}} \tag{3-24}$$

Equation **(3-23)** can then be seen to be the same as **(3-20)**. This boundary condition is redundant and we need not use it.

4. From the Maxwell's equation containing $\nabla\times\mathbf{H}$, we see that the tangential component of **H** is continuous if there are no surface currents. The tangent component of **H** can be written in terms of the electric field

$$\mathbf{H}\times\hat{\mathbf{n}} = \frac{\mathbf{B}}{\mu}\times\hat{\mathbf{n}} = \frac{\sqrt{\mu\epsilon}}{\mu k}\mathbf{k}\times\mathbf{E}\times\hat{\mathbf{n}} \tag{3-25}$$

The boundary condition is then written

$$\left[\frac{1}{k_i}\sqrt{\frac{\epsilon_i}{\mu_i}}(\mathbf{k}_i\times\mathbf{E}_i + \mathbf{k}_r\times\mathbf{E}_r) - \frac{1}{k_t}\sqrt{\frac{\epsilon_t}{\mu_t}}(\mathbf{k}_t\times\mathbf{E}_t)\right]\times\hat{\mathbf{n}} = 0$$

$$(\mathbf{k}\times\mathbf{E})\times\hat{\mathbf{n}} = \left[(E_yk_z - E_zk_y)\,\hat{\mathbf{i}} + (E_zk_x - E_xk_z)\,\hat{\mathbf{j}} + (E_xk_y - E_yk_x)\,\hat{\mathbf{k}}\right]\times\hat{\mathbf{k}}$$

$$= (E_zk_x - E_xk_z)\,\hat{\mathbf{i}} - E_yk_z\,\hat{\mathbf{j}} = 0 \tag{3-26}$$

Each vector component must equal zero. The x component gives

$$\sqrt{\frac{\epsilon_i}{\mu_i}}\left[E_{Pi} + E_{Pr}\right] = \sqrt{\frac{\epsilon_t}{\mu_t}}E_{Pt} \tag{3-27}$$

If we apply Snell's law **(3-24)** to **(3-18)**, we obtain the same equation as **(3-27)**. Therefore, we do not need boundary condition 1.

The y component of **(3-26)** gives

$$\sqrt{\frac{\epsilon_i}{\mu_i}}\,(E_{Ni} - E_{Nr})\cos\theta_i = \sqrt{\frac{\epsilon_t}{\mu_t}}\,E_{Nt}\cos\theta_t \tag{3-28}$$

Of the four boundary conditions of Maxwell's equations only two are needed to obtain the relationships between the incident, reflected, and transmitted waves; the conditions utilized are that the tangential components of **E** and **H** are continuous across the boundary. The boundary conditions place independent requirements on the polarizations parallel to and normal to the plane of incidence and generate two pair of equations that are treated separately. There are three unknowns but only two equations for each polarization; thus, the amplitudes of the reflected and transmitted light can only be found in terms of the incident amplitude.

σ Case (Perpendicular Polarization)

For this component of the polarization, **E** is perpendicular to the plane of incidence, that is, the x, z plane. This means that **E** is everywhere normal to $\hat{\mathbf{n}}$ and parallel to the boundary surface between the two media. [This case could be labeled the *transverse electric field* (TE) case; we do not use this notation because it implies that the wave may not be a transverse electromagnetic wave (a TEM wave); instead, we use the subscript N. The TE notation will be reserved for inhomogeneous waves in a guiding structure, discussed in Chapter 5.]

The second and fourth boundary conditions provide in **(3-20)** and **(3-28)** relationships between the various normal electric fields. The amplitude of reflected and transmitted light will be found, using these equations, in terms of the incident amplitude. We use Snell's law **(3-24)** to modify **(3-28)**

$$E_{Ni} - E_{Nr} = \frac{\mu_i \sin\theta_i}{\mu_t \cos\theta_i}\cdot\frac{\cos\theta_t}{\sin\theta_t}\cdot E_{Nt} = \frac{\mu_i \tan\theta_i}{\mu_t \tan\theta_t}E_{Nt} \tag{3-29}$$

Adding **(3-29)** to **(3-20)** yields

$$2E_{Ni} = \left(1 + \frac{\mu_i \tan\theta_i}{\mu_t \tan\theta_t}\right)E_{Nt}$$

$$\frac{E_{Nt}}{E_{Ni}} = \frac{2}{1 + \dfrac{\mu_i \tan\theta_i}{\mu_t \tan\theta_t}} \tag{3-30}$$

As was shown in Chapter 2, for the majority of optical materials, $\mu_i \approx \mu_t$ and the equation simplifies to

$$t_N = \frac{E_{Nt}}{E_{Ni}} = \frac{2\sin\theta_t\cos\theta_i}{\sin(\theta_i + \theta_t)} \tag{3-31}$$

The amplitude ratio t_N is called the *amplitude transmission coefficient for perpendicular polarization*.

Now **(3-30)** can be substituted back into **(3-20)** to obtain the amplitude ratio for the reflected light.

$$E_{Ni} + E_{Nr} = \frac{2E_{Ni}}{1 + \dfrac{\mu_i \tan \theta_i}{\mu_t \tan \theta_t}}$$

$$\frac{E_{Nr}}{E_{Ni}} = \frac{1 - \left(\dfrac{\mu_i \tan \theta_i}{\mu_t \tan \theta_t}\right)}{1 + \left(\dfrac{\mu_i \tan \theta_i}{\mu_t \tan \theta_t}\right)} \tag{3-32}$$

Again when $\mu_i \approx \mu_t$, we obtain the ratio of reflected amplitude to the incident amplitude, which is called the *amplitude reflection coefficient for perpendicular polarization.*

$$r_n = \frac{E_{Nr}}{E_{Ni}} = \frac{-\sin(\theta_i - \theta_t)}{\sin(\theta_i + \theta_t)} \tag{3-33}$$

π Case (Parallel Polarization)

For this component of polarization, **E** is everywhere parallel to the plane of incidence; however, **B** and **H** are everywhere normal to **n̂** and parallel to the boundary between the two media. [This case could be labeled the *transverse magnetic field* (TM) case, but we will use the subscript P for the same reason we did not use the TE notation. See Chapter 5, where the TM notation is utilized for inhomogeneous waves.] The second boundary condition provides **(3-21)** that can be written

$$E_{Pi} - E_{Pr} = \frac{\cos \theta_t}{\cos \theta_i} E_{Pt} \tag{3-34}$$

Applying Snell's law **(3-24)** to **(3-27)** yields

$$E_{Pi} + E_{Pr} = \sqrt{\frac{\mu_i \epsilon_t}{\mu_t \epsilon_i}}\, E_{Pt} = \frac{\mu_i \sin \theta_i}{\mu_t \sin \theta_t} E_{Pt} \tag{3-35}$$

Adding this equation to **(3-34)** yields the desired ratio of amplitudes

$$\frac{E_{Pt}}{E_{Pi}} = \frac{2 \cos \theta_i \sin \theta_t}{\cos \theta_t \sin \theta_t + \dfrac{\mu_i}{\mu_t} \cos \theta_i \sin \theta_i} \tag{3-36}$$

For the usual situation of $\mu_i \approx \mu_t$, the *amplitude transmission coefficient for parallel polarization* is

$$t_P = \frac{E_{Pt}}{E_{Pi}} = \frac{2 \cos \theta_i \sin \theta_t}{\sin(\theta_i + \theta_t) \cos(\theta_i - \theta_t)} \tag{3-37}$$

Substituting **(3-37)** into **(3-35)** produces the amplitude reflection ratio

$$\frac{E_{Pr}}{E_{Pi}} = \frac{\left(\dfrac{\mu_i}{\mu_t}\right) \sin 2\theta_i - \sin 2\theta_t}{\sin 2\theta_t + \left(\dfrac{\mu_i}{\mu_t}\right) \sin 2\theta_i} \tag{3-38}$$

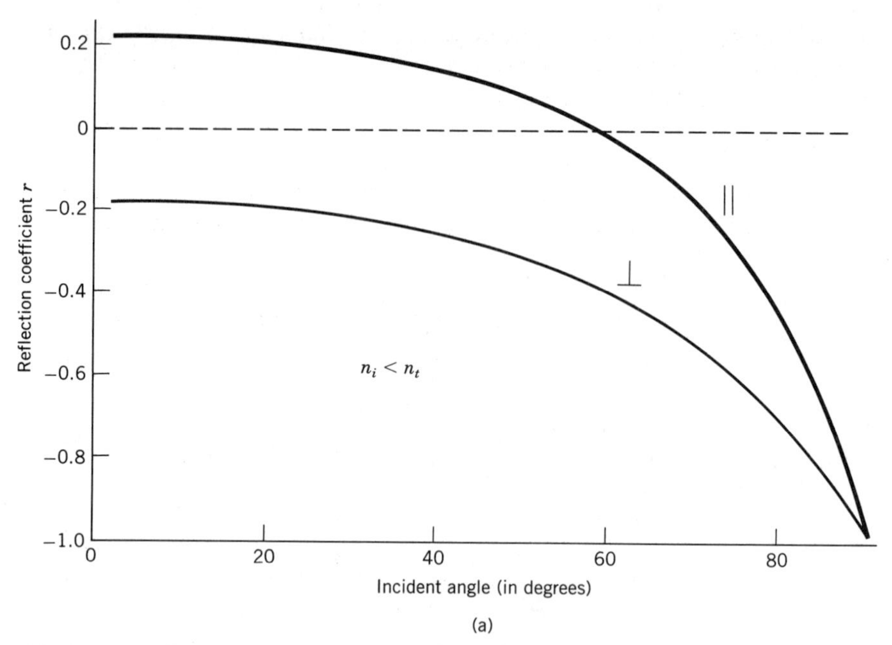

FIGURE 3-5*a*. Reflection coefficient for $n_i = 1.0$ and $n_t = 1.5$, i.e., the ratio of index of refraction is 1.5.

The assumption of $\mu_i \approx \mu_t$ (from now on this assumption will be used) results in the *amplitude reflection coefficient for parallel polarization* of

$$r_P = \frac{E_{Pr}}{E_{Pi}} = \frac{\tan(\theta_i - \theta_t)}{\tan(\theta_i + \theta_t)} \tag{3-39}$$

The amplitude reflection coefficients are plotted as a function of the incident angle in Figure 3-5. Figure 3-5*a* corresponds to the condition of

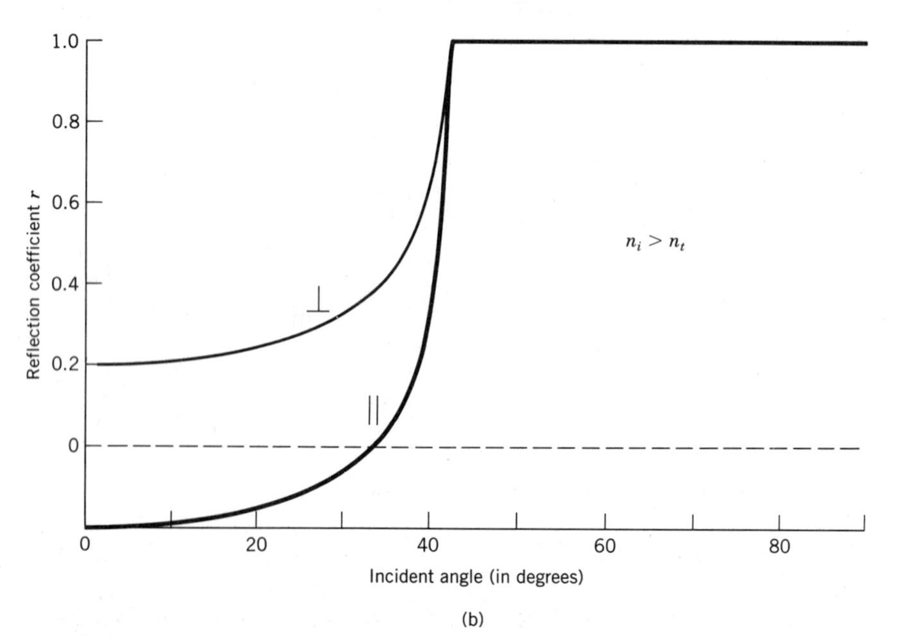

FIGURE 3-5*b*. Reflection coefficient for $n_i = 1.5$ and $n_t = 1$, i.e., ratio of index of refraction is 0.67.

$n_i < n_t$ and Figure 3-5*b* corresponds to the condition of $n_i > n_t$. As can be seen by examining the two figures, the behavior of the reflection coefficient is quite different for these two conditions.

The plots of the reflection coefficients shown in Figure 3-5*a* and 3-5*b* demonstrate that a sign change occurs for r_P, labeled || in the figure, for a range of angles that depend on the relative index of refraction. This phase change is important because r_P must pass through zero for the phase change to occur.

We will discuss in detail the behavior of the amplitude reflection coefficient when $\theta \approx 0°$, $r_P = 0$, and $r = 1$, which occurs if $n_i > n_t$, as shown in Figure 3-5*b*.

REFLECTED AND TRANSMITTED ENERGY

The fraction of the incident amplitude reflected and transmitted at a surface is not experimentally available. The parameter that can be measured is the energy. At first, we might think that we could simply square the ratios we have derived to obtain the energies but this would lead to erroneous results. The correct way to proceed is to use the average Poynting vector incident on a unit area, given by

$$\langle \mathbf{S} \rangle \cdot \hat{\mathbf{n}} = |\langle \mathbf{S} \rangle| \cos \theta_i$$

where the expression for the average Poynting vector is obtained from **(2-26)**. (We assume that $\mu = \mu_0$, resulting in

$$\mu c = \sqrt{\frac{\mu_0}{\epsilon_0}}$$

which is defined as the impedance of vacuum.) The energy flow across the boundary can be obtained from the following equations:

$$\upsilon_i \langle U_i \rangle = \langle S_i \rangle \cos \theta_i = \frac{n_i}{2}\sqrt{\frac{\epsilon_0}{\mu_0}}\, |E_i|^2 \cos \theta_i \qquad (3\text{-}40)$$

$$\upsilon_i \langle U_r \rangle = \langle S_r \rangle \cos \theta_i = \frac{n_i}{2}\sqrt{\frac{\epsilon_0}{\mu_0}}\, |E_r|^2 \cos \theta_i \qquad (3\text{-}41)$$

$$\upsilon_t \langle U_t \rangle = \langle S_t \rangle \cos \theta_t = \frac{n_t}{2}\sqrt{\frac{\epsilon_0}{\mu_0}}\, |E_t|^2 \cos \theta_t \qquad (3\text{-}42)$$

Each of these three equations apply separately to the normal and parallel components of polarization, resulting in six equations for the description of reflected and transmitted energy at a boundary.

We define *reflectivity* as

$$R = \frac{\langle U_r \rangle}{\langle U_i \rangle} = \frac{|E_r|^2}{|E_i|^2} \qquad (3\text{-}43)$$

and *transmissivity* as

$$T = \frac{\upsilon_t \langle U_t \rangle}{\upsilon_i \langle U_i \rangle} = \left(\frac{n_t}{n_i}\right)\frac{\cos \theta_t}{\cos \theta_i} \cdot \frac{|E_t|^2}{|E_i|^2} \qquad (3\text{-}44)$$

[The quantities defined by **(3-43)** and **(3-44)** are ratios of the Poynting vectors and therefore assume a wave of known frequency and phase. Experimentally, the ratios of the incident flux to the reflected and transmit-

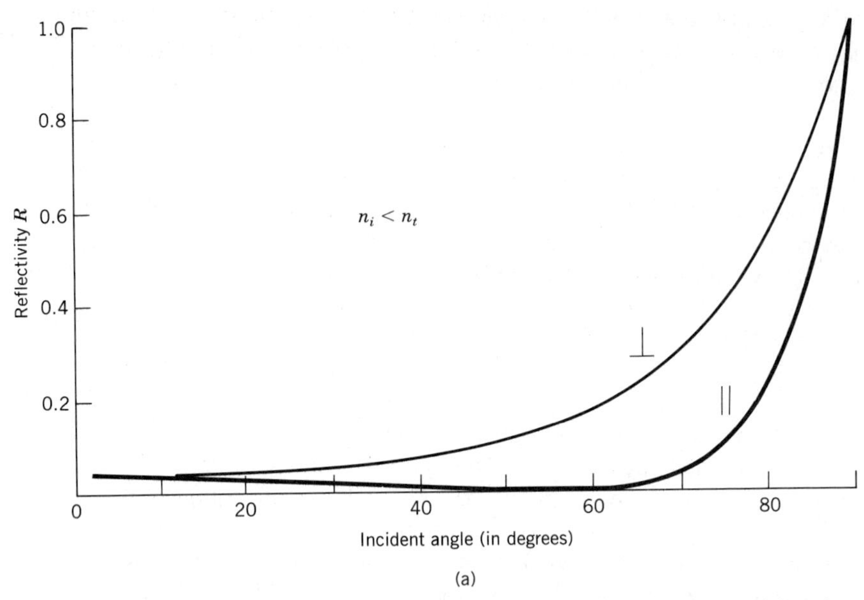

FIGURE 3-6a. Reflectivity as a function of incident angle for the case of $n_i = 1.0$ and $n_t = 1.5$.

ted flux are determined using radiometric units called the *reflectance* and *transmittance*.] The expressions for the reflectivity and the transmissivity can be used to demonstrate that energy is conserved, $T + R = 1$, when light encounters a boundary (see Problem 3-7).

Figure 3-6 displays the dependence of the reflectivity on the angle of incidence for both polarizations. Here, as was done for the amplitude reflection coefficient, the conditions of $n_i < n_t$ (Figure 3-6*a*) and $n_i > n_t$ (Figure 3-6*b*) are shown. The special cases when $r_N^2 = r_P^2 = 1$ and $r_P^2 = 0$

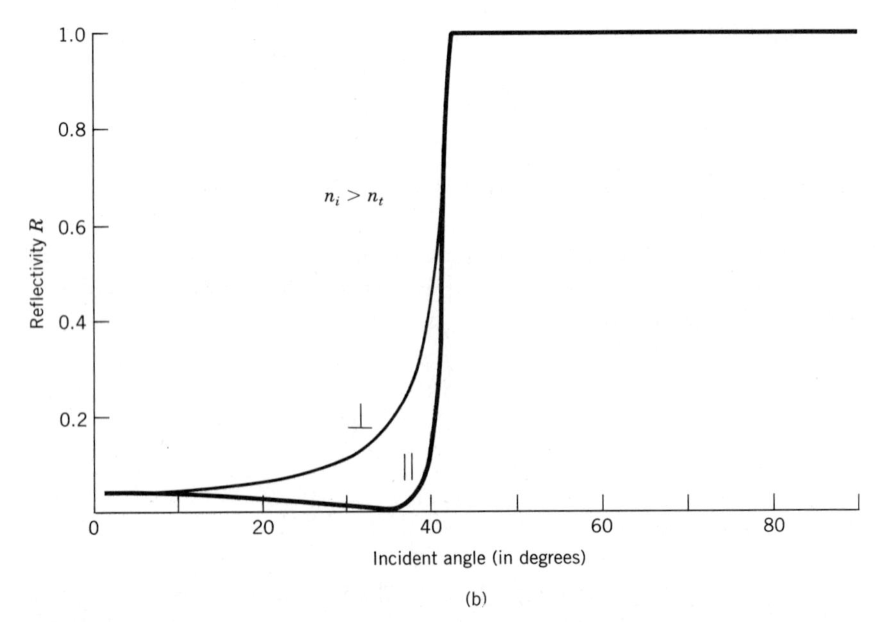

FIGURE 3-6b. Reflectivity as a function of incident angle for the case of $n_i = 1.5$ and $n_t = 1$.

are quite noticeable. After first discussing reflection and transmission when a light wave is incident normal to the boundary, we will examine these special cases.

NORMAL INCIDENCE

The meaning of a plane of incidence is lost for $\theta \approx 0$, since the two vectors that are supposed to define the plane are parallel and thus cannot define a plane. On examining Figure 3-6, it can be seen that $R_N = R_P$ when $\theta < 10°$. For these reasons, at normal incidence, that is, when $\theta_i \approx 0$, the distinction between the normal and parallel polarizations is lost.

The reflection and transmission coefficients at $\theta = 0°$ can be found if the limit of the coefficients is taken as $\theta \to 0°$. The sine found in the denominator of **(3-37)** can be eliminated by using Snell's law and the two trigonometric identities

$$\sin(\theta_i + \theta_t) = \sin\theta_i \cos\theta_t + \cos\theta_i \sin\theta_t$$

$$\cos(\theta_i - \theta_t) = \cos\theta_i \cos\theta_t + \sin\theta_i \sin\theta_t$$

$$\frac{E_{Pt}}{E_{Pi}} = \frac{2\cos\theta_i \sin\theta_t}{\sin\theta_t\left[\left(\frac{n_t}{n_i}\right)\cos\theta_t + \cos\theta_i\right]\left[\cos\theta_i\cos\theta_t + \sin\theta_i\sin\theta_t\right]}$$

Sin θ_t cancels out of the numerator and denominator so that the limit of the ratio as θ_i and $\theta_t \to 0$ is found to be

$$\frac{E_{Pt}}{E_{Pi}} = \frac{2n_i}{n_t + n_i} \tag{3-45}$$

In the same way, the limit of **(3-31)** as $\theta \to 0$ is

$$\frac{E_{Nt}}{E_{Ni}} = \frac{2n_i}{n_t + n_i}$$

yielding the same transmission coefficient for the two polarizations.

The limit of **(3-33)** as $\theta \to 0$ can also be obtained using the above identities

$$\frac{E_{Nr}}{E_{Ni}} = \frac{-\sin\theta_t\left[\left(\frac{n_t}{n_i}\right)\cos\theta_t - \cos\theta_i\right]}{\sin\theta_t\left[\left(\frac{n_t}{n_i}\right)\cos\theta_t + \cos\theta_i\right]} = \frac{-(n_t - n_i)}{n_t + n_i} \tag{3-46}$$

The limit of **(3-39)** is a little more difficult to obtain because terms that go to zero as θ approaches zero do not cancel out of the equation

$$\frac{E_{Pr}}{E_{Pi}} = \frac{\dfrac{\sin(\theta_i - \theta_t)}{\cos(\theta_i - \theta_t)}}{\dfrac{\sin(\theta_i + \theta_t)}{\cos(\theta_i + \theta_t)}}$$

Near normal incidence, $\theta_i \approx \theta_t \approx 0$ and we have $\cos\theta_i = \cos\theta_t \approx 1$ and $\sin\theta_t \approx \theta_t$, leading to

$$\sin\theta_i = \left(\frac{n_t}{n_i}\right)\sin\theta_t \approx \left(\frac{n_t}{n_i}\right)\theta_t$$

We can write

$$\frac{\sin(\theta_i \pm \theta_t)}{\cos(\theta_i \pm \theta_t)} \approx \frac{\sin\ \theta_i \cos\ \theta_t \pm \cos\ \theta_i \sin\ \theta_t}{\cos\ \theta_i \cos\ \theta_t - \sin\ \theta_i \sin\ \theta_t}$$

$$\approx \frac{\left(\frac{n_t}{n_i}\right)\theta_t \pm \theta_t}{1 - \left(\frac{n_t}{n_i}\right)\theta_t^2}$$

Keeping only terms linear in θ_t gives

$$\frac{E_{Pr}}{E_{Pi}} = \frac{n_t - n_i}{n_t + n_i} \tag{3-47}$$

The reflection coefficient is directly proportional to the difference between the propagation velocities of the two media forming the interface, a result identical to the one obtained for waves on a string **(3-5)**. Experimental proof of this result is shown in Figure 3-7. A glass rod, with an index of refraction of about 1.5, is seen passing through air ($n = 1.0$), xylene ($n = 1.5$), and water ($n = 1.3$). The rod is easily seen in air and water because of light reflected from its surface. The glass rod is nearly invisible in the xylene because the indices of refraction of the rod and the liquid are nearly equal.

The reflection coefficient for the two polarizations has the same magnitude; the difference in sign between **(3-46)** and **(3-47)** is due to the geometry and sign convention used in Figure 3-4. (Upon reflection, the normal polarization along the y axis is not affected by the change in propagation direction, but the parallel polarization in the x, z plane has its x component reversed in sign).

FIGURE 3-7. A glass rod rests in a beaker containing water with xylene floating on its surface. The rod is invisible in the xylene because no light is reflected from its surface.

Most optical glasses have an index of refraction of about 1.5 and air has an index of 1.0; therefore, $n_t/n_i = 1.5$ and

$$R = \left(\frac{n_t - n_i}{n_t + n_i}\right)^2 = \left(\frac{\frac{n_t}{n_i} - 1}{\frac{n_t}{n_i} + 1}\right)^2 = 0.04$$

This means that at each glass-air interface, 4% of the incident light is reflected. The efficiency of solar collectors is affected by this loss and a trade-off between thermal conduction loss to the surrounding air, reduced by multiple glass panes, and reflective loss must be made when designing solar collectors. Early camera lenses used optical designs that required few elements because of the problems created by reflections at air-glass interfaces.

If we calculate the transmissivity for a glass-air interface, we find that $T = 0.96$; thus, $T + R = 1.0$ and energy is conserved. It can be shown in general that this is true (see Problem 3-15).

In electromagnetic theory, the ability to transmit energy from a source, say a generator or oscillator, to a load is affected by the impedance of the source, the transmission medium, and the load. The same formalism can be used to describe the transmission of light. The equations for reflection and refraction can be reformulated in terms of the impedance of each medium (see Problem 3-10), yielding new definitions for the reflectivity and transmissivity

$$R = \left(\frac{Z_t - Z_i}{Z_t + Z_i}\right)^2 \tag{3-48}$$

$$T = \frac{4Z_iZ_t}{(Z_t + Z_i)^2} \tag{3-49}$$

The choice of using the impedance formulation or the index of refraction formulation for optics problems is one of personal preference.

POLARIZATION BY REFLECTION

In Figure 3-8 we replot the reflectivity of a wave polarized in the plane of incidence to emphasize the region where the amplitude of the reflected wave goes to zero. For the component of polarization parallel to the plane of incidence, the angle of incidence for which there is no reflected wave is named after **Sir David Brewster (1781–1868)**, the inventor of the kaleidoscope.

From the equation for the reflection coefficient **(3-39)**, we see that the reflection coefficient is zero whenever $\tan(\theta_i + \theta_t) = \infty$. This occurs if the sum of the angles is 90°, that is, when $\theta_i + \theta_t = \pi/2$. To calculate *Brewster's angle*, we use Snell's law

$$\frac{n_t}{n_i} = \frac{\sin\theta_i}{\sin\theta_t} = \frac{\sin\theta_i}{\sin\left(\frac{\pi}{2} - \theta_i\right)} = \frac{\sin\theta_i}{\cos\theta_i} = \tan\theta_B \tag{3-50}$$

Brewster's angle is therefore given by

$$\theta_B = \tan^{-1}\left(\frac{n_t}{n_i}\right)$$

In Figure 3-8, when $n_t > n_i$, the ratio of indices is $n_t/n_i = 1.5$ and

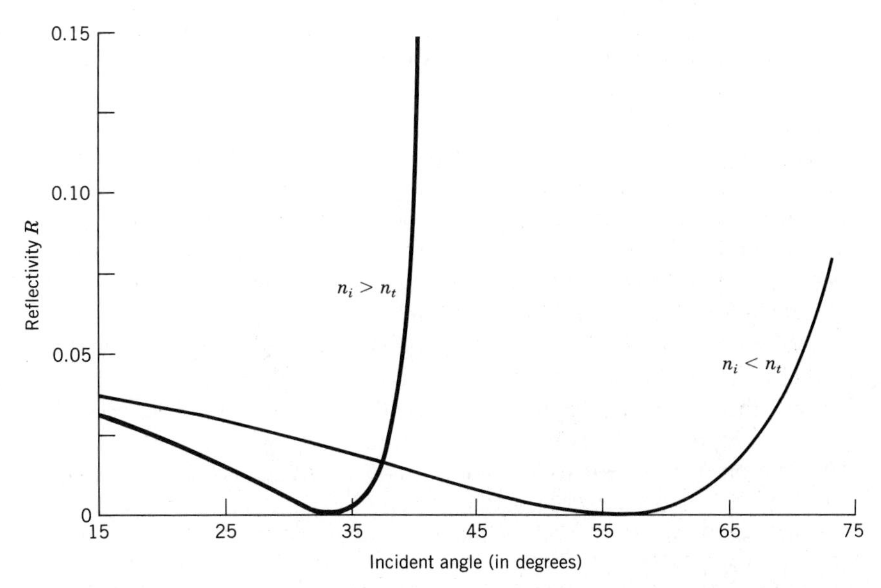

FIGURE 3-8. The reflectivity of a wave polarized in the plane of incidence as a function of the incident angle around the Brewster's angle. The two indices are 1.0 and 1.5.

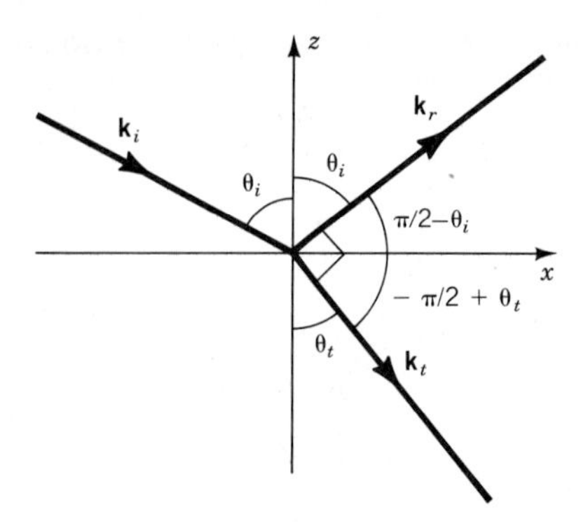

FIGURE 3-9. Geometry used to explain the occurrence of Brewster's angle.

Brewster's angle is $\theta_B = 56.3°$. When $n_i > n_t$, the ratio of the indices is $n_t/n_i = 0.67$ and Brewster's angle is the complement of the previous angle, $\theta_B = 33.7°$. As can be seen in Figure 3-8, the reflectivity remains near zero over a large range of angles, making the effect quite easy to observe.

Some laser designs use windows or crystal surfaces set at the Brewster's angle to reduce the reflective losses in the laser cavity for one polarization. Some brands of sunglasses are designed to take advantage of the fact that light of one polarization is not reflected strongly from dielectric surfaces. These sunglasses use a material (polaroid) that transmits light polarized in one direction and absorbs light polarized normal to that direction (see Chapter 13 for a discussion of polaroid).

A simple physical explanation can be given for Brewster's angle.[10] In Figure 3-9, the angle between $\mathbf{k}_r$ and $\mathbf{k}_t$ is

$$\left(\frac{\pi}{2} - \theta_i\right) - \left(-\frac{\pi}{2} + \theta_t\right) = \pi - (\theta_i + \theta_t)$$

At Brewster's angle $(\theta_i + \theta_t) = \pi/2$, so that $\mathbf{E}_t$ is parallel to $\mathbf{k}_r$. When light is incident on a medium, the electric field causes the electrons to vibrate in the direction of the field of the transmitted wave $\mathbf{E}_t$. The vibrating electrons radiate an electromagnetic wave that propagates back into the first medium; this is the origin of the reflected wave. There is no radiation produced in the direction of vibration[3] of the electrons. Thus, when the reflected and transmitted waves are propagating at right angles to each other, the reflected wave does not receive any energy from oscillations in the plane of incidence. The geometry of the waves at Brewster's angle, shown in Figure 3-9, demonstrates that no radiation will be observed. Magnetic materials complicate the problem[11] but the mathematics needed to describe the effects are a straightforward extension of this derivation.

TOTAL REFLECTION

From Figure 3-6*a*, it is apparent that when light is incident on a more dense (larger value of index n) medium from a less dense one, then the reflectivity for the perpendicular polarization is a monotonically increasing function of the incident angle. The reflectivity for the parallel polarization decreases to zero at the Brewster's angle and then exhibits the same behavior as the reflectivity for the perpendicular polarization.

When light is incident on a less dense medium from a more dense medium, Figure 3-6*b* shows that the reflectivity is 1 beyond a critical angle. Beyond this critical angle, the light is said to undergo *total reflection*. A schematic representation of refraction and total reflection is shown in Figure 3-10. When $n_i > n_t$, Snell's law states that $\theta_t > \theta_i$ because for θ between 0° and $\pi/2$, $\sin\theta$ is a monotonically increasing function. The angle θ_t reaches $\pi/2$ and $\sin\theta_t$ reaches 1 when the incident angle is equal to $\theta_c (< \pi/2)$ obtained from Snell's law

$$\theta_c = \sin^{-1}\left(\frac{n_t}{n_i}\right) \tag{3-51}$$

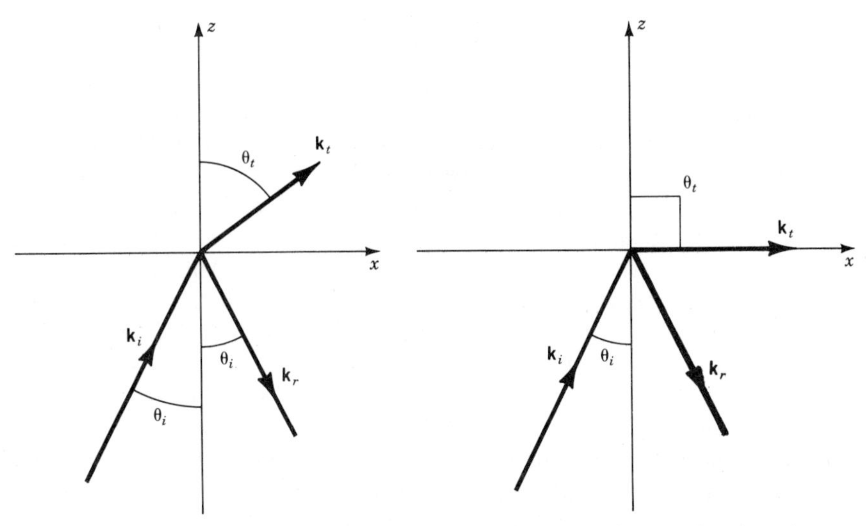

FIGURE 3-10. Example of a wave incident at an angle well below the critical angle is shown on the left; this is normal refraction. On the right is shown a wave incident at the critical angle when total internal reflection occurs.

Figure 3-11 shows an experimental demonstration of total reflection. In the experiment, a water/xylene interface is formed by floating a layer of xylene on water. When light is incident at an angle less than the critical angle

FIGURE 3-11. Total reflection occurring at a water/xylene interface. Light from a HeCd laser is incident on the interface from the xylene. The water contains the dye Rhodamine 6G in solution. If the blue laser light propagates in the water, it is absorbed by the Rhodamine, which then produces a red fluorescence. At angles of incidence greater than the critical angle, no radiation propagates into the water and no red light is visible. At angles of incidence less than the critical angle, light travels into the water and red radiation is observed. See color insert.

$$\theta_c \approx \sin^{-1}\left(\frac{1.3}{1.5}\right) = 62.7^\circ$$

then conventional reflection and refraction take place. When light is incident at an angle greater than the critical angle, total reflection takes place, as is shown in Figure 3-11. The water contains the dye Rhodamine 6G in solution. The dye strongly absorbs the blue light, from the HeCd laser used to produce the beams in the figures, and reemits light in the yellow to red region of the spectrum. The dye complicates the calculation of θ_c because n_t is now complex; however, it makes a dramatic demonstration. At angles less than θ_c, a rapidly decaying beam of orange light is seen propagating in the water; see Figure 2-11 and the color insert. For angles equal to or greater than θ_c, the orange beam disappears and only blue light is seen.

Total reflection occurs when $\theta_c < \theta_i \le \pi/2$. Beyond the critical angle θ_c, the angle θ_t becomes imaginary. To understand how the angle becomes imaginary and the physical significance of the imaginary angle, recall that

$$\cos\theta_t = \sqrt{1 - \sin^2\theta_t}$$

Snell's law can be used to rewrite this identity

$$\cos\theta_t = \sqrt{1 - \left(\frac{n_i}{n_t}\right)^2 \sin^2\theta_i} = \sqrt{1 - \left(\frac{\sin\theta_i}{\sin\theta_c}\right)^2}$$

Over the interval $0 \le \theta \le \pi/2$, we have $0 \le \sin\theta \le 1$ so that $\sin\theta_i/\sin\theta_c > 1$ when $\theta_c < \theta_i$. This leads to the conclusion that $\cos\theta_t$ is an imaginary function

$$\cos\theta_t = i\sqrt{\left(\frac{\sin\theta_i}{\sin\theta_c}\right)^2 - 1}, \qquad \theta_c < \theta_i \le \frac{\pi}{2}$$

For notational convenience, select the negative root and set it equal to α so that $-i\alpha = \cos\theta_t$. The transmitted wave, represented in Figure 3-10 by its propagation vector $\mathbf{k}_t$, can be written in this notation as

$$E_t \propto e^{-i\mathbf{k}_t \cdot \mathbf{r}} = e^{-ik_t(x\sin\theta_t + z\cos\theta_t)}$$

$$E_t = e^{-k_t\alpha z} e^{-ik_t x\sqrt{1+\alpha^2}}$$

The transmitted wave propagates parallel to the surface (i.e., along the x axis) and is attenuated exponentially in the z direction, i.e., normal to the propagation direction. This is called an inhomogeneous plane wave. As we will see in a moment, the wave is not transverse but has a component of the field parallel to $\mathbf{k}$.

The reflectivities for the two polarizations are

$$R_P = \frac{\tan^2(\theta_i - \theta_t)}{\tan^2(\theta_i + \theta_t)} = \frac{\tan^2(\theta_i - \frac{\pi}{2})}{\tan^2(\theta_i + \frac{\pi}{2})} = \frac{\operatorname{ctn}^2\theta_i}{\operatorname{ctn}^2\theta_i} = 1$$

$$R_N = \frac{\sin^2(\theta_i - \theta_t)}{\sin^2(\theta_i + \theta_t)} = \frac{\sin^2(\theta_i - \frac{\pi}{2})}{\sin^2(\theta_i + \frac{\pi}{2})} = \frac{\cos^2\theta_i}{\cos^2\theta_i} = 1$$

Since the reflectivities are 1, there should be no energy flow across the boundary. Calculation of the z component of the Poynting vector will verify this.

To calculate the Poynting vector, we use the geometry of Figure 3-10 and assume that the wave is incident with perpendicular polarization, i.e., **E** is parallel to the y axis

$$E_{yt} \propto e^{-k_t \alpha z} e^{-ik_t x\sqrt{1+\alpha^2}}$$

$$\mathbf{H} \propto \mathbf{k}\times\mathbf{E} = -E_y\, k_z\, \hat{\mathbf{i}} + E_y\, k_x\, \hat{\mathbf{k}}$$

$$H_{zt} \propto \sqrt{1+\alpha^2}\sqrt{\frac{\epsilon_t}{\mu_t}}\, E_{yt}$$

$$H_{xt} \propto i\alpha E_{yt}\sqrt{\frac{\epsilon_t}{\mu_t}}$$

The Poynting vector is

$$\mathbf{S} = \mathbf{E}\times\mathbf{H} = E_y\, H_z\, \hat{\mathbf{i}} - E_y\, H_x\, \hat{\mathbf{k}}$$

$$S_x \propto E_{yt}^2\sqrt{1+\alpha^2}\sqrt{\frac{\epsilon_t}{\mu_t}}$$

$$S_z \propto -E_{yt}^2\, i\alpha\sqrt{\frac{\epsilon_t}{\mu_t}}$$

Energy flows along the x direction but not in the z direction because the Poynting vector in the z direction is imaginary. An inhomogeneous wave attached to the surface and decaying exponentially from the surface, such as this one, is called an *evanescent wave*.

We can define a penetration depth $z = 1/\gamma$ as the value of z where the amplitude of the wave drops to $1/e$ of its original value; γ is called the decay constant

$$\frac{1}{\gamma} = \frac{1}{k_t\alpha}$$

We can detect this exponentially decaying wave by bringing another high index material within a few penetration depths of the first material. This is called *frustrated total reflection* and is analogous to quantum mechanical tunneling.[12] Figure 3-12 shows two applications of frustrated total reflection, a variable attenuator and a prism coupler for optical waveguide. The latter device uses frustrated total reflection to couple light into an optical wave guide.[13]

While the amplitude of the wave is not modified when the light undergoes total reflection, the phase of the light is modified. We can examine this phase change by looking at the reflection coefficients. The phase will be important in Chapter 5 in which the propagation of light in an optical fiber is discussed. We will rewrite **(3-33)** and **(3-39)** to get

$$r_N = -\frac{\sin\theta_i\cos\theta_t - \cos\theta_i\sin\theta_t}{\sin\theta_i\cos\theta_t + \cos\theta_i\sin\theta_t}$$

$$= \frac{\sqrt{1+\alpha^2}\cos\theta_i + i\alpha\sin\theta_i}{\sqrt{1+\alpha^2}\cos\theta_i - i\alpha\sin\theta_i} \tag{3-52}$$

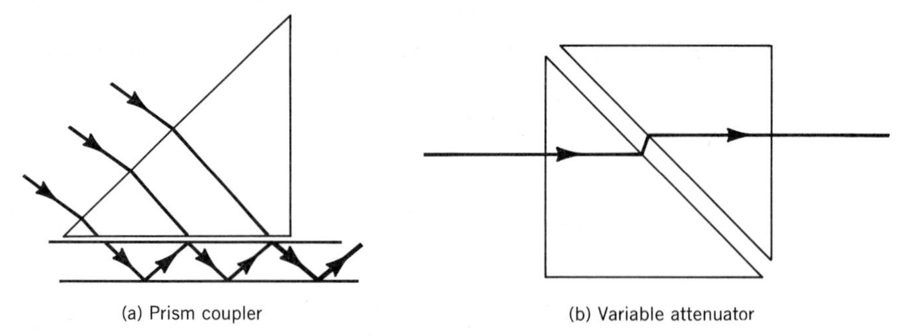

FIGURE 3-12. Two examples of frustrated total reflection. (a) The evanescent wave from a light beam undergoing total internal reflection in a prism is coupled into a mode of an optical waveguide. (b) Two identical prisms are used to make a variable attenuator by varying the amount of evanescent wave coupled into the second prism.

$$r_P = \frac{\sin\theta_i \cos\theta_i - \cos\theta_t \sin\theta_t}{\sin\theta_i \cos\theta_i + \cos\theta_t \sin\theta_t} = \frac{\sin\theta_i \cos\theta_i + i\alpha\sqrt{1+\alpha^2}}{\sin\theta_i \cos\theta_i - i\alpha\sqrt{1+\alpha^2}} \tag{3-53}$$

Both r_N and r_P are complex numbers of the form $(a + ib)/(a - ib)$. These complex variables can be written as

$$\frac{ae^{i\varphi}}{ae^{-i\varphi}} = e^{2i\varphi}$$

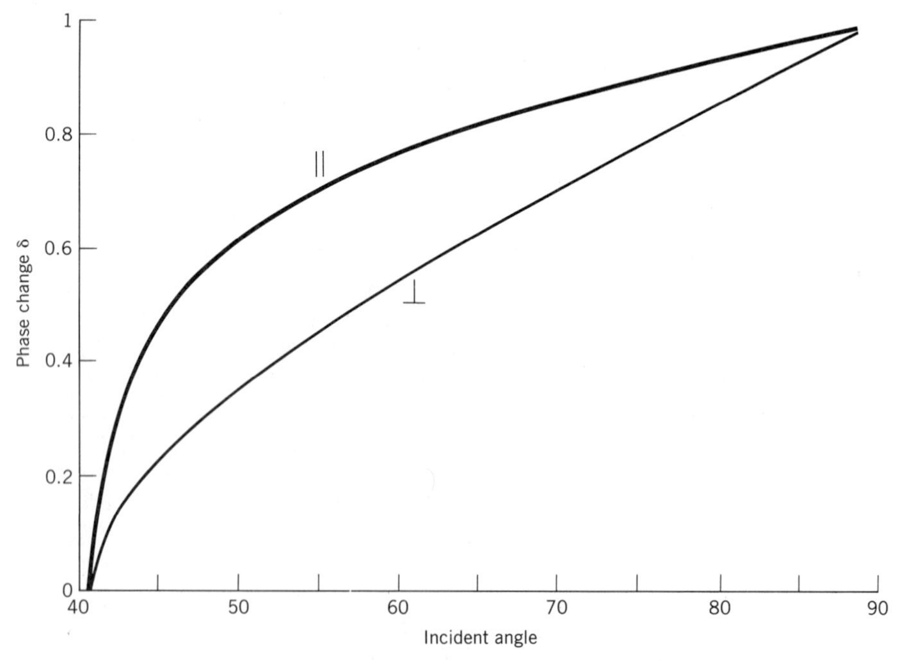

FIGURE 3-13. The phase change upon total reflection for each of the two polarizations. For this calculation, we assumed that the index of the dense medium was 1.5 and the index of the less dense medium was 1.0.

We then define $\delta/2 = \varphi$ and $\tan(\delta/2) = b/a$. This leads to phase shifts of

$$\tan\left(\frac{\delta_N}{2}\right) = \frac{\alpha \sin\theta_i}{\sqrt{1+\alpha^2}\cos\theta_i} \tag{3-54}$$

$$\tan\left(\frac{\delta_P}{2}\right) = \frac{\alpha\sqrt{1+\alpha^2}}{\sin\theta_i \cos\theta_i} \tag{3-55}$$

The phase shifts as a function of incident angle are shown in Figure 3-13 for the case of $n_i = 1.5$ and $n_t = 1.0$.

REFLECTION FROM A CONDUCTOR

The results derived for reflection and transmission from a boundary apply to materials with a complex index of refraction as well as a material with a real index. To use the derived equations for materials with complex indices, simply replace the real index by the equivalent complex one. We will demonstrate the procedure by finding the reflectivity of a metal that has a complex index of refraction. We will limit our discussion to the near-normal reflection from an air/metal interface where

$$n_t = \mathcal{N}$$

(a complex index of refraction) and $n_i = 1$. If we replace the real index in **(3-46)** by a complex index

$$R = \frac{(\mathcal{N}-1)(\mathcal{N}^*-1)}{(\mathcal{N}+1)(\mathcal{N}^*+1)} = \frac{(n-1+in\kappa)(n-1-in\kappa)}{(n+1+in\kappa)(n+1-in\kappa)} = \frac{(n-1)^2+(n\kappa)^2}{(n+1)^2+(n\kappa)^2}$$

$$R = 1 - \frac{4n}{(n+1)^2+(n\kappa)^2} \tag{3-56}$$

The association of R and the absorption in a material can be observed in Figure 3-14, where $n\kappa$ and R are plotted as a function of wavelength for the

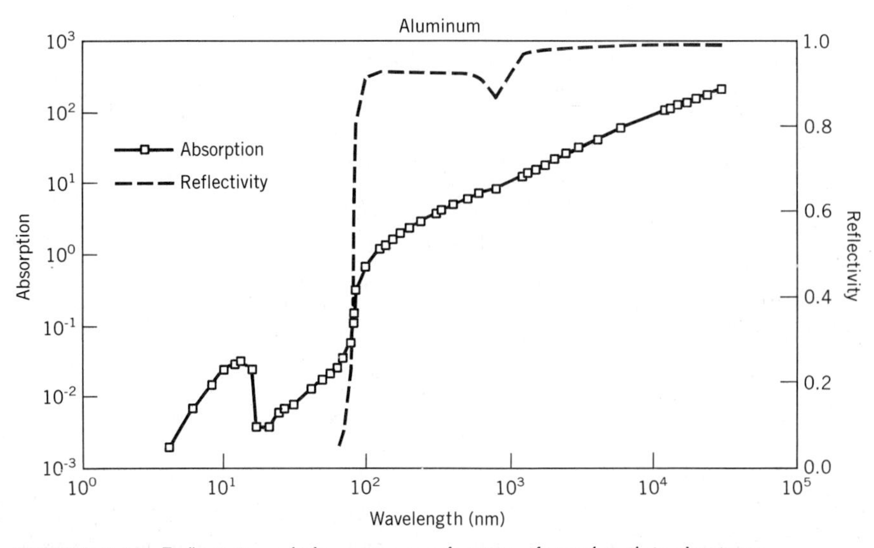

FIGURE 3-14. Reflectivity and absorption as a function of wavelength in aluminum.

metal aluminum. The reason for the dependence of $n\kappa$ on wavelength will be discussed in Chapter 7.

If the index of refraction were purely imaginary, then

$$\mathcal{N} = in\kappa$$

$$R = \frac{(in\kappa - 1)(-in\kappa - 1)}{(in\kappa + 1)(-in\kappa + 1)} = \frac{-(in\kappa - 1)(in\kappa + 1)}{-(in\kappa + 1)(in\kappa - 1)} = 1$$

and the material would be a perfect reflector.

SUMMARY

By requiring that the phase of a wave be continuous across the boundary between media with different wave propagation velocities, the laws of reflection and refraction were derived. The media's propagation velocities are used to define the index of refraction of the medium containing the incident wave n_i and the transmitted wave n_t.

Law of Reflection

1. The reflected wave's propagation vector lies in the plane defined by the incident wave's propagation vector and the normal to the boundary at the point where the incident wave intersects the boundary. The plane is called the plane of incidence.

2. The angle between the reflected wave's propagation vector and the normal to the boundary θ_r is equal to the angle between the incident wave's propagation vector and the normal θ_i

$$\sin\theta_i = \sin\theta_r$$

Law of Refraction

1. The propagation vector of the wave transmitted across the boundary lies in the plane of incidence.

2. The angle between the transmitted wave's propagation vector and the normal to the boundary θ_t is given by Snell's law

$$\sin\theta_t = \frac{n_i}{n_t}\sin\theta_i$$

Maxwell's equations and their boundary conditions were used to derive the amplitude of the reflected and transmitted wave for the electric field in the plane of incidence

$$t_P = \frac{E_{Pt}}{E_{Pi}} = \frac{2\cos\theta_i\sin\theta_t}{\sin(\theta_i + \theta_t)\ \cos(\theta_i - \theta_t)}$$

$$r_P = \frac{E_{Pr}}{E_{Pi}} = \frac{\tan(\theta_i - \theta_t)}{\tan(\theta_i + \theta_t)}$$

and normal to the plane of incidence

$$t_N = \frac{E_{Nt}}{E_{Ni}} = \frac{2\sin\theta_t\cos\theta_i}{\sin(\theta_i + \theta_t)}$$

$$r_N = \frac{E_{Nr}}{E_{Ni}} = -\frac{\sin(\theta_i - \theta_t)}{\sin(\theta_i + \theta_t)}$$

The reflection from a surface under several special conditions was discussed. The condition of normal incidence reduces the two equations to a single equation

$$R = \left(\frac{n_t - n_i}{n_t + n_i}\right)^2$$

which demonstrates that the reflectivity depends on the relative propagation velocities in the two media.

There is an angle for which a wave, polarized parallel to the plane of incidence, is not reflected. The angle is called the Brewster's angle and is given by

$$\theta_B = \tan^{-1}\left(\frac{n_t}{n_i}\right)$$

When the incident wave strikes a boundary with a medium having an index of refraction less than the index of the incident wave's medium, then when the angle of incidence exceeds an angle given by

$$\theta_c = \sin^{-1}\left(\frac{n_t}{n_i}\right)$$

the incident wave will be totally reflected. The totally reflected wave will experience a phase shift given by

$$\tan\left(\frac{\delta_N}{2}\right) = \frac{\alpha \sin\theta_i}{\sqrt{1+\alpha^2}\cos\theta_i}$$

$$\tan\left(\frac{\delta_P}{2}\right) = \frac{\alpha\sqrt{1+\alpha^2}}{\sin\theta_i \cos\theta_i}$$

PROBLEMS

3-1. An unpolarized beam of light is one whose Stokes vectors are $s_1 = s_2 = s_3 = 0$. If such a beam in air is incident at an angle of 30° on glass with an index of 1.50, what is the percentage of light energy refracted in the normal and parallel polarized components? What is the degree of polarization?

3-2. Light is incident on a medium of index $n = 1.682$ from air. What is the Brewster's angle?

3-3. Assume we have two strings joined at the origin. The velocity of the wave in the string on the left is $v_1 = 20$ m/sec. The velocity for the string on the right is $v_2 = 10$ m/sec. The string on the right has a wave with an amplitude of 3 cm and a wavelength of 1 m moving toward the junction.
(a) What is the amplitude of the transmitted and reflected wave and the wavelength of the transmitted wave?
(b) What is the ratio of the power transmitted to the power reflected?

3-4. A Gaussian displacement traveling along a string

$$s(x,t) = A\exp\left\{-\left[\frac{t-(x/v_1)}{t_1}\right]^2\right\}$$

is incident from the left onto a junction at $x = 0$ where the linear mass density of the string changes from μ_1 to μ_2 and the velocity of propagation of a wave along the string changes from v_1 to v_2. Write down s_{trans} and s_{refl}.

3-5. Calculate the critical angle and Brewster's angle for water ($n = 1.33$) and dense flint glass ($n = 1.75$) when light propagating in the medium is incident on an interface with air.

3-6. Use the computer to generate a plot $R_p(R_\pi)$ and $R_N(R_\sigma)$ for an air-glass interface ($n_1 = 1$ and $n_2 = 1.5$) for the following two cases: (a) when the light is incident from the air side; (b) when the light is incident from the glass side.

3-7. A tank of water is covered with a 1 cm thick layer of oil ($n_0 = 1.48$); above the oil is air. If a beam of light originates in the water, what angle must the light beam make at the water-oil interface if no light is to escape into the air?

3-8. As the sun rises over a still pond, an angle will be reached when its image on the water's surface is completely linearly polarized in a plane parallel to the water's surface. What is the incident angle?

3-9. Use **(3-43)** and **(3-44)** to prove that energy is conserved at the interface between two dielectrics, i.e., $T + R = 1$.

3-10. Derive **(3-48)** and **(3-49)** using the definitions of impedance, found in the discussion following **(2-20)**, of reflectivity and of transmissivity.

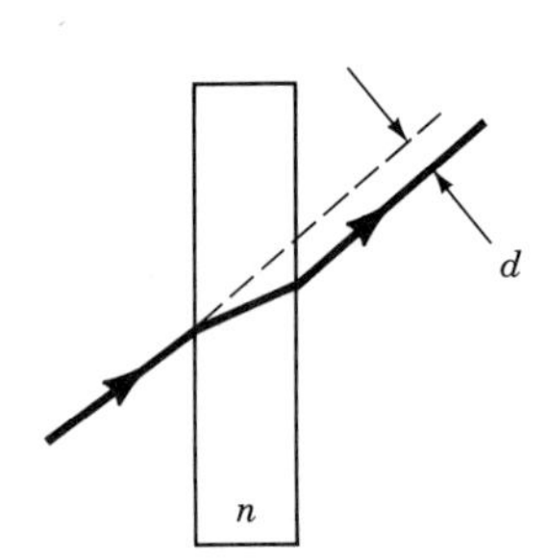

FIGURE 3-15. Displacement of light after passing through dielectric medium.

3-11. Derive an expression for the displacement of a light wave's propagation vector after it passes through a plane parallel glass plate of index n and thickness t as is shown in Figure 3-15. Show that for small n, the displacement is

$$d \approx t \sin\theta \left(1 - \frac{1}{n}\right)$$

where d is defined in Figure 3-15.

3-12. Derive an expression for the transmittance of light normally through a stack of N glass plates, each separated by a small air space. Assume no absorption and that all the plates have an index of refraction n.

3-13. Derive Brewster's law for the case when we cannot assume $\mu_1 = \mu_2 = \mu_0$. Carry out the derivation for both polarization states.

3-14. The index of refraction of Germanium at $\lambda = 500$ nm is $3.47 - i(1.4)$. What is the reflection coefficient of a polished germanium surface at normal incidence?

3-15. Use the equations for transmissivity and reflectivity for normal incidence to prove that energy is conserved.

3-16. Determine the value of the amplitude reflection coefficients for light incident at 30° on an air-glass interface where $n_i = 1.0$ and $n_t = 1.5$. Write the Stokes vector for the reflected light.

3-17. Mirrors used in x-ray optics are often based on total reflection. The index of refraction for x-rays of wavelength 0.15 nm in a vacuum is 1.0. For silver, the index is 0.99998. What is the angle for total reflection? Sketch the experimental arrangement needed to take advantage of total reflection.

3-18. What is the refractive index of a glass plate that polarizes light reflected at 57.5°?

3-19. A beam of light passes normally through one side and is totally reflected off the hypotenus of a 45°–90°–45° glass prism of index 1.6. What is the decay constant for the evanescent wave at the point of total reflection, assuming the prism is in air? What happens if the prism is in water?

3-20. A glass container of index 1.65 is filled with carbon tetrachloride with an index of 1.46. If a beam of light strikes the container normally, how much light is transmitted through the container?

Interference

The superposition principle, introduced in Chapter 1, states that the sum of solutions to the wave equation is also a solution to the wave equation. In this chapter, a general approach to the addition of waves of the same frequency will be developed so that the superposition principle may be applied to a large number of three-dimensional waves traveling in different directions.

The physical consequence of the superposition principle is the observation of bright and dark bands of light called *fringes* when a number of waves coexist in a region in space. These fringes are commonly observed in soap bubbles or on oil films on a wet roadway. (Discussion of the factors that determine when the fringes can be observed must be delayed until Chapter 8 when we will examine the addition of waves of different frequencies.) The bright regions occur when a number of waves add together to produce an intensity maximum of the resultant wave; this is called *constructive interference*. *Destructive interference* occurs when a number of waves add together to produce an intensity minimum of the resultant wave.

Collectively, the distribution of fringes is called an *interference pattern*. In this chapter, techniques for adding waves will be used to develop a theoretical explanation of the interference patterns produced by a number of simple physical systems. The systems to be treated are the interference between waves originating from two slits illuminated by a single, small light source (called Young's two-slit experiment) and reflections from the front and back surfaces of a thin dielectric layer. The interference of waves reflected from a dielectric layer is used as a model to explain the operation of a Michelson interferometer and a Fabry–Perot interferometer. The results for the dielectric layer are also used in Appendix 4-A to describe the design of dielectric mirrors and antireflection coatings.

The first observation of interference was made by **Robert Boyle (1627–1691)** in 1663, who saw what are now called Newton's rings. **Robert Hooke (1635–1703)** was a co-discoverer of the rings, but it is Newton who has his name associated with the interference pattern because he performed a number of experiments on the effect. The first experiments involving interference effects in light were performed by **Thomas Young (1773–**

1829) in 1802. His experiments disagreed with the then accepted particle theory developed by Newton and were rejected by most of the scientific community. (Young must have become discouraged for he left optics to decipher the Rosetta Stone.) Young's experiments were not decisive in establishing wave theory because the observed intensity patterns could have originated from processes other than interference. Ten years after Young's experiments, Fresnel performed experiments that confirmed Young's results and eliminated all other possible sources of the observed patterns. These experiments led to the rejection of the particle theory of Newton and its replacement by the wave theory of light.

ADDITION OF WAVES

In the discussion of the reflection of a wave on a vibrating string in Chapter 3, a simple algebraic technique was used to combine multiple waves. The addition of these waves was accomplished for each instant in time by adding the amplitudes of the waves algebraically, at each value of x, to obtain the resultant wave. There are two additional ways to add waves of the same frequency that are sometimes easier to apply to a general three-dimensional wave. We will demonstrate the three approaches by applying them to the addition of two or more scalar plane waves of the form

$$y = Y\cos(\omega t - \mathbf{k}\cdot\mathbf{r} + \varphi)$$

All of the waves will have the same frequency ω and all of the sources of phase differences between the various waves will occur in the phase

$$\phi = -\mathbf{k}\cdot\mathbf{r} + \varphi$$

The differences in ϕ could arise from differences in propagation paths, reflection, etc.

Trigonometric Approach

The two scalar waves

$$y_1 = Y_1 \cos(\omega t + \phi_1), \qquad y_2 = Y_2 \cos(\omega t + \phi_2) \tag{4-1}$$

can be added algebraically to produce the resultant wave

$$\begin{aligned} y &= y_1 + y_2 \\ &= (Y_1 \cos \phi_1 + Y_2 \cos \phi_2) \cos \omega t - (Y_1 \sin \phi_1 + Y_2 \sin \phi_2) \sin \omega t \end{aligned}$$

By defining two new variables Y and δ through the equations

$$Y \cos \delta = Y_1 \cos \phi_1 + Y_2 \cos \phi_2 \tag{4-2}$$

$$Y \sin \delta = Y_1 \sin \phi_1 + Y_2 \sin \phi_2 \tag{4-3}$$

the sum of the two waves can be rewritten as

$$\begin{aligned} y &= Y \cos \delta \cos \omega t - Y \sin \delta \sin \omega t \\ &= Y \cos (\omega t + \delta) \end{aligned} \tag{4-4}$$

The resultant **(4-4)**, obtained by adding together two sinusoidal waves of frequency ω, is a sinusoidal wave of the same frequency, ω. To solve for Y, add the square of **(4-2)** and **(4-3)** together to obtain

$$Y^2 = (Y_1 \cos \phi_1 + Y_2 \cos \phi_2)^2 + (Y_1 \sin \phi_1 + Y_2 \sin \phi_2)^2$$

To solve for δ, divide **(4-3)** by **(4-2)**

$$\tan\,\delta = \frac{Y_1 \sin\,\phi_1 + Y_2 \sin\,\phi_2}{Y_1 \cos\,\phi_1 + Y_2 \cos\,\phi_2}$$

This wave addition technique is the one used in Chapter 3 to generate Figure 3-2 and to obtain the equation for a standing wave **(3-6)**. The origin of the phase difference in ϕ between the incident and reflected wave for that example was the reflection.

This technique is not useful for adding a large number of waves because of the bookkeeping required when manipulating a large number of trigonometric functions; therefore, two other techniques will be introduced.

Complex Approach

If we start with N waves of the form given by **(4-1)**, they can be rewritten in complex notation as

$$y_1 = Y_1 e^{i(\omega t + \phi_1)}, \ldots, y_i = Y_i e^{i(\omega t + \phi_i)}, \ldots, y_N = Y_N e^{i(\omega t + \phi_N)} \tag{4-5}$$

To simplify the discussion, assume that the phase of each wave is ϕ_0 larger than the preceding wave and that the amplitude of each wave is identical and equal to the value Y_0. (Waves with these properties will be encountered when discussing multiple reflections in a dielectric film.) The sum of N waves of this type is written as

$$y = \sum_{j=1}^{N} y_j = Y_0 e^{i\omega t}\left[e^{i\phi_0} + e^{2i\phi_0} + \ldots + e^{Ni\phi_0}\right] = Y_0 e^{i\omega t} \sum_{j=1}^{N} e^{j(i\phi_0)} \tag{4-6}$$

and can be shown (Problem 4-16) to be equal to

$$y = Y_0 e^{i(\omega t + \phi_0)} \frac{1 - e^{Ni\phi_0}}{1 - e^{i\phi_0}} \tag{4-7}$$

To obtain the real part of y, **(4-7)** must be placed in standard form by eliminating the complex term from the denominator

$$y = Y_0 e^{i(\omega t + \phi_0)} \frac{e^{-i\phi_0} - e^{i(N-1)\phi_0} - 1 + e^{iN\phi_0}}{-2(1 - \cos\,\phi_0)}$$

Now the real part of y can be separated from the imaginary part and, through the use of some trigonometric identities, placed in the form

$$\mathcal{Re}\{y\} = \frac{Y_0 \sin\, N\dfrac{\phi_0}{2}}{\sin\,\dfrac{\phi_0}{2}} \cos\left[\omega t + (N+1)\,\frac{\phi_0}{2}\right] \tag{4-8}$$

Again, the sum of a number of harmonic waves of the same frequency ω leads to a resultant wave that is also a sinusoidal wave with the frequency ω and with an amplitude and phase given by

$$Y = Y_0 \frac{\sin\, N\dfrac{\phi_0}{2}}{\sin\,\dfrac{\phi_0}{2}}, \qquad \delta = (N+1)\,\frac{\phi_0}{2}$$

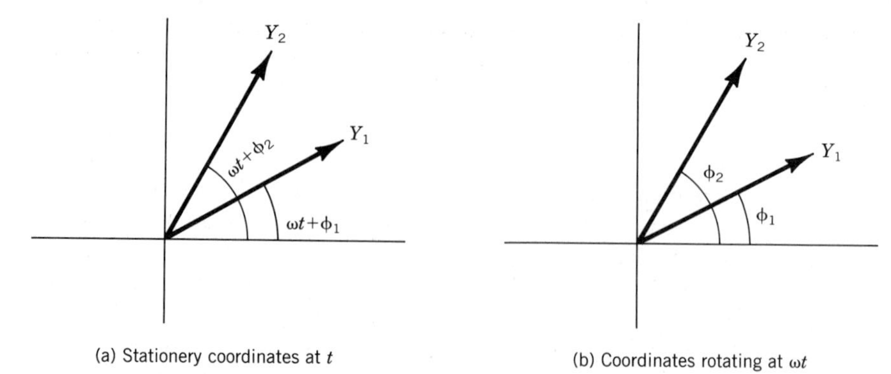

FIGURE 4-1. Vector representation of two waves with the same frequency and different phase. The coordinate system on the left is stationary, whereas the one on the right is rotating at an angular frequency equal to the waves' frequency so that the vectors appear stationary in this frame.

The results obtained by applying the trigonometric approach are identical to these if we set $N = 2$, $Y_1 = Y_2 = Y_0$, $\phi_1 = \phi_0$, and $\phi_2 = 2\phi_0$.

Vector Approach

The two waves given in **(4-1)** can be represented, at a time t_1, by vectors. The length of the vectors is equal to the maximum amplitude of the wave and the angle the vector makes with the abscissa is given by the phase $\omega t_1 + \phi$. If these vectors are drawn in a coordinate system rotating about the axis normal to the plane of the figure at a frequency ω, the vectors will appear stationary. Figure 4-1 shows the vector representation of the waves in a fixed coordinate system on the left and in a coordinate system rotating at an angular velocity ω on the right.

Vectors are added graphically by placing the vectors tail to head as shown in Figure 4-2. The resultant wave is then drawn from the tail of the first wave to the head of the last wave. The resultant vector shown in Figure 4-2 is stationary in the rotating frame and rotating at a frequency ω in the fixed coordinate system; therefore, the resultant wave is a harmonic wave of

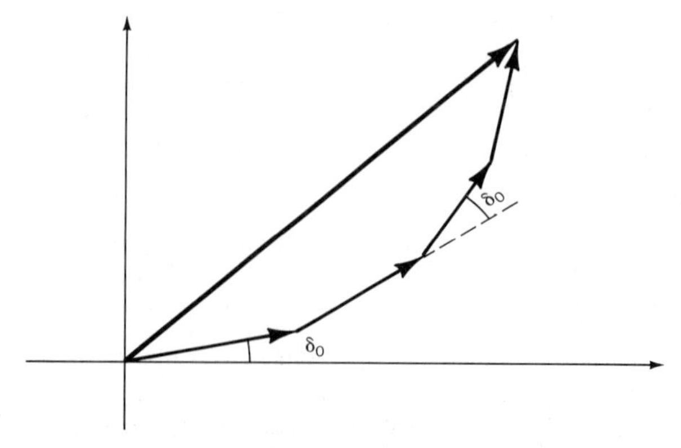

FIGURE 4-2. Vector representation of four waves with equal amplitudes and phases in an arithmetic progression starting at δ_0. The coordinate system is rotating at a frequency ω.

frequency ω. This approach is the graphic implementation of the complex approach. The advantage of the vector approach is the visual insight of the interference process it provides.

INTERFERENCE

To add light waves, attention must be paid to the fact that light wave displacements are vector quantities. We need to evaluate what effect this complication has on the techniques developed for adding scalar waves. We will represent the addition of two electromagnetic plane waves in its most general form. We will then discover, if we limit our attention to plane electromagnetic waves propagating in free space, that orthogonal polarizations will not interfere due to the fact that light waves are transverse.

We will represent two electromagnetic waves by

$$\mathbf{E}_1 = \mathcal{A}e^{i\omega t} \quad \text{and} \quad \mathbf{E}_2 = \mathcal{B}e^{i\omega t}$$

As we stated earlier in this chapter, we will assume that the waves all have the same frequency and will need to only concern ourselves with the *complex amplitudes* of the waves. The components of the complex amplitudes of the two waves are

$$\mathcal{A}_x = a_1 e^{-ig_1}, \qquad \mathcal{A}_y = a_2 e^{-ig_2}, \qquad \mathcal{A}_z = a_3 e^{-ig_3} \tag{4-9}$$

$$\mathcal{B}_x = b_1 e^{-ih_1}, \qquad \mathcal{B}_y = b_2 e^{-ih_2}, \qquad \mathcal{B}_z = b_3 e^{-ih_3} \tag{4-10}$$

where g and h are of the form $(\mathbf{k}\cdot\mathbf{r} - \phi)$. The vector $\mathbf{r}$ represents the position of the point in space where the wave amplitudes are added. Since we can only measure the intensity of the light wave, the time average of the square of the sum of the two waves must be calculated before the theoretical result can be compared with experiment,

$$\langle \mathbf{E}^2 \rangle = \langle (\mathbf{E}_1 + \mathbf{E}_2) \cdot (\mathbf{E}_1 + \mathbf{E}_2) \rangle = \langle \mathbf{E}_1^2 \rangle + \langle \mathbf{E}_2^2 \rangle + 2\langle \mathbf{E}_1 \cdot \mathbf{E}_2 \rangle \tag{4-11}$$

Equation **(2-25)** defines the intensity of the wave as proportional to the square of the electric field of the light wave. We are only interested in relative intensities so we will replace $\langle \mathbf{E}^2 \rangle$ by I and neglect the constants of proportionality.

The resultant wave's intensity is given by

$$I = I_1 + I_2 + 2\langle \mathbf{E}_1 \cdot \mathbf{E}_2 \rangle$$

I_1 and I_2 are the intensities of each wave, independent of the other. All information about interference is contained in the third term of this equation. If the third term is zero at all positions, the waves do not interfere and are said to be *incoherent* or *noncoherent*. (Details of this concept will be given in Chapter 8.) We will evaluate the third interference term using the complex notation of **(2-28)**; the third term becomes

$$2\langle \mathbf{E}_1 \cdot \mathbf{E}_2 \rangle = \mathcal{Re}\{\mathbf{E}_1 \cdot \mathbf{E}_2^*\} = \tfrac{1}{2}(\mathbf{E}_1 \cdot \mathbf{E}_2^* + \mathbf{E}_1^* \cdot \mathbf{E}_2)$$

$$= a_1 b_1 \cos(g_1 - h_1) + a_2 b_2 \cos(g_2 - h_2) + a_3 b_3 \cos(g_3 - h_3)$$

(This general result will be of use not only in our discussion of interference but also in the discussion of holography in Chapter 12).

Assume that the two waves to be added together are plane waves of the same frequency, orthogonally polarized, and propagating parallel to one another along the z axis. One wave is assumed to have its $\mathbf{E}$ vector located in the x, z plane so that $a_2 = 0$ and the other is assumed to have its **E** vector in the y, z plane so that $b_1 = 0$. With these assumptions, the interference term is

$$2\langle \mathbf{E}_1 \cdot \mathbf{E}_2 \rangle = a_3 b_3 \cos(g_3 - h_3)$$

Up to now the electromagnetic properties of the waves have been ignored. Now we will use the physical fact that light waves are transverse in free space. The transverse nature of light requires that $a_3 = b_3 = 0$ for the waves represented by $\mathbf{E}_1$ and $\mathbf{E}_2$ and results in the interference term being equal to zero.

Maxwell's equations provide the key result for interference that light waves propagating in free space and polarized at right angles to each other will not interfere. The key assumptions leading to this result are that the medium is isotropic and free of charge; thus, the result applies to many simple dielectrics. (There are materials in which an electromagnetic wave can have a longitudinal component and for those special cases, the above result must be modified.)

Since only parallel polarized waves interfere with one another, we can simplify the notation without loss of generality by assuming that all light waves are linearly polarized in the y direction and the waves are propagating in the x, z plane

$$a_1 = a_3 = b_1 = b_3 = 0$$

$$I_1 = \frac{a_2^2}{2}, \qquad I_2 = \frac{b_2^2}{2}$$

$$2\langle \mathbf{E}_1 \cdot \mathbf{E}_2 \rangle = a_2 b_2 \cos(g_2 - h_2) = 2\sqrt{I_1 I_2} \cos \delta$$

allowing the relation for the intensity of the resultant wave to be written as

$$I = I_1 + I_2 + 2\sqrt{I_1 I_2} \cos \delta \tag{4-12}$$

where

$$\delta = \left[(\mathbf{k}_1 \cdot \mathbf{r} - \mathbf{k}_2 \cdot \mathbf{r}) + (\phi_1 - \phi_2) \right] = \Delta + \Delta\phi \tag{4-13}$$

Two electromagnetic waves have been added using the complex approach and have been used to demonstrate that only parallel components of the electric field contribute to the interference term. The interference term has been shown to be a function of the amplitudes of the two waves and a harmonic function of the phase difference δ between the two waves.

In **(4-13)**, we have separated the phase difference δ into two components $\Delta\phi = \phi_1 - \phi_2$ equal to the phase difference due to differences in the epoch angles of the oscillators, producing the waves and a component, $\Delta = (\mathbf{k}_1 \cdot \mathbf{r}) - (\mathbf{k}_2 \cdot \mathbf{r})$ equal to the phase difference due to propagation path differences. The propagation paths for the two waves are measured in units of wavelength since $|\mathbf{k}| = 2\pi/\lambda$.

We know from the discussion of reflection and refraction in Chapter 3 that the wavelength of light depends on the propagation velocity in the medium; see Figure 3-2. In order to allow the light to propagate along paths in media with different indices of refraction and still evaluate their phase differences, all path lengths are converted to an equivalent path length in a vacuum. The equivalent path length or the *optical path length* between points A and B in a medium with an index of refraction n is defined as the distance a wave in a vacuum would travel during the time it took light to travel from A to B in the actual medium. If the distance between A and B is r and the velocity of propagation in the medium is v, then the time to travel from A to B is $\tau = r/v$. The distance light would travel in a vacuum in the time τ, that is, the optical path length, is given by

$$c\tau = \frac{cr}{v} = nr \tag{4-14}$$

This equation provides the mathematical definition of an optical path length.

The phase difference Δ can now be expressed in terms of the optical path length. Both waves were assumed to have the same frequency; therefore, the propagation constant of wave i is

$$|\mathbf{k}_i| = \frac{n_i 2\pi}{\lambda_0}$$

where λ_0 is the wavelength of the waves in a vacuum and n_i is the index of refraction associated with the path of the ith wave. With the definitions just introduced, it is apparent that Δ is the difference in optical path lengths of the two waves. This may be made obvious by assuming the index of refraction is the same for both propagation paths; then,

$$\Delta = \mathbf{k}_1 \cdot \mathbf{r} - \mathbf{k}_2 \cdot \mathbf{r} = (\mathbf{k}_{01} - \mathbf{k}_{02}) \cdot n\mathbf{r}$$

where the subscript 0 denotes the propagation vector in a vacuum.

When the two waves have the same amplitude, **(4-12)** can be rewritten as

$$I = 4I_1 \cos^2 \frac{\delta}{2} \tag{4-15}$$

Figure 4-3 shows a plot of the intensity distribution that is described by **(4-15)**.

The importance of the time average in the equations for interference cannot be overemphasized. Temporal variations of δ must be small during the period T, over which we take the temporal average, if the interference is to be observed. From an operational viewpoint, the detector used to observe interference must respond in a time T that is faster than the time for the intensity variations of the interference process to occur if interference is to be observed. We can define a time τ_c that characterizes the temporal variations of the phase; if $\tau_c > T$, then the waves are said to be *coherent*, interference effects are observable and the interference term is nonzero. (We will have more to say about the subject of coherence in Chapter 8; in this chapter, all temporal effects will be assumed to be negligible with $\tau_c >> T$.)

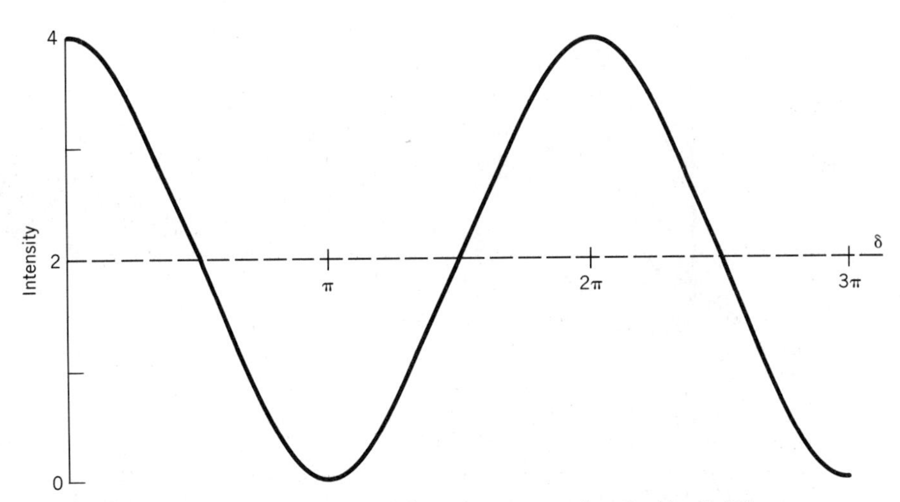

FIGURE 4-3. Interference of two waves of equal intensity as described by **(4-15)**.

The intensity is $4I_1$ when constructive interference occurs because the phase difference between the two waves is $\delta = m \cdot 2\pi$, and the intensity is zero when destructive interference occurs because the phase difference between the two waves is $\delta = (2m + 1)\pi$. Thus, constructive interference occurs when the wave amplitudes add to produce a maximum, and destructive interference occurs when the wave amplitudes add to produce a minimum in the observed light distribution. Conservation of energy requires the energy in the combined beams to be $2I_1$ if the energy in each beam is I_1. Is conservation of energy violated when the observed intensity reaches a value of $4I_1$? The answer is no; the energy has had its spatial distribution modified by interference but over the entire cross section of the resultant wave the energy is still $2I_1$ (the average of $\cos^2 \delta$ yields 1/2).

YOUNG'S INTERFERENCE

There are a number of ways to produce two waves that exhibit interference. One way is to mix samples of a single wavefront. This approach was first used by **Francesco Maria Grimaldi (1618–63)** who used an extended source to illuminate a pair of pinholes. The variations in the light distribution, observed by Grimaldi, could be explained not only by the theory of interference, but also by the theory of diffraction (we will discuss the theory of diffraction in Chapters 9–11). Thomas Young attempted to discriminate between the two theories by placing a pinhole between the source and the pair of pinholes. Although Young's experiment was an improvement over Grimaldi's, it did not remove the possibility that diffraction created the observed light distribution. Fresnel and Lloyd each developed an experimental arrangement (see Figure 4-4) that removed all questions as to the origin of the observed dark and light bands. The interference pattern from a Lloyd's mirror experiment is shown in Figure 4-5.

We will examine the experiment of Young's in detail because of its simplicity. The experiments of Fresnel and Lloyd can be explained by the theory of Young's experiment and the law of reflection. All discussion

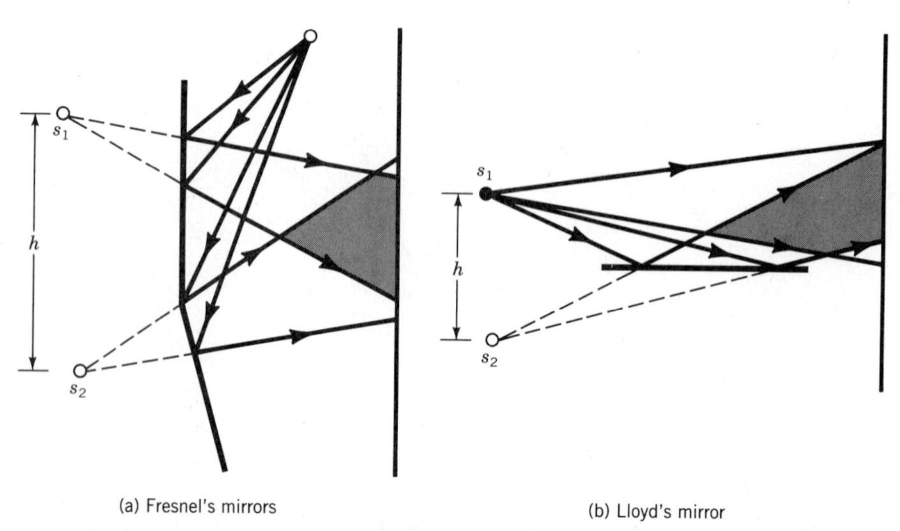

FIGURE 4-4. Fresnel's mirrors and Lloyd's mirror.

of diffraction in Young's experiment will be delayed until we develop diffraction theory in Chapter 9.

The experimental geometry of Young's experiment is shown in Figure 4-6. The origin of the x, z coordinate system is centered between two slits, labeled s_1 and s_2, separated by a distance h and extending out of the sheet of paper. (Slits make the interference pattern much easier to observe than the interference pattern produced by two pinholes by increasing the length of the interference fringes.) These slits are illuminated by a light source s that produces a single frequency and whose extent is limited by a pinhole. The light upon reaching s_1 and s_2 differs in phase by $\phi_2 - \phi_1$. (This phase would be nonzero if the pinhole at the source was not centered with respect to the two slits). The two slits obtain samples of the wavefront that is emitted from the pinhole. We observe the interference produced by the two samples after they have propagated through the distance D. We will depend on diffraction, the subject of later chapters, to overlap the two waves; however, a lens could be used to image the source pinhole on the observation screen and thereby overlap the waves from s_1 and s_2.

FIGURE 4-5. Interference fringes from a Lloyd's mirror experiment. A HeNe laser was focused onto a pinhole and the light transmitted through the pinhole illuminated a flat mirror. The band at the plane established by the mirror surface is a dark band.

The light from s_1 travels a path r_1 and the light from s_2 travels the path r_2 to the observation point P in the observation plane, a distance D from the slit plane. The intensity at P is determined by the total phase difference

$$\delta = \phi_2 - \phi_1 + k(r_2 - r_1)$$

The distances r_1 and r_2 are not easy to measure; however, the angles are experimentally more accessible.

The angles shown in Figure 4-6 are defined by the equations

$$\sin\theta_1 = \frac{r - r_1}{\frac{h}{2}}, \qquad \sin\theta_2 = \frac{r_2 - r}{\frac{h}{2}}$$

The angles are defined in terms of distances that are not easily measured. To define the angles in terms of easily measured parameters, we assume that the distance from the slits to the observation screen D is much larger than the height of the observation point above the z axis, i.e., $x < D$. With this assumption, $\sin\theta_1 \approx \tan\theta_1$, and we may write the distances r_1 and r_2 in terms of quantities that are easy to measure

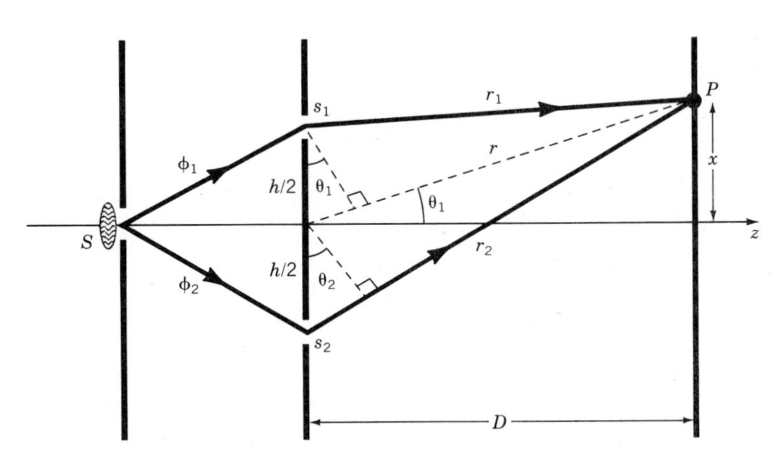

FIGURE 4-6. Young's two-slit experiment.

$$\sin\theta_1 = \frac{r - r_1}{\frac{h}{2}} \approx \tan\theta_1 = \frac{x}{D}$$

and

$$\sin\theta_2 = \frac{r_2 - r}{\frac{h}{2}} \approx \tan\theta_2 = \frac{x + \frac{h}{2}}{D}$$

We further assume that $h < x < D$ so that we may neglect terms that involve h^2

$$r_2 - r \approx \frac{xh}{2D}$$

Now adding $(r_2 - r) + (r - r_1)$, we obtain the difference in the path lengths from s_1 to P and s_2 to P in terms of easily measured experimental parameters

$$r_2 - r_1 = \frac{xh}{D}$$

The interference term

$$2\sqrt{I_1 I_2}\cos\delta$$

from **(4-12)** can now be evaluated by using

$$\delta = \phi_2 - \phi_1 - k(r_2 - r_1) = \Delta\phi - \frac{2\pi xh}{\lambda D} \qquad \text{(4-16)}$$

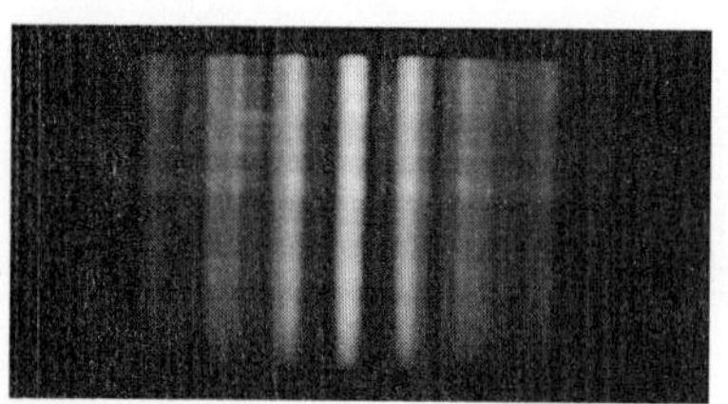

FIGURE 4-7. Fringes from a Young's two-slit experiment made with white light. See the color insert for a color rendition of this interference pattern.

The interference pattern that results consists of bright and dark bands perpendicular to the plane of Figure 4-6. An experimental example of Young's fringes is shown in Figure 4-7.

Whenever the argument of the cosine changes by 2π, the intensity on the screen passes through one cycle in intensity, say maximum to minimum and back to maximum,

$$\left(\frac{2\pi x_1 h}{\lambda D} - \Delta\phi\right) - \left(\frac{2\pi x_2 h}{\lambda D} - \Delta\phi\right) = 2\pi$$

A single cycle in intensity is called a *fringe* and corresponds to a spacing between two bright or two dark bands. The spatial period of a fringe, i.e., its width, is given by

$$x_1 - x_2 = \frac{\lambda D}{h} \qquad \text{(4-17)}$$

From **(4-17)** we see that the spacing of the fringes on the viewing screen is inversely proportional to the slit spacing. Note also that the fringe spacing is a linear function of the wavelength. The fringes in Figure 4-7 were produced with white light and the wavelength dependence of these fringes can be seen in the color insert.

The fringes are generated by adding waves originating from two samples of a wavefront obtained by using the two slits. For this reason, the term *interference by division of wavefront* is often used to describe this type of interference. Another view of the interference process is that it performs a comparison of waves and provides a measure of the similarity of the two waves; here, the comparison would be between two different parts of the same wavefront. (We will discuss this view of interference during the discussion of coherence in Chapter 8.)

DIELECTRIC LAYER

Nature has constructed an interference experiment that we encounter in our everyday lives. It is the interference produced by waves reflecting from two surfaces of a thin, dielectric film. In contrast to the Young's experiment, the two interfering waves come from the same position on a wavefront, resulting in this type of interference being labeled *interference by division of amplitude*.

The geometrical model that will be used to discuss interference by reflection from a thin, dielectric layer is shown in Figure 4-8. In Figure 4-8, light enters a thin, dielectric film of thickness d at point A. A portion of the amplitude of the wave is reflected toward D, remaining in the medium with index n_1, while another portion proceeds from A to B and then to C in the medium with index n_2. A lens is assumed to gather the waves at C and D and bring them together on a viewing screen, as shown in Figure 4-9.

The paths taken from C and D to the viewing screen are the same for both waves; therefore, this portion of the two paths will not contribute to the phase difference δ between the two waves and may be ignored. If we use the geometry in Figure 4-8, the optical path lengths for the waves from A to D and from A to C via B will be obtained. These optical path lengths will then be used to find δ for use in **(4-13)**.

The optical path, as light travels from A to B in the thin film, is

$$\frac{k_2 d}{\cos\,\theta_t} = \frac{\omega n_2 d}{c\,\cos\,\theta_t}$$

Light is reflected off the back surface of the dielectric film at point B and travels an equivalent optical path from B to C. Thus, the optical path from A to B to C in the medium with index n_2 is

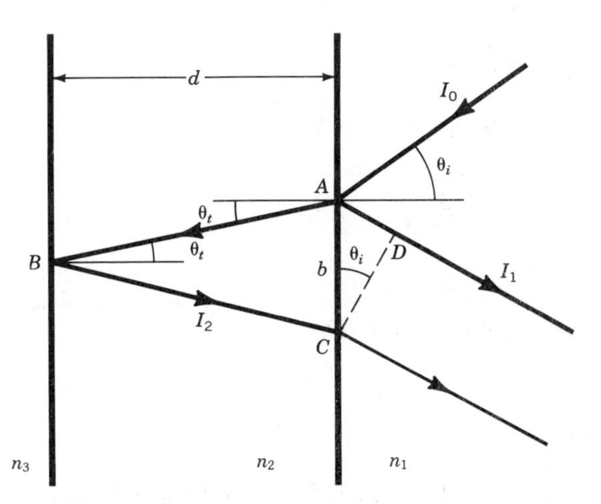

FIGURE 4-8. Dielectric layer of index n_2 and thickness d separating a medium of index n_1 and a medium of index n_3.

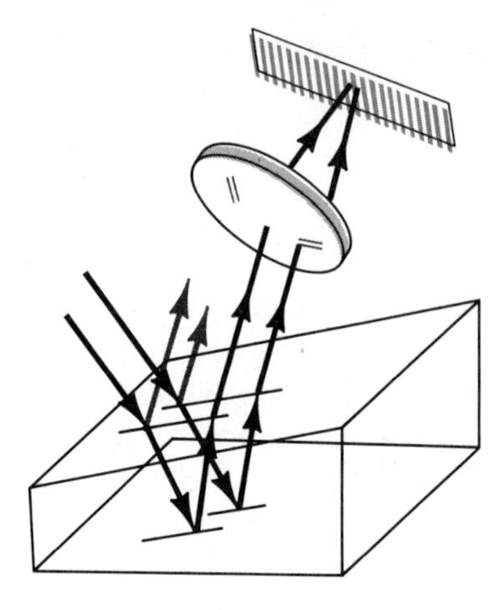

FIGURE 4-9. Use of a lens to produce fringes from a dielectric layer.

$$\overline{ABC} = \frac{2\omega n_2 d}{c \cos \theta_t} \tag{4-18}$$

The optical path length from A to D in the medium with index n_1 is

$$\begin{aligned}\overline{AD} &= k_1 2b \sin \theta_i = 2k_1 d \tan \theta_t \sin \theta_i \\ &= 2\omega n_1 d \frac{\sin \theta_i \sin \theta_t}{\cos \theta_t}\end{aligned} \tag{4-19}$$

Applying Snell's law, we can rewrite **(4-19)** as

$$\overline{AD} = 2\omega d n_2 \frac{\sin^2 \theta_t}{c \cos \theta_t}$$

Since the source for the two waves is identical, $\Delta\phi = 0$ and the phase difference between the waves at points C and D will determine the value of the interference term

$$\delta = \overline{ABC} - \overline{AD} = \frac{4\pi n_2 d}{\lambda_0} \cos \theta_t \tag{4-20}$$

There is an additional phase change of π upon reflection from a dense medium (see Figure 3-5 and remember that we earlier assumed the polarization will be normal to the plane of incidence for all interference problems). If $n_2 > n_1$, the reflection at A will experience this additional phase change, and if $n_3 > n_2$, the reflection at B will experience a phase change. We will assume $n_3 < n_2 > n_1$ for our discussion so that the only phase change due to reflection occurs at A. A bright band will occur when $\delta = 2\pi m$

$$\delta = 2\pi m = \frac{2n_2 d}{\lambda_0} 2\pi \cos \theta_t + \pi$$

Three of the parameters in this equation could independently cause δ to vary and produce a set of interference fringes. These parameters are d, λ_0, and θ_t.

Fizeau Fringes

If we illuminate the dielectric layer with a plane wave, with a single frequency, then $\cos \theta_t$ is constant over the layer and each bright or dark band will show the region of constant optical thickness. The thickness associated with a particular bright band is given by

$$d = \frac{\lambda_0(2m - 1)}{4n_2 \cos \theta_t} \tag{4-21}$$

These are called *Fizeau fringes* or fringes of constant thickness. The geometry that could produce Fizeau fringes is shown schematically in Figure 4-10*a* where a plane wave is assumed to illuminate a dielectric film with a nonuniform thickness. The fringes, as observed by the eye, are produced by reflections of the wave in the film. Figure 4-10*b* displays fringes in a vertical soap film illuminated by sodium light (see Problem 4-13).

Color Fringes

The thickness of the dielectric film could be a constant and the illumination could be from a source producing a range of wavelengths. For example, consider an oil film floating on a wet street, illuminated by sunlight (Figure 4-11*a*), or vertical soap film as in Figure 4-11*b*, illuminated by an incandescent bulb. Images of these figures can be found in the color insert. Those regions of the film with the same thickness display the same color band; the color pattern repeats when the thickness changes by one wavelength. The wavelength for a bright band is given by

$$\lambda_0 = \frac{4n_2 d \cos \theta_t}{2m - 1} \tag{4-22}$$

Haidinger's Fringes

If the thickness and wavelength of the illumination are constant, then fringes of constant inclination or *Haidinger's fringes* will be observed. The fringes will be concentric circles if the line of sight from the viewing point to the

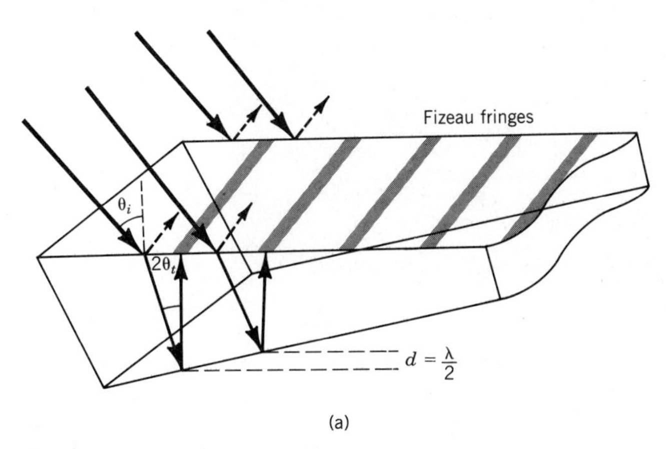

(a)

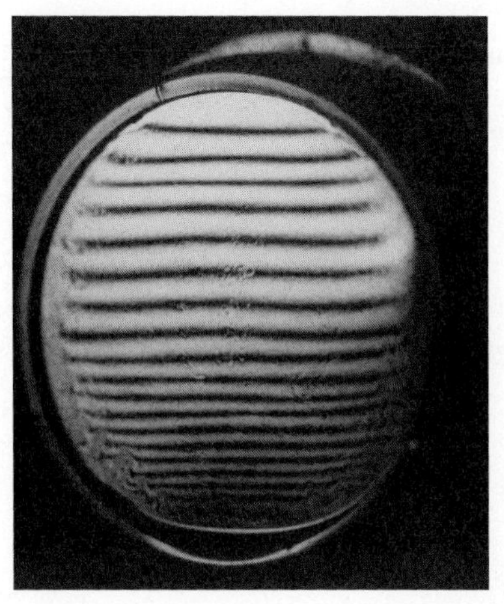

(b)

FIGURE 4-10. (a) The production of Fizeau fringes in a dielectric layer with a nonuniform thickness. The fringes appear to be located at the dielectric layer, but originate because of the interference of light collected by a lens such as our eye. (b) Fizeau fringes produced by illuminating a vertical soap film with sodium light. Gravity creates a nearly linear thickness change in the film.

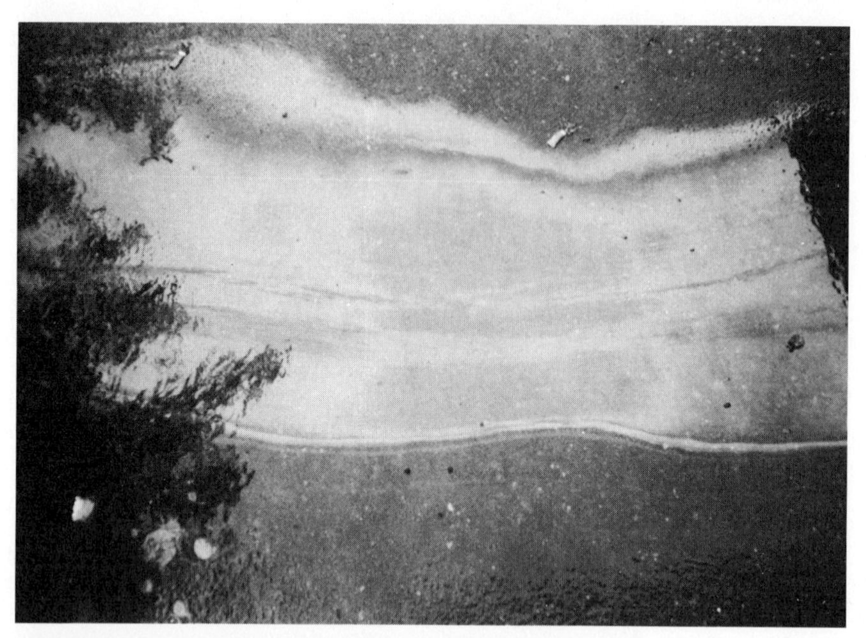

FIGURE 4-11*a*. An oil film on a wet street.

FIGURE 4-11*b*. A soap film similar to the one shown in Figure 4-10*b*, but here the soap film is illuminated by white light. The color pattern repeats when the thickness changes by one wavelength, thus, each cycle in color corresponds to one order m in the interference pattern. A reproduction of these photos is found in the color insert.

dielectric layer is normal to the surface and if the incident wave is nearly parallel to the line of sight. Bright bands will occur at an angle given by

$$\cos\,\theta_t = \frac{(2m - 1)\,\lambda_0}{4n_2 d}$$

We can rewrite this in terms of the angle of incidence by using Snell's law

$$\sin\,\theta_i = \frac{1}{n_1}\sqrt{n_2^2 - \frac{n_2(2m-1)\,\lambda_0}{4d}} \tag{4-23}$$

An example of these fringes is shown in Figure 4-13*b*.

Antireflection Coating

The interference resulting from light reflected by the surfaces of a dielectric layer can be used to reduce the reflectivity of the substrate supporting the layer. This is accomplished by selecting a layer thickness that results in destructive interference between the wave reflected by the front and back surface. Assume that light is incident normal to the surface of the dielectric so that $\theta_t \approx 0$ and $\cos(\theta_t) \approx 1$; also assume that $n_1 < n_2 < n_3$ so that there is an equal phase change for reflections from the two surfaces of the dielectric film. To observe a dark band, we require **(4-20)** to fulfill the condition

$$\delta = (2m+1)\,\pi = \frac{2n_2 d{\bullet}2\pi\,\cos\,\theta_t}{\lambda_0}$$

Thus, the film thickness required to produce destructive interference for light incident normal to the surface is

$$d = \frac{2m+1}{4}\cdot\frac{\lambda_0}{n_2} = (2m+1)\,\frac{\lambda}{4}$$

This equation is identical to **(4-21)** but now destructive interference produces low intensity. This change in intensity is due to the assumption made about the relative size of the various indices.

The dielectric film thickness must be an odd multiple of $\lambda/4$ for there to be destructive interference between the light reflected from the front and back surfaces of the layer. When this condition is met, there will be no light reflected from the surface and the dielectric layer will be an antireflection coating. The index n_2 of the film appears in the equation to correct the wavelength for the velocity of propagation in the medium.

We have said nothing about the amount of light reflected at the surfaces. If equal amounts of light are not reflected by the two surfaces, then total cancellation of the reflected component will not occur. Fresnel's formula can be used to determine the condition for which equal fractions of light will be reflected from the two surfaces. If we require that the reflectivity of the two surfaces are equal, then for normal incidence **(4-46)** yields

$$\frac{n_2 - n_1}{n_2 + n_1} = \frac{n_3 - n_2}{n_3 + n_2}$$

We can assume $n_1 = 1$ since most optical surfaces are used in air. With this assumption, the equation reduces to

$$n_2 = \sqrt{n_3}$$

For some flint glasses

$$n = 1.7 \quad \text{and} \quad \sqrt{n} = 1.305$$

Magnesium fluoride has an index of $n = 1.38$ that is reasonably close to the desired index n_2 and should act as a good antireflection coating for glasses with $n = 1.7$. For many years, magnesium fluoride was used as an antireflection coating for glasses of all indices because it was easy to apply and relatively rugged. (Additional details on the design of antireflection coatings and high reflectivity mirrors are contained in Appendix 4A.)

Newton's Rings

The dielectric film need not have an index of refraction larger than its surroundings. Newton's rings are an example of interference fringes produced by a dielectric film whose index of refraction is less than its surroundings. Newton's rings are Fizeau fringes or fringes of constant thickness and can be obtained by placing a convex lens in contact with a flat glass plate, forming an air wedge. Circular interference fringes are formed

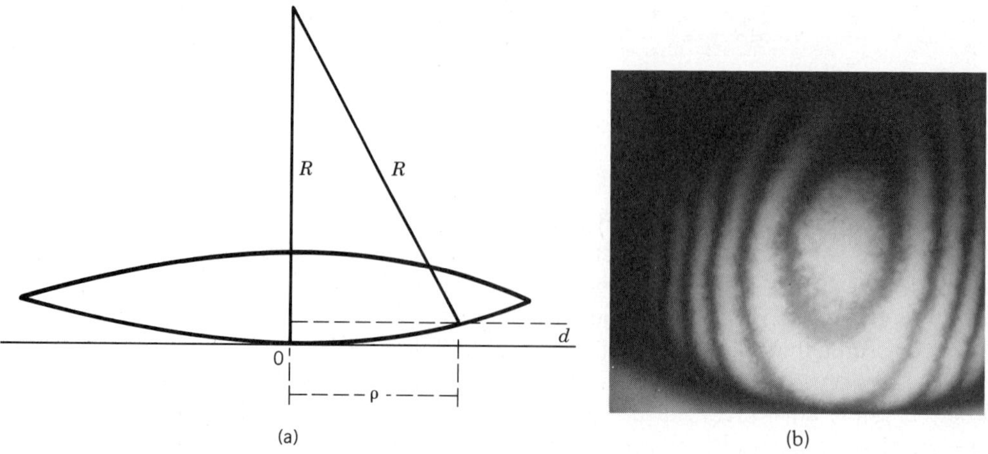

FIGURE 4-12. (a) Experimental set-up for the observation of Newton's rings. (b) Newton's rings produced by the air film between an optical flat and a convex lens.

about the point of contact, point O in Figure 4-12*a*. The radius of a dark or light band of the fringes can be calculated in terms of the radius of curvature of the lens.

In Figure 4-12*a*, the radius of curvature of the lens is R and the radius of a band making up one of Newton's rings is ρ. These two parameters can be related by the Pythagorean theorem

$$R^2 = \rho^2 + (R - d)^2$$

where d is the thickness of the air gap at the fringe.

$$\rho^2 = d(2R - d)$$

Since $d << R$, we can write

$$d = \frac{\rho^2}{2R} \tag{4-24}$$

We assume normal incidence so that $\cos(\theta_t) \approx 1$ and see from **(4-21)** that the dark bands will be rings with a radius of

$$\rho^2 = \frac{m\lambda_0 R}{n_2}, \qquad m = 0, 1, 2, \ldots \tag{4-25}$$

and the bright bands will be rings with a radius of

$$\rho^2 = \frac{\left(m + \frac{1}{2}\right)\lambda_0 R}{n_2}, \qquad m = 0, 1, 2, \ldots \tag{4-26}$$

Usually, the gap between the test plate and the lens is air so that $n_2 = 1$. The rings can also be viewed in transmission as is shown in Figure 4-12*b*. Bands that are dark on reflection are bright on transmission.

Modern lens grinders use these fringes to monitor their work.[14] Newton's rings can provide not only a qualitative idea of the quality of the lens, but also a quantitative measure by providing the lens designer with a measurable parameter that will relate to the radius of curvature of the lens under construction. This parameter is the radius of a Newton's ring. Through the use of either **(4-25)** or **(4-26)**, the optician can measure the radius of a Newton's ring to determine if the lens being ground has reached the proper curvature. If the Newton's rings are not symmetric, then errors in the figure of the lens can be identified and corrected. (The *figure* of an optical surface refers to how well the actual shape of the surface conforms to the desired shape.)

MICHELSON INTERFEROMETER

Albert Abraham Michelson (1852–1931), the first American Scientist to win the Nobel Prize in physics, developed the design for an interferometer that has found a number of useful applications. The interferometer will work with a much larger source than one can use with the two-slit arrangement of Young's experiment, giving much brighter interference fringes. A beam splitter (a semitransparent mirror) is used to divide the light into two beams; see Figure 4-13*a*.

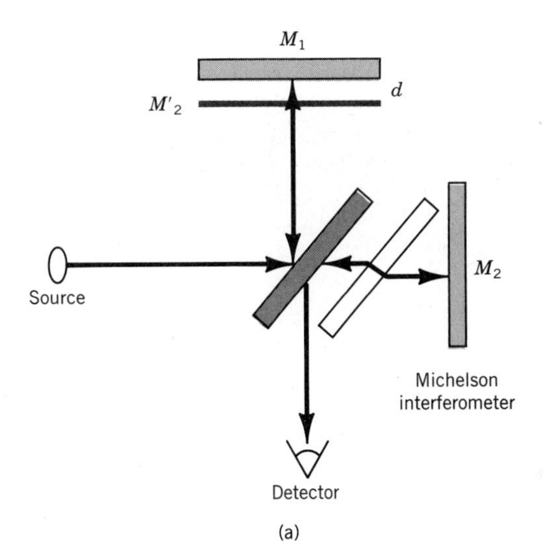

FIGURE 4-13*a*. Michelson interferometer. The shaded diagonal plate is a beam splitter with one surface covered by a partially reflective coating. It creates two beams traveling toward mirrors M_1 and M_2. The second diagonal plate is called the compensating plate. The compensating plate is made of the same material and thickness as the beam splitter. Its purpose is to equalize the optical path length of the two arms of the interferometer.

The two beams are directed along orthogonal paths, usually called the arms of the interferometer, where they strike two mirrors, M_1 and M_2, and then return to the beam splitter where they interfere. Looking at the beam splitter from the detector, we see an image of mirror M_2 near mirror M_1. The image M_2' and the mirror M_1 form a dielectric layer of thickness d. The interferometer is assumed to be in the air so that the dielectric layer has an index of refraction $n_2 = 1$. Also, light reflected from M_1 and M_2 experiences the same phase change upon reflection, thus, no additional phase shift needs to be added to δ in calculating the phase difference between the two waves. The total phase difference for a bright band is

$$\delta = 2\pi m = \frac{2d \cdot 2\pi \cos \theta_t}{\lambda_0}$$

The source shown in Figure 4-13*a* is nearly a point source, generating spherical waves that produce fringes of constant inclination **(4-23)**. The fringes are circularly symmetric, as shown in Figure 4-13*b*, if d is a constant across the aperture. The maximum value of m, the order of the fringe, occurs at the center of the set of fringe rings where $\theta_t = 0$

$$m_{\max} = \frac{2d}{\lambda_0} \tag{4-27}$$

and the order of the fringes decreases as we move out from the center fringe in the viewing plane. The order of the fringe is shown by **(4-27)** to be equal to the difference in lengths of the two interferometer arms expressed in the number of wavelengths of light contained in d.

As d increases, a bright band will move out from the center of the aperture and a new bright band of higher order will take its place at the center. If we placed a detector at the position of the center fringe and monitored the intensity as we moved one of the mirrors, and thereby changed d, we would see

$$I = I_1 + I_2 + 2\sqrt{I_1 I_2} \cos\left(\frac{2\omega d}{c}\right) \tag{4-28}$$

where we have substituted

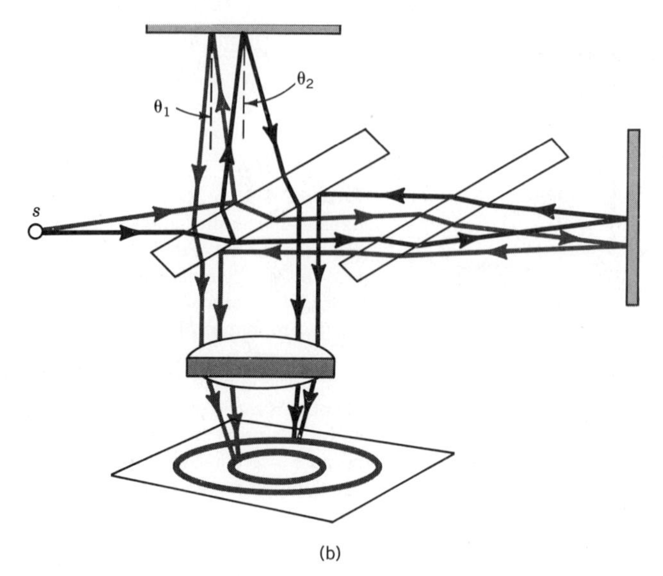

(b)

FIGURE 4-13*b*. Haidinger's fringes produced in a Michelson interferometer. Both mirror surfaces are assumed to be normal to the axis of the interferometer. The light source is a point source producing spherical waves. The gray line has a smaller angle of incidence on the mirrors than has the black line, $\theta_1 < \theta_2$.

$$\frac{\omega}{c} = \frac{2\pi}{\lambda_0}$$

If the beam splitter is a 50:50 splitter, then the intensity of light in the two arms will be the same and we can write

$$I = I_0\left[1 + \cos\left(\frac{2\omega d}{c}\right)\right] \tag{4-29}$$

If we use the physical arrangement shown in Figure 4-13, the light in the M_1 arm of the interferometer travels an extra distance $2d$ to reach the detector. This means that two waves are added together that originated at different times; the difference in time between the origination of the two waves is

$$\tau = \frac{2d}{c} \tag{4-30}$$

which is called the *retardation time* (the wave in arm M_1 has been delayed or retarded). Using the definition **(4-30)**, we can rewrite **(4-29)** as

$$I = I_0[\,1 + \cos\,\omega\tau] \tag{4-31}$$

The signal is made up of a constant term plus an oscillatory term. The oscillatory term will provide information about the coherence properties of the light as we will see in Chapter 8.

The Michelson interferometer has a number of shortcomings that have been overcome by modifying its design. The Twyman–Green interferometer (Figure 4-14*a*) is a Michelson interferometer that utilizes a plane wave through the use of a collimating lens. The simple fringe pattern makes the Twyman–Green interferometer useful for the evaluation of optical components (see Figure 4-14*b*). Without the collimating lens, a large number of fringes

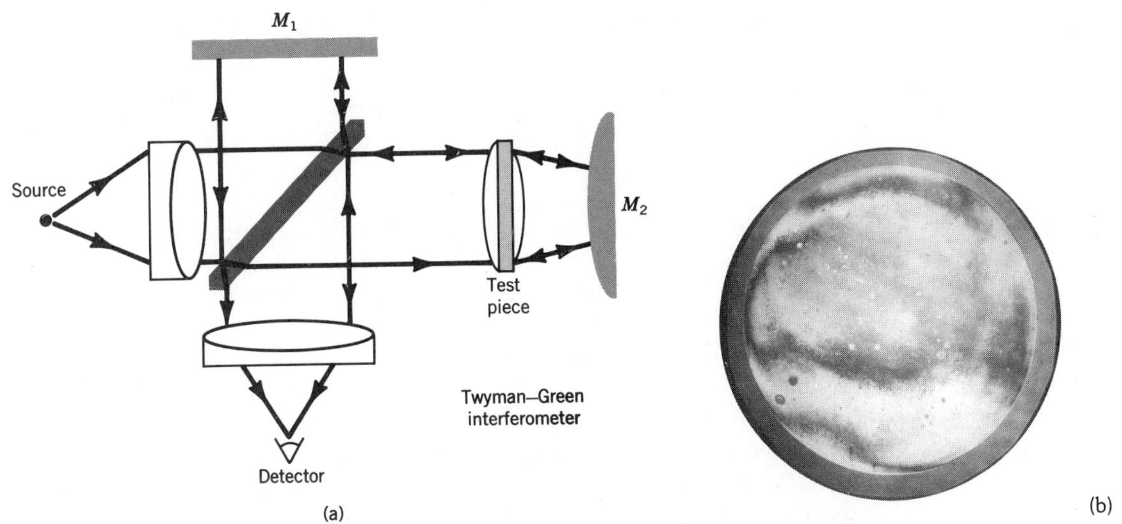

FIGURE 4-14. (a) An improvement on the Michelson interferometer for the testing of optical components. If the test piece is perfect, the waves interfering on the beam splitter are plane waves. M_2 can be either a plane mirror or spherical mirror. The spherical mirror has several experimental advantages. (b) The Twyman–Green interferometer can be used to test a mirror surface by replacing the "test piece" and M_2, shown in (a), by the test mirror. Here is shown a mirror in just such a configuration. The mirror was designed for use at 10 μm, but the wavelength used in the interferometer was 0.44 μm. Thus, the fringes in the interferogram correspond to 1/25 of a fringe at the mirror's operating wavelength.

(Haidinger's fringes) would be present, making interpretation of the data difficult. By the addition of a collimating lens, θ_t is converted from a variable to a constant. This means that since the angle of incidence is a constant, the fringes are those of equal thickness (Fizeau fringes) and **(4-21)** applies.

The Mach–Zehnder interferometer (Figure 4-15) also uses collimated light but further simplifies the understanding of the observed fringes by passing light through the test area only once. By separating the two optical paths, very large

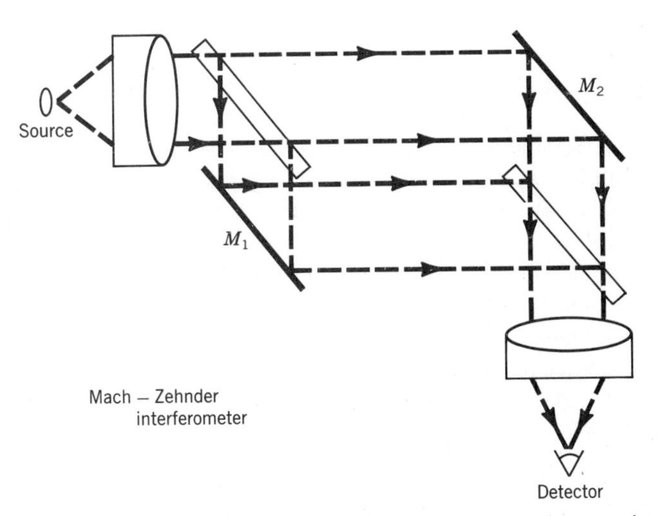

FIGURE 4-15. A Mach-Zehnder interferometer configured to use plane waves. Either of the arms can be used as the test area.

objects such as wind tunnels can be tested. Either arm of the interferometer can contain the test area.

Measurements using interferometers are expressed in fringes per unit length. One fringe spacing equals the distance between adjacent dark bands.

INTERFERENCE BY MULTIPLE REFLECTION

In our previous discussion of a thin dielectric film, we neglected multiple reflections. We now wish to consider what effect multiple reflections have on interference. We will find that the dielectric layer forms a resonant cavity. If the losses at each reflection are not too great, a set of standing waves is created in the dielectric layer, similar to a set of standing waves that forms on a guitar string. The mathematics become somewhat involved, but once completed, they demonstrate that the dielectric layer acts like a wavelength filter, transmitting some frequencies and rejecting others.

Before looking at the optical problem, consider a guitar string fixed, at each end, to rigid mounts, the bridges. The boundary conditions for the wave equation become $y = 0$ at $x = 0$ and $x = L$. The boundary conditions are met if

$$\frac{2\pi L}{\lambda} = kL = m\pi, \qquad m = 1, 2, 3, \ldots \tag{4-32}$$

The solutions to the wave equation become

$$y(x, t) = Y_m \sin\left(\frac{m\pi x}{L}\right) \cos \omega_m t \tag{4-33}$$

where

$$\omega_m = \frac{\pi mc}{L}$$

and c is the wave velocity. The solutions **(4-33)** are called the normal modes of vibration of the string and are standing waves of the form **(3-6)**. The spatial dependence of a standing wave is

$$y(x) = Y_m \sin\left(\frac{m\pi}{L}\right) x$$

These functions are the eigenmodes of the string and the frequencies ω_n are the eigenvalues.

In a dielectric layer (see Figure 4-16), the light waves reflecting back and forth between the boundaries of the dielectric layer will result in solutions of the same form as **(4-33)**, causing the dielectric layer to behave as a resonant structure.

We will ignore the fields in the dielectric layer and direct our attention toward the light transmitted through the layer. Assume that a wave of amplitude A is incident on the layer. We let f_1 be the fraction of A that is reflected from the surface, f_2 the fraction transmitted into the layer, and f_3 the fraction of the wave transmitted out of the film. (If the wave were incident normal to the layer, Fresnel's equations would give

$$f_2 = \frac{2n_1}{n_1 + n_2}, \qquad f_3 = \frac{2n_2}{n_1 + n_2}, \qquad f_1 = \frac{\pm(n_2 - n_1)}{n_1 + n_2}$$

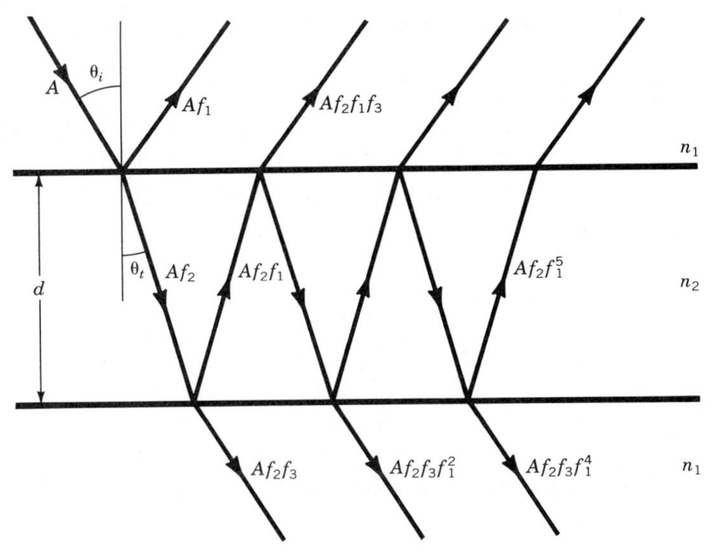

FIGURE 4-16. Reflection of a plane wave in a plane parallel dielectric layer.

but we will retain the current notation to allow the results to be applied to any illumination angle. A second method of handling the notation required for multiple reflections is introduced in Appendix 4-A.)

We will determine the interference of the transmitted waves by calculating the amplitude and phase of each transmitted wave. The waves will then be added using the complex approach. As shown in Figure 4-9, a lens will be used to collect all of the transmitted waves and bring them together at a viewing screen. All of the waves are assumed to travel the same optical path length upon exit from the dielectric so that the wave-combining process can be neglected.

We will represent the first transmitted wave by

$$\mathrm{E} = Af_2 f_3 e^{i(\omega t - \mathbf{k}\cdot\mathbf{r})}$$

The reflection coefficient of the dielectric layer is defined as $r = f_1$ and the reflectivity of the layer is defined as r^2. The transmitted amplitudes for the various reflected waves form a geometric progression in terms of the reflectivity:

$$\mathrm{E},\ \mathrm{E}r^2,\ \mathrm{E}r^4,\ \mathrm{E}r^6,\ \ldots$$

Each transmitted wave will have a constant phase difference, relative to its neighbor, given by **(4-20)**

$$\delta = \frac{4\pi d n_2 \cos\theta_t}{\lambda_0} \tag{4-34}$$

The general nth transmitted wave is of the form

$$\mathrm{E}r^{2n}e^{i\delta_n} = r^{2n}Af_2 f_3 e^{i(\omega t - \mathbf{k}\cdot\mathbf{r})}e^{in\delta} \tag{4-35}$$

where $n = 0, 1, 2, \ldots$. To add all of the transmitted waves, we use the complex method (the vector approach is used in Problem 4-19). In adding the transmitted waves, we can factor E outside the summation because all of the transmitted waves are at the same frequency and all are propagating in a parallel direction

$$A = E\,(1 + r^2e^{i\delta} + r^4e^{i2\delta} + r^6e^{i3\delta} + \ldots)$$

$$= E\,(1 + r^2\cos\delta + r^4\cos 2\delta + \ldots)$$

$$+ iE\,(r^2\sin\delta + r^4\sin 2\delta + r^5\sin 3\delta + \ldots) \qquad (4\text{-}36)$$

To evaluate **(4-36)**, we use the identity

$$\frac{1}{1+x} = 1 - x + x^2 - x^3 + x^4 - \ldots, \qquad -1 < x < 1$$

to write

$$\frac{1}{1 - r^2e^{i\delta}} = 1 + r^2e^{i\delta} + r^4e^{i2\delta} + r^6e^{i3\delta}\ldots, \qquad r^2 < 1$$

The right side of this equation is identical to **(4-36)** if we multiply by E; the left side can be rewritten

$$\frac{1}{1 - r^2e^{i\delta}} = \frac{1}{1 - r^2e^{i\delta}}\,\frac{1 - r^2e^{-i\delta}}{1 - r^2e^{-i\delta}} = \frac{1 - r^2e^{-i\delta}}{1 - r^2\left(e^{i\delta} + e^{-i\delta}\right) + r^4}$$

using the identity

$$e^{\pm i\delta} = \cos\delta \pm i\sin\delta$$

$$\frac{1}{1 - r^2e^{i\delta}} = \frac{1 - r^2\cos\delta + ir^2\sin\delta}{1 - 2r^2\cos\delta + r^4} \qquad (4\text{-}37)$$

Equating real and imaginary parts of **(4-36)** and **(4-37)**, we obtain for the real part

$$\frac{1 - r^2\cos\delta}{1 - 2r^2\cos\delta + r^4} = 1 + r^2\cos\delta + r^4\cos 2\delta + \ldots \qquad (4\text{-}38)$$

and for the imaginary part

$$\frac{r^2\sin\delta}{1 - 2r^2\cos\delta + r^4} = r^2\sin\delta + r^4\sin 2\delta + r^6\sin 3\delta + \ldots \qquad (4\text{-}39)$$

The intensity of the transmitted light is

$$I = \mathcal{A}\mathcal{A}^*$$

which is the sum of the squares of the real and the imaginary terms we have just calculated

$$I \propto \frac{\mathcal{E}^2\left[(1 - r^2\cos\delta)^2 + r^4\sin^2\delta\right]}{(1 - 2r^2\cos\delta + r^4)^2} = \frac{\mathcal{E}^2}{1 - 2r^2\cos\delta + r^4} \qquad (4\text{-}40)$$

To obtain the form of the transmitted wave normally used, the denominator of **(4-40)** is rewritten

$$1 - 2r^2\cos\delta + r^4 = (1 - 2r^2\cos\delta + r^4) - 2r^2 + 2r^2$$

$$= 2r^2(1 - \cos\delta) + (1 - 2r^2 + r^4)$$

This new expression allows **(4-40)** to be written in the form

$$I \propto \frac{\mathcal{E}^2}{(1 - r^2)^2 + 4r^2\sin^2\frac{\delta}{2}} \qquad (4\text{-}41)$$

When $\delta = 0,\ 2\pi,\ 4\pi, \ldots$

$$I_{max} \propto \frac{\mathcal{E}^2}{(1 - r^2)^2} \tag{4-42}$$

When $\delta = \pi,\ 3\pi,\ 5\pi, \ldots$

$$I_{min} \propto \frac{E^2}{(1 + r^2)^2} \tag{4-43}$$

We define the fringe visibility as

$$v = \frac{I_{max} - I_{min}}{I_{max} + I_{min}} \tag{4-44}$$

The fringe visibility is a useful characterization of interference; the larger the value of this parameter, the easier it is to observe the fringe pattern. The maximum value of v is 1 and occurs when $I_{min} = 0$; the minimum value is zero and occurs when $I_{max} = I_{min}$. Using **(4-42)** and **(4-43)** in **(4-44)**, we obtain the fringe visibility in terms of the reflectivity of the film surface

$$v = \frac{2r^2}{1 + r^4}$$

As the reflectivity r^2 approaches 1, the fringe visibility approaches its maximum value of 1.

We can write the intensity transmitted through the dielectric layer in terms of I_{max}

$$I = \frac{I_{max}}{1 + \dfrac{4r^2}{(1 - r^2)^2} \sin^2 \dfrac{\delta}{2}}$$

The factor

$$\frac{4r^2}{(1 - r^2)^2} = F$$

is called the *contrast* F and its use allows the relative transmission of the dielectric layer to be expressed in a format called the *Airy function*

$$\frac{I}{I_{max}} = \frac{1}{1 + F \sin^2 \dfrac{\delta}{2}} \tag{4-45}$$

If we plot the Airy function for different values of the reflectivity r^2 we obtain Figure 4-17 that displays a periodic maximum in the transmission of the dielectric layer as δ is varied. The peak in transmission occurs when d is equal to a multiple of $\lambda/2$, where λ is the illuminating wavelength. Thus, the maximum transmission of the film occurs when the film thickness will support standing waves in the layer, that is, the transmission maxima occur for eigenvalues of the layer.

The peak in transmission in Figure 4-17 narrows as the reflectivity r^2 increases. The value of δ over which I goes from I_{max} to $I_{max}/2$ is a measure of the fringe sharpness and can be obtained by noting $I = I_{max}$ when $\delta = 0$. When $I = I_{max}/2$,

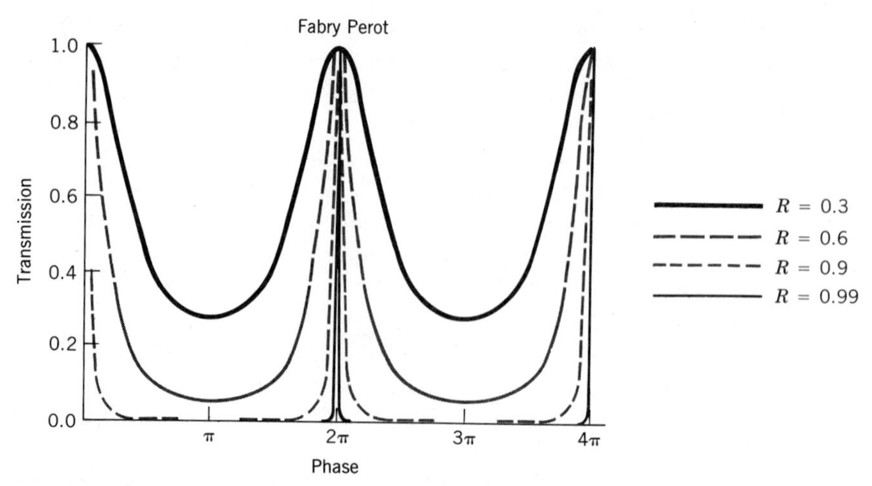

FIGURE 4-17. This is a plot of the fraction of transmitted light as a function of the optical path length *nd* for reflectivities of 0.3, 0.6, 0.9, and 0.99. The fringes become narrower as the reflectivity increases.

$$\frac{4r^2 \sin^2 \frac{\delta}{2}}{(1 - r^2)^2} = 1$$

The fringe sharpness is given by

$$\delta_{1/2} = 2 \sin^{-1}\left(\frac{1 - r^2}{2r}\right) = 2 \sin^{-1}\left(\frac{1}{\sqrt{F}}\right) \tag{4-46}$$

The sharpness of the fringes increases and $\delta_{1/2}$ decreases with increasing reflectivity. If the reflectivity is $r^2 = 0.9$, then

$$\delta_{1/2} = 0.211 \text{ rad}$$

The multiple reflections are the cause of the sharp fringes, as we can see by comparing the fringe sharpness of the dielectric layer with multiple reflections to a Michelson interferometer that was modeled as a dielectric layer with one reflection. If we rewrite **(4-29)** as

$$I = I_{max} \cos^2 \frac{\delta}{2}$$

the sharpness of the fringes for a Michelson interferometer can be obtained by noting that $I = I_{max}$ whenever $\delta = 0$ and $I = I_{max}/2$ whenever

$$\delta_{1/2} = 2 \cos^{-1}\left(\frac{1}{\sqrt{2}}\right) = \frac{\pi}{2} = 1.57 \text{ rad}$$

This value for the fringe sharpness is much larger than the value obtained for the dielectric layer, and thus, the fringes of a Michelson interferometer are much broader.

It is obvious that a device constructed as a dielectric layer could be used to accurately measure wavelength. Such a device was first constructed by **Marie Paul Auguste Charles Fabry (1867–1945)** and **Jean Baptiste Gaspard Gustave Alfred Perot (1863–1925)**.

FABRY–PEROT INTERFEROMETER

The Fabry–Perot interferometer is constructed using two highly reflective surfaces usually separated by air. In Figure 4-18 is displayed a typical experimental arrangement. Two plane glass plates separated by a distance d have reflective dielectric mirrors on their facing surfaces. (We will limit our attention here to the interferometers constructed using plane, parallel plates, but interferometers are also constructed using spherical mirrors and we will discuss some of their properties in Chapter 5.) The waves exiting the plates, after multiple reflections, are collected by the lens and imaged on an observation screen. Only one propagation vector, incident at an angle θ, is followed through the system in Figure 4-18. Other incident propagation vectors will result in a bright band if $\delta = 2\pi m$. The incident angle of the propagation vectors forming the bright bands must satisfy the equation

$$m\lambda = 2nd \cos \theta$$

The observed fringes, such as those in Figure 4-19, are circularly symmetric if the illumination is symmetric about the symmetry axis of the optical system.

To use the Fabry–Perot as a wavelength-measuring instrument, a detector behind an aperture in the screen measures the light passing through the screen as the optical path length $n_2 d$ is varied by either varying the pressure of the air surrounding the interferometer or by using piezoelectric translators to move one mirror relative to the other. If a single wavelength illuminates the mirrors, then the signal recorded by the detector will look like one of the curves in Figure 4-17.

The accuracy with which an interferometer can measure the wavelength of its illumination is called *chromatic resolving power*, $\mathcal{R}$, and is defined as $\lambda/\Delta\lambda$, where λ is the mean wavelength of the illumination and $\Delta\lambda$ is the wavelength difference that can be resolved. We need a criterion for resolution, and the one we will use assumes that two wavelengths λ_1 and λ_2 of equal intensity are present. The criterion for resolution of the two wavelengths states that the two wavelengths are just resolved if the half-maximum intensity of a fringe produced by λ_1 falls on the half-maximum intensity of a fringe produced by λ_2. When this occurs, the transmitted intensity is a constant as d is varied from the resonant condition of λ_1 to the resonant condition of λ_2. The phase shift in going from the intensity maximum for λ_1 to the intensity maximum for λ_2 is then $\Delta\delta = 2\delta_{1/2}$. The fringes are assumed to be narrow so that this is a small value and we may make the approximation

$$\sin \delta_{1/2} = \sin \frac{\Delta\delta}{2} \approx \frac{\Delta\delta}{2}$$

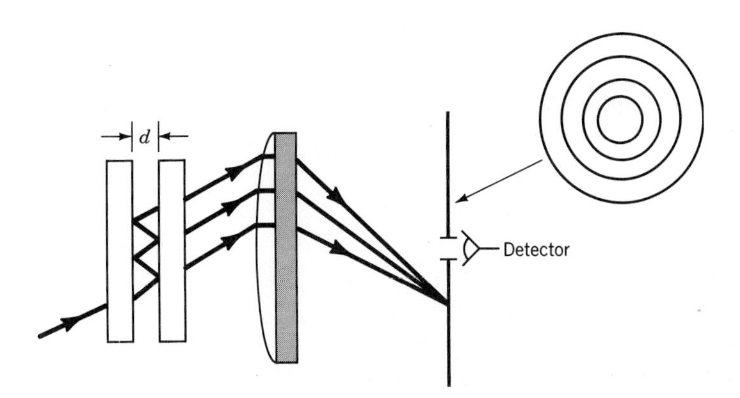

FIGURE 4-18. The experimental arrangement of a Fabry–Perot interferometer. Shown on the right is a schematic drawing of the interference pattern observed on the screen; Figure 4-19 shows an actual fringe pattern.

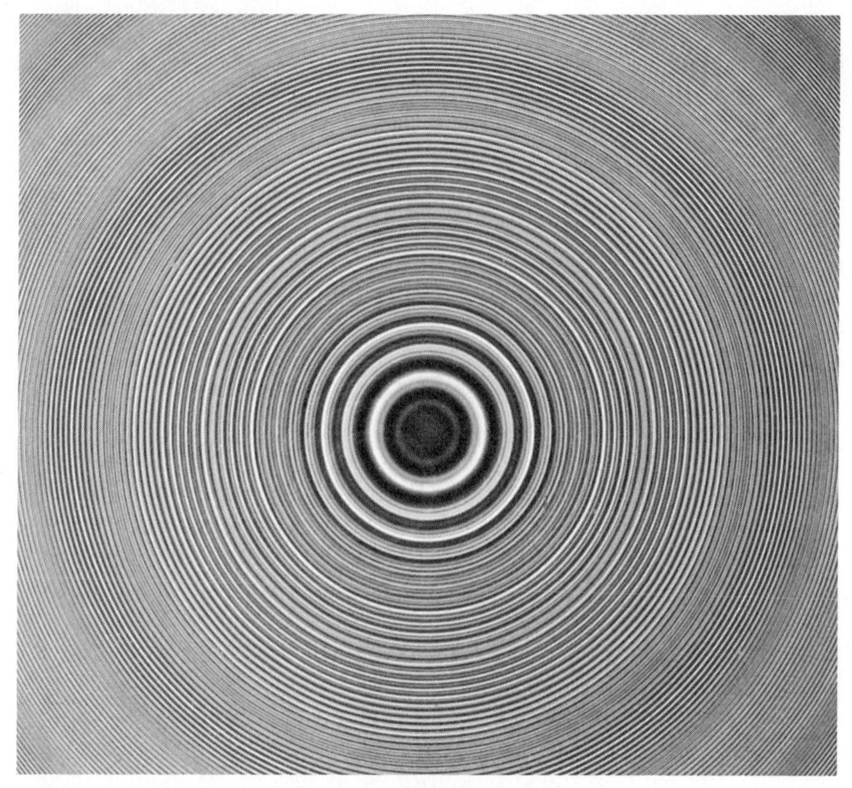

FIGURE 4-19. The output fringes from a Fabry–Perot interferometer. Several sets of fringes due to multiple colors are present in this photo, as can be seen in the print in the color insert. Courtesy of Fredrick L. Roesler, University of Wisconsin.

and

$$\sin\left(\frac{\delta_{1/2}}{2}\right) = \frac{\Delta\delta}{4}$$

Using **(4-46)** yields

$$\frac{\Delta\delta}{4} \approx \frac{1}{\sqrt{F}} = \frac{1-r^2}{2r} \tag{4-47}$$

Differentiating **(4-34)** to get a relationship between $\Delta\delta$ and $\Delta\theta$ yields

$$\Delta\delta = -4\pi n_2 d \sin\,\theta_t \frac{\Delta\theta_t}{\lambda_0} \tag{4-48}$$

A bright band will occur whenever

$$2n_2 d \cos\,\theta_t = m\lambda \tag{4-49}$$

If we differentiate this equation, we will obtain a relationship between $\Delta\theta$ and $\Delta\lambda$

$$-2n_2 d \sin\,\theta_t \Delta\theta_t = m\Delta\lambda$$

$$-\sin\,\theta_t \Delta\theta_t = \frac{m\Delta\lambda}{2n_2 d} \tag{4-50}$$

Using **(4-50)** in **(4-48)** and then equating with **(4-47)**, we get

$$\Delta\delta = 2\pi m \frac{\Delta\lambda}{\lambda} = \frac{2(1-r^2)}{r}$$

The resolving power is thus

$$\mathcal{R} = \frac{\lambda}{\Delta\lambda} = \frac{m\pi r}{1-r^2} = \frac{m\pi}{2}\sqrt{F} \tag{4-51}$$

We see from **(4-49)** that the order number m has a maximum value whenever $\cos\theta_t = 1$, that is, m is a maximum at the center of the Fabry–Perot fringe pattern and this maximum is given by

$$m_{max} = \frac{2n_2 d}{\lambda_0} \tag{4-52}$$

The peaks in the transmitted intensity shown in Figure 4-17 occur when the spacing d is a multiple of $\lambda/2$, where λ is the illuminating wavelength. Every time d is changed by $\lambda/2$, another peak in the intensity corresponding to the wavelength λ is recorded by the detector. The wavelength difference $(\Delta\lambda)_{SR}$ corresponding to a change in d of $\lambda/2$, or a change in m of one order, is called the *free spectral range* of the interferometer. This parameter is the maximum wavelength difference that can be unambiguously measured by the interferometer. When we change the order m by 1 we change the phase $\Delta\delta = 2\pi$. We have just derived the relationship

$$\Delta\delta = 2\pi m \frac{\Delta\lambda}{\lambda} \tag{4-53}$$

This allows us to write for the free spectral range

$$(\Delta\lambda)_{SR} = \frac{\lambda}{m}$$

Substituting the maximum value for m into the equation will give us the minimum free spectral range

$$(\Delta\lambda)_{SR} = \frac{\lambda^2}{2n_2 d} \tag{4-54}$$

or in terms of frequency

$$(\Delta\nu)_{SR} = \frac{c}{2n_2 d}$$

If we substitute **(4-52)** into **(4-51)**, we see that increasing the separation d increases the resolving power of the Fabry–Perot, but accompanying that increase is a decrease in the free spectral range as shown by **(4-54)**. If two wavelengths are separated by more than the free spectral range, we will obtain an incorrect value for their wavelength difference. This occurs because the order m in Fabry–Perot interferometers is very large and there is no tag labeling the order of the fringe. We cannot discriminate between, say, the fringe associated with $(m+1)\lambda_1$ and the fringe associated with $(m-1)\lambda_2$. This means that the separation between fringes yields a multiple of the true wavelength separation. To insure that the multiple is 1, another wavelength selective device is often used as a prefilter of the light incident on the Fabry–Perot.

As an example of the chromatic resolving power of a Fabry–Perot interferometer, assume that $r^2 = 0.9$; the reflecting surfaces are separated by $d = 1$ cm, the dielectric between the two reflecting surfaces is air, $n_2 = 1.0$; and the illuminating wavelength is $\lambda_0 = 500$ nm. The resolving power is

$$\frac{\lambda}{\Delta\lambda} = 1.2 \times 10^6$$

and the smallest wavelength difference that can be measured by the instrument is

$$\Delta\lambda = 4.2 \times 10^{-4} \text{ nm}$$

The ratio of the free spectral range to the minimum resolvable wavelength is called the *finesse* $\mathcal{F}$. Using **(4-51)** and **(4-54)**, we can write

$$\mathcal{F} = \frac{(\Delta\lambda)_{SR}}{\Delta\lambda} = \frac{\pi}{2}\sqrt{F} \tag{4-55}$$

The finesse is the key measure of performance of the interferometer, and as it should be, it is independent of the spacing d.

In terms of frequency, the finesse is

$$\mathcal{F} = \frac{(\Delta\nu)_{SR}}{\Delta\nu}$$

Since the finesse is proportional to the reciprocal to the minimum resolvable bandwidth of the instrument, the finesse can be thought of as proportional to the decay time of the Fabry–Perot, i.e., the time taken for the optical fields associated with the standing waves in the Fabry–Perot to fall from their steady-state value to zero after light is removed from the interferometer. The relaxation time is given by

$$\tau_D = \frac{1}{2\pi\Delta\nu}$$

This allows us to associate the finesse of the Fabry–Perot cavity with the Q of a classical oscillator; see **(1A-11)**.

Two other interpretations that can be given to the finesse are that it is the effective number of interfering beams involved in forming the interference fringe, or that it is a measure of the photon lifetime.

It would appear that by increasing the reflectivity, and thus the contrast F, we could continue to increase the finesse without limit. This is not the case; the figure on the mirrors, that is, the flatness of the mirrors, will place an ultimate limit on the resolving power of the interferometer. To see why, return to the expression for the phase difference of each transmitted wave, **(4-35)**, and assume the index in the interferometer is $n_2 = 1$, the operating wavelength is the vacuum wavelength $\lambda = \lambda_0$, and the illumination is a plane wave so that $\cos\,\theta_t = 1$; then,

$$\delta = \frac{4\pi d}{\lambda}$$

The variation of δ with mirror spacing d is

$$\Delta\delta = \frac{4\pi\Delta d}{\lambda}$$

We assume that because the mirrors are not perfectly flat, d varies across the Fabry–Perot's aperture by a fraction of a wavelength given by $\Delta d = \lambda/M$. The phase variation due to the mirror *figure* (the variation in surface flatness) is given by

$$\Delta\delta = \frac{4\pi}{M}$$

Using the relation, **(4-53)**, we can write the resolving power in terms of the mirror figure

$$\mathcal{R} = \frac{\lambda}{\Delta\lambda} = \frac{m}{2}M$$

In calculating the resolution due to mirror reflectivity, we found that the variation in phase due to the mirror's reflectivity is

$$\Delta\delta < \frac{2(1 - r^2)}{r}$$

If we have

$$\frac{1}{M} < \frac{1 - r^2}{2\pi r} = \frac{1}{\pi}\sqrt{F} = \frac{1}{2\mathcal{F}}$$

then the mirror figure, not its reflectivity, will determine the wavelength resolution. We define a new finesse called the figure finesse to characterize the performance of a Fabry–Perot limited by mirror flatness

$$\mathcal{F}_F = \frac{M}{2}$$

The size a of the pinhole used in Figure 4-18 to limit the view of the detector can also determine the wavelength resolution. There is a pinhole finesse, defined as

$$\mathcal{F}_P = \frac{4\lambda f^2}{a^2 d}$$

where f is the focal length of the lens. (The origin of this relationship is diffraction.)

The net finesse due to the pinhole size, mirror figure, and reflectivity is called the *instrument finesse* $\mathcal{F}_I$

$$\frac{1}{\mathcal{F}_I^2} = \frac{1}{\mathcal{F}^2} + \frac{1}{\mathcal{F}_F^2} + \frac{1}{\mathcal{F}_P^2}$$

SUMMARY

The addition of overlapping electromagnetic waves led to the result that the intensity of the resultant wave was the sum of the intensities of the individual waves unless

$$\langle \mathbf{E}_1 \cdot \mathbf{E}_2 \rangle \neq 0$$

The term is zero if the polarizations of the two waves are orthogonal and the propagation medium is isotropic and free of charge. When the term is nonzero, the waves are said to interfere. Two experimental realizations of interference analyzed in this chapter were Young's two-slit experiment and the thin dielectric layer.

The geometry associated with Young's two-slit experiment is shown in Figure 4-6. The result of the analysis demonstrates that if the intensity passed by each slit was equal to I_0, then the intensity a distance D from the two slits and a height x above the optical axis was

$$I = 4I_0 \cos^2\left(\frac{\pi x h}{\lambda D}\right)$$

where h is the spacing between the slits. The distance between two bright bands was found to be

$$\Delta x = \frac{\lambda D}{h}$$

The geometry associated with the dielectric layer model is shown in Figure 4-8. The spatial intensity distribution of the light reflected from the dielectric layer was found to be a function of the incident angle, the layer's thickness, and the wavelength of the illumination. When λ and θ_t were held constant, the fringes were called Fizeau fringes or fringes of constant thickness. When θ_t and d were held constant, then color fringes were created. Finally, when d and λ were held constant, Haidinger's fringes or fringes of equal inclination were generated. The phase difference between a wave reflected off the front and back surface of the layer was shown to be given by

$$\delta = \left(\frac{4\pi n_2 d}{\lambda_0}\right) \cos \theta_t$$

To this phase shift must be added any phase change that occurred upon reflection from an interface. The intensity of the light reflected from the dielectric layer depended on the phase shift produced by the layer

$$I = I_1 + I_2 + 2\sqrt{I_1 I_2} \cos \delta$$

where the two intensities were due to the light reflected from each interface of the dielectric layer.

The dielectric layer was found to be a useful model to explain antireflection coatings, Newton's rings, the Michelson interferometer, and the Fabry–Perot interferometer.

By using the dielectric layer model, the Michelson interferometer's output intensity was found to be

$$\frac{I}{I_0} = 1 + \cos\left(\frac{2\omega d}{c}\right)$$

where the Michelson interferometer was assumed to have a 50:50 beam splitter so that the intensity in each arm of the interferometer was equal. The variable ω in this equation is the angular frequency of the illuminating radiation and d is the difference in length in the two arms of the interferometer.

The Fabry–Perot interferometer was explained by allowing multiple reflections to occur in the dielectric layer. The transmission of light through the interferometer was found to be

$$\frac{I}{I_{max}} = \frac{1}{1 + F\sin^2\left(\frac{\delta}{2}\right)}$$

where

$$F = \frac{4r^2}{(1 - r^2)^2} = \left(\frac{2\mathcal{F}}{\pi}\right)^2$$

The parameter $\mathcal{F}$ is called the reflective finesse of the Fabry–Perot interferometer. The instrument finesse $\mathcal{F}_I$ was defined as the ratio of the free spectral range and the minimum resolvable wavelength of the instrument. The reflective finesse is not the only contributor to the instrument finesse. Other contributions to the instrument finesse, such as the mirror figure, add in the reciprocal

$$\frac{1}{\mathcal{F}_I^2} = \sum_i \frac{1}{\mathcal{F}_i^2}$$

The finesse was shown to be a measure of the resolving power of the Fabry–Perot

$$\mathcal{R} = \frac{2n_2 d}{\lambda_0}\mathcal{F}$$

where $\mathcal{R}$ was defined as the ratio $(\lambda/\Delta\lambda)$.

PROBLEMS

4-1. Eight interference fringes are spread over 2 cm on a screen 100 cm away from a double slit with 0.2 mm separation. What is λ?

4-2. When a glass plate with $n = 1.517$ is placed in one of the arms of a Michelson interferometer, 18.5 fringes are displaced. What is the glass thickness if $\lambda = 589.3$ nm?

4-3. Light with a wavelength of $\lambda = 514.53$ nm is incident normal to a soap film ($n = 4/3$). The first order is strongly reflected. What is the film thickness? What is λ in the film?

4-4. Two flat-glass plates touch at one edge and are separated by a hair at the other. Light with $\lambda = 632.8$ nm is incident normal and 9 fringes are visible. What is the hair thickness?

4-5. The equivalent of a Lloyd's mirror experiment occurs in radar when the receiver R and the source S in Figure 4-20 are close to the ground. In radar, the observed interference pattern is called "multipath." Assume a source with a frequency of 12 GHz located 1 m above the ground, as shown in Figure 4-20, and the receiver is 30 m from the source. How high above the ground will the receiver see the first maximum in received signal?

4-6. If we coat glass ($n = 1.5$) with another material ($n = 2.0$), what thicknesses give maximum reflection? Minimum reflection? Assume that $\lambda = 500$ nm.

4-7. A solid Fabry–Perot is made of an uncoated 2 cm thick slab of material ($n = 4.5$). Relying only on the reflectivities of the air/material interface, what is the fringe contrast and the resolving power at $\lambda = 500$ nm? A glass slab such as this, often with reflective coatings, is called a *solid etalon* and is used as a wavelength selective filter.

4-8. Assume we move one mirror of a Michelson interferometer through a distance of 3.142×10^{-4}m and we see 850 bright fringes pass by. What is the illuminating wavelength?

4-9. What is the smallest thickness that a film of index 1.455 may have if it is to generate a minimum in reflected light under normal illumination, $\lambda = 500$ nm? The film has air on both sides. What would the transmission of this film be like at $\lambda = 500$ nm?

4-10. A glass chamber 25 mm long filled with air is placed in front of one of the slits in the Young's two-slit experiment. The air is removed and replaced by a

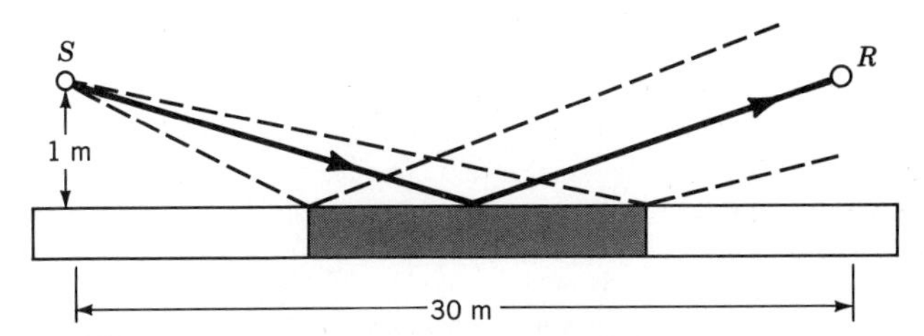

FIGURE 4-20. Source is located 1 m above the ground and 30 m from a receiver of variable height above the ground.

gas. On comparing the two fringe systems, we find that when the gas is present, the fringes are displaced by 21 bright bands toward the side with the chamber. If $\lambda = 656.2816$ nm and n for air is 1.000276, what is the index of refraction of the gas?

4-11. Assume the mirrors of a Michelson interferometer are not perfectly aligned. We see as an output a 3×3 cm illuminated field containing 24 vertical bright fringes. What is the angle the two mirrors make with respect to each other? Assume $\lambda = 632.8$ nm.

4-12. Looking into a Michelson interferometer, we see a dark central disk surrounded by concentric light and dark rings. One mirror is 2 cm farther from the beam splitter than the other and $\lambda = 500$ nm. What is the order of the central disk and the 6th dark ring?

4-13. A metal ring is dipped into a soap solution ($n = 1.34$) and held in a vertical plane so that a wedge-shaped film forms under the influence of gravity. At near-normal illumination with blue-green light ($\lambda = 488$ nm) from an argon laser, one can see 12 fringes per cm. Determine the wedge angle of the soap film. Compare this result to what is seen in Figure 4-10*b*.

4-14. A positive lens with a radius of curvature of 20 cm rests on an optical flat and is illuminated, with the incident wave normal to the surface by sodium *D* light ($\lambda = 589.29$ nm). The gap between the two surfaces is then filled with carbon tetrachloride ($n = 1.461$). What is the radius of the 23rd dark band before introducing the liquid? After introducing the liquid?

4-15. Two narrow parallel slits illuminated by yellow sodium light ($\lambda = 589.29$ nm) are found to produce fringes with a separation of 0.5 mm on a screen 2.25 m away. What is the distance between the slits?

4-16. Prove that the sum shown in **(4-6)** is given by **(4-7)**.

4-17. Assume two plane waves of the same wavelength, $\lambda = 632.8$ nm, are polarized with their electric fields parallel to the y axis. The propagation vectors for the two waves are shown in Figure 4-21. Calculate the separation of intensity maxima along the x and z-axes.

4-18 A plane wave is incident at an angle of θ with respect to the normal of a thin parallel slab of glass with an index of refraction $n = 1.5$. What glass thicknesses d will result in maximum transmission? With the derived equation, find the incident angle that will yield maximum transmission of a wave with wavelength 514.5 nm for a plate that exhibits a maximum in transmission for light of wavelength of 632.8 nm at an angle of incidence of 30°. The variation in transmission with the incident angle is often used to tune dielectric filters or etalons.

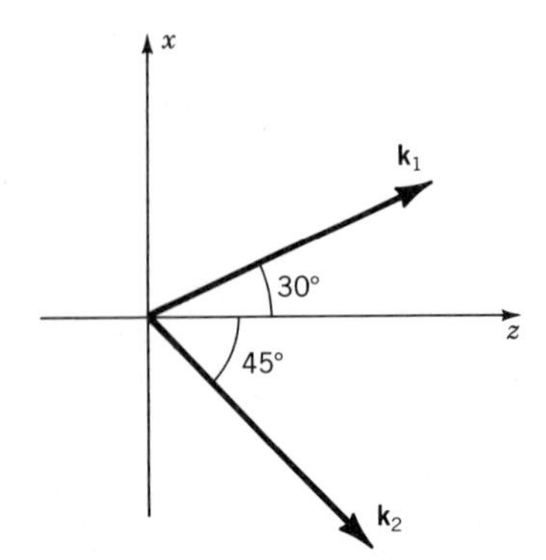

FIGURE 4-21. Propagation vectors of waves to be added together as described in Problem 4-17.

4-19. Add 7 harmonic waves together graphically. The waves have the same wavelength and amplitude but each differs in phase from the next by 20°. Assuming equal phase differences between each wave and its neighbor, for what value of the phase difference would the resultant wave have zero amplitude? What would happen to the resultant amplitude if the phase of each wave differed by 2π? By extending this problem to N waves, we could develop a theory of the Fabry–Perot interferometer using the vector approach to wave addition.

4-20. The fringes shown in Figure 4-22 were produced using a wavelength of 589 nm. The fringes are due to an air gap between two flat glass plates created by placing a scrap of tissue between the plates, along one edge. What is the angle of the air gap?

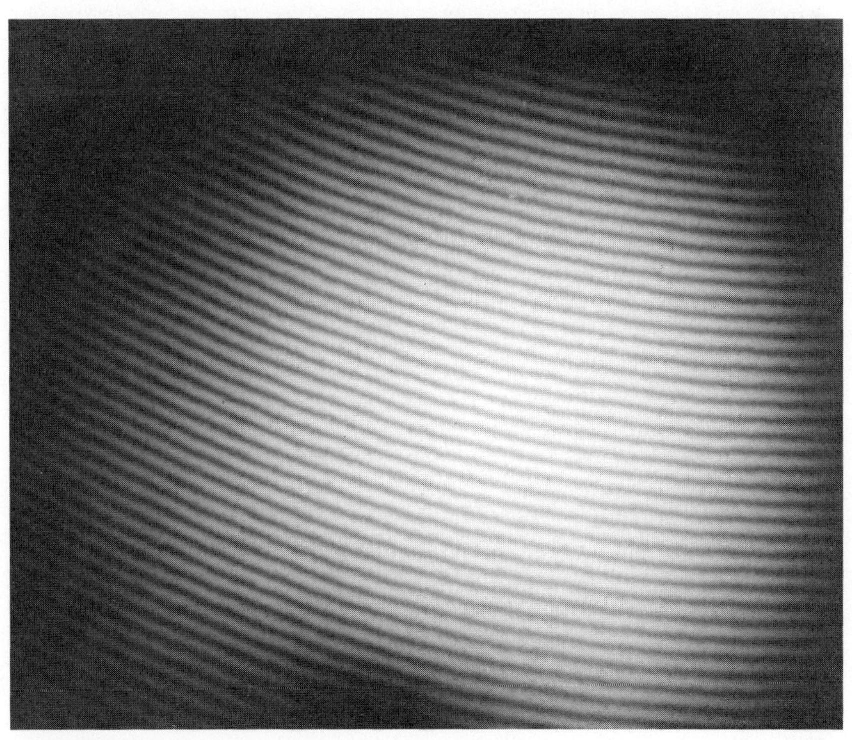

FIGURE 4-22. Fizeau fringes to be used in Problem 4-20.

Appendix 4-A

MULTILAYER DIELECTRIC COATINGS

Interference between waves reflected from the interfaces of a dielectric film can be used to reduce the reflection from an optical surface. This concept can be extended to a multilayer dielectric coating to produce any desired reflection property, and in this appendix, several procedures for designing multilayer coatings will be described.

Fraunhofer produced antireflection layers on glass surfaces by acid etching in 1817 but it was not until 1891 that **Dennis Taylor** associated the reduced reflectivity with an increased transparency. Full utilization of the concept of interference filters had to wait until the late 1940s and early 1950s when techniques were developed that allowed the production of rugged dielectric films. Cold mirrors designed to reflect the visible wavelengths and transmit the infrared wavelengths were one of the first products of this technology and today are found in every dentist's lamp. In the 1970s the multilayer coating technology had developed to a point that allowed the mass production of laser mirrors with very low absorption. Coating technology is now being used to produce durable mirrors for copy machines and conductive coatings to provide frost-free aircraft windshields.

We have introduced the theory of multilayer interference filters in Chapters 3 and 4. This theory will now be used to develop several methods for designing a filter. Only the reflectivity of the filter will be considered, but enough information will be given to allow transmission to also be calculated. We need a notation that will allow the manipulation of a large number of dielectric layers. To see how the notation is generated, we rewrite **(3-33)** as

$$r_N = \frac{E_{Nr}}{E_{Ni}} = \frac{n_i \cos \theta_i - n_t \cos \theta_t}{n_i \cos \theta_i + n_t \cos \theta_t}$$

and **(3-39)** as

$$r_P = \frac{E_{Pr}}{E_{Pi}} = \frac{\dfrac{n_i}{\cos \theta_i} - \dfrac{n_t}{\cos \theta_t}}{\dfrac{n_i}{\cos \theta_i} + \dfrac{n_t}{\cos \theta_t}}$$

First, we defined the ith medium as the incident medium and the $(i + 1)$ medium as the transmitted medium. The two reflection coefficients can be written as one coefficient by defining an effective index. The effective index for normal polarization is

$$N_j = \mathcal{N}_j \cos \theta_j$$

and for parallel polarization is

$$N_j = \frac{\mathcal{N}_j}{\cos\,\theta_j}$$

where the notation indicates that the dielectric films can have a complex index of refraction. This notation allows generalized equations for the amplitude reflection coefficient and transmission coefficient to be defined at the interface between medium i and $i + 1$

$$r_i = \frac{N_i - N_{i+1}}{N_i + N_{i+1}}, \qquad r'_i = -r_i = \frac{N_{i+1} - N_i}{N_i + N_{i+1}} \tag{4A-1a}$$

$$t_i = \frac{2N_i}{N_i + N_{i+1}}, \qquad t'_i = \frac{2N_{i+1}}{N_i + N_{i+1}} \tag{4A-1b}$$

[When $\theta_j = 0$, the generalized equation **(4A-1)** agrees with **(3-45)** and **(3-46)**.] The subscripts allow an unlimited number of interfaces. We will start at $j = 0$ and for m layers there will be $m + 2$ indices of refraction n_0, $n_1, \ldots, n_m, n_{m+1}$, where n_0 is the refractive index in the incident medium and n_{m+1} is the refractive index of the substrate medium, that is, the medium on which the dielectric layers will be deposited. This notation has allowed the design and construction of multilayer stacks as large as $m > 100$.

Vector Approach

The vector approach to designing multilayer interference filters provides only an approximate determination of the reflectivity, but gives the designer a means for visualizing the operation of the multilayer stack.

We begin by revisiting the problem of interference by multiple reflections discussed in Chapter 4. In Chapter 4, we formulated the problem in terms of the transmitted wave; here, we wish to reformulate the effect in terms of the reflected wave and limit our attention to the first reflection from the front and back surface of the film. Redrawing Figure 4-11 for this problem, we produce Figure 4A-1.

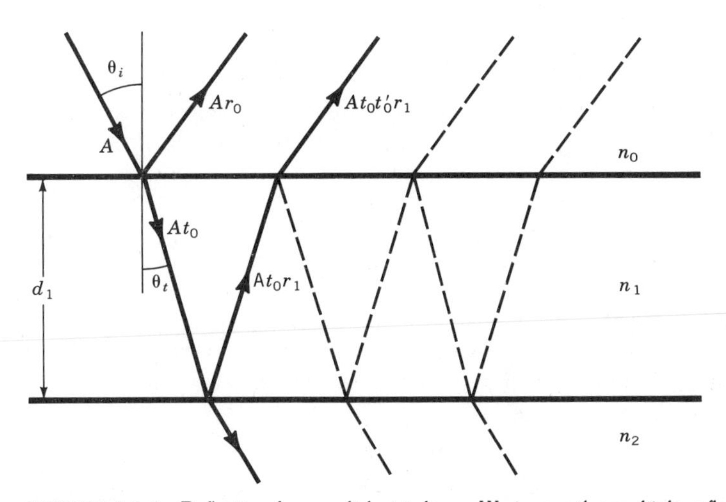

FIGURE 4A-1. Reflection from a dielectric layer. We ignore the multiple reflections, some of which are indicated by the dotted lines.

To add the reflected waves, we must include the phase shift for the wave that is reflected from the back surface. From Figure 4A-1, the reflection coefficient is

$$\rho = \frac{1}{A}(Ar_0 + At_0 t_0' r_1 e^{-i\delta_1})$$

From energy conservation, we can write

$$t_0 = 1 - r_0, \qquad t_0' = 1 - r_0' = 1 + r_0$$

Thus,

$$t_0 t_0' = 1 - r_0^2$$

and we may rewrite the reflection coefficient as

$$\rho = r_0 + r_1 e^{-\delta_1} - r_0^2 r_1 e^{-i\delta_1}$$

Only the first two terms of this expansion will be retained, that is, we will limit our analysis to a single reflection from each interface

$$\rho = r_0 + r_1 e^{-i\delta_1} \tag{4A-2}$$

From **(4-34)**, we have

$$\delta_1 = \frac{4\pi d_1 n_1 \cos\theta_1}{\lambda_0}$$

It is quite simple to extend **(4A-2)** to multilayers. An m layer stack is

$$\rho = r_0 + \sum_{j=1}^{m} r_j \exp\{\Sigma_{k=1}^{j} \delta_k\} \tag{4A-3}$$

We can perform the summation graphically according to the procedure shown in Figure 4-2. In Figure 4A-2, a graphic addition of the reflection coefficients for a two layer stack is shown.

As an example, we will design antireflection coatings using one and two layers. Table 4A-1 lists some of the coating materials we can select for our design. The index of refraction of many of the films depends on how the films were prepared; Table 4A.1 gives average values.

First, we will graphically design a single-layer antireflection coating for a glass of index 1.70 using MgF_2. The glass is to be used in air so $n_0 = 1.0$, $n_1 = 1.38$, and $n_2 = 1.7$.

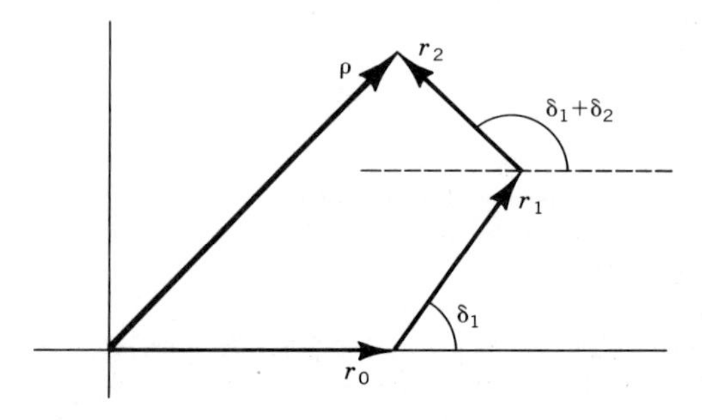

FIGURE 4A-2. Graphic formulation of Equation **(4A-3)**.

TABLE 4A.1 Thin-Film Materials

Material	Index of Refraction	Wavelength Range, μm
Cryolite(Na_3AlF_6)	1.35	0.15–14
Magnesium fluoride (MgF_2)	1.38	0.12–8
Silicon dioxide (SiO_2)	1.46	0.17–8
Thorium fluoride (ThF_4)	1.52	0.15–13
Aluminum oxide (Al_2O_3)	1.62	0.15–6
Silicon monoxide (SiO)	1.9	0.5–8
Zirconium dioxide (ZrO_2)	2.00	0.3–7
Cerium dioxide (CeO_2)	2.2	0.4–16
Titanium dioxide (TiO_2)	2.3	0.4–12
Zinc sulfide (ZnS)	2.3	0.4–12
Zinc selenide (ZnSe)	2.44	0.5–20
Cadmium telluride (CdTe)	2.69	1.0–30
Silicon (Si)	3.5	1.1–10
Germanium (Ge)	4.05	1.5–20
Lead telluride (PbTe)	5.1	3.9–20+

$$r_0 = \frac{n_0 - n_1}{n_0 + n_1} = \frac{1 - 1.38}{1 + 1.38} = -0.16, \qquad r_1 = \frac{1.38 - 1.7}{1.38 + 1.7} = -0.1$$

We know from our discussion in Chapter 4 that we should make $n_1 d_1 = \lambda/4$. We will design for the wavelength $\lambda = 500$ nm, resulting in $d_1 = 910$ nm. The phase change on passing through the film will be calculated, if we assume normal incidence, for $\lambda = 500$ nm and $\lambda = 650$ nm to allow an estimate of the bandpass of the antireflection coating

$$\delta_1(500) = \frac{4\pi d_1 n_1}{\lambda_0} = \frac{4\pi(9.1 \times 10^{-7})(1.38)}{5 \times 10^{-7}} = 180°$$

$$\delta_1(650) = \frac{4\pi(9.1 \times 10^{-7})(1.38)}{6.5 \times 10^{-7}} = 140°$$

The vector diagram for these two cases, following the procedure for vector addition outlined in Figure 4-2, is shown in Figure 4A-3.

When the film's thickness is set to produce a phase shift of 180°, as it was for the wavelength of $\lambda = 500$ nm, then the vectors associated with the reflection coefficients r_0 and r_1 are antiparallel. For this case, it is possible

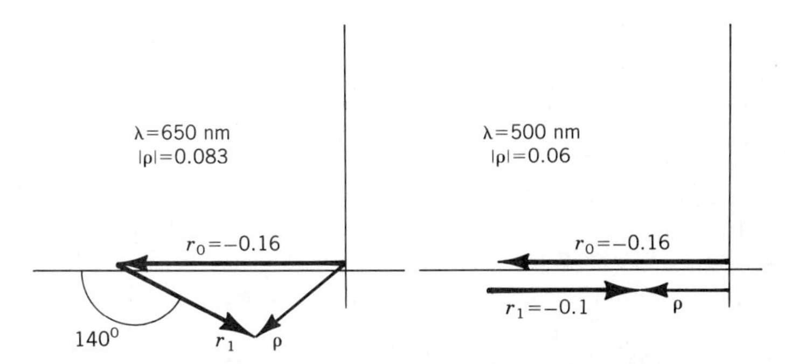

FIGURE 4A-3. Design of an antireflection coating for a glass of index 1.7 using a film of index 1.38. The design wavelength that was used to select the film thickness was 500 nm.

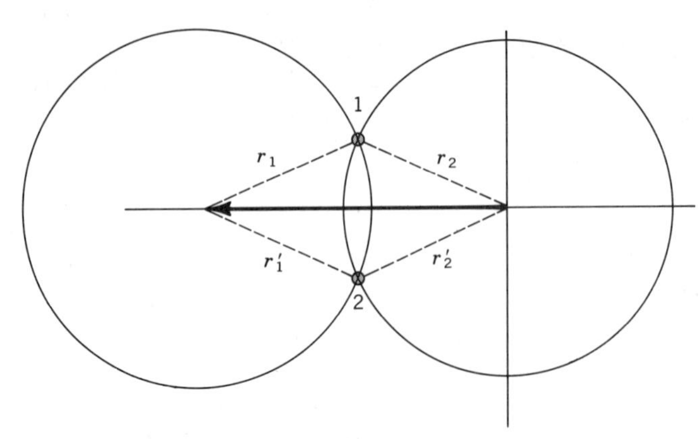

FIGURE 4A-4. Design of a two-layer antireflection coating on a glass of index 1.517 using films with indices 1.46 and 2.3.

to make the resultant reflection coefficient ρ zero if the reflection coefficients from the two interfaces can be made equal. When the phase shift produced by propagation through the dielectric film is not 180°, then there is no possibility of obtaining a resultant reflection coefficient that is zero.

A number of other graphical means of designing multilayer filters are used.[15] One that electrical engineering students would recognize is the use of the Smith chart. The Smith chart is a graphical calculator used in microwave design for the solution of impedance-matching problems. The problem of reflection reduction (or enhancement) can be approached as an impedance-matching problem as was pointed out by the introduction of **(3-48)** and **(3-49)**. The design equations for thin films are similar to those used by the microwave engineer, and thus the Smith chart is found useful and is used extensively by thin-film designers. We will not discuss the Smith chart approach as it offers little advantage over the vector approach and a discussion of its use would take us far afield from optics.

A little more challenging design is a two-layer antireflection coating. The glass will be used in air ($n_0 = 1$), the top-most layer will be SiO_2, ($n_1 = 1.46$), the second layer will be TiO_2, ($n_2 = 2.3$), and the films are to be deposited on a glass substrate called BK-7 ($n_3 = 1.517$). Using these index values in **(4A-1)** yields

$$r_0 = -0.187, \qquad r_1 = -0.213, \qquad r_2 = 0.195$$

We do not know the proper thickness for the two layers so we will draw two circles of radii r_1 and r_2 as shown in Figure 4A-4.

As shown in Figure 4A-4, we center the circle of radius $|r_1|$ at the head of r_0 and the circle of radius $|r_2|$ at the tail of r_0. The intersections of the two circles, labeled 1 and 2, represent two solutions to the design problem. Point 1 has

$$\delta\,(SiO_2) = 118°, \qquad \delta\,(TiO_2) = 26.5°$$

and Point 2 has

$$\delta'\,(SiO_2) = 61.5°, \qquad \delta'\,(TiO_2) = 153.5°$$

The solution associated with Point 2 is easier to manufacture because it is thicker, but the solution associated with Point 1 is less sensitive to wavelength changes.

Matrix Approach

A quantitative approach to designing multilayer filters that allows the computer generation of the spectral characteristics of the filters is the use of matrices. The approach is called the *method of resultant waves* or the *E^+, E^- matrix method*.[16] The boundary conditions associated with Maxwell's equations, as stated in Chapter 4, are placed into a matrix equation format. To accomplish the reformulation, the boundary conditions are manipulated so that the information about the angle of incidence and the polarization are

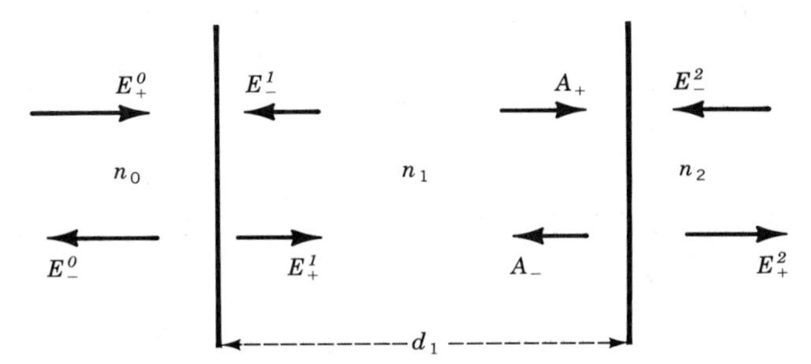

FIGURE 4A-5. Geometry for waves in a dielectric film.

placed into an effective index of refraction as defined in **(4A-1)**. The fields on each side of the boundary can then be represented by plane waves incident normal to the interface as shown in Figure 4A-5.

Each dielectric layer has two interfaces but the two interfaces, as shown in Figure 4A-5, are formally identical so we need only configure the problem for a general interface and repeat the calculation m times for the m interfaces of a dielectric stack $m - 1$ layers high.

We use the boundary conditions **(3-20, 3-21, 3-27, 3-28)** from Chapter 3 to produce the boundary conditions at a generalized interface.

$$n_i \cos \theta_i (E_{N+}^i - E_{N-}^i) = n_{i+1} \cos \theta_{i+1} (E_{N+}^{i+1} - E_{N-}^{i+1}) \tag{4A-4}$$

$$E_{N+}^i + E_{N-}^i = E_{N+}^{i+1} + E_{N-}^{i+1} \tag{4A-5}$$

$$(E_{P+}^i - E_{P-}^i) \cos \theta_i = (E_{P+}^{i+1} - E_{P-}^{i+1}) \cos \theta_{i+1} \tag{4A-6}$$

$$(E_{P+}^i + E_{P-}^i) n_i = (E_{P+}^{i+1} + E_{P-}^{i+1}) n_{i+1} \tag{4A-7}$$

For the normal components,

$$E_{N+}^i = \left(\frac{n_i \cos \theta_i + n_{i+1} \cos \theta_{i+1}}{2n_i \cos \theta_i} \right) E_{N+}^{i+1} + \left(\frac{n_i \cos \theta_i - n_{i+1} \cos \theta_{i+1}}{2n_i \cos \theta_i} \right) E_{N-}^{i+1} \tag{4A-8}$$

$$E_{N-}^i = \left(\frac{n_i \cos \theta_i - n_{i+1} \cos \theta_{i+1}}{2n_i \cos \theta_i} \right) E_{N+}^{i+1} + \left(\frac{n_i \cos \theta_i + n_{i+1} \cos \theta_{i+1}}{2n_i \cos \theta_i} \right) E_{N-}^{i+1} \tag{4A-9}$$

For the parallel components,

$$E_{P+}^i = \left(\frac{n_i \cos \theta_{i+1} + n_{i+1} \cos \theta_i}{2n_i \cos \theta_i} \right) E_{P+}^{i+1} + \left(\frac{n_{i+1} \cos \theta_i - n_i \cos \theta_{i+1}}{2n_i \cos \theta_i} \right) E_{P-}^{i+1} \tag{4A-10}$$

$$E_{P-}^i = \left(\frac{n_{i+1} \cos \theta_i - n_i \cos \theta_{i+1}}{2n_i \cos \theta_i} \right) E_{P+}^{i+1} + \left(\frac{n_i \cos \theta_{i+1} + n_{i+1} \cos \theta_i}{2n_i \cos \theta_i} \right) E_{P-}^{i+1} \tag{4A-11}$$

Using the definitions in **(4A-1a)** and **(4A-1b)**, we can simplify our notation and reduce **(4A-9)** through **(4a-12)** to the following two equations:

$$E^i_+ = \frac{E^{i+1}_+ + r_i E^{i+1}_-}{t_i} \tag{4A-12}$$

$$E^i_- = \frac{r_i E^{i+1}_+ + E^{i+1}_-}{t_i} \tag{4A-13}$$

We may combine **(4A-12)** and **(4A-13)** into a matrix equation for the interface

$$\begin{pmatrix} E^i_+ \\ E^i_- \end{pmatrix} = \begin{pmatrix} \frac{1}{t_i} & \frac{r_i}{t_i} \\ \frac{r_i}{t_i} & \frac{1}{t_i} \end{pmatrix} \begin{pmatrix} E^{i+1}_+ \\ E^{i+1}_- \end{pmatrix}$$

Normally, this is written in a more compact notation

$$\boldsymbol{E}^i = \boldsymbol{I}_i \boldsymbol{E}^{i+1}$$

where $\boldsymbol{I}_i$ is the ith interface matrix

$$\boldsymbol{I}_i = \begin{pmatrix} \frac{1}{t_i} & \frac{r_i}{t_i} \\ \frac{r_i}{t_i} & \frac{1}{t_i} \end{pmatrix} \tag{4A-14}$$

The problem of finding the values of A_+ and A_- in Figure 4A-5 is a simple propagation problem. The fields A_+ and E_- must be modified by the phase shift they experience after propagating through the dielectric layer, here labeled 1

$$A_+ = e^{i\delta_1} E^1_+, \qquad E^1_- = e^{i\delta_1} A_-, \qquad \text{or} \qquad A_- = e^{-i\delta_1} E^1_-$$

These equations can be combined into a matrix equation

$$\boldsymbol{A} = \boldsymbol{T}_1 \boldsymbol{E}^1$$

which can be generalized for the ith dielectric layer by defining a transmission matrix of the form

$$\boldsymbol{T}_i = \begin{pmatrix} e^{i\delta_i} & 0 \\ 0 & e^{-i\delta_i} \end{pmatrix} \tag{4A-15}$$

The effect of an m layer dielectric film can be described by the matrix equation

$$\begin{pmatrix} E^0_+ \\ E^0_- \end{pmatrix} = \boldsymbol{M} \begin{pmatrix} E^f_+ \\ E^f_- \end{pmatrix} \tag{4A-16}$$

where

$$\boldsymbol{M} = \boldsymbol{I}_0 \cdot \boldsymbol{T}_1 \cdot \boldsymbol{I}_1 \cdot \boldsymbol{T}_2 \cdot \cdots \cdot \boldsymbol{I}_{m-1} \cdot \boldsymbol{T}_m \cdot \boldsymbol{I}_m$$

and the E^f are the fields in the final medium. The reflection coefficient of the stack is

$$\rho = \frac{E^0_-}{E^0_+} \tag{4A-17}$$

To simplify the problem without affecting the desired ratios, we normally assume $E^f_+ = 1$ and $E^f_- = 0$.

It is simple to program a personal computer to explore the properties of various combinations of films. We will sketch out a very simple example. The objective is to design a multilayer stack made up of alternating layers of index n_1 and n_2 with the light incident normal to the stack. The thickness of each layer is selected to be one-quarter wavelength thick at the design wavelength. A schematic representation of the film stack would have the following index layers:

$$n_0 \mid n_1 \mid n_2 \mid n_1 \mid \ldots \mid n_1 \mid n_g$$

The first step in the solution of the problem is to form three matrices. The front layer matrix is

$$\boldsymbol{F} = \boldsymbol{I}_0\boldsymbol{T}_1 = \frac{i}{2}\begin{bmatrix} 1 + n_1 & -(1 - n_1) \\ 1 - n_1 & -(1 + n_1) \end{bmatrix}$$

The back layer matrix is

$$\boldsymbol{B} = \boldsymbol{T}_{N+1}\boldsymbol{I}_{N+1} = \frac{i}{2n_1}\begin{bmatrix} n_1 + n_g & n_1 - n_g \\ -(n_1 - n_g) & -(n_1 + n_g) \end{bmatrix}$$

There are N pairs of layers of index n_1 and n_2 and the matrix for one pair is

$$\boldsymbol{M} = \boldsymbol{I}_1\boldsymbol{T}_2\boldsymbol{I}_2\boldsymbol{T}_1 = -\frac{1}{2}\begin{pmatrix} \dfrac{n_1}{n_2} + \dfrac{n_2}{n_1} & \dfrac{n_1}{n_2} - \dfrac{n_2}{n_1} \\[2ex] \dfrac{n_1}{n_2} - \dfrac{n_2}{n_1} & \dfrac{n_1}{n_2} + \dfrac{n_2}{n_1} \end{pmatrix}$$

For N pair of the $n_1 \,|\, n_2$ layers,

$$\boldsymbol{M}^N = \frac{(-1)^N}{2}\begin{pmatrix} \left(\dfrac{n_1}{n_2}\right)^N + \left(\dfrac{n_2}{n_1}\right)^N & \left(\dfrac{n_1}{n_2}\right)^N - \left(\dfrac{n_2}{n_1}\right)^N \\[2ex] \left(\dfrac{n_1}{n_2}\right)^N - \left(\dfrac{n_2}{n_1}\right)^N & \left(\dfrac{n_1}{n_2}\right)^N + \left(\dfrac{n_2}{n_1}\right)^N \end{pmatrix}$$

We will assume that $n_g = 1$, which allows us to write the transmission **(4A-15)** for this problem as

$$\begin{pmatrix} E^0_+ \\ E^0_- \end{pmatrix} = \frac{(-1)^{N+1}}{8}\begin{pmatrix} \left(\dfrac{n_1}{n_2}\right)^N + \left(\dfrac{n_2}{n_1}\right)^N & \left(\dfrac{n_1}{n_2}\right)^N - \left(\dfrac{n_2}{n_1}\right)^N \\[2ex] \left(\dfrac{n_1}{n_2}\right)^N - \left(\dfrac{n_2}{n_1}\right)^N & \left(\dfrac{n_1}{n_2}\right)^N + \left(\dfrac{n_2}{n_1}\right)^N \end{pmatrix}\begin{pmatrix} E^f_+ \\ E^f_- \end{pmatrix}$$

We can now write the reflection coefficient of the stack as

$$\rho = \frac{\left(\dfrac{n_1}{n_2}\right)^N - \left(\dfrac{n_2}{n_1}\right)^N}{\left(\dfrac{n_1}{n_2}\right)^N + \left(\dfrac{n_2}{n_1}\right)^N} \tag{4A-18}$$

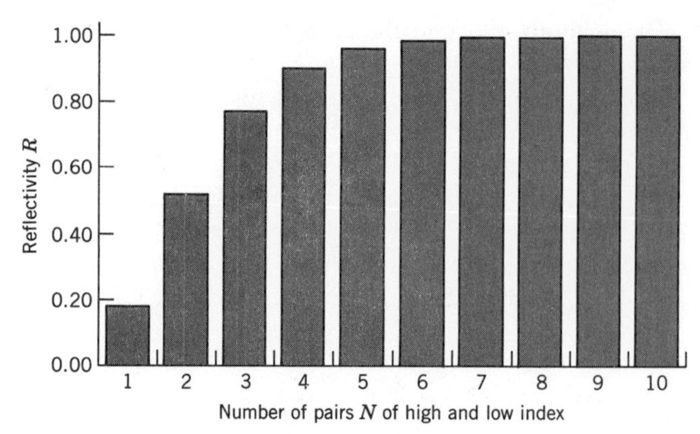

FIGURE 4A-6. The reflectivity as a function of the number of pairs of low-index and high-index films N is shown for a stack with a low-index film of 1.46 and a high-index film of 2.3.

We know that one of the indices is larger than the other. For large N, one of the two ratios will dominate and the reflectivity will approach 1, as shown in Figure 4A-6 where the reflectivities for $N = 1$ to 10 are displayed. As the difference between n_1 and n_2 increases, so does the wavelength range over which the reflectivity can be made nearly 1. A simple computer model using **(4A-18)** generated the curve in Figure 4A-7, which demonstrates the dependence of the reflectivity on the index ratio.

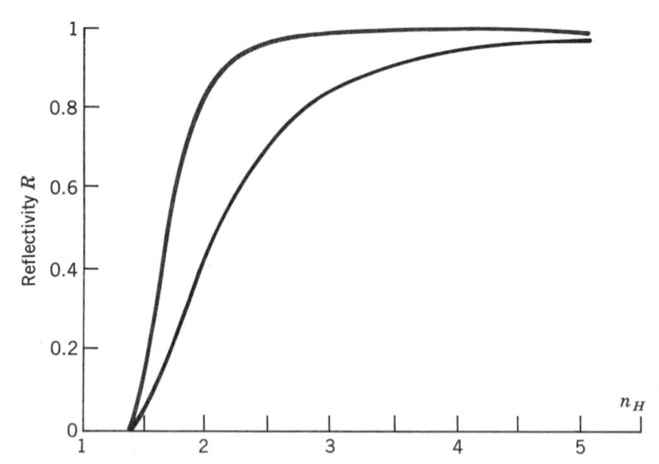

FIGURE 4A-7. The reflectivity of a dielectric stack as a function of the ratio of the index of refraction of the two materials used to construct the stack. The low index of refraction layer was set at 1.35 and the high index of refraction layer was allowed to range from 1.38 to 5.1. Two different size stacks were used, one with $N = 2$ and the other with $N = 4$.

Geometrical Optics

The wavelength of light is on the order of 10^{-7} m. In terms of the size of objects in our daily lives, this is a very small dimension. The contribution to our observations from such a small wavelength should be negligible much of the time. In this chapter, we will assume that the wavelength is zero, which will lead to a geometrical theory of optics.

It is common to observe light rays, such as sunlight filtering through trees or headlights shining through the rain, that appear to travel in straight lines. These common experiences are the origin of the assumption that light propagation is rectilinear. Based on the assumption of rectilinear propagation, geometric constructions using straight lines are used to represent the propagation path. Early man must have used the geometric concepts of light to construct his optical instruments.

One of the oldest optical instruments is a converging lens of rock crystal that was found in Nineveh. There are historical references to the use of magnifying lenses and mirrors to start fires in Greece at least as early as 424 B.C.; see *The Clouds*, Act II, by Aristophanes. The first actual references to a geometrical theory of light were by the Greeks and Romans. The Platonic school taught the theory of rectilinear propagation of light and **Cleomedes**, a Roman about the time of Augustus, discussed atmospheric refraction from a geometrical optics' viewpoint.

Optical instruments such as the telescope and microscope were designed using geometrical constructions. A number of nations claim the inventor of the telescope, one of the first complex refractive optical systems. **Hans Lippershey** of Wesel, Holland constructed one of the first telescopes in 1608. He used crystal rather than glass for the refractive elements. Upon application for a patent on October 2, 1608, he was told to make a binocular telescope, which he quickly did, suggesting that he had a good grasp of geometrical optics. **Galileo Galilei (1564–1642)** was the first to use the telescope to make important scientific discoveries. His astronomical discoveries were not popular. In fact, those who were invited to look through the telescope either refused to believe what they saw or refused to look through the instrument. Another complex optical instrument, the microscope, was probably invented by **Zacharias Joannides** some time before 1610.

The theoretical foundation of geometric optics was established by **Pierre de Fermat** (1601–1665) who deduced the laws of reflection and refraction from the assumption (now called Fermat's principle) that light travels from a point in one medium to a point in another medium in the least time. Fermat did not publish, but instead wrote his thoughts into the margins of other books for his own enjoyment. Fermat's notes were published by his son five years after Fermat's death.

The success of Fermat's principle in optics led to its application in classical mechanics by **William Rowan Hamilton** in 1831. This, in turn, led to the use of the same mathematical formalism for both areas of physics. The formal theory is called the Lagrangian formulation of optics and it allows the calculation of the ray trajectories when the index of refraction is a function of position.

The frequent observation of light traveling in straight lines caused Newton to reject the wave theory of Huygens because of its apparent inability to explain rectilinear motion. Today, the geometrical theory of light can be formally connected to the wave theory, removing the objections of Newton. When the wavelength is allowed to approach zero, the wave equation is found to take on the form of an equation for the trajectory of a classical particle. As was mentioned earlier, narrow beams of light called *light rays* are often observed and we associate the derived trajectories with these light rays. For example, we will discover that when the index of refraction is independent of position (a homogeneous medium), the predicted trajectory is a straight line.

The equation for a ray trajectory, obtained from the wave equation, leads to the statement of Fermat's principle. Fermat's principle is then used to derive the laws of reflection and refraction. These laws are used to develop the matrix formalism of geometrical optics. A matrix, called the *ABCD* matrix, is derived that can be used to describe any optical system. To demonstrate the use of the *ABCD* matrix, the operating parameters of a Fabry–Perot resonator are calculated. The connection between the *ABCD* matrix and the more traditional formulations of geometrical optics based on the thin-lens equation is presented in Appendix 5-A.

The theory of geometrical optics as presented here is associated with rays that propagate at very small angles with respect to the optical axis. This theory is called the paraxial theory and it fails to describe the actual performance of an optical system. The departures of the paraxial predictions from actual performance are called aberrations. An introduction to aberrations is given in Appendix 5-B.

The use of geometrical optics and the concept of total reflection are used to derive a simple theory of guided waves. Some of the basic properties of guided waves are obtained by using this simple theory. The Lagrangian formulation of geometrical optics is used to prove that rectilinear propagation is obtained when the propagation medium is homogeneous and to derive the propagation in an optical fiber with an index of refraction that varies in the radial direction.

EIKONAL EQUATION

In this section, we will derive the effect on the wave equation caused by allowing the wavelength to approach zero. We will find that, in the limit as $\lambda \to 0$, the wave equation leads to an expression that describes the path taken by the normal to the wave surface of constant phase. This normal is called the *optical ray* and its equation of motion is called the *eikonal equation*. This discussion will provide a justification for geometrical optics but is not needed to understand or utilize the ray approach to optics.

It is the spatial rather than the temporal behavior of the light wave that is of interest in geometrical optics. For this reason, we assume that the light has a well-defined frequency and look at the behavior of the complex amplitude, which must obey the Helmholtz equation **(2-14)**

$$(\nabla^2 + k^2)E(\mathbf{r}) = 0$$

where

$$k = \frac{\omega}{c} = \omega\sqrt{\epsilon\mu} = \frac{2\pi}{\lambda}$$

If we simply allow $\lambda \to 0$, then $k \to \infty$ and the Helmholtz equation becomes indeterminate. We can draw quantitative conclusions about the equation by being more careful about taking the limit.

Assume that the solution of the Helmholtz equation is of the form

$$\mathcal{E} = Ae^{ik_0\mathcal{S}} \tag{5-1}$$

where

$$k_0 = \omega\sqrt{\epsilon_0\mu_0} = \frac{2\pi}{\lambda_0}$$

is the propagation constant in a vacuum. Comparing $\mathcal{S}$ to the function $\mathcal{S}(\mathbf{r})$ introduced in Chapter 1 suggests that $\mathcal{S}$ can be interpreted as the function describing the phasefront of the wave. The parameters A and $\mathcal{S}$ will be assumed to be slowly varying functions of x, y, and z that do not go to infinity with k_0. Substituting the assumed solution into the Helmholtz equation requires that we obtain the second spatial derivative of **(5-1)**. For the x derivative,

$$\frac{\partial \mathcal{E}}{\partial x} = ik_0\mathcal{E}\frac{\partial \mathcal{S}}{\partial x} + e^{ik_0\mathcal{S}}\frac{\partial A}{\partial x} = ik_0\mathcal{E}\frac{\partial \mathcal{S}}{\partial x} + \frac{\mathcal{E}}{A}\frac{\partial A}{\partial x}$$

$$\frac{\partial \mathcal{E}}{\partial x} = \left(ik_0\frac{\partial \mathcal{S}}{\partial x} + \frac{\partial \ln A}{\partial x}\right)\mathcal{E}$$

$$\frac{\partial^2 \mathcal{E}}{\partial x^2} = \left[-k_0^2\left(\frac{\partial \mathcal{S}}{\partial x}\right)^2 + 2ik_0\left(\frac{1}{2}\frac{\partial^2 \mathcal{S}}{\partial x^2} + \frac{\partial \ln A}{\partial x}\frac{\partial \mathcal{S}}{\partial x}\right) + \frac{\partial^2 \ln A}{\partial x^2}\right]\mathcal{E}$$

We get similar results for y and z. Substituting these results into the Helmholtz equation yields

$$(\nabla^2 + k^2)\mathcal{E} = -k_0^2\left[\left(\frac{\partial \mathcal{S}}{\partial x}\right)^2 + \left(\frac{\partial \mathcal{S}}{\partial y}\right)^2 + \left(\frac{\partial \mathcal{S}}{\partial z}\right)^2 - \frac{k^2}{k_0^2}\right]\mathcal{E} \tag{5-2}$$

$$+2ik_0\left(\frac{\nabla^2\mathcal{S}}{2} + \nabla \ln A \cdot \nabla\mathcal{S}\right)\mathcal{E} + \left[\nabla^2 \ln A + (\nabla \ln A)^2\right]\mathcal{E}$$

The third term on the right in **(5-2)** does not contain k_0 and thus creates no problem when we let $k_0 \to \infty$. The first two expressions will not create any problems if the quantities in parentheses are zero and if k/k_0 remains finite and equal to the index of refraction as k_0 approaches the limit.

Since

$$\nabla \mathcal{S} \cdot \nabla \mathcal{S} = |\nabla \mathcal{S}|^2$$

setting the first bracket equal to zero yields

$$|\nabla \mathcal{S}|^2 = \left(\frac{k}{k_0}\right)^2 = n^2 \tag{5-3}$$

Surfaces of constant $\mathcal{S}$ are surfaces of constant phase, i.e., phasefronts. The normals to these surfaces are given by $\nabla \mathcal{S}$ in **(5-3)** and represent the rays of geometric optics. The equation

$$\nabla \mathcal{S} = n\hat{\mathbf{s}} \tag{5-4}$$

is the *eikonal equation* and S is called the *eikonal*. We define $\hat{\mathbf{s}}$ as a unit vector, normal to the phasefronts and tangent to the light ray. The curve in Figure 5-1 represents the wavefront, $\mathbf{r}$ is the position vector of the wavefront, and the unit vector $\hat{\mathbf{s}}$ is defined as shown in Figure 5-1.

If n varies in space, i.e., the dielectric constant ϵ is a function of position, then the rays are curved. However, if n is independent of position, then

$$\mathcal{S} = n(\alpha x + \beta y + \gamma z)$$

When the direction cosines obey the equation

$$\alpha^2 + \beta^2 + \gamma^2 = 1$$

the surface described by $\mathcal{S}$ is a plane. For this case, the rays are straight lines in the direction given by the direction cosines

$$\nabla \mathcal{S} = n(\alpha\hat{\mathbf{i}} + \beta\hat{\mathbf{j}} + \gamma\hat{\mathbf{k}}) = n\hat{\mathbf{s}}$$

By taking the limit of the Helmholtz equation when $\lambda \to 0$, as we have done here, we implicitly assume that the electromagnetic wave, in the region around the ray, is a plane wave whose normal is the ray. We follow this normal to determine the propagation of the wave.

The curl of a gradient is zero [see **(2A-10)**]; thus, if we take the curl of **(5-4)**, it is equal to zero

$$\nabla \times \nabla \mathcal{S} = \nabla \times (n\hat{\mathbf{s}}) = 0$$

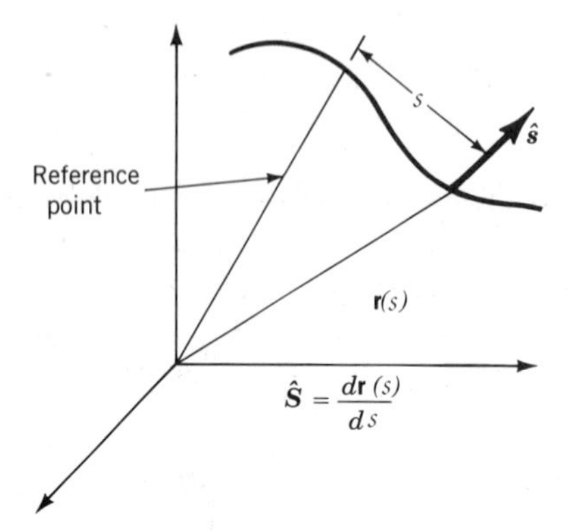

FIGURE 5-1. Geometrical definition of the unit vector normal to the phasefront and parallel to the ray.

If we integrate over a surface A, we obtain

$$\oint_A \nabla \times (n\hat{\mathbf{s}})\, da = 0$$

where the incremental area in rectangular coordinates is $da = dx\, dy$. We can now apply the Stokes theorem from vector calculus to give

$$\oint_C n\hat{\mathbf{s}} \cdot d\mathbf{r} = 0 \tag{5-5}$$

where the notation indicates a line integral over a closed path, C. This is called *Lagrange's integral invariant* and has a formal mathematical connection to Fermat's principle, defined in the next section.

Setting the second bracket of **(5-2)** equal to zero yields an equation

$$\nabla \ln A \cdot \nabla \mathcal{S} = \frac{\nabla^2 \mathcal{S}}{2} \tag{5-6}$$

which describes, through a hydrodynamic model, how energy propagates. The energy is found to flow along the geometrical rays, in a manner similar to a fluid flowing along flow lines, and the Poynting vector is everywhere parallel to the geometrical ray. Problem 5-17 asks the reader to show that the use of

$$\mathcal{E} = E_0 e^{ik_0\mathcal{S}}, \qquad \mathcal{B} = Be^{ik_0\mathcal{S}}$$

in Maxwell's equations will demonstrate this fact by proving that $\nabla\mathcal{S}$ is proportional to the Poynting vector. We will not concern ourselves further with **(5-6)**.

FERMAT'S PRINCIPLE

When we allowed $\lambda \to 0$, we obtained from the Helmholtz equation the result **(5-5)**. This is a mathematical statement that the integral

$$\int_1^2 n\hat{\mathbf{s}} \cdot d\mathbf{r}$$

is independent of the path of integration. It can be shown[5] that the surface over which such an integral exists is the optical wavefront and the normal to this surface is obtained by applying Fermat's principle.

Fermat's Principle states that

Light travels the path which takes the least time.

To discuss Fermat's principle, we must use the concept of optical path length introduced in Chapter 4. The optical path length is the distance light would travel in a vacuum during the same time it took to go the distance ℓ in the medium with index of refraction n. In the last chapter, we used this concept to aid in the calculation of phase differences between two beams of light that start from the same point but travel different paths before they again meet. The definition of Fermat's principle is a statement of minimization; the optical path length will be the quantity that will be minimized.

In a homogeneous medium with index n, light travels a distance

$$\ell = \int_1^2 ds$$

Historically, geometrical optics developed from Fermat's principle and not from the wave theory as the presentation in this chapter might imply. It was *Cataoptrica*, a book written by **Hero of Alexandria** (ca. 100 A.D.), that inspired Fermat to apply the concept of least time to refraction. Hero was an Egyptian or Greek who proved the law of reflection from the premise that the rays of light take the shortest path between any two points.

in a time ℓ/v. In a vacuum, during this same time, light would travel

$$\frac{\ell}{v}c = n\ell = \Delta = S(\mathbf{r}_1) - S(\mathbf{r}_2)$$

If light is traveling over a path C, then the optical path length is the path integral

$$\Delta = \int_C n\,ds$$

where n is the index of refraction and ds is an incremental path length. In Figure 5-2, we see that the optical path length is the separation between the wavefronts at positions $\mathbf{r}_1$ and $\mathbf{r}_2$ along the optical ray.

A question that arises is how does the light know to take the proper path? The answer lies in the fact that these rays are, in reality, waves. The waves traverse all possible paths. When the various waves reach their destination, they interfere. The phase differences for those waves that travel over the "wrong" paths are such that they destructively interfere, while those waves traveling over the "correct" paths constructively interfere. We can prevent light from traversing all possible paths by the use of obstructions; when we do, we observe light in regions that geometrical optics say should be dark. This behavior is called diffraction, a property of waves introduced in Chapter 9. Theoretical support of this interpretation of Fermat's principle will be given in Chapter 11.

Fermat's principle states that the optical path length of an actual ray between any two points 1 and 2 is a minimum, i.e., any curve we choose that joins these points and lies in a "neighborhood" of the proper path has the same optical path length. The time required to traverse any of the paths in this neighborhood is the same. (A physical interpretation of the "neighborhood" of the optical path will be given in Chapter 11.) Mathematically, this path is found through the calculus of variations (see Chapter 19 of the book by Feynman[17] for an excellent introduction to this subject). The mathematical statement of Fermat's Principle is

$$\delta\Delta = \delta\int_{P_1}^{P_2} n\,ds = 0$$

where the symbol δ is to indicate that the variation has been taken.

Actually, the statement that Δ is a minimum is not entirely true; rather, Δ is a stationary value. To understand how a stationary optical path length can occur, consider an elliptical reflector (Figure 5-3), such as is used in some optically pumped laser designs. We wish to apply Fermat's principle to determine the optical path taken by light traveling from one focus P_1 of the elliptical reflector to the other focus P_2.

The actual optical path length for a reflection from point B in the elliptical optical reflector is given by the line P_1BP_2. This path results in an angle of reflection that equals the angle of incidence. If the angle ϕ is varied, then light from P_1 will strike the elliptical reflector at points to the left or right of point B. For the elliptical surface, the distance from P_1 to surface b and then to P_2 is unchanged. This behavior is shown in Figure 5-4, curve b.

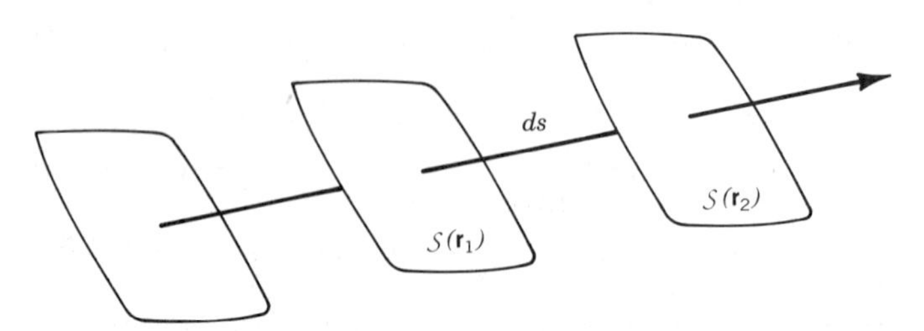

FIGURE 5-2. The integral along the optical ray between the wavefronts at $\mathbf{r}_1$ and $\mathbf{r}_2$ is the optical path length.

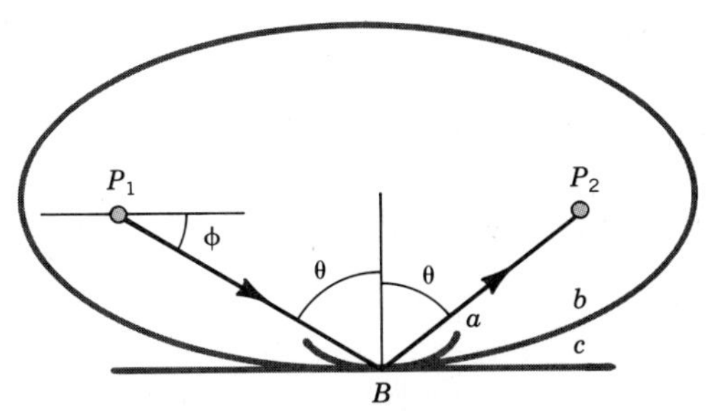

FIGURE 5-3. An elliptical reflector, curve b, with a light source at one focus of the ellipse and a detector at the other. Light is reflected at point B and the angle of incidence and reflection at point B are shown. Two other surfaces with the same surface normal at B are shown. Curve a is assumed to be a sphere with a larger curvature than curve b, and curve c is a plane, and therefore has a smaller curvature than curve b.

Now consider a more tightly curved surface a, here a spherical surface, and the lesser curved surface of a plane c. Each surface is positioned so that it has the same surface normal as surface b at point B. If the shape of the reflector is modified so it now has the curvature of surface a, then as ϕ is varied, the optical path length decreases in size from the correct path length. See Figure 5-4, curve a. The proper optical path is the maximum path length.

Finally, replace the elliptical reflector with a flat mirror, surface c. As already stated, the normal to surface c is the same as the normal of the elliptical reflector at point B. For this surface, as ϕ is varied, the optical path length increases, as is shown in Figure 5-4, curve c. Here, the proper optical path is the minimum path length.

Fermat's principle can be expressed in terms of the time taken to traverse the optical path.

$$\delta \Delta = \delta \int_{P_1}^{P_2} n\, ds = c\delta \int_{t_1}^{t_2} dt = 0 \tag{5-7}$$

From **(5-7)**, we see that minimizing the optical path length is equivalent to minimizing the propagation time. In the following examples, we will minimize

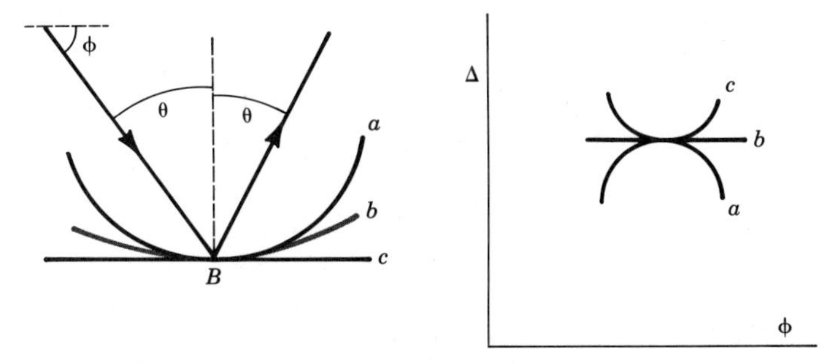

FIGURE 5-4. An enlarged picture of the reflection point from Figure 5-3 is shown on the left. On the right is shown the change in optical path length as we vary the angle of light leaving the origin at point P_1. The labels a, b, c refer to the three surfaces, spherical, elliptical and planar, discussed in the text and shown in Figure 5-3. Fermat's principle predicts an optical path that is the same for all three surfaces.

the propagation time in the application of Fermat's principle. From these examples, we will learn that Fermat's principle requires the time taken to traverse an optical path to remain constant, to first order, for incremental changes in the path length. Mathematically, this means that the first derivative of the time with respect to the optical path length must equal zero.

APPLICATIONS OF FERMAT'S PRINCIPLE

The principle of Fermat's makes some of the laws of optics self-evident. For example, the *principle of reciprocity* says that if light travels from point P_1 to P_2 over a path, then it will travel over the same path in going from P_2 to P_1. Since light travels at the same speed in both directions, the path of least time must be the same between P_1 and P_2, no matter which direction the light travels. The laws of reflection and refraction can be quickly derived using Fermat's principle.

Law of Reflection

Consider the optical path taken by light originating at point P_1, a distance a above the mirror M, and reflected by the mirror to point P_2, a distance c from point P_1 and a distance b above the mirror, as shown in Figure 5-5. The time for light to travel from P_1 to P_2 over the path P_1OP_2 of Figure 5-5 is

$$t = \frac{\overline{P_1O} + \overline{OP_2}}{v} = \frac{1}{v}\left[\sqrt{b^2 + x^2} + \sqrt{a^2 + (c - x)^2}\right]$$

For this time to be stationary, the first derivative of the time with respect to the path length must be zero, where a, b, and c are constants since P_1 and P_2 are fixed

$$\frac{dt}{dx} = \frac{1}{v}\left[\frac{x}{\sqrt{b^2 + x^2}} - \frac{c - x}{\sqrt{a^2 + (c - x)^2}}\right] = 0$$

$$\frac{x}{\sqrt{b^2 + x^2}} = \frac{c - x}{\sqrt{a^2 + (c - x)^2}}$$

$$\sin i = \sin r$$

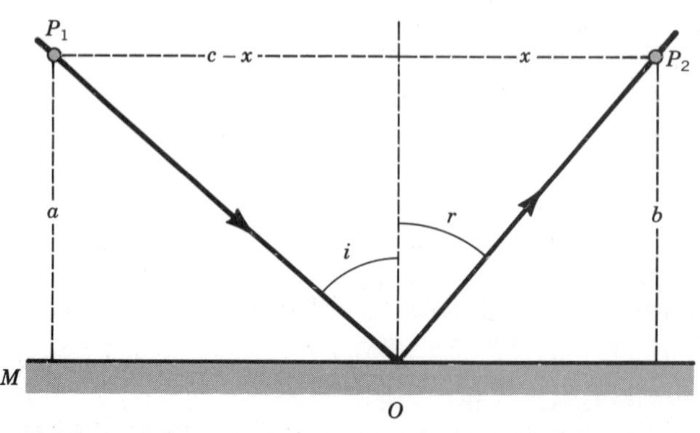

FIGURE 5-5. Geometry for use of Fermat's principle to derive the law of reflection.

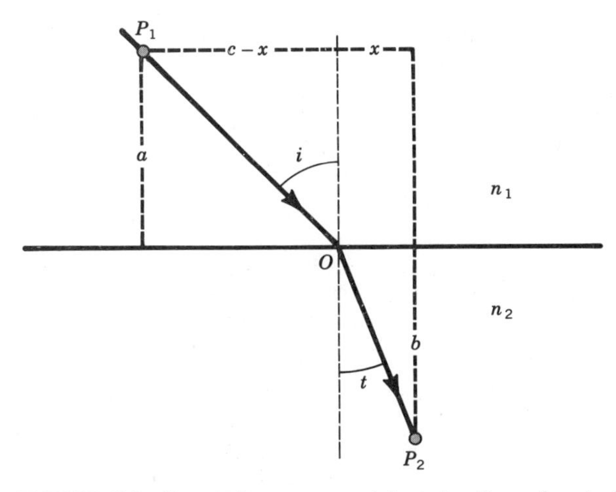

FIGURE 5-6. Geometric arrangement for using Fermat's principle to derive the law of refraction.

Thus, the angle of incidence equals the angle of reflection. The second derivative d^2t/dx^2 is greater than zero so the function is a minimum, as Hero stated 1800 years ago.

Law of Refraction

To obtain the law of refraction using Fermat's principle, consider the path of a ray of light P_1OP_2 in Figure 5-6. All of the paths under consideration start at P_1, a distance a above the interface, and end at P_2, a distance b below the interface, so that a, b, and c are constants. The time to travel from P_1 to P_2 is

$$t = \frac{\overline{P_1O}}{v_1} + \frac{\overline{OP_2}}{v_2} = \frac{1}{v_1}\sqrt{a^2 + (c - x)^2} + \frac{1}{v_2}\sqrt{b^2 + x^2}$$

where v_1 is the velocity of propagation of light in the upper medium of Figure 5-6 and v_2 is the velocity of propagation of the lower medium. The first derivative of the time with respect to the path length is

$$\frac{dt}{dx} = \frac{-(c - x)}{v_1\sqrt{a^2 + (c - x)^2}} + \frac{x}{v_2\sqrt{b^2 + x^2}} = 0$$

$$\frac{\sin t}{v_2} = \frac{\sin i}{v_1}$$

$$\frac{\sin i}{\sin t} = \frac{v_1}{v_2} = \frac{n_2}{n_1}$$

which is Snell's law.

Propagation Through an Optical System

Assume a spherical wave diverging from a source s enters an optical system, represented by the shaded area in Figure 5-7. The optical system forms an image of s at point P; thus, upon emerging from the optical system, the wave

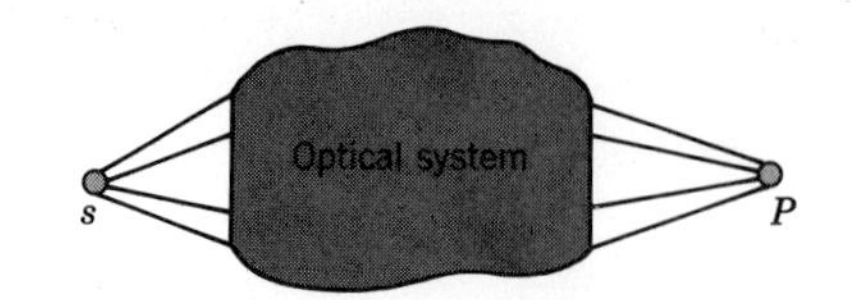

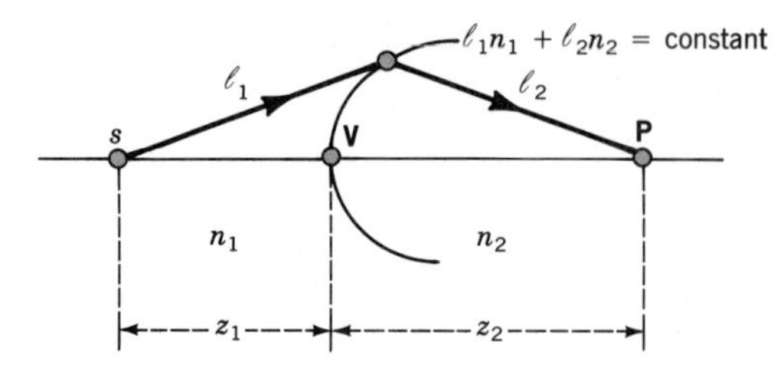

FIGURE 5-7. Fermat's principle for an imaging system requires that the optical path lengths for all rays connecting the object s and the image P must be equal. For a single surface, this requirement results in a Cartesian oval.

converges to an image of the source at point P. Fermat's principle states that the light will traverse the minimum optical path length in going from s to P; however, the light from s to P in Figure 5-7 travels a large number of paths. For agreement with Fermat's principle to occur, all the rays, from the object s to the image P, must traverse identical optical path lengths—no matter how complicated the optical system.

In the simplest application of this fact, Fermat's principle can be used to generate a single surface that will image point s onto point P as shown in Figure 5-7. Fermat's principle also can be used to prove that ellipsoid and hyperbolic surfaces will convert spherical waves into plane waves.

LENS DESIGN AND MATRIX ALGEBRA

The lens designer uses geometric optics to trace rays through an optical system to determine the system's performance before it is constructed. Ideally, the optical system would collect light from an object point and focus it to an image point. In practice, the image point is blurred. The lens designer adjusts the lens' material, shape, and position in an attempt to reduce the blur to an acceptable size. Before the computer, this was a labor-intensive process (see, for example, A.E. Conrady, *Applied Optics and Optical Design*, Dover Publications, New York, 1957).

The ray-tracing formulas used by the lens designer are developed by applying the laws of refraction and ray propagation, as obtained from Fermat's principle. The resulting equations are placed in a matrix formalism because matrix algebra allows an easily understood formalism that can be used to treat the propagation of light through any optical system, from an optical fiber to a complex zoom lens.

In this section, we will, in a stepwise fashion, follow a ray through two surfaces of a lens. It is possible to formulate an exact matrix formulation involving sine, cosine, and tangent functions;[18] we will not do so. Our goal is to become familiar with the general process used by the lens designer but not to study the procedure [Reference 18(a) provides an excellent introduction to lens design, including Fortran programs based on the matrix formalism. Reference 18(b) introduces the same material utilizing the more traditional techniques]. For this reason, we will restrict the discussion to an approximation that allows the sine and tangent functions to be replaced by their angle in radians.

The sine, cosine, and tangent of a small angle γ can be approximated by the Taylor series

$$\sin\gamma = \gamma - \frac{\gamma^3}{3!} + \frac{\gamma^5}{5!} - \frac{\gamma^7}{7!} + \cdots$$

$$\cos\gamma = 1 - \frac{\gamma^2}{2!} + \frac{\gamma^4}{4!} - \frac{\gamma^6}{6!} + \cdots$$

$$\tan\gamma = \gamma + \frac{\gamma^3}{3} + \frac{2\gamma^5}{15} + \frac{17\gamma^7}{315} + \cdots$$

The *paraxial approximation* assumes that γ is small so that only the first term of each expansion is needed. The paraxial approximation is also called the first-order theory; retention of higher-order terms lead to the third-, fifth-, seventh-, etc., order theories. We will use the paraxial approximation to construct a matrix that describes the propagation of a ray through an optical system.

Before we begin, we will establish a sign convention. (The sign convention used here was selected because it is easy to remember. There is no "best" sign convention and a large number of other conventions are used in geometrical optics. Moving from one convention to another can be a source of confusion.) We assume that the optical surfaces will all have spherical curvature with a radius of curvature R. The position of the origin of the x, z coordinate system of Figure 5-8 is such that the z axis lies along the axis of rotational symmetry of the optical components called the *optical axis*. The optical surface under consideration is assumed to intersect the optical axis at the point V, called the *vertex* of the optical surface. The origin of the x, z coordinate system is positioned at the vertex of the optical surface. The sign convention will be the same as that used in a Cartesian system. As is shown in Figure 5-8, all distances along z to the right of V and all distances along x above the optical axis (the z axis) are positive, as are angles measured counterclockwise from the positive z axis, as shown in Figure 5-8. (Normally, the sign of the incident, reflected, and refracted angles are not specified but assume the sign dictated by the ray-tracing procedure.) The radius of

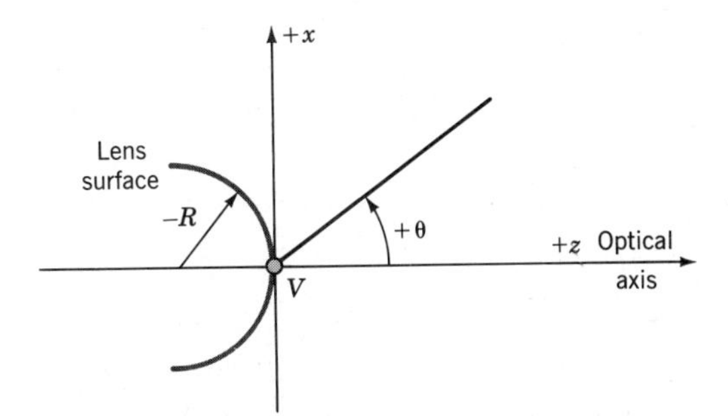

FIGURE 5-8. Coordinate system that establishes the sign convention used for geometrical optics. Distances to the right of V and above the z axis are positive. Angles measured counterclockwise from the positive z axis are positive. The optical surface under consideration intersects the z axis at the position V and is rotationally symmetric about the z axis. The z axis is defined as the optical axis. The radius of curvature of the optical surface is defined as being negative if the center of curvature is to the left of V.

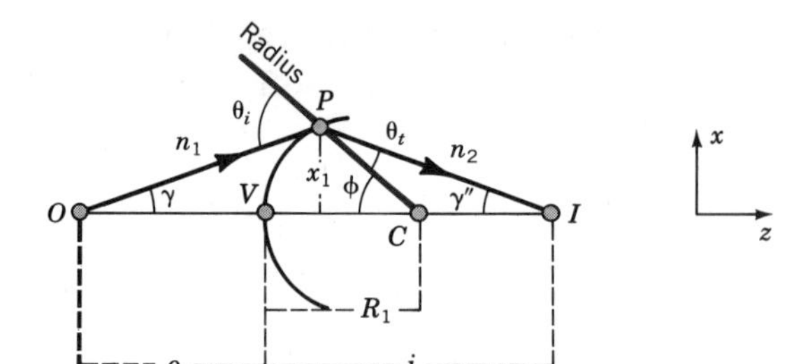

FIGURE 5-9. Geometry used for the development of the lens design equations.

curvature of the optical surface is measured from the surface to the center of curvature and is negative if the center of curvature of the surface is to the left of the vertex *V*. The index of refraction to the left of the optical surface has a lower subscript than the index to the right of the surface, i.e., the indices are ordered n_1, n_2, n_3, etc.

We will analyze a general lens by considering each surface independently. This will require that a new coordinate system be constructed at each surface. Figure 5-9 contains the first optical surface to be considered, with the previously defined coordinate system constructed at its vertex, labeled *V* in Figure 5-9. Light will travel from the object point *O* in Figure 5-9 toward the right, hitting the image point *I* after refraction at the optical surface a distance x_1 above the optical axis. Any ray, such as the one just defined, that has as its origin the object/optical axis intersection is called an *axial ray*.

The first surface has a positive radius of curvature R_1 centered on point *C* of Figure 5-9. The radius of curvature through the point where the light ray intersects the optical surface makes an angle ϕ with the optical axis

$$\sin\phi = \frac{x_1}{R_1}$$

We will use the paraxial approximation that $\sin\phi \approx \phi$.

In the paraxial theory, all rays leaving the object point arrive at the image point; however, in the exact theory, this is not the case. The failure of all the light rays from an object to converge to a single image point after passing through an optical system is called optical *aberration*. To treat the aberrations mathematically, we extend the paraxial theory to third, fifth, seventh, etc. order by including additional terms in the expansion of the sine function (a treatment of aberrations consistent with the present analysis can be found in Reference 18 where computer routines are provided for calculating the various aberrations of an optical system. Appendix 5B provides a brief introduction to the subject).

Snell's law is applied using the angle, shown in Figure 5-10, between the incident ray and the normal to the curved surface, the line labeled radius at the point *P*, where the ray intersects the optical surface. By using the first-order approximation

$$n_1\theta_i = n_2\theta_t$$

the angle of incidence θ_i and the angle of transmission θ_t can be written in terms of known angles using triangles from Figure 5-9. The triangles are redrawn in Figure 5-10 with their boundaries highlighted for clarity. Figure

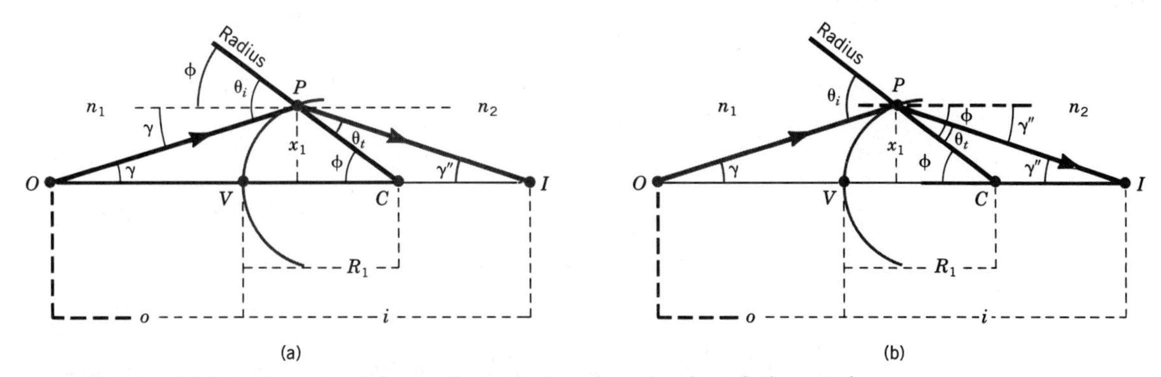

FIGURE 5-10. (a) Triangle constructed using the incident ray from the object 0, the optical axis, from point 0 to point C, and the radius of curvature of the optical surface. (b). Triangle constructed using the transmitted ray traveling toward the image point I, the optical axis and the radius of curvature of the optical surface.

5-10*a* is the triangle used to find the angle of incidence at point *P*, where ϕ is measured from the horizontal to the normal

$$\theta_i = \gamma - \phi$$

Figure 10*b* is the triangle used to find the angle of transmission at point *P*, where again ϕ is measured from the horizontal to the normal

$$\theta_t = -\phi - (-\gamma'')$$

$$\theta_t = \gamma'' - \phi$$

Note from Figure 5-9 that γ is the angle the ray from the object *O* makes with the optical axis and γ'', which is negative in this sign convention, is the angle the ray, traveling toward the image *I*, makes with the optical axis. From Figure 5-9, we have

$$\sin\phi = -\frac{x_1}{R_1}$$

The paraxial approximation allows us to write

$$\phi \approx -\frac{x_1}{R_1}$$

Snell's law can now be written

$$n_1\left(\gamma + \frac{x_1}{R_1}\right) = n_2\left(\gamma'' + \frac{x_1}{R_1}\right)$$

Solving this equation for γ'' yields

$$\gamma'' = \frac{n_1}{n_2}\gamma + \frac{n_1 x_1}{n_2 R_1} - \frac{x_1}{R_1}$$

$$\gamma'' = \frac{n_1}{n_2}\gamma + \frac{x_1(n_1 - n_2)}{n_2 R_1} \qquad \text{(5-8a)}$$

The quantity

$$\Phi = \frac{n_2 - n_1}{R_1}$$

is called the *power* of the surface, with units of *diopters* when R_1 is in meters. If the lens under consideration is located in air, then $n_1 = 1$ and the power for this surface is a positive number when the radius is positive

$$\Phi_1 = \frac{n_2 - 1}{R_1} > 0$$

There is no change in the ray's height above the optical axis x_1 across the surface so that

$$x_1 = x_1'' \tag{5-8b}$$

where the primes denote that we are now across the boundary into the material with index n_2. Each ray is characterized by its slope and its distance from the optical axis. These can be combined into a vector, allowing **(5-8)** to be written in matrix form

$$\begin{pmatrix} 1 & 0 \\ \dfrac{n_1 - n_2}{n_2 R_1} & \dfrac{n_1}{n_2} \end{pmatrix} \begin{pmatrix} x_1 \\ \gamma \end{pmatrix} = \begin{pmatrix} x_1'' \\ \gamma'' \end{pmatrix}$$

The matrix multiplying the ray vector describes refraction across the interface and is called the *refraction matrix*.

$$\boldsymbol{R}_1 = \begin{pmatrix} 1 & 0 \\ \dfrac{n_1 - n_2}{n_2 R_1} & \dfrac{n_1}{n_2} \end{pmatrix} \tag{5-9}$$

This matrix defines the ray's new direction after it has undergone refraction. The determinant of $\boldsymbol{R}_1$ is

$$\|\boldsymbol{R}_1\| = (1)\left(\frac{n_1}{n_2}\right) - (0)\left(\frac{n_1 - n_2}{n_2 R_1}\right) = \frac{n_1}{n_2}$$

[Most lens design texts modify **(5-8)** so that the index of the medium is combined with the ray slope, i.e., γn_2. Also the ray tracing is usually done in the y, z plane rather than the x, z plane. The equations are easily modified by the reader who prefers the other convention.]

We now must consider the translation from the front surface to the back surface in the lens medium of index n_2. We will denote in Figure 5-11 the entry point into the lens as point A and the exit point from the lens as point B. We will retain the same coordinate system, with its origin at V_1, for this part of the ray-tracing problem.

Referring to Figure 5-11, we can write

$$x_2'' - x_1'' = \Delta x = d \tan \gamma_2'' \approx d\gamma_2''$$

where d is a positive number because the light ray is traveling in the positive z direction, from left to right $[d = (z_2 - z_1)]$. As can be seen in Figure 5-11 because of rectilinear propagation, the angle γ'' does not change from A to B but the height above the optical axis changes from x_1'' to x_2''

$$x_2'' = x_1'' + d\gamma_2'' \tag{5-10a}$$

$$\gamma'' = \gamma_2'' \tag{5-10b}$$

These equations can be rewritten in matrix form

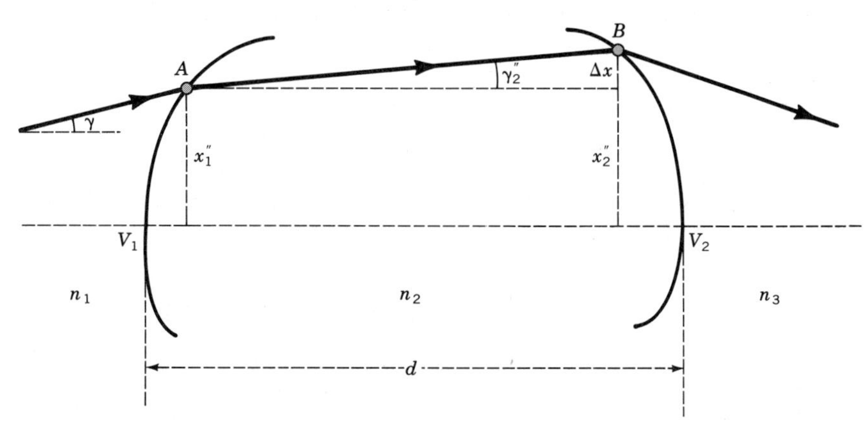

FIGURE 5-11. Geometric construction for propagation of a light ray in the lens. The first surface has a radius of curvature of R_1 and the second surface has a radius of curvature of R_2, a negative quantity in this drawing.

$$\begin{pmatrix} 1 & d \\ 0 & 1 \end{pmatrix} \begin{pmatrix} x_1'' \\ \gamma_2'' \end{pmatrix} = \begin{pmatrix} x_2'' \\ \gamma'' \end{pmatrix}$$

The matrix that describes the propagation from one surface to another is called the *transfer matrix*

$$\boldsymbol{T} = \begin{pmatrix} 1 & d \\ 0 & 1 \end{pmatrix} \tag{5-11}$$

The determinant of $\boldsymbol{T}$ is

$$\|\boldsymbol{T}\| = 1$$

To follow the ray across the surface at B, we must construct another refraction matrix using the same procedure as used to construct **(5-9)**. The origin of the coordinate system is moved to the vertex of the second surface V_2 in Figure 5-11, and the following matrix equation is constructed:

$$\begin{pmatrix} 1 & 0 \\ \dfrac{n_2 - n_3}{n_3 R_2} & \dfrac{n_2}{n_3} \end{pmatrix} \begin{pmatrix} x_2'' \\ \gamma'' \end{pmatrix} = \begin{pmatrix} x_2' \\ \gamma' \end{pmatrix}$$

The refraction matrix for the second surface is, therefore,

$$\boldsymbol{R}_2 = \begin{pmatrix} 1 & 0 \\ \dfrac{n_2 - n_3}{n_3 R_2} & \dfrac{n_2}{n_3} \end{pmatrix} \tag{5-12}$$

The radius of curvature of this second surface is negative according to the sign convention we have established. If the lens is in air, then $n_3 = 1$ and the power for this surface is

$$\Phi_2 = \frac{n_3 - n_2}{R_2} = \frac{1 - n_2}{-R_2} > 0$$

The power of the lens is $\Phi = \Phi_1 + \Phi_2$.

A lens with a positive power such as the one shown in Figure 5-11 is called a *converging lens.* If a plane wave is incident onto a converging lens, the lens will focus the plane wave to a spot; the distance from the vertex to the spot defines the *focal length* of the lens, and the plane normal to the optical axis at this distance is called the *focal plane.* If the lens is a converging lens, then the focal distance is positive.

A *diverging lens* will modify a plane wave so that the wave will appear to originate from a position in front of the lens, i.e., on the object side, a distance equal to the focal length of the lens. The focal distance of a diverging lens is negative as is the power.

The product of the three matrices **(5-9)**, **(5-11)**, and **(5-12)** is called the *system matrix*

$$\boldsymbol{S} = \boldsymbol{R}_2 \boldsymbol{I} \boldsymbol{R}_1$$

or the *ABCD matrix*

$$\begin{pmatrix} 1 & 0 \\ \dfrac{n_2 - n_3}{n_3 R_2} & \dfrac{n_2}{n_3} \end{pmatrix} \begin{pmatrix} 1 & d \\ 0 & 1 \end{pmatrix} \begin{pmatrix} 1 & 0 \\ \dfrac{n_1 - n_2}{n_2 R_1} & \dfrac{n_1}{n_2} \end{pmatrix} = \begin{pmatrix} A & B \\ C & D \end{pmatrix} \tag{5-13}$$

The elements of the system matrix are known as the *Gaussian constants*

$$A = 1 + \frac{d(n_1 - n_2)}{n_2 R_1}, \qquad B = \frac{d n_1}{n_2}$$

$$C = \frac{1}{n_3}\left[\frac{n_2 - n_3}{R_2} + \frac{n_1 - n_2}{R_1} + \frac{d(n_1 - n_2)(n_2 - n_3)}{n_2 R_1 R_2}\right],$$

$$D = \frac{n_1}{n_3}\left[1 + \frac{d(n_2 - n_3)}{n_2 R_2}\right]$$

The determinant of the system matrix is

$$\|\boldsymbol{S}\| = \|\boldsymbol{R}_2\| \bullet \|\boldsymbol{T}\| \bullet \|\boldsymbol{R}_1\| = AD - BC = \frac{n_1}{n_3}$$

If the index of refraction is the same at the beginning and end of the ray path, the determinant of the system matrix is 1.

The use of the paraxial *ABCD* matrix to evaluate an optical system is the first step taken in optical design. It can be used to determine the size of the optical system, the general properties of the optical components, and to estimate the light throughput for the optical system. Additional information about this matrix, such as its connection to the thin-lens equation and to the principal planes of the optical system, is given in Appendix 5-A. Here, we will apply the matrix to the Fabry–Perot resonator used in the construction of lasers.

GEOMETRIC OPTICS OF RESONATORS

As an example of the use of the *ABCD* matrix, we will evaluate a Fabry–Perot resonator used in laser designs and identical to the Fabry–Perot interferometer studied in Chapter 4. The resonator consists of two mirrors that may have spherical curvature, separated by a distance d, as is shown in Figure 5-12. The resonator performs a role similar to a resonant cavity in a microwave system by producing large internal fields with small power input. It does this by trapping energy in the cavity. The trapped energy is distributed over the eigenmodes of the cavity (also called the resonant modes of the cavity).

The *ABCD* matrix will be used to determine the importance of the three parameters R_1, R_2, and d on the design of the resonator. We will discover that the parameters must meet a "stability condition" for a resonant mode to exist in the resonator. The stability condition is associated with the trapping of a ray inside the resonator; the trapped ray establishes a resonator mode. This calculation will tell us when a mode will exist but the physical process

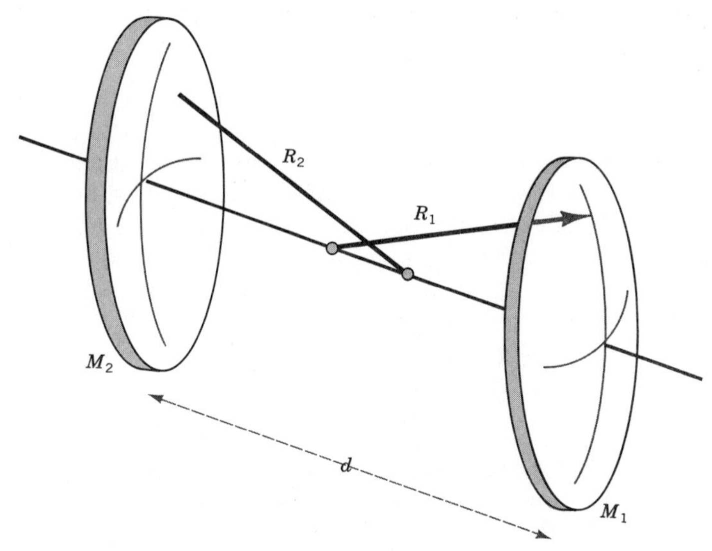

FIGURE 5-12. A Fabry–Perot resonator made of mirror M_1 with radius of curvature R_1 separated by a distance d from mirror M_2 with radius of curvature R_2.

behind the loss of stability can only be discussed through the use of the diffraction theory developed in Chapter 9.[19]

Before the system can be evaluated, an *ABCD* matrix for a mirror must be obtained. The refraction matrix **(5-9)** can be used to obtain the reflection matrix by setting $n_2 = -n$. A negative index of refraction $-n$ can be interpreted physically as associated with a ray traveling in the opposite direction from the ray associated with the positive index n. (If this modification is used on the equation for the power of a surface, we find that a mirror with a positive power must have a negative radius of curvature, which is in agreement with the knowledge that a concave mirror will focus sunlight.) The reflection matrix for a mirror of radius R, in a medium of index n is

$$\boldsymbol{M} = \begin{pmatrix} 1 & 0 \\ \dfrac{2n}{-nR} & \dfrac{n}{-n} \end{pmatrix} = \begin{pmatrix} 1 & 0 \\ \dfrac{-2}{R} & -1 \end{pmatrix}$$

The element M_{22} is negative because the angle of reflection is measured in the opposite sense to the angle of incidence. (A summary of the various *ABCD* matrices we have used is found in Figure 5A-4.)

To follow a light ray through a Fabry–Perot resonator, we start the light ray at mirror M_1 traveling toward mirror M_2. Let the origin of the coordinate system be at the vertex of M_2. Mirror M_1 acts as an object that is to be imaged by M_2 and the ray is followed through its reflection at mirror M_2. The transfer matrix describing the passage of the ray through the Fabry–Perot cavity is

$$\boldsymbol{T}_1 = \begin{pmatrix} 1 & -d \\ 0 & 1 \end{pmatrix}$$

The reflection takes place at mirror M_2, which has a positive spherical curvature R_2. The reflection matrix is

$$\boldsymbol{M}_2 = \begin{pmatrix} 1 & 0 \\ -\dfrac{2}{R_2} & -1 \end{pmatrix}$$

where we assume that the index between the two mirrors is 1. In the next step of this ray-tracing problem, mirror M_2 is treated as an object and we follow the ray from this new object through its reflection at mirror M_1. The light propagates through the cavity toward mirror M_1 and the second transfer matrix becomes

$$\boldsymbol{T}_2 = \begin{pmatrix} 1 & d \\ 0 & 1 \end{pmatrix}$$

Finally, the light completes its roundtrip by reflecting off of mirror M_1. Mirror M_1 has a negative radius of curvature since the distance from the mirror surface to the center of curvature is a negative quantity. The reflection matrix for mirror M_1 is

$$\boldsymbol{M}_1 = \begin{pmatrix} 1 & 0 \\ \dfrac{2}{R_1} & -1 \end{pmatrix}$$

The *ABCD* matrix of the system is obtained by multiplying the matrices together in the matrix equation

$$\begin{pmatrix} x_1 \\ \gamma_1 \end{pmatrix} = \begin{pmatrix} 1 & 0 \\ \dfrac{2}{R_1} & -1 \end{pmatrix} \begin{pmatrix} 1 & d \\ 0 & 1 \end{pmatrix} \begin{pmatrix} 1 & 0 \\ -\dfrac{2}{R_2} & -1 \end{pmatrix} \begin{pmatrix} 1 & -d \\ 0 & 1 \end{pmatrix} \begin{pmatrix} x_0 \\ \gamma_0 \end{pmatrix}$$

To simplify the notation, replace the mirror's radius of curvature with its focal length $f_i = R_i/2$

$$\begin{pmatrix} x_1 \\ \gamma_1 \end{pmatrix} = \begin{bmatrix} 1 - \dfrac{d}{f_2} & d\left(\dfrac{d}{f_2} - 2\right) \\ \dfrac{1}{f_1} + \dfrac{1}{f_2}\left(1 - \dfrac{d}{f_1}\right) & 1 - \dfrac{2d}{f_1} - \dfrac{d}{f_2}\left(1 - \dfrac{d}{f_1}\right) \end{bmatrix} \begin{pmatrix} x_0 \\ \gamma_0 \end{pmatrix} \tag{5-14}$$

In equation form,

$$x_1 = Ax_0 + B\gamma_0 \tag{5-15a}$$

$$\gamma_1 = Cx_0 + D\gamma_0 \tag{5-15b}$$

where

$$A = 1 - \frac{d}{f_2} \qquad B = d\left(\frac{d}{f_2} - 2\right)$$

$$C = \frac{1}{f_1} + \frac{1}{f_2}\left(1 - \frac{d}{f_1}\right) \qquad D = 1 - \frac{2d}{f_1} - \frac{d}{f_2}\left(1 - \frac{d}{f_1}\right)$$

Rewriting the first ray equation, **(5-15a)** yields

$$\gamma_0 = \frac{dx_0}{dz} = \frac{1}{B}(x_1 - Ax_0)$$

If we pass through the resonator a second time, we obtain

$$\gamma_1 = \frac{1}{B}(x_2 - Ax_1) \tag{5-16}$$

From **(5-15b)**, we obtain

$$\gamma_1 = \frac{dx_1}{dz} = Cx_0 + \frac{D}{B}(x_1 - Ax_0) \tag{5-17}$$

Equating **(5-16)** and **(5-17)** yields

$$\frac{1}{B}(x_2 - Ax_1) = Cx_0 + \frac{D}{B}(x_1 - Ax_0)$$

$$x_2 - (A + D)x_1 + (AD - BC)x_0 = 0 \tag{5-18}$$

The quantity $(AD - BC)$ is the determinant of the $ABCD$ matrix. (It is easy to show that the determinant of the resonator's system matrix, $AD - BC$, is equal to 1, as it should be, since the index of refraction of the origin and endpoint is the same).

The difference equation **(5-18)** is equivalent in form to the differential equation for a harmonic oscillator **(1A-1)**. The solution of the differential equation for the harmonic oscillator, using complex notation, is

$$x = x_0 e^{\pm i\sqrt{(s/m)}\,x}$$

Therefore, a reasonable guess for a solution to **(5-18)** is a function of the form

$$x_n = x_0 e^{in\theta}$$

where n is the number of trips the light has made through the cavity. Substituting this function into **(5-18)** yields

$$e^{i2\theta} - (A + D)e^{i\theta} + 1 = 0$$

$$e^{\pm i\theta} = \frac{1}{2}\left[(A + D) \pm \sqrt{(A + D)^2 - 4}\right]$$

$$= \frac{A + D}{2} \pm i\sqrt{1 - \frac{(A + D)^2}{4}} \tag{5-19}$$

Using the Euler formula **(1B-4)**, we can write

$$\cos\theta = \frac{A + D}{2} \tag{5-20}$$

The general solution of the difference equation is therefore

$$x_n = x_0(e^{in\theta} + e^{-in\theta}) = 2x_0 \cos n\theta$$

The function x_n equals the height above the optical axis where the ray strikes the mirror on pass n through the cavity. It cycles periodically through a set of values, retracing its path after one cycle but never exceeding $\pm 2x_0$. To obtain this harmonic behavior, the radical in **(5-19)** must be imaginary.

$$-1 \leq \frac{A + D}{2} \leq 1$$

If the stability condition, **(5-24)** is not met, then **(5-19)** is no longer complex. The identification of $(A + D)/2$ with a cosine function **(5-20)** is no longer proper, and, analogous with the differential equation, the solution of **(5-18)** becomes a function of two exponentials

$$x_n = C_1 e^{a_+ x} + C_2 e^{a_- x}$$

$$e^{a_\pm x} = \frac{A+D}{2} \pm \sqrt{\left(\frac{A+D}{2}\right)^2 - 1}$$

At some value of n, one of the two terms of x_n will exceed the mirror's radius and the beam will escape from the resonator.

The harmonic behavior of x is associated with the stability of the cavity modes. For a mode to be stable, the ray associated with the mode must never miss a mirror as it reflects to and fro in the cavity. If the ray missed the mirror after a number of reflections, then the fields in the Fabry–Perot would not build up to large values and we would not have a resonator. If we use the relationships for A and D, the stability condition can be rewritten as

$$-1 \le \frac{d^2}{2f_1 f_2} - d\left(\frac{1}{f_1} + \frac{1}{f_2}\right) + 1 \le 1 \tag{5-21}$$

The inequality can be simplified by using the identity

$$2\left(1 - \frac{d}{2f_1}\right)\left(1 - \frac{d}{2f_2}\right) = \frac{d^2}{2f_1 f_2} - d\left(\frac{1}{f_1} + \frac{1}{f_2}\right) + 2$$

to allow us to rewrite **(5-21)**. The left side of the inequality is

$$0 \le \frac{d^2}{2f_1 f_2} - d\left(\frac{1}{f_1} + \frac{1}{f_2}\right) + 2$$

$$0 \le \left(1 - \frac{d}{2f_1}\right)\left(1 - \frac{d}{2f_2}\right) \tag{5-22}$$

The right side of the inequality can be rewritten

$$\frac{d^2}{2f_1 f_2} - d\left(\frac{1}{f_1} + \frac{1}{f_2}\right) + 2 \le 2$$

$$\left(1 - \frac{d}{2f_1}\right)\left(1 - \frac{d}{2f_2}\right) \le 1 \tag{5-23}$$

Combining these two inequalities, we obtain the stability condition for the Fabry–Perot resonator

$$0 \le \left(1 - \frac{d}{2f_1}\right)\left(1 - \frac{d}{2f_2}\right) \le 1 \tag{5-24}$$

In applying this stability condition, the fact that the sign convention requires facing mirrors to have radius of curvatures of opposite sign must be kept in mind.

GUIDED WAVES

In the discussion of the Fabry–Perot interferometer, a light wave was assumed to undergo multiple reflections at the boundaries of a lossless dielectric; see Figure 4-11. The wave trapped in the dielectric layer slowly died away because of the light transmitted out of the dielectric layer at each reflection from a boundary. If the incident angle of the wave in the layer met the requirement for total reflection **(3-51)**, then the wave would not experience loss but would propagate along the dielectric layer; such a wave, confined to the layer, is called a *guided wave* and the dielectric layer is called a *waveguide*.

In this section, we will use geometric optics to discover the properties of guided waves. The theory to be used is not a "pure" geometric theory because we borrow the concepts of interference and phase change upon total reflection from the wave theory. The composite theory is useful because it provides an intuitive understanding of guided waves.

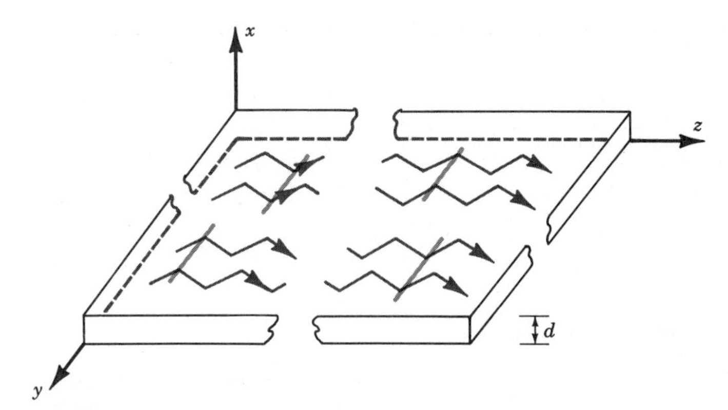

FIGURE 5-13. Geometry of planar waveguide. Light is propagated along the z direction.

We will discuss the requirements placed on the angle of incidence in the waveguide and briefly review techniques for introducing guided waves into the waveguide. The discussion will be limited to a one-dimensional problem by assuming that the waveguide, Figure 5-13, has a thickness d in the x direction and extends to infinity in the y and z directions with light propagating along the z direction. The results for the dielectric slab of Figure 5-13 will, in general, not apply to optical fibers that have cylindrical symmetry. However, the results derived here will apply to those rays in a fiber that pass through the cylindrical axis of the fiber. The rays that cross the fiber axis as they propagate are called *meridional rays.* (In geometrical optics, any ray that lies in a plane containing the optical axis is a meridional ray; all other rays are called *skew rays.*)

Tyndall demonstrated that light could be guided by a water-jet in 1870, but a theoretical investigation of guided waves was not conducted until the publication of an analysis by **Peter Debye** in 1910. The use of guided waves was not practical until the idea of cladding a core glass of index n_2 by another glass of index $n_1 < n_2$ was introduced in 1954. The guiding structures were suggested for use as endoscopes in medical diagnosis but this application did not trigger much interest in guided-wave technology. Research in guided-wave optics did not blossom into a full-scale research area until proposals were made in the mid-1960's for the use of guided waves in communications.

Consider a three-layer sandwich of dielectrics with indices $n_2 > n_3 > n_1$. When light coming from the region of index n_3 strikes the two interfaces, we can have the three possible outcomes, shown schematically in Figure 5-14.

When θ_3 in Figure 5-14*a* is small and the rays meet the inequalities

$$0 \le \theta_1 \le \frac{\pi}{2}, \qquad \theta_3 < \sin^{-1}\left(\frac{n_1}{n_3}\right), \qquad \theta_2 < \sin^{-1}\left(\frac{n_1}{n_2}\right) \tag{5-25}$$

the light is transmitted across the two interfaces, as shown in Figure 14*a*. We will call rays that meet the inequalities in **(5-25)** *radiation modes* or air modes because medium 1 is often air. The discussions in Chapter 3 on reflection and refraction were descriptions of the behavior of radiation modes. If the thickness d of the dielectric layer is on the order of a wavelength, the interfaces are very flat, or the illuminating light wave has a high degree of

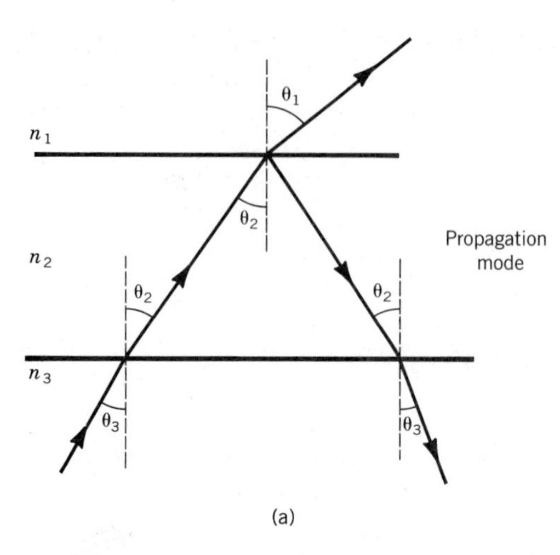

FIGURE 5-14*a*. The rays shown here are called radiation modes, air modes, or propagation modes. They occur when the inequalities of **(5-25)** are met.

coherence (see Chapter 8); then, interference, discussed in Chapter 4, would be observed.

When θ_3 increases to a large enough value, θ_2 will exceed the critical angle for total reflection and light rays will not enter the region with index n_1; see Figure 5-14*b*. For this condition to occur, θ_2 and θ_3 are limited by the inequalities

$$\sin^{-1}\left(\frac{n_1}{n_2}\right) < \theta_2 < \sin^{-1}\left(\frac{n_3}{n_2}\right), \qquad \sin^{-1}\left(\frac{n_1}{n_3}\right) < \theta_3 < \frac{\pi}{2} \tag{5-26}$$

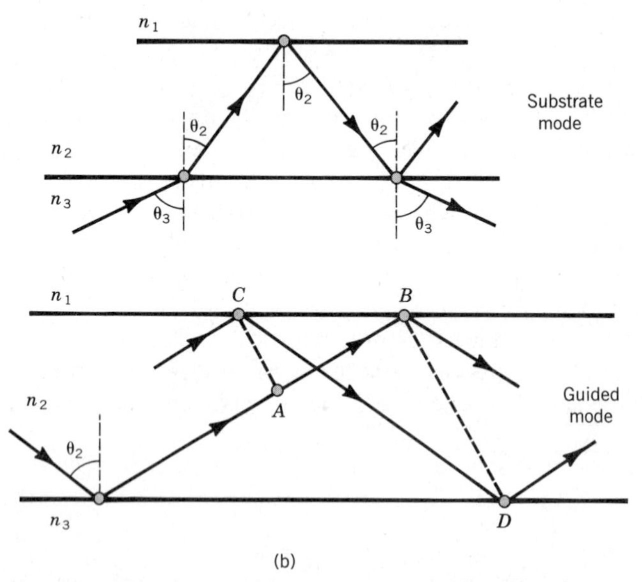

FIGURE 5-14*b*. The rays shown here are the substrate modes of a dielectric guide using the zigzag-wave or geometric optical model of propagation. As in (a), these modes are propagation modes but the rays do not enter the medium with index n_1. For these modes to exist, the inequalities of **(5-26)** must be met. (bottom) The waveguide modes of a dielectric layer are illustrated using the zigzag-wave or geometric optical model of propagation. The dotted lines *AC* and *BD* represent the wavefront of a plane wave represented by the rays at two different positions during the propagation down the dielectric layer.

We call rays that satisfy **(5-26)** *substrate modes* because the medium in which they propagate, with index of refraction n_3, is the *substrate* or supporting medium for the dielectric layer with index n_2. (These modes are propagating waves and thus could also be called radiation modes.) We treated the waves associated with the rays obeying **(5-26)** in Chapter 3 during the discussion of total reflection.

Finally, it is possible to introduce a ray into the region of index n_2 so that the angle of incidence of the ray satisfies the inequality

$$\sin^{-1}\left(\frac{n_3}{n_2}\right) < \theta_2 < \frac{\pi}{2} \tag{5-27}$$

These rays, called *waveguide modes*, are associated with waves trapped in the region of index n_2 that we will call the waveguide region or the guiding layer; see Figure 5-14*c*. To obtain the guided-wave modes of Figure 5-14*c*, the incident angle θ_3 must exceed $\pi/2$. Thus, a waveguide mode cannot be excited by simply illuminating the guide surface. Figure 3-12 illustrated one method of coupling energy into a guide. Discussion of the use of prism coupling can be found in any text on guided waves. We will discuss a second technique for coupling energy into the waveguide modes in some detail in the next section.

End Coupling

One technique for launching a waveguide mode is to couple the light into the polished end of the dielectric layer, along the z direction and perpendicular to the x, y plane. This technique is called *end-fired coupling*.

As an example, we will consider a step-index waveguide, Figure 5-15, and use geometric optics to follow the light, introduced into a flat end surface of the guide, through the guide. In Figure 5-15, the x, y plane of the step-index guide and its index of refraction profile, a step function, are shown; a fiber geometry is shown because we will derive the limits placed on the incident angle for end-fired coupling into a fiber. For a fiber, the high-index

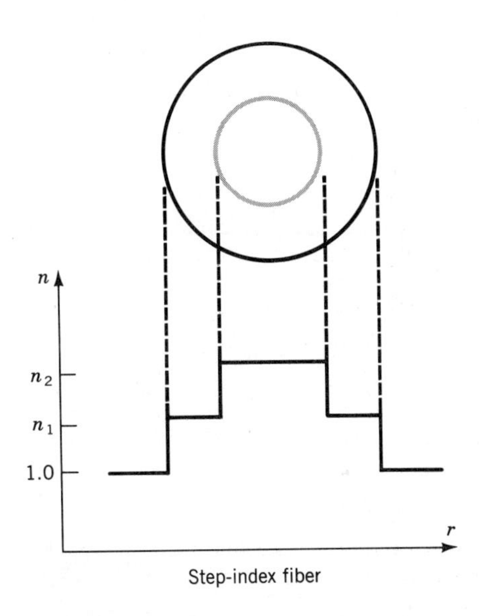

FIGURE 5-15. Step-index fiber. The fiber is shown end on and the index of refraction profile is displayed.

region at the center is called the core and the lower-index region surrounding the core is called the cladding.

The guide will accept rays that are incident onto the end face of the guide at an angle less than a critical angle that we will denote by θ_{NA} and call the *acceptance angle*. The value of this maximum angle is obtained by using the fact that rays trapped in the guide must be incident on the core/cladding interface at an angle θ_2, which exceeds the critical angle for total reflection. When this requirement is met, the light is trapped in the guide until it emerges from the other end.

The maximum incident angle θ_{NA} that will propagate as a guided wave can be obtained by using the geometry in Figure 5-16. The angle θ_2 in Figure 5-16 must exceed the critical angle

$$\theta_2 \geq \theta_c = \sin^{-1}\frac{n_1}{n_2}$$

The angle θ_2 can be written in terms of the sides of the triangle drawn in Figure 5-16

$$a^2 = b^2 + c^2$$

$$1 = \left(\frac{b}{a}\right)^2 + \left(\frac{c}{a}\right)^2$$

$$\sin\theta_2 = \frac{b}{a}$$

$$\left(\frac{c}{a}\right)^2 = 1 - \left(\frac{b}{a}\right)^2 \leq 1 - \left(\frac{n_1}{n_2}\right)^2$$

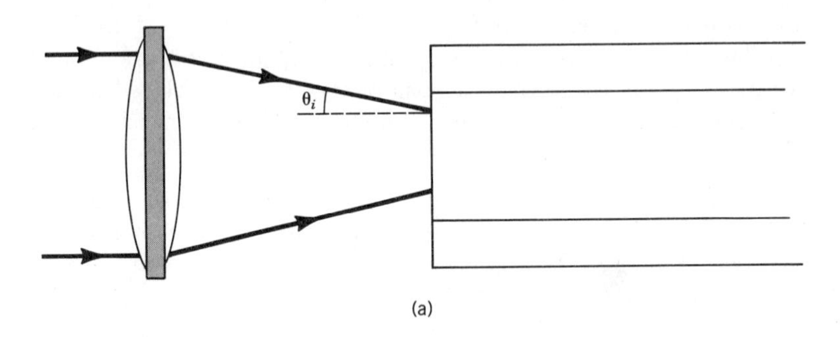

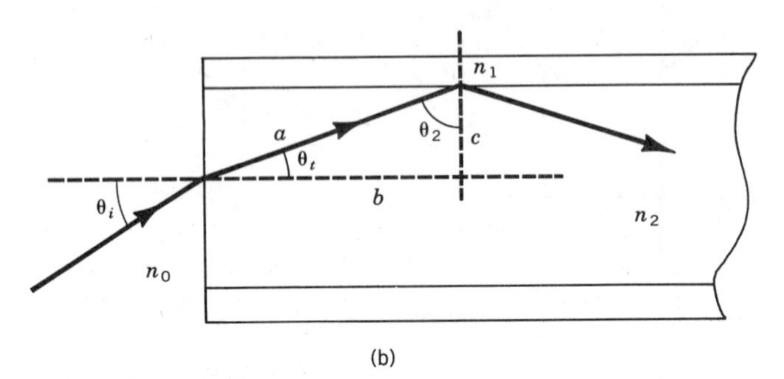

FIGURE 5-16. (a) Edge-coupled guide. A lens is used to focus light into the polished end of a guided-wave structure. (b) Coordinates for determining the numerical aperture of a step-index optical guide.

The sides of the triangle in Figure 5-16 can also be related to the transmission angle θ_t at the front surface of the guide

$$\frac{c}{a} = \sqrt{1 - \left(\frac{n_1}{n_2}\right)^2} = \sin\,\theta_t$$

Snell's law relates the transmission angle to the angle of incidence on the fiber face

$$\frac{n_2}{n_0} = \frac{\sin\,\theta_i}{\sin\,\theta_t}$$

The acceptance angle can be written in terms of the above relations

$$\sin\,\theta_i \le \sin\,\theta_{NA} = \frac{n_2}{n_0}\sqrt{1 - \left(\frac{n_1}{n_2}\right)^2}$$

We define the *numerical aperture* of the guide in terms of the acceptance angle as

$$\begin{aligned} NA &= n_0\,\sin\,\theta_{NA} = n_2\,\sin\,\theta_t \\ &= n_2\sqrt{1 - \left(\frac{n_1}{n_2}\right)^2} \\ &= \sqrt{n_2^2 - n_1^2} \qquad (5\text{-}28) \end{aligned}$$

The numerical aperture is proportional to the largest incident angle that the guide can accept and transmit. The results just derived will apply to a planar guide or the meridional rays of a fiber, but must be modified for other rays in a fiber.

In addition to meridional rays, a fiber will allow the propagation of skew rays. As mentioned earlier, skew rays travel the length of the fiber without ever crossing the cylindrical axis of the fiber. Figure 5-17 shows the a skew ray from two perspectives.

Figure 5-17*b* is an end view of the fiber showing the skew ray propagating down the fiber without crossing the axis of the fiber. The size of the vectors representing the rays in Figure 5-17 is used to denote the depth of the rays propagating into the fiber; smaller vectors represent deeper rays. The angle between each pair of rays is 2φ, where φ is defined in Figure 5-17*a*.

Figure 5-17*a* displays the geometry associated with the skew ray. The skew ray propagates in a plane that makes an angle φ with the plane containing the cylindrical axis of the fiber. The component of the ray propagating in the core parallel to the normal vector of the core/cladding interface is

$$k_t\,\cos\,\varphi\,\sin\,\theta_t$$

The normal vector of the core/cladding interface is the radius vector from the cylindrical axis of the fiber to the point where the ray strikes the interface and, with the ray, establishes the plane of incidence. The component of the ray along the normal is given, in terms of the angle of incidence θ_2, as

$$k_t\,\cos\,\theta_2 = k_t\sqrt{1 - \sin^2\,\theta_2}$$

If we assume that θ_2 is equal to the critical angle for total reflection and use Snell's law to replace θ_t by θ_i

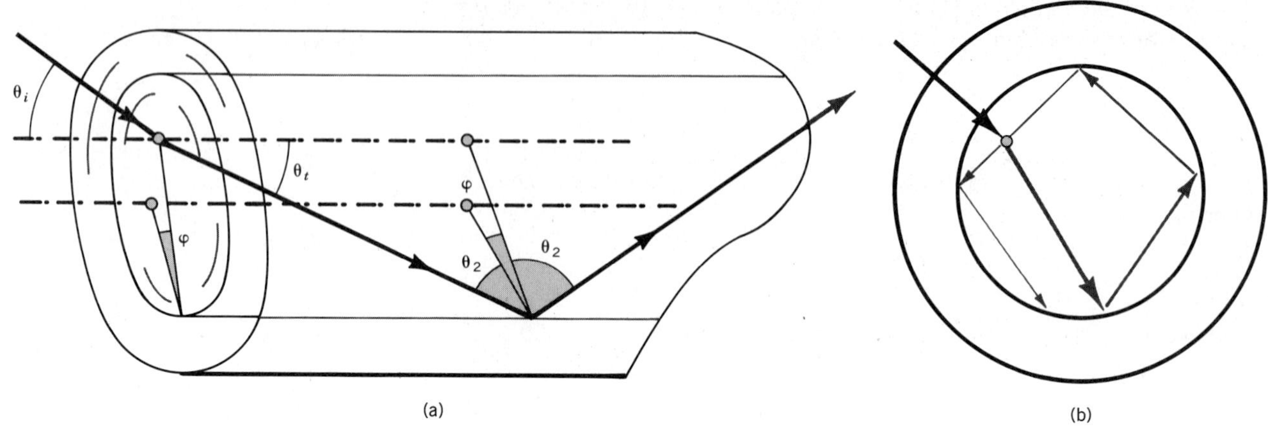

FIGURE 5-17. Propagation of a skew ray in an optical fiber. The ray propagates in a helical path. (a) The geometry used to evaluate the numerical aperture for a skewed ray. (b) The propagation of a skew ray as viewed from the end of the fiber. As the ray progresses down the fiber, the arrows representing the rays decrease in thickness.

$$\cos\varphi \frac{n_0}{n_2} \sin\theta_i = \sqrt{1 - \left(\frac{n_1}{n_2}\right)^2}$$

We thus discover that the numerical aperture for a skew ray, propagating in the plane defined by φ, is given by

$$(NA)_s = \frac{NA}{\cos\varphi}$$

where the NA represents the numerical aperture for a meridional ray given by **(5-28)**. We see that the maximum acceptance angle of a meridional ray is the minimum acceptance angle for skew rays. The skew rays thus act to increase the light-gathering capability of a fiber.

As an example of the application of **(5-28)**, assume $n_0 = 1$, $n_1 = 1.5$, and $n_2 = 1.53$. The critical angle is then

$$\theta_c = \sin^{-1}\left(\frac{n_1}{n_2}\right) = \sin^{-1}\left(\frac{1.5}{1.53}\right) = 78.6°$$

The numerical aperture is 0.303, which corresponds to a maximum incident angle of $\theta_{NA} = 17.6°$. If a skew ray is propagating in this fiber with $\varphi = 50°$, then the maximum incident angle for this skew ray is 28.1°.

The symmetry associated with the propagation of a ray in an optical fiber predicts that the angle of the output ray from a fiber will be equal to the incident angle. Because the meridional ray is confined to a plane, the ray will always emerge at a position determined by the input ray. The skew ray, however, will exit a fiber at a position that is a function of the number of reflections it has undergone, rather than upon the input ray. The skew ray will thus smooth the output distribution of light along a circumference of a circle of radius r, where r is the position of the input ray at the entrance to the fiber. An example of this smoothing is shown in Figure 5-18. In Figure 5-18*a*, an array of slits is shown distributed along the radius of a fiber from the center (on the right of Figure 5-18*a*) to the edge of the core (on the left of Figure 5-18*a*). The light distribution on exiting the fiber is shown in Figure 5-18*b*.

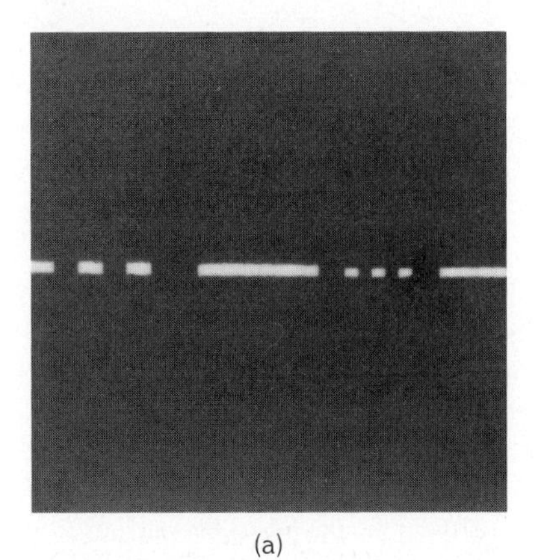

(a)

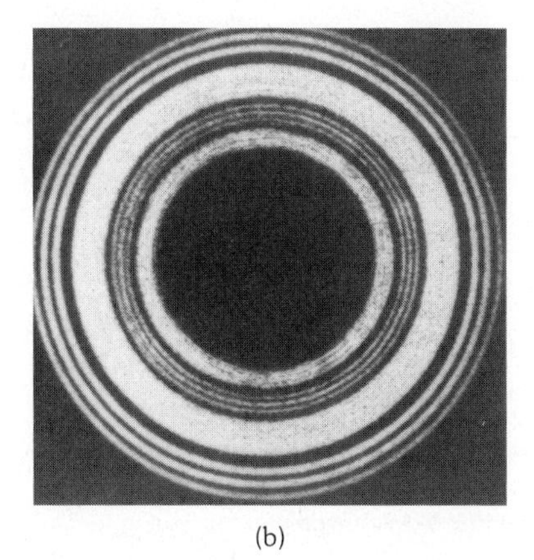

(b)

FIGURE 5-18. (a) A set of slits forms a one-dimensional pattern. This pattern is positioned at the fiber input with the rightmost edge of the pattern centered on the fiber axis. (b) The output of the multimode fiber with (a) as an input displays the intensity smoothing produced by skew rays. (Courtesy of A. Ti and A. Friesem, Environmental Research Institute of Michigan.)

Guided Modes

To analyze how the waveguide modes propagate in the waveguide region of index n_2, we use the geometry shown in Figure 5-14*c*. We have developed all of the necessary theory to use the wave model for this problem, but we will instead use the more pictorial geometric model. This ray model, sometimes called the *zigzag model*, will demonstrate the application of geometric optics and further will develop an intuitive understanding of guided optics.

In free space, we proved (Chapter 2) that the electric and magnetic vectors are perpendicular to the propagation vector of the electromagnetic wave. It is customary to denote such waves as *transverse electromagnetic* (TEM) waves in guided-wave theory. In Chapter 4, we found that for the case of total reflection, we no longer had transverse waves; therefore, we must modify the notation for the polarization of waves undergoing total reflection. The new notation defines *TE waves* as waves with a longitudinal component of the magnetic field (these waves correspond closely to the E_N polarization for free space waves) and *TM waves* as waves with a longitudinal component of the electric field. We will limit our discussion to the TE waves because the results are qualitatively the same for the two polarizations, and the purpose of our discussion is to develop a qualitative understanding of guided waves.

Using Figure 5-19, we follow a plane wave propagating through the waveguide of index n_2 and thickness d. We have denoted the wavefronts associated with the rays, which are assumed to be planar, by heavy dotted lines in Figure 5-19. One ray travels from C to D and undergoes two reflections, while a second ray goes from A to B and experiences no reflection. Since the wavefronts are surfaces of constant phase, we must have the phase difference associated with the two propagation paths equal to a multiple of 2π. We can derive the length of the two propagation paths, solid lines AB

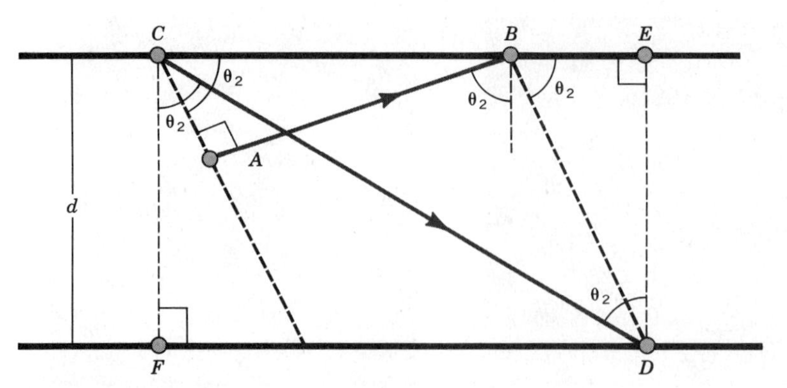

FIGURE 5-19. Geometry for finding the guided modes using the ray model of optics.

and *CD*, using the geometry of Figure 5-19, which is a redrawn version of Figure 5-14*c*.

The optical path lengths associated with the distances $\overline{CD}$ and $\overline{AB}$ can be determined using the triangles drawn in Figure 5-19. Triangle *CDF* of Figure 5-19 yields

$$\frac{d}{\overline{CD}} = \cos\,\theta_2$$

Triangle *ABC* yields

$$\frac{\overline{AB}}{\overline{CB}} = \sin\,\theta_2$$

Triangle *CDE* yields

$$\frac{\overline{CE}}{d} = \tan\,\theta_2$$

Finally, triangle *BDE* yields

$$\frac{d}{\overline{BE}} = \tan\,\theta_2$$

The distance along the guide axis is

$$\overline{CE} = \overline{CB} + \overline{BE}$$

Solving for $\overline{CB}$ and substituting for $\overline{CE}$ and $\overline{BE}$ yield

$$\overline{CB} = d\left(\tan\,\theta_2 - \frac{1}{\tan\,\theta_2}\right)$$

The phase difference between the two paths is given by

$$n_2k(\overline{CD} - \overline{AB}) + \delta_1 + \delta_3$$

where δ_1 and δ_3 are phase shifts due to total reflection and whose values can be obtained from **(3-54)**. We require that the total phase difference be equal to an integer multiple of 2π

$$n_2k(\overline{CD} - \overline{AB}) + \delta_1 + \delta_3 = m{\cdot}2\pi \tag{5-29}$$

$$\frac{n_2kd}{\cos\theta_2} - n_2kd\sin\theta_2\left(\tan\theta_2 - \frac{1}{\tan\theta_2}\right) + \delta_1 + \delta_3 = m\cdot 2\pi$$

$$\frac{n_2kd}{\cos\theta_2}(1 - \sin^2\theta_2 + \cos^2\theta_2) + \delta_1 + \delta_3 = m\cdot 2\pi$$

$$\frac{n_2kd}{\cos\theta_2}\cdot 2\cos^2\theta_2 + \delta_1 + \delta_3 = m\cdot 2\pi$$

$$2n_2kd\cos\theta_2 + \delta_1 + \delta_3 = m\cdot 2\pi \qquad (5\text{-}30)$$

This relation is called the *dispersion relation* for TE guided waves. We have neglected any effect on the plane waves used in this derivation arising from the finite size of the guide. Thus, this is an infinite plane wave result.

We see that θ_2 can only have certain values θ_m determined by the integer values of m in **(5-30)**. The rays associated with the discrete values of θ_m are the waveguide modes. Figure 5-20 displays a number of discrete modes propagating in a dielectric layer. Each mode in Figure 5-20 can be characterized by the number of bounces it makes as it propagates through the layer.

Remember that to have total reflection, the angle θ_2 must satisfy the inequality **(5-27)**

$$\sin\theta_m > \frac{n_3}{n_2} \qquad \text{or} \qquad \cos\theta_m < \sqrt{1 - \left(\frac{n_3}{n_2}\right)^2}$$

This means that there is a maximum number of modes m_{max} that can propagate in the guide. This maximum value is given by

$$m_{max} \leq \frac{2dn_2}{\lambda_0}\sqrt{1 - \left(\frac{n_3}{n_2}\right)^2} + \frac{\delta_1 + \delta_3}{2\pi} \qquad (5\text{-}31)$$

We define the quantity

$$V = kdn_2\sqrt{1 - \left(\frac{n_3}{n_2}\right)^2} = kd\sqrt{n_2^2 - n_3^2} \qquad (5\text{-}32)$$

called the *normalized film thickness* for the case of planar waveguides and *normalized frequency* or simply the *V number* for optical fibers. The inequality **(5-31)** can be written in terms of the V number

$$m_{max} \leq \frac{V}{\pi} + \frac{\delta_1 + \delta_3}{2\pi}$$

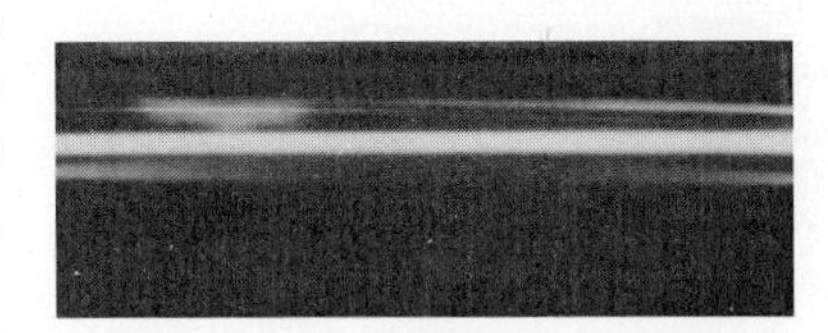

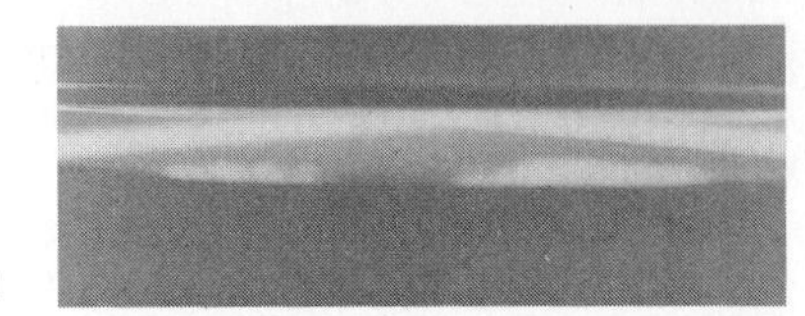

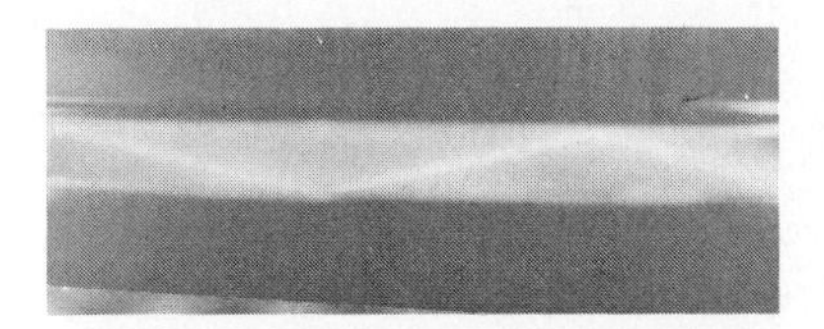

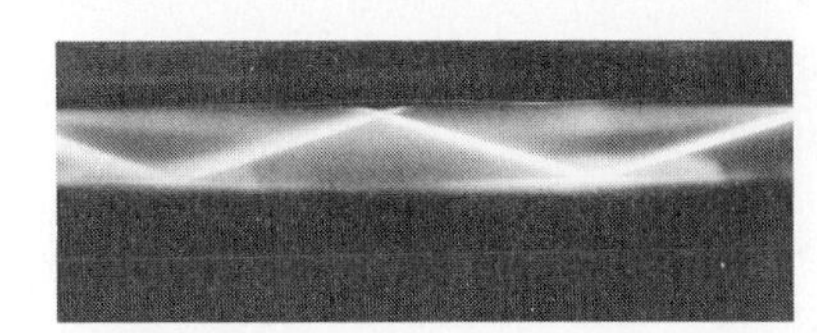

FIGURE 5-20. Several modes are shown propagating in a guiding layer of xylene bounded above by air and below by water containing a dye.

As an example of how the maximum number of modes and the normalized film thickness are calculated, assume that a waveguide 100 μm thick with an index of $n_2 = 1.53$ is formed between a medium with index $n_1 = 1.0$ and a medium with index $n_3 = 1.5$. If the wavelength of the propagating light is 1 μm, the normalized film thickness is

$$V = \frac{2\pi}{1\times 10^{-6}}\cdot(1\times 10^{-4})\sqrt{(1.53)^2 - (1.5)^2} = 189.4$$

The maximum number of modes in this guide is

$$m_{max} \leq \frac{189.4}{\pi} + 1 = 61.3$$

yielding a maximum number of modes of about 60. We assume that the phase shift upon reflection should not exceed 180°.

We will discover in a moment [in **(5-33)** and **(5-34)**] that δ_1 and δ_3 are negative. Since m can only be a positive integer or zero, if $V < \delta_1 + \delta_3$, then no mode can propagate in the dielectric layer. This condition is called the *cut-off condition* and the V number, where $V_c = \delta_1 + \delta_3$ is defined as the cut-off value for the waveguide V_c. We will return to this cut-off value of V in a moment.

Since the number of modes in a guide is a function of V and $\delta_1 + \delta_3$ is limited in magnitude to be no larger than 2π, we can evaluate the contributions to the number of waveguide modes by the basic guide parameters from **(5-32)**. The number of modes in the guide increase with increasing guide index n_2, decreasing substrate index n_3, increasing guide thickness, and decreasing wavelength.

We derived the functional form of δ_1 and δ_3 during our discussion of total reflection **(4-54)** and **(4-55)**. It is these parameters that contain the effects of polarization. For the TE waves, we can rewrite **(4-54)** as

$$\delta_1 = -2 \tan^{-1} \frac{\sqrt{\sin^2 \theta_2 - (n_1/n_2)^2}}{\cos \theta_2} \tag{5-33}$$

$$\delta_3 = -2 \tan^{-1} \frac{\sqrt{\sin^2 \theta_2 - (n_3/n_2)^2}}{\cos \theta_2} \tag{5-34}$$

Substituting these relationships into **(5-30)** yields a transcendental equation that must be solved to determine the values of θ_2 associated with the waveguide modes. A graphical solution to this transcendental equation will be presented after the introduction of several new variables that are often used in the wave formalism of guided-wave theory to describe the observed propagation in a guide. If we assume that the ray path is described by the propagation vector of the wave, then this notation can be incorporated into our ray theory.

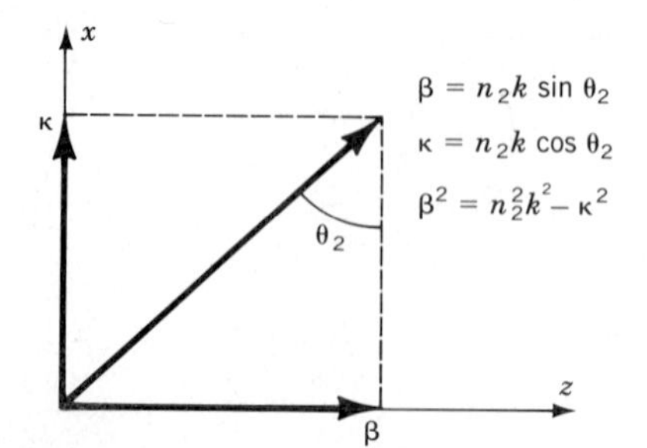

FIGURE 5-21. Definition of the propagation constant of a guided wave.

Propagation Vector Formalism

The geometric ray is parallel to the propagation vector **k**. Thus, there are a discrete set of propagation vectors associated with the set of rays predicted by **(5-30)**. We can decompose any one of these propagation vectors into two components, as shown in Figure 5-21.

To the external observer, the light associated with a particular waveguide mode appears to propagate through the guide with an effective propagation constant β, defined in Figure 5-21. The bounds on β for waveguide modes follow directly from **(5-27)**

$$kn_3 < \beta < kn_2$$

We can use the effective propagation constant to define an *effective guide index of refraction*

$$N = \frac{\beta}{k}$$

The limits on this effective index are

$$n_3 < N < n_2$$

The effect of introducing the new terms β and κ in Figure 5-21 is to divide the wave, in the guide, into two components: a standing wave that exists between the two interfaces and a wave propagating along the z axis at a velocity ω/β.

Using the newly defined variables, we may rewrite **(5-33)** and **(5-34)**

$$\delta_{1,3} = -2\tan^{-1}\left[\frac{\sqrt{\beta^2 - (n_{1,3}k)^2}}{\kappa}\right] \tag{5-35}$$

The radical in **(5-35)** is equal to the decay constant of an evanescent wave associated with total reflection, introduced in Chapter 3

$$\sqrt{\beta^2 - n^2k^2} = kn\alpha = \gamma$$

We introduce a decay constant for each of the guide boundaries

$$\gamma_1^2 = (kn_1\alpha_1)^2 = \beta^2 - n_1^2k^2 = (n_2^2 - n_1^2)k^2 - \kappa^2 \tag{5-36a}$$

$$\gamma_3^2 = (kn_3\alpha_3)^2 = \beta^2 - n_3^2k^2 = (n_2^2 - n_3^2)k^2 - \kappa^2 \tag{5-36b}$$

Combining all of the new variables we have introduced, we may rewrite the dispersion relation **(5-30)** as

$$\kappa d - \tan^{-1}\left(\frac{\gamma_1}{\kappa}\right) - \tan^{-1}\left(\frac{\gamma_3}{\kappa}\right) = m\pi$$

Solution for Asymmetric Guide

To solve the dispersion relation graphically, we first take the tangent of both sides of the dispersion equation to get

$$\tan(\kappa d - m\pi) = \tan\left[\tan^{-1}\left(\frac{\gamma_1}{\kappa}\right) + \tan^{-1}\left(\frac{\gamma_3}{\kappa}\right)\right]$$

We now apply the trigonometric identity

$$\tan(a \pm b) = \frac{\tan a \pm \tan b}{1 - \tan a \tan b}$$

to both sides of the equation

$$\frac{\tan \kappa d - \tan m\pi}{1 + \tan \kappa d \tan m\pi} = \frac{(\gamma_1/\kappa) + (\gamma_3/\kappa)}{1 - (\gamma_1\gamma_3/\kappa^2)}$$

$$\tan \kappa d = \frac{\kappa(\gamma_1 + \gamma_3)}{\kappa^2 - \gamma_1\gamma_3} \tag{5-37}$$

This is the mode equation for the TE waveguide modes and is the form of

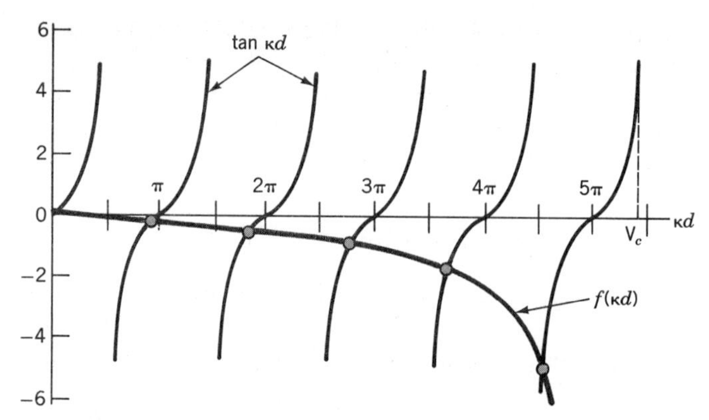

FIGURE 5-22. Using $n_1 = 1.0$, $n_2 = 1.62$, and $n_3 = 1.515$, the values for a sputtered glass wave guide, along with $d = 3\mu$m and $\lambda = 0.63\mu$m, we plot the two sides of **(5-37)** as separate functions. The intersections of the curves are the solutions of **(5-37)**. Table 5-1 lists the angle of incidence for the modes identified by the points of intersection. The cut off value for the V number is the asymptote of the f(κd) curve.

the dispersion relation usually encountered in the wave formalism of guided-wave optics. We can solve this equation by plotting, in Figure 5-22, the functions tan(κd) and

$$f(\kappa d) = \frac{\kappa d(\gamma_1 d + \gamma_3 d)}{(\kappa d)^2 - (\gamma_1 d)(\gamma_3 d)}$$

In Figure 5-22, the intersections of the $f(\kappa d)$ curve with the various tan(κd) curves are the solutions of **(5-37)**. The value of κd at each intersection allows the calculation of the guided ray's incident angle for that mode

$$\cos\,\theta_2 = \frac{(\kappa d)\lambda}{2\pi n_2 d}$$

Table 5-1 list the angle of incidence of a guided ray for the modes shown in Figure 5-22. It is important to note that the intersection of $f(\kappa d)$ and tan(κd) at the origin in Figure 5-22, is not a physically realizable solution.

The curve $f(\kappa d)$ in Figure 5-22 does not extend without limit toward larger and larger values of κd, but rather has a maximum value of κd, its asymptote, beyond which, one of the γ's becomes imaginary. From **(5-36)**,

$$\gamma_1 \rightarrow \text{imaginary when } (n_2^2 - n_1^2)k^2 < \kappa^2$$

$$\gamma_3 \rightarrow \text{imaginary when } (n_2^2 - n_3^2)k^2 < \kappa^2$$

Our initial conditions assumed $n_3 > n_1$ so γ_3 will become imaginary first. The point of interest to us is when $\gamma_3 = 0$, i.e., just before γ_3 becomes

TABLE 5.1 Incident Angle of Guided Mode

Mode	κd	θ_2
0	0.9π	86.6°
1	1.8π	83.3
2	2.75π	79.7
3	3.7π	76.1
4	4.55π	72.8

imaginary. Physically, it is at this point that substrate modes are produced and the guided modes disappear. Remember if $\gamma_3 = 0$, the evanescent wave extends infinitely far into the substrate because the penetration depth of the evanescent wave $1/\gamma$ becomes infinite. At this point, θ_2 no longer meets the condition for total reflection at the boundary between medium 2 and 3.

When $\gamma_3 = 0$,

$$\left(n_2^2 - n_3^2\right) k^2 = \kappa^2$$

We take the square root and multiply both sides by the guide thickness

$$kd\sqrt{n_2^2 - n_3^2} = \kappa d$$

From **(5-32)**,

$$V_c = \kappa d \tag{5-38}$$

where V_c is the cut-off value of the normalized film thickness. For a given guide thickness d, there is a minimum wavelength that will propagate in the guide. This cut-off limit occurs for rays that have incident angles that are less than the critical angle.

By examining Figure 5-22, we see that not only is there a maximum κd, beyond which $f(\kappa d)$ will not intersect $\tan(\kappa d)$, but there is also a minimum value of κd, $\kappa d = \pi/2$, below which there are no solutions possible in an asymmetric waveguide, i.e., no guided mode can be supported. This means that the asymmetric guide can only support modes in a band of wavelengths established by $\pi/2 \le \kappa d \le V_c$. (The guide is said to act as a bandpass filter of frequency.)

Solution for Symmetric Guide

A symmetric guide will exhibit propagation characteristics that differ from those of an asymmetric guide. If we rewrite **(5-38)**,

$$\tan V_c = \tan \kappa d = \frac{\gamma_1}{\kappa}$$

where we use the fact that at cut-off, $\gamma_3 = 0$ and a propagating wave penetrates the interface. Using this relationship, we can obtain a new expression for the cut-off condition

$$\tan V_c = \frac{\sqrt{\left(n_2^2 - n_1^2\right) k^2 - \kappa^2}}{\kappa}$$

$$\tan V_c = \sqrt{\frac{n_3^2 - n_1^2}{n_2^2 - n_3^2}} \tag{5-39}$$

This new expression for the cut-off condition leads to the conclusion that for a *symmetric guide*, where $n_1 = n_3$, there is no minimum value for κd, because V_c will equal zero. (The symmetric guide would act as a low-pass filter of frequency, because there is a minimum propagating wavelength but no maximum wavelength, $0 \le \kappa d \le V_c$.) To understand this behavior, we will solve the dispersion equation for a symmetric guide.

Because the indices on either side of the guide are the same, **(5-36a)** and **(5-36b)** reduce to a single equation

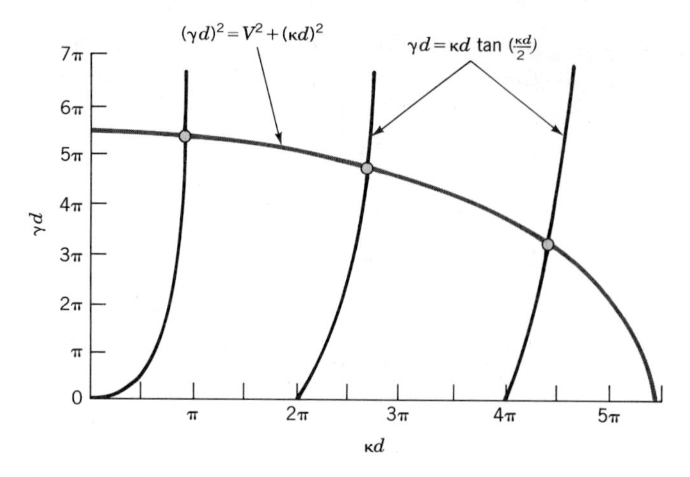

FIGURE 5-23. Using $n_1 = n_3 = 1.515$ and $n_2 = 1.62$, along with $d = 3\mu$m and $\lambda = 0.63\mu$m, we plot **(5-40)**. The intersections of the curves are the solutions of **(5-40)** for even modes. The lowest mode is never cut-off for this symmetric guide.

$$\gamma^2 = (n_2^2 - n_3^2)\, k^2 - \kappa^2$$

$$(\gamma d)^2 = V^2 - (\kappa d)^2 \tag{5-40a}$$

which is the equation of a circle. The dispersion relationship for the symmetric guide is

$$\kappa d - m\pi = 2 \tan^{-1}\left(\frac{\gamma d}{\kappa d}\right)$$

$$\gamma d = \kappa d \tan \frac{1}{2}(\kappa d - m\pi)$$

This dispersion relationship can be expressed in two forms

$$\gamma d = \begin{cases} \kappa d \tan \dfrac{\kappa d}{2}, & m \text{ even} \\ -\kappa d \operatorname{ctn} \dfrac{\kappa d}{2}, & m \text{ odd} \end{cases} \tag{5-40b}$$

The TE modes are determined by the simultaneous solution of **(5-40a)** and **(5-40b)**. The graphic solution for the even modes of a symmetric guide with properties similar to the asymmetric guide just discussed are shown in Figure 5-23.

An intersection between **(5-40a)** and **(5-40b)** at the origin is a physically realizable solution for a symmetric guide and results in a minimum κd of zero. It corresponds to $\theta_2 = \pi/2$, which is a ray propagating along the z axis, parallel to the interfaces of the guide. The maximum κd, and thus the value of V_c, occurs at the value of κd where the circle intersects the abscissa.

The term "cut-off" describes the inability of a waveguide to support a given propagation mode. The high-frequency cut-off value was defined as $V_c = \delta_1 + \delta_3$, based on the mathematical requirement that the mode number be a positive integer. The high-frequency cut-off condition was given a more physical interpretation as the incident angle for which $\gamma_3 = 0$ and the evanescent field becomes a propagating field (since $n_3 > n_1$, γ_3 reaches zero before γ_1 and is thus the limiting parameter). When performing a graphic solution for the guided modes, we found that an asymmetric guide of thickness d and with indices n_1, n_2, and n_3 had a low-frequency cut-off also. The asymmetric guide could only support guided waves for V numbers larger than $\pi/2$. This means that an asymmetric guide cannot support guided modes for wavelengths larger than

$$\lambda \geq 4d\sqrt{n_2^2 - n_3^2}$$

[In optical fibers, the propagation constants of the various TE modes are obtained by graphic solutions of equations involving Bessel functions rather than the tangent functions of **(5-38)**; therefore, for fibers, the low-frequency cut-off condition is determined by the first zero of the Bessel function, $J_0(\kappa d)$, i.e., $2.405 \leq \kappa d \leq V_c$.]

The cut-off value of the V number was given by **(5-39)**. This relationship demonstrated that a symmetric guide had no maximum cut-off wavelength. A graphic solution for even modes of a symmetric guide illustrated this fact. The physical origin of the existence of a low-frequency cut-off can be understood by considering the procedure used to analyze a waveguide through the use of Maxwell's equations.

To obtain wave solutions, we apply the boundary conditions introduced in Chapter 3, during the derivation of Fresnel's formula. In a symmetric guide, the boundary conditions in a plane normal to the guide interfaces would be the same at each interface. For an asymmetric guide, this would not be the case. For this reason, it is impossible for a plane wave to propagate through an asymmetric guide with its propagation vector parallel to the two interfaces. Such a wave can exist in a symmetric guide because the boundary conditions can be satisfied on the wavefront at both interfaces.

PARAMETRIC CHARACTERIZATION OF A LIGHT GUIDE

It would be very painful if repeated graphic solutions had to be obtained for waveguides whenever the thickness, wavelength, or indices were changed. Two new variables make it necessary to calculate the properties of a guide only once. Once solutions involving these parameters are obtained, parametric curves relating these new parameters can be used to find the properties of any guide.

Normalized Index

$$\mathbf{b} = \frac{N^2 - n_3^2}{n_2^2 - n_3^2}$$

where N is the effective guide index of refraction. By substituting the limits on N into the equation for $\mathbf{b}$, we find that $0 < \mathbf{b} < 1$.

Asymmetry Parameter

$$\boldsymbol{a} = \frac{n_3^2 - n_1^2}{n_2^2 - n_3^2}$$

Because of the initial assumption that $n_2 > n_3 > n_1$, we see that $\boldsymbol{a} > 0$.

These new parameters allow the phase shift δ_1 **(5-33)** to be rewritten

$$\delta_1 = -2 \tan^{-1} \sqrt{\frac{N^2 - n_1^2}{n_2^2 \cos^2 \theta_2}}$$

$$= -2 \tan^{-1} \sqrt{\frac{N^2 - n_1^2}{n_2^2 - N^2}}$$

$$= -2 \tan^{-1} \sqrt{\frac{N^2 - n_3^2 + n_3^2 - n_1^2}{n_2^2 - N^2 - n_3^2 + n_3^2}} \sqrt{\frac{n_2^2 - n_3^2}{n_2^2 - n_3^2}}$$

$$= -2 \tan^{-1} \sqrt{\frac{\boldsymbol{b} + \boldsymbol{a}}{1 - \boldsymbol{b}}}$$

If we use the same procedure, the phase shift δ_3 **(5-34)** can be rewritten as

$$\delta_3 = -2 \tan^{-1} \sqrt{\frac{\boldsymbol{b}}{1 - \boldsymbol{b}}}$$

The dispersion relation **(5-30)** can be expressed in terms of these new parameters as

$$2kdn_2 \cos \theta_2 - 2 \tan^{-1} \sqrt{\frac{\boldsymbol{b} + \boldsymbol{a}}{1 - \boldsymbol{b}}} - 2 \tan^{-1} \sqrt{\frac{\boldsymbol{b}}{1 - \boldsymbol{b}}} = 2\pi m$$

The first term of this expression can be written in terms of the V number

$$2kdn_2 \cos \theta_2 = 2kd\sqrt{n_2^2 \cos^2 \theta_2}$$

$$= 2kd\sqrt{n_2^2 - N^2}$$

$$= 2V\sqrt{1 - \boldsymbol{b}}$$

The dispersion relation can now be written as

$$V\sqrt{1 - \boldsymbol{b}} - \tan^{-1} \sqrt{\frac{\boldsymbol{b} + \boldsymbol{a}}{1 - \boldsymbol{b}}} - \tan^{-1} \sqrt{\frac{\boldsymbol{b}}{1 - \boldsymbol{b}}} = m\pi$$

Figure 5-24 is a graphic representation of this normalized dispersion relationship for several values of a and m. Dispersion curves such as those shown in Figure 5-24 can be used to analyze any slab waveguide design.

The parameters used in the graphic solutions discussed earlier can be simply expressed in terms of these new normalized parameters

$$\kappa d = V\sqrt{1 - \boldsymbol{b}}, \qquad \gamma_1 d = V\sqrt{\boldsymbol{b} + \boldsymbol{a}}, \qquad \gamma_3 d = V\sqrt{\boldsymbol{b}}$$

This new formulation of the dispersion relation makes it easy to evaluate the cut-off condition for any mode m in a waveguide. Several examples will demonstrate the use of this relationship.

When the incident angle of a ray in the guide θ_2 is equal to the critical angle, we have

$$\sin \theta_2 = \frac{n_3}{n_2}$$

and the effective index is

$$N = n_2 \sin \theta_2 = n_3$$

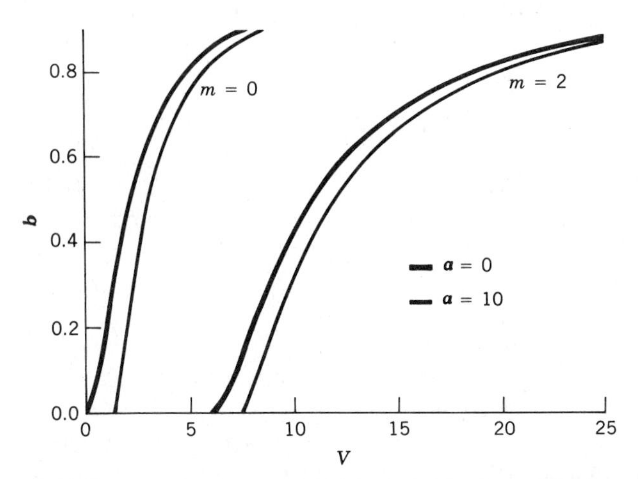

FIGURE 5-24. Normalized $\boldsymbol{b} - V$ diagram for the first and third modes for a symmetric guide and an asymmetric guide with an asymmetry parameter of 10.

At an incident angle equal to the critical angle, the normalized index becomes zero, $\boldsymbol{b} = 0$, and the dispersion relation becomes

$$V_{mc} = \tan^{-1} \sqrt{\boldsymbol{a}} + m\pi$$

where V_{mc} is the cut-off value of the V number for the mth mode. We will first examine the cut-off values of V numbers for symmetric guides.

When $n_1 = n_3$, a guide's asymmetry parameter is zero, $\boldsymbol{a} = 0$, and the guide is symmetric. From the dispersion relationship, we find that the cut-off value of the V number for a symmetric guide is $V_{mc} = m\pi$. For the zero mode of a symmetric guide, $m = 0$ and the cut-off value of the V number is zero. Using **(5-32)**, we can write the V number in terms of the guide parameters

$$\frac{2\pi d}{\lambda_0} \sqrt{n_2^2 - n_3^2} = 0$$

From this equation, we conclude that for the zero mode in a symmetric guide, where n_2, λ, and $n_3 = n_1$ are specified, there is no minimum value for the guide thickness d. Likewise, for a given n_2, d, and $n_3 = n_1$, there is no maximum wavelength λ_0, beyond which a wave will not propagate. The zero mode is never cut-off in a symmetric guide!

Continuing the discussion of the symmetric guide, when $m = 1$, we see that

$$V_{1c} = \pi$$

$$d = \frac{\lambda_0}{2\sqrt{n_2^2 - n_3^2}}$$

If we set $\lambda_0 = 632.8$ nm, $n_2 = 1.62$, and $n_3 = n_1 = 1.47$, then the smallest guide thickness that will allow the $m = 1$ mode to propagate is $d = 0.47$ μm. If the guide is thinner, only the zero mode will propagate and the guide is said to be a *single-mode guide.*

Consider now an asymmetric guide where $\lambda_0 = 632.8$ nm, $n_2 = 1.62$, $n_3 = 1.47$, and $n_1 = 1.0$ (air)

$$\mathbf{a} = 2.5$$

$$V_{0c} = \tan^{-1}\sqrt{2.5} = 1.007$$

$$V_{1c} = \tan^{-1}\sqrt{2.5} + \pi = 4.149$$

For the zero mode, the minimum guide thickness is $d = 0.15\ \mu\text{m}$, and for the mode $m = 1$, it is $d = 0.62\ \mu\text{m}$. If an asymmetric guide, with the parameters we have just specified, has a thickness d in the range

$$0.15 \leq d \leq 0.62\ \mu\text{m}$$

then the guide will only propagate one mode $m = 0$ at a wavelength of 632.8 nm. This guide would be called a single-mode waveguide. All thicker guides would be called *multimode guides.*

The number of modes in the multimode guide is simply $m + 1$, where m is the largest mode that can propagate in the guide

$$m + 1 = 1 + \frac{2d}{\lambda_0}\sqrt{n_2^2 - n_3^2} - \frac{1}{\pi}\tan^{-1}\sqrt{\mathbf{a}}$$

For example, if $n_2 = 1.5$, $d = 1$ mm, $\lambda_0 = 632.8$ nm, and $n_3 = n_1 = 1.0$, we have a symmetric guide with $\mathbf{a} = 0$ and the number of modes is given by $m + 1 = 3534$. If we put a cladding on this waveguide so that $n_3 = n_1 = 1.49$, then the number of modes decreases to $m + 1 = 547$.

LAGRANGIAN FORMULATION OF OPTICS

Because of the success of Fermat's principle in optics, the concept was extended to classical mechanics by Hamilton. The connection between geometrical optics and particle trajectories made by Hamilton brought him very close to the invention of quantum mechanics. He failed, however, to make the connection between wave optics and mechanics. Failure to make this connection should not be criticized; there were no experiments to suggest the need for such a connection.

The mathematical expression in classical mechanics, equivalent to Fermat's principle, is called *Hamilton's principle.* It states that the trajectory of a particle between times t_1 and t_2 is such that

$$\delta\int_{t_1}^{t_2} (T - V)\, dt = 0 \tag{5-41}$$

where T is the kinetic energy, V is the potential energy of the particle, and the symbol δ indicates that the variation must be taken. The energy quantity

$$\mathcal{L} = T - V$$

is called the Lagrangian in classical mechanics.

If the forces acting on the particle are conservative, i.e., if the total energy $T + V$ is a constant during the motion, then Hamilton's principle leads to the equations of motion of the particle being of the form

$$\frac{d}{dt}\left(\frac{\partial \mathcal{L}}{\partial \dot{q}_k}\right) = \frac{\partial \mathcal{L}}{\partial q_k} \tag{5-42}$$

This is called the Lagrangian equation; the q's are called generalized coordinates. In rectangular coordinates, $q_k \Rightarrow (q_1, q_2, q_3) \Rightarrow (x, y, z)$. For a simple pendulum, we would formulate the problem so that $q_k \Rightarrow q_1 \Rightarrow \theta$, where θ is the angle the pendulum makes with the vertical. In our shorthand notation,

$$\dot{q} = \frac{\partial q}{\partial t}$$

and

$$\frac{\partial \mathcal{L}}{\partial \dot{q}}$$

is a generalized momentum. For example, if $V = 0$, then

$$\mathcal{L} = \frac{mv^2}{2} = \frac{m(\dot{x})^2}{2}$$

and

$$\frac{\partial \mathcal{L}}{\partial \dot{x}} = m\dot{x}$$

Hamilton placed optics and classical mechanics under a common mathematical formalism. We will use that formalism to derive the equation of motion for a light ray and then use the equation to derive rectilinear propagation and the law of refraction.

In optics, we can write the path length of a ray as

$$ds = \sqrt{(dx)^2 + (dy)^2 + (dz)^2}$$

$$= dz\sqrt{1 + \left(\frac{dx}{dz}\right)^2 + \left(\frac{dy}{dz}\right)^2}$$

Fermat's principle can be rewritten

$$\delta\int n\,ds = \delta\int n(x, y, z)\sqrt{1 + \left(\frac{dx}{dz}\right)^2 + \left(\frac{dy}{dz}\right)^2}\,dz = 0 \tag{5-43}$$

Comparing **(5-43)** to Hamilton's principle, we can define an optical Lagrangian

$$\mathcal{L} = n(x, y, z)\sqrt{1 + \left(\frac{dx}{dz}\right)^2 + \left(\frac{dy}{dz}\right)^2} \tag{5-44}$$

where we are allowing z to play the same role in optics as time t does in classical mechanics.

The optical equations of motion are

$$\frac{d}{dz}\left(\frac{\partial \mathcal{L}}{\partial x'}\right) = \frac{\partial \mathcal{L}}{\partial x}, \qquad \frac{d}{dz}\left(\frac{\partial \mathcal{L}}{\partial y'}\right) = \frac{\partial \mathcal{L}}{\partial y} \tag{5-45}$$

where

$$x' = \frac{dx}{dz} \quad \text{and} \quad y' = \frac{dy}{dz}$$

We can substitute **(5-44)** into **(5-45)** to get the functional form of the optical equation of motion. For the x component, where we assume that n is independent of x',

$$\frac{\partial \mathcal{L}}{\partial x'} = \frac{nx'}{\sqrt{1 + x'^2 + y'^2}}, \qquad \frac{\partial \mathcal{L}}{\partial x} = \sqrt{1 + x'^2 + y'^2}\,\frac{\partial n}{\partial x}$$

The x component of the equation of motion is thus

$$\frac{d}{dz}\left(\frac{nx'}{\sqrt{1 + x'^2 + y'^2}}\right) = \sqrt{1 + x'^2 + y'^2}\,\frac{\partial n}{\partial x}$$

We may use the identity

$$\frac{d}{ds} = \frac{1}{\sqrt{1 + x'^2 + y'^2}}\,\frac{d}{dz}$$

to rewrite the equation of motion as

$$\frac{d}{ds}\left(n\,\frac{dx}{ds}\right) = \frac{\partial n}{\partial x}$$

If we carry out the same procedure for the y and z components, we obtain the vector equation

$$\frac{d}{ds}\left(n\,\frac{d\mathbf{r}}{ds}\right) = \nabla n\,(x, y, z) \tag{5-46}$$

This is called the *ray equation*; the vector $\mathbf{r}$ represents the position of any point on the ray. The derivative $d\mathbf{r}/ds$ is the unit vector along the tangent to the ray.

Rectilinear Propagation

A very simple application of the ray equation **(5-46)** is the calculation of the light trajectory in a homogeneous medium, where the index of refraction in independent of position $\nabla n = 0$

$$\frac{d^2\mathbf{r}}{ds^2} = 0$$

The solution of this equation is

$$\mathbf{r} = \boldsymbol{a}s + \boldsymbol{b}$$

which is the equation of a straight line. Thus, in a homogeneous medium, light rays travel in straight lines.

Law of Refraction

We can derive the law of refraction using the ray equation by assuming that the index of refraction is only a function of the x coordinate as shown in Figure 5-25. In the shaded region, the index changes in some unknown manner in the x direction from n_1 to n_2, but there is no index variation in the z direction. The z component of the ray equation is

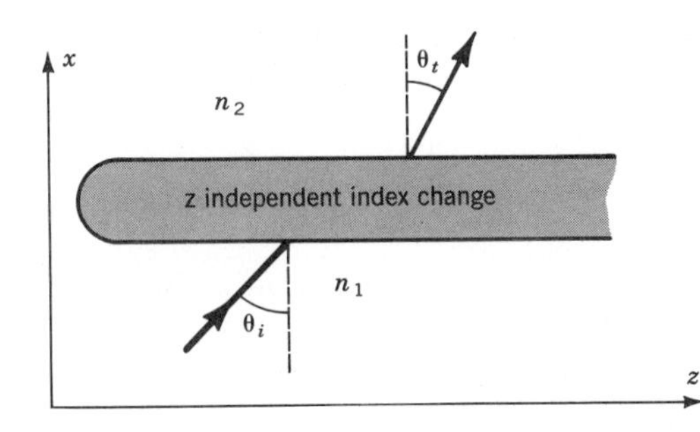

FIGURE 5-25. Refraction by a boundary between two media with different indices of refraction.

FIGURE 5-26. A guided wave lens has been formed in a guiding layer. Two narrow beams of light from a HeNe laser propagate through the lens and are focused by the lens. In a "good" guiding structure, the two beams could not be seen. The guide shown here was intentionally created with a large amount of scattering to allow the photo to be made. (Courtesy of Carl Verber, Georgia Institute of Technology.)

$$\frac{d}{ds}\left(n\frac{dz}{ds}\right) = \frac{\partial n}{\partial z} = 0 \qquad (5\text{-}47)$$

Integrating, we find

$$n\frac{dz}{ds} = \text{constant}$$

From the geometry of Figure 5-25, we see that $dz/ds = \sin\theta$ (remember that ds is measured along the ray path), which leads to

$$n \sin\theta = \text{constant} \qquad (5\text{-}48)$$

Thus, by assuming only that the index of refraction did not vary in the z direction, we obtain the law of refraction

$$n_1 \sin\theta_i = n_2 \sin\theta_t$$

An application of this result is seen in Figure 5-26 where a lens has been created in a guided-wave structure. It was not necessary to control the detailed way the index changed but only that an index change occurred.

PROPAGATION IN A GRADED INDEX OPTICAL FIBER

Generally, the ray equation **(5-46)** is difficult to solve. To reduce the complexity of the problem, the assumption is made that the light rays propagate in the general direction of the z axis and those that travel at an angle to the z axis depart by only small angles. This is another statement of the paraxial ray approximation and allows us to write $ds \approx dz$ (equivalent to the approximation $\sin\theta \approx \tan\theta$). This approximation results in the paraxial ray equation

$$\frac{d}{dz}\left(n\frac{d\mathbf{r}}{dz}\right) = \nabla n \qquad (5\text{-}49)$$

We will apply the paraxial approximation to evaluate the problem of an optical fiber whose index of refraction varies in the direction normal to, but not along, the axis of the fiber, i.e., the z axis. Before applying the paraxial approximation, we will rewrite the general ray equation in cylindrical coordinates. Although we could use rectangular coordinates in this problem, it seems more appropriate to use cylindrical polar coordinates that are the symmetry coordinates of the fiber. The z axis of the coordinate system is aligned along the cylindrical axis of the fiber and the cylindrical coordinates are related to the rectangular coordinates through the equations

$$r = \sqrt{x^2 + y^2}$$

$$\phi = \tan^{-1} \frac{y}{x}$$

$$z = z$$

We must rewrite the partial derivatives in x and y of n in terms of the cylindrical coordinates

$$\frac{\partial n}{\partial x} = \frac{\partial n}{\partial r}\frac{\partial r}{\partial x} + \frac{\partial n}{\partial \phi}\frac{\partial \phi}{\partial x}$$

$$= \frac{\partial n}{\partial r} \cos \phi - \frac{\partial n}{\partial \phi}\frac{\sin \phi}{r}$$

$$\frac{\partial n}{\partial y} = \frac{\partial n}{\partial r}\frac{\partial r}{\partial y} + \frac{\partial n}{\partial \phi}\frac{\partial \phi}{\partial y}$$

$$= \frac{\partial n}{\partial r} \sin \phi + \frac{\partial n}{\partial \phi}\frac{\cos \phi}{r}$$

The derivatives of x and y with respect to s are

$$\frac{dx}{ds} = \frac{dr}{ds} \cos \phi - r \frac{d\phi}{ds} \sin \phi$$

$$\frac{dy}{ds} = \frac{dr}{ds} \sin \phi + r \frac{d\phi}{ds} \cos \phi$$

With these newly formulated derivatives, we can multiply the x component of the ray equation by x and the y component by y, then add to generate the r component of the ray equation

$$\frac{d}{ds}\left(n \frac{dr}{ds}\right) - n\, r\left(\frac{d\phi}{ds}\right)^2 = \frac{\partial n}{\partial r} \qquad (5\text{-}50)$$

If we multiply the x component of the ray equation by y and the y component by x and then add, we generate the ϕ component of the ray equation

$$\frac{d}{ds}\left(nr^2 \frac{d\phi}{ds}\right) = \frac{\partial n}{\partial \phi} \qquad (5\text{-}51)$$

The z component of the ray equation is

$$\frac{d}{ds}\left(n \frac{dz}{ds}\right) = 0$$

Because the index of refraction is independent of the z coordinate, we do not need to make the paraxial approximation to solve this equation; we need only perform the integration. After integration, we get a result analogous to **(5-48)**

$$n\frac{dz}{ds} = n \sin \theta = \frac{\beta}{k} = \text{constant} \qquad (5\text{-}52)$$

Not only is the index of refraction in this fiber independent of z, but it is also independent of ϕ so that

$$\frac{d}{ds}\left(nr^2\frac{d\phi}{ds}\right) = 0$$

$$nr^2\frac{d\phi}{ds} = \text{constant} = K_1 \tag{5-53}$$

To proceed farther, we must use the paraxial approximation. To apply the paraxial approximation, we must replace ds by dz in **(5-50)** and **(5-53)**. The paraxial approximation requires that the variation of $n(r, \phi)$ with r be small, allowing us to replace $n(r, \phi)$ by a spatial average of the guide's index $n_2 \approx < n(r, \phi) >$, where the $< \ldots >$ here denotes the spatial average. If we use the spatial average, the ray equations become

$$\frac{d^2r}{dz^2} - r\left(\frac{d\phi}{dz}\right)^2 = \frac{1}{n_2}\frac{\partial n}{\partial r} \tag{5-54}$$

$$r^2\frac{d\phi}{dz} = \frac{K_1}{n_2} \tag{5-55}$$

We cannot obtain an exact solution for an arbitrary refractive index profile so we will assume a parabolic-shaped profile; see Figure 5-27.

$$n(r) = n_2\left[1 - \left(\frac{r}{a}\right)^2\Delta\right], \qquad r < a \tag{5-56}$$

where we have introduced the quantity Δ that is defined as

$$\Delta = \frac{n_2^2 - n_1^2}{2n_2^2}$$

In practice, the index of the cladding is almost the same as the core, resulting in an approximation for Δ of

$$\Delta = \frac{(n_2 - n_1)(n_2 + n_1)}{2n_2^2} \approx \frac{n_2 - n_1}{n_2}$$

Equation **(5-56)** describes the variation in the index of refraction illustrated in Figure 5-27. The core of the fiber has an index of refraction that varies from n_2 to n_1 along the radius of the fiber from the center outward. At the

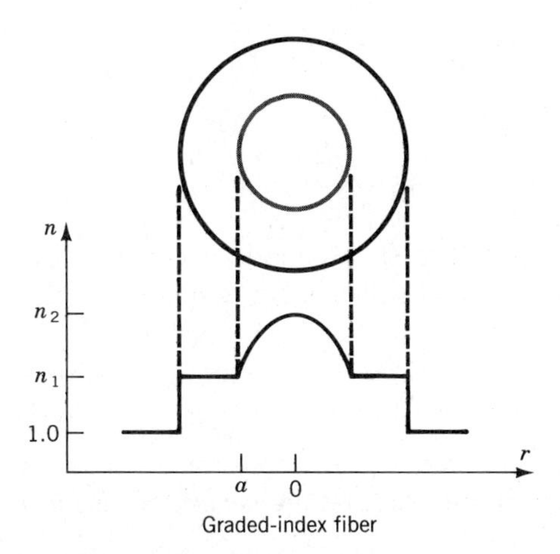

FIGURE 5-27. Profile of the index of refraction variation as a function of position in a graded-index optical fiber with a parabolic variation in index of refraction.

boundary between the core and the cladding, the index of refraction is n_1 and remains constant throughout the cladding, as shown in Figure 5-27. The assumption that n_1 is a constant throughout the cladding is not important for this problem because, as we will learn, the guided rays do not travel in the region $r > a$ and we can ignore completely any index variation in this region.

The derivative of the index along the fiber radius is

$$\frac{\partial n}{\partial r} = -2n_2\Delta\frac{r}{a^2}$$

We can substitute this derivative into **(5-54)** along with **(5-55)** where we introduce a new constant $K_2 = K_1^2/n_2^2$ for brevity

$$\frac{d^2r}{dz^2} - \frac{K_2^2}{r^3} = -\frac{2r\Delta}{a^2} \tag{5-57}$$

The objective is to find a solution to **(5-57)** that will provide an equation for the ray's coordinates (r, ϕ) as a function of position z along the fiber. If we multiply both sides of **(5-57)** by dr/dz, we create the derivative

$$\frac{d}{dz}\left[\frac{1}{2}\left(\frac{dr}{dz}\right)^2 + \frac{r^2\Delta}{a^2} + \frac{K_2}{2r^2}\right] = 0$$

This can be integrated to give

$$\frac{1}{2}\left(\frac{dr}{dz}\right)^2 + \frac{r^2\Delta}{a^2} + \frac{K_2}{2r^2} = K_3$$

We may integrate a second time to get

$$z - Z_0 = \frac{a}{2\sqrt{2\Delta}}\sin^{-1}\left[\frac{4\Delta(r/a)^2 - 2K_3}{\sqrt{4K_3^2 - 8\Delta(K_2/a^2)}}\right] \tag{5-58}$$

where Z_0 is yet another constant of integration whose value is determined by the origin of the coordinate system.

This is the solution we sought but its form is difficult to interpret. We can simplify the expression and make it easier to understand the physical significance of the result by introducing the variables

$$\Omega = \frac{\sqrt{2\Delta}}{a}, \qquad b^2 = K_2\left(\frac{\Omega}{K_3}\right)^2, \qquad A = \frac{\sqrt{K_3}}{\Omega}$$

The r coordinate of the ray's trajectory is

$$r(z) = A\sqrt{1 + \sqrt{1 - b^2}\cdot\sin[2\Omega(z - Z_0)]} \tag{5-59}$$

The ray's r coordinate is shown by **(5-59)** to be a periodic function of the distance along the cylindrical axis of the fiber oscillating between a maximum and minimum value

$$A\sqrt{1 - \sqrt{1 - b^2}} < r < A\sqrt{1 + \sqrt{1 - b^2}} \tag{5-60}$$

along the z axis. Thus, the ray moves through the fiber constrained between the two extremes defined by **(5-60)**. These extremes are called the *classical turning points* of the ray.

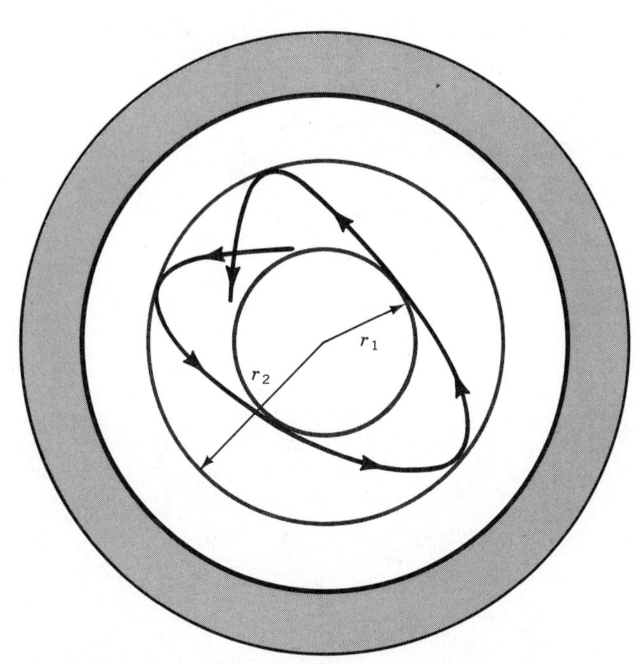

FIGURE 5-28. A ray projection showing the path of a general ray in the core of a graded-index fiber. The path is contained between the classical turning points. The cladding is denoted by the shaded region.

Figure 5-28 shows the propagation of a general ray in the core of a graded-index fiber; the cladding of the fiber is shown as the shaded region in this figure. The radii r_1 and r_2 in Figure 5-28 correspond to the maximum and minimum values of **(5-60)**.

By substituting **(5-59)** into **(5-55)** and integrating, we find the angular coordinate of the ray's trajectory to be

$$\phi = \phi_0 + \tan^{-1} \frac{\sqrt{1-b^2} + \tan \Omega(z - Z_0)}{b} \tag{5-61}$$

To understand the trajectories of the rays as they move between the turning points, we will evaluate two limiting cases. The first occurs when we set the first integrating constant from **(5-55)** equal to zero, i.e., $K_1 = 0$. This is equivalent to assuming that the azimuth angle ϕ is a constant and the ray is confined to a plane. Whenever $K_1 = 0$, we have $b = 0$, allowing the use of **(5-61)** to demonstrate that the azimuth angle is a constant

$$\phi = \phi_0 + \frac{\pi}{2}$$

The inequality in **(5-60)** becomes

$$0 < r < a\sqrt{2\frac{K_3}{\Delta}}$$

and the equation for $r(z)$, **(5-59)**, becomes

$$r(z) = A\sqrt{\cos^2 \Omega(z - Z_0) + \sin^2 \Omega(z - Z_0) + 2 \sin \Omega(z - Z_0) \cos \Omega(z - Z_o)}$$

$$= A\sqrt{2} \sin\left[\Omega(z - Z_0) + \frac{\pi}{4}\right]$$

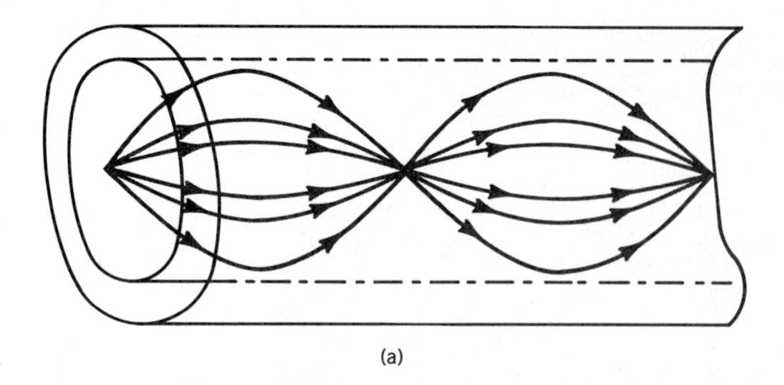

(a)

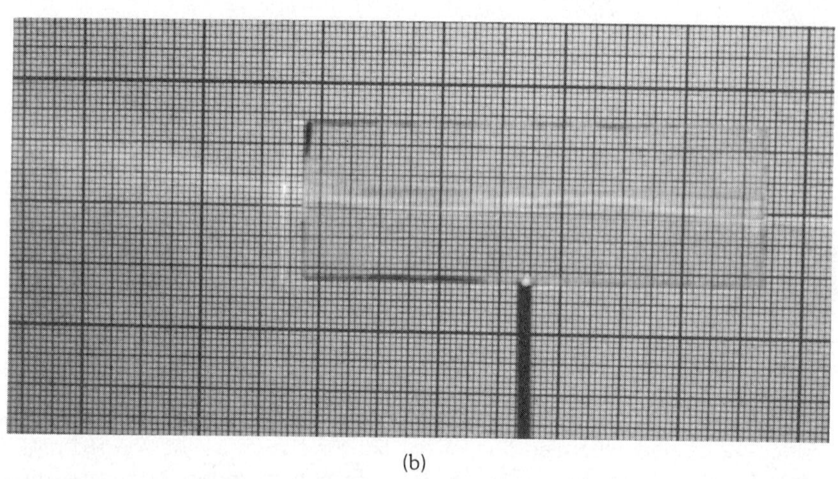

(b)

FIGURE 5-29. (a) Meridional rays propagating in a graded-index fiber with a parabolic index profile. (b) Beam of light propagating along a graded index guide. (Figure *b* courtesy of David Hamblen, Kodak Research Laboratory.)

The ray is a meridional ray, confined to the meridional plane defined by ϕ. The ray moves in a sinusoidal trajectory, periodically crossing the symmetry axis (the z axis) with a period of

$$\frac{1}{\Omega} = \frac{a}{\sqrt{\Delta}}$$

Several trajectories of this type, each corresponding to a different mode, are shown in Figure 5-29*a*. A small beam of light is shown propagating along a graded-index guide in Figure 5-29*b*, demonstrating the sinusodial propagation path predicted by our result.

The graded index guide shown in Figure 5-29*b* can be described by an *ABCD* matrix. To generate the *ABCD* matrix, it is better to return to **(5-57)** which is modified by the assumption that $K_1 = 0$ to read

$$\frac{d^2r}{dz^2} + \frac{2r\Delta}{a^2} = 0$$

The solution of this differential equation is

$$r(z) = \frac{\gamma_0}{\Omega} \sin \Omega z + r_0 \cos \Omega z$$

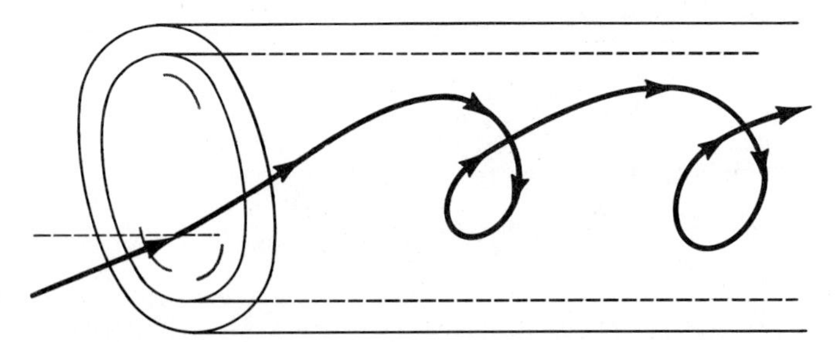

FIGURE 5-30. Propagation of a skew ray in a graded-index fiber.

We have assumed that the fiber has its entrance face at $Z_0 = 0$ and at that position the ray's parameters are r_0 and γ_0, where we have defined the ray's angle by

$$\gamma = \frac{dr}{dz}$$

The *ABCD* matrix for the fiber is

$$\begin{pmatrix} \sin \Omega z & \dfrac{1}{\Omega} \cos \Omega z \\ -\Omega \cos \Omega z & \sin \Omega z \end{pmatrix}$$

By adjusting the length of this type of fiber, we can construct the equivalent of a lens and use the fiber to focus, collimate, or image the meridional rays.

A second limiting case occurs when $b = 1$. In this case, r is a constant, equal to

$$a\sqrt{\frac{K_3}{2\Delta}}$$

and the azimuth angle is a linear function of the z coordinate

$$\phi(z) = \phi_0 + \Omega\,(z - Z_0)$$

These are called helical rays because they travel on a helical path about the z axis a fixed distance A from the axis. A typical helical ray is shown in Figure 5-30.

A general ray has a trajectory that falls between these two limiting cases.

It is possible to attack the problem of a wave propagating in a fiber by using the same mathematical method used to treat an electron in a potential well. In quantum mechanics, we solve the Schrödinger equation of an electron

$$\frac{d^2\Psi}{dx^2} + \frac{2m}{\hbar}(E - V)\,\Psi = 0$$

In optics, we want to solve the Helmholtz equation of a wave

$$(\nabla^2 + k^2 n_2^2)\,E = 0$$

In a dielectric slab, where the index in the guide is uniform and

$$\frac{\partial^2 E}{\partial y^2} = 0, \qquad \frac{\partial^2 E}{\partial z^2} = -\beta^2 E$$

the Helmholtz equation becomes

$$\frac{d^2 E}{dx^2} + [(k^2 n_1^2 - \beta^2) - (k^2 n_1^2 - k^2 n_2^2)]\,E = 0$$

The Schrödinger and Helmholtz equations have the same mathematical form. In quantum mechanics, we use the WKB method to solve the Schrödinger equation in the classical limit, where the change of potential energy within the particle's wavelength (a deBroglie wavelength) is small compared with the kinetic energy.[20] In optics, we apply the same WKB method[21] to solve the Helmholtz equation in the geometrical optic limit when the index of refraction changes slowly over the distance of a wavelength.

The result of the application of the WKB method is a dispersion equation of the form

$$\int_0^{r_t} \sqrt{n^2(r)k_0^2 - \beta^2}\, dr = \delta_1 + \delta_3 + m\pi$$

where $n(r)$ is the refractive index profile that might have the form of **(5-56)**. The turning point r_t is the upper limit of the integration and found by solving the equation

$$n_2^2(r)k_0^2 - \beta^2 = 0$$

The dispersion equation can be used to perform an analysis, similar to the one carried out for planar guides, to determine the cut-off condition and an estimate of the number of modes in a fiber.

SUMMARY

With an objective of demonstrating the fact that geometrical optics is a special case of the wave theory, the wavelength was allowed to approach zero and the solution of the Helmholtz equation was evaluated. The solution was found to predict the motion of the normal to the phasefront of a wave. Geometrical optics was found to treat a light wave at a point as if it were a plane wave and to obtain the wave's propagation path by following the path of the normal to the plane at that point. No application of this result was made, but it was pointed out that the result led mathematically to a statement of Fermat's principle.

The time it takes a light ray to travel an optical path between any two points is a minimum, i.e., the optical path of a light ray between any two points is equal to the optical length of any other curve that joins these points and lies in a neighborhood of the proper path.

The optical path length was defined as the distance light would travel in a vacuum during the time it traveled a distance ℓ in a medium with index n. Thus, the optical path length was

$$\Delta = n\ell$$

Fermat's principle was used to derive the principle of reciprocity, the law of refraction, and the law of reflection.

The law of refraction and the concept of rectilinear propagation were used to generate a matrix that describes the propagation of light through an optical element. The resulting matrix equation was of the form

$$\begin{pmatrix} x' \\ \gamma' \end{pmatrix} = \begin{pmatrix} A & B \\ C & D \end{pmatrix} \begin{pmatrix} x \\ \gamma \end{pmatrix}$$

where x is the height above the optical axis of the ray on entering the optical system and γ is the angle the ray makes with the optical axis. The primes denote the same parameters on leaving the optical system.

The *ABCD* matrix was derived after making the paraxial approximation. This approximation assumes that the first terms of the expansions of the trigonometric functions are all that are needed to represent the functions, for example, $\sin \theta \approx \theta$.

Refraction at a spherical surface with radius of curvature R_1 separating two media with index of refractions n_1 and n_2 was shown to be described by the matrix

$$\boldsymbol{R}_1 = \begin{pmatrix} 1 & 0 \\ \dfrac{n_1 - n_2}{n_2 R_1} & \dfrac{n_1}{n_2} \end{pmatrix} \tag{5-9}$$

The propagation through a medium with an index of refraction n_2 is described by the matrix

$$\boldsymbol{T} = \begin{pmatrix} 1 & d \\ 0 & 1 \end{pmatrix} \tag{5-11}$$

where d is the path length traveled by a ray in the medium.

The elements of the *ABCD* matrix for a complete lens were found by combining the refraction and transfer matrices.

$$A = 1 + \frac{d\,(n_1 - n_2)}{n_2 R_1}, \qquad B = \frac{d n_1}{n_2}$$

$$C = \frac{1}{n_3}\left[\frac{n_2 - n_3}{R_2} + \frac{n_1 - n_2}{R_1} + \frac{d\,(n_1 - n_2)\,(n_2 - n_3)}{n_2 R_1 R_2}\right], \qquad D = \frac{n_1}{n_3}\left[1 + \frac{d\,(n_2 - n_3)}{n_2 R_2}\right]$$

The relationship between these matrix elements and other more familiar representations of geometrical optics have been made in Appendix 5-A. In this chapter, the *ABCD* matrix was used to find the stability condition for a Fabry–Perot resonator. The stability condition establishes the requirements upon the mirror curvatures R and separation d that result in the trapping of a light ray in a Fabry–Perot cavity. The result of the analysis was a stability condition of the form

$$0 \le \left(1 - \frac{d}{2f_1}\right)\left(1 - \frac{d}{2f_2}\right) \le 1 \tag{5-24}$$

where $f = R/2$ is the focal length of a mirror with spherical curvature R.

Geometrical optics was also used to determine some of the properties of guided waves. It was learned that light could be coupled into the end of an optical waveguide, but that there was a maximum acceptance angle for the light given by

$$NA = \sqrt{n_2^2 - n_1^2} \tag{5-28}$$

Rays, guided by the waveguide, propagate by undergoing total reflection. The angles of reflection can only assume discrete values determined by a dispersion equation. Several expressions for the dispersion equation were derived. The first

$$2 n_2 k d \cos \theta_2 + \delta_1 + \delta_3 = m \cdot 2\pi \tag{5-30}$$

was derived by assuming that the ray was associated with an infinite plane wave whose phase front propagated through the guide without distortion. The second representation

$$\tan \kappa d = \frac{\kappa(\gamma_1 + \gamma_3)}{\kappa^2 - \gamma_1\gamma_3} \tag{5-37}$$

was constructed by anticipating the results that would have been obtained if a wave analysis had been carried out. It was included to connect the geometrical results derived here with the results found in the guided-wave literature. The final representation

$$V\sqrt{1-\boldsymbol{b}} - \tan^{-1}\sqrt{\frac{\boldsymbol{b}+\boldsymbol{a}}{1-\boldsymbol{b}}} - \tan^{-1}\sqrt{\frac{\boldsymbol{b}}{1-\boldsymbol{b}}} = m\pi$$

was introduced as an aid in the interpretation of the waveguide cut-off parameter V_c.

The V number of a waveguide was defined as

$$V = kd\sqrt{n_2^2 - n_3^2}$$

The parameter $\boldsymbol{b}$ called the normalized index

$$\boldsymbol{b} = \frac{N^2 - n_3^2}{n_2^2 - n_3^2}$$

That mode for which $\boldsymbol{b}$ is equal to zero is said to be cut-off. At cut-off, the V number is obtained from the dispersion relation

$$V_{mc} = \tan^{-1}\sqrt{\boldsymbol{a}} + m\pi$$

where the asymmetry parameter is

$$\boldsymbol{a} = \frac{n_3^2 - n_1^2}{n_2^2 - n_3^2}$$

The definition of the V number can be used to find out the values of the physical parameters when cut-off occurs. The V number also provides information about the maximum number of modes that can propagate in a guide

$$m + 1 = 1 + \frac{2d}{\lambda_0}\sqrt{n_2^2 - n_3^2} - \frac{1}{\pi}\tan^{-1}\sqrt{\boldsymbol{a}}$$

PROBLEMS

5-1. What length plane mirror must you purchase to see your full height when it is mounted in a vertical position? Do not assume that you have eyes on top of your head.

5-2. Prove that the reflected ray from a plane mirror turns through an angle 2θ when the mirror rotates through θ.

5-3. Consider a series of plane interfaces, all parallel. At the first, the index changes from n_0 to n_1; at the second from n_1 to n_2; at the mth from n_{m-1} to n_m, etc. If θ_m is the angle of refraction and θ_{m-1} is the angle of incidence at the mth interface, show

$$n_0 \sin \theta_0 = n_m \sin \theta_m$$

5-4. A fish appears to be 2 m below the surface of a pond ($n = 1.33$). What is its actual depth?

5-5. A collimated laser beam shines on a tank of water. Part reflects off the surface (beam A) and part reflects off the bottom and exits the water as beam B, as shown in Figure 5-31. Show that beams A and B are parallel.

5-6. The Selfoc fiber has a graded index of refraction given by

$$n^2(x, y, z) = n_0^2[1 - \alpha^2(x^2 + y^2)]$$

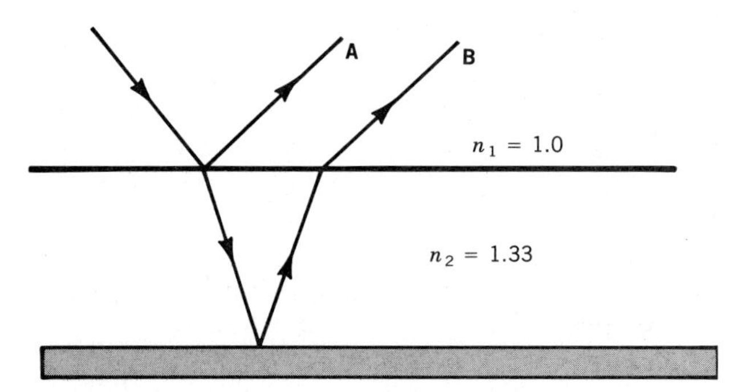

FIGURE 5-31. Prove beams A and B are parallel.

where n_0 and α are material constants. Find the path of a ray in the fiber without assuming a paraxial ray.

5-7. To analyze a Fabry–Perot cavity with mirrors of radii R_1 and R_2, we can replace the mirror with a set of lenses of focal lengths $f_1 = R_1/2$ and $f_2 = R_2/2$, alternately spaced by the cavity length. Rederive the $ABCD$ matrix for this lens system. What effect does this $ABCD$ matrix have on the stability condition **(5-24)**?

5-8. Assume that we have a step index fiber with $n_2 = 1.6$ and $n_1 = 1.46$. Considering only meridional rays, what is the smallest value that the angle of incidence in the fiber θ_2 can have? What is the numerical aperture when $d = 0.6\ \mu m$ and $\lambda = 1\ \mu m$?

5-9. Using the parameters in Problem 5-8, how many modes can propagate in the fiber? What happens if we increase d to $3\ \mu m$? If we increase λ to $1.3\ \mu m$?

5-10. What is the maximum radius that the fiber of Problem 5-8 can have if single-mode operation is desired?

5-11. Calculate the phase shifts, introduced on total reflection inside a plane guide, with $n_2 = 1.53$, $n_3 = 1.5$, and $n_1 = 1.0$. Assume the angle of incidence is $\theta_2 = 85°$.

5-12. Calculate the object-image matrix (see Appendix 5-A) for the lens system shown in Figure 5-32, where the lens on the left has a focal length f_1 and the one on the right has a focal length f_2. Prove that conjugate planes occur for $s = -f_1$ and $s' = f_2$. This latter matrix is sometimes called the focal plane matrix.

5-13. Using the object-image matrix from Problem 5-12, (a) calculate the positions of the cardinal points when $s = -f_1$ and $s' = f_2$. (b) What is the effective

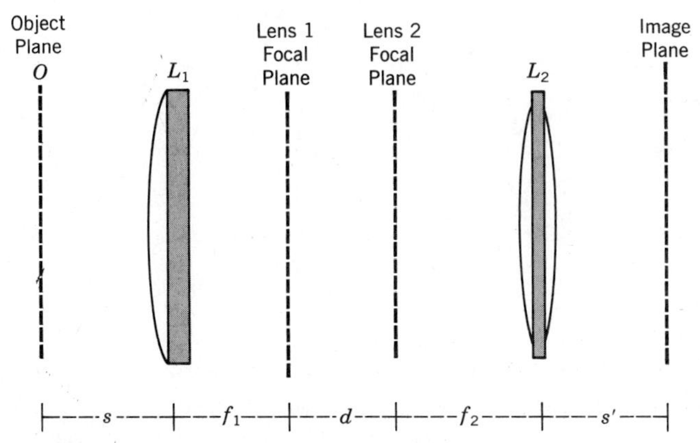

FIGURE 5-32. General two-lens optical system.

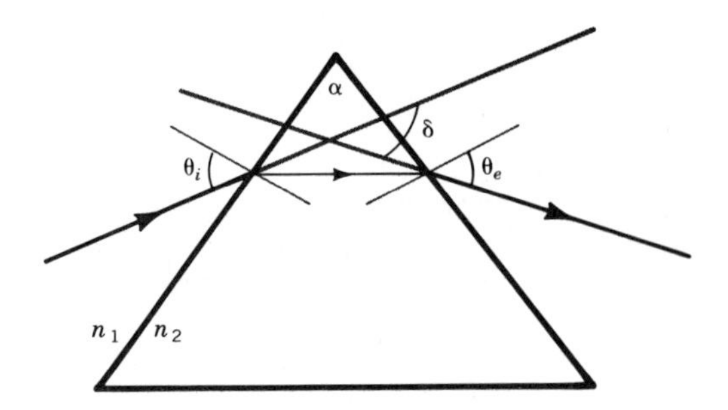

FIGURE 5-33. Prism with apex angle equal to α.

focal length of this optical system if the two lenses are in contact with each other?

5-14. Using the results of Problem 5-13, assume $f_1 = 2f_2$ and sketch out the optical system, including the positions of the principal planes for (a) a Huygens eyepiece, $d = -1.5f_2$; (b) an astronomical telescope, $d = 0$; and (c) a compound microscope, $d = 2f_2$. The condition of $d < 0$ occurs when the distance between the two lenses is less than $f_1 + f_2$.

5-15. Using the results of Problem 5-12, assume $f_1 = -2f_2$ and sketch out the optical system, including the positions of the principal planes for (a) a Galilean telescope, $d = 0$; and (b) a telephoto lens, $d = -0.5f_2$.

5-16. Trace a ray through the prism shown in Figure 5-33. The angle between the initial and final ray is called the *angle of deviation*, denoted by δ in Figure 5-33. Derive an equation in terms of the apex angle α in Figure 5-33 for the angle of deviation. Assume the prism has an index of refraction n_2 and is in a medium of index n_1. What is the minimum angle of deviation? Prove that the angle of minimum deviation occurs when $\theta_i = \theta_e$.

5-17. Represent the electric and magnetic fields of a light wave by

$$\mathbf{E} = E_0 e^{ik_0 \mathcal{S}}, \qquad \mathbf{B} = B_0 e^{ik_0 \mathcal{S}}$$

where $\mathcal{S}$ is the eikonal. Prove that the Poynting vector is proportional to $\nabla \mathcal{S}$.

5-18. Assume a plane wave in incident normal to the polished end of a step-index fiber, shown in Figure 5-34. Derive an equation for the minimum radius that

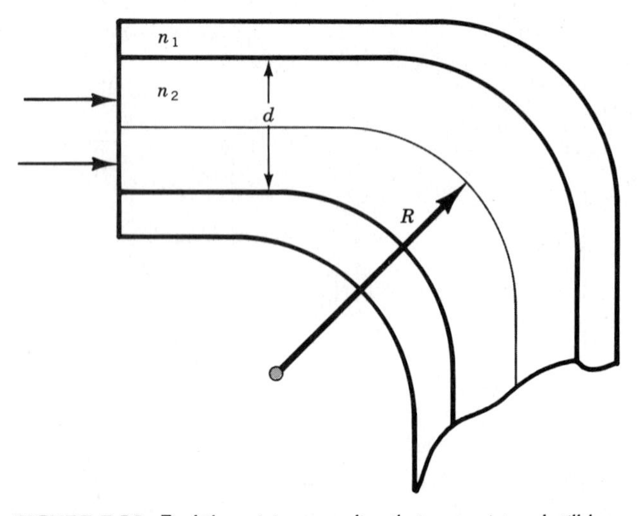

FIGURE 5-34. Find the minimum radius that can exist and still have guided waves.

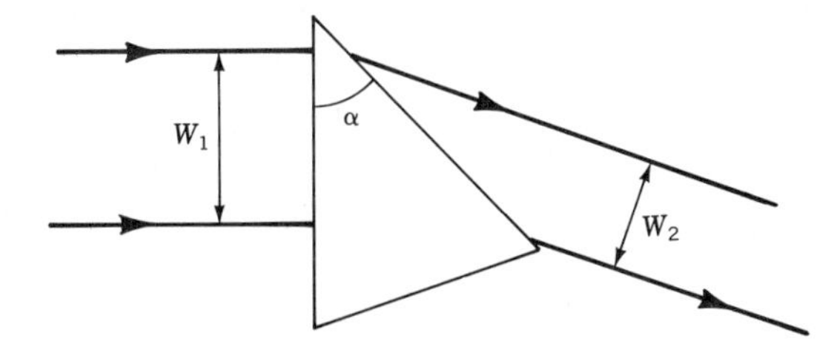

FIGURE 5-35. A prism used to magnify or demagnify a parallel beam of light.

a fiber can be bent before light is lost from the guided mode. Assume that the index of the core is 1.66 and the index of the cladding is 1.52. Using the equation just derived, what is the minimum radius?

5-19. If a parallel beam of width W_1 is incident on a prism such as the one in Figure 5-35, find an equation that gives the width of the beam exiting the prism in terms of the incident angle and the apex angle α. Sir David Brewster, in 1835, first suggested the use of this type of magnification. Today, a component based on this concept is found in the CinemaScope optical system. Can you guess why this optical component is used?

5-20. In Figure 5-36, the bottom trace shows the initiation of a laser diode pulse (λ = 850 nm) at the entrance of a step-index optical fiber and the top trace shows the arrival of that pulse at the end of 1100 m of the fiber. Calculate the index of refraction of the core of the fiber. The scope time base is 1 μsec/division. The core of the fiber is 100 μm and the cladding is 140 μm.

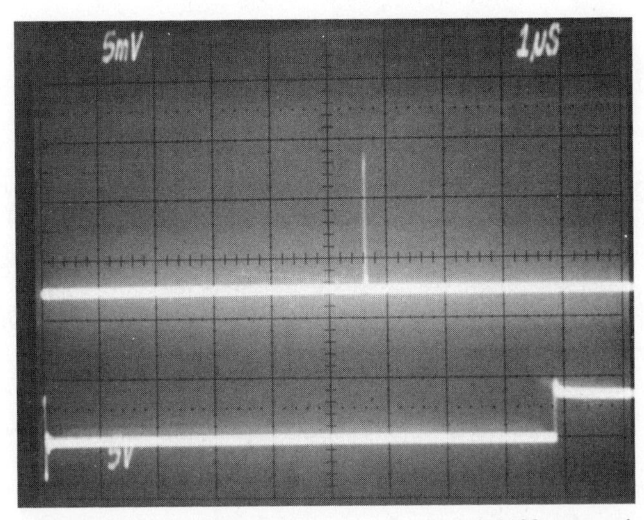

FIGURE 5-36. An optical pulse propagating through 1100 meters of optical fiber. The bottom trace indicates the initiation of the optical pulse at the start of the fiber and the top trace indicates the arrival of the pulse at the end of the fiber. The time scale is 1 μsec/division.

Appendix 5-A

ABCD MATRIX

In Chapter 5, we derived the general form of the *ABCD* matrix and generated the matrix for two refracting surfaces. Now we would like to relate the elements of the *ABCD* matrix to some of the traditional parameters of geometrical optics.

If the optical element is thin, we can assume $d \to 0$ and because most imaging systems will be used in air, we can also assume that $n_1 = n_3 = 1$. The system matrix **(5-13)**, for the thin lens, becomes

$$\boldsymbol{S} = \begin{bmatrix} 1 & 0 \\ (n_2 - 1)\left(\dfrac{1}{R_2} - \dfrac{1}{R_1}\right) & 1 \end{bmatrix} \tag{5A-1}$$

For this case, the determinant of the system matrix is

$$\|\boldsymbol{S}\| = 1$$

The *lateral magnification* is defined as $\beta \equiv x_1'/x_1$. Because $A = 1$ and $B = 0$, the values of the x coordinate at the vertices V_1 and V_2 are unchanged by the system matrix $\beta = 1$. This is equivalent to saying that the front and back vertices of the thin lens define planes of unit magnification. The planes of unit magnification are important reference planes for both thick ($d \neq 0$) and thin ($d = 0$) lenses. We will introduce a technique for locating the planes of unit magnification for a thick lens later in this appendix.

The *ABCD* matrix for the general imaging condition can be shown to be equivalent to the more familiar thin-lens equation. To obtain the equivalence, we first apply the transfer matrix from the object point *O* in Figure 5A-1 to the lens, then apply the lens' system matrix, and finally use a second transfer matrix to reach the image point *I*.

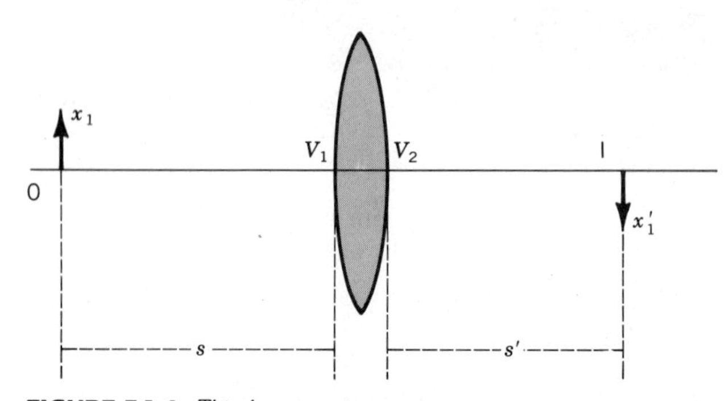

FIGURE 5A-1. Thin-lens imaging system.

$$\begin{pmatrix} 1 & s' \\ 0 & 1 \end{pmatrix} \begin{bmatrix} 1 & 0 \\ (n_2 - 1)\left(\dfrac{1}{R_2} - \dfrac{1}{R_1}\right) & 1 \end{bmatrix} \begin{pmatrix} 1 & -s \\ 0 & 1 \end{pmatrix} \begin{pmatrix} x_1 \\ \gamma \end{pmatrix} = \begin{pmatrix} x_1' \\ \gamma' \end{pmatrix}$$

A negative s is used in the propagation matrix for the region to the left of the lens because of the sign convention; s is a negative quantity and the negative sign insures that the propagation matrix contains a positive quantity for propagation from left to right.

$$\begin{bmatrix} 1 + s'(n_2 - 1)\left(\dfrac{1}{R_2} - \dfrac{1}{R_1}\right) & s' - s - ss'(n_2 - 1)\left(\dfrac{1}{R_2} - \dfrac{1}{R_1}\right) \\ (n_2 - 1)\left(\dfrac{1}{R_2} - \dfrac{1}{R_1}\right) & 1 - s(n_2 - 1)\left(\dfrac{1}{R_2} - \dfrac{1}{R_1}\right) \end{bmatrix} \begin{pmatrix} x_1 \\ \gamma \end{pmatrix} = \begin{pmatrix} x_1' \\ \gamma' \end{pmatrix}$$

This is called the object-image matrix, $\boldsymbol{O}$.

Thin-Lens Equation

If x_1' is an image of x_1, then all rays, regardless of their value of γ, must arrive at x_1' according to Fermat's principle. This means x_1' is independent of γ and we must have the object-image matrix element $B = 0$. This requirement, called the *imaging requirement*, leads to

$$s' - s - ss'(n_2 - 1)\left(\frac{1}{R_2} - \frac{1}{R_1}\right) = 0$$

which can be rewritten in the form of the familiar *thin-lens equation*

$$\frac{1}{s'} - \frac{1}{s} = (n_2 - 1)\left(\frac{1}{R_1} - \frac{1}{R_2}\right) \tag{5A-2}$$

If we allow the index of refraction in the object and image space to differ, then **(5A-2)** is modified to read

$$\frac{n_3}{s'} - \frac{n_1}{s} = \frac{n_2 - n_1}{R_1} - \frac{n_3 - n_2}{R_2}$$

The elements $ABCD$ of the object-image matrix can be identified with some simple parameters of the thin lens. From our definition of the lateral magnification of the imaging system, we find that the element A of the object-image matrix is

$$\begin{aligned} \beta &= \frac{x_1'}{x_1} = A \\ &= 1 + s'(n_2 - 1)\left(\frac{1}{R_2} - \frac{1}{R_1}\right) \end{aligned} \tag{5A-3}$$

Optical Invariant

The determinants of both transfer matrices and the lens system matrix have determinants that are equal to 1,

$$\|\boldsymbol{T}\| = \|\boldsymbol{S}\| = 1$$

therefore, the object-image matrix has a determinant equal to one,

$$\|\mathbf{O}\| = \|\mathbf{T}\|\cdot\|\mathbf{S}\|\cdot\|\mathbf{T}\| = 1$$

The determinant is equal to the product

$$\|\mathbf{O}\| = AD - BC = 1$$

For the imaging condition, we must have $B = 0$, requiring that

$$\|\mathbf{O}\| = AD = 1$$

If we now introduce the concept of an *angular magnification* as

$$\beta_\gamma = \frac{\gamma'}{\gamma} = D = 1 - s(n_2 - 1)\left(\frac{1}{R_2} - \frac{1}{R_1}\right) \tag{5A-4}$$

we are led to an important invariant of an optical system. This definition leads to the requirement that

$$\|\mathbf{O}\| = \beta\beta_\gamma = 1$$

which can be rewritten as

$$x_1'\gamma' = x_1\gamma$$

A more general form of this invariant can be obtained by recalling that when the index of refraction differs in the object and image space, then the object-image determinant is

$$\|\mathbf{O}\| = \beta\beta_\gamma = \frac{n_1}{n_3}$$

The *optical invariant* is then

$$\mathcal{H} = n_1x_1\gamma = n_3x_1'\gamma' \tag{5A-5}$$

This relationship is also called the *Lagrange invariant* or *Smith-Helmholtz* equation. The invariant was derived using ray parameters from object and image space and can be applied to any two conjugate (object/image) planes of the optical system.

Lens Maker's Equation

When the indices are the same in object and image space, the object-image matrix now simplifies to

$$\mathbf{O} = \begin{pmatrix} \beta & 0 \\ C & \dfrac{1}{\beta} \end{pmatrix}$$

We still have one constant C that has not been identified with any simple property of a thin lens. That connection can be made by defining the focal point of a lens.

When we put the object at infinity, the image point, located a distance f_2 from the lens vertex V_2, is defined as the *focal point* of the lens F_2; see Figure 5A-2.

$$\frac{1}{s'} = \frac{1}{f_2} = (n_2 - 1)\left(\frac{1}{R_1} - \frac{1}{R_2}\right) \tag{5A-6}$$

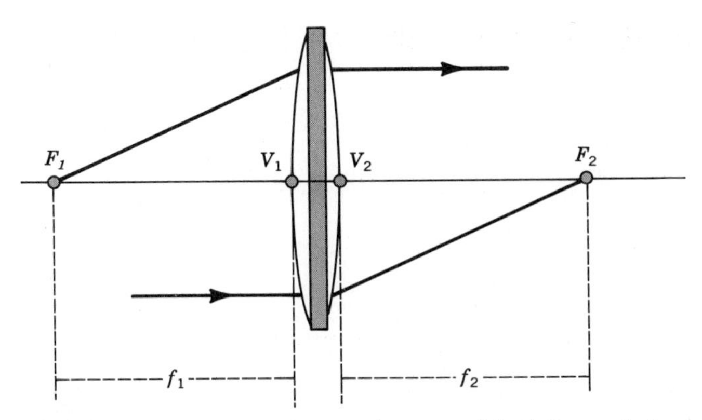

FIGURE 5A-2. Definition of the focal points and focal planes of a lens.

This relationship for the focal length is called the *lens maker's equation*. From the lens maker's equation, we learn that the same focal length (or power) can be obtained by using a variety of indices and lens curvatures. Comparing **(5A-6)** with the element C of the object-image matrix, we discover that $C = -1/f_2$, i.e., C is equal to the negative of the power of the lens and the object-image matrix is further simplified

$$\boldsymbol{O} = \begin{pmatrix} \beta & 0 \\ -\dfrac{1}{f_2} & \dfrac{1}{\beta} \end{pmatrix} \tag{5A-7}$$

Gaussian Formalism

If the object were located at infinity to the right of the lens, then the image would be located at a distance equal to the focal distance f_1 from V_1 at the front of the lens. This distance defines a second focal point F_1. If the index of refraction is the same on both sides of the lens, then the focal length, i.e., the distance from each focal point to its vertex, is $-f_1 = f_2 = f$. Using the definition of the focal length given by **(5A-6)**, we can rewrite **(5A-2)** as

$$\frac{1}{s'} - \frac{1}{s} = \frac{1}{f} \tag{5A-8}$$

This is called the *Gaussian form* of the thin-lens equation. The sign of the focal length f used in **(5A-8)** is given by f_2 in **(5A-6)**. If f_2 is positive, the lens is called a *positive lens*; it will focus a plane wave to a point located a distance f_2 from the lens vertex. If f_2 is negative, the lens is called a *negative lens*; it will cause a plane wave to diverge, as if it originated from a point located a distance $-f_2$ from the lens vertex.

Newtonian Formalism

The pair of focal points F_1 and F_2 is the first of three pairs of points called the *cardinal points* of the lens. The two focal points can be used to establish reference planes as replacements for the vertices of the lens. The object and image distances are then measured with respect to the focal planes.

The lens equation **(5A-8)** is modified to reflect the new reference points by substituting

$$Z = s - f_1, \qquad Z' = s' - f_2$$

yielding the *Newtonian form* of the thin-lens equation

$$ZZ' = -f_1 f_2 \tag{5A-9}$$

Principal Planes

The thin lens results in an *ABCD* matrix that is very simple. It is possible to retain this simplicity for very complex optical systems by defining a pair of cardinal points, called the principal points, to replace the vertices V_1 and V_2 as reference points of the lens; see Figure 5A-3. These new principal points define planes of unit magnification, just as the vertices did, but they may not lie anywhere near the outer surfaces of the optical system. With these new reference points, the *ABCD* matrix for any complex lens assumes the simple form of the thin lens. These reference points are said to be *conjugate points* because the points are related through the optical system as an *object-image pair.*

We need to replace the *ABCD* system matrix for a thin lens defined using the vertices V_1 and V_2 by the more general *ABCD* system matrix referenced to the principal planes. The new matrix will describe the propagation of light from the principal point H_1 to the principal point H_2. By combining this new *ABCD* system matrix with transfer matrices from the object to H_1 and from H_2 to the image, an object-image matrix for any complex lens can be generated.

To create the new *ABCD* matrix, we begin by using a ray-trace procedure similar to the one used in Chapter 5 to produce the system matrix. The system matrix is then combined with two transfer matrices that describe the ray propagation from H_1 through the optical system and onto H_2.

$$\mathbf{M} = \begin{pmatrix} 1 & h_2 \\ 0 & 1 \end{pmatrix}\begin{pmatrix} A & B \\ C & D \end{pmatrix}\begin{pmatrix} 1 & -h_1 \\ 0 & 1 \end{pmatrix} \tag{5A-10}$$

Here, as with the thin-lens image-object matrix, the propagation matrix for

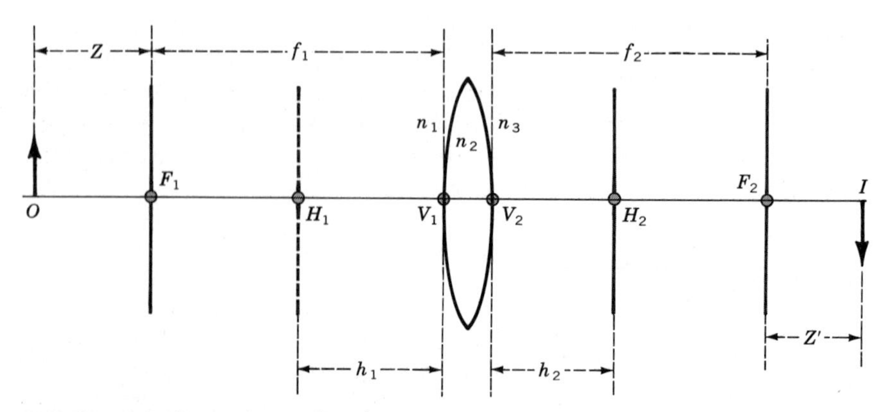

FIGURE 5A-3. Principal points for a lens.

the region to the left of the lens contains a negative distance because of the sign convention selected for geometric optics.

$$\mathbf{M} = \begin{pmatrix} A + h_2C & B - h_1A + h_2(D - h_1C) \\ C & D - h_1C \end{pmatrix} \tag{5A-11}$$

The principal planes H_1 and H_2 located at h_1 and h_2 are the equivalent of the thin-lens vertices that require they be planes of unit magnification. Because the principal planes are planes of unit magnification

$$\mathbf{M}_{11} = 1$$

We can therefore write

$$A + Ch_2 = 1$$

$$h_2 = \frac{1 - A}{C}$$

Because the two planes are defined by conjugate points, the imaging condition must hold between them. This means

$$\mathbf{M}_{12} = 0$$

Since

$$\|\mathbf{M}\| = \frac{n_1}{n_3}$$

we must have

$$\mathbf{M}_{22} = \frac{n_1}{n_3}$$

$$D - Ch_1 = \frac{n_1}{n_3}$$

$$h_1 = \frac{D - n_1/n_3}{C}$$

When $n_3 = n_1$, then

$$h_1 = \frac{D - 1}{C}$$

We now have defined the location of the principal points with respect to the lens' vertices. For these equations to be meaningful, we must have $C \neq 0$ (it occurs in the denominator of h_1 and h_2). In **(5A-6)** we associated C with $-1/f$; here, C is equal to the reciprocal of the *effective focal length*, $\mathbf{f}_2$, measured from the principal plane H_2, i.e.,

$$\mathbf{C} = -\frac{1}{\mathbf{f}_2}$$

There is a special optical system for which $C = 0$: the *afocal system*. This optical system produces an image at infinity of an object located at infinity. Because an astronomical telescope performs this optical operation, an afocal system is often called a *telescopic system*. Because the object and the image are both located at infinity, it is more useful to specify the size of the object by the angle it subtends. We, therefore use angular magnification **(5A-4)** for the telescopic system, as a measure of its performance

$$\beta_\gamma = \frac{\gamma'}{\gamma} = D$$

Nodal Points

The third set of cardinal points are called the *nodal points*, $\mathbf{n}_1$ and $\mathbf{n}_2$, and the pair of planes normal to the optical axis, that contain these points are called *nodal planes*. The nodal points are defined as those points for which

$$\gamma' = \gamma$$

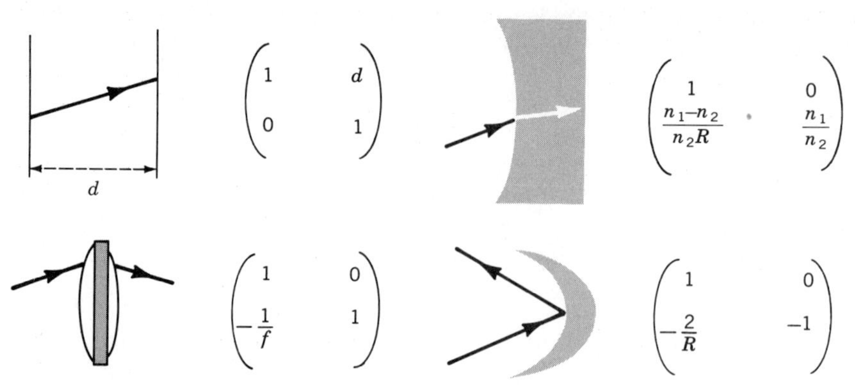

FIGURE 5A-4. The *ABCD* matrix for translation in a uniform medium and refraction at an interface. Also shown are the *ABCD* matrices for a thin lens and a mirror.

or equivalently, the points of unit angular magnification

$$\beta_\gamma = 1$$

The image of an object will not move if the imaging lens is rotated a small amount about an axis passing through a nodal point perpendicular to the optical axis. Thus, it is quite easy to locate the nodal points of a lens experimentally. If the index of refraction on each side of an optical system is the same, $n_1 = n_3$, then the nodal points coincide with the principal points and we need only deal with four cardinal points. The *ABCD* matrices for some common optical elements and media are shown in Figure 5A-4.

Aperture Stop and Pupils

The finite size of an optical system limits the bundle of axial rays leaving the object point O on the optical axis of Figure 5A-5. The obstruction that

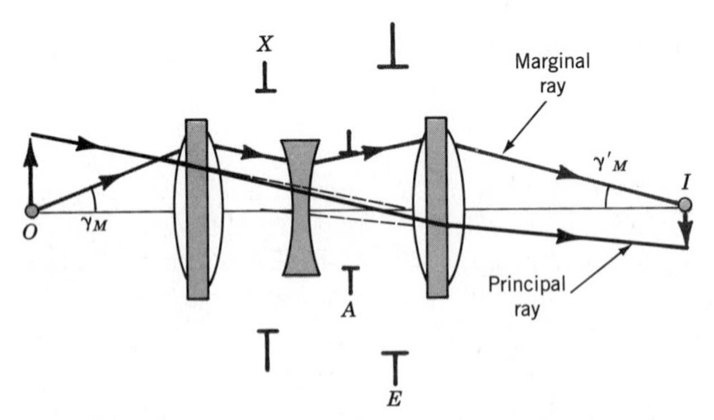

FIGURE 5A-5. A complex lens system containing a stop A that acts as the aperture stop for the lens. The rays that just skim pass the aperture stop are called *marginal rays*. Any ray that passes through the center of the aperture stop is called a *principal ray*. The image of the aperture stop as seen from the object is called the *entrance pupil*, labeled E in the figure. The image of the aperture stop as seen from the image is called the *exit pupil*, labeled X. Because the entrance and exit pupils are conjugates of the aperture stop, the principal ray will also pass through the center of these pupils, or as in this figure, they will appear to do so as indicated by the dotted lines.

defines the maximum angle γ_M of a ray from the object to the image is called the *aperture stop*. The rays that leave the object at angles nearly equal to γ_M are called *marginal rays*. In Figure 5A-5 we show a marginal ray leaving the object at γ_M and just skimming the aperture stop labeled **A**.

To locate the aperture stop is quite easy. In conventional ray tracing, we trace a single ray through the system, as illustrated in Figure 5A-6. We then calculate the ratio of aperture radius to ray height

$$\left| \frac{r_j}{x_j} \right|$$

The smallest ratio indicates the aperture stop. With the matrix technique, we create an *ABCD* matrix from the object *O*, on the optical axis, to each stop in the optical system, as illustrated in Figure 5A-6. Because the object height is zero, the matrix equation for each ray height would then be

$$\begin{pmatrix} x_j \\ \gamma_j \end{pmatrix} = \begin{pmatrix} A & B \\ C & D \end{pmatrix} \begin{pmatrix} 0 \\ \gamma \end{pmatrix}$$

From this equation, the value of x_j in each aperture is given by

$$x_j = B\gamma$$

If we set each x_j equal to the radius of the stop at that point r_j, then the object angle for the marginal ray of that stop is

$$\gamma_j = \frac{r_j}{B}$$

Thus, to find the aperture stop, we simply calculate the matrix element B for the optical elements to the left of each stop. The stop with the smallest ratio of the stop radius to matrix element B is the aperture stop.

The maximum angle γ_M determines the numerical aperture of the lens

$$NA = n_3 \sin \gamma'_M \qquad \text{(5A-12)}$$

and the $f/\#$ of the lens

$$f/\# = \frac{1}{2 \sin \gamma'_M} \qquad \text{(5A-13)}$$

The image of the aperture stop formed by the optical system to its left, i.e., the image of the aperture stop as seen from the object position, is called the *entrance pupil* and is labeled *E* in Figure 5A-5. The image of the aperture

The equation normally used to define and calculate the $f/\#$ is a special case of **(5A-13)**. If a plane wave fully illuminates the aperture of the optical system, then γ'_M is given by

$$\gamma' = \frac{r_M}{f_2}$$

where f_2 is the effective focal length and r_M is the aperture radius. Applying the paraxial approximation allows us to write the normally encountered definition of the $f/\#$

$$f/\# = \frac{f_2}{D}$$

where D is the diameter of the aperture stop.

FIGURE 5A-6. Geometry for finding the aperture stop of an optical system.

stop formed by the optical system to its right, i.e., the image of the aperture stop as seen from the image position, is called the *exit pupil* and is labeled X in Figure 5A-5.

A ray passing through the center of the aperture stop is called the *principal ray*. (Sometimes, the term *chief ray* is used in place of principal ray and other times the term chief ray is reserved for the center ray of a ray bundle. Most of the time, the two rays are identical.) Because the entrance and exit pupils are conjugates (images) of the aperture stop, principal rays pass through the centers of these pupils also. In Figure 5A-5, the entrance and exit pupils are virtual images; thus, a principal ray appears to originate at the centers of these pupils.

Other stops exist in an optical system; they may be placed there to control aberrations, act as lens mounts, or reduce scattered light (glare stops). The stop that limits the maximum oblique principal ray that can propagate through the optical system is called the *field stop*. This stop will occur at an object or image plane and is therefore conjugate to the final image plane and limits the size of the image.

If a ray bundle is created by the principal ray and a marginal ray for an object point on the optical axis, we find that this size ray bundle cannot propagate through the system for object points located off the optical axis. Another way to state this fact is that oblique ray bundles will not fill the aperture stop. This means that the flux density that can pass from the object to the image is a function of the distance off the optical axis. This effect is called *vignetting*. Vignetting is not always a defect. Sometimes lens designers "hide" aberrations in the low-light region produced by vignetting.

Appendix 5-B

ABERRATIONS

A perfect optical system, from a geometric optics point of view, is one that redirects all light rays from an object point through a single conjugate (image) point. For a real image, the wavefront associated with each of the rays leaving the optical system must be spherical, centered on the image point. Fermat's principal states that the optical path length of each ray will be identical.

If the perfect optical system were a positive lens, then a plane wave incident on the lens would be focused to a point. If we inspected the light distribution in planes on either side of the focal plane, we would expect to see uniformly illuminated disks of light that decreased in diameter as we approached focus and increased in diameter as we moved away from focus.

As we already know, light is a wave and geometric optics is only an approximate theory. We expect to see a departure from the predictions of geometric optics when dimensions are on the order of a wavelength. Small dimensions are found near the focal plane, and as shown in Figure 5B-1, there is a departure from the predictions of geometric optics.

The light distributions are quite complicated with the light intensity on the optical axis passing through zero periodically. We do, however, observe identical distributions in planes symmetrically positioned about the focal plane. These light distributions are due to diffractive effects and a critical analysis of them has been made.[6] These diffractive effects will not be dealt with in this book. It was necessary to point out the effects, however, because they are an obvious component of the images we will present during our discussion of aberrations.

The objective of the lens designer is to produce an optical system that is perfect from the geometrical optics viewpoint, but the designer is usually prevented from doing so by cost. The most economical materials available to the lens designer are glasses with a uniform index of refraction. To form these materials into optical components, grinding and polishing techniques are used, again because of their economy. When two materials are placed in contact, separated only by maybe some abrasive material, and moved with respect to one another in a great number of different relative positions, the points of contact between the materials will be worn away until the two surfaces form plane or spherical surfaces, touching at every point. Thus, the lowest-cost optical surface has a spherical shape (a plane is viewed as a spherical surface with an infinite radius of curvature).

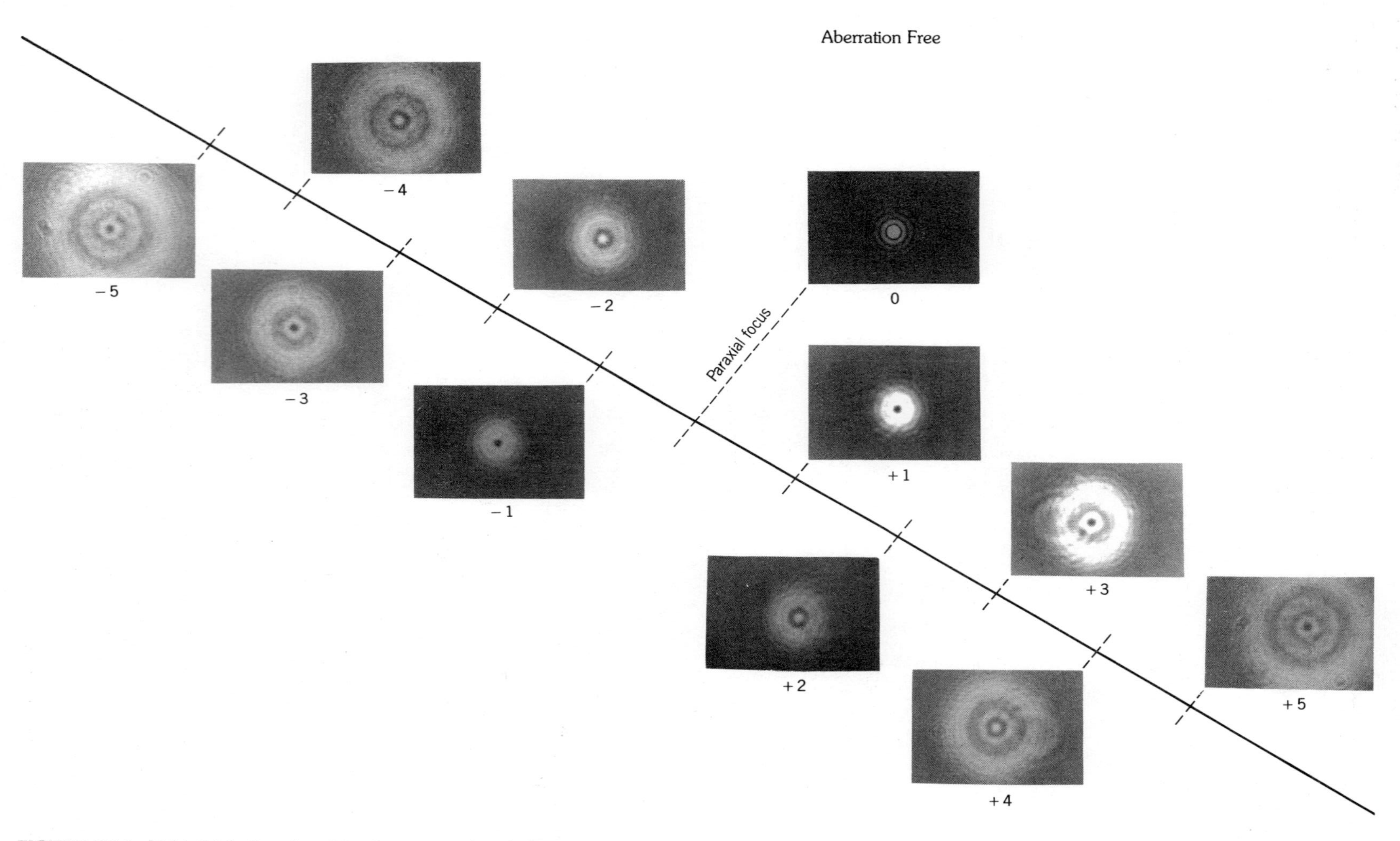

FIGURE 5B-1. Light distributions found in planes normal to the optical axis and equally spaced along the optical axis of a nearly perfect lens. The planes were spaced about 1mm apart and the focal length of the lens was about 500 mm.

Economy requires that the optical designer attempt to produce a perfect optical system using optical components with a uniform index of refraction and having spherical surfaces. Normally, these components will be axially centered and rotationally symmetric. Paraxial theory predicts the formation of a perfect image by these components. However, when the exact theory is used, we discover that a spherical component can only produce an imperfect image. The departure from the desired image is called aberration. The goal of the lens designer is to combine a number of spherical components in such a way as to reduce the aberrations.

The first aberration corrected lenses were designed by **Chester Moor-Hall (1704–1771)** and **John Dollond (1706–1761)** around 1729 for use in astronomy. Much later in 1840 **Josef Max Petzval (1807–1891)** designed a portrait camera lens. These designs were accomplished without the benefit of a formal theory. **Ludwig Philipp von Seidel (1821–1896)** established a rigorous theory of third-order aberrations in 1856. **Ernst Abbe (1840-1905)** developed and implemented the analytical methods of optical design.

Optimization techniques are used to guide the lens designer in reducing the aberrations. The optimization techniques require a method of quantifying the departure from a perfect image. In this appendix, we will introduce the methods used to quantify aberrations. The approach taken measures the departure of the actual image, as predicted by application of the laws of refraction and reflection (called *finite ray tracing*), from the predictions of paraxial theory. Here, we will only discuss the differences between paraxial theory and the third-order approximation to the exact theory. The third-order approximation to aberration theory leads to what are called the *primary* aberrations. (In England, the aberrations we will discuss are viewed as the *first-order* corrections to the paraxial theory, whereas in the United States, they are viewed as the *third-order* correction to the first-order theory.)

There are a two basic ways to characterize the aberrations:

1. In terms of the departure of the wavefront from a spherical surface, called the reference sphere.
2. By the distance that a ray misses the paraxial image point.

To quantify these characterizations, a set of rays is followed through the optical system. To select those rays, the entrance pupil of the system is divided into an array of points, see Figure 5B-2, and rays are traced from the object to the entrance pupil, through the system to the exit pupil, and finally from the exit pupil to the image plane.

In Figure 5B-2, the various coordinate planes used in the following discussion are defined. The vector **h** defines the location of the Gaussian ray's intercept in the plane of interest, i.e., the ray point determined by paraxial theory; here, the point shown is the conjugate point of the object. The second point shown in the image plane is the point determined by finite ray tracing to be the actual image point. In the exit pupil, it is sometimes convenient to use polar coordinates (ρ, θ) and to divide the pupil into *zones*. For other aberrations, a rectangular coordinate system (ξ, η) is used and the aberrations are described in terms of rays propagating in two perpendicular planes, the *tangential* and the *sagittal planes*.

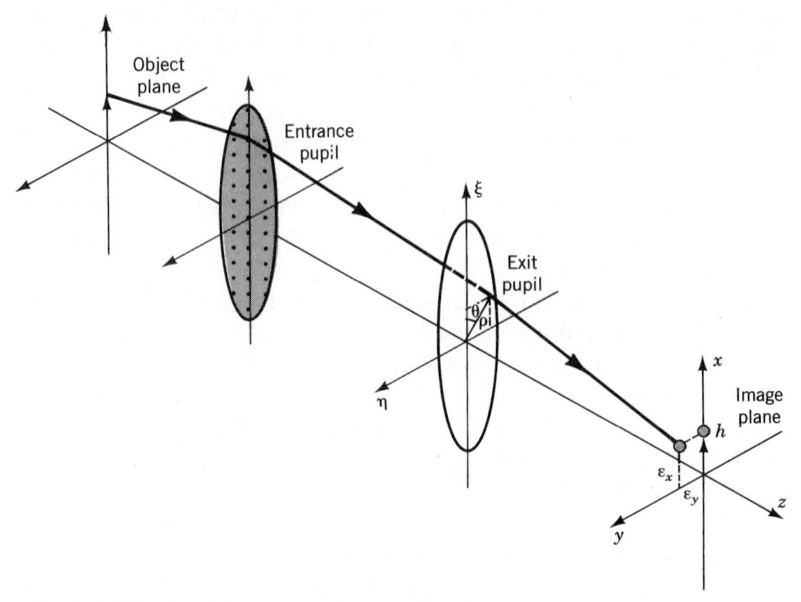

FIGURE 5B-2. Coordinate system for aberration theory.

The *tangential plane* is another name for the meridional plane and is defined as any plane containing the object point and the optical axis of the system; see Figure 5B-3. In our discussion, the tangential plane has always been the x, z plane. In the exit pupil of Figure 5B-2, it is defined as the ξ, z plane. All rays confined to this plane are called tangential rays.

The plane at right angles to the tangential plane and containing the principal ray is called the *sagittal plane*; see the right-hand side of Figure 5B-3. The sagittal plane changes at each surface, but at the pupils and aperture stop, the sagittal plane is also the η, z plane. This is because the principal ray crosses the optical axis at these positions. It should be pointed out that the principal ray lies in both the tangential and sagittal planes: It is the only ray that does so.

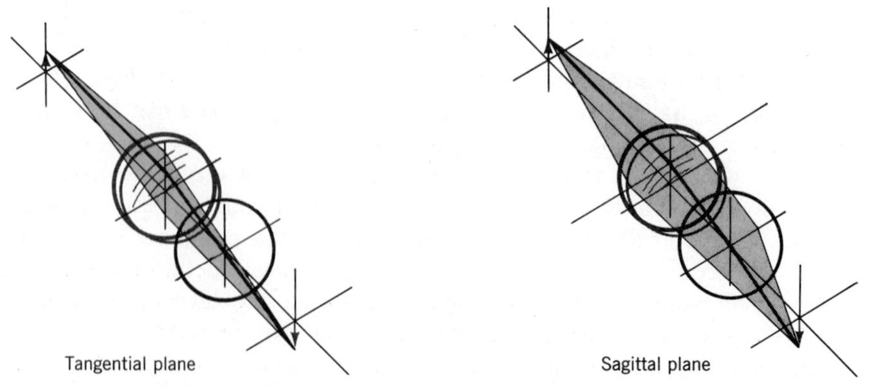

FIGURE 5B-3. Definition of tangential plane, on left, and sagittal plane, on right, of a simple optical system. The double circle represents a lens and the single circle represents the aperture stop. The dark line is the principal ray.

WAVEFRONT ABERRATION COEFFICIENTS

One technique for characterizing aberrations was introduced by Hamilton while he was in his early twenties. This technique expresses the aberration of an optical system in terms of the departure of the geometric wavefront from a spherical shape. Paraxial theory predicts that an optical system will image a point source into a point image. All rays from the point source will travel to the point image. The geometrical wavefront is constructed by generating the normal to each of the rays and is a sphere for a point image. This spherical surface, predicted by the paraxial theory, is called the reference sphere.

If we calculate the distance between the reference sphere $\mathcal{S}(\xi, \eta, z)$ and the actual wavefront, $\Sigma(\xi, \eta, z)$ in the exit pupil, we generate the wave aberration function Δ

$$\Delta(\xi, \eta, z) = n_2[\Sigma(\xi, \eta, z) - \mathcal{S}(\xi, \eta, z)] \tag{5B-1}$$

This function defines the optical path difference between the two surfaces and is measured along the radius of the reference sphere.

We will not develop the theory, but simply state that we may find the wavefront surface, usually in the exit pupil, and then measure the departure of this surface from a reference sphere positioned around the image point. The departure at each ray is given as an optical path length and is a function of the image position $\mathbf{h}$ and the coordinates of the point in the exit pupil, specified by the vector $\boldsymbol{\rho}$.

Because the optical system is axially symmetric, Δ must not change when we simultaneously rotate the vectors $\boldsymbol{\rho}$ and $\mathbf{h}$. This means that we should be able to describe the functional dependence of Δ in terms of scalar quantities. The simplest independent scalar relationships involving the variables are

$$\rho^2, \qquad \boldsymbol{\rho}\cdot\mathbf{h} = \rho h \cos\theta, \qquad h^2$$

If we expand Δ as a power series in the scalar variables, we obtain

$$\begin{aligned}\Delta(h^2, \rho^2, \rho h \cos\theta) = \Delta_{000} & \\ &+\Delta_{020}\rho^2 + \Delta_{111}h\rho\cos\theta \\ &+\Delta_{040}\rho^4 + \Delta_{131}h\rho^3\cos\theta + \Delta_{222}h^2\rho^2\cos^2\theta \\ &+\Delta_{220}h^2\rho^2 + \Delta_{311}h^3\rho\cos\theta \end{aligned} \tag{5B-2}$$

The indices of the coefficients of the expansion Δ_{ijk} indicate the power of each of the variables h, ρ, and $\cos\theta$. The constant term Δ_{000} must be zero because by definition the reference sphere and wavefront are in contact at the optical axis; see Figure 5B-4. The second and third terms are associated with a shift in the location of the Gaussian image $\Delta_{020}\rho^2$ and a small change in the magnification of the image $\Delta_{111}h\rho\cos\theta$. These terms do not effect image quality and thus are not really aberrations. The last five terms correspond to the lowest-order aberrations.

The wavefront aberration function **(5B-2)** can be used to determine the interference pattern that will be produced by the optical system in an interferometer such as the Twyman–Green interferometer (Figure 4-14). The interferogram is represented by constructing contours of constant optical path difference.

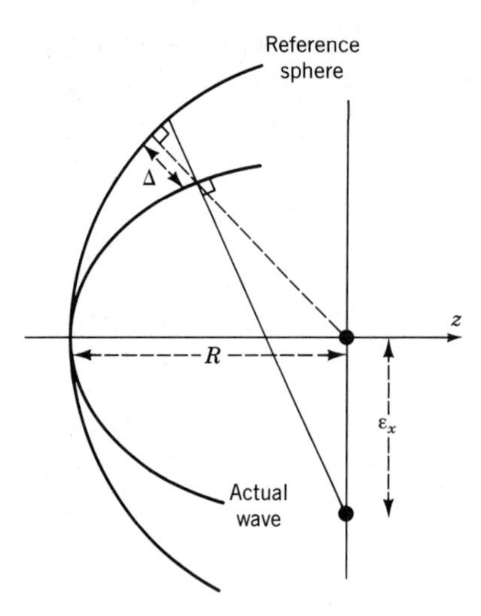

FIGURE 5B-4. Reference sphere and actual wavefront are shown in the exit pupil of a lens. Also, the optical path difference and the ray intercept in the image plane are shown.

Optical Path Difference (OPD)

It is often convenient to have a single number to characterize aberration. One technique used to generate such a number is to set ρ and h equal to their maximum values; then, the Δ_{ijk} are equal to the optical path difference contribution from each of the aberrations. The OPDs are usually quoted in wavelengths.

An optical system with an OPD of $\lambda/4$ is considered a perfect system from a geometrical optics point of view. This criterion is often called the *Rayleigh limit* (proposed by Lord Rayleigh in 1879). A system that meets the Rayleigh limit has its performance controlled by the wave properties of light and is called a *diffraction-limited system*. In such systems it is necessary to use an additional criterion to specify its performance.

As we will see in Chapter 10, a diffraction-limited system produces as an image of a point a small disk surrounded by a set of rings (see the paraxial focal plane in Figure 5B-1, where we see the ring pattern known as the Airy pattern. We will give a mathematical description of this pattern in Chapter 10.) The fraction of energy contained in the central disk is used to characterize the performance of a nearly perfect, diffraction-limited system. The characterization is made through use of the *Strehl ratio*, which is the ratio of light intensity at the peak of the diffraction pattern of an aberrated image relative to the peak

TABLE 5B.1 Characterization of a Diffraction Limited Optical System

OPD	Fraction of the Rayleigh Limit	Intensity	Strehl Ratio
0	Perfect lens	84%	1.0
$\lambda/16$	1/4	83	0.988
$\lambda/8$	1/2	80	0.952
$\lambda/4$	1	68	0.8095

in a perfect image. A perfect lens, with an OPD of 0, should contain 84% of the energy in the central disk of the Airy pattern at paraxial focus. Table 5B.1 compares several measures of lens perfection.

TRANSVERSE RAY COEFFICIENTS

To study the performance of a lens design, rays are traced through the optical system. These rays are then used to generate the wavefront and the wavefront aberration coefficients. Instead of obtaining the wavefront surface, we may instead use the rays to determine the miss distance of the ray from the paraxial image point in the image plane. The two approaches are equivalent; they simply represent a change of the reference plane where the aberration measurement is made. The use of the ray intercept is of more immediate interest to the designer since it represents the effect of the aberration on the image.

As seen in Figure 5B-2, the ray intercept is represented by its coordinates ϵ_x and ϵ_y. In the image plane, the aberrations are specified by expanding ϵ_x and ϵ_y as a power series in h, ρ, and θ.

$$\epsilon_x = \sigma_1 \rho^3 \cos\theta + \sigma_2 \rho^2 h (2 + \cos 2\theta) + (3\sigma_3 + \sigma_4)\rho h^2 \cos\theta + \sigma_5 h^3$$

$$\epsilon_y = \sigma_1 \rho^3 \sin\theta + \sigma_2 \rho^2 h \sin 2\theta + (\sigma_3 + \sigma_4)\rho h^2 \sin\theta \qquad \text{(5B-3)}$$

The coefficients of this expansion σ_i are called *the Seidel coefficients* (sometimes, the term *third-order coefficients* is used.) The two descriptions of aberrations are geometrically related and the coefficients of expansions **(5B-2)** and **(5B-3)** are simply related through paraxial ray-trace parameters. We can obtain the intercepts by calculating the normal to the wave surface

$$\epsilon_x = -R\frac{\partial \Delta}{\partial x}, \qquad \epsilon_y = -R\frac{\partial \Delta}{\partial y} \qquad \text{(5B-4)}$$

The normals yield the direction of the ray and R is the distance from the exit pupil to the image plane, approximately equal to the radius of the reference sphere.

The third-order aberration coefficients can all be calculated from paraxial ray-trace data by tracing two rays: a marginal and a principal ray. (See W.T. Welford, *Aberrations of the Symmetrical Optical System*, Academic Press, New York, 1974 for a systematic development of the theory.) We will examine qualitatively each of the terms in the expansions introduced above, but will make no attempt to calculate the coefficients of the expansion nor develop a detailed theory of any of the aberrations.

One of the aberrations, spherical aberration, can be considered alone because it is the only term in **(5B-2)** and **(5B-3)** that is independent of h. Therefore on axis, it is the only aberration that is not identically zero. When we move off axis, all of the aberrations are present and it is difficult to independently analyze a single aberration with a simple optical system. Spherical aberration is also symmetric about the optical axis so we can perform all ray tracing in the tangential plane. We will use the ray tracing of a very simple optical system to study spherical aberration. We have not developed the formalism for tracing skew rays[18] and therefore can only provide a qualitative discussion of the off-axis aberrations. In this qualitative discussion, we will assume that only the aberration under discussion is present; all other aberrations will be assumed to be zero.

SPHERICAL ABERRATIONS

The first term in the two expansions introduced above is associated with *spherical aberration*.

$$\text{Wave} \Leftrightarrow \{\Delta_{040}\rho^4, \qquad \text{Seidel} \Leftrightarrow \begin{cases} \sigma_1\rho^3 \cos\theta \\ \sigma_1\rho^3 \sin\theta \end{cases} \tag{5B-5}$$

The aberration depends on the ray position in the exit pupil but is independent of the position in the image plane. To understand this aberration, we will present an example that illustrates spherical aberration and demonstrates the technique used by the lens designer to obtain a measure of the aberration. Our example will involve a single refracting surface. Such an analysis is not wasted because each surface contributes to the total aberration of an optical system as if it were alone. The total aberration of a system is simply the sum of the contributions from all surfaces.

The example we wish to examine is the lens shown in Figure 5B-5*a*. This is not a contrived lens, but is often found in front of the photosensor of a sunlight activated street or yard light. The lens found in the yard light has a lower index of refraction and therefore has a different length.

From paraxial theory given in Chapter 5,

The optical surface we are considering is spherical. Fermat's principle requires that the surface be Cartesian, see Figure 5-7. For an object at infinity, the required Cartesian surface is an ellipsoid with an excentricity of 1/2. Because our surface is not the surface required, we should expect to encounter aberrations.

$$n_2\gamma'' = n_1\gamma - x_1\frac{n_2 - n_1}{R}$$

$$x_1 = x_1'',$$

$$x_2'' = x_1'' + 2R\gamma_2''$$

$$\gamma'' = \gamma_2''$$

We want the incoming ray to strike the optical axis at the back face of the lens, located $2R$ from the front vertex

$$x_2'' = 0$$

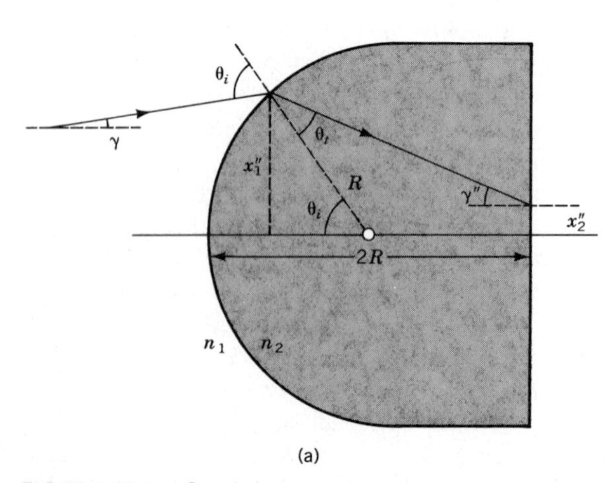

(a)

FIGURE 5B-5*a*. Simple lens with only one refractive surface.

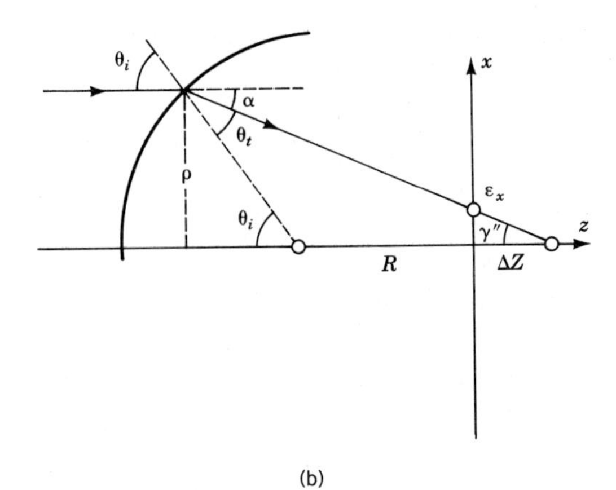

FIGURE 5B-5*b*. Geometry for finite lens trace of lens in (a).

We assume the object is at infinity

$$\gamma = 0$$

The index of refraction required to focus the plane wave from the object onto the back of the lens is obtained by solving the above equations for n_2

$$x_1 = -2R\gamma''$$

$$n_2\gamma'' = -\frac{x_1(n_2 - n_1)}{R}$$

$$n_2 = 2n_1$$

If n_1 is 1 (the lens is located in air), then $n_2 = 2$.

We will use Snell's law to perform a finite ray trace on this optical system and discover what aberrations exist in this lens. The geometry used in the ray trace is shown in Figure 5B-5*b*.

The aperture stop is the diameter of the lens and the entrance and exit pupils are at the same location because there are no optical components to the left or right of the stop. The radius of the ray intercept in the exit pupil is ρ

$$\rho = R \sin \theta_i$$

Snell's law allows us to write the transmitted angle in terms of the incident angle

$$\sin \theta_t = \frac{\sin \theta_i}{2}$$

The angle the refracted ray makes with the optical axis is given by

$$\gamma'' = \alpha = \theta_i - \theta_t$$

The law of sines allows us to calculate the z axis intercept

$$\Delta z = \frac{R \sin \theta_i}{2 \sin(\theta_i - \theta_t)} - R$$

This is called the *longitudinal spherical aberration*. We see that the optical

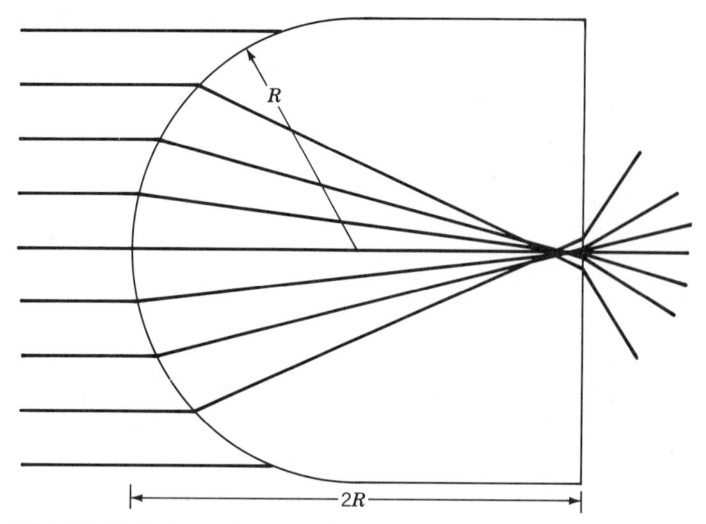

FIGURE 5B-6. A fan of rays are shown propagating through the lens defined in Figure 5B-5. (Courtesy of David Hamblen, Kodak Research Laboratory.)

The difficulty with the longitudinal intercept is that it has a large range of possible values $(-\infty, \infty)$. This makes it hard to specify improvement criteria or to check for convergence. Skew rays may never intercept the optical axis. This means a separate code would have to be written to handle skew rays. With transverse intercept the range of values can be restricted and a single code can handle skew and meridional rays by making measurements with respect to the principal rays.

system will have a focus that varies with the height of the ray in the aperture, $R \sin \theta_i$. This variation is illustrated in Figure 5B-6, where the paths of a fan of rays through the lens are shown. When spherical aberration causes the focal length to decrease with increasing zone radius, as it does in Figure 5B-6, the aberration is said to be *undercorrected*. When the marginal rays have a larger focal length than the paraxial rays, the aberration is said to be *overcorrected*. We can plot the displacement of the intercept from the paraxial intercept along the z direction as a function of the ray height; see Figure 5B-7.

The set of images equivalent to those shown in Figure 5B-1, using an optical system with spherical aberration, is shown in Figure 5B-8. We observe here that rays through the outer zones focus nearer the lens than the paraxial rays, as predicted by Figure 5B-6. The images suggest that if we

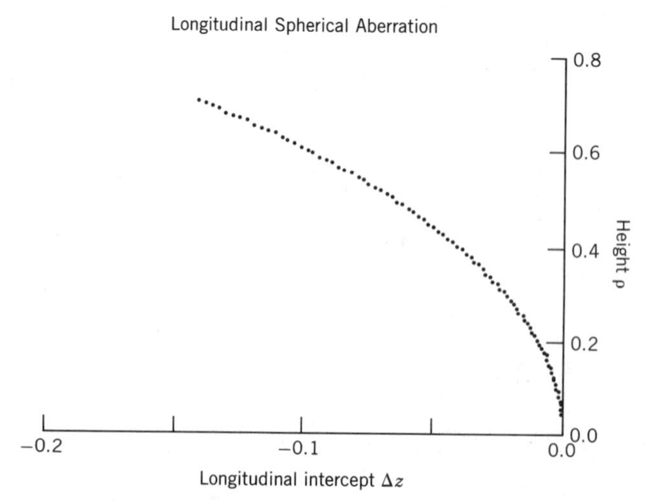

FIGURE 5B-7. Longitudinal spherical aberration of a single spherical refracting surface. This is undercorrected spherical aberration.

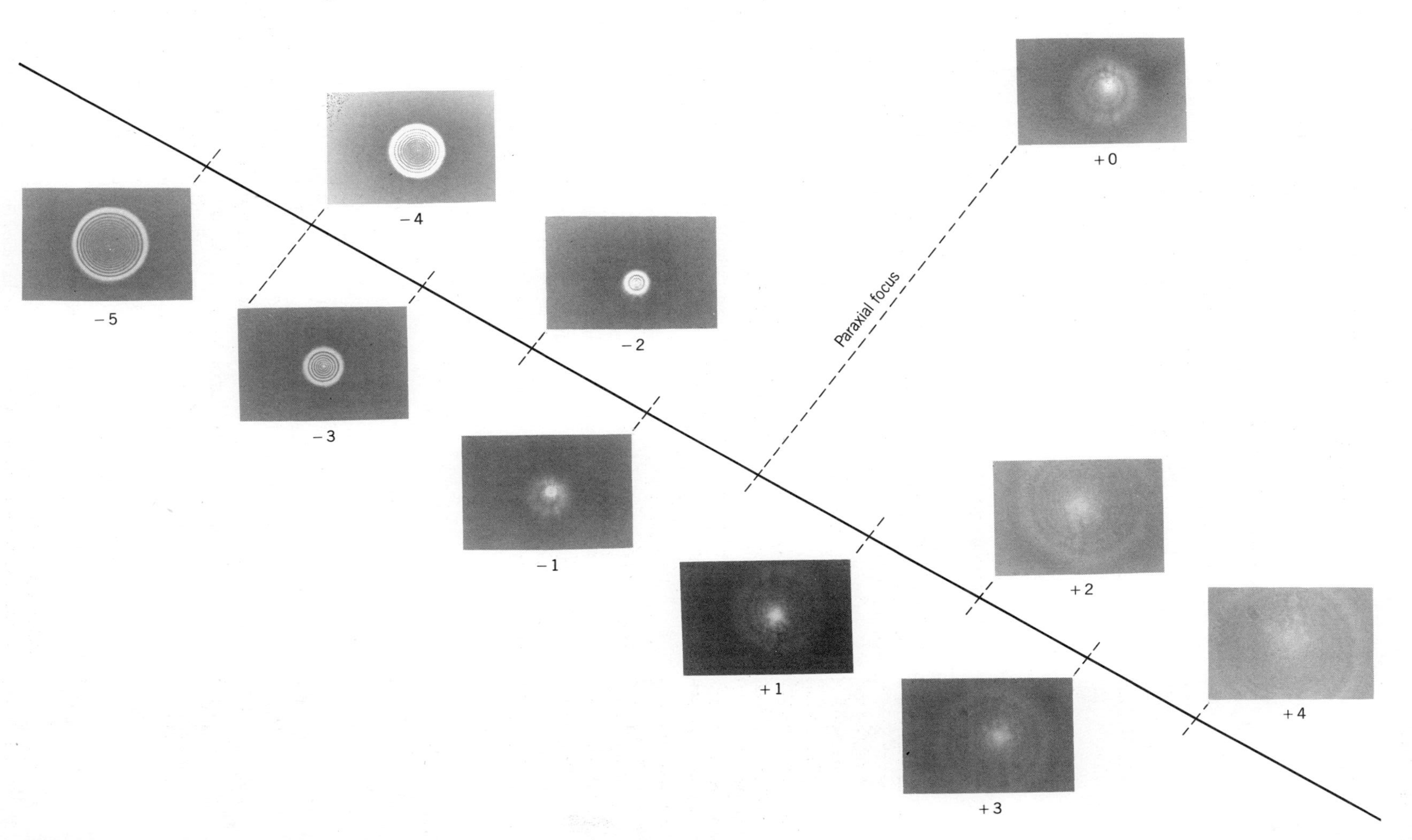

FIGURE 5B-8. Images of a point source recorded in planes spaced equally on either side of the paraxial focal plane of an optical system with undercorrected spherical aberration

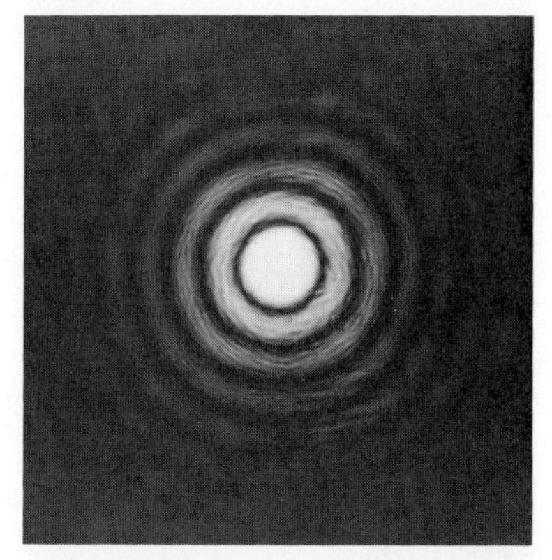

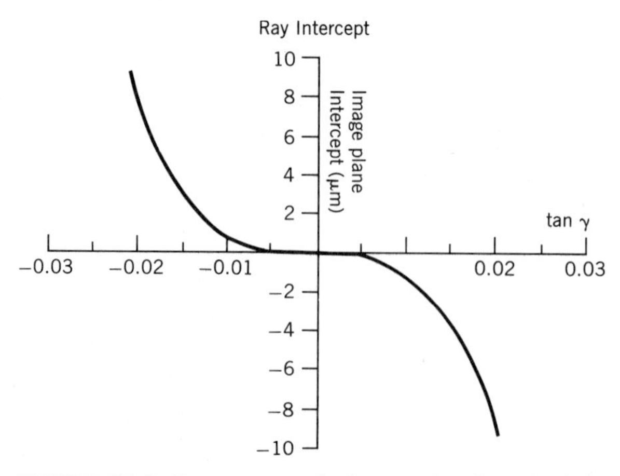

FIGURE 5B-9. Ray–intercept plot for a single spherical refracting surface.

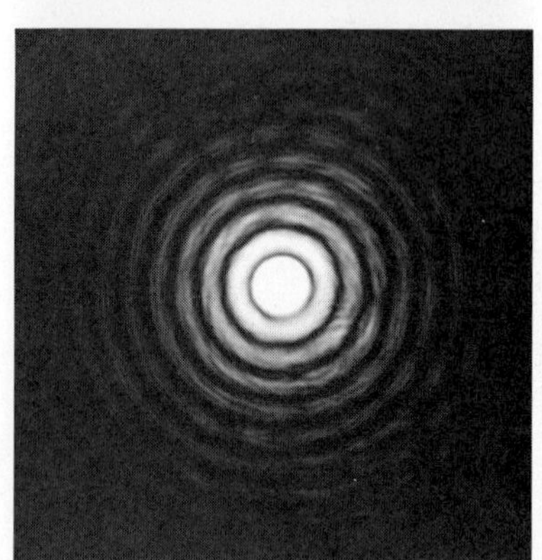

moved the focal plane inward from the paraxial plane toward the lens, we could improve the image. This is called defocus and is used by lens designers to minimize the effects of aberrations. It does not, however, remove the aberrations; in fact, spherical aberration cannot be completely eliminated from a single spherical lens.

The characterization of spherical aberration by the longitudinal displacement of the ray is not as useful to the lens designer as is the use of the image plane intercept ϵ_x, because it is more difficult to construct an optimization procedure with the longitudinal displacement.

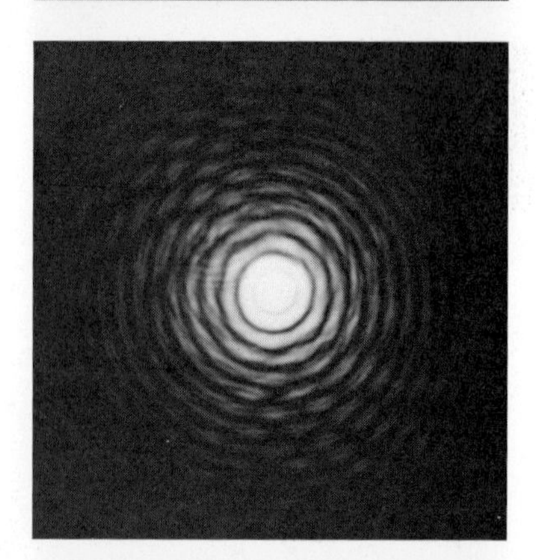

Ray Intercept Plot

The image plane intercept ϵ_x is a measure of the *transverse spherical aberration* and can be calculated by using the simple trigonometric relationship suggested by Figure 5B-5*b*

$$\epsilon_x = \Delta z \tan \gamma'' = R\left[\frac{\sin \theta_i}{2 \cos(\theta_i - \theta_t)} - \tan(\theta_i - \theta_t)\right]$$

Because

$$\tan \gamma \propto \rho$$

the ray intercept curve in Figure 5B-9 is a cubic function, as predicted by **(5B-5)**. We find that as the aperture of the optical system increases, so does the spherical aberration and more and more energy appears outside the paraxial focus. In Figure 5B-10, we have recorded the image of a point source in the paraxial focal plane for an optical system with increasing amounts of spherical aberration present. We increased the spherical aberration by increasing the aperture of the optical system. At first, diffraction dominates, and as the aperture size increases, the Airy disk decreases in size; see Chapter 10 for an explanation of this effect. Quickly, however, the spherical aberration dominates and we see a spreading of energy in the paraxial focal plane.

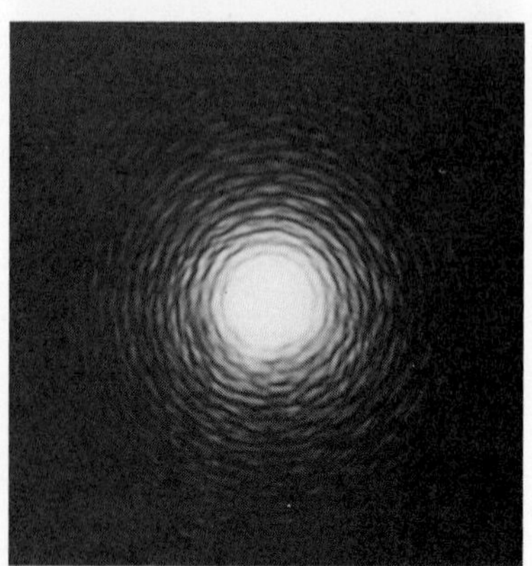

FIGURE 5B-10. Images of a point source produced by an optical system with increasing amounts of spherical aberration.

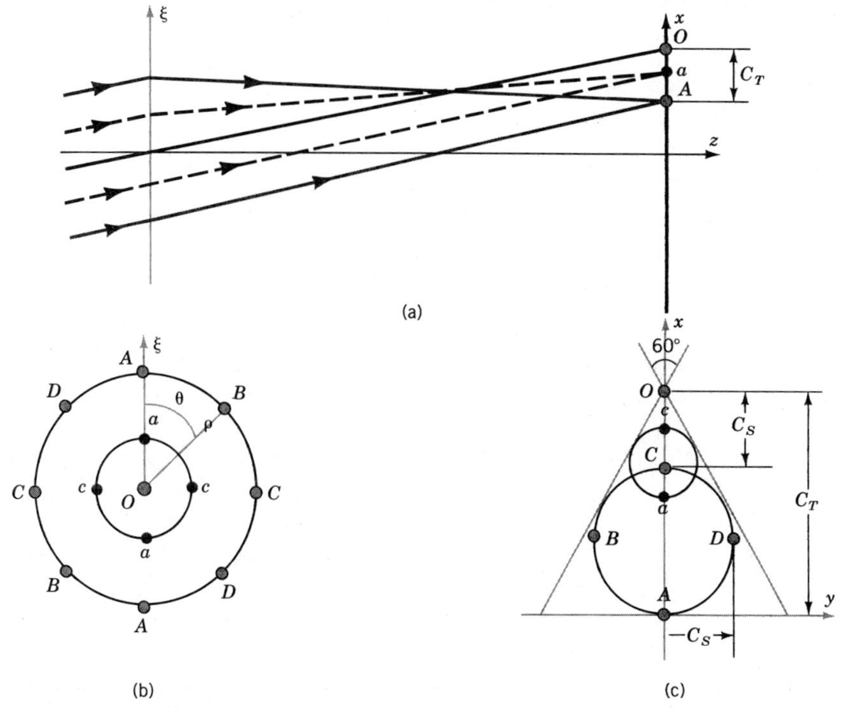

FIGURE 5B-11. Coma is due to aperture zones having different magnification. (a) Off-axis rays are focused at different heights above the optical axis with the height proportional to the square of the zone radius, i.e., each aperture zone has different magnification. (b) We select rays from two different zones in the exit pupil. (c) At the image plane, the tangential rays A and a focus at a distance C_T from the principal ray O, while the sagital rays C and c focus a distance C_s from the principal ray. The envelope of the circular images created by rays passing through individual zones form a 60°wedge that means $C_T = 3C_s$.

COMA

The first off-axis aberration we wish to discuss is called coma. It is associated with the terms

$$\text{Wave} \Leftrightarrow \left\{ \Delta_{131}\rho^3 h \cos\theta, \right. \qquad \text{Seidel} \Leftrightarrow \left\{ \begin{array}{l} \sigma_2\rho^2 h\,(2 + \cos 2\theta) \\ \sigma_2\rho^2 h \sin 2\theta \end{array} \right. \tag{5B-6}$$

from **(5B-2)** and **(5B-3)**. The image of a spot in the presence of coma is a comet shape (Figure 5B-11); thus, we arrive at the name coma. The aberration is measured by measuring the distance between the ray intercept of the principal ray, the head of the comet, and a zonal ray in the comet tail. Physically, the magnification of the object rays passing through marginal zones differs from rays passing through zones near the optical axis.

To describe the formation of the comet, we will follow several skew rays. To treat the skew rays, we decompose each ray into a sagittal component and a tangential component; the tangential component is shown in the top image of Figure 5B-11.

Because of the change in magnification with ray height, rays passing through a marginal zone are focused at a different height above the optical axis than the chief ray, labeled O in Figure 5B-11. To generate the pattern formed by this aberration, assume that the image height h and the zone radius ρ are fixed. The fixed zone radius defines a large circle in the exit pupil shown at the left of Figure 5B-11. The value of h defines the position

TABLE 5B.2 Ray Intercepts for Coma

Label	θ	Tangential	Sagittal
A	0°	$3\sigma_2^2\rho^2 h$	0
B	45°	$2\sigma_2^2\rho^2 h$	$\sigma_2^2\rho^2 h$
C	90°	$\sigma_2^2\rho^2 h$	0
D	135°	$2\sigma_2^2\rho^2 h$	$-\sigma_2^2\rho^2 h$
A	180°	$3\sigma_2^2\rho^2 h$	0

of O in the image plane and scales the size of the comet. Table 5B.2 lists the ray intercepts in the sagittal and tangential planes for rays passing through the points labeled A, B, C, D, A.

From the results in Table 5B.2, we see that the coma is given by

$$C_T = 3\sigma_2^2\rho^2 h, \qquad C_S = \sigma_2^2\rho^2 h \tag{5B-6}$$

About 55% of the light falls in the region between O and C in Figure 5B-11 so that the sagittal coma C_S is a useful measure of the image quality.

An important relationship discovered simultaneously in 1873 by Abbe and Helmholtz is the *optical sine theorem*,

$$n_1 x_1 \sin \gamma_1 = n_3 x_3 \sin \gamma_3 \tag{5B-7}$$

It can be expressed in terms of the magnification of the optical system

$$\beta = \frac{x_3}{x_1}$$

where the optical system is assumed to be in air, $n_1 = n_3 = 1$

$$\beta = \frac{\sin \gamma_1}{\sin \gamma_3}$$

Satisfying this *Abbe's sine condition* is a necessary condition for no coma and if spherical aberration is absent, it is also a sufficient condition. Coma arises from a variation in the magnification with ray angle. By meeting the sine condition, i.e., by requiring the magnification of the optical system to be a constant for all ray angles, we remove comatic aberration.

Lens designers have developed a parameter to measure the departure from the sine condition called *offense against sine condition* (OSC). This term

$$\mathrm{OSC} = \frac{C_S}{h}$$

was developed to reduce the calculation load in lens design. With computers, its original need has disappeared but lens designers continue to find it useful.

The optical sine theorem **(5B-7)**, looks like a generalization of the Lagrange invariant **(5A-5)**, but that is not completely true. The Lagrange invariant is true for all rays but only for angles that satisfy the paraxial approximation. The optical sine theorem is true for all angles but only for rays in the sagittal plane.

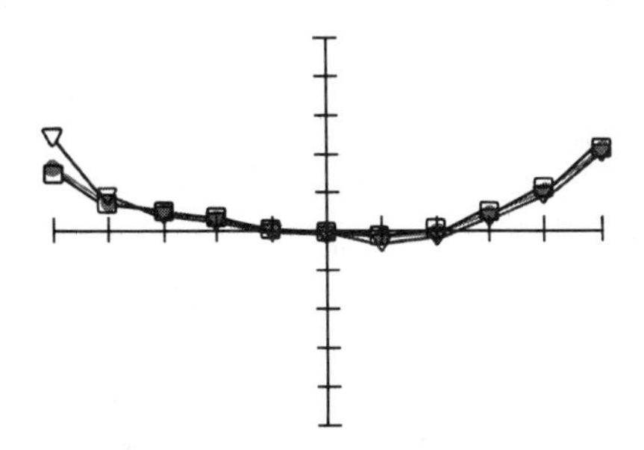

FIGURE 5B-12. A ray intercept curve of a lens whose dominate aberration is coma. The three curves are generated by performing ray traces at three wavelengths; See Appendix 7-A.

Because of the quadratic dependence of C_S on ρ **(5B-6)** we would expect the ray intercept curve to be parabolic in shape. Figure 5B-12 shows a ray intercept curve for a Cooke triplet lens. Coma, in this lens design, is the dominate aberration so the general shape of the curve is parabolic.

Spot Diagram

Much can be learned from the general shape of the image of a point source. For that reason, a useful method of displaying aberrations is the *spot diagram*. To produce a spot diagram, a number of rays are traced through the optical system to the image plane. The intersection of these rays with the image plane is represented by small circles or squares. A spot diagram of the optical system in Figure 5B-5*a* is shown in Figure 5B-13. Our simple lens produces a not so simple off-axis spot diagram. This is due to the fact that in addition to coma the other off-axis aberrations are present. With a more complicated system, the other aberrations can be reduced.

In Figure 5B-14, a set of images of a point source produced by an optical system with all off-axis aberration other than coma minimized is shown. The images were made with increasing amounts of coma. Geometrical optics was used to generate the spot diagram in Figure 5B-13 so there is no indication of the structure that diffraction effects produce in the actual image, but

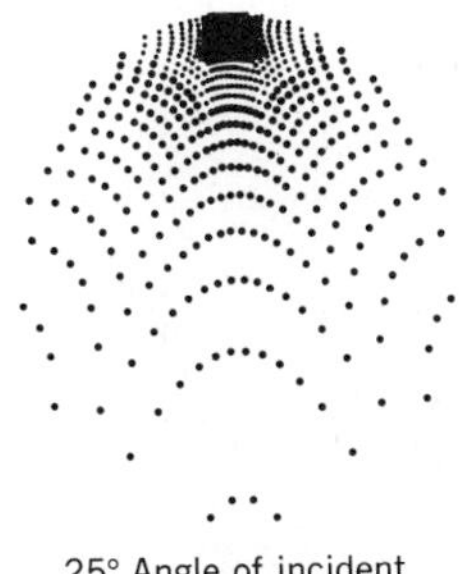

FIGURE 5B-13. Spot diagrams of the lens in Figure 5B-5. (a) Incident rays are parallel to the optical axis. These rays are called axial rays because they originate from the foot of an object at infinity. (b) Incident rays that make an angle of 25° with respect to the optical axis. (Courtesy David Hamblen, Kodak Research Laboratory.)

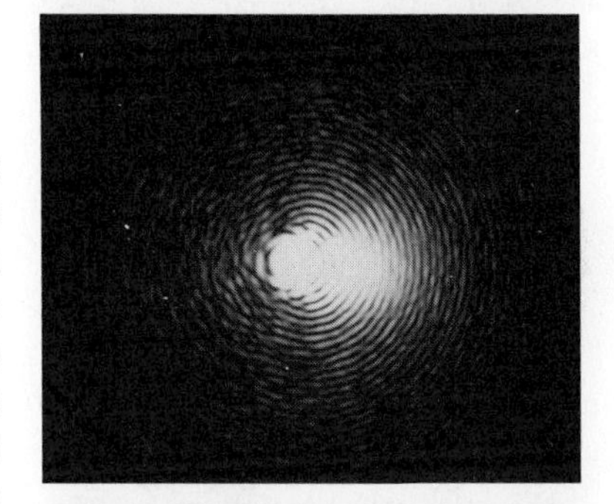

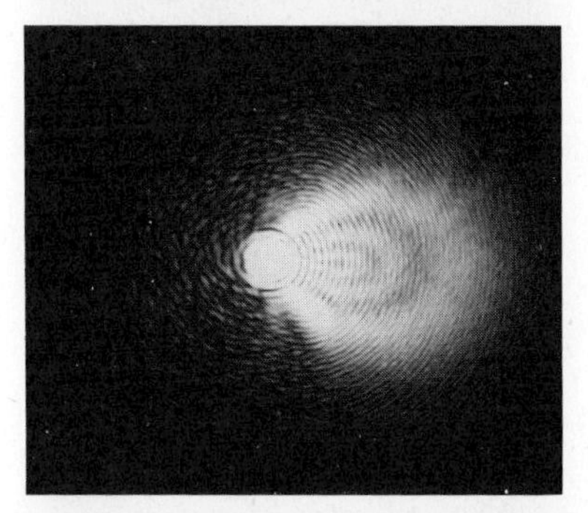

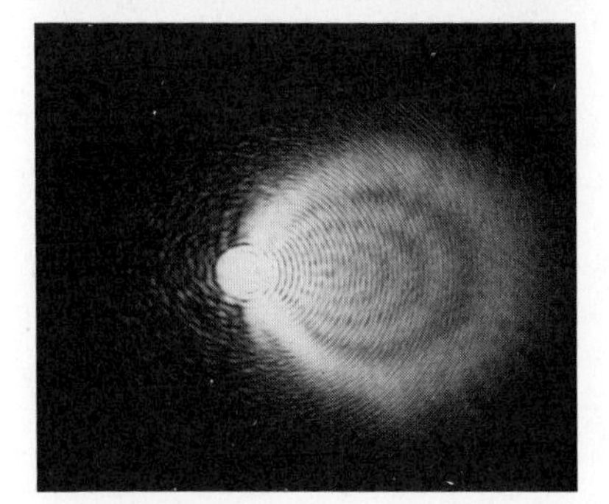

FIGURE 5B-14. Images of a point source produced by an optical system with increasing amounts of coma.

the general shape is as we expect. The modern view of an optical system utilizes the response of an optical system to a point object as the means of characterizing the system. For this reason, the spot diagram and images of a point source produced by the optical system are a useful representation of the performance of the optical system.

ASTIGMATISM

In the absence of all other aberrations the tangential and sagittal rays from an off axis point do not focus in the same plane. This behavior is predicted by the following terms from **(5B-2)** and **(5B-3)**:

$$\text{Wave} \Leftrightarrow \left\{ \Delta_{222}\rho^2 h^2 \cos^2\theta, \right. \qquad \text{Seidel} \Leftrightarrow \begin{cases} 3\sigma_3 \rho h^2 \cos\theta \\ \sigma_3 \rho h^2 \sin\theta \end{cases} \tag{5B-8}$$

This aberration is due to the lens exhibiting a different power in the tangential plane and the sagittal plane. If we look at the wavefront expansion term we see that there is no aberration in the sagittal plane $\theta = \pi/2$, but there is a quadratic distortion of the wavefront in the tangential plane. The additional curvature brings the tangential component of the wave to a focus earlier than predicted by paraxial theory. A spherical lens with astigmatism behaves for off-axis rays as if a cylindrical lens were in contact with the spherical lens. Figure 5B-15 displays the focal properties of such a lens.

The rays in the tangential plane $\theta = 0$ come to a focus in the tangential image plane, called the *tangential field*. The tangential field is defined by

$$\epsilon_x = 0$$

$$\epsilon_y = \sigma_3 \rho h^2$$

The tangential rays continue on from the focus to form a line image of the point source, called the *sagittal focal line*. The sagittal focal line occurs where

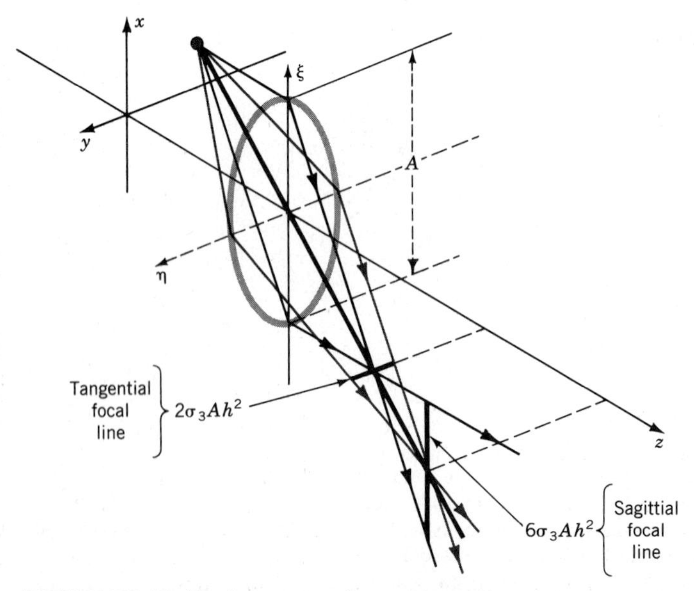

FIGURE 5B-15. Focal properties of an astigmatic lens.

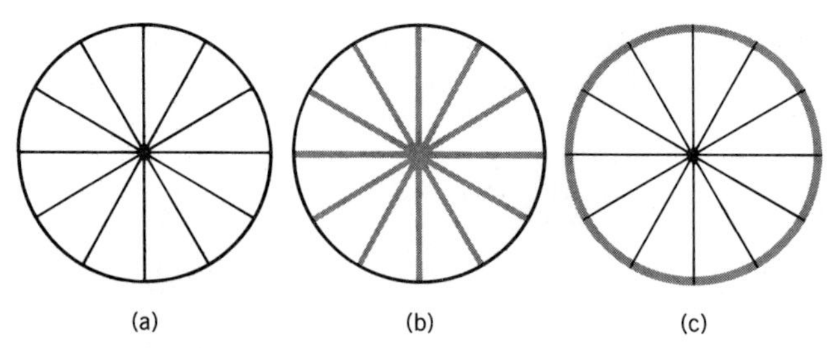

FIGURE 5B-16. (a) Object to be imaged by optical system with astigmatism. (b) Image of wheel in tangential focal plane. (c) Image of wheel in sagittal focal plane.

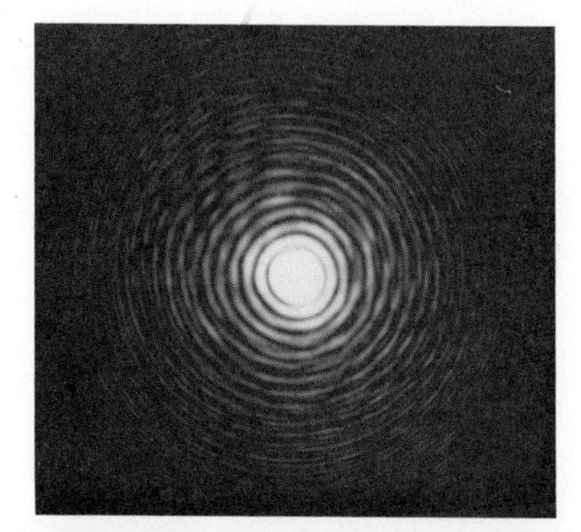

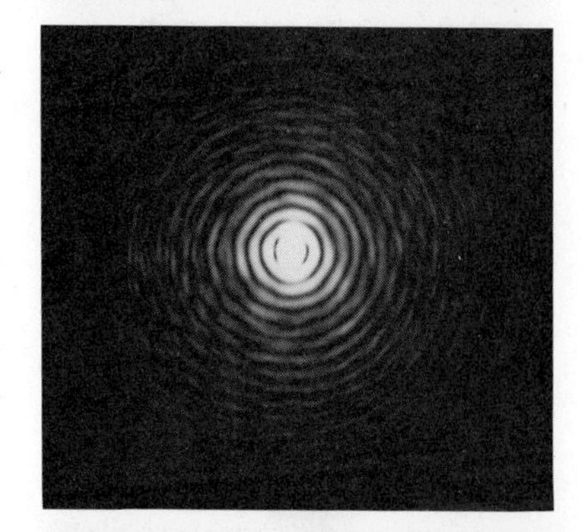

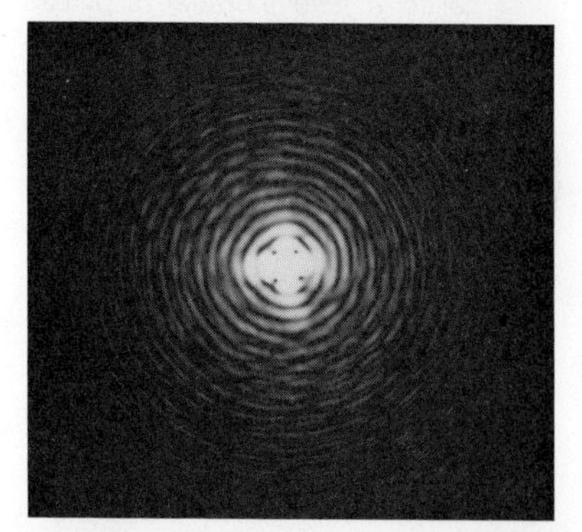

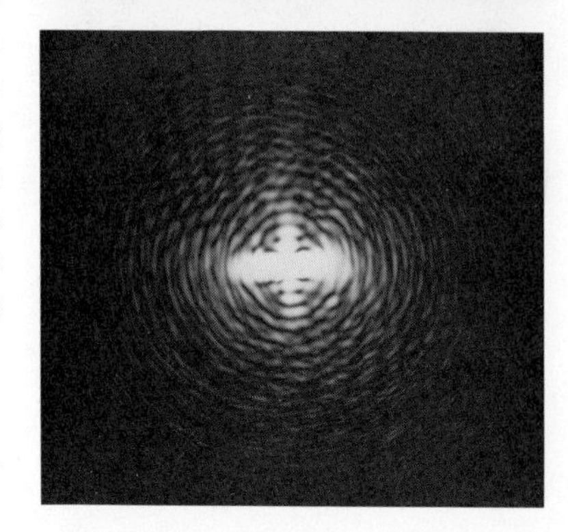

$$\epsilon_x = 3\sigma_3 \rho h^2$$

$$\epsilon_y = 0$$

These relationships also define the plane where the rays in the sagittal plane $\theta = \pi/2$ come to focus. This focal plane is called the sagittal image plane or the *sagittal field*. Before reaching their focal plane, the sagittal rays form a line image, called the *tangential focal line*. It is a little confusing when stated, but obvious by examining Figure 5B-15, that the sagittal focal line, which defines the sagittal field, lies in the tangential plane and the tangential focal line, which defined the tangential field, lies in the sagittal plane.

Because of the axial symmetry of the optical systems, we know that the tangential plane represents an infinite number of planes about the optical axis over the angular range

$$0 \leq \theta \leq 2\pi$$

If we place the wheel shown in Figure 5B-16*a* in the object plane of the optical system, then the image in the tangential field should have the appearance of Figure 5B-16*b*. The name tangential now assumes some significance. The points along the circumference of the wheel are all associated with tangential rays and are in focus.

The spokes of the wheel are normal to the circumference of the wheel and are associated with rays in the sagittal plane that is defined as the plane normal to the tangential plane. All of the spokes are thus imaged in the sagittal field, as shown in Figure 5B-16*c*.

An optical system with astigmatism produces two line images of a point source. Astigmatic images of a point source, recorded in the paraxial plane, are shown in Figure 5B-17. A ray intercept curve is not as useful for astigmatism because the curve only displays intercepts for rays in one plane. Instead of a ray intercept plot, the z coordinate of the tangential and sagittal image points is plotted as a function of h. The z coordinate is found by dividing the intercept value, ϵ_x or ϵ_y, by the tangent of the ray angle γ_3. The result is a pair of curves such as those shown in Figure 5B-18. This is a plot of the image surfaces for the lens shown in Figure 5B-5. The shift of the off-axis image point along the optical axis is called the *sagitta* or *sag* of the lens. The curves are parabolic in shape due to the dependence on h^2.

FIGURE 5B-17. Images of a point source produced by an optical system with increasing amounts of astigmatism. The images were produced in the paraxial image plane.

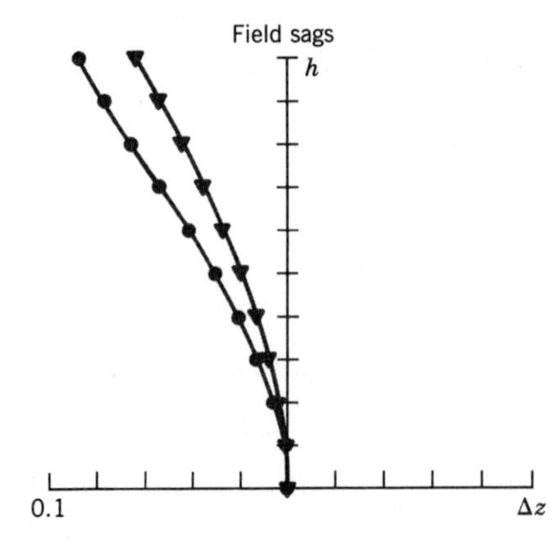

FIGURE 5B-18. A plot of the image surfaces of the lens shown in Figure 5B-5. These curves are called the *field sags* by lens designers. (Courtsey David Hamblen, Kodak Research Laboratory.)

FIELD CURVATURE

There is another aberration term with a functional form similar to the astigmatism aberration term

$$\text{Wave} \Leftrightarrow \left\{ \Delta_{220}\rho^2 h^2, \qquad \text{Seidel} \Leftrightarrow \begin{cases} \sigma_4 \rho h^2 \cos\theta \\ \sigma_4 \rho h^2 \sin\theta \end{cases} \right. \tag{5B-9}$$

Since the curvature is proportional to the sum of the powers of the optical elements, we must combine positive and negative lenses to reduce the field curvature. Often, a negative lens called a *field flattener* is placed at or near an image plane. It has little effect on image quality but reduces the field curvature by reducing the Petzval sum.

If we compare the wavefront expansion component associated with an image plane shift, $\Delta_{020}\rho^2$, with this wavefront aberration term, we discover that the aberration must be associated with an image plane shift that changes with image height h^2, i.e., the term is associated with an image plane curvature. This aberration is called *field curvature* or *Petzval curvature.*

The field curvature is measured in the same way as astigmatism, by the displacement along the optical axis of the image point as a function of the image height. If there is no astigmatism, this displacement, the sag, is given by the *Petzval sum*

$$\mathcal{H}^2 \sum_i \frac{1}{n_i f_i}$$

where $\mathcal{H}$ is the Lagrange invariant **(5A-5)**.

DISTORTION

The final aberration is associated with the terms

$$\text{Wave} \Leftrightarrow \left\{ \Delta_{311}\rho h^3 \cos\theta, \qquad \text{Seidel} \Leftrightarrow \begin{cases} \sigma_5 h^3 \\ 0 \end{cases} \right. \tag{5B-10}$$

These terms are similar to the term involving a transverse displacement in focus, $\Delta_{111}h\rho \cos\theta$, introduced early in our discussion. Because it is associated with a change in image height, this shift can also be viewed as a change in magnification. The cubic dependence on h causes the magnification to increase with image height. This term thus leads to a variation in image scale over the image plane. An object in the shape of a square would be imaged as one of the two images shown at the right of the square in Figure 5B-19.

Distortion is measured by measuring the fractional change in image position, shown by the arrows in Figure 5B-19. A distortion plot for the lens in Figure 5B-5 is shown in Figure 5B-20. The distortion is easy to control because it is zero for optical systems symmetric about an aperture stop.

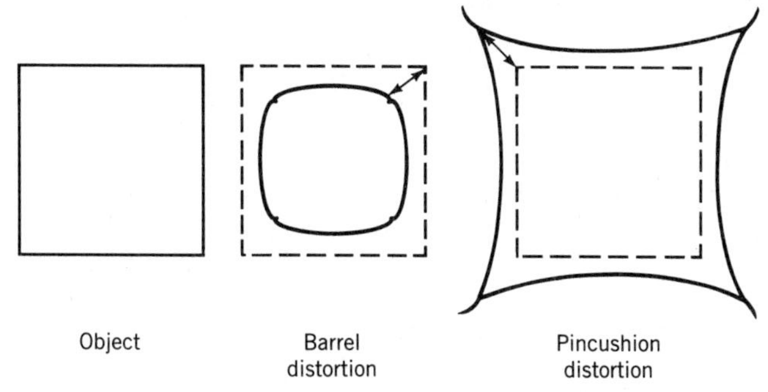

FIGURE 5B-19. If the square shown at the left is an object of an optical system with distortion as its primary aberration, then one of the two images on the right will be produced. The image labeled barrel distortion is associated with positive distortion and the image labeled pincushion distortion is associated with negative distortion. The percent distortion is obtained by making the measurement shown by the arrows.

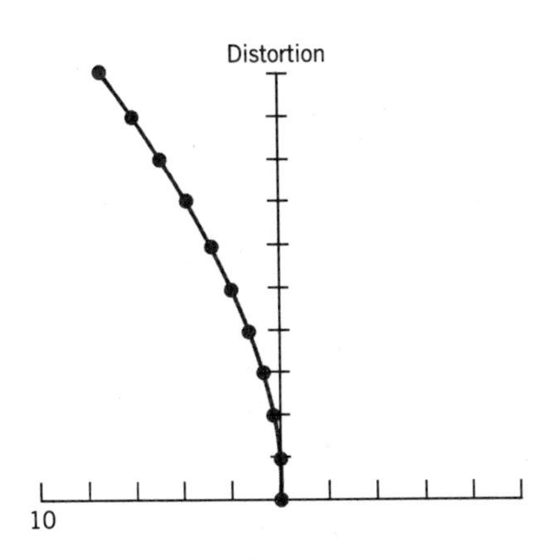

FIGURE 5B-20. Distortion shown as percent change in image height as a function of image height. This curve was generated for the lens in Figure 5B-5. (Courtesy David Hamblen, Kodak Research Laboratory.)

ABERRATION REDUCTION

Seldom do economics allow one to design an optical system from scratch, minimizing the aberrations and optimizing the performance for the job to be done. What is usually done is to combine a number of off-the-shelf lenses to accomplish the desired operation. It is still possible to control the aberrations in these "store bought" systems; to see how we must introduce two new parameters.[5]

Coddington Shape Factor

$$\sigma = \frac{R_1 + R_2}{R_2 - R_1} = \frac{c_1 + c_2}{c_1 - c_2}$$

$$c_i = \frac{1}{R_i}$$

The shape factors for a number of lenses with the same absolute value of the focal length are shown in Figure 5B-21. The positive and negative lenses shown in Figure 5B-21 each have the same power. The process of increasing or decreasing the Coddington shape factor is called *bending*. Bending is the basic tool of lens design.

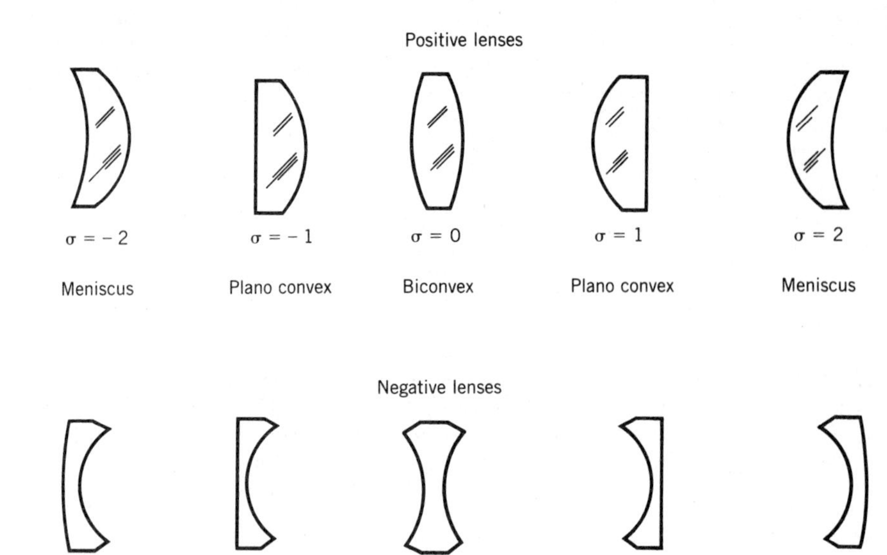

FIGURE 5B-21. The Coddington shape factors for a number of simple positive and negative lenses are shown. All of the lenses have the same absolute value of power. The process of increasing or decreasing the shape factor is called "bending."

Coddington Position Factor

$$\pi = \frac{s' + s}{s' - s}$$

$$= \frac{1 + \beta}{1 - \beta}$$

where β is the magnification. A collimating lens, with a point object at the front focal point, allowing the lens to produce a plane wave, has a position factor $\pi = +1$. A condensing lens that focuses a plane wave to a point has a position factor of $\pi = -1$. A lens arranged to produce unit magnification, the object positioned at $2f$ in front of the lens and the image at $2f$ behind the lens, has a position factor of $\pi = 0$.

Third-order aberration theory allows coma and spherical aberration to be formulated in terms of these Coddington factors.[5] Plots of spherical aberration as a function of σ, with π as a parameter, is shown in Figure 5B-22.

The optimum shape factor for minimizing spherical aberration as a function of the position factor is

$$|\sigma| = \frac{2(n^2 - 1)}{n + 2} |\pi|$$

The minimums obtained from this equation can be compared to those shown in Figure 5B-22.

For coma, third-order theory produces a linear relationship between the shape factor and the aberration. The position factor simply changes the curve's intercept. Coma is minimum whenever

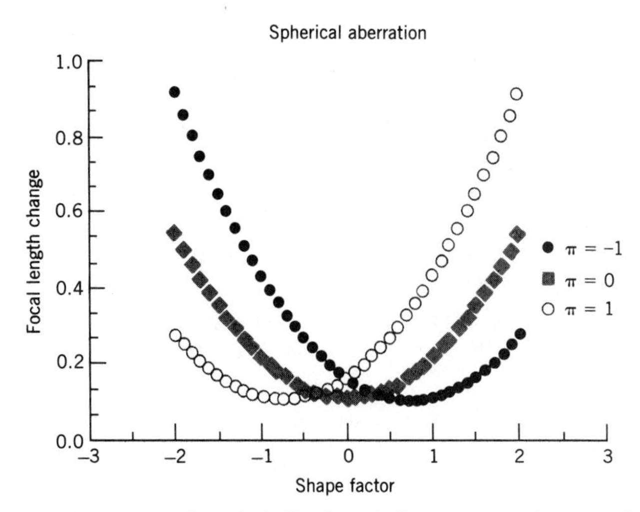

FIGURE 5B-22. Longitudinal spherical aberration as a function of the Coddington shape factor is shown for three values of the Coddington position factor.

$$\left| \sigma \right| = \frac{(2n + 1)(n - 1)}{n + 1} \left| \pi \right|$$

These results suggest that to minimize coma and spherical aberration, we need only to purchase lenses with the shape factor dictated by the imaging job to be done. This is true; however, we could quickly go broke trying to buy lenses for every new imaging configuration.

A solution to our problem is to perform *lens splitting*. The concept is shown in Figure 5B-23. Assume that in the laboratory we wish to transfer an image from one point to another, without a change in magnification. From the curves in Figure 5B-21, we would select a bioconvex lens with a power

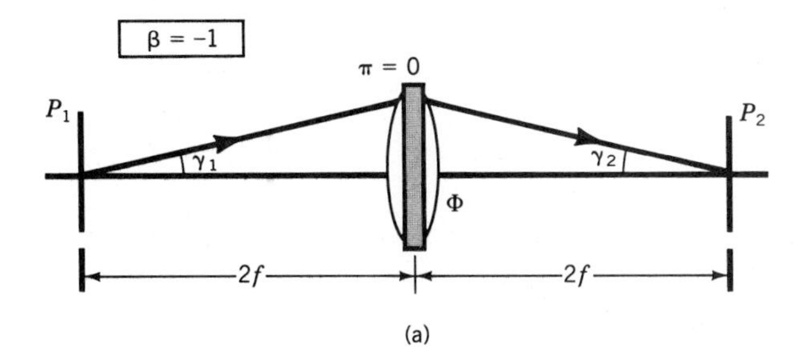

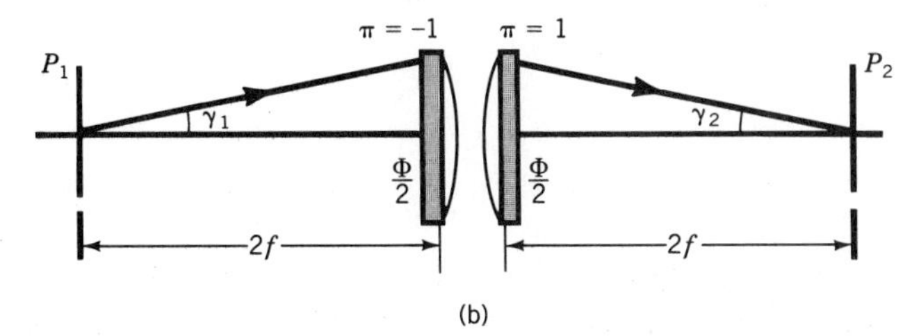

FIGURE 5B-23. The use of lens splitting to reduce the aberrations in a lens system.

to allow the transfer over the distance $4f$. If we split the lens into two lenses with half the power, we still accomplish the desired task.

No matter what magnifcation β we desire, we will benefit if we split a lens. We will add an additional restriction to the splitting procedure. We require the position parameter, after splitting, to be

$$|\pi| = 1$$

By making this restriction, we reduce the number of lens shapes needed to only one, a planoconvex lens. This is because the minimum of that position parameter is, from Figure 5B-21, ± 0.8 that is close enough to one for most applications. With this added restriction we find that the powers of the split lenses are given by

$$\Phi_1 = \frac{\beta \Phi}{\beta - 1}, \qquad \Phi_2 = \frac{\Phi}{1 - \beta}$$

The splitting procedure provides a number of benefits:

1. The relative apertures of the lenses have been reduced by a factor of two. This reduces spherical aberration, coma, astigmatism and curvature of field.
2. Spherical aberration and coma have been minimized by using the best lens shapes.
3. The numbers of lenses we must have in our stock has been reduced. Now we need purchase only planoconvex lenses.

Fourier Analysis

In the foregoing theoretical development of light, we have assumed that only one frequency of light was present. In nature this never occurs; thus, we need to expand our discussion to allow for multiple frequencies. We will introduce several mathematical techniques in this chapter that will help in handling multiple frequencies. The techniques we will discuss were developed by **Jean Baptiste Joseph Baron de Fourier (1768–1830)**. (The technique was developed to aid Fourier in the solution of heat flow problems. His first paper on the subject was rejected because Lagrange did not believe the series would converge. In the 18th century, mathematicians did not view a function as an infinite series of terms and the approach presented by Fourier required a major modification in the thinking of mathematicians. Fourier did not discover any of the principal results of the theory that bears his name. Dirichlet was one of the key contributers to the development of the theory, establishing some of the convergence criteria for the series.) The Fourier theory states that a Fourier series, a sum of sinusoidal functions, can be used to describe any periodic functions and the Fourier transform, an integral transform, can be used to describe nonperiodic functions.

Our discussions concerning light waves have also been limited to plane wavefronts. We will learn in later chapters that the Fourier theory, developed to handle multiple frequencies, can be used to describe an arbitrary wavefront in terms of combinations of plane waves. The mathematical techniques for handling multiple frequencies and arbitrary wavefronts, based on the Fourier theory, form the foundations of the modern approach to physical optics. Applications of the Fourier theory will be found in Chapter 8 in the discussions concerning coherence and Chapter 10 in the discussions concerning diffraction.

The Fourier theory allows the representation of a function in terms of its frequency or temporal characteristics and permits one to easily move between the two representations. The ability to move from a temporal to frequency representation and back, provided by the Fourier theory, allows the theory of optics to be developed using single frequencies and simple waveforms. The resultant theory can then be applied to more general waves through application of the Fourier series or transform.

In this chapter, we will first discuss the Fourier series for the representation of periodic functions and then introduce the Fourier transform, as an extension of the series, to handle nonperiodic functions. The use of a Fourier series to describe a square wave is discussed as is the use of a Fourier transform to describe a sinusoidal wave of finite duration.

The measurement process used to obtain information about continuous functions found in nature is accomplished by making discrete measurements (called samples) of the functions. The development of the Fourier theory presents the opportunity to justify the experimental approach, by examining the processes of replication and sampling. The opportunity to discuss this important concept is taken in this chapter even though the theory will not be directly applied in this book.

In the mid-1940s the concepts developed for electrical communication systems based on linear system theory and dependent on the use of the Fourier method were introduced by **P.M. Duffieux**[22] and **R.K. Luneberg**[23] for the analysis of optical imaging systems. In this chapter, the concepts of impulse response (also called the Green's function) and convolution integrals are introduced and their use in the description of a linear system operating on an arbitrary input is discussed. The linear system approach to optics will be associated with the Fresnel formulation of diffraction in Chapters 9 and 10. This approach to optics has resulted in the development of the application of optical signal processing (Appendix 10-B) and has led to many of the advances in the area of medical imaging.

The mathematical concepts and examples in this chapter will often be presented without immediate association with the physical observations that require their use. The physical observations will be introduced in later chapters after the development of the mathematical tools.

FOURIER SERIES

We wish to examine the use of a trigonometric expansion of sines and cosines called the Fourier series to describe periodic functions. The possibility of such an expansion was known to Euler, but it was not until the derivation and use of the expansion by Fourier that the usefulness of such an expansion was recognized.

The Fourier theorem as stated and proved by Dirichlet is this

If a function f(t) is periodic, has a finite number of points of ordinary discontinuity, and has a finite number of maxima and minima in the interval representing the period, then the function can be represented by a Fourier series

$$f(t) = \frac{a_0}{2} + \sum_{\ell=1}^{\infty} a_\ell \cos(\ell\omega t) + \sum_{\ell=1}^{\infty} b_\ell \sin(\ell\omega t) \tag{6-1}$$

The requirements on the function are all met by physically realizable functions.

We will not prove the theorem but simply show that it is plausible by proving that the right side of **(6-1)** is periodic. We have required the left side of **(6-1)**, $f(t)$, to be periodic, i.e., $f(t) = f(t+T)$, where $T = 2\pi/\omega$; thus, the right side of **(6-1)** must also be periodic.

$$\frac{a_0}{2} + \sum_{\ell=1}^{\infty} a_\ell \cos \ell\omega t + \sum_{\ell=1}^{\infty} b_\ell \sin \ell\omega t$$

$$= \frac{a_0}{2} + \sum_{\ell=1}^{\infty} a_\ell \cos \ell\omega(t + T) + \sum_{\ell=1}^{\infty} b_\ell \sin \ell\omega(t + T)$$

For all values of ℓ, we must have

$$a_\ell \cos \ell\omega t = a_\ell \cos (\ell\omega t + 2\pi\ell)$$

$$b_\ell \sin \ell\omega t = b_\ell \sin (\ell\omega t + 2\pi\ell)$$

which are true if ℓ is an integer.

Examination of **(6-1)** shows that the expansion is in terms of sine and cosine functions that are harmonics of the frequency $\omega = 2\pi/T$, where T is the period of the periodic function $f(t)$. Each harmonic ℓ of the fundamental frequency ω is multiplied by a coefficient, and the task of applying the Fourier theorem reduces to the problem of finding the coefficients a_ℓ and b_ℓ. The steps needed to derive the expressions for the coefficients are quite simple, as are the resulting equations for determining the coefficients. We will derive the expressions used to determine the coefficients of the harmonics making up the Fourier series and discuss two special cases that result in a shortcut in applying the Fourier series to certain classes of functions.

dc Term

The coefficient associated with $\ell = 0$ is called the *dc term* because it is associated with zero frequency. (There is no b_0 coefficient because the sine of zero frequency is zero.) To determine the constant a_0, we multiply both sides of **(6-1)** by dt and integrate over one period $(-\pi/\omega < t < \pi/\omega)$

$$\int_{-\pi/\omega}^{\pi/\omega} f(t)\, dt = \int_{-\pi/\omega}^{\pi/\omega} \frac{a_0}{2}\, dt + \sum_{\ell=1}^{\infty} \int_{-\pi/\omega}^{\pi/\omega} a_\ell \cos \ell\omega t\, dt + \sum_{\ell=1}^{\infty} \int_{-\pi/\omega}^{\pi/\omega} b_\ell \sin \ell\omega t\, dt$$

The integral of a sine or a cosine function over one period is zero; thus,

$$a_0 = \frac{\omega}{\pi} \int_{-\pi/\omega}^{\pi/\omega} f(t)\, dt \tag{6-2}$$

We see that a_0 is the average value of $f(t)$ over one period. If $f(t)$ is symmetric about the abscissa, then $a_0 = 0$.

Cosine Series

To obtain the coefficients of the cosine series a_ℓ, we multiply both sides of **(6-1)** by $\cos n\omega t$, where n represents a preselected harmonic of the series

$$\int_{-\pi/\omega}^{\pi/\omega} f(t) \cos n\omega t\, dt = \int_{-\pi/\omega}^{\pi/\omega} \frac{a_0}{2} \cos n\omega t\, dt$$

$$+ \sum_{\ell=1}^{\infty} \int_{-\pi/\omega}^{\pi/\omega} a_\ell \cos \ell\omega t \cos n\omega t\, dt + \sum_{\ell=1}^{\infty} \int_{-\pi/\omega}^{\pi/\omega} b_\ell \sin \ell\omega t \cos n\omega t\, dt$$

We now use the trigonometric identities

$$\cos(\ell\omega t)\cos(n\omega t) = \frac{1}{2}[\cos(\ell+n)\omega t + \cos(\ell-n)\omega t]$$

$$\sin(\ell\omega t)\cos(n\omega t) = \frac{1}{2}[\sin(\ell+n)\omega t + \sin(\ell-n)\omega t]$$

to evaluate the two summations. The first summation contains terms of the form

$$\int_{-\pi/\omega}^{\pi/\omega} a_\ell \cos \ell\omega t \cos n\omega t \, dt$$

$$= \frac{1}{2}\int_{-\pi/\omega}^{\pi/\omega} a_\ell \cos(\ell+n)\omega t \, dt + \frac{1}{2}\int_{-\pi/\omega}^{\pi/\omega} a_\ell \cos(\ell-n)\,\omega t \, dt$$

When $\ell \neq n$, both of the integrals are zero (see Problem 6-14). When $\ell = n$, the first integral is zero but the second integral is $(\pi/\omega)a_n$. The second summation contains terms of the form

$$\int_{-\pi/\omega}^{\pi/\omega} a_\ell \sin \ell\omega t \cos n\omega t \, dt$$

$$= \frac{1}{2}\int_{-\pi/\omega}^{\pi/\omega} a_\ell \sin(\ell+n)\omega t \, dt + \frac{1}{2}\int_{-\pi/\omega}^{\pi/\omega} a_\ell \sin(\ell-n)\,\omega t \, dt$$

which are zero for all values of ℓ. (The fact that the integrals involving sines and cosines are zero except when $\ell = n$ defines a property of sinusoids known as orthogonality.) Therefore,

$$\int_{-\pi/\omega}^{\pi/\omega} f(t) \cos n\omega t \, dt = \frac{1}{2}\int_{-\pi/\omega}^{\pi/\omega} a_n \, dt = \frac{\pi a_n}{\omega}$$

The coefficients of the cosine series are obtained by using the integral

$$a_n = \frac{\omega}{\pi}\int_{-\pi/\omega}^{\pi/\omega} f(t) \cos n\omega t \, dt \qquad \text{(6-3)}$$

Sine Series

An integral similar to **(6-3)** can be derived for the coefficients b_ℓ of the sine series if we multiply both sides of **(6-1)** by $\sin(n\omega t)$ and make use of the identity

$$\sin(\ell\omega t) \cdot \sin(n\omega t) = \frac{1}{2}[\cos(\ell-n)\omega t - \cos(\ell+n)\omega t]$$

We find that the coefficients of the sine series are given by

$$b_n = \frac{\omega}{\pi}\int_{-\pi/\omega}^{\pi/\omega} f(t) \sin n\omega t \, dt \qquad \text{(6-4)}$$

The equations for the Fourier coefficients **(6-3)** and **(6-4)** are sometimes called Euler's formulas, in recognition of Euler's early involvement with the expansion.

The sine and cosine series can individually be used to represent certain classes of functions. For example, suppose f is an even function

$$f(t) = f(-t)$$

then $f(t)$ can be represented by a series of cosines [a_0 is included in this series as the coefficient of cos(0)]. This occurs because the integral, over one period about zero (from $-\pi/\omega$ to π/ω), of an even function is nonzero, but the integral of an odd function over the same interval is zero (see Problem 6-15). Using this fact, we can determine that **(6-4)** will be zero whenever $f(t)$ is an even function. [To understand why **(6-4)** is zero for an even function, remember that the sine is an odd function and the product of an odd function and an even function is an odd function.]

If $f(t)$ is an odd function

$$f(t) = -f(-t)$$

then it can be represented by a series of sine terms. If $f(t)$ is neither odd nor even (for example, $f(t) = e^t$) then both the sine and cosine series are required.

Exponential Representation

The representation of the Fourier series given in **(6-1)** is convenient for analyzing real functions, but for extending our discussion to Fourier transforms, we will find it useful to express the Fourier series as an exponential series.

The first step in reformulating the Fourier series is to use the identities

$$\cos \ell\omega t = \frac{1}{2}(e^{i\ell\omega t} + e^{-i\ell\omega t})$$

$$\sin \ell\omega t = \frac{-i}{2}(e^{i\ell\omega t} - e^{-i\ell\omega t})$$

to rewrite **(6-1)** as

$$f(t) = \frac{a_0}{2} + \frac{1}{2}\sum_{\ell=1}^{\infty}(a_\ell - ib_\ell)\, e^{i\ell\omega t} + \frac{1}{2}\sum_{\ell=1}^{\infty}(a_\ell + ib_\ell)\, e^{-i\ell\omega t} \tag{6-5}$$

where the coefficients in the summations are given by

$$\alpha_{\mp\ell} = a_\ell \pm ib_\ell = \frac{\omega}{\pi}\int_{-\pi/\omega}^{\pi/\omega} f(t)(\cos \ell\omega t \pm i \sin \ell\omega t)\, dt$$

$$= \frac{\omega}{\pi}\int_{-\pi/\omega}^{\pi/\omega} f(t)e^{\pm i\ell\omega t}\, dt$$

This allows **(6-5)** to be rewritten as a summation over positive and negative values of ℓ

$$f(t) = \sum_{\ell=-\infty}^{\ell=\infty} \alpha_\ell e^{i\ell\omega t} \tag{6-6}$$

where

$$\alpha_\ell = \frac{\omega}{2\pi}\int_{-\pi/\omega}^{\pi/\omega} f(t)e^{-i\ell\omega t}\, dt \tag{6-7}$$

We can establish some general properties of α_ℓ given by **(6-7)** by replacing t by $-t$

$$\alpha_\ell = \frac{\omega}{2\pi}\int_{-\pi/\omega}^{\pi/\omega} f(-t)e^{i\ell\omega t}\,dt$$

Since $f(-t) = f(t)$ for an even function and $f(-t) = -f(t)$ for an odd function, we can make the following statement about the coefficients:

$\alpha_\ell = \alpha_{-\ell}$	$f(t)$	even
$\alpha_\ell = -\alpha_{-\ell}$	$f(t)$	odd
$\alpha_\ell \neq \alpha_{-\ell}$	$f(t)$	neither odd nor even
α's complex	$f(t)$	neither odd nor even

PERIODIC SQUARE WAVE

As an example of how the Fourier series is applied, we will evaluate the function

$$f(t) = \begin{cases} 1, & -\dfrac{T}{k} \le t \le \dfrac{T}{k} \\ 0, & \dfrac{T}{k} \le t \le T - \dfrac{T}{k} \end{cases} \tag{6-8}$$

The graphical representation of **(6-8)**, shown in Figure 6-1, consists of a periodic array of rectangular pulses called a square wave. The process of calculating the Fourier coefficients of the square wave is called *harmonic analysis*.

The coefficients of the Fourier series in exponential form are given by

$$\alpha_\ell = \frac{1}{T}\int_{-T/k}^{T/k} e^{-i\ell\omega t}\,dt$$

For $\ell \neq 0$, we have

$$\alpha_\ell = -\frac{1}{2\pi\ell i}\left(\exp\{-i\ell(2\pi/k)\} - \exp\{-i\ell(2\pi/k)\}\right) = \frac{2}{k}\frac{\sin(2\pi\ell/k)}{2\pi\ell/k} \tag{6-9}$$

For α_0, the integral is

$$\alpha_0 = \frac{1}{T}\int_{-T/k}^{T/k} dt = \frac{1}{T}\left(\frac{T}{k} + \frac{T}{k}\right) = \frac{2}{k} \tag{6-10}$$

One of the mathematical dividends provided by the Fourier series is that it can be used to evaluate infinite series. Although it does not impact on our study of optics, it is interesting to see an example of this application of the Fourier series. We demonstrate this use of the Fourier series by evaluating the square wave at $t = 0$ in Figure 6-1 to obtain

$$f(0) = 1 = \frac{1}{2} + 2\left(\frac{1}{\pi} - \frac{1}{3\pi} + \frac{1}{5\pi} - \frac{1}{7\pi} + \cdots\right)$$

By rewriting this relationship, we obtain the sum of Gregory's series

$$\frac{\pi}{4} = 1 - \frac{1}{3} + \frac{1}{5} - \frac{1}{7} + \cdots$$

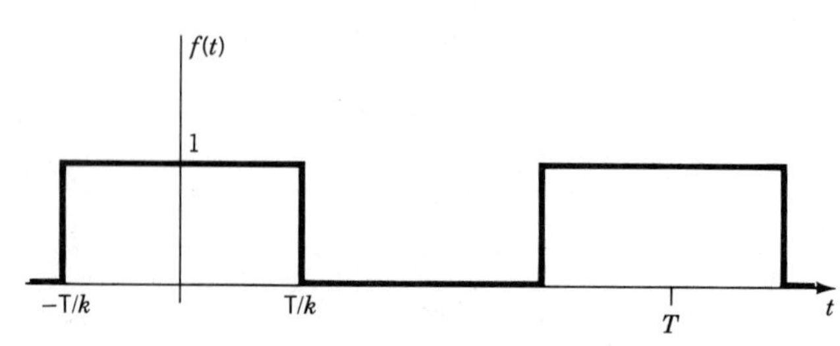

FIGURE 6-1. Generalized square wave where k is a constant.

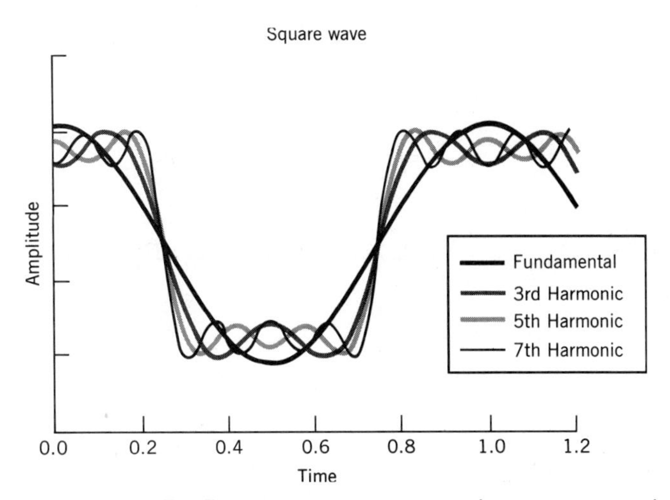

FIGURE 6-2. The Fourier series approximation of a square wave with the series terminated after the fundamental, third, fifth, and seventh harmonic.

As an example, let $k = 4$. The Fourier series is given by

$$f(t) = \frac{1}{2} + 2\left(\frac{\cos \omega t}{\pi} - \frac{\cos 3\omega t}{3\pi} + \frac{\cos 5\omega t}{5\pi} \cdots\right)$$

where we have combined the positive and negative exponents of **(6-6)** in order to express the expansion in terms of cosine functions. As you can see in Figure 6-1, the square wave defined by **(6-8)** is an even function; from our previous comments concerning even and odd functions, we are not surprised to find that the Fourier series is a cosine series.

In Figure 6-2 we plot the Fourier series for $f(t)$ with the series terminated at $\ell = 1, 3, 5$, and 7. Each additional term adds another odd harmonic to the previous estimate of the function. As we include more and more terms, the series becomes a better approximation of the square wave we are attempting to represent.

Increasing the value of k is equivalent to increasing the period of the square wave. If we think of each positive going part of $f(t)$ in Figure 6-1 as a pulse, then the width of the pulse decreases as k increases, and the time between pulses increases. We can easily calculate the coefficients of the harmonics for three examples of square waves with $k = 4, 8$, and 16. The results of the calculation are shown in Table 6.1. A convenient way of displaying these results is to plot the size of the coefficients α_ℓ as a function of $\ell\omega$. This plot is called the *frequency spectrum* and is shown in Figure 6-3. Each spectrum displays the coefficients, that is, the amplitudes, of each of the harmonic waves in the Fourier series of a square wave with different values of k.

TABLE 6.1 Fourier Coefficients for a Square Wave

k	α_0	α_1	α_2	α_3	α_4	α_5	α_6	α_7	α_8
4	0.5	0.318	0	−0.106	0	0.064	0	−0.045	0
8	0.25	0.225	0.159	0.075	0	−0.045	−0.053	−0.032	0
16	0.125	0.122	0.113	0.098	0.080	0.059	0.038	0.017	0

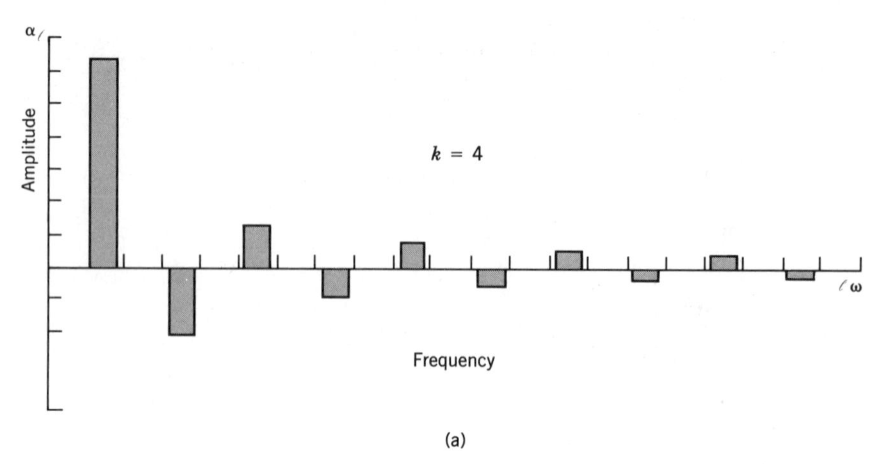

FIGURE 6-3*a*. The coefficients of the Fourier series of a square wave of width T/2.

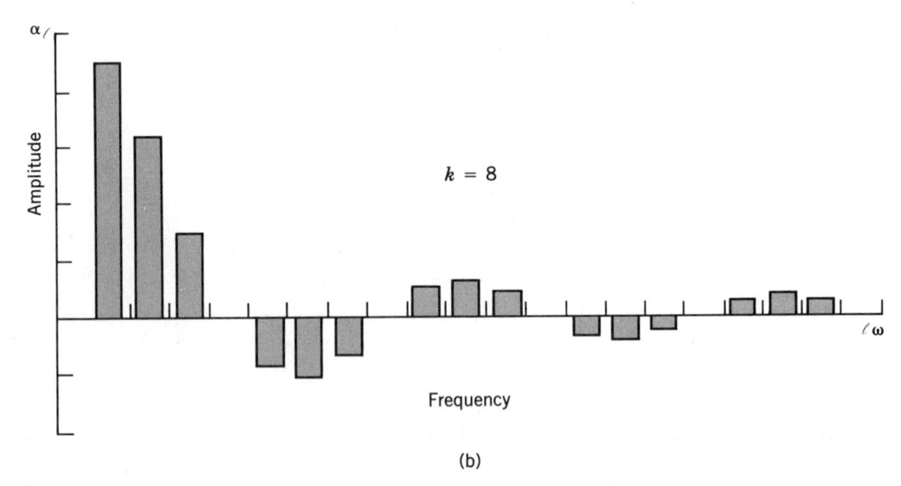

FIGURE 6-3*b*. Coefficients of the Fourier series of a square wave of width T/4.

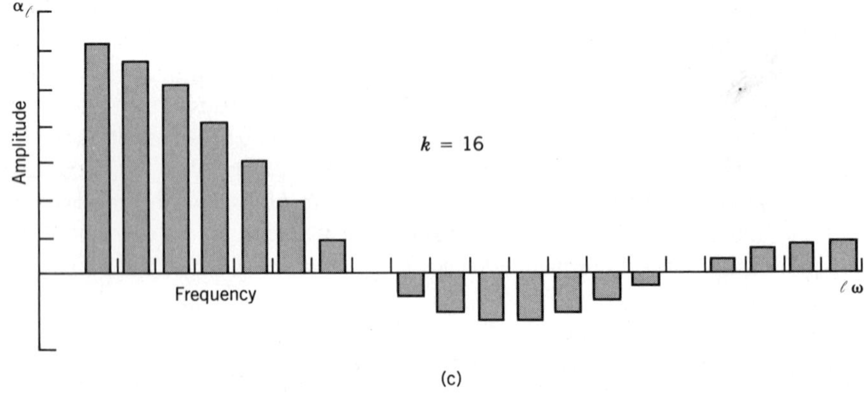

FIGURE 6-3*c*. Coefficients of the Fourier series of a square wave of width T/8.

The discrete spectra in Figure 6-3 are symmetric about zero because $f(t)$ is symmetric. For this reason, we only display the positive values of ℓ, that is, $\ell > 0$. As we decrease the width of the square pulse, that is, increase the value of k, there is the suggestion that a smooth curve could be drawn through the α's in Figure 6-3. On examining the interval between zero frequency and the frequency of the first occurrence of a zero coefficient, we find that the number of coefficients contained in this interval increases as the width of the pulse decreases. If we measure the position of the first zero coefficient in terms of the harmonic ℓ associated with the zero, we see that the frequency $\ell\omega$ at which the zero occurs increases as the width of the pulse decreases. We will find this reciprocal relationship between frequency and time is a fundamental property of Fourier series and transforms and will be repeatedly encountered both in mathematics and optics.

THE FOURIER INTEGRAL

In the discussion of Fourier series, we have required that $f(t)$ be periodic. We now wish to expand the theory to handle nonperiodic functions. We can apply a Fourier expansion to nonperiodic functions by recognizing that a nonperiodic function is really a periodic function whose period is infinite. Allowing the period of a periodic function to approach infinity is an extrapolation of the procedure used to generate Figure 6-3, that is, k increases until the width of the pulse is an infinitesimal fraction of the period T. Since $\omega = \pi/T$, we have $\omega \to 0$ as $T \to \infty$ and in the limit as the fundamental frequency approaches zero, the summation over discrete harmonics of the fundamental frequency becomes a definite integral over a continuous distribution of frequencies.

In taking the limit, we first define the fundamental frequency as $\Delta\omega$ and rewrite **(6-9)** in terms of the frequency $\Delta\omega$

$$f(t) = f\left(t + \frac{2\pi}{\Delta\omega}\right) = \frac{1}{2\pi}\sum_{\ell=-\infty}^{\ell=\infty}\left\{\int_{-\pi/\Delta\omega}^{\pi/\Delta\omega} f(t)e^{-i\ell\Delta\omega t}\, dt\right\} e^{i\ell\Delta\omega t}\Delta\omega \qquad (6\text{-}11)$$

The limit is now taken as $\Delta\omega \to 0$. The harmonics making up the distribution become infinitely close to one another and, in the limit, we replace the discrete set of harmonics with a continuous function

$$\lim_{\Delta\omega\to 0} (\ell\Delta\omega) = \omega$$

Also as the limit is taken, the period approaches infinity

$$\lim_{\Delta\omega\to 0} (T) = \lim_{\Delta\omega\to 0}\left(\pm\frac{\pi}{\Delta\omega}\right) = \pm\infty$$

Taking the limit of **(6-11)** yields

$$f(t) = \frac{1}{2\pi}\int_{-\infty}^{\infty}\int_{-\infty}^{\infty} f(\tau)e^{i\omega(t-\tau)}d\tau\, d\omega \qquad (6\text{-}12)$$

We define the function $F(\omega)$ as the *Fourier transform* of $f(t)$

$$\mathcal{F}\{f(t)\} \equiv F(\omega) \equiv \int_{-\infty}^{\infty} f(\tau)e^{-i\omega t}\, d\tau \qquad (6\text{-}13)$$

The transformation from a temporal to a frequency representation given by

(6-13) does not destroy information; thus, the inverse transform can also be defined by simply substituting the definition **(6-13)** into **(6-12)**

$$\mathcal{F}^{-1}\{F(\omega)\} = f(t) = \frac{1}{2\pi}\int_{-\infty}^{\infty} F(\omega)e^{i\omega t}\,d\omega \tag{6-14}$$

[$f(t)$ and $F(\omega)$ are called a Fourier transform pair and will be denoted by lower- and uppercase letters.] The nonperiodic function $f(t)$ is represented by an infinite number of sinusoidal functions with angular frequencies infinitely close together. $F(\omega)$ measures the spectral density, that is, the fractional contribution of frequency ω to the representation of the function. The absolute value of $F(\omega)$ is called the spectrum of the function $f(t)$.

In other books, slightly different definitions of the Fourier transform are used. In some books, the transform **(6-13)** and its inverse **(6-14)** are defined in a symmetric fashion

$$\mathcal{F}\{f(t)\} = F(\omega) = \frac{1}{\sqrt{2\pi}}\int_{-\infty}^{\infty} f(\tau)e^{-i\omega\tau}\,d\tau$$

$$\mathcal{F}^{-1}\{F(\omega)\} = f(t) = \frac{1}{\sqrt{2\pi}}\int_{-\infty}^{\infty} F(\omega)e^{i\omega t}\,d\omega$$

In other books, the constants

$$\frac{1}{\sqrt{2\pi}}$$

are absent and the integrals are expressed in terms of ν rather than $\omega\,(=2\pi\nu)$. Sometimes, the positive and negative exponentials in **(6-13)** and **(6-14)** are interchanged. The definition one selects is somewhat arbitrary.

We have written the relationships using time and frequency but we could replace time by a space variable, say, x. The transform or conjugate variable must have reciprocal units; thus, when a space variable is used, the conjugate units would be "distance" and its reciprocal 1/"distance". The conjugate variable to the space variable is called *spatial frequency* and in optics is the propagation constant **k**. Another example of conjugate variables are the periodic lattice and the reciprocal lattice, which are members of a three-dimensional Fourier transform pair used in crystallography.

There are validity conditions, called Dirichlet conditions, placed on $f(t)$ for $F(\omega)$ to exist. These are the same conditions we placed on $f(t)$ for the Fourier series to exist. They state that $f(t)$ must

1. Be single valued.
2. Have a finite number of maxima and minima in any finite interval.
3. Have a finite number of finite discontinuities but no infinite discontinuities in any finite interval.
4. Lead to a finite frequency spectrum.

$$\int_{-\infty}^{\infty} f(\tau)e^{-i\omega\tau}\, d\tau < \infty$$

(The approach we have used to obtain the Fourier transform would not be satisfactory to a mathematician. It would be more correct to consider the Fourier series as a special case of the Fourier transform. In this case, the validity conditions for the series follow naturally from a statement of the conditions for the transform.) These conditions are met by all physically occurring functions but not by such useful functions as constants and periodic functions. Techniques involving the use of limits allow these useful functions to be included. The difficulty also disappears when the theory of generalized functions is used.[24] (We will discuss an example of a generalized function, the Dirac delta function, in this chapter.)

Evaluation of the Fourier Transform

It is not immediately obvious how the Fourier transform defined by **(6-13)** is to be carried out. By expressing the transform in terms of its real and imaginary components, we see that

$$F(\omega) = \int_{-\infty}^{\infty} f(\tau) \cos \omega\tau\, d\tau - i\int_{-\infty}^{\infty} f(\tau) \sin \omega\tau\, d\tau$$

If $f(\tau)$ is a real function, then the Fourier transform can be obtained by calculating the *cosine transform*

$$\int_{-\infty}^{\infty} f(\tau) \cos \omega\tau\, d\tau \tag{6-15a}$$

and the *sine transform*

$$\int_{-\infty}^{\infty} f(\tau) \sin \omega\tau\, d\tau \tag{6-15b}$$

If $f(\tau)$ is not only real-valued but also even, we need only calculate the cosine transform **(6-15a)**. If $f(\tau)$ is complex, it can be expressed as $f(\tau) = \eta(\tau) + i\xi(\tau)$ and the Fourier transform is

$$F(\omega) = \int_{-\infty}^{\infty} \eta(\tau) \cos \omega\tau\, d\tau + \int_{-\infty}^{\infty} \xi(\tau) \sin \omega\tau\, d\tau$$
$$-i\left[\int_{-\infty}^{\infty} \xi(\tau) \cos \omega\tau\, d\tau - \int_{-\infty}^{\infty} \eta(\tau) \sin \omega\tau\, d\tau\right]$$

which demonstrates that to calculate the Fourier transform of a general function, we must evaluate the sine and cosine transforms of both the real and imaginary components of $f(\tau)$. The calculation of the Fourier transform on a digital computer makes use of an algorithm developed by James W. Cooley and J.W. Tukey in 1965.[24] Subroutines based on this algorithm are now standard components in computer software packages.

In general, the Fourier transform is a complex function and to display the Fourier transform

$$F(\omega) = f(\omega)e^{-i\phi(\omega)}$$

we plot the amplitude spectrum $f(\omega)$ and the phase spectrum $\phi(\omega)$. If the original function $f(t)$ is real and even, then $\phi(\omega)$ is a constant and we ignore it.

We will now examine three applications of Fourier transforms; Appendix 6-A contains a few additional Fourier transform pairs along with some of the important properties of the Fourier transform.

RECTANGULAR PULSE

To understand the Fourier transform, the transform of the real, even function

$$f(\tau) = \mathbf{rect}(\tau) = \begin{cases} 1, & |\tau| \leq \dfrac{1}{2} \\ 0, & \text{all other } \tau \end{cases} \tag{6-16}$$

will be calculated. This function is a rectangular pulse and is the result of allowing $k \rightarrow \infty$ in the expression for a square wave **(6-8)**. (It might be easier to think of the process of obtaining the single pulse as one in which we keep the pulse width constant and allow the period $T \rightarrow \infty$.) To calculate the Fourier transform, we use **(6-15a)** that reduces to

$$F(\omega) = \int_{-1/2}^{1/2} \cos\, \omega\tau\, d\tau = \frac{1}{\omega}(\sin\ \omega\tau)_{-1/2}^{1/2} = \frac{\sin \dfrac{\omega}{2}}{\dfrac{\omega}{2}} \tag{6-17}$$

We interpret this equation as follows: $\cos\, \omega\tau$ is a weighting function called the kernel. The shape and duration of the weighting function determine the time average of $f(t)$ calculated by **(6-15a)**. In Figure 6-4, the cosine weighting function is plotted as a two-dimensional surface in ωt space.

The function $\mathbf{rect}(t)$ slices the weighting function perpendicular to the ω axis. The profile of each slice is modified by the cosine weighting function; the frequency of the weighting function is determined by the position of the slice on the ω axis. The extent of the weighting function in the t direction is determined by $\mathbf{rect}(t)$. The value of $F(\omega)$ at each frequency is the area under the cosine curve. Figure 6-4 displays a few representative points. The Fourier transform of the rectangular pulse **(6-16)** is the continuous frequency spectrum shown in Figure 6-5 and is given by a function of the form

$$\text{sinc}(x) = \frac{\sin\, x}{x} \tag{6-18}$$

where x may represent ω or k, for example. This function is encountered so often it has been given its own name: the *sinc function*. It has zeroes whenever $x = n\pi$.

In some books, the sinc function is defined as

$$\text{sinc}(x) = \frac{\sin\, \pi x}{\pi x}$$

The only advantage of this alternate definition is that the zeroes occur at integer values of x.

At $x = 0$, the sinc function takes on the indeterminate form 0/0 and we must apply L'Hospital's rule to determine the value of the function

$$\lim_{x \to 0} \frac{\dfrac{d}{dx}(\sin\, x)}{\dfrac{d}{dx}(x)} = \lim_{x \to 0} \cos\, x = 1$$

Comparing a plot of **(6-18)** in Figure 6-5 with the envelope of the coefficients of the Fourier series in Figure 6-3c, we find that they are equivalent. We can

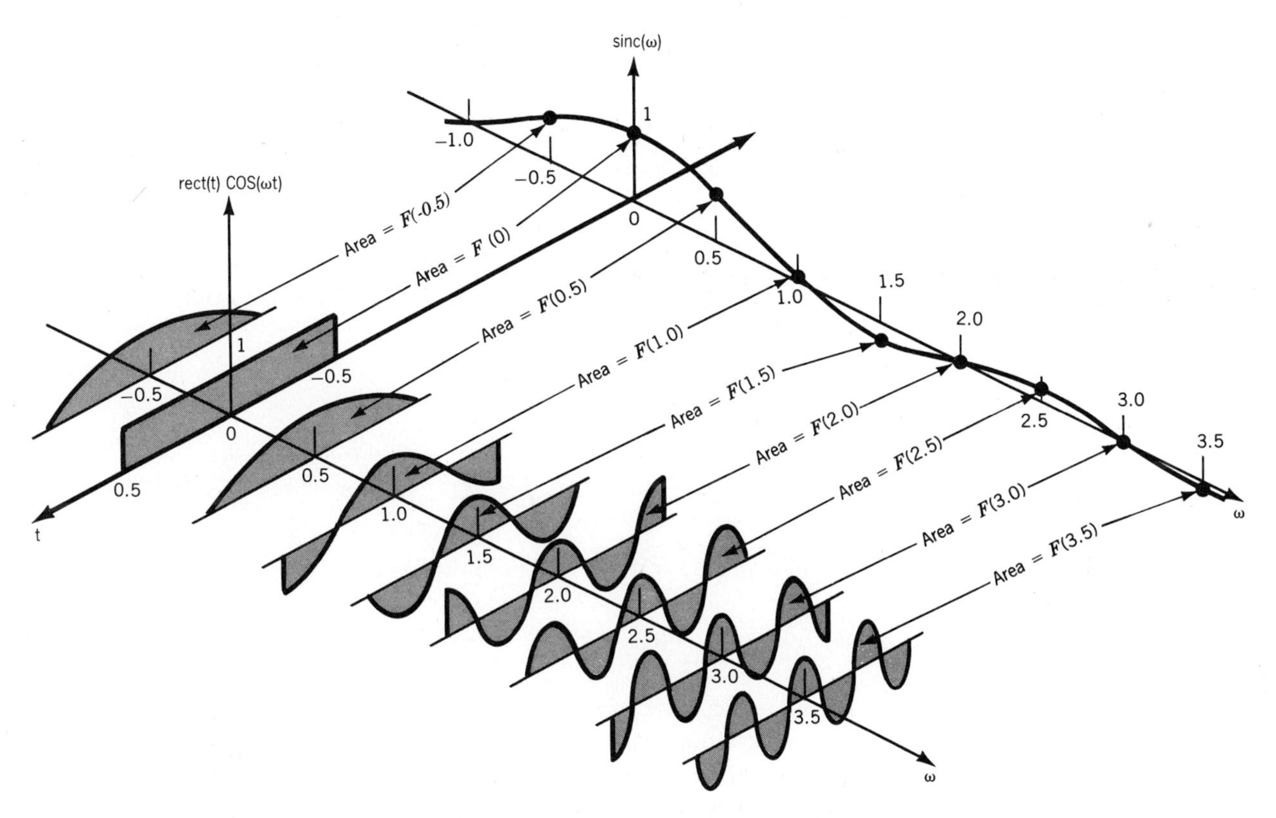

FIGURE 6-4. Geometrical construction of the Fourier transform integral of a rectangular pulse. (Jack D. Gaskill, *Linear Systems, Fourier Transforms and Optics*, Wiley, New York, 1978.)

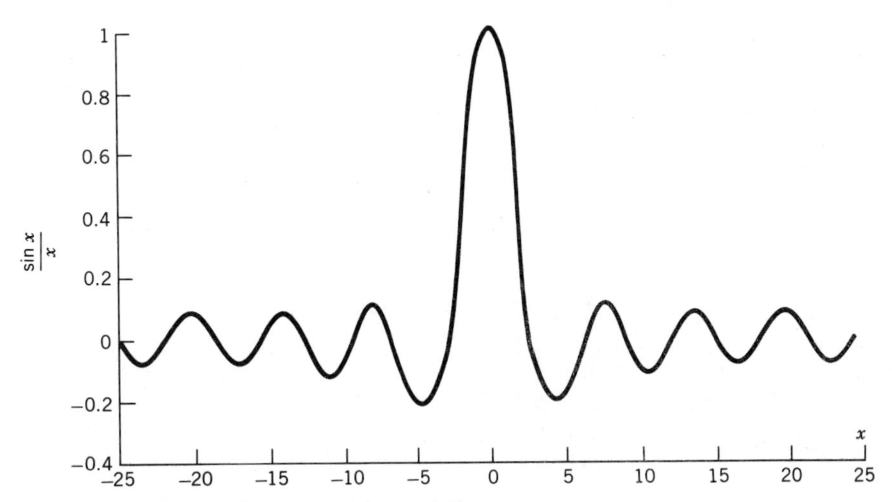

FIGURE 6-5. The sinc function $\text{sinc}(x) = \sin(x)/x$.

also compare **(6-9)** with **(6-18)** to see that **(6-9)** is a discrete representation of **(6-18)**. Equation **(6-9)** is said to be a *sampled* version of **(6-18)**.

PULSE MODULATION-WAVE TRAINS

The Fourier transform provides us with a tool to evaluate any wave of finite duration. As an example, consider a wave of frequency ω_0 that is turned on at time $-t_1$ and off at time t_1 (see Figure 6-6). The wave shown in Figure 6-6 has its amplitude modulated by a rectangular pulse of width $2t_1$. Because the wave is symmetric about the time origin, we need only calculate the cosine transform

$$F(\omega) = \int_{-t_1}^{t_1} A \cos \omega_0 \tau \cos \omega \tau \, d\tau$$

$$= \int_{-t_1}^{t_1} A[\cos(\omega_0 + \omega)\tau + \cos(\omega_0 - \omega)\tau] \, d\tau$$

The Fourier transform of the pulse-modulated wave contains two terms

$$F(\omega) = A\left[\frac{\sin(\omega_0 + \omega)t_1}{\omega_0 + \omega} + \frac{\sin(\omega_0 - \omega)t_1}{\omega_0 - \omega}\right] \tag{6-19}$$

The frequency spectrum given by **(6-19)** is shown in Figure 6-7. There are two identical frequency spectra, centered at ω_0 and $-\omega_0$, where ω_0 is called the *carrier frequency*. The small peaks to the sides of each large central peak are called *side lobes*. The first term of **(6-19)** is associated with the negative frequency distribution in Figure 6-7. It appears to contain redundant information but we must retain the negative frequencies if we wish to recover the original signal. If the conjugate variables were x and k, the negative values of k would have physical significance, as we will see later in the discussion of diffraction.

The major contribution to $F(\omega)$ occurs from the central peak (in fact, the first side lobe's peak is only 21.7% of the center peak); thus, the spectrum can be evaluated without excessive error by considering only the central peak. The width of the central peak can be defined as twice the distance from the carrier frequency ω_0 to the frequency where $F(\omega) = 0$. The

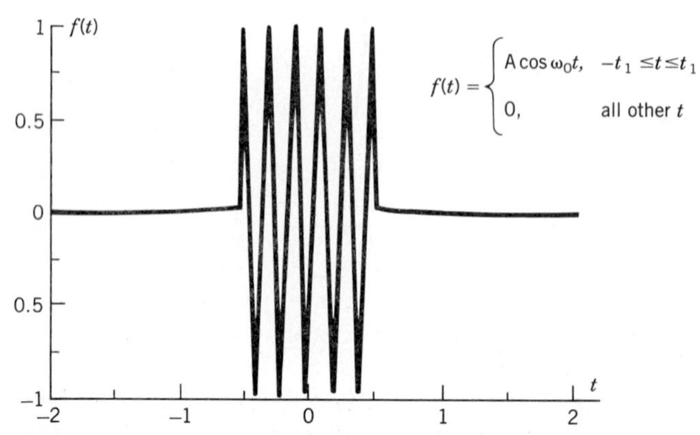

FIGURE 6-6. A wave of frequency ω_0 whose amplitude is modulated by a rectangular pulse of duration $2t_1$.

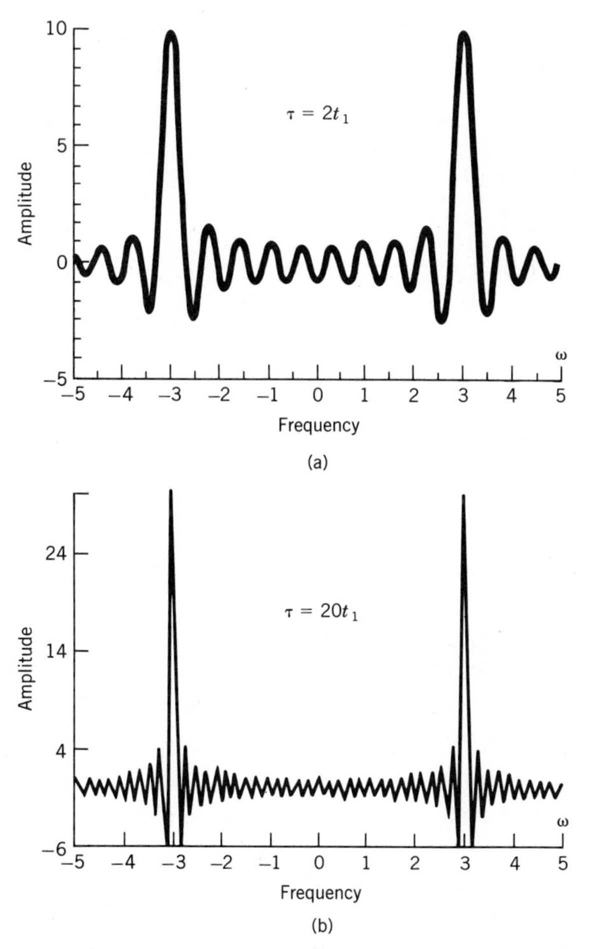

FIGURE 6-7. The frequency spectrum of a pulse of width (a) $2t_l$ and (b) $20t_l$ and carrier frequency $\omega_0 = 3$. Note that the wider pulse results in a narrower frequency spectrum.

frequency spectrum of the pulse-modulated wave $F(\omega)$ is equal to zero when $\sin(\omega_0 - \omega)t_1 = 0$ and $\omega_0 \neq \omega$. The zeroes occur when

$$\omega = \omega_0 \pm \frac{n\pi}{t_1}, \qquad n = 1, 2, \ldots$$

The width of the central peak

$$2(\omega_0 - \omega) = \frac{\pi}{t_1}$$

is inversely proportional to the pulse width t_1. Here, we see a reciprocal relationship between conjugate variables similar to what we observed in the Fourier series of Figure 6-3.

As an example, suppose $\omega_0 = 10^6$ Hz and $t_1 = 10$ μsec; then, the width of the frequency spectrum would be 600 kHz (from 700 kHz to 1.3 MHz). If the 1 MHz signal remained on for 1 sec, then the width of the spectral distribution would be about 6 Hz.

The Fourier spectra at two different values of pulse width t_1 are shown in Figure 6-7. There are two ways to interpret the frequency spectra of Figure 6-7:

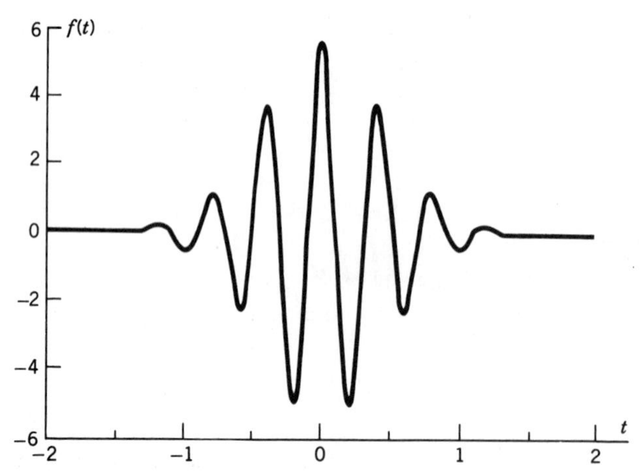

FIGURE 6-8. A wave of frequency ω_0 whose amplitude is modulated by a Gaussian pulse.

1. The classical viewpoint treats the frequency plot as a display of the actual frequencies contained in the pulse.
2. The quantum viewpoint treats the frequency plot as a display of the uncertainty in assigning a particular frequency to the pulse. Another way to state this viewpoint is that the frequency spectrum is the probability that a given frequency is present in the pulse.

A second pulse shape that will be analyzed is a pulse with a Gaussian profile shown in Figure 6-8 and described mathematically in **(6-20)**

$$f(t) = A\sqrt{\frac{\pi}{\alpha}}e^{-t^2/4\alpha}\cos\omega_0 t \tag{6-20}$$

We can rewrite this real function using complex notation by applying **(2B-6)**

$$f(t) = \frac{A}{2}\sqrt{\frac{\pi}{\alpha}}e^{-t^2/4\alpha}\left(e^{i\omega_0 t} + e^{-i\omega_0 t}\right)$$

The Fourier transform of this Gaussian modulated wave is

$$F(\omega) = \frac{A}{2}\sqrt{\frac{\pi}{\alpha}}\left\{\int_{-\infty}^{\infty} e^{-[(\tau^2/4\alpha)+i(\omega-\omega_0)\tau]}\,d\tau + \int_{-\infty}^{\infty} e^{-[(\tau^2/4\alpha)+i(\omega+\omega_0)\tau]}\,d\tau\right\}$$

This integral can be solved by completing the squares in the exponent

$$-\frac{\tau^2}{4\alpha} - i(\omega \pm \omega_0)\tau$$

$$= -\alpha(\omega \pm \omega_0)^2 - \left[\frac{\tau^2}{4\alpha} + i(\omega \pm \omega_0)\tau - \alpha(\omega \pm \omega_0)^2\right]$$

$$= -\alpha(\omega \pm \omega_0)^2 - \left[\frac{\tau}{2\sqrt{\alpha}} + i\sqrt{\alpha}(\omega \pm \omega_0)\right]^2$$

Now by substituting

$$u_{\pm}^2 = \left[\frac{\tau}{2\sqrt{\alpha}} + i\sqrt{\alpha}\left(\omega \pm \omega_0\right) \right]^2, \qquad du_{\pm} = \frac{1}{2\sqrt{\alpha}}\, d\tau$$

we can solve the integrals to get

$$F(\omega) = A\pi \left[e^{-\alpha\,(\omega - \omega_0)^2} + e^{-\alpha\,(\omega + \omega_0)^2} \right]$$

The Fourier transform of a Gaussian is another Gaussian. The widths of the transform pair are conjugate variables and are thus inversely proportional to each other.

DIRAC DELTA FUNCTION

Consider the envelope of the two pulses we have just examined using Fourier transforms. We can widen the temporal pulse, and as we do, the frequency spectrum narrows until in the limit of a cw signal, only one frequency exists in frequency space. In this limit, the frequency spectrum becomes the *Dirac delta function* (sometimes called the impulse function). The Dirac delta function was the first generalized function to be defined and is the only one we will discuss[25] (the generalized function is also called a singularity function, functional, or distribution).

The definition of the delta function usually encountered is as follows:

$$\delta\,(t - t_0) = 0, \qquad t \neq t_0 \tag{6-21}$$

i.e., the function is zero everywhere except at the point t_0. The integral of the delta function is

$$\int_{-\infty}^{\infty} \delta\,(t - t_0)\, dt = 1 \tag{6-22}$$

i.e., the delta function has a finite area contained beneath it.

A mathematically more precise definition of the delta function, based on distribution theory, is obtained by using the *sifting property* of the delta function

$$\int_{-\infty}^{\infty} f\,(t)\delta\,(t - t_0)\, dt = f\,(t_0) \tag{6-23}$$

A distribution is not an ordinary function, but rather it is a method of assigning a number to a function. The assignment is expressed formally by an integral of the form of **(6-23)**, where the delta function located at t_0 assigns the value $f\,(t_0)$ to the function $f\,(t)$. It should be emphasized that it is not the delta function itself but rather the assignment operation that is defined.

The Fourier transform of the delta function is easily obtained using **(6-23)**

$$D(\omega) = \int_{-\infty}^{\infty} \delta\,(t - t_0) e^{-i\omega t}\, dt = e^{-i\omega t_0} \tag{6-24}$$

The function $D(\omega)$ has a constant amplitude but a phase that varies linearly with ω. If $t_0 = 0$, that is, the delta function, is centered at the origin $t = 0$, then the delta function is an even function and the Fourier transform is given

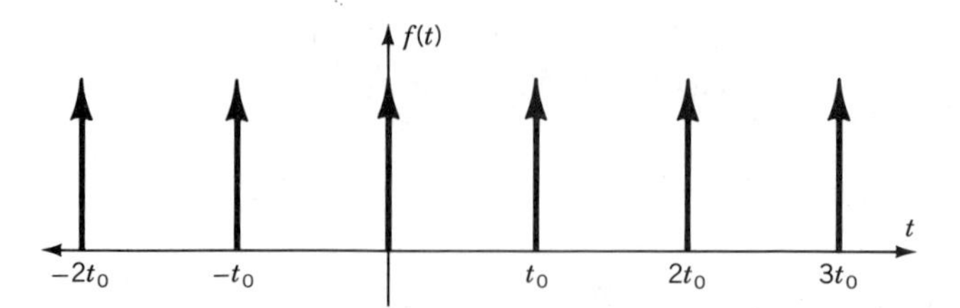

FIGURE 6-9. The comb function consisting of delta functions spaced by t_0.

by the cosine transform. The transform of the delta function located at the origin is a constant $[D(\omega) = \cos 0 = 1]$.

A series of equally spaced delta functions, called the *Dirac series* or sometimes the *comb function*, is written

$$\mathbf{comb}(t) = \sum_{n=-N}^{N} \delta(t - t_n) \tag{6-25}$$

where $t_n = nt_0$; see Figure 6-9. It is useful because it performs a sampling operation on another function, as we will see in a moment. The Fourier transform of the comb function is

$$\mathcal{F}\{\mathbf{comb}(t)\} = COMB(\omega) = \sum e^{-\omega t_n} \tag{6-26}$$

If there are two delta functions at t_0 and $-t_0$ (see Figure 6-10), then the Fourier transform is a cosine function of frequency $1/t_0$

$$C(\omega) = e^{-i\omega t_0} + e^{i\omega t_0} = 2\cos \omega t_0$$

as shown in the lower half of Figure 6-10.

If we have a series of $2N + 1$ delta functions equally spaced about the origin, we can write their sum as a geometric series

$$\mathcal{F}\left\{\sum_{n=-N}^{N} \delta(t - nt_0)\right\} = \sum_{n=-N}^{N} \mathcal{F}\{\delta(t - nt_0)\} = \sum_{n=-N}^{N} e^{-i\omega nt_0} \tag{6-27}$$

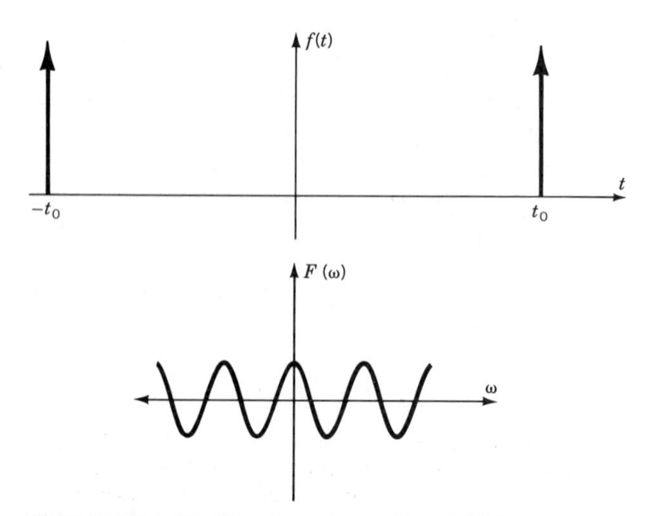

FIGURE 6-10. The Fourier transform of two delta functions positioned at $\pm t$ is the cosine function with a frequency of $1/t_0$.

Since this is the sum of a geometric series, we can write

$$\mathcal{F}\left\{\sum_{n=-N}^{N} \delta(t - nt_0)\right\} = \left(\frac{e^{-iN\omega t_0} - 1}{e^{-i\omega t_0} - 1}\right) + \left(\frac{e^{iN\omega t_0} - 1}{e^{i\omega t_0} - 1}\right) - 1$$

$$\mathcal{F}\left\{\sum_{n=-N}^{N} \delta(t - nt_0)\right\} = \frac{\cos (N-1)\omega t_0 - \cos N\omega t_0}{2 \sin^2 \dfrac{\omega t_0}{2}}$$

$$\mathcal{F}\left\{\sum_{n=-N}^{N} \delta(t - nt_0)\right\} = \frac{\sin \frac{1}{2}[(2N+1)\omega t_0]}{\sin\left(\dfrac{\omega t_0}{2}\right)} \tag{6-28}$$

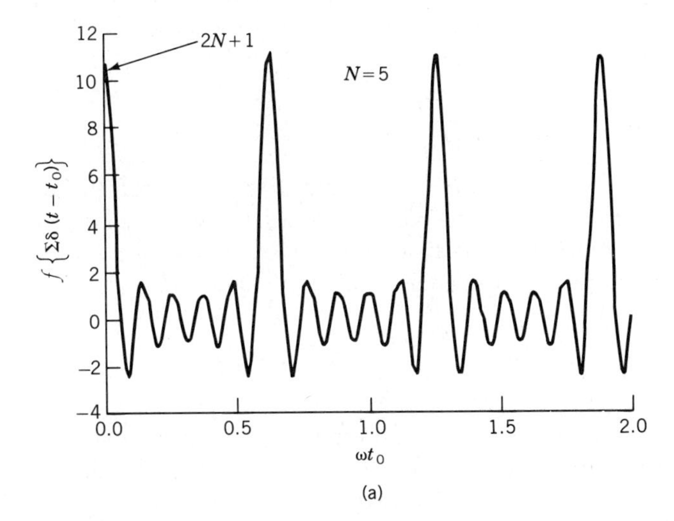

(a)

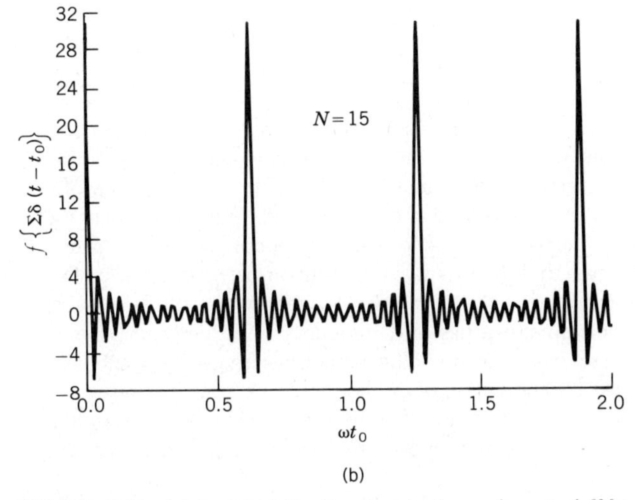

(b)

FIGURE 6-11. (a) A plot of the Fourier transform of a set of $2N+1$ equally spaced delta functions where $N = 5$. The maximum value of the Fourier transform is $2N + 1$ and the first zero is inversely proportional to $(2N + 1)$. (b) A plot of the Fourier transform of a set of $2N + 1$ equally spaced delta functions where $N = 15$. Note that the width of the primary peaks narrows as N increases.

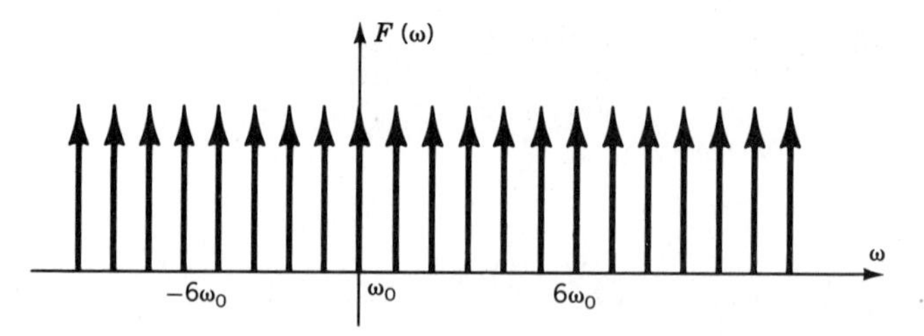

FIGURE 6-12. The Fourier transform of the infinite comb function shown in Figure 6-9.

A plot of **(6-28)** is shown in Figure 6-11 for two values of N : $N = 5$ and $N = 15$.

As can be seen in Figure 6-11, **(6-28)** is a periodic function made up of large primary peaks surrounded by secondary peaks that decrease in amplitude as you move away from the primary peak. The amplitude of the primary peak is $(2N + 1)$ and the first zero (a measure of the width of the primary peak) is given by

$$\omega = \frac{n\pi}{(2N + 1)t_0}$$

In the limit as $N \rightarrow \infty$, Figure 6-11a, and b suggest that **(6-28)** approaches a delta function; this can be proved formally.[26] Thus, the Fourier transform of the comb function in the time domain, for $N \rightarrow \infty$, is a similar comb function in the frequency domain, as shown in Figure 6-12

$$\mathcal{F}\left\{\sum_{n=-\infty}^{\infty} \delta(t - nt_0)\right\} = \frac{1}{t_0}\sum_{n=-\infty}^{\infty} \delta(\omega - n\omega_0)$$

In the limit as $N \rightarrow \infty$, the comb function becomes a periodic function and the coefficients of the Fourier series of the periodic function can be shown to be equal to the values of the Fourier integral at $n\omega_0 = 2\pi n/t_0$, which is the location of the delta functions in the frequency domain (see Figure 6-12).

A second approach to evaluating the spectral content of a nonperiodic function is to assume that the function, over the interval of interest, is one period of a periodic function. In making this assumption, we treat the function as if it were replicated over all time; the period of the replication would be equal to the length of the interval of interest. We will look at an example of the application of this replication process and then treat the replication process formally. The result of the formal treatment will be the demonstration that the process of replication in the time domain results in a frequency spectrum consisting of discrete frequencies, a sampled version of the continuous frequency distribution that would be obtained by application of the Fourier transform. This result will justify the statement of an important theorem from communication theory that specifies the number of samples of a function that are needed to represent the function.

As an example of the application of replication, a Fourier series is used to represent a straight line over the interval $-1 \leq t \leq 1$. The details are left for Problem 6-8; here, we will only display, in Figure 6-13, the first three terms of the series over the interval $-2 < x < 2$. In the interval of interest, the series approaches the straight line. Outside the interval, the fit is poor. The curve

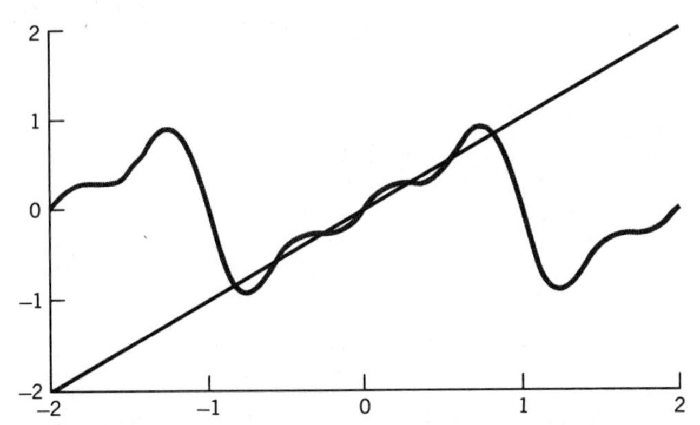

FIGURE 6-13. Fourier series approximation of the function $g(t) = t$ over the interval $-1 \leq t \leq 1$ using the first three terms of the series. We display the interval $-2 \leq t \leq 2$ to show the failure of the approximation outside the desired interval.

for the Fourier series (the gray curve) demonstrates that the function has been replicated.

To treat the replication process formally, assume we have a nonperiodic function $g(t)$ defined over the interval $-t_0 < t < t_0$, such as the function $g(t) = t$ shown in Figure 6-13. We replicate $g(t)$ $2N$ times, creating the function

$$g_N(t) = \sum_{n=-N}^{N} g(t - nt_0) \tag{6-29}$$

shown in Figure 6-14. We can use a property of a Fourier transform called *the shifting property* **(6A-5)** to write

$$\mathcal{F}\{g_N(t)\} = \sum_{n=-N}^{N} G(\omega)e^{-in\omega t_0} \tag{6-30}$$

We have already found the sum of the geometric progression of this form in **(6-27)** and **(6-28)**

$$\mathcal{F}\{g_N(t)\} = G(\omega)\sum_{n=-N}^{N} e^{-in\omega t_0} = G(\omega)\frac{\sin\left[\frac{\omega t_0}{2}(2N+1)\right]}{\sin\left(\frac{\omega t_0}{2}\right)}$$

This equation can be rewritten as a function centered on the point $\omega t_0 = 2\pi n$ and defined over the frequency region one-half a period on either side of these points, i.e., we replace ω by $\omega \pm 2\pi n/t_0$

$$\mathcal{F}\{g_N(t)\} \approx G(\omega)\frac{\sin\left[\left(\frac{\omega t_0}{2} - n\pi\right)(2N+1)\right]}{\sin\left(\frac{\omega t_0}{2} - n\pi\right)}$$

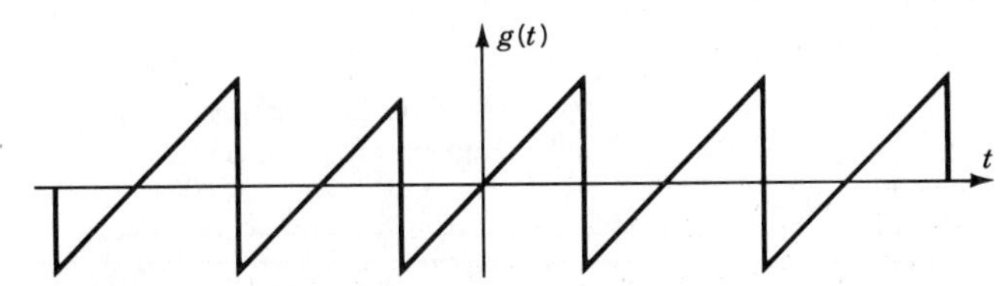

FIGURE 6.14. Replication of the function $g(t) = t$ shown in Figure 6-13.

As was mentioned earlier, in the limit as $N \to \infty$, this becomes a series of delta functions. We can use the definition $\omega_0 = 2\pi/t_0$ to write each delta function in terms of frequency

$$\delta\left(\frac{\omega t_0}{2} - n\pi\right) = \frac{2}{t_0}\delta\left(\omega - \frac{2n\pi}{t_0}\right) = \frac{2}{t_0}\delta(\omega - n\omega_0)$$

When $\omega t_0 \neq 2\pi n$, then $\mathcal{F}\{g_N(t)\} = 0$. Because there are a periodic array of these delta functions (see Figure 6-12), the Fourier transform is an infinite sum

$$\mathcal{F}\{g_N(t)\} = \frac{1}{t_0}\sum_{n=-\infty}^{n=\infty} G(\omega)\delta(\omega - n\omega_0) \tag{6-31}$$

We have shown mathematically what can be surmised by inspecting Figures 6-3 and 6-5. The frequency spectrum of a rectangular pulse is shown in Figure 6-5. If we replicate the rectangular pulse, we generate a square wave. The frequency spectrum of the square wave, shown in Figure 6-3, is a sampled version of Figure 6-5. By replication in the time domain, a function that is sampled in the frequency domain is obtained (the comb function, discussed in the previous section, performs the sampling).

The converse is also true. We could measure the spectrum in Figure 6-5 at discrete, equally spaced, frequency intervals and obtain the same spectrum shown in Figure 6-3. If we took the inverse transform of the discrete frequency samples, we would not get a square pulse but instead would generate the function that led to Figure 6-3; namely, a periodic square wave whose period equals the original pulse width.

A natural question is how should $F(\omega)$ be sampled if the resulting periodic function is to truly represent the desired function over one period? The answer is called the *sampling theorem* and was developed by Claude Shannon to determine the amount of information that can be transmitted in a communication channel.[27] The sampling theorem states that

> if the Fourier transform $\boldsymbol{F(\omega)}$ of the function $\boldsymbol{f(t)}$ is zero above some cut-off frequency
>
> $$F(\omega) = 0, \qquad \omega > \omega_c$$
>
> then $\boldsymbol{f(t)}$ is uniquely determined from its values measured at a set of times
>
> $$t = nt_0 = n\frac{\pi}{\omega_c}$$
>
> Thus, at a minimum, *we must sample twice in one period of the highest frequency present in a waveform.*

Experimentally, the sampling theorem is very important because the normal procedure for measuring a temporal signal is to sample the signal at a number of points in a time interval. The sampled data are then plotted or put into a computer for data analysis. An example of the use of sampled temporal data is found in the use of digital audio recordings. These recordings are made by sampling the audio signal and storing the sampled data in digital form. The sampling theory states that if frequencies above a certain value are unimportant, 20 kHz for human hearing, then samples need only be taken at a temporal spacing of $t_0 = 1/2\nu$, 25 μsec for audio signals. (The actual sampling frequency used in digital audio recording is 44.1 kHz, corresponding to a temporal sampling of 22.7 μsec. The frequency used is slightly higher than required in order to be compatible with television.)

CORRELATION

We often find it necessary to compare functions. With light, we effectively compare two waves by interfering them. Where the two waves are alike, we see a bright band and when they are dissimilar, we see a black band. We will show an example of this type of comparison after discussing the methodology. The method for calculating the similarity of two functions is called the *correlation integral* and the resulting function is called the *correlation function*, $h(\tau)$. If we wish to compare $a(t)$ and $b(t)$, where $a(t)$ and $b(t)$ are different functions, the integral is called the *cross correlation function*

$$h(\tau) = a(t) \oplus b(t) = \int_{-\infty}^{\infty} a(t)b^*(t-\tau)\,dt \tag{6-32}$$

If $a(t)$ and $b(t)$ are the same function, then the correlation integral is called the *autocorrelation function*. It is useful to normalize the correlation functions, by dividing by the root mean square average of the two functions, to allow comparison with other correlations. The normalized correlation function is

$$h(\tau) = a(t) \oplus b(t) = \frac{\int_{-\infty}^{\infty} a(t)b^*(t-\tau)\,dt}{\left[\int_{-\infty}^{\infty} a(t)a^*(t)\,dt\right]^{1/2}\left[\int_{-\infty}^{\infty} b(t)b^*(t)\,dt\right]^{1/2}} \tag{6-33}$$

If $a(t)$ and $b(t)$ were light waves, the integrals in the denominator would be the average intensity of each wave; thus, the name average energy is usually associated with these integrals.

To develop a physical intuition about the correlation function, we will calculate the autocorrelation function of $A(t)$, a square pulse, defined as

$$A(t) = \begin{cases} A, & -t_0 \le t \le t_0 \\ 0, & \text{all other } t \end{cases}$$

We will use this example to discover that the autocorrelation function is always an even function and that $h(0)$ of the autocorrelation function is the average energy of the function. The function, a construction showing the correlation value for $t = \tau$, and the normalized autocorrelation function are shown in Figure 6-15.

To calculate the correlation function, we simply slide one function across the second, calculating the overlapping area for each displacement τ. The autocorrelation function at τ is the overlap area of the function and its clone, the area shaded in Figure 6-15. For $A(t)$, the area of overlap equals the area of the two pulses $(A{\cdot}2t_0 + A{\cdot}2t_0)$, minus the area of each pulse not overlapped $(A\tau + A\tau)$. The area is thus

$$4At_0 - 2A\tau$$

We divide by the area of the pulses to normalize, yielding

$$h(\tau) = \begin{cases} 1 - \dfrac{|\tau|}{2t_0}, & |\tau| \le 2t_0 \\ 0, & |\tau| > 2t_0 \end{cases}$$

If we plot $h(\tau)$, we obtain a triangle whose base is twice the width of the pulse; this is the autocorrelation of the square pulse $A(t)$.

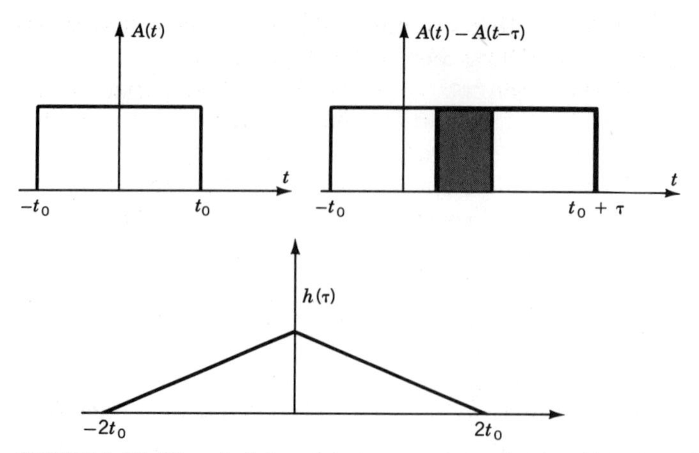

FIGURE 6-15. The calculation of the autocorrelation function $h(\tau)$, of the function $A(t)$. We simply slide one $A(t)$ over another copy of $A(t)$ and record the overlapping area, shown here as the shaded area.

A negative shift of $A(t)$ with respect to its clone (leftward shift in Figure 6-15) is equivalent to a positive shift between the two functions. We can easily demonstrate this fact and thereby discover that the autocorrelation is an even function. Mathematically, the use of a negative shift to generate the autocorrelation function is written as

$$h(-\tau) = \int_{-\infty}^{\infty} A(t)A(t + \tau)\, dt$$

Let $t + \tau = \gamma$ and $dt = d\gamma$ so that the correlation integral can be rewritten

$$h(-\tau) = \int_{-\infty}^{\infty} A(\gamma - \tau)A(\gamma)\, d\gamma = h(\tau)$$

This means that the autocorrelation is always an even function.

The maximum value of the autocorrelation occurs when the two identical functions are aligned and $\tau = 0$, where the autocorrelation is given by

$$h(0) = \int_{-\infty}^{\infty} [A(t)]^2\, dt$$

This integral is equal to the average energy of $A(t)$.

If the two functions are identical but one leads the other by a time Γ, then the maximum value of what should now be called a cross correlation occurs at $\tau = \Gamma$. As an example of this property, we will calculate the cross correlation function of two periodic functions with the same period but different epoch angles

$$a(t) = A \cos(\omega_0 t + \theta)$$

$$b(t) = B \cos(\omega_0 t + \phi)$$

The cross correlation function is

$$h(\tau) = \frac{AB}{2} \cos(\omega_0 \tau + \theta - \phi)$$

The peak of this correlation function is periodic and the location of the maximum allows the determination of the relative phase difference between

$a(t)$ and $b(t)$, i.e., how much $a(t)$ leads or lags $b(t)$. This result is the mathematical representation of an optical interference experiment.

In summary, the peak value of a correlation function as well as the value of the relative displacement τ measure the degree of similarity and the relative temporal position of the functions.

One of the properties of the Fourier transform is that the correlation integral is given by the Fourier transform of the product of the Fourier transforms of the two functions

$$h(\tau) = a(t) \oplus b(t) = \mathcal{F}\{A(\omega)B^*(\omega)\} \tag{6-34}$$

CONVOLUTION INTEGRALS

Another class of integrals we will find useful is called *convolution integrals*

$$g(\tau) = a(t) \otimes b(t) = \int_{-\infty}^{\infty} a(t)b(\tau - t)\, dt \tag{6-35}$$

In German, this integral is called the *faltung* or folding integral because the function $b(t)$ is folded over the ordinate before the integral is performed. The weighting function $b(\tau - t)$, called the convolution kernel, can be thought of as a window that moves in time and through which we observe the function $a(t)$. The convolution is the time average of the temporal function $a(t)$ viewed through this window.

The convolution function is easily confused with the correlation function **(6-33)**, but they are not the same. In general, the correlation operation does not commute

$$a(t) \oplus b(t) \neq b(t) \oplus a(t)$$

while the convolution does (see Appendix 6A)

$$a(t) \otimes b(t) = b(t) \otimes a(t)$$

There is a simple relationship between the convolution and correlation functions

$$a(t) \oplus b(t) = a(t) \otimes b^*(-t) \tag{6-36}$$

We see that the correlation and convolution functions are identical if the weighting function $b(t)$ is a real, even function. If we look back at **(6-12)** we can now recognize it as a convolution integral.

We will evaluate the convolution of the two functions in Figures 6-16*a* and *b*. Figures 6-16*c* and *d* display graphically the evaluation of the integral. The convolution and correlation for the two functions shown in Figures 6-16*a* and *b* are listed in Table 6.2. The differences between the convolution and correlation for these two functions are not large even though the functions are

The two functions in Figures 6-16*a* and *b* are called functions with compact support. This means that both are identically zero outside some finite interval. For this type of function, the width of the convolution is equal to the sum of the widths of the two functions, of compact support, being convolved. Figures 6-16 and 6-17 verify this statement. For functions that do not have compact support, the relationship between the widths is only approximate.

TABLE 6.2 Comparison of Correlation and Convolution

Correlation $h(\tau)$		Convolution $g(\tau)$	
0	$\tau < -3$	0	$\tau < -1$
$(1/3)(\tau + 3)^2$	$-3 < \tau < 0$	$(1/3)(\tau + 1)^2$	$-1 < \tau < 2$
3	$0 < \tau < 1$	3	$2 < \tau < 3$
$3 - (1/3)(\tau - 1)^2$	$1 < \tau < 4$	$3 - (1/3)(\tau - 3)^2$	$3 < \tau < 6$
0	$\tau > 4$	0	$\tau > 6$

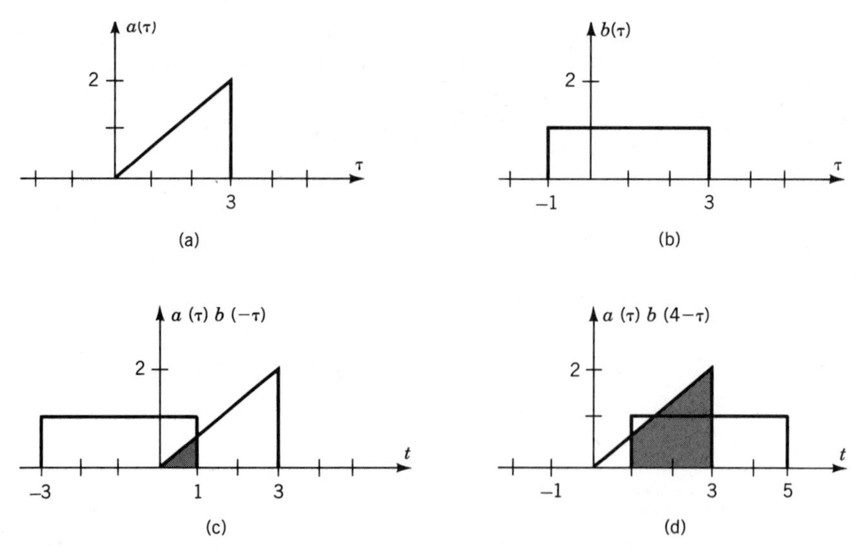

FIGURE 6-16. The calculation of the convolution integral involving the functions $a(\tau)$ and $b(\tau)$, shown in (a) and (b), respectively, is obtained by the operation shown in (c) and (d). We reflect $b(\tau)$ through the ordinate and then slide the reflected $b(-t)$ over $a(t)$, respectively, as we did for the correlation function.

not symmetric about the origin; one of the problems at the end of the chapter involves two functions that produce larger differences in the correlation and convolution.

Figure 6-17 displays a plot of the convolution and correlation functions for our example. Note that the convolution operation is a smoothing operation, i.e., sharp peaks are rounded and steep slopes are reduced. Because of the smoothing process, the convolution is often referred to as filtering. The amount of smoothing depends on the nature of the two functions. For example, if we replaced $b(t)$ in the above example with a delta function, then the convolution of $a(t)$ with $\delta(t)$ would be

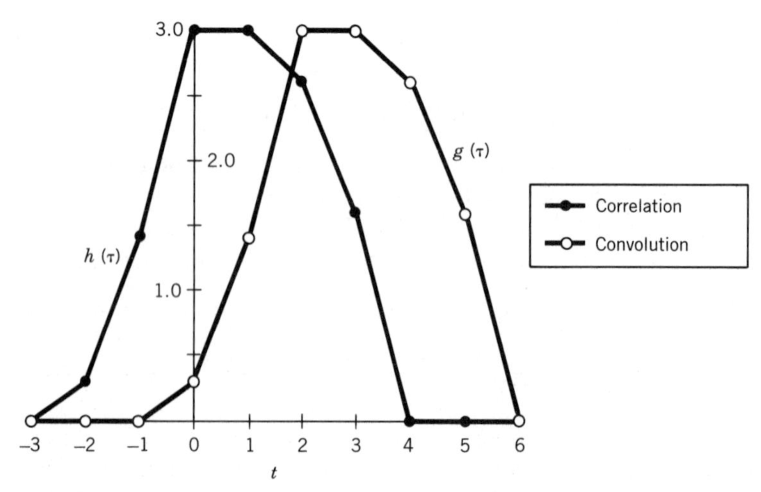

FIGURE 6-17. Convolution and correlation functions for the two functions shown in Figure 6-16.

$$a(t) \otimes \delta(t) = \int_{-\infty}^{\infty} a(t)\delta(\tau - t)\, dt = a(\tau) \tag{6-37}$$

Thus, the convolution of an arbitrary function $a(t)$ with a delta function reproduces the value of the function $a(t)$ at the delta function position. If we move the delta function over $a(t)$, the convolution function produced is identical to the original function. If the function $b(t)$ is allowed to change from the delta function to a rectangular pulse of increasing width, then the resulting convolution becomes an increasingly smoothed version of $a(t)$. The amount of smoothing is directly proportional to the width of the rectangular pulse.

A Fourier transform property allows us to write the Fourier transform of the convolution as the product of the Fourier transforms of the two functions involved in the convolution

$$\mathcal{F}\{a(t) \otimes b(t)\} = \mathcal{F}\left\{\int_{-\infty}^{\infty} a(t)b(\tau - t)\, dt\right\} = A(\omega)B(\omega) \tag{6-38}$$

LINEAR SYSTEM THEORY

Why are we interested in the convolution? It is an important function in the theory of linear systems and we will find it useful to treat optical systems as linear systems. To define a linear system, we use an operational definition. We then use the operational definition to prove that we can characterize a linear system by determining its response to a delta function input. The output of the linear system to an arbitrary input function will be shown to be the convolution of the input function and the delta function response.

To define a linear system, assume that the system is a black box that may contain an optical, electrical, or mechanical system. The black box uniquely maps any input onto an output but not necessarily in a one-to-one manner. We will represent the operation of the black box by the mathematical operator $\mathcal{T}$, which maps the input function $f(t)$ onto the output function $g(t)$

$$\mathcal{T}\{f_1(t)\} \Rightarrow g_1(t), \qquad \mathcal{T}\{f_2(t)\} \Rightarrow g_2(t)$$

The box (system) has the *homogeneous property* if

$$\mathcal{T}\{af_1(t)\} \Rightarrow ag_1(t)$$

It has *linearity* if it obeys the principle of superposition

$$\mathcal{T}\{af_1(t) + bf_2(t)\} \Rightarrow ag_1(t) + bg_2(t)$$

It has *stationarity* or is *shift invariant* if

$$\mathcal{T}\{f_1(t - t_0)\} \Rightarrow g_1(t - t_0)$$

If the box is linear and stationary (invariant), then we will be able to develop a number of useful relationships between the input and output of the system that form the foundation of linear system theory. The relationships are based on the principle of superposition that allows the decomposition of a complex input into a linear combination of simple functions. Theory allows the calculation of the effect of the linear system on the simple functions. The modified versions of the simple functions are then recombined to form the response to the complex input.

The simple functions selected for characterization of a system are the eigenfunctions of the linear, invariant system. These eigenfunctions are exponentials of the form $e^{i\omega t}$. The linear system modifies the phase and amplitude of the eigenfunctions but the eigenfunctions retain their form, i.e., if $f(t) + ig(t)$ is an eigenfunction of the linear system, then the output will be $c_1 f(t) + ic_2 g(t)$. The constants c_1 and c_2 are called eigenvalues of the system. The problem of finding the output of a linear system to a complex input is therefore reduced to a problem of properly decomposing the input into a set of eigenfunctions, then modifying and recombining these eigenfunctions into the output function.

To prove that the exponential $e^{i\omega t}$ is an eigenfunction, we denote the operation of the system on the exponential by

$$\mathcal{T}\left\{e^{i\omega t}\right\} = e(t)$$

Since the system is invariant

$$e(t + t_1) = \mathcal{T}\left\{e^{i\omega(t+t_1)}\right\} = \mathcal{T}\left\{e^{i\omega t}e^{i\omega t_1}\right\}$$

Because the system is homogeneous, this can be written

$$\mathcal{T}\left\{e^{i\omega(t+t_1)}\right\} = \mathcal{T}\left\{e^{i\omega t}\right\}e^{i\omega t_1} = e^{i\omega t_1}e(t)$$

At $t = 0$, we have

$$e(t + t_1)\big|_{t=0} = e(t_1) = e(0)e^{i\omega t_1}$$

but t_1 is arbitrary so we can replace t_1 by t and rewrite this result as

$$e(t) = e(0)e^{i\omega t}$$

The multiplier of the exponent $e(0)$ is a constant, possibly complex, demonstrating that the exponential is an eigenfunction.

When we put an impulse function (a delta function) into the input of the linear system, we obtain

$$\mathcal{T}\{\delta(t)\} \Rightarrow s(t)$$

where $s(t)$ is called the *impulse response* [in mathematics, $s(t)$ is called the *Green's function* and in optics, it is called the *point spread function*]. Because of the assumed properties of linearity and stationarity,

$$\mathcal{T}\{f(t_1)\delta(t - t_1) + f(t_2)\delta(t - t_2)\} \Rightarrow f(t_1)s(t - t_1) + f(t_2)s(t - t_2)$$

where $f(t_1)$ and $f(t_2)$ are eigenfunctions of the linear operation $\mathcal{T}$. For a large set of impulse responses,

$$\mathcal{T}\left\{\sum_{n=1}^{N} f(t_n)\delta(t - t_n)\right\} \Rightarrow \sum_{n=1}^{N} f(t_n)s(t - t_n) \tag{6-39}$$

We can extrapolate the result given by **(6-39)** to a continuous distribution by using the sifting property of the delta function

$$f(t_1) = \int f(t')\delta(t' - t_1)\,dt'$$

to decompose the input function

$$\mathcal{T}\left\{\int f(t')\delta(t-t')\,dt'\right\}$$

We now use the linearity of the system and the fact that $f(t')$ is an eigenfunction of $\mathcal{T}$ to write

$$\mathcal{T}\left\{\int f(t')\delta(t-t')\,dt'\right\} \Rightarrow \int f(t')\mathcal{T}\{\delta(t-t')\}\,dt' \Rightarrow \int f(t')s(t-t')\,dt'$$

The integral

$$\int f(t')s(t-t')\,dt' \tag{6-40}$$

is a convolution integral (sometimes, this integral is called the superposition integral and the result just obtained explains why). Our result demonstrates the fact that a linear system is completely characterized by its response to an impulse. To obtain the output from a linear system for a complex input, we need only convolve the input with the impulse response of the system.

The Fourier transform of $s(t)$ is $S(\omega)$ and is called the *transfer function frequency response*. The frequency spectrum of the system's output is the product of the input spectrum (the Fourier transform of the input function) and the transfer function $S(\omega)F(\omega)$. The output of the system is the Fourier transform of this product, as stated mathematically by **(6-38)**.

Another interpretation of the impulse response $s(t)$ emphasizes its role as a weighting function in the convolution integral **(6-40)**. The impulse response can be viewed as a measure of the ability of the system to remember past events. This is in keeping with the earlier interpretation of the weighting function as a window through which a time average is performed. The window determines how much of the past history of the function can be seen when the time average is performed.

FOURIER TRANSFORMS IN TWO DIMENSIONS

We have limited our discussion to one-dimensional temporal functions but in optics, we will need to perform transforms of functions with two spatial coordinates. We can define a two-dimensional Fourier transform by making a simple extension of the one-dimensional definition **(6-13)**

$$F(\xi, \eta) = \iint_{-\infty}^{\infty} f(x, y)e^{-i(\xi x + \eta y)}\,dx\,dy$$

If $f(x, y)$ is separable in x and y, we can write

$$F(\xi, \eta) = \iint_{-\infty}^{\infty} f(x)g(y)e^{-i\xi x}e^{-i\eta y}\,dx\,dy$$

$$F(\xi, \eta) = \int_{-\infty}^{\infty} f(x)e^{-i\xi x}\,dx\int_{-\infty}^{\infty} g(y)e^{-i\eta y}\,dy \tag{6-41}$$

$$= F(\xi)G(\eta) \tag{6-42}$$

For separable functions, our previous discussions are easily extended to two dimensions; however, performing the integration of the two-dimensional transform can become very difficult if the function is not separable.

In optics, most of the functions we wish to consider have circular symmetry, and it is appropriate to make a change of variables to polar format

$$x = r \cos\theta, \qquad y = r \sin\theta$$

$$\xi = \rho \cos\Theta, \qquad \eta = \rho \sin\Theta$$

In polar coordinate, the circularly symmetric function is not only separable but because it has circular symmetry, it is independent of θ; thus,

$$f(x, y) \Rightarrow f(r, \theta) = f(r)g(\theta) \doteq f(r)$$

$$\mathcal{F}\{f(r, \theta)\} = F(\rho, \Theta) = F(\rho)$$

$$F(\rho, \Theta) = \int_0^{2\pi} d\theta \int_0^{\infty} f(r) e^{-i\rho r(\cos\theta\cos\Theta + \sin\theta\sin\Theta)}\, r\, dr$$

$$= \int_0^{\infty} f(r)\, r\, dr \int_0^{2\pi} e^{-i\rho r\cos(\theta-\Theta)}\, d\theta \tag{6-43}$$

The second integral belongs to a class of functions called the Bessel function defined by the integral

$$\mathbf{J}_n(r\rho) = \int_0^{2\pi} e^{\,i[r\rho\sin(\theta - n\theta)]}\, d\theta$$

The integral in **(6-43)** corresponds to the $n = 0$, zero-order Bessel function. Using this definition, we can write **(6-43)** as

$$F(\rho) = \int_0^{\infty} f(r)\, \mathbf{J}_0(r\rho) r\, dr \tag{6-44}$$

This transform is called the Fourier–Bessel transform or the *Hankel zero-order transform*. We now apply **(6-44)** to a simple circular symmetric function, sometimes called the *top-hat* function

$$f(x, y) = \begin{Bmatrix} 1, & \sqrt{x^2 + y^2} \le 1 \\ 0, & \text{all other } x, y \end{Bmatrix} = f(r, \theta) = f(r) = \begin{cases} 1, & r \le 1 \\ 0, & \text{all other } r \end{cases}$$

The transform of the top-hat function is

$$F(\rho) = \int_0^1 \mathbf{J}_0(r\rho) r\, dr$$

We use the identity

$$x\mathbf{J}_1(x) = \int_0^x \alpha \mathbf{J}_0(\alpha)\, d\alpha$$

to obtain

$$F(\rho) = \frac{\mathbf{J}_1(\rho)}{\rho} \tag{6-45}$$

The Bessel functions are important in optics because most optical systems have circular symmetry. A whole family of Bessel functions exist, and as in the case for sines and cosines, they may be calculated using a series expansion. The series expansion is

$$\mathbf{J}_n(\rho) = \sum_{k=0}^{\infty} \frac{(-1)^k \rho^{n+2k}}{2^{n+2k} k!(n+k)!} \tag{6-46}$$

The values of Bessel functions have been tabulated and are found in most collections of mathematical tables. We will discuss the Bessel function of order 1 when we discuss diffraction by a circular aperture.

SUMMARY

In this chapter, we have introduced a number of mathematical tools that will be needed to interpret the optical observations presented in later chapters.

Fourier Series To describe a periodic function $f(t)$, the series

$$f(t) = \frac{a_0}{2} + \sum_{\ell=1}^{\infty} a_\ell \cos \ell\omega t + \sum_{\ell=1}^{\infty} b_\ell \sin \ell\omega t$$

can be used. The coefficients of the two summations are obtained by carrying out the integrals

$$a_n = \frac{\omega}{\pi} \int_{-\pi/\omega}^{\pi/\omega} f(t) \cos n\omega t \, dt$$

$$b_n = \frac{\omega}{\pi} \int_{-\pi/\omega}^{\pi/\omega} f(t) \sin n\omega t \, dt$$

Fourier Transform A nonperiodic function can be represented by the integral

$$\mathcal{F}\{f(t)\} = F(\omega) = \int_{-\infty}^{\infty} f(\tau) e^{-i\omega\tau} \, d\tau$$

which transforms $f(t)$ from a temporal representation to the frequency representation $F(\omega)$. The inverse transform can also be performed

$$\mathcal{F}^{-1}\{F(\omega)\} = f(t) = \frac{1}{2\pi} \int_{-\infty}^{\infty} F(\omega) e^{i\omega t} \, d\omega$$

Correlation The correlation function is a useful integral for comparing the similarity between two functions

$$h(\tau) = a(t) \oplus b(t) = \int_{-\infty}^{\infty} a(t) b^*(t - \tau) \, dt$$

It can be thought of as the calculation of the area of overlap of two functions as one of the functions slides over the other. We suggested that optical interference was related to this mathematical function.

The Fourier transform provides another way of calculating the correlation function. The correlation function is the Fourier transform of the product of the Fourier transforms of the two functions to be correlated

$$h(\tau) = a(t) \oplus b(t) = \mathcal{F}\{A(\omega) B^*(\omega)\}$$

Convolution A second integral of use in linear system theory is the convolution integral. It is sometimes called the smoothing operation because if the function $a(t)$ has any sharp peaks, they will be rounded, or if $a(t)$ has any steep slopes, they will be reduced. The amount of smoothing depends on the nature of $a(t)$ and $b(t)$. The convolution integral of $a(t)$ and $b(t)$ is defined as

$$g(\tau) = a(t) \otimes b(t) = \int_{-\infty}^{\infty} a(t)b(\tau - t)\, dt$$

As was the case with the correlation, the Fourier transform can be used in the calculation of the convolution. The Fourier transform of the convolution is the product of the Fourier transforms of the two functions to be convolved

$$\mathcal{F}\left\{a(t) \otimes b(t)\right\} = A(\omega)B(\omega)$$

Linear Systems In the discussion of linear systems, the delta function was found to be useful in the description of the response of a linear system. The delta function is defined by the integral

$$\int_{-\infty}^{\infty} f(t)\delta(t - t_0)\, dt = f(t_0)$$

If the delta function is the input function to the linear system, then the output is $s(t)$, the impulse response of the linear system. The impulse response can be used in the convolution integral to predict the output of the linear system. For an arbitrary input $f(t)$, the output of a linear system is given by

$$\int f(t')s(t - t')dt'$$

Again, the Fourier transform can be used to calculate this information. The Fourier transform of the impulse response is called the transfer function of the linear system $S(\omega)$. If $F(\omega)$ is the frequency spectrum (Fourier transform) of the input function $f(t)$, then the output frequency spectrum of the linear system is given by $S(\omega)\bullet F(\omega)$.

Two-dimensional Fourier Transforms If the two-dimensional function under study $h(x, y)$ is separable in its dependence on the spatial coordinates $h(x, y) = f(x)g(y)$, then the two-dimensional Fourier transform is

$$F(\xi, \eta) = \int_{-\infty}^{\infty} f(x)e^{-i\xi x}\, dx \int_{-\infty}^{\infty} g(y)e^{-i\eta y}\, dy$$

For circularly symmetric functions,

$$f(x, y) \Rightarrow f(r, \theta) = f(r)g(\theta) = f(r)$$

the Fourier transform is

$$\mathcal{F}\left\{f(r, \theta)\right\} = F(\rho, \Theta) = F(\rho)$$

and is given by the Hankel transform

$$F(\rho) = \int_{0}^{\infty} f(r)\, \mathbf{J}_0(r\rho)r\, dr$$

Sampling Theorem The sampling theorem states that if the Fourier transform $F(\omega)$ of the function $f(t)$ is zero above some cut-off frequency

$$F(\omega) = 0, \qquad \omega \geq \omega_c$$

then $f(t)$ is uniquely determined by the values of $f(t)$ measured at a set of times calculated using the formula

$$t = nt_0 = n\frac{\pi}{\omega_c}$$

This means that we must take two sample points in every period t_0, where $1/t_0$ is the highest frequency contained in the function $f(t)$.

PROBLEMS

6-1. Prove the linearity theorem of Fourier transforms

$$\mathcal{F}\{ag(x) + bh(x)\} = a\mathcal{F}\{g(x)\} + b\mathcal{F}\{h(x)\}$$
$$= aG(k) + bH(k)$$

where

$$\mathcal{F}\{g(x)\} = \int_{-\infty}^{\infty} g(x)e^{-ikx}\,dx = G(k)$$

6-2. Prove the similarity theorem of Fourier transforms if

$$\mathcal{F}\{g(x)\} = G(k)$$

then

$$\mathcal{F}\{g(ax)\} = \frac{1}{|a|}G\left(\frac{k}{a}\right)$$

6-3. Prove the shift theorem of Fourier transforms if

$$\mathcal{F}\{g(x)\} = G(k)$$

then

$$\mathcal{F}\{g(x-a)\} = G(k)e^{-ika}$$

6-4. Prove the convolution theorem of Fourier transforms if

$$\mathcal{F}\{g(x)\} = G(k)$$

and

$$\mathcal{F}\{h(x)\} = H(k)$$

then

$$\mathcal{F}\left\{\int_{-\infty}^{\infty} g(\xi)h(x-\xi)\,d\xi\right\} = G(k)H(k)$$

6-5. Prove the autocorrelation theorem of Fourier transforms

$$\mathcal{F}\left\{\int_{-\infty}^{\infty} g(\xi)g^*(\xi-x)\,d\xi\right\} = |G(k)|^2$$

6-6. Find the Fourier series for the function $f(x) = x^2$ over the range $-a \leq x \leq a$.

6-7. Find the Fourier series representation of the periodic function

$$f(t) = \begin{cases} 1, & 0 < t < \dfrac{T}{2} \\ -1, & \dfrac{T}{2} < t < T \end{cases}$$

6-8. Find the Fourier transform of

$$f(x) = \begin{cases} e^{-ax}, & x > 0 \\ 0, & x < 0 \end{cases}$$

6-9. Compare the convolution and correlation of the following two functions:

$$a(t) = \begin{cases} 1, & 0 \le t \le 1 \\ 0, & 0 > t > 1 \end{cases}$$

$$b(t) = \begin{cases} \delta(t) - e^{-t}, & t \ge 0 \\ 0, & t < 0 \end{cases}$$

6-10. We can perform repeated convolutions, and as we do, the final convolution will tend toward a Gaussian function. To demonstrate this fact, calculate the convolutions of

$$f_n(t) = \begin{cases} 1, & |\tau| \le \tau_n \\ 0, & \text{all other } t \end{cases}$$

for

$$\tau_1 = 1, \quad \tau_2 = 2, \quad \tau_3 = \tfrac{1}{2}, \quad \tau_4 = \tfrac{3}{2}$$

Plot the results for each convolution

$$f_1 \otimes f_2, \quad f_1 \otimes f_2 \otimes f_3, \quad \text{and} \quad f_1 \otimes f_2 \otimes f_3 \otimes f_4$$

6-11. Assume that the function

$$f(t) = \begin{cases} t, & -1 \le t \le 1 \\ 0, & \text{all other } t \end{cases}$$

is periodic and find the Fourier series over the interval $-1 < t < 1$.

6-12. Evaluate the infinite series derived in Problem 6-11 when $t = \pi/3$.

6-13. Using the results of Problem 6-11, write a computer program to verify Figure 6-13.

6-14. Show that

$$\int_{-\pi/\omega}^{\pi/\omega} \cos(m - n)\omega t \, dt = 0$$

unless $m = n$.

6-15. Show that

$$\int_{-\pi/\omega}^{\pi/\omega} f(t)\, dt = \begin{cases} 0, & f(t) = -f(-t) \\ \text{nonzero}, & f(t) = f(-t) \end{cases}$$

6-16. Assume that the Fourier transform of $f(t)$ is $F(\omega)$. What is the Fourier transform of $f(t + t') + f(t - t')$?

6-17. Assume that the Fourier transform of $f(t)$ is $F(\omega)$. What is the Fourier transform of $f(t)\ \sin(\omega'/2)t$?

6-18. The Dirac delta function is the unit operator for convolutions, just as zero is for addition and one is for multiplication. Prove that this statement is true.

6-19. Find the Fourier transform of the function

$$f(t) = \begin{cases} 1 + \cos \omega_0 t, & -\dfrac{T}{2} \le t \le \dfrac{T}{2} \\ 0, & \text{all other } t \end{cases}$$

6-20. Use the shifting property of the Fourier transform **(6A-5)** to rewrite **(6-29)** into the form shown in **(6-30)**.

6-21. Use **(6-34)** to find the Fourier transform of

$$f(t) = \begin{cases} 1 - \left|\dfrac{t}{t_o}\right| & |t| \le t_0 \\ 0, & |t| > t_0 \end{cases}$$

Appendix 6-A

FOURIER TRANSFORM PROPERTIES

Some of the important properties of the Fourier transform are given below. Their proof is left to the reader as problems. We use the following definitions:

$$F(\omega) = \mathcal{F}\{f(t)\} = \int_{-\infty}^{\infty} f(\tau)e^{-i\omega\tau}\, d\tau \tag{6A-1}$$

$$G(\omega) = \mathcal{F}\{g(t)\} = \int_{-\infty}^{\infty} g(\tau)e^{-i\omega\tau}\, d\tau \tag{6A-2}$$

We also let a and b be constants.

Linearity

$$\mathcal{F}\{af(t) + bg(t)\} = aF(\omega) + bG(\omega) \tag{6A-3}$$

Scaling

$$\mathcal{F}\{f(at)\} = \frac{1}{|a|} F\left(\frac{\omega}{a}\right) \tag{6A-4}$$

Shifting

$$\mathcal{F}\{f(t - t_0)\} = e^{-i\omega t_0}F(\omega) \tag{6A-5}$$

Conjugation

$$\mathcal{F}\{f^*(t)\} = F^*(-\omega) \tag{6A-6}$$

Differentiation

$$\mathcal{F}\left\{\frac{d^n f(t)}{dt^n}\right\} = (i\omega)^n F(\omega) \tag{6A-7}$$

Convolution

$$\mathcal{F}\left\{\int_{-\infty}^{\infty} f(\tau)g(t - \tau)\, d\tau\right\} = F(\omega)G(\omega) \tag{6A-8}$$

Parseval's Theorem

$$\int_{-\infty}^{\infty} f(t)g^*(t)dt = \int_{-\infty}^{\infty} F(\omega)G(\omega)\, d\omega \tag{6A-9}$$

Correlation

$$\mathcal{F}\left\{\int_{-\infty}^{\infty} f(t)g^*(t-\tau)\, dt\right\} = F(\omega)G^*(\omega) \tag{6A-10}$$

A few common Fourier transform pairs are listed below. Some of these have been derived in the chapter and others are the subject of problems at the end of Chapter 6. See A. Papoulis[26] for a more complete listing, as well as more details on the subject of Fourier transforms.

$f(t)$	$F(\omega)$	
$\begin{cases} 1, & -t_0 \le t \le t_0 \\ 0, & \text{all other } t \end{cases}$	$\dfrac{\sin \omega t_0}{\omega t_0}$	(6A-11)
$\begin{cases} 1 - \left\lvert \frac{t}{t_0} \right\rvert, & -t_0 \le t \le t_0 \\ 0, & \text{all other } t \end{cases}$	$\operatorname{sinc}^2 \omega t_0$	(6A-12)
e^{-t}	$\dfrac{1}{1+\omega^2}$	(6A-13)
comb t	comb ω	(6A-14)

CONVOLUTION PROPERTIES

Commutative

$$f(x) \otimes g(x) = g(x) \otimes f(x) \tag{6A-15}$$

Distributive

$$[af(x) + bg(x)] \otimes w(x) = a[f(x) \otimes w(x)] + b[g(x) \otimes w(x)] \tag{6A-16}$$

Associative

$$[f(x) \otimes g(x)] \otimes w(x) = f(x) \otimes [g(x) \otimes w(x)] \tag{6A-17}$$

Identity

$$f(x) \otimes \delta(x) = f(x) \tag{6A-18}$$

Shift-Invariant

$$f(x - x_0) \otimes g(x) = f(x) \otimes g(x - x_0) \tag{6A-19}$$

Dispersion

In the derivation of the propagation equation from Maxwell's equations, the material constants σ, μ, and ϵ were assumed to be independent of frequency. The material constants σ, μ, and ϵ are compared for dielectrics in Table 7-1 and for metals in Table 7-2. The measurements were made near dc (usually around 60 Hz) and at visible light frequencies (usually at the sodium D line). The dielectrics listed in Table 7-1 do not show the agreement between the index n and the dielectric constant $\sqrt{\epsilon/\epsilon_0}$ displayed in a similar list in Table 2-2.

In the discussion of propagation in a conducting medium (Chapter 2), we found from **(2-63)** that

$$2n^2\kappa = c^2\frac{\mu\sigma}{\omega}$$

This equality is not supported by the experimentally obtained values in Table 7-2. The disagreement is obvious when comparing the last two columns of Table 7-2.

The disagreement between the measurements made at low frequencies and at light frequencies arises because the electrons and nuclei cannot respond instantaneously to an applied electromagnetic field at light frequencies. The result of the sluggish response of atoms and molecules to light waves is *dispersion*, i.e., the variation of the index of refraction with frequency. A proper treatment of the subject requires the use of quantum mechanics, but we can obtain an equation of the same form by using a classical model of a molecule. The classical model is useful in visualizing the optical response of materials but cannot be used to calculate some experimentally observable parameters.

In this chapter, the propagation of a wave on a stiff string will be discussed to develop a familiarity with waves displaying frequency-dependent propagation parameters. Having demonstrated that such waves can exist using this simple model, we will discuss the effect of dispersion on the propagation of two closely spaced frequencies. It will be shown that the two waves,

TABLE 7.1 Static Dielectric Constant and Index of Refraction

Dielectric Material	Yellow light, n	static, $(\epsilon/\epsilon_0)^{1/2}$
CH_3OH (methyl alcohol)	1.34	5.7
C_2H_3OH (ethyl alcohol)	1.36	5.0
Water	1.33	9.0
Fused silica	1.458	1.94
NaCl	1.50	2.37

TABLE 7.2 Conductivity and Optical Attenuation

Metal	n	$n\kappa$	$n^2\kappa$	$c^2\mu\sigma/\omega$
Sodium	0.044	2.42	0.213	867
Silver	0.2	3.44	1.38	2564
Cadmium	1.130	5.01	11.3	490
Aluminum	1.44	5.23	15.1	1335
Gold	0.47	2.83	2.7	1546
Copper	0.62	2.57	3.19	2228
Nickel	1.58	3.42	10.8	520
Lead	2.01	3.48	14.0	192
Iron	1.51	1.63	4.9	415

propagating at slightly different velocities, will alternately undergo constructive and destructive interference, exhibiting the phenomena of beats. The results derived for two frequencies will then be extended to a Gaussian distribution of frequencies and a discussion of phase and group velocities of a wave will be given.

To explain the origin of material dispersion, a simple classical model will be developed. The model will first treat a system of charges that do not interact. The complexity of the theory will be increased by adding an unspecified force between two charges so that they experience a restoring force when acted on by an electromagnetic field. The final addition to the theory will allow pairs of charges to interact. The objective is to obtain an equation for the dielectric constant (or equivalently, the index of refraction) from the model. The resulting theory will explain the origin of a complex dielectric constant and the observation of resonant absorption. The theory derived is quite simplistic but is, in general, supported when compared to optical experiments.

The dispersion theory predicts that the index of refraction will decrease with wavelength, except in the region of a resonant absorption where it will increase with wavelength. The increase of the index of refraction with an increase in wavelength is called *anomalous dispersion*.

Anomalous dispersion can yield both a phase and a group velocity greater than the speed of light. The final topic of this chapter will discuss the theory developed to explain this apparent contradiction of the theory of relativity.

The first observations of dispersion were recorded by **Marcus (1648)** and **Grimaldi (1665)**. Newton attempted to explain his observations of dispersion, using a glass prism, by his particle model but was not successful. **Augustin Louis Cauchy (1789–1857),** the French mathematician famous for his contributions to calculus and complex variables, derived the dispersion relationship that bears his name

$$n = a + \frac{b}{\lambda^2} + \frac{d}{\lambda^4} + \frac{f}{\lambda^6} + \cdots$$

using the mechanistic concept of an elastic solid. Later observations of anomalous dispersion by **P. Le Roux (1832–1907)** pointed out the limited applicability of Cauchy's equation.

Maxwell suggested that the atoms and molecules may act as oscillators and possess natural frequencies, but he did not follow up his idea. **Sellmeier (1871)** independently used the idea of natural resonances to derive the Sellmeier relation

$$n^2 = 1 + \frac{a\lambda^2}{\lambda^2 - \lambda_0^2}$$

to explain simple cases of anomalous dispersion.

A more rigorous classical theory was derived first by **Paul Karl Ludwig Drude (1863–1906)** and **W. Voigt (1850–1919)**. The theory was extended to its present form by **Hendrik Antoon Lorentz (1853–1928)**, a Dutch physicist who shared the Nobel Prize with Zeeman for the study of the effects of magnetism on light. Before we look at the theory of Lorentz, we will examine a string model that includes dispersion.

STIFF STRINGS

The simple wave equation, discussed in Chapter 1 for a wave on a string, was derived by assuming that a pulse could propagate along the string without changing shape. The discussion of Fourier transforms in Chapter 6 leads us to view this pulse as a distribution of frequencies. By assuming that the pulse's shape does not change, we are effectively stating that the frequency distribution must not change in either amplitude or relative phase as the wave propagates along the string. All of the frequencies must propagate at the same velocity if they are to retain the same relative phase. Thus, our discussion of waves in Chapter 1 was limited to nondispersive waves. We can prove that this is true by substituting a harmonic wave into the wave equation. This will allow the derivation of the *dispersion equation*. The dispersion equation relates the phase velocity of a wave to its frequency and propagation constant for the medium modeled by the wave equation.

The wave equation is given by **(1-6)** as

$$\frac{\partial^2 f}{\partial x^2} - \left(\frac{1}{c^2}\right)\frac{\partial^2 f}{\partial t^2} = 0$$

Using the expressions **(1-21)** and **(1-22)** to rewrite the simple wave equation, we obtain

$$(-ik)^2 f = \left(\frac{i\omega}{c}\right)^2 f$$

leading to the dispersion equation

$$c^2 = \left(\frac{\omega}{k}\right)^2 \tag{7-1}$$

For the simple wave equation and a harmonic wave, we find that the phase velocity ω/k is a constant equal to c, i.e., the medium is nondispersive and all frequencies travel at the same velocity.

We now will introduce a model of wave propagation that includes dispersion. The model selected has additional interest in that it has a frequency region for which no wave can propagate and only evanescent waves exist. The simple model used in Chapter 1 for the propagation of a wave on a string did not recognize any force other than the tension in the string. A real guitar string is stiff and tends to straighten out even when it is not strung. We will extend the string model to recognize that the string is stiff. Rather than introduce a force dependent on the length of the string, as might be the case for a real guitar string, we will assume that the vibrating string experiences a restoring force proportional only to the string's displacement, i.e., a simple elastic resorting force. The wave equation for this stiff string is

$$\frac{\partial^2 f}{\partial x^2} - sf = \frac{1}{c^2}\frac{\partial^2 f}{\partial t^2} \tag{7-2}$$

where s is the force constant. (In quantum mechanics, this equation is called the Klein–Gordon equation.) By substituting a harmonic wave into **(7-2)**, we obtain

$$(-ik)^2 f - sf = \left(\frac{i\omega}{c}\right)^2 f$$

leading to the dispersion equation

$$\frac{\omega}{k} = c\sqrt{1 + \frac{s}{k^2}} \tag{7-3}$$

The wave equation is now dispersive; the phase velocity ω/k depends on k, i.e., on the wavelength. Figure 7-1 shows a plot of the two dispersion equations, **(7-1)** and **(7-3)**.

The stiff string has a minimum, or cut-off frequency $\omega_c = c\sqrt{s}$, below which there are no propagating constants associated with the frequencies. We can discover the significance of the cut-off frequency by examining **(7-3)**. Below ω_c the propagation constant k is imaginary

$$k = \frac{i}{c}\sqrt{\omega_c^2 - \omega^2}, \qquad \omega_c > \omega \tag{7-4}$$

The significance of a complex propagation constant was discovered in Chapter 1 during the study of the propagation of a wave in a medium with loss.

By attaching a perfect string to a stiff string, total reflection could be observed at the connection point when the wave propagating in the perfect string was incident onto the connection point and had a frequency below the cut-off frequency of the stiff string. If the stiff string consisted of a short segment joined at either end by perfect strings, then the wave propagating in the first perfect string toward the stiff string segment could tunnel through the barrier imposed by the stiff string segment and continue propagating, at a reduced amplitude, along the second perfect string.

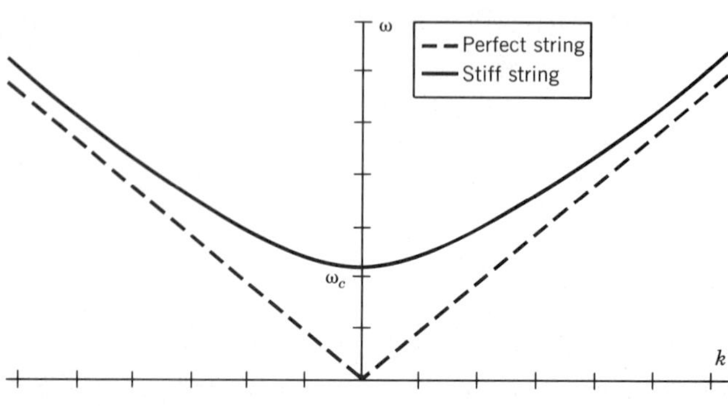

Figure 7-1. Dispersion curve for a nondispersive wave (the perfect string) and a dispersive wave (the stiff string). There is a range of frequencies $\omega < \omega_c = cs^{1/2}$, for which there are no real values of k for the stiff string.

Here, the propagation constant has no real component. To discover the significance of a purely imaginary propagation constant, we define

$$k = -i\kappa$$

and replace k in the complex representation of the one-dimensional wave by $-i\kappa$. The result is

$$f(x,\ t) = Ae^{i(\omega t + i\kappa x + \phi)} = Ae^{-\kappa x}e^{i(\omega t + \phi)} \tag{7-5}$$

The actual wave is the real part of **(7-5)**

$$f(x,\ t) = Ae^{-\kappa x}\cos(\omega t + \phi)$$

This is a nonpropagating, evanescent wave whose amplitude decreases exponentially with increasing values of x. Thus, if the propagation constant is purely imaginary, a wave cannot propagate in the medium.

GROUP VELOCITY

In the previous section, we pointed out that when a distribution of frequencies needed to describe a pulse shape propagated at different velocities, then the shape of the pulse, described by the distribution, would change as it propagated. Fourier theory will here be used to prove that statement.

We will first evaluate the propagation of two waves with slightly different frequencies. We will add the two waves to obtain, as the resultant wave, an amplitude-modulated wave. We will use this wave to define the group velocity that describes the propagation of the amplitude modulation. The procedure will then be extended to first, the addition of $2n$ waves and finally, a continuous Gaussian distribution of frequencies. Through the description of combining waves, first for a few discrete frequencies and then for a continuous distribution of frequencies, we will demonstrate how an arbitrary waveform can be produced: an immediate application of the Fourier theory of Chapter 6.

The initial analysis will be in terms of the temporal properties of the wave because it can easily be related to the material of Chapter 6. After having developed an understanding of the temporal properties, the discussion will be extended to an evaluation of the spatial properties of the wave, for it is the spatial properties of a wave that will allow a pulse shape to be followed through a dispersive medium. The details of the mathematics will not be emphasized but rather the emphasis will be on the development of a qualitative understanding of the effect of dispersion on the shape of the propagating pulse.

Assume two copropagating waves of equal amplitude but different frequencies and propagating constants

$$\omega_1 = \omega + \Delta\omega, \qquad k_1 = k + \Delta k$$

$$\omega_2 = \omega - \Delta\omega, \qquad k_2 = k - \Delta k$$

The resulting wave is obtained by simple algebraic addition of the two waves

$$f(x,\ t) = A\cos(\omega_1 t - k_1 x) + A\cos(\omega_2 t - k_2 x)$$

$$f(x,\ t) = A\cos[(\omega t - kx) + (\Delta\omega t - \Delta kx)] + A\cos[(\omega t - kx) - (\Delta\omega t - \Delta kx)]$$

If we use the trig identity

$$\cos(\alpha \pm \beta) = \cos\alpha\cos\beta \mp \sin\alpha\sin\beta$$

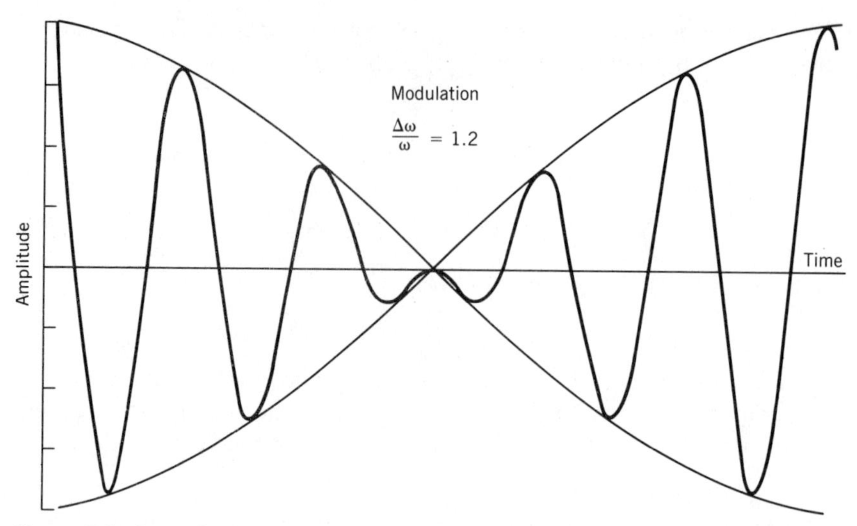

Figure 7-2. A traveling wave group with a ratio of the carrier frequency to the modulation frequency of 1.2. We could also identify this wave as an amplitude-modulated wave or as an example of beating between two frequencies. All of these descriptions are equivalent.

the resultant wave becomes

$$f(x, t) = 2A \cos(\Delta\omega t - \Delta k x) \cos(\omega t - kx)$$

This wave is shown in Figure 7-2.

Because the two waves are at slightly different frequencies, their relative phase difference is a function of time. The time-dependent phase results in the two waves alternately adding constructively and destructively, creating an amplitude modulation that is a periodic series of maxima due to constructive interference. We will call these maxima *groups*.

The modulation is on a carrier wave of frequency

$$\omega = \tfrac{1}{2}(\omega_1 + \omega_2)$$

and propagation constant k traveling at a velocity

$$v = \frac{\omega}{k} \tag{7-6}$$

This velocity we recognize as the phase velocity **(1-14)**.

The amplitude modulation of the carrier wave is a wave of frequency $\Delta\omega$, called the *beat frequency*, and propagation constant Δk propagating at a velocity of

$$u = \frac{\Delta\omega}{\Delta k} \tag{7-7}$$

This velocity is called the *group velocity*. In a nondispersive medium, the group and phase velocity are identical.

The simple example of the addition of two harmonic waves of different frequencies can now be extended to $2n$ waves. If the waves are uniformly distributed from frequency ω to frequency $2n\omega$, the temporal behavior of the wave is found to be

$$f(0, t) = \frac{m(0, t)}{2}(e^{i\omega_0 t} + e^{-i\omega_0 t}) \tag{7-8}$$

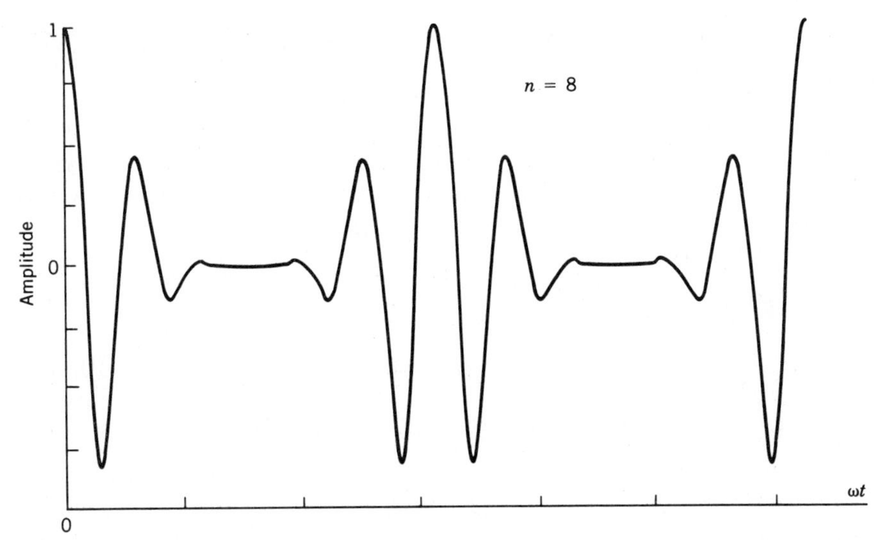

Figure 7-3. The resultant waveform given by **(7-8)** when $n = 8$. In this plot, the frequencies were distributed equally about ω_0, each separated by $0.1\ \omega_0$.

As an example, the result of adding $2n = 16$ waves of frequency $m\omega$, where $1 \leq m \leq 16$, is shown in Figure 7-3. The carrier's frequency is $\omega_0 = n\omega = 8\omega$ and the modulation envelope is proportional to

$$m(0,\ t) \propto \cos^4(\omega/2)t$$

We can use the Fourier series to describe $f(0,\ t)$ of **(7-8)**

$$\begin{aligned}\alpha_\ell &= \frac{\omega'}{4\pi} \cdot \int_{-\infty}^{\infty} m(0,\ t)(e^{in\omega t} + e^{-in\omega t})e^{-i\ell\omega t}\, dt \\ &= \frac{\omega'}{4\pi}\left\{ \int_{-\infty}^{\infty} m(0,\ t)\exp[-i(\ell - n)\omega t]\, dt \right. \\ &\qquad \left. + \int_{-\infty}^{\infty} m(0,\ t)\exp[-i(\ell + n)\omega t]\, dt \right\}\end{aligned} \tag{7-9}$$

where ω' is the average modulation frequency. The spectrum of the modulated wave consists of two copies of the spectrum of $m(0,\ t)$, one centered a distance $n\omega$ above and one centered a distance $n\omega$ below zero frequency.

Although the temporal behavior of these waves is interesting, the objective of this discussion is to learn what happens to a complex waveform as it propagates in a dispersive medium. The complex waveform can be analyzed as a harmonic carrier wave modulated in both space and time by a modulation function $m(x,\ t)$. To see the effects of dispersion on the spatial properties of the modulation, the modulation must be described as a function of the propagation constant rather than frequency.

By analogy to **(6-14)** and the discussion of the temporal properties of a modulated wave, the x coordinate and the propagation constant k can be used to describe a modulated wave as

$$f(x,\ t) = \int M(k)e^{i(\omega t - kx)}\, dk \tag{7-10}$$

The modulated waves used as examples have all involved periodic amplitude modulation. This procedure can be extended to a nonperiodic modulation; an example is found in Chapter 6. The modulation envelope $m(0,\ t)$ is equal to **(6-16)** and the waveform $f(0,\ t)$ is shown in Figure 6-6. The frequency spectrum for the pulse modulation, shown in Figure 6-7, exhibits the same characteristics seen in **(7-9)**, i.e., two copies of the spectrum of $m(0,\ t)$ centered on the carrier frequency and the negative carrier frequency.

where $M(k)$ is the Fourier transform of $m(x, 0)$, the spatial dependence of the modulation. The dispersion equation can be used to eliminate ω from **(7-10)**, allowing the modulated wave to be written in terms of a harmonic carrier wave with a spatially dependent modulation. The resulting expression will be used to follow a Gaussian wave through a medium with a simple dispersion relation.

The dispersion equation for the propagation medium can be written as a Taylor series, treating k as the independent variable

$$w(k) = \omega_0 + \left(\frac{d\omega}{dk}\right)_{k=k_0}(k-k_0) + \left(\frac{d^2\omega}{dk^2}\right)_{k=k_0}(k-k_0)^2 + \cdots$$

This expansion of the dispersion relation can be used to rewrite **(7-10)** as

$$f(x, t) = e^{i(\omega_0 t - k_0 x)} \int M(k)e^{i(k-k_0)(ut-x)} \exp\left[i(k-k_0)^2 wt\right] dk \qquad \text{(7-11)}$$

where we have made use of the following definitions to simplify the expression:

$$u = \left(\frac{d\omega}{dk}\right)_{k_0}, \qquad w = \left(\frac{d^2\omega}{dk^2}\right)_{k_0}$$

These definitions are not arbitrarily selected; if $w = 0$, then the surface of constant phase $ut - x$ propagates with the velocity

$$u = \left(\frac{d\omega}{dk}\right)_{k_0}$$

Thus, u is the group velocity, agreeing with the definition used for two waves of different frequencies **(7-7)**. To display the relationship between group and phase velocity in terms of spatial parameters, the definition of group velocity is rewritten as

$$u = v - \lambda\left(\frac{dv}{d\lambda}\right) \qquad \text{(7-12)}$$

To obtain a physical interpretation of **(7-11)**, a Gaussian wave **(6-20)**

$$f(x, t) = A\sqrt{\frac{\pi}{\alpha}}e^{-(x^2/4\alpha)} \cos(\omega_0 t - k_0 x)$$

will be followed as it propagates through a dispersive medium. In Chapter 6, the Fourier transform of a Gaussian wave from a temporal to a frequency representation was derived. By analogy, the Fourier transform from a spatial representation (x variable) to a spatial frequency representation (k variable) is

$$M(k) = \int_{-\infty}^{\infty} f(x, 0)e^{-ikx}\, dx$$

$$= A\pi\left[e^{-\alpha(k-k_0)^2} + e^{-\alpha(k+k_0)^2}\right]$$

We will neglect the component $e^{-\alpha(k+k_0)^2}$ and only follow the positive spatial frequency components.

Inserting the spatial transform into **(7-11)** yields

$$f(x, t) = A\pi e^{i(\omega_0 t - k_0 x)} \int_{-\infty}^{\infty} e^{i(ut-x)(k-k_0)}e^{-i(\alpha - iwt)(k-k_0)^2}\, d(k-k_0)$$

To perform this integration, we observe that the integral is of the form

$$\int e^{ax-bx^2}\,dx = \sqrt{\frac{\pi}{b}}\,e^{(a^2/4b)}$$

where

$$a = i(ut - x) \qquad \text{and} \qquad b = (\alpha - iwt)$$

We can therefore write

$$f(x, t) = A\pi e^{i(\omega_0 t - k_0 x)}\sqrt{\frac{\pi}{\alpha - iwt}}\exp[-(ut - x)^2/4(\alpha - iwt)]$$

The second exponent contains a complex quantity in the denominator that needs to be removed to allow the separation of the real and imaginary parts of the waveform

$$f(x, t) = A\sqrt{\frac{\pi^3(\alpha + iwt)}{\alpha^2 + (wt)^2}}$$
$$\exp[-(ut - x)^2(\alpha + iwt)\,/\,4(\alpha^2 + w^2t^2)]\,\exp[i(\omega_0 t - k_0 x)]$$

The imaginary part of this function is a phase term that varies slowly with time and is not of interest in a discussion of the shape of the waveform. The real part of the function describes the envelope of the waveform, which is a Gaussian.

If $w = 0$, then the medium is nondispersive and

$$f(x, t) = A\sqrt{\frac{\pi^3}{\alpha}}\exp\left[-(ut - x)^2\,/\,4\alpha\right]\cos(\omega_0 t - k_0 x)$$

This is a Gaussian pulse moving at velocity u. The half-width $\sqrt{\alpha}$ of the Gaussian pulse is independent of time; the pulse retains its Gaussian shape and its width is unchanged as it propagates in the nondispersive media. This is the propagation character assumed for the pulse in Chapter 1.

If $w \neq 0$, then the width of the Gaussian pulse is

$$\sqrt{\frac{\alpha^2 + w^2t^2}{\alpha}}$$

This width increases with time. If the medium has a large enough dispersion, the pulse will deform rapidly and the concept of a group, propagating with a group velocity u, will lose its physical significance.

To further evaluate the propagation of the Gaussian pulse, we will assume that the dispersion is given by

$$w(k) = \omega_p\left(1 + \frac{\mathbf{a}^2k^2}{2}\right) \tag{7-13}$$

where ω_p is a constant with units of frequency and $\mathbf{a}$ is a constant with units of length. The two parameters from the Taylor expansion of the dispersion relation are

$$u = \omega_p\mathbf{a}^2k_0, \qquad w = \omega_p\mathbf{a}^2$$

If the pulse has propagated in the dispersive medium a long distance, i.e., if the time is large, the width of the Gaussian pulse is

$$\sqrt{\alpha + \frac{(\omega_p \mathbf{a}^2 t)^2}{\alpha}} \approx \frac{\omega_p \mathbf{a}^2 t}{\sqrt{\alpha}}$$

The width changes faster for a narrow pulse (small α) than for a broad pulse.

Through the use of an assumed dispersion relationship, we have demonstrated that the effect on a light pulse by a dispersive medium is a spatial broadening of the pulse as it propagates. An analogous derivation can be used to demonstrate that there will also be a temporal broadening of a pulse as it propagates.

The dispersive medium is one that does not respond in the same way to all frequencies. As we try to propagate narrower and narrower pulses through a dispersive medium, Fourier theory states that the medium must pass a wider and wider range of frequencies (Figure 6-7 displays the spectral content of pulses differing in width by a factor of 10.) In a later section, a model will be formulated to describe the response of a dispersive medium based on a treatment of molecules as classical oscillators driven by electromagnetic radiation. On a microscopic scale, the oscillators absorb and reradiate the electromagnetic wave and this process modifies the propagation velocity of the wave. The oscillators have resonant frequencies associated with the binding forces in the molecules. Near and below resonance, the oscillators absorb and reemit radiation, whereas far above the resonant frequency, the oscillator will not absorb energy from the radiation. The absorption and reemission process slows the propagation of the low-frequency components. The wider the range of frequencies contained in a pulse, the more noticeable is the failure of the oscillators to respond to the high-frequency components. This failure is exhibited by a rapid loss of the high-frequency components and a resulting broadening of the pulse. The high frequencies, those far above the oscillator resonances, are lost because they run away from the lower frequencies.

DISPERSION OF GUIDED WAVES

Dispersion, through a broadening of pulses as the pulses propagate, causes distortion and reduces the allowable data rates in optical fiber communication. Because of its importance, we will briefly review the sources of dispersion in fibers. This topic is interesting because dispersion in fibers can be due to processes other than the response of molecules to light.

The effects on the propagation of light that are due to the wavelength dependence of the refractive index are called *material dispersion*. Material dispersion will be treated in the next section by a classical model based on the interaction of light and molecules. Material dispersion in silica-based optical fibers is negligible at wavelengths around 1.3 μm. Most applications, requiring high data rates, attempt to take advantage of this fact by the proper choice of materials in fiber construction and the proper choice of light sources.

There are two additional sources of dispersion in a guided-wave system. If the guide has a gradient index of refraction, then the dependence of the refractive index on wavelength could cause a different profile for different wavelengths, i.e., Δ in **(5-56)** could be a function of the wavelength. This type of dispersion is called *profile dispersion*.

The final source of dispersion in a guided wave is called *mode dispersion*. If a light pulse introduced into a guide excites several modes, then the energy in each mode will appear to travel at a different velocity and thus the modes will arrive at the end of the fiber at different times. If we

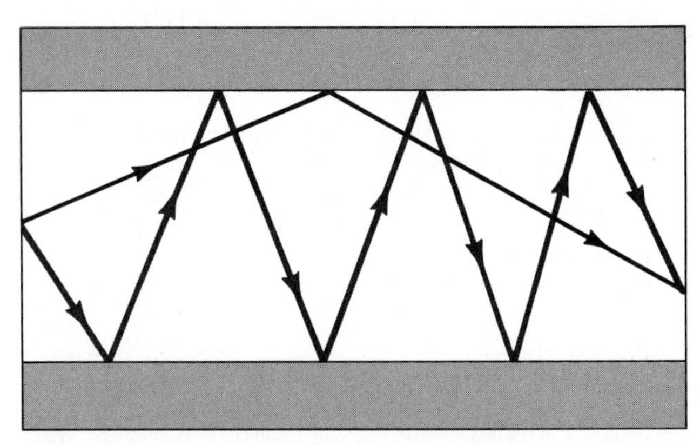

Figure 7-4. Two waveguide modes propagating in an optical guide. The different path-lengths show the origin of mode dispersion in optical guides.

retain the geometric view of guided waves used in Chapter 5, we see that the physical reason that the modes appear to travel at different velocities is that rays associated with each mode travel different optical path lengths. See Figure 7-4.

A simple relation can be derived for a fiber if skew rays are ignored. For meridional rays, an optical fiber can be viewed as a symmetric guide with a velocity of propagation given by ω/β, where

$$\beta = kn_2 \sin \theta_2$$

The limits on θ_2 for guided waves are given by **(5-27)**. Using the two limits of θ_2, we find that the maximum difference in propagation time for a fiber of length L is

$$\Delta t = \frac{Ln_2}{c}\left(\frac{1}{\sin \theta_c} - \frac{1}{\sin (\pi/2)}\right) = L\frac{n_2(n_2 - n_1)}{n_1 c} \tag{7-14}$$

If we code information as pulses and propagate the pulses down the fiber, we must not have any overlapping pulses. The pulse width at the start of the fiber is τ and after propagating along the fiber it is $\tau + \Delta t$. The minimum pulse spacing that can be used is thus Δt. Energy would be wasted if the pulse width exceeded Δt so we set the pulse width at Δt. We saw in Chapter 6 that we need two samples for the highest frequency transmitted; thus, the maximum bandwidth of the fiber is

$$B = \omega_c \propto \frac{\pi}{\Delta t}$$

Mode dispersion is the major difference between the various classes of optical fibers. The allowable communication bandwidth of a fiber shown in Table 7.3 indicates the effect of mode dispersion.

TABLE 7.3 Optical Fiber Bandwidth

Fiber Type	Bandwidth
Multimode, step-index	20 MHz · km
Multimode, graded-index	1 GHz · km
Single-mode, step-index	100 GHz · km

Mode dispersion in graded-index fibers is lower than step-index fibers because rays traveling at small angles relative to the fiber axis travel almost exclusively in the high-index region of the fiber and therefore at slow propagation speeds. However, rays traveling at large angles travel in the low-index region and have high propagation velocities. The gradient in the index of refraction partially compensates for the difference in geometrical path lengths by modifying the optical path lengths of the various rays. Single-mode fibers only transmit one mode; thus, mode dispersion does not exist.

MATERIAL DISPERSION

Although the data presented in Tables 7.1 and 7.2 at the beginning of this chapter demonstrated that the electromagnetic properties of materials were dependent on frequency, we still have no way to estimate that dependence. We saw that a relationship for dispersion was needed to calculate the effect of dispersion on the shape of a light pulse. A model could provide such a relationship and also an understanding of such optical properties as the origin of color and the reflectivity of metals. The model we will develop, based on the treatment of molecules as classical oscillators, will provide all of these benefits. Rather than develop one all-inclusive model whose properties might be difficult to interpret, we will develop three models of increasing complexity. We will identify the physical significance of the results of each of the models as we complete its development.

We assume that the medium in which the wave is propagating is made up of classical molecules consisting of positive and negative charges with a charge of magnitude q. If we expose a material composed of these molecules to a plane harmonic wave $\mathrm{E} = \mathbf{E}e^{i\omega t}$, then the field inside the material will be due to the following:

1. The applied field

$$\mathrm{E} = \mathbf{E}e^{i\omega t}$$

2. A contribution from any molecules that may have a net charge. These are called free charges and their density ρ is what is used in Maxwell's equation

$$\nabla \cdot \mathbf{D} = \rho$$

   We will assume $\rho = 0$, i.e., there are equal numbers of positive and negative charges in the material.

3. A contribution arising from the induced dipole created when the positive and negative charges are separated by the applied field (we will ignore the added complication of polar molecules that have permanent as well as induced dipole moments). The result of separating the positive and negative charges is an electric dipole moment

$$\mathbf{p} = \sum_i q_i \mathbf{r}_i - \sum_j q_j \mathbf{r}_j$$

   where the sum over i is over the positive charges multiplied by their displacement $\mathbf{r}$, relative to the center of mass, and the sum over j is over the negative charges multiplied by their displacement.

To take the contribution due to "bound charge" into account, we sum over all the electric dipole moments $\mathbf{p}$ in a microscopically large volume with respect to the elemental charge, and macroscopically small volume ΔV, with respect to the illuminating wavelength. The resulting vector **P** is called the polarization vector.

$$\mathbf{P} = \frac{1}{\Delta V}\sum_i \mathbf{p}_i \tag{7-15}$$

We use the electric displacement to describe this internal field

$$\mathbf{D} = \epsilon \mathrm{E} = \epsilon_0 \mathrm{E} + \mathbf{P} \tag{7-16}$$

To keep the math simple, we will assume a one-dimensional problem, with each molecule containing one positive and one negative charge of equal magnitude. This allows us to define the electric dipole moment as

$$\mathbf{p} = \sum_i q_i \mathbf{r}_i - \sum_j q_j \mathbf{r}_j = q(x_+ + x_-) = qx$$

We will also assume the initial velocity is equal to zero and the initial position of the charges is the equilibrium position.

Conductive Gas

If the negative charges of mass m are not bound to the positive charges and the positive charges are too heavy to respond to the light field, then the equation of motion is simply

$$F = ma = m\frac{d^2x}{dt^2} = qEe^{i\omega t} \tag{7-17}$$

where the only force experienced by the charges is assumed to be the applied light wave. The solution of **(7-17)** is

$$x = -\frac{qEe^{i\omega t}}{m\omega^2} \tag{7-18}$$

which gives the displacement of the charges due to the light wave.

If we use the separation between the positive and negative charge obtained from the solution to the equation of motion **(7-18)**, the polarization for the medium is

$$P = Nqx = -\frac{Nq^2E}{m\omega^2} \tag{7-19}$$

where N is the number density ($\#/m^3$) of the charges. Using this polarization in **(7-16)** we obtain

$$\epsilon E = \epsilon_0 E - \frac{Nq^2E}{m\omega^2}$$

We now have an equation for the dielectric constant, or equivalently the index of refraction, of the material

$$\frac{\epsilon}{\epsilon_0} = n^2 = 1 - \frac{Nq^2}{\epsilon_0 m\omega^2} \tag{7-20}$$

We define the *plasma frequency* as

$$\omega_p = \sqrt{\frac{Nq^2}{\epsilon_0 m}} \tag{7-21}$$

In terms of the plasma frequency, the index of refraction is

$$n^2 = 1 - \frac{\omega_p^2}{\omega^2} \tag{7-22}$$

The plasma frequency is the natural frequency of the collective motion of the charges. The charges move much like a liquid. To understand the origin of the plasma frequency, consider a cloud of electrons confined to move in one dimension by a very large magnetic field. The Coulomb repulsion between the electrons causes each electron to assume an equilibrium position between its neighbors. If one of the electrons is displaced from its equilibrium position, it will begin to oscillate. The neighboring electrons will experience perturbing forces due to the oscillating electron. In a short time, the initial displacement of a single electron will cause the whole cloud to oscillate. The frequency of this collective oscillation is the plasma frequency.

From this simple equation, originally derived by Drude in 1900, we can describe the optical behavior of electrons in metals. The conduction electrons in a metal are not restrained (at least to first order) by the core ions in the metal and thus fit our model. Better agreement is obtained, however, if collisions of the electrons with other particles are included by the addition of a damping term $-\gamma(dx/dt)$ to the equation of motion. (This damping term is a function of the charge's velocity because the damping should depend inversely on the time between collisions.) The theory is also appropriate for the description of plasmas, as long as they are not too dense.

Equation **(7-22)** can be rewritten in the form of a dispersion relation

$$\omega = \omega_p \sqrt{1 + \left(\frac{c}{\omega_p}\right)^2 k^2}$$

$$\approx \omega_p \left[1 + \frac{1}{2}\left(\frac{c}{\omega_p}\right)^2 k^2\right]$$

which is identical to **(7-13)** if $\mathbf{a} = (c/\omega_p)$. Thus, the dispersion relationship assumed during the discussion of pulse propagation is one appropriate for the description of an electromagnetic wave propagating in a plasma.

When $\omega < \omega_p$, **(7-22)** is negative and the index of refraction is imaginary

$$\mathcal{N} = \frac{\sqrt{\omega^2 - \omega_p^2}}{\omega} = i\frac{\sqrt{\omega_p^2 - \omega^2}}{\omega} = in\kappa$$

For this situation, a wave incident on the medium would be completely reflected, as we found out in Chapter 3 by calculating the reflectivity

$$R = \frac{(\mathcal{N} - 1)(\mathcal{N}^* - 1)}{(\mathcal{N} + 1)(\mathcal{N}^* + 1)} = 1$$

Experimental verification of this result is found in Figure 3-14.

As a verification of the qualitative agreement of this theory with experiment, the plasma frequency ω_p will be calculated for some materials. We continue to make the realistic assumption that because the nucleus (positive charge) is so massive, only the electrons (negative charge) respond to the high frequency ω. (It is not a necessary assumption since m could be the reduced mass and q the effective charge.) The assumption allows the use of the fundamental constants associated with electrons to calculate the plasma frequency.

$$q = 1.6 \times 10^{-19} \text{ C (coulomb)}$$

$$\epsilon_0 = 8.85 \times 10^{-12} \frac{\text{C}^2 \text{sec}^2}{\text{kg}\cdot\text{m}^2}$$

$$m = 9.11 \times 10^{-31} \text{ kg}$$

$$\omega_p = \sqrt{\frac{N(1.6 \times 10^{-19})^2}{(8.85 \times 10^{-12})(9.11 \times 10^{-31})}} = \sqrt{N(3.175 \times 10^3)}$$

1. For a gas at room temperature and a pressure of 2.5 Torr, $N = 10^{23}/\text{m}^3$, $\omega_p = 1.78 \times 10^{13}$ rad/sec, and $\nu_p = 2.84 \times 10^{12}$ Hz. For $\omega > \omega_p$, n is real; thus, from this result, we would expect to find most gases at low pressure to be transparent to visible light.

2. In the ionosphere (the Earth's atmosphere at altitudes above 70 km), $N = 10^{11}/\text{m}^3$, $\omega_p = 1.78 \times 10^7$ rad/sec, and $\nu_p = 2.84$ MHz, resulting in the prediction that the ionosphere should reflect frequencies below about 3 MHz. This result agrees with the experience of amateur radio operators, who know that the E layer (100–200 km) reflects frequencies around 3 MHz. The higher-altitude F layers (200–400 km) reflect higher frequencies, and the lower-altitude D layers (50–100 km) reflect lower frequencies, as they should if ω_p is proportional to the number density of charged particles.

3. In a metal, $N \approx 2.5 \times 10^{28}/\text{m}^3$, $\omega_p = 8.91 \times 10^{15}$ rad/sec, $\nu_p = 1.42 \times 10^{15}$ Hz. This plasma frequency yields a plasma wavelength of $\lambda_p = 212$ nm. Experimentally, the plasma wavelength is observed to be $\lambda_p = 210$ nm for Na. From the result for sodium, we would expect a metal like potassium to exhibit high reflectivity in the visible and infrared but be transparent in the uv. As is shown in Figure 7-5, potassium is relatively transparent below 310 nm but is reflective above 400 nm.

4. The number of free charges in a semiconductor can be modified by changing the impurity doping. The change in plasma frequency with

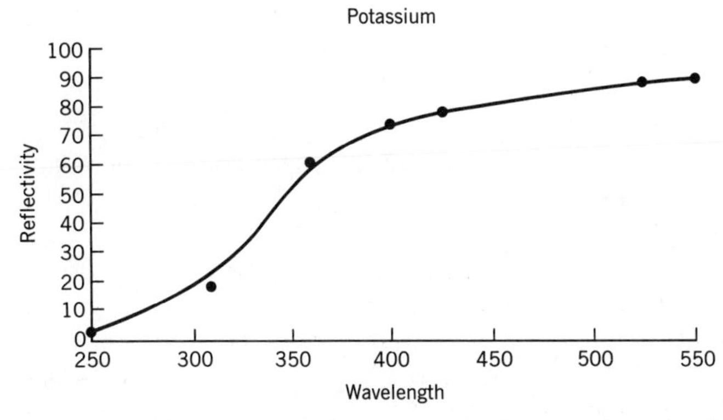

Figure 7-5. The reflectivity of potassium as a function of wavelength near the transition associated with the plasma frequency. [H.E. Ives and H.B. Briggs, *J. Opt. Soc. Am.* **26**: 238 (1936).]

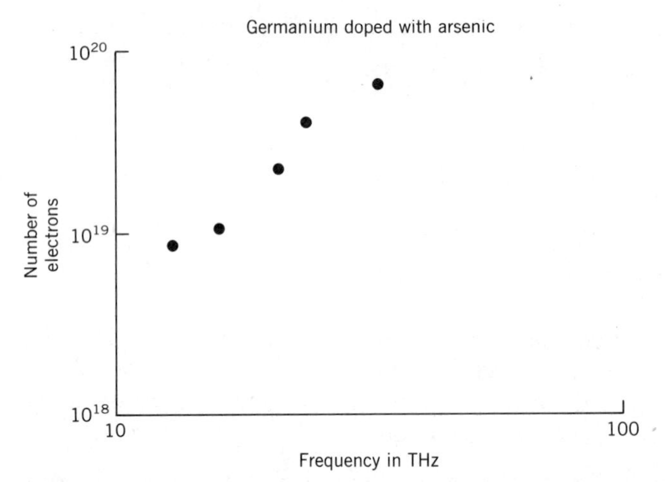

Figure 7-6. The plasma frequency as a function of carrier concentration. [J.I. Pankove, "Properties of Heavily Doped Germanium," *Progress in Semiconductors*, Vol. 9, A.F. Gibson and R.E. Burgess, eds. Heywood (1965), p.67.]

doping is shown in Figure 7-6, where the plasma frequency of germanium is plotted as a function of the concentration of arsenic.

Molecular Gas

If the charges in the molecules are bound by some internal force that acts to bring them back to their equilibrium position, then we must modify the equation of motion **(7-17)** by the addition of a restoring force $F(x)$

$$m\frac{d^2x}{dt^2} = -F(x) \tag{7-23}$$

(The solution of this model is identical to the solution of the harmonic oscillator in Appendix 1-A and only the highlights will be given here.) The exact form of the restoring force need not be known; we only require the displacement from equilibrium be small so we can write the force as the first nonzero term in a Taylor series, that is, we define the spring constant

$$\left\{\frac{\partial F}{\partial x}\right\}_{x=0} = s$$

(See Appendix 1-A.) By approximating the restoring force with the first term of a Taylor series expansion, we can rewrite **(7-23)** as the equation of motion of a harmonic oscillator **(1A-1)**

$$m\frac{d^2x}{dt^2} + sx = 0 \tag{7-24}$$

The solution of **(7-24)** is

$$x = x_0 e^{i\omega_0 t} \tag{7-25}$$

where

$$\omega_0^2 = \frac{s}{m}$$

We would have a more realistic model if we included a damping term

$$\frac{b}{m}\frac{dx}{dt} = \gamma\frac{dx}{dt}$$

[see **(1A-8)**], modifying **(7-24)** to read

$$\frac{d^2x}{dt^2} + \gamma\frac{dx}{dt} + \omega_0^2 x = 0 \tag{7-26}$$

We will assume $\gamma << \omega_0$, which will allow the solution of this equation to be written as **(1A-10)**

$$x = x_0 e^{-(\gamma/2)t} e^{i\omega_0 t} \tag{7-27}$$

If we illuminate the molecule with light of frequency ω, then we must add a driving term to **(7-26)**

$$\frac{d^2x}{dt^2} + \gamma\frac{dx}{dt} + \omega_0^2 x = \frac{q}{m}\mathcal{E} \tag{7-28}$$

Rather than solve the equation for x, we will replace x by the polarization using the definition of the polarization

$$P = Nqx \tag{7-29}$$

where the number density is N molecules per unit volume. Substituting the polarization into **(7-28)** gives

$$\frac{d^2P}{dt^2} + \gamma\frac{dP}{dt} + \omega_0^2 P = \frac{Nq^2}{m}\mathcal{E}$$

We assume that the polarization P and the electric field are at the same frequency but could have a different phase. The applied field and the induced polarization are thus

$$\mathcal{E} = E_0 e^{i\omega t}, \qquad P = \chi E_0 e^{i\omega t}$$

where χ is a complex constant. We can use the relations

$$\frac{dP}{dt} = i\omega P, \qquad \frac{d^2P}{dt^2} = -\omega^2 P$$

to reformulate the equation of motion (this formulation is sometimes called Drude's formula)

$$\chi(-\omega^2 + i\gamma\omega + \omega_0^2) = \frac{Nq^2}{m}$$

The assumption about the form of P allows

$$\epsilon\mathcal{E} = \epsilon_0\mathcal{E} + P$$

to be rewritten

$$\epsilon = \epsilon_0 + \frac{P}{\mathcal{E}} = \epsilon_0 + \chi$$

Drude's formula can be substituted into this equation to yield

$$\epsilon = \epsilon_0 + \chi = \epsilon_0 + \frac{Nq^2/m}{\omega_0^2 - \omega^2 + i\gamma\omega}$$

We continue with the assumption that $\mu \approx \mu_0$, so $n^2 = \epsilon/\epsilon_0$ and

$$\frac{\epsilon - \epsilon_0}{\epsilon_0} = n^2 - 1 = \frac{Nq^2/m\epsilon_0}{\omega_0^2 - \omega^2 + i\gamma\omega} \tag{7-30}$$

Using the definition of the plasma frequency yields

$$n^2 - 1 = \frac{\omega_p^2}{\omega_0^2 - \omega^2 + i\gamma\omega} \tag{7-31}$$

To separate the real and imaginary parts of the index of refraction, we use the assumption that the material is a dilute gas, so $n \approx 1$

$$n^2 - 1 = (n + 1)(n - 1) \approx 2(n - 1)$$

An equivalent approach is to assume

$$\frac{Nq^2/m\epsilon_0}{\omega_0^2 - \omega^2 + i\gamma\omega} < 1$$

and use the approximation that

$$\sqrt{1 + \delta} \approx 1 + \frac{\delta}{2}$$

This second approach corresponds physically to the assumption that the number of molecules N is small, or equivalently that $\omega_0^2 - \omega^2$ is large. Either approximation leads to

$$n = 1 + \frac{\frac{\omega_p^2}{2}(\omega_0^2 - \omega^2)}{(\omega_0^2 - \omega^2)^2 + \gamma^2\omega^2} \tag{7-32}$$

$$n\kappa = \frac{\frac{\omega_p^2}{2}\bullet\omega\gamma}{(\omega_0^2 - \omega^2)^2 + \gamma^2\omega^2} \tag{7-33}$$

When $\omega_0^2 >> \omega^2$, we can neglect ω^2 and **(7-32)** becomes

$$n \approx 1 + \frac{\omega_p^2}{2\omega_0^2} = \text{constant} \tag{7-34}$$

Since we are assuming $\omega \approx 0$, **(7-33)** yields $n\kappa = 0$. For frequencies well below the resonant frequency of the oscillator, the index of refraction is real and independent of the frequency. There is no dispersion in this frequency region. As an example, in argon gas, $n = 1.000516$ for all frequencies below 1 MHz.

As ω approaches ω_0 from below, $\omega_0^2 - \omega^2$ decreases. We assume that ω is far enough from ω_0 so that $\omega_0 - \omega > \gamma$. (The physical significance of this limit can be found in Figure 7-8 where we see the assumption means that the applied frequency must be outside the curve labeled absorption.) These limits lead to the conclusion that

$$(\omega_0^2 - \omega^2) = (\omega_0 + \omega)(\omega_0 - \omega) > \omega_0(\omega_0 - \omega) > \gamma\omega_0 > \gamma\omega$$

$$(\gamma\omega)^2 < (\gamma\omega_0)^2 < (\omega_0^2 - \omega^2)^2$$

We can therefore neglect $(\gamma\omega)^2$ and rewrite **(7-32)** and **(7-33)** as

$$n = 1 + \frac{\omega_p^2}{2(\omega_0^2 - \omega^2)} \tag{7-35}$$

$$\begin{aligned} n\kappa &= \frac{\omega_p^2 \omega \gamma}{2(\omega_0^2 - \omega^2)^2} \\ &= (n-1)\left(\frac{\omega\gamma}{\omega_0^2 - \omega^2}\right) \end{aligned} \tag{7-36}$$

As ω nears ω_0, we find from **(7-35)** that n increases with increasing frequency. This is called *normal dispersion*. We also find from **(7-36)** that

$$n\kappa << |n - 1|$$

and we may consider the index to be real if $\omega_0^2 - \omega^2$ is not too close to zero, that is, if

$$\omega_0^2 - \omega^2 > \omega\gamma$$

Examples of the behavior predicted by **(7-35)** can be found among the inert gases. Consider argon; we can fit **(7-35)** to experimental data for the index of refraction of argon over the wavelength region from 240 to 600 nm, shown in Figure 7-7 as points. The smooth curve is predicted by **(7-35)**, where we let ω_p and ω_0 be adjustable parameters. The equation of the smooth curve shown in Figure 7-7 is given by

$$n - 1 = \frac{19.732 \times 10^{28}}{(7.0876 \times 10^{32}) - \omega^2}$$

The numerator in this equation is proportional to the plasma frequency equal to

$$\omega_p^2 = \frac{19.732 \times 10^{28}}{2}$$

$$\omega_p = 4.4 \times 10^{14} \text{ rad/sec}$$

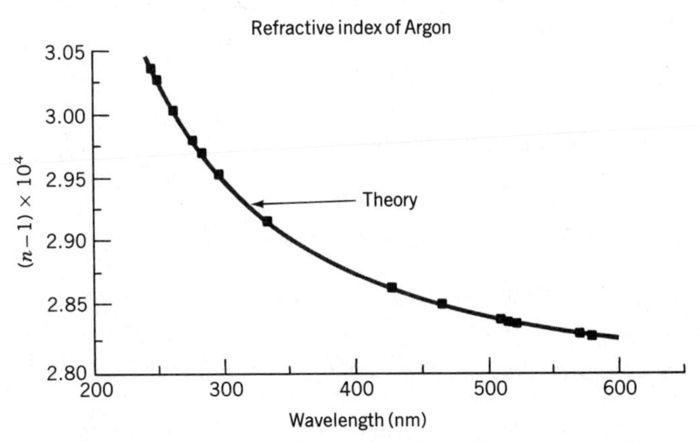

Figure 7-7. Dispersion in argon in the visible region of the spectrum. [S.A. Korff, "Optical Dispersion," *Rev. Mod. Phys.*, **4**: 471–503 (1932).]

The plasma wavelength is

$$\lambda_p = 4.2\ \mu\text{m}$$

If the charged particles are assumed to be electrons, the number density can be obtained from the plasma frequency

$$N = \frac{\omega_p^2}{3.175 \times 10^3} = 3.1 \times 10^{25}/\text{m}^3$$

which is roughly the density of a gas at one atmosphere, corresponding to the experimental conditions associated with the data.

The equation for the index predicts a static dielectric constant of $n = 1.000278$; the measured dc value is 1.000516. The resonant frequency ω_0 in the equation corresponds to a wavelength of 71 nm. The atomic spectral data do not support this result. Whereas the other constants correspond to realistic values, the classical theory is not able to predict the correct resonant frequency. To make an accurate prediction, we need the quantum theory.

When $\omega \approx \omega_0$, ω can be replaced by ω_0 in **(7-32)** and **(7-33)** except where the difference $\omega_0^2 - \omega^2$ occurs. This difference can be approximated by

$$\omega_0^2 - \omega^2 = (\omega_0 + \omega)(\omega_0 - \omega) \approx 2\omega_0(\omega_0 - \omega)$$

allowing **(7-32)** and **(7-33)** to be rewritten as

$$n = 1 + \frac{\omega_p^2 \dfrac{\omega_0 - \omega}{4\omega_0}}{(\omega_0 - \omega)^2 + (\gamma/2)^2} \tag{7-37}$$

$$n\kappa = \frac{\dfrac{\omega_p^2 \gamma}{8\omega_0}}{(\omega_0 - \omega)^2 + (\gamma/2)^2} \tag{7-38}$$

The previous approximation concerning $\omega\gamma$ is no longer correct and we retain the damping term in **(7-37)** and **(7-38)**.

The parameters obtained by fitting **(7-35)** to the experimental data in Figure 7-7 cannot be immediately applied to this new approximate result of our model. Near the resonance, these parameters lead to $n >> 1$ and

$$\omega_p > \omega_0^2 - \omega^2 + i\gamma\omega$$

in contradiction to the previous assumption that

$$\frac{Nq^2/m\epsilon_0}{\omega_0^2 - \omega^2 + i\gamma\omega} = \frac{\omega_p^2}{\omega_0^2 - \omega^2 + i\gamma\omega} < 1$$

We need only make one modification to the parameters to remove this contradiction. The number density can be reduced to $N = 10^{23}/\text{m}^3$, corresponding to a gas pressure of about 2 Torr. This assumption about the number density will lower the plasma frequency to

$$\omega_p = 1.27 \times 10^{13}\ \text{rad/sec}$$

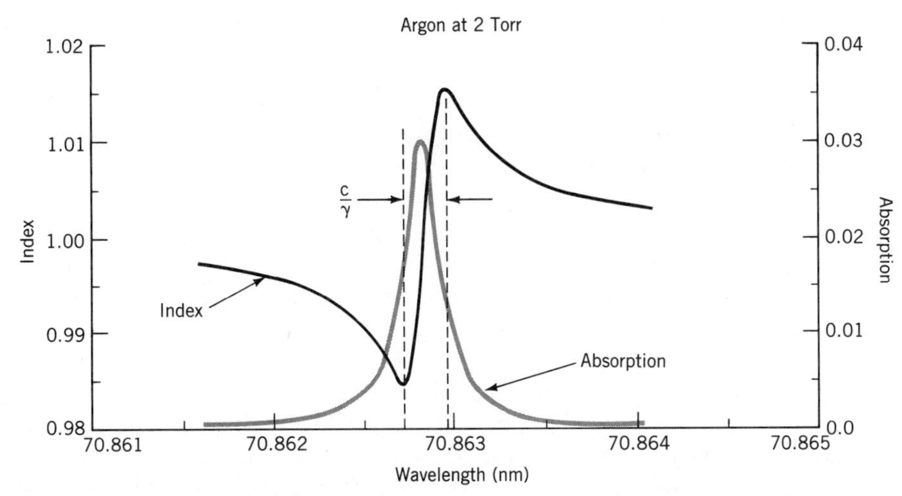

Figure 7-8. A plot of the index n **(7-37)**, and the absorption $n\kappa$, **(7-38)**, calculated with the parameters used to calculate the curve shown in Figure 7-7 modified to apply to a pressure of 2 Torr.

With the new plasma frequency, the assumption stated above is correct and the other parameters from Figure 7-6 can be used to obtain the curves shown in Figure 7-8.

As we stated before, the classical model does not correctly predict the details near a resonance, but the general behavior shown in Figure 7-8 is what is observed experimentally. This agreement with experiment is obtained by introducing into the model parameters obtained from experiments made far from the resonance condition.

The curve labeled absorption in Figure 7-8, is a plot of $n\kappa$ using **(7-38)**. The curve can be described mathematically by a Lorentzian function. The curve is often called the absorption line or the resonance line in the literature. It is easy to show that its maximum occurs at ω_0 and its width is γ.

As the resonant condition is approached from longer wavelengths, the index curve in Figure 7-8, plotted using **(7-37)**, increases; this is called normal dispersion. Within γ of the resonant frequency ω_0, the index n decreases from its maximum to its minimum value. This region is called the region of anomalous dispersion because the index decreases with increasing frequency. The occurrence of anomalous dispersion can be demonstrated by evaluating n at three points

$$\omega_0 - \omega = \frac{\gamma}{2} \rightarrow n = 1 + \frac{\omega_p^2}{4\omega_0\gamma}$$

$$\omega_0 - \omega = 0 \rightarrow n = 1$$

$$\omega_0 - \omega = -\frac{\gamma}{2} \rightarrow n = 1 - \frac{\omega_p^2}{4\omega_0\gamma}$$

At wavelengths shorter than the resonant wavelength, the index is less than unity but approaches the value of 1 as the wavelength decreases.

Finally when $\omega >> \omega_0$, we return to **(7-31)** to obtain a result identical to **(7-22)** modified by the addition of a damping term

$$n \approx 1 - \frac{\omega_p^2}{\omega^2 - i\gamma\omega}$$

The second term on the right of this equation will approach zero as the frequency ω becomes much larger than ω_p. Therefore, for very high frequencies (short wavelengths), the index will approach 1 from values less than 1 and the medium appears to the wave as if it were a vacuum. This behavior can be seen in Figure 7-15, where the index of refraction for aluminum is plotted. For wavelengths below the plasma frequency, the index of refraction of aluminum is less than 1 and approaches the value of 1 as the wavelength decreases.

If an absorption, such as the one shown in Figure 7-8, occurs in the visible region of the spectrum, then the material will be colored. Resonant frequencies occurring in the ultraviolet and visible regions of the spectrum are due to the electrons in atoms and molecules. The vibrations of the atoms making up a molecule result in resonances in the infrared. Molecules can also rotate and the resonances associated with the reorientation of molecules are found throughout the far infrared and into the microwave regions of the spectrum. In each of these wavelength regions, the data suggest that a number of oscillators with different spring constants s_j exist. In order to model a material containing a collection of resonances, the assumption is made that a fraction f_j of the oscillators in a material will have the resonant frequency ω_j and a loss term γ_j. With this assumption, **(7-31)** is modified to read

$$n^2 - 1 = \omega_p^2 \sum_{j=1}^{n} \frac{f_j}{\omega_j^2 - \omega^2 + i\gamma_j\omega}$$

where f_j is called the *oscillator strength*.

Dense Dielectric

In the treatment of dispersion up to now, we have neglected the fact that the electric dipoles induced by the applied field will interact, through the long-range field produced by one dipole at the site of another. One would think that it would be impossible to take into account the modification of the applied field at the molecule due to its many neighbors, but this is not the case. Lorentz solved the problem by noting that

1. Molecules have a minimum separation that is twice the molecular radius.
2. The distribution of the neighboring molecules can be treated as isotropic for a large number of materials.
3. The molecular dimensions are small with respect to the wavelength of light so that the electric field variables can be treated as constants with respect to spatial variables and propagation times can be ignored.

Using these assumptions, Lorentz derived what is now called the Lorentz field, $\mathbf{E}_L = \mathbf{E} + a\mathbf{P}$, where $\mathbf{E}$ is the applied field. For nonpolar dielectrics with certain symmetries such as cubic, $a = 1/3\epsilon_0$ (the derivation of this result can be found in a number of solid-state physics books[28] and will not be given here). We will assume this value for a in our discussion

$$\mathbf{E}_L = \mathbf{E} + \frac{\mathbf{P}}{3\epsilon_0} \tag{7-39}$$

Much of what we earlier derived can be modified to apply to a dense medium by using the Lorentz field in place of the applied field. The modified one-dimensional equation of motion is

$$\frac{d^2P}{dt^2} + \gamma\frac{dP}{dt} + \omega_0^2 P = \frac{Nq^2}{m}\left(E + \frac{P}{3\epsilon_0}\right)$$

and the dielectric constant is

$$\epsilon = \epsilon_0 + \frac{Nq^2/m}{\omega_0^2 - \omega^2 - (Nq^2/3\epsilon_0 m) + i\gamma\omega} \tag{7-40}$$

The index of refraction is

$$n^2 = 1 + \frac{\omega_p^2}{\omega_0^2 - (\omega_p^2/3) - \omega^2 + i\gamma\omega} \tag{7-41}$$

This equation can be put into the same mathematical form as **(7-31)**, derived for a dilute dielectric medium, by defining a new resonant frequency

$$\omega_m^2 = \omega_0^2 - \frac{\omega_p^2}{3}$$

The effect of having a dense medium, i.e., a medium in which the individual charges are allowed to interact, is simply to lower the resonant frequency from that of a free molecule. In quantum mechanics, this shift in resonant frequency is called *renormalization*. Physically, the shift can be understood by noting that the neighboring molecules will be oscillating together. The electric field from the neighbors acts against the restoring force of the oscillator, reducing the effective "spring constant" s. The difference between ω_m and ω_0 leads to the Clausius–Mossotti relation **(7-47)** by solving **(7-40)** for ω_p^2.

Figure 7-9 shows the variation in the index of refraction for a solid, the semiconductor GaP. The region of the spectrum over which a material exhibits low absorption and high transmission is the region that the material can serve as an optical material. The useful wavelength span of any optical

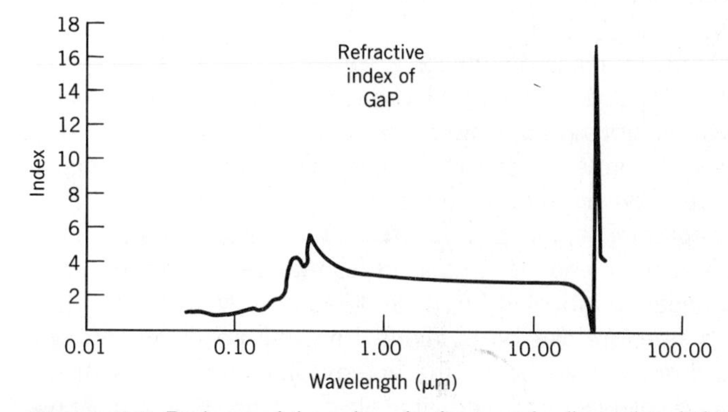

Figure 7-9. Real part of the index of refraction of gallium phosphide. The wavelength is shown in units of μm.

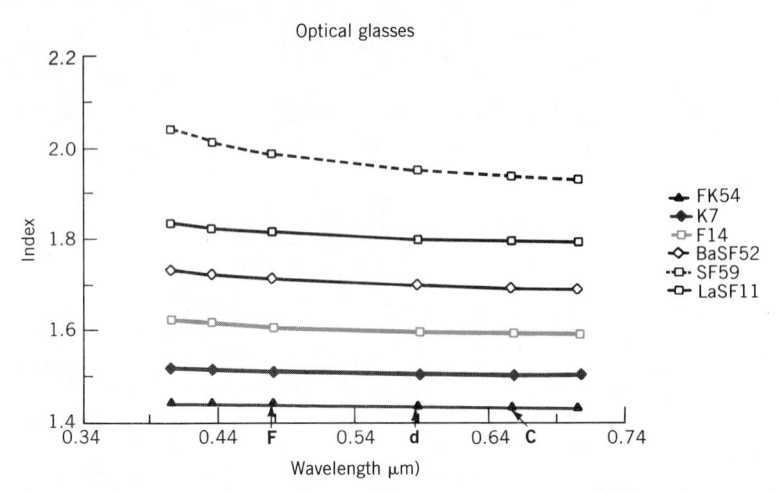

Figure 7-10. The index of refraction as a function of wavelength for a few optical glasses. The materials were selected to show the range of indices available for visible optics. Outside of the visible region, there are few indices available to the optical designer. The wavelengths labeled **C**, **d**, and **F** are Fraunhofer lines used to characterize the dispersion in optical glasses; see Appendix 7-A. (Data from Schott Optical Glass Catalog No. 3111.)

material is limited on the low- and high-frequency ends by an increase in absorption. The spectrum of GaP contains the general characteristics of all optical materials. It is transparent between 600 nm and 4.55 μm, with strong absorptions in the visible and far infrared region, making it a useful infrared optical material. Because of the availability of ultraviolet and infrared data on GaP, we will look at it in some detail. All optical materials will exhibit the same general behavior regardless of their useful range of wavelengths. For example, Figure 7-10 displays the index of refraction for a number of optical glasses. (The labels **C**, **d**, and **F** on the wavelength axis of Figure 7-10 denote the Fraunhofer solar spectrum lines used by optical designers to characterize dispersion in glass; see Appendix 7-A.).

In the visible region, the optical glasses shown in Figure 7-10 exhibit the same type of dispersion as exhibited by GaP in the near infrared, i.e. a smooth, nearly linear curve with a negative slope. On either side of the low-absorption region, strong absorptions will appear. For example, a nearby resonance in the ultraviolet is the origin of the rise in the index of SF59 glass at short wavelengths. (The data shown were obtained from the Schott Optical Glass Catalog No. 3111.)

In Figure 7-11, the visible and ultraviolet regions of the spectrum shown in Figure 7-9 are expanded. If the small variations are ignored, then the index does show anomalous dispersion in the region of maximum absorption and we do see the index approaching the value of 1 at short wavelengths (high frequencies). However, the curve is a large departure from Figure 7-8. This departure is due to a failure of the simple classical model. The allowable electronic energies in a solid fall into bands and this band structure is not adequately modeled by the simple harmonic oscillator. Evidence of the band structure can be obtained from the imaginary component of the index. For energies below the energy difference, between the conduction band and valence band, very little absorption is observed. At the frequency corresponding to the energy gap between these two bands, a

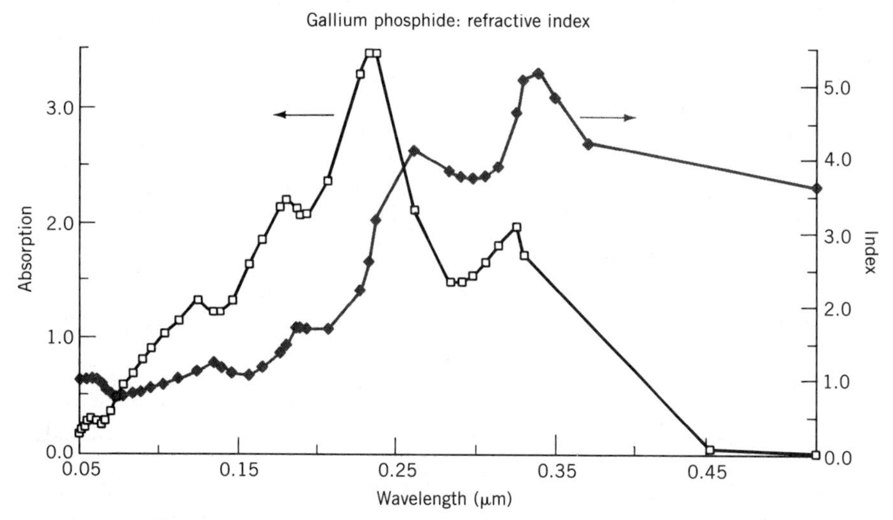

Figure 7-11. The complex index of refraction of gallium phosphide. The wavelength is shown in units of μm.

rapid rise in absorption is observed. In Figure 7-11, the rise starts around 450 nm.

This rapid increase in the absorption, below 450 nm, is due to the light wave exciting electrons, located in the valence band, into the conduction band of the GaP. For this to happen, the frequency of the light must exceed the bandgap frequency of GaP, which occurs at 2.25 eV. The conversion between energy in electron volts and wavelength is

$$E\ (\text{energy in electron volts}) = \frac{1.239}{\lambda(\mu\text{m})}$$

(Details of interpreting the absorption spectrum of a semiconductor can be found in the book by J.I. Pankove.[29])

All optical materials have strong absorptions at short wavelengths. Insulators such as the glasses for visible optics, shown in Figure 7-9, have much larger bandgaps than GaP, usually greater than three times the energy, in electron volts. Because of the large bandgap, insulators exhibit a rapid rise in absorption at a much shorter wavelength than the GaP. The index of refraction of the glass SF59 in Figure 7-9 displays the expected rapid rise at short wavelengths in an insulator.

Optical materials also exhibit absorption bands in the infrared and GaP is no exception. In Figure 7-12, we expand the infrared region of the spectrum shown in Figure 7-9 to display both the real and imaginary component of the index of refraction. The structure of this resonance is different from the structure of the ultraviolet resonance. For the infrared resonance, the model leading to **(7-40)** applies. The charges making up the oscillator associated with this resonance are the Ga and P atoms in the solid. As can be seen in Figure 7-12, the optical behavior of a solid is complicated, even when the model applies; there is no way to produce simple expressions for n and $n\kappa$ from **(7-40)**.

Rather than work with the complicated functions obtained for n and $n\kappa$ that would describe the curves in Figure 7-12, the dielectric constant is used to characterize the material

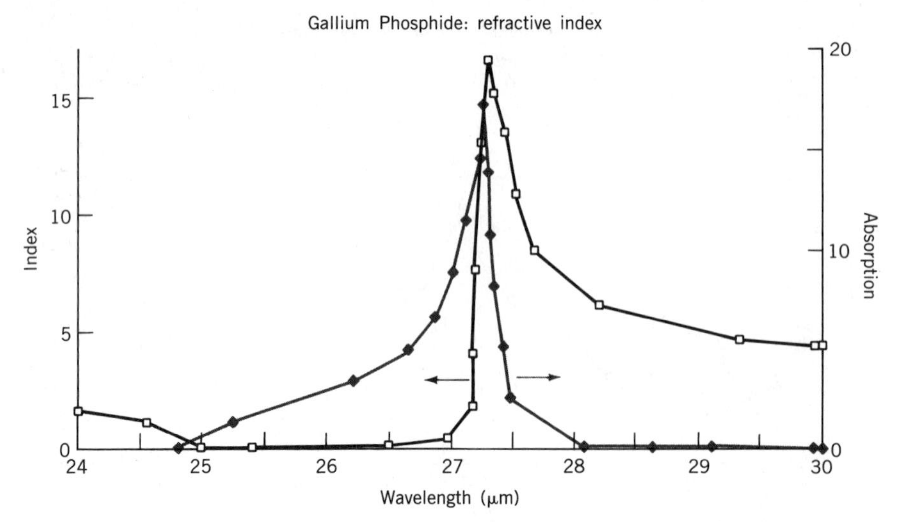

Figure 7-12. The complex index of refraction of gallium phosphide in the infrared.

$$\mathcal{Re}\{\epsilon\} = n^2(1 - \kappa^2) = 1 + \omega_p^2\left[\frac{\omega_m^2 - \omega^2}{(\omega_m^2 - \omega^2)^2 + \gamma^2\omega^2}\right] = c^2\mu\epsilon \quad (7\text{-}42)$$

$$\mathcal{Im}\{\epsilon\} = 2n^2\kappa = \omega_p^2\left[\frac{\gamma\omega}{(\omega_m^2 - \omega^2)^2 + \gamma^2\omega^2}\right] = \frac{c^2\mu\sigma}{\omega} \quad (7\text{-}43)$$

The functional forms of the real and imaginary parts of the dielectric constant are the same as were obtained for n and $n\kappa$ by assuming a dilute medium.

The similarity between the dielectric constant results of **(7-42)** and **(7-43)** and the index results for a dilute medium given by **(7-37)** and **(7-38)** can be demonstrated by making a normalized plot of **(7-42)** and **(7-43)**. The universal plot is generated by first defining the following normalized parameters:

$$a^2 = \left(\frac{\omega_p}{\omega_m}\right)^2, \qquad g = \frac{\gamma}{\omega_m}$$

$$x = \frac{\omega^2 - \omega_m^2}{\omega_m^2}, \qquad y = \frac{n^2(1 - \kappa^2)}{a^2}, \qquad z = \frac{2n^2\kappa}{a^2}$$

[Comparing the definition of g with **(1A-11)** shows that $g = 1/Q$.] Substituting into **(7-42)** yields a general function for the real part of the dielectric constant

$$y = \frac{1}{a^2} - \frac{x}{x^2 + g^2(1 + x)}$$

and substituting into **(7-43)** yields a general function for the imaginary part of the dielectric constant

$$z = \frac{g\sqrt{1 + x}}{x^2 + g^2(1 + x)}$$

Figure 7-13 shows a plot of $y(x)$ and $z(x)$.

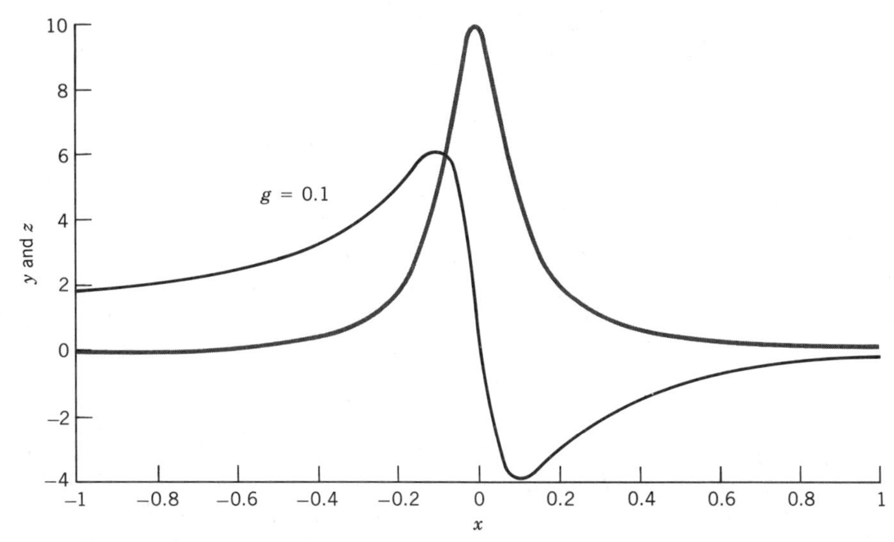

Figure 7-13. Normalized plots of the real (dark curve) and imaginary (light curve) parts of the dielectric constant.

We have implicitly assumed that only one resonant frequency is present. Normally, there are many resonances present. If each resonance is very narrow, i.e., if γ is small, then the multiple resonant frequencies can be treated independent of each other, according to the theory just developed. However, if γ is large for some resonant frequency, then that frequency will affect its neighboring resonances. To incorporate the effects of nearby strong absorptions, an average index of refraction $\langle n \rangle$ can be introduced, where $\langle n \rangle$ is experimentally determined. This is done to generate an experimentally based dispersion relation

$$y = \frac{\langle n \rangle}{a^2} - \frac{x}{x^2 + g^2(1 + x)}$$

Metals

The derived results for the dielectric constant of a solid can be applied to a metal and are consistent with the results we obtained for an electron gas. In a metal, the conduction electrons are not bound tightly to the core metal ions; thus, for these electrons, the resonant frequency can be neglected, $\omega_0 << \omega$. Not all electrons are free; for those electrons that are bound, the resonant frequency ω_0 is used (within a metal, **E** and not **E** + **P**/3 is the proper internal field). By using this information **(7-42)** and **(7-43)** can be rewritten as

$$c^2\mu\epsilon \approx \frac{\epsilon}{\epsilon_0} = 1 - \frac{\omega_p^2}{\omega^2 + \gamma^2} + \omega_p^2\left[\frac{\omega_0^2 - \omega^2}{(\omega_0^2 - \omega^2)^2 + \gamma^2\omega^2}\right] \tag{7-44}$$

$$\frac{c^2\mu\sigma}{\omega} \approx \frac{\sigma}{\epsilon_0\omega} = \frac{\omega_p^2\gamma}{\omega(\omega^2 + \gamma^2)} + \omega_p^2\left[\frac{\gamma\omega}{(\omega_0^2 - \omega^2)^2 + \gamma^2\omega^2}\right] \tag{7-45}$$

It is the bound electrons that lead to the color of some metals. If we neglect the effect of the bound electrons, we discover that this result is in agreement with the analysis of the dispersion of free charges made earlier.

Note that in **(7-44)**, ϵ becomes negative when

$$\omega^2 \le \omega_p^2 - \gamma^2$$

Since

$$n = \sqrt{\frac{\epsilon}{\epsilon_0}},$$

the index becomes imaginary and the material reflects all the incident light; see **(3-56)** and Figure 3-14. These results are identical to the results derived by assuming a free charge, as can be seen by setting $\gamma \approx 0$ and comparing the resultant equation to **(7-19)**.

Lorenz–Lorentz Law

If we have a transparent substance, where the index of refraction is real, and we can neglect the damping constant γ, we find

$$\frac{1}{\rho}\left(\frac{n^2-1}{n^2+2}\right) = \text{constant} \tag{7-46}$$

where ρ is the density. (Remember that the density is proportional to N, the number density.) In this form, **(7-46)** is called the *Lorenz–Lorentz* law. (For a gas, $n \approx 1$ and the Lorenz-Lorentz law can be rewritten as

$$\frac{n-1}{\rho} = \text{constant}$$

which is sometimes called the *law of Gladstone and Dale*.)

The constant in **(7-46)** is called the *atomic refractivity*. For molecular gases and liquids of molecular weight M, **(7-46)** is modified to read

$$\frac{n^2-1}{n^2+2}\cdot\frac{M}{\rho} = \text{constant}$$

where the constant is now called the *molar refractivity*. From this result, we can obtain an equation that allows the calculation of the index of refraction of a mixture of noninteracting substances by using the index of refraction of each of the components of the mixture. For a mixture of N materials that do not interact,

$$\left(\frac{n^2-1}{n^2+2}\right)\bullet\frac{W}{\rho} = \sum_{j=1}^{N}\left(\frac{n_j^2-1}{n_j^2+2}\right)\bullet\left(\frac{W_j}{\rho_j}\right)$$

where W is the weight of the mixture and W_j the weight of the jth component of the mixture. By dividing through by W, the weight fraction of each component can be used in the summation on the right.

In the limit as we approach very low frequencies, the Lorenz–Lorentz equation is written in terms of the dielectric constant

$$\frac{1}{\rho}\left(\frac{(\epsilon/\epsilon_0)-1}{(\epsilon/\epsilon_0)+2}\right) = \text{constant} \tag{7-47}$$

and is called the *Clausius–Mossotti relation*. The constant in the Clausius-Mossotti relation is called the *molar polarization*. If **(7-46)** and **(7-47)** are correct, we should be able to use the index of refraction of a gas at a given wavelength to predict the dielectric constant of a liquid at that same wavelength. For nonpolar molecules in the visible region, we can use the relations to an accuracy of a few percent (see Table 7.4 for the molar refractivity of a few simple molecules).

The Clausius-Mossotti equation fails to predict the molar refractivity of polar liquids or gases (for example, water). The cause of the failure has as its origin the failure of the Lorentz theory to include permanent dipole moments. In the treatment by Lorentz, he assumed that all of the dipole moments were parallel, aligned by the field that induced them. Permanent dipole moments would be oriented at random.[28]

We use **(7-41)** to rewrite **(7-46)**

TABLE 7.4 Molecular Refractivity

	n		Molar Refractivity	
Material	Liquid	Gas	Liquid	Gas
O_2	1.221	1.000271	2.00	2.01
HCl	1.245	1.000447	6.88	6.62
H_2O	1.334	1.000244	3.71	3.70
CS_2	1.628	1.00147	21.33	21.78

$$\frac{n^2 - 1}{n^2 + 2} = \frac{\omega_p^2/3}{\omega_m^2 - \omega^2 + i\gamma\omega} \tag{7-48}$$

Up to now, we have assumed that the solid contains N identical oscillators. If we instead suppose that a molecule contains several types of oscillators, we may write, as we did for a molecular gas,

$$\frac{n^2 - 1}{n^2 + 2} = \frac{\omega_p^2}{3} \sum_j \frac{f_j}{\omega_j^2 - \omega^2 + i\gamma_j\omega} \tag{7-49}$$

Drude called the quantity f_j the number of dispersion electrons per atom. **W. Pauli (1900–1958)** gave f_j its present name of oscillator strength. Quantum mechanics leads to an expression equivalent to **(7-48)**. However, the oscillator strengths f_j are not introduced in an ad hoc fashion as was done here; instead, the f_j are found to be proportional to intra-molecular dipole transition probabilities, i.e., the probability that a transition between two quantum levels will take place and a single photon produced. The frequencies ω_j are those associated with transitions between pairs of quantum states and the damping term γ_j is found to be proportional to $1/\tau$, where τ is the lifetime of the upper quantum state, i.e., the one at the higher energy.

SIGNAL VELOCITY

The definition of phase velocity

$$v = \frac{\omega}{k}$$

and group velocity

$$u = \frac{d\omega}{dk} = v - \lambda\frac{dv}{d\lambda}$$

generate an unusual result in the region of anomalous dispersion. The wave velocity decreases with increasing wavelength, i.e.,

$$\frac{dv}{d\lambda} < 0$$

Because this derivative is negative, the group velocity must be greater than the phase velocity, $u > v$. On the high-frequency side of the absorption peak (see Figure 7-7), the index of refraction is less than 1, which implies that $v > c$. We are, therefore, led to the conclusion that the group velocity is greater than the velocity of light. This result is in contradiction to a basic premise of the theory of relativity. The dilemma, posed by this result, was raised during the years immediately following Einstein's publication of the special theory of relativity, as a demonstration that the basic assumption of Einstein was in error.

The apparent conflict between the propagation velocity of light near a resonant absorption and the theory of relativity was resolved by **Sommerfeld**[30] and **Brillouin**.[31] The original papers and a detailed discussion of the topic can be found in the book by Brillouin[32] on the subject of wave propagation. We will not perform the derivation of the theories here, but will rather discuss the conclusions that can be drawn from the theory.

The conclusions of the theory developed by Sommerfeld and Brillouin are

1. No signal can propagate faster than the speed of light in a vacuum.
2. The leading edge of a pulse propagating in a dispersive medium travels at the vacuum velocity of light. This component of the pulse, called the *Sommerfeld precursor*, has a very small amplitude that grows from zero to a maximum and then decays back to zero. The frequencies associated with the precursor are very high at its front edge but decrease behind the pulse front. The frequencies are not a function of the incident wave but rather a function of the plasma frequency of the medium and the distance traveled

$$\frac{\omega_p^2 z_0}{c}$$

   where z_0 is the distance propagated.
3. At a later time in the pulse's temporal evolution, a second precursor, the *Brillouin precursor,* arrives. The amplitude and frequency of this precursor grow and then decay. The frequency of this precursor is not equal to the incident wave or the first precursor. The second precursor is much lower in frequency than the first precursor and the difference in frequency between the two precursors increases with propagation distance.
4. The main part of the pulse finally arrives and the wave assumes the "steady-state" values dictated by the incident wave. The time needed to reach this final state is associated with the signal velocity in the medium, which is limited by the theory of relativity to a value no higher than the velocity of light.

Figure 7-14 displays the general shape of this propagating pulse with the time of arrival plotted from left to right.

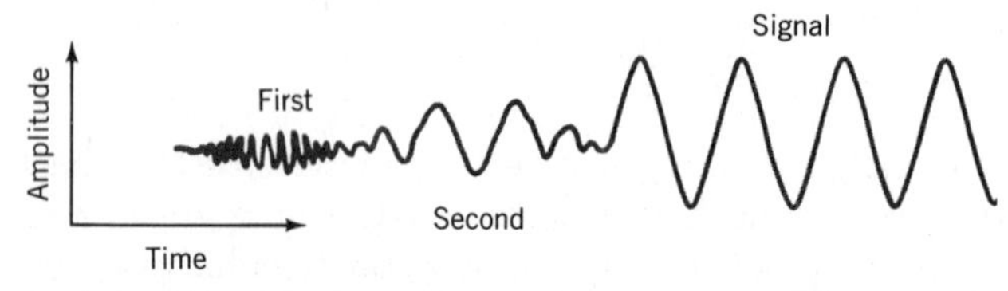

FIGURE 7-14. The first and second precursors and the signal shown with time of arrival increasing to the right. The origin and thus front edge of the pulse is on the left.

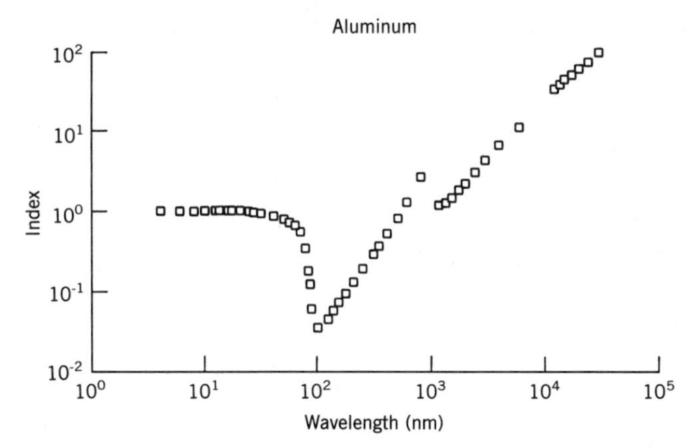

Figure 7-15. The index of refraction of aluminum from soft x-ray wavelengths to submillimeter wavelengths.

The classical model of dispersion can provide a qualitative understanding of the propagation of light. As light propagates in a dispersive medium, some of it is absorbed and reradiated by the electrons in the medium. Because of the inertia of the oscillators, the reradiated energy is out of phase with the incident wave. It is this lag in the reradiated energy that results in a slower propagation velocity.

To describe the shape of the leading edge of the propagating wave, we must use a number of frequencies. A number of very high frequencies will be needed if the front of the wave is very steep. The free and bound electrons will not follow these very high frequencies because of inertia. Thus, the very high frequencies will pass through the medium without seeing the electrons, i.e., as if they were propagating in a vacuum. The waves associated with the highest frequencies make up the first precursor (see Figure 7-14) and travel at the velocity of light in a vacuum.

Verification that very high frequencies propagate through a medium as if it were a vacuum can be obtained from the index of refraction of materials at very high frequencies (very short wavelengths). Figure 7-15 shows the index of refraction of aluminum, which at very short wavelengths is approximately equal to 1.

The second precursor and the final arrival of the signal component of the pulse can also be understood in a qualitative manner as the variation in the response of the classical oscillators to the Fourier components making up the pulse.

We are now in a position to understand the significance of the phase velocity and why it often exceeds the speed of light. The phase velocity establishes the phase of a steady-state wave propagating in a medium. We display this for two cases in Figure 7-16. We assume that the wave is a single frequency starting at $t = 0$ and $z = 0$ and propagating in the positive z direction. The wave cannot reach z_0 until after the time $t_0 = z_0/c$. Ignoring the transient behavior at the leading edge of this wave, after time $t_0 = z_0/c$, we have a wave of a single frequency. The phase of this wave is established by the phase velocity v in the relation $t = z_0/v$. In Figures 7-16*a*, we have $v < c$ and in 16*b*, we have $v > c$. In Figure 7-16*b*, causality prevents us from observing a wave at time z_0/v, but we use this time to establish the phase of the wave that we can first observe at $t_0 = z_0/c$.

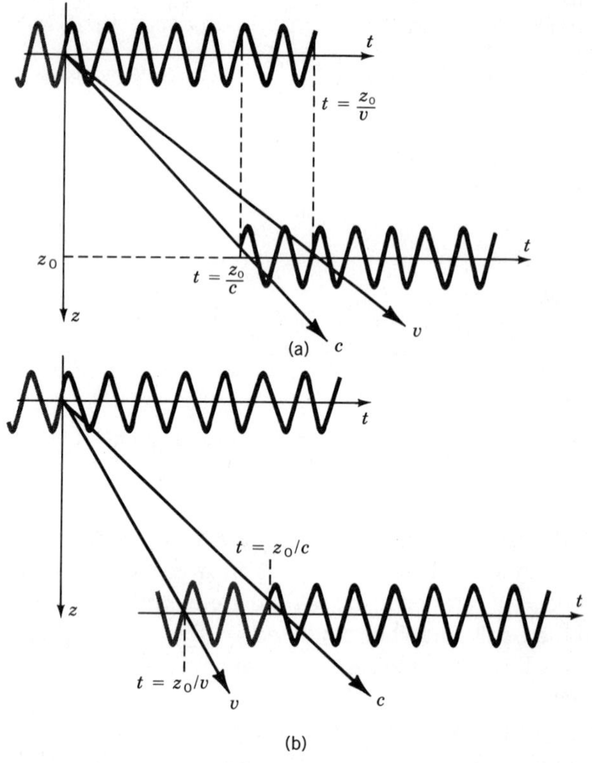

FIGURE 7-16. The use of phase velocity to establish the phase of a propagating wave. (a) $v < c$, (b) $v > c$.

SUMMARY

The original model, used in Chapter 1 to discuss wave propagation, was based on the assumption that the wave would propagate without change. This assumption is equivalent to assuming that all frequencies propagate at the same velocity. If a wave model allowed a frequency-dependent wave velocity, that is, dispersion, we found that a pulse would change shape as it propagated. By assuming a functional form for the relationship between the phase velocity and wavelength, we were able to demonstrate that a pulse would increase in both temporal and spatial breadth as it propagated.

It was necessary to introduce a new velocity to describe the propagation of a pulse in a dispersive medium. This velocity was called the group velocity

$$u = \frac{d\omega}{dk}$$

A wave of a single frequency would propagate at the phase velocity

$$v = \frac{\omega}{k}$$

and a wave made up of a group of frequencies would travel at the group velocity

$$u = v - \lambda\left(\frac{dv}{d\lambda}\right)$$

A model was developed to explain dispersion in an optical material based on a classical harmonic oscillator. The model was used to calculate the index of refraction for three cases.

Plasma A neutral gas of unbound charges has no resonant frequency. The index of refraction of a neutral gas, or plasma, was found to be given by

$$n^2 = 1 - \frac{\omega_p^2}{\omega^2}$$

where the plasma frequency is defined as

$$\omega_p = \sqrt{\frac{Nq^2}{\epsilon_0 m}}$$

If a loss term is included in the model for a conductive gas, then the index of refraction is complex. The real component of the index is

$$n^2(1 - \kappa^2) = 1 - \frac{\omega_p^2}{\omega^2 + \gamma^2}$$

and the imaginary component is

$$2n^2\kappa = \frac{\omega_p^2 \gamma}{\omega(\omega^2 + \gamma^2)}$$

Molecular Gas By assuming a collection of noninteracting harmonic oscillators, the index of refraction for a gas was calculated to be

$$n^2 = 1 + \frac{\omega_p^2}{\omega_0^2 - \omega^2 + i\gamma\omega}$$

To compare this result to experiment, the real and imaginary parts of the index were separated by assuming that the gas was dilute. The real part of the index could then be written

$$n = 1 + \frac{(\omega_p^2/2)(\omega_0^2 - \omega^2)}{(\omega_0^2 - \omega^2)^2 + \gamma^2\omega^2}$$

The imaginary part of the complex index of refraction of a dilute gas was found to be

$$n\kappa = \frac{(\omega_p^2/2)\omega\gamma}{(\omega_0^2 - \omega^2)^2 + \gamma^2\omega^2}$$

Dense Dielectric The final case evaluated allowed the individual oscillators to interact. The interaction lowered the resonant frequency from that of a noninteracting oscillator to

$$\omega_m^2 = \omega_0^2 - \frac{\omega_p^2}{3}$$

The expression for the index of refraction could no longer be separated easily into its real and imaginary components. Instead, the expression for the dielectric constant was used to describe the dispersion. The real part of the dielectric constant was found to be

$$\mathcal{Re}\{\epsilon\} = 1 + \omega_p^2\left[\frac{\omega_m^2 - \omega^2}{(\omega_m^2 - \omega^2)^2 + \gamma^2\omega^2}\right]$$

The imaginary part of the dielectric constant had the form

$$\mathcal{I}m\{\epsilon\} = \omega_p^2 \left[\frac{\gamma\omega}{(\omega_m^2 - \omega^2)^2 + \gamma^2\omega^2} \right]$$

For a transparent, nonpolar material, which has only a real dielectric constant, the index of refraction was shown to obey the Lorenz–Lorentz law

$$\left(\frac{n^2 - 1}{n^2 + 2}\right)\left(\frac{M}{\rho}\right) = \text{constant}$$

This relationship allows the use of the index of refraction of a gas to predict the dielectric constant of a liquid.

The final topic covered in this chapter concerned the propagation of a pulse of light through a dispersive medium. When the carrier frequency of the pulse is in the region of anomalous dispersion, the resultant group velocity seemed to contradict the postulate that information cannot be transmitted faster than the speed of light. We learned that the leading edge of a pulse travels no faster than the speed of light and the information or signal velocity is always limited to a value that does not exceed this velocity.

PROBLEMS

7-1. In the ionosphere, the lowest layer (called the D layer) has an electron density that reaches its maximum of $N \approx 10^9/\text{m}^3$ at midday and rapidly disappears after sunset. The highest layer (called the F_2 layer) has a density of $10^{12}/\text{m}^3$. What are the plasma frequencies of these two layers? How does the behavior of the D level affect radio transmission?

7-2. At a wavelength of $\lambda = 589.3$ nm, the index of N_2 at STP is $n = 1.000297$ and the index of liquid nitrogen at $-196°$ C is $n = 1.205$. The gas density is $\rho = 1.251 - 10^{-3}$ g/m^3 and the liquid density $\rho = 0.808$ g/m^3. Use this data to see how well the Lorenz–Lorentz equation agrees with experiment.

7-3. Calculate the index of refraction for acetone and ether in the gas phase at STP using the indices from the liquids.

TABLE 7.5 Problem 7-3

Material	ρ	n
Acetone	0.7899	1.3588
Ethyl ether	0.7138	1.3526

Compare with the observed values of 1.00108 for acetone and 1.00152 for the ether. *Hint*: Remember from chemistry that the volume occupied by one molecular weight of a gas is 22.4 Liters.

7-4. Derive the Clausius–Mossotti equation **(7-47)** using the equation

$$\epsilon = \epsilon_0 + \frac{\epsilon_0\omega_p^2}{\omega_0^2 - \dfrac{\omega_p^2}{3} - \omega^2 - i\gamma\omega}$$

7-5. In the case of normal dispersion in the visible region, in optical materials with the nearest absorption in the ultraviolet, the index of refraction can be approximated by Cauchy's equation

$$n = A + \frac{B}{\lambda^2} + \frac{C}{\lambda^4}$$

Derive Cauchy's equation from **(7-31)**.

7-6. The following measurements were made for air:

TABLE 7.6 Problem 7-6

λ	$n - 1$
6.563×10^{-7} m	2.916×10^{-4}
5.184	2.940
4.308	2.966
3.441	3.016
2.948	3.065

Fit Cauchy's equation to this data, using only the first two terms of the equation.

7-7. From the equation for $n\kappa$ for a molecular gas

$$n\kappa = \frac{\omega\gamma(\omega_p^2/2)}{(\omega_0^2 - \omega^2)^2 + \gamma^2\omega^2}$$

find the relationship between the width of the absorption peak at half-maximum and γ. What about n?

7-8. From the equations for the real components of the dielectric constant for dense dielectrics,

$$\epsilon = 1 + \frac{\omega_p^2(\omega_0^2 - \omega^2)}{(\omega_0^2 - \omega^2)^2 + \gamma^2\omega^2}$$

find the maximum, minimum, and width.

7-9. If we have a step-index fiber with a numerical aperture of 0.16, $n_2 = 1.45$, and a core diameter of 90 μm, how much intermodal dispersion should we expect?

7-10. Calculate the group velocity of light at a wavelength of 500 nm when the index of refraction is given by

$$n = 1.577 + \frac{1.78 \times 10^{-14}}{\lambda^2}$$

7-11. The experimentally derived dispersion relation for hydrogen in the wavelength range from $0.4 < \lambda < 9\ \mu$m is

$$n^2 - 1 = 2.721 \times 10^{-4} + \frac{2.11 \times 10^{-18}}{\lambda^2}$$

Using the results of Problem 7-5, find the resonant frequency.

7-12. Find the group velocity for a radio wave propagating in the ionosphere. The dispersion of the phase velocity is

$$v = \sqrt{c^2 + b^2\lambda^2}$$

7-13. Write a computer program to generate the plots shown in Figure 7-9. Vary the constant g, which corresponds to the loss term, and see what effect it has on the shape of the curves.

7-14. Starting with **(7-45)**, show that the dc conductivity of a metal is given by

$$\sigma_0 = \frac{Nq^2}{m\gamma}$$

7-15. Derive the Sellmeier's equation

$$n^2 = 1 + \frac{a\lambda^2}{\lambda^2 - \lambda_0^2}$$

using **(7-31)** as the starting point. Under what conditions will Sellmeier's equation reduce to Cauchy's equation?

7-16. The following data have been gathered for the index of refraction of water-ethanol mixtures:

TABLE 7.7 Problem 7-16

Weight Fraction of Ethanol	ρ	n
0.0000	0.998	1.337
0.2075	0.97	1.3508
0.4089	0.939	1.3616
0.5998	0.899	1.367
0.7999	0.854	1.3693
1.0000	0.805	1.3676

Use this data to evaluate the Lorenz–Lorentz law.

7-17. Derive **(7-12)** from the definition of the group velocity

$$u = \frac{d\omega}{dk}$$

7-18. Find the radii of curvature for a cemented achromatic double with a focal length of 10 cm using BK7 as the crown and SF6 as the flint. Assume that the crown is equiconvex (the radius of curvature of both surfaces is the same) and set one curvature of the flint equal to the crown so they may be glued together.

7-19. Redesign the lens in Problem 7-18 but now make the flint an equiconvex lens.

7-20. Redesign the lens in Problem 7-18 using SK4 as the crown and F4 as the flint.

Appendix 7-A

CHROMATIC ABERRATIONS

The dispersion in optical glasses, illustrated in Figure 7-10, produces wavelength-dependent paraxial properties of optical systems. These wavelength effects are historically called *primary chromatic aberrations*. Wavelength-dependent variations of the aberrations introduced in Appendix 5-B also occur. To display the wavelength dependence of aberrations, ray-intercept curves are generated for three wavelengths as illustrated in Figure 7A-1.

The index of refraction variation with wavelength causes a variation in focal length, or equivalently the power, of a lens

$$\frac{1}{f} = \Phi = (n-1)\left(\frac{1}{R_1} - \frac{1}{R_2}\right)$$

$$\delta\Phi = \delta n\left(\frac{1}{R_1} - \frac{1}{R_2}\right) = \frac{\delta n}{n_i - 1}\Phi = \frac{\Phi}{\upsilon_i} \tag{7A-1}$$

where υ_i is the *Abbe value* of the glass (In optics literature, the Abbe value is also known as the *V number*, *constringence*, or *reciprocal dispersive power*.) The change in the focal length is called *longitudinal chromatic aberration* or *longitudinal color*. The magnification of a lens also changes with wavelength. This change is called *lateral chromatic aberration* or *lateral color*.

The dispersion properties of glasses are measured at standard wavelengths; see Table 7A.1. These wavelengths are prominent lines in the solar spectrum that were labeled with letters by Joseph von Fraunhofer. These

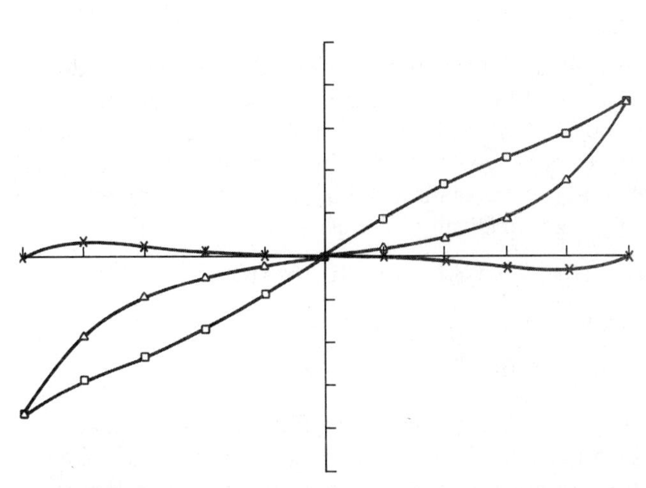

FIGURE 7A-1. Ray intercept curves for a Fraunhofer objective (the type analyzed in Problem 7-18) at *C*, *d*, and *F* wavelengths.

TABLE 7A.1 Standard Wavelengths

Fraunhofer Designation	Wavelength, nm	Element
r	706.5188	He
C	656.2725	H
C'	643.8469	Cd
D_1	589.5923	Na
D_2	588.9953	Na
d	587.5618	He
e	546.0740	Hg
F	486.1327	H
F'	479.9914	Cd
g	435.8343	Hg
h	404.6561	Hg

letter labels are used in all discussions of dispersion in the visible region of the spectrum.

The Abbe number, defined in **(7A-1)**, is used to characterize the dispersion of a particular glass. Its definition suggests an infinitesimal change, but in practice it is defined as

$$\nu_d = \frac{n_d - 1}{n_F - n_C} \tag{7A-2}$$

The numerator $n_d - 1$ is called the *refractivity* and the denominator $n_F - n_C$ is called the *partial dispersion*. The partial dispersion spans the visible spectrum with the refractivity positioned roughly in the center.

Glasses are grouped into two types: *crown* and *flint*. A very long time ago, the crowns were window glass composed of soda, lime, and silica. The glass was formed by blowing a globe and whirling it into a disk. The appearance of the disk led to the name crown. Flint glass was composed of alkalis, lead oxide, and silica and used for cut glass tableware. Today, the adjectives separate glasses into low-dispersion, low-index crowns that fulfill the following inequalities:

$$n_d > 1.60, \qquad \nu_d > 50$$

$$n_d < 1.60, \qquad \nu_d > 55$$

and high-dispersion, high-index flints that do not fulfill the above inequalities.

Glass types are identified by an alphanumeric such as BK7 and by a numerical index such as 517642. The letter B indicates a borosilicate and the letter K indicates a "crown" type of glass. The numerical index provides much more information about the glass type but is not easily remembered. The first three digits are the refractivity of the glass

$$n_d - 1 = (1.51680 - 1) \times 10^3 \approx 517$$

and the last three digits are the Abbe number

$$\nu_d = (64.17) \times 10 \approx 642$$

A lens where $\delta\Phi = 0$ is called an achromat. From **(7A-1)**, we see that a single lens cannot be an achromat unless it has zero power. In the solution of Problem 5-13, it was discovered that the focal length of two lenses in contact is given by

$$\frac{1}{f} = \frac{1}{f_1} + \frac{1}{f_2} \tag{7A-3}$$

where f_1 and f_2 are the focal lengths of the individual lenses. In terms of the power of a lens, this result is written

$$\Phi = \Phi_1 + \Phi_2 \tag{7A-4}$$

The variation in the power of the lens pair with wavelength is given by

$$d\Phi = \delta\Phi_1 + \delta\Phi_2 = \frac{\Phi_1}{\upsilon_{d1}} + \frac{\Phi_2}{\upsilon_{d2}} \tag{7A-5}$$

By proper choice of optical glasses, we can keep the power required by **(7A-4)** constant while reducing the variation in the power predicted by **(7A-5)** to zero.

$$d\Phi = 0$$

$$\frac{\Phi_1}{\upsilon_{d1}} = -\frac{\Phi_2}{\upsilon_{d2}} \tag{7A-6}$$

We can solve **(7A-6)** for the focal lengths to obtain

$$-\frac{f_1}{f_2} = \frac{\upsilon_{d2}}{\upsilon_{d1}} \tag{7A-7}$$

Because all optical glasses have normal dispersion (see Figure 7-10), the Abbe value must be positive. Equation **(7A-7)** therefore implies that an achromat must be a combination of a positive and negative lens.

Simultaneous solutions of **(7A-3)** and **(7A-6)** yield

$$\Phi_1 = \frac{\upsilon_{d1}}{\upsilon_{d1} - \upsilon_{d2}}\Phi, \qquad \Phi_2 = \frac{\upsilon_{d2}}{\upsilon_{d2} - \upsilon_{d1}}\Phi \tag{7A-8}$$

We see that the powers of the two lenses making up the achromat are inversely proportional to the difference in the Abbe values $\Delta\upsilon$ of their glasses. High-power, short-focal-length lenses are more expensive, have a smaller maximum aperture, and have larger aberrations than low-power lenses. For this reason, glasses that yield a $\Delta\upsilon > 20$ are usually selected in an achromat design. The properties of a few of the optical glasses found in the Schott Optical Glass Catalog (No. 3111) are listed in Table 7A.2.

Assume we wish to construct an achromatic doublet using BK7 as the crown and F4 as the flint component. From **(7A-8)** and the data in Table 7A.2, we have

$$\Phi_1 = \frac{64.17}{64.17 - 36.63}\Phi = 2.33\Phi$$

$$\Phi_2 = \frac{36.63}{36.63 - 64.17} = -1.33\Phi$$

The power of the achromat is 43% of the power of the crown element and the flint's power is 57% of the crown. We see that the positive element overshoots the desired power and the overshoot is subtracted out by the negative element. Along with the reduction in power is a cancellation of the dispersion effects for two wavelengths.

Newton believed that all glasses exhibited the same dispersion, making it impossible to produce an achromat. His belief led him to invent the reflecting telescope. A lawyer and amateur astronomer, Chester Moor Hall proved Newton wrong by discovering that crown and flint glass could be combined into a doublet with no chromatic aberration. He attempted to keep his discovery secret by using different London lensmakers to produce the two components of his doublet. Unfortunately, both lensmakers were swamped with work and subcontracted the job to the same individual, John Dollond. Dollond recognized the purpose of the two lenses and applied for a patent on the idea. The court awarded Dollond the patent even though Moor Hall was the inventor. The judge ruled that the person who made an invention public for the general good deserved the patent.

TABLE 7A.2 Optical Glasses

Type	n_d	ν_d	$n_F - n_C$	n_C
BK1 510635	1.51009	63.46	0.008038	1.50763
BK7 517642	1.51680	64.17	0.008054	1.51432
SK4 613586	1.61272	58.63	0.010451	1.60954
F2 620364	1.62004	36.37	0.017050	1.61503
F4 617366	1.61659	36.63	0.016834	1.61164
SF6 805254	1.80518	25.43	0.031660	1.79609

The method of producing an achromat just described insures that the focal length of the doublet is the same at two wavelengths, C and F light. At other wavelengths, this is not the case. The chromatic aberration still present is called the *secondary spectrum*. If we wished to make the focal length the same at C, F, and d light, then we must make the variation in power equal to zero.

$$\begin{aligned}\delta\Phi_{d,C} &= (n_{1d} - n_{1C})\frac{\Phi_1}{n_{1d} - 1} + (n_{2d} - n_{2C})\frac{\Phi_2}{n_{2d} - 1} \\ &= \left(\frac{n_{1d} - n_{1C}}{n_{1F} - n_{1C}}\right)\frac{\Phi_1}{\upsilon_{1d}} + \left(\frac{n_{2d} - n_{2C}}{n_{2F} - n_{2C}}\right)\frac{\Phi_2}{\upsilon_{2d}}\end{aligned} \tag{7A-9}$$

The fractions in the parentheses are called *relative partial dispersions*. Only if the partial dispersions of the two glasses are equal can we reduce the secondary spectrum by achromatizing the lens at three wavelengths. Abbe was the first to note that if you plot the relative partial dispersion vs υ_d for the various optical glasses, a nearly linear relation is observed between the two parameters. This means that we cannot meet the condition for reduced secondary spectrum and also fulfill the requirement that the Abbe values for the two glasses have a large difference. It is for this reason that it is impossible to reduce the chromatic aberration below that obtained by the simple calculations above.

Coherence

Everyday experience suggests that the observation of interference must be subject to very restrictive conditions. The intensity distribution from a number of light sources is normally quite uniform, with no evidence of the fringe patterns discussed in Chapter 4. When interference is observed, it is usually associated with small dimensions; for example, two slits separated by only a few millimeters, or thin films much less than a millimeter thick, are required to produce interference fringes in natural light. The need for small dimensions is not, however, associated with the wavelength of light. When laser illumination is used, interference is quite easy to produce. In fact, interference is a major noise source in images produced with laser illumination (see Appendix 10-C for a brief discussion of the generation of a random interference pattern called Speckle noise). The ability to observe interference is associated with a property of a wave called coherence. In this chapter, the measurement of coherence will be discussed. (An excellent review of coherence from the point of view of interference is found in the article by H. Paul.[33])

The terms coherent and incoherent were introduced in Chapter 4 in association with the presence or absence of interference. Two waves were said to be coherent if the interference term

$$\langle \mathbf{E}_1 \cdot \mathbf{E}_2 \rangle = \sqrt{I_1 I_2} \cos \delta$$

was nonzero within the region occupied by both $\mathbf{E}_1$ and $\mathbf{E}_2$. Simply restated, the two waves are coherent if $\mathbf{E}_1$ and $\mathbf{E}_2$ have a constant phase relation δ. The measurement of the property of coherence is, thus, a comparison of the relative phase of two light waves or two samples of the same wave. The comparison can be made in terms of the waves' temporal or spatial dependence, making coherence a function of both space and time. In this chapter, we will treat the temporal and spatial dependence separately.

The mathematical implementation of a comparison of two quantities is a correlation operation. We will develop the connection between coherence theory and the correlation operation and discuss the experimental techniques that duplicate the correlation operation. We will discover that a temporal correlation of a wave with itself can be performed using a Michelson interferometer and that a spatial correlation can be performed by using Young's two-slit experimental configuration.

Interference of waves of different frequencies will be analyzed using the amplitude-modulated wave, derived in Chapter 7 by the addition of two waves of different frequency. An association will be established between the beat frequency of an amplitude-modulated wave and coherence that will demonstrate the need for a quantified measurement of coherence. The connection between the concepts of modulation and interference led **A.T. Forrester** to develop the idea of photoelectric mixing that is now the basis of a number of applications, including light-beating spectroscopy.

The Michelson interferometer's operation will be examined when the input wave contains two frequencies. The conclusions of this examination will then be extended to a continuous distribution of frequencies. The output intensity will be associated with the measurement of coherence and a coherence function will be defined. The coherence function will be calculated for a typical spectral distribution found in emission or absorption spectroscopy. This example will demonstrate one of the important applications of coherence theory, Fourier transform spectroscopy.

Fourier theory provides a connection between the spectral analysis view of the coherence function and the concept of correlation that will be formally established in this chapter. The correlation viewpoint will be used to discuss spatial coherence. The spatial coherence of a line source will be calculated to demonstrate the application of the concept of spatial coherence. The theoretical analysis of a line source will aid in understanding the use of Michelson's stellar interferometer to measure the spatial extent of a star.

Stellar interferometry has been extended to allow the utilization of intensity interferometry. A brief discussion of this application will conclude the chapter.

One of the first experiments on coherence was conduced by **E. Verdet**, who measured the spatial coherence length of an extended light source in 1865. Michelson expanded the work of Verdet during the 1890s by connecting the source distribution and the source's spectral content with the visibility of interference fringes. Coherence was successfully treated classically by **F. Zernike (1885-1966)**,[34] with **H.H. Hopkins**[35] placing Zernike's ideas into an easier-to-understand framework.

A key event in the development of coherence theory was the introduction of a new interferometric technique by **R. Hanbury Brown** and the derivation of a theoretical basis for the technique by **Richard Twiss**. The technique involved intensity correlation between signals collected at two points in space. It could be explained quite easily by classical theory but seemed to contradict quantum theory. The contradiction led to a vigorous controversy during the early 1960s, which was resolved by **R.J. Glauber** who placed the quantum theory of coherence on a firm theoretical foundation.

A detailed review of both classical and quantum theories of coherence can be found in a review article by L. Mandel and E. Wolf[36] and in a textbook on the subject, with extensive references, written by Arvind S. Marathay.[37]

PHOTOELECTRIC MIXING

In Chapter 7, two harmonic waves of different frequencies, but equal amplitudes, were added together to produce a resultant, harmonic, carrier wave with a frequency equal to the average of the frequencies of the two waves. The carrier wave's amplitude was found to be modulated by a harmonic wave whose frequency equaled the beat frequency, i.e., it was equal to one-half of the difference between the frequencies of the two primary waves. In this section, we will calculate the intensity of the amplitude-modulated wave at an arbitrary point in space. We will compare this result to the intensity produced by interference between two waves of equal frequency. The result of this comparison will allow us to establish under what conditions waves of different frequencies will interfere.

Assume initially that we have two, one-dimensional waves of identical amplitudes but different frequencies. The ith wave is

$$E_i(x, t) = E_0 \cos(\omega_i t - kx + \phi_i), \qquad i = 1, 2 \tag{8-1}$$

In Chapter 7, we added two such waves and obtained a resultant, amplitude-modulated wave of the form

$$E(x,t) = 2E_0 \cos(\omega t - kx) \cos(\Delta\omega t - \Delta kx)$$

where

$$\Delta\omega = \frac{\omega_2 - \omega_1}{2}$$

is the beat frequency between the two waves. Because the time average of the Poynting's vector is the quantity measured, the time average of E^2 must be found. This is accomplished by using (**2-23**) to obtain the intensity given by (**2-27**)

$$I(x) = |\langle \mathbf{S} \rangle| \propto \langle |\mathbf{E}|^2 \rangle = 4E_0^2 \langle \cos^2(\omega t - kx) \cos^2(\Delta\omega t - \Delta kx) \rangle \tag{8-2}$$

We wish to compare this intensity distribution to the intensity distribution obtained in Chapter 4 for the interference between two waves (**4-11**)

$$I(x) = |\langle \mathbf{S} \rangle| \propto \langle |\mathbf{E}_1|^2 \rangle + \langle |\mathbf{E}_2|^2 \rangle + 2\langle |\mathbf{E}_1 \cdot \mathbf{E}_2| \rangle$$

When the two waves are of identical frequency, the intensity distribution given by this relationship was found to be

$$\frac{I}{I_0} = 2 + 2\cos\delta$$

where from (**2-24**),

$$I_0 = \langle S \rangle \propto \frac{1}{T} \int_{t_0}^{t_0+T} E_i^2 \, dt \propto \frac{|E_0|^2}{2}$$

and T is the measurement period.

Performing the temporal average of (**8-2**) yields

$$\frac{I(x)}{I_0} = 2 + 2\langle \cos 2(\Delta\omega t - \Delta kx) \rangle \tag{8-3}$$

where the assumption that $\omega T >> 1$ is used, as it was in Chapter 2, to simplify the temporal average (ω is at optical frequencies $\approx 10^{14}$ Hz).

The temporal average of (**8-3**) yields the desired relationship

$$\frac{I(x)}{I_0} = 2 + \frac{2}{\Delta\omega T}\left\{\cos\left[\Delta\omega(2t_0 + T) - 2\Delta kx\right] \sin\Delta\omega T\right\} \tag{8-4}$$

If $\Delta\omega T >> 1$, then the resultant intensity at the position x is equal to the sum of the intensities of the individual waves

$$I \propto \langle E^2 \rangle = \langle E_1^2 \rangle + \langle E_2^2 \rangle = |E_0|^2 \propto 2I_0$$

If a detector with a 100 kHz bandwidth $T \approx 10^{-5}$ is used to measure the intensity and the beat frequency is $\Delta\omega > 10^6$, or in terms of the wavelength $\Delta\lambda \approx 10^{-4}$ nm, then the time-bandwidth product is $\Delta\omega T \approx 10$. Under this condition, a 10% error is made when the second term in (**8-4**) is neglected. When the second term can be neglected, the source is said to be incoherent.

This result can be extended to the addition of N waves at N different frequencies. When the assumption that $\Delta\omega T >> 1$ holds, the resultant intensity is given by the sum of N individual intensities

$$\langle E^2 \rangle = \sum_{i=1}^{N} \langle E_i^2 \rangle$$

Real sources contain a continuous distribution of frequencies. For an incoherent source with an intensity distribution $I(\omega)$, the observed intensity is given by an integral over the frequency distribution of the intensity of the source

$$I_0 \propto \langle E^2 \rangle = \int_{\omega_1}^{\omega_2} I(\omega)\, d\omega \tag{8-5}$$

where $\omega_2 - \omega_1$ is the frequency bandwidth of the detector or the source, whichever is smaller. Equation (**8-5**) describes any source for which the beat frequency is not observable because of the detector's frequency response.

If $\Delta\omega T < 1$ so that $\sin\Delta\omega T \approx \Delta\omega T$, then (**8-4**) is written

$$\frac{I}{I_0} = 2 + 2\cos\left[\Delta\omega(2t_0 + T) - 2\Delta kx\right]$$

Because the detector's response time will allow the observation of the beat frequency, the wave is now considered to be coherent. To obtain the resultant intensity for the coherent waves, we must add amplitudes of the individual waves before calculating the intensity. (Since we know that interference can occur, we need to develop some measure that will tell us that $\omega_2 - \omega_1$ is small enough to allow interference.)

The view of interference as a function of a detector's response time led Forrester to conduct an experiment to demonstrate the production of beats from light waves. This first demonstration of photoelectric mixing, in 1955, involved two light waves with discrete frequencies separated by about 10 GHz. (In the derivation presented here, the wave's amplitude E_0 has been treated as a constant, but in the original experiment by Forrester, this was not the case. The noise introduced by the temporal variations of the amplitude made the experimental observation of beats very difficult.) Later, the concept was extended first to beating between the frequency components of a continuous distribution of light, and later, to beating between light produced by two different lasers. The former type of mixing, usually called *homodyne*, is associated with the application of light-beating spectroscopy and the second type of mixing, usually called *heterodyne*, is associated with the application of laser radar.

If the beat frequency is small enough, we may neglect it to obtain from (**8-4**),

$$\frac{I}{I_0} = 2 + 2\cos(2\Delta kx) = 2 + 2\cos\delta$$

which is identical to the result obtained in Chapter 4 during the analysis of interference between two waves of equal amplitude and frequency.

We have demonstrated that interference between waves of different frequencies is possible if the difference in frequency between the waves is sufficiently "small," but we have not quantified small. If we can quantify small, we will have a measure of the coherence of light. As the first step in developing that measure, we will look at the response of a Michelson interferometer to waves of different frequencies.

INTERFERENCE SPECTROSCOPY

The Michelson interferometer can be used to obtain a characterization of the coherence of a wave. To discover the origin of this characterization, we will calculate the form of the interference produced by a Michelson interferometer when the input wave contains two frequencies. The results of this analysis will then be extended to a continuous distribution of frequencies. We will assume that the detector cannot respond to the beat frequencies produced on the detector surface by the two waves. Because of this assumption, the output intensity of the interferometer is obtained by adding the intensities associated with each frequency contained in the input. We will find that measurement of the interferometer's output intensity, as we move one mirror, will generate information about the spectral content of the source.

We will begin the analysis by assuming that the light wave incident on the Michelson interferometer contains two monochromatic (single frequency) waves of frequencies ω_1 and ω_2 and of intensities I_1 and I_2. Because the beat frequency $(\omega_2 - \omega_1)/2$ is too high to be detected, the output intensity of the Michelson interferometer is calculated by adding the intensities associated with the interference pattern created by each frequency. Each frequency will produce an interference pattern at the detector whose intensity is given by (**4-31**). We add the two intensities

$$\begin{aligned} I_d &= I_{d1} + I_{d2} \\ &= I_1 + I_2 + I_1\cos\omega_1\tau + I_2\cos\omega_2\tau \qquad (8\text{-}6) \\ &= I_0(1 + \gamma) \end{aligned}$$

Just as in (**4-31**), the resultant intensity contains a constant term

$$I_0 = I_1 + I_2$$

and an oscillating term $I_0\gamma(\tau)$. For the current example of two frequencies, the normalized oscillating term is

$$\gamma(\tau) = \frac{I_1}{I_0}\cos\omega_1\tau + \frac{I_2}{I_0}\cos\omega_2\tau$$

If the two arms of the Michelson interferometer are the same length, we have $\tau = 0$ and γ has its maximum value

$$\gamma(\tau) = \frac{I_1 + I_2}{I_0} = 1$$

The minimum value that γ can have is $\gamma = 0$. Whenever $\gamma = 0$, there is no interference and the resultant intensity is

$$I_d = I_1 + I_2 = I_0$$

Examples of the output intensity of the interferometer is given in Figure 8-1 for three representative values of γ. The intensities plotted in Figure 8-1 are measured on axis as one mirror is moved to vary d.

By measuring γ, we can measure the separation between the two wavelengths present in the source illumination. To see how this measurement is accomplished, first note that the two wavelengths of the source can prevent the observation of interference. The loss of interference fringes is due to a filling in of the minimum intensity band associated with one frequency by the maximum intensity band of another frequency. The loss of visible fringes occurs when

$$\gamma(\tau) = \frac{I_1}{I_0} \cos \omega_1 \tau + \frac{I_2}{I_0} \cos \omega_2 \tau = 0$$

If

$$I_1 = I_2 = \tfrac{1}{2} I_0$$

the function $\gamma(\tau)$ for the two frequencies can be rewritten as

$$\gamma(\tau) = \cos(\frac{\omega_1 + \omega_2}{2})\tau \cos(\frac{\omega_1 - \omega_2}{2})\tau$$

which equals zero whenever

$$\frac{\omega_2 - \omega_1}{2}\tau = \pi(\nu_2 - \nu_1)\tau = \frac{\pi}{2}$$

$$\Delta\nu = \frac{1}{2\tau}$$

Physically, γ is zero when the maximum intensity of the fringe associated with one frequency occurs at the same mirror separation as the minimum intensity of the fringe associated with the second frequency.

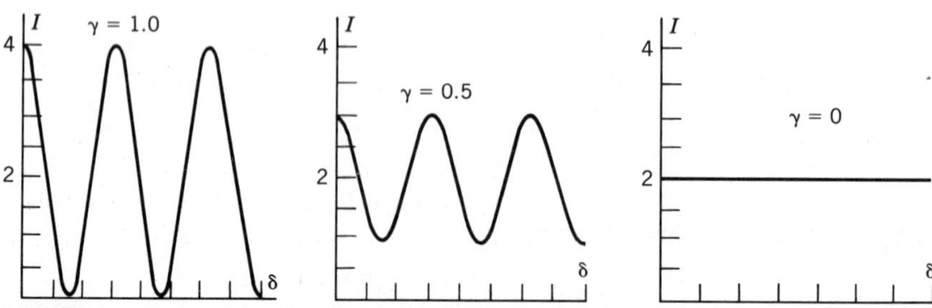

FIGURE 8-1. The intensity output of a Michelson interferometer as a function of mirror spacing for three values of the degree of coherence function.

We now have a technique for measuring the separation, in wavelength, of two monochromatic sources. We begin by adjusting the two arms of a Michelson interferometer so that they have equal optical path lengths. Light from the two monochromatic sources is combined and introduced into the Michelson interferometer. A detector monitors the central band at the output of the interferometer as one mirror is moved, increasing d. When the difference in the optical path length of the two arms becomes large, the output intensity will no longer vary as d changes ($\gamma = 0$). The value of d where the variation in intensity ceases can be used to calculate the wavelength separation between the two sources

$$\Delta\nu = \frac{c}{4d}.$$

The spectral resolving power needed to discriminate between two wavelengths, or frequencies, is defined as

$$\mathcal{R} = \left|\frac{\lambda_a}{\Delta\lambda}\right| = \left|\frac{\nu_a}{\Delta\nu}\right|$$

where

$$\lambda_a = \frac{\lambda_1 + \lambda_2}{2}$$

is the average wavelength and

$$\Delta\lambda = \lambda_2 - \lambda_1$$

is the resolvable wavelength separation. The maximum resolving power of a Michelson interferometer is given by

$$\mathcal{R}_{\text{max}} = 2\nu_a\tau_{\text{max}} = \frac{2d_{\text{max}}}{\lambda/2}$$

The length

$$d_{\text{max}} = \frac{c\tau_{\text{max}}}{2}$$

is the maximum displacement, measured from the position of zero optical path difference, that can be obtained with the Michelson interferometer.

Naturally occurring sources contain a distribution of frequencies; therefore, the theory must be expanded to include a continuous distribution of frequencies.

FOURIER TRANSFORM SPECTROSCOPY

We can extend the theory for interference produced by a source with two frequencies to a continuous distribution of frequencies. This extension will demonstrate that the coherence function of a light source can be measured using a Michelson interferometer. We will also learn that by taking the Fourier transform of the measured coherence function, we can obtain the spectral distribution of a light source. We will use a commonly occurring spectral distribution to learn how to measure the coherence function.

The resultant intensity at the output of a Michelson interferometer for a source with a continuous spectral distribution of intensity $I(\omega)$ is obtained by analogy with the two frequency results. The form of the resultant intensity, as was the case for (**8-6**), is given by

$$I_d(\tau) = \int_0^\infty I(\omega)(1 + \cos \omega\tau)\, d\omega \tag{8-7}$$

This integral contains the sum of constant and oscillatory terms. The constant term is

$$I_0 = \int_0^\infty I(\omega)\, d\omega$$

and the oscillatory term is

$$\int_0^\infty I(\omega) \cos \omega\tau\, d\omega$$

By defining a normalized intensity distribution of frequencies called the *spectral distribution function* or the *power spectrum* of the light source

$$P(\omega) = \frac{I(\omega)}{\int_0^\infty I(\omega)\, d\omega} = \frac{I(\omega)}{I_0} \tag{8-8}$$

the oscillating term can be written in a form independent of the incident intensity

$$\gamma(\tau) = \int_0^\infty P(\omega) \cos \omega\tau\, d\omega \tag{8-9}$$

The oscillatory function (**8-9**) called the *degree of coherence* is the cosine transform of $P(\omega)$. [Here, the degree of coherence is shown as the real part of the Fourier transform of $P(\omega)$. We would expect, in general, to find $\gamma(\tau)$ to be a complex function. See (**8-26**) for a general expression of $\gamma(\tau)$.]

As was the case for two frequencies, if the two arms of the Michelson interferometer are the same length $\tau = 0$ and we have the maximum value for γ,

$$\gamma(0) = \int_0^\infty P(\omega)\, d\omega = 1$$

If there is no interference, the resultant intensity (**8-7**) is obtained by integrating over $I(\omega)$. From (**8-8**), we know that $I(\omega)$ equals the fraction of light contained in an interval between ω and $\omega + d\omega$, $P(\omega)$, multiplied by the total intensity I_0. The integrand is therefore the spectral intensity distribution. Performing the integration over this function produces a resultant intensity that equals the incident intensity

$$I_d(\tau) = \int_0^\infty I(\omega)\, d\omega = \int_0^\infty I_0 P(\omega)\, d\omega = I_0$$

We see that for $\gamma = 0$, the output intensity of the interferometer is a constant, independent of the difference in length between the two interferometer arms $d = \tau c/2$. For this condition, the wave is said to be incoherent. We find that the degree of coherence $\gamma(\tau)$ has a total range of values

$$|\gamma(\tau)| \le 1$$

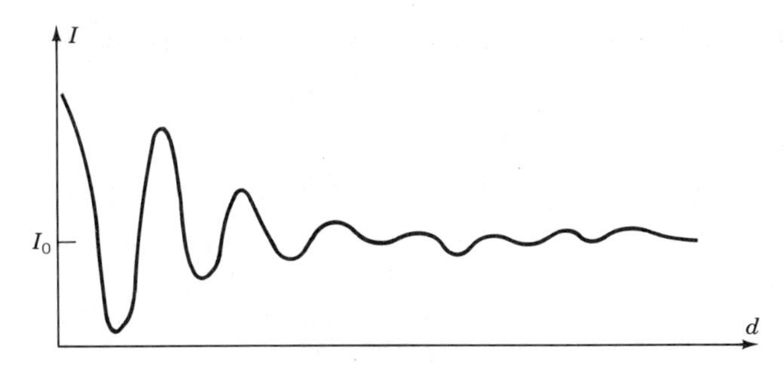

FIGURE 8-2. A typical interferogram from a Michelson interferometer.

Because γ describes the interference part of the intensity distribution of the Michelson interferometer, it is the quantitative measure of coherence we desire. Our discussion suggests that we should be able to measure the coherence function of any light source with the Michelson interferometer. From (**8-9**), we infer that by taking the Fourier transform of the measured coherence function, the spectral distribution of a light source can be recovered. Michelson used this procedure to determine the shape of a number of spectral emission lines such as the hydrogen line at $\lambda = 656.3$ nm. The approach developed by Michelson was ignored for many years because the Fabry–Perot interferometer was much easier to use. Now, with the availability of small computers and the fast Fourier transform algorithm, the technique is very popular in the far-infrared region of the spectrum.

A typical output from a Michelson interferometer might look like Figure 8-2. Upon examining Figure 8-2, it is not clear what quantity, associated with this output, should be measured to obtain the coherence function. We will calculate the output of a Michelson interferometer for a commonly occurring spectral distribution to discover the form of the degree of coherence. In performing this calculation, we will discover the parameter that must be measured to obtain the degree of coherence.

Gaussian Spectral Distribution

We will calculate the coherence function of an atomic emission line from a gas at low pressure. The spectral distribution of this line is Gaussian rather than the Lorentzian shape of Figure 7-7. The physical origin of this spectral distribution is the atomic motion in the gas.

When an atom or a molecule in a low-pressure gas emits radiation, a sharp quasimonochromatic line at frequency ν_0 is emitted *in the atom's rest frame.* If the atom is moving at a relative velocity v toward or away from the viewer, then the observer will see a frequency

$$\nu = \nu_0\left(1 \pm \frac{v}{c}\right) \tag{8-10}$$

This effect is called the *Doppler effect* after **Christian Doppler (1803–1853)**, who explained the frequency shift for sound waves in 1842. The explanation of the effect is quite simple. As the source of light moves toward the observer, the wave crests emitted at times t_0 and $t_0 + \Delta t$ are closer to each other than would be the case for an at rest source because the source has moved forward. More wave crests arrive, in a unit time, at the observer due to the motion, and the frequency, measured by the observer, is higher than if the source were at rest. [**Armand H.L. Fizeau(1819–1896)** made the first correct application of the Doppler effect to starlight and showed how it could be used to measure stellar velocities.]

If the molecules are in thermal equilibrium, classical statistical mechanics states that they will have a Maxwellian velocity distribution. The spectral distribution function, resulting from the motion of the molecules in thermal equilibrium, has the form

$$P(\omega) = \sqrt{\pi \tau_d} \exp\left[-(\omega - \omega_0)^2 \left(\frac{\tau_d}{2}\right)^2\right] \tag{8-11}$$

This is a Gaussian line centered at the frequency ω_0 with a width of

$$\Delta\omega = \frac{1}{\tau_d}$$

From the kinetic theory of gases, the dependence of the linewidth on the physical properties of the gas is

$$\tau_d^2 = \frac{c^2 m}{8\omega_0^2 kT}$$

(The Lorentzian spectral distribution of Figure 7-7 is observed if the duration of the oscillations is shorter than the time τ_d, see Problem 8-1. We associate the interruption of the oscillations with atomic collisions. Thus, a Lorentzian distribution is said to be pressure broadened while the Gaussian distribution is said to be thermally broadened).

The Gaussian spectral distribution (**8-11**) for atomic or molecular emission from a low-pressure gas can be used to calculate $\gamma(\tau)$ by calculating the Fourier transform of $P(\omega)$. $P(\omega)$ has the same form as the transform of (**6-20**). We can therefore immediately write

$$\gamma(\tau) = e^{-(\tau/\tau_d)^2} \cos \omega_0 \tau \tag{8-12}$$

$$= v(\tau) \cos \omega_0 \tau \tag{8-13}$$

The degree of coherence is made up of a rapidly varying term $\cos(\omega_0 \tau)$ and a slowly varying term $v(\tau)$ that modulates the amplitude of the rapidly varying term. When $\tau << \tau_d$, only the rapidly varying component

$$\gamma(\tau) \approx \cos \omega_0 \tau$$

is important. This rapidly varying term, equivalent to (**4-31**), is the result expected for a monochromatic wave. When τ is on the order of τ_d, then the effects of the finite frequency spread, associated with (**8-11**), can be observed in the slower varying term

$$v(\tau) \approx e^{-(\tau/\tau_d)^2}$$

The combined terms in (**8-13**) are due to the fact that the spectral distribution is centered about ω_0. If the center of the spectral distribution were shifted to zero from ω_0, then $v(\tau)$ would equal the Fourier transform of the shifted function.

FRINGE CONTRAST AND COHERENCE

The slowly varying function $v(\tau)$ discussed above is quite easy to measure. [Remember from (**4-30**) that $\tau = 2d/c$, where d is the difference in length of the two arms of the Michelson interferometer; thus, we normally have no trouble accessing all values of τ.] We will show that it is equivalent to a parameter called the fringe contrast by Michelson.

If we illuminate the Michelson interferometer with light from a Doppler-broadened transition, described by (**8-11**), then the intensity output of the interferometer, as the mirror in one arm is moved, would be

$$I_d(\tau) = I_1 + I_2 + 2\sqrt{I_1 I_2}\; v(\tau) \cos \omega_0 \tau \tag{8-14}$$

We define a parameter, called the *fringe visibility*, by the equation

$$\mathcal{V} = \frac{I_{\text{max}} - I_{\text{min}}}{I_{\text{max}} + I_{\text{min}}} \tag{8-15}$$

The maximum intensity observed at the output of the Michelson interferometer is

$$I_{\text{max}} = I_1 + I_2 + 2\sqrt{I_1 I_2}\; v(\tau)$$

and the minimum intensity is

$$I_{\text{min}} = I_1 + I_2 - 2\sqrt{I_1 I_2}\; v(\tau)$$

The fringe visibility is thus

$$\mathcal{V} = \frac{2\sqrt{I_1 I_2}}{I_1 + I_2}\, v(\tau) \tag{8-16}$$

If $I_1 = I_2$, then $\mathcal{V} = v(\tau)$; the visibility of the fringes is equal to the degree of coherence. [If we used complex valued field distributions, then $v(\tau)$ would be complex and the fringe visibility would be equal to the real part of $v(\tau)$.] We now have an experimental method of obtaining the coherence of a light source that is a general result and not limited to the Gaussian distribution used in the example.

TEMPORAL COHERENCE TIME

The arms of the Michelson interferometer act as delay lines, allowing light waves that were generated at different times to interfere. Because the interfering light waves differ in their time origin, the coherence measured by the Michelson interferometer is called *temporal coherence*.

The degree of coherence function completely characterizes the temporal coherence of a source but is too complicated for general use. We need a simple parameter that can be used to specify the amount of coherence present in a light source. The parameter used to characterize temporal coherence will be a characteristic time, called the *coherence time*; it specifies the duration of time that a source maintains its phase.

The wave groups, shown in Figures 7-2 and 7-3, were created by the addition of a number of waves of different frequencies. The temporal duration of a group is equal to the time that the different source frequencies, contributing to the group, remain in a fixed phase relationship and can therefore be interpreted as the coherence time of the source. To calculate the group duration, and thus the coherence time, we find the time between zeroes in the amplitude modulation at a fixed position in space

$$(\Delta\omega t_2 - \Delta k x_1) - (\Delta\omega t_1 - \Delta k x_1) = \Delta\omega(t_2 - t_1)$$

$$= (2m + 1)\frac{\pi}{2} - (2m - 1)\frac{\pi}{2}$$

We find that the coherence time is equal to

$$\tau_c = t_2 - t_1 = \frac{\pi}{\Delta\omega}$$

In this section, a more formal approach will be used to obtain the coherence time. We will discover that the coherence time for a Gaussian spectral distribution is proportional to the reciprocal of the spectral width of the source $\tau_c \propto 1/\Delta\omega$, agreeing with the simple definition of coherence time just introduced. The reciprocal relationship between coherence time and spectral linewidth should not be much of a surprise; it has already been demonstrated that the spectral distribution and the coherence function are a Fourier transform pair.

One characterization of a well-behaved function such as γ^2 would be the root-mean-square width of the function

$$\tau_c^2 = \frac{\int_{-\infty}^{\infty} (\tau - \langle\tau\rangle)^2 \gamma^2(\tau)\, d\tau}{\int_{-\infty}^{\infty} \gamma^2(\tau)\, d\tau} \tag{8-17}$$

The average value of τ is zero since $\gamma^2(\tau)$ is an even function; thus, (8-17) can be rewritten

$$\tau_c^2 = \frac{\int_{-\infty}^{\infty} \tau^2 \gamma^2(\tau)\, d\tau}{\int_{-\infty}^{\infty} \gamma^2(\tau)\, d\tau} \tag{8-18}$$

Another equally adequate definition of the coherence time, used extensively in the coherence theory literature, can be obtained by approaching coherence theory from a statistical point of view. By treating the lack of interference as arising from intensity fluctuations, the definition of the coherence time becomes

$$\tau_c = 4\int_0^{\infty} \gamma^2(\tau)\, d\tau \tag{8-19}$$

[Coherence will be associated with amplitude fluctuations in the next section, but a detailed treatment that included a discussion of intensity fluctuations and a derivation of (**8-19**) would take us too far astray. The interested reader can find a discussion of this subject in the literature.[38]]

We will apply the definition given in (**8-19**) to find the coherence time for a Doppler-broadened line with a spectral distribution given by (**8-11**) and a degree of coherence (or a coherence function) given by (**8-12**). If we use (**8-12**), the coherence time can be quickly calculated

$$\tau_c = 4\int_0^{\infty} \cos^2 \omega_0\tau e^{-2(\tau/\tau_d)^2} d\tau$$

$$= 2\int_0^{\infty} e^{-2(\tau/\tau_d)^2} d\tau + 2\int_0^{\infty} \cos 2\omega_0\tau e^{-2(\tau/\tau_d)^2} d\tau$$

$$\tau_c = \tau_d\sqrt{\frac{\pi}{2}}\{1 + \exp[-(\omega_0\tau_d)^2/2]\}$$

The frequency ω_0^2 is a very large number (somewhere around 10^{28} for visible light); thus, the second term in the equation for τ_c is very small and may be neglected. By neglecting the second term, the coherence time becomes

$$\tau_c \approx \tau_d \sqrt{\frac{\pi}{2}}$$

The line width of the Gaussian line is $\Delta\omega \propto 1/\tau_d$; thus, $\tau_c \propto 1/\Delta\omega$, agreeing with the earlier result. In fact, it is easy to show that for a Gaussian line, $\tau_c \Delta\omega > 1/4\pi$. We have proven that very narrow spectral lines are very coherent, as we expected at the beginning of the analysis.

If the wave was frozen in time, the interference could be thought of as occurring between the wave at two different positions, in space, along the direction of propagation. For this reason, temporal coherence is also called *longitudinal coherence*. If we use this spatial concept of temporal coherence, the coherence time is equivalent to a *coherence length* through the relationship

$$\ell_\ell = c\tau_c = \frac{c}{\Delta\nu} \tag{8-20}$$

In a Michelson interferometer, the coherence length is equal to twice the difference in optical path length between the arms of the Michelson interferometer, measured at the position when the fringe visibility goes to zero. Applying the concept of coherence length to the wave group discussion above leads to a view of the coherence length as the spatial extent of a wave group, frozen in time.

We can also express ℓ_ℓ in terms of wavelength,

$$\frac{\Delta\nu}{\nu} = \frac{|\Delta\lambda|}{\lambda}$$

$$\ell_\ell = \frac{\lambda^2}{\Delta\lambda} \tag{8-21}$$

For a sodium lamp, $\lambda \approx 500$ nm and $\Delta\lambda = 0.1$ nm (see Table 7A-2 for the wavelengths of the two visible lines produced by a sodium lamp), which gives $\ell_\ell \approx 2$ mm. For a laser, typical values of ℓ_ℓ are 10 m for a 1 mW HeNe laser and 4 cm for an argon laser. Larger coherence lengths can be produced in the visible region of the spectrum but only with effort. Laser sources with coherence lengths on the order of kilometers can be produced in the infrared. This is a result of the fact that changes in dimensions due to thermal expansion and vibration are a small fraction of a wavelength at infrared wavelengths, making it easier to stabilize the infrared laser.

AUTOCORRELATION FUNCTION

Another way to view the loss of interference in a Michelson interferometer is that it is a result of the random fluctuations, in time, of the light wave's amplitude and/or phase. Up to now, all waves were assumed to have amplitudes A and phases $(-kx + \phi)$ that were independent of time. By approaching interference in terms of fluctuating phases and amplitudes, we find that interference can be associated with the mathematical operation of correlation. If we use this mathematical formalism, the coherence function can be shown to be equal to the autocorrelation of a wave. From this viewpoint, the concept of coherence is seen to be a statement concerning the similarity of two waves. We will discover that this mathematical view of correlation is consistent with the view of coherence in terms of beat frequencies or the Fourier transform of the power spectrum.

The complex representation of the light wave is

$$\mathcal{E}(t) = A(t) \exp\{ i[\omega_0 t + \phi(t)] \}$$

where we are allowed to suppress the spatial dependence by the experimental configuration of a Michelson interferometer. The wave in one arm of the interferometer occurred at time t and the wave in the other arm occurred at the earlier time $(t-\tau)$, where $\tau = 2d/c$. The amplitudes in the two arms are added and the detected interference signal is the temporal average of the resultant wave

$$I_d = \frac{1}{T}\int_{t_0-(T/2)}^{t_0+(T/2)} \left[\mathcal{E}(t) + \mathcal{E}(t-\tau)\right]\left[\mathcal{E}(t) + \mathcal{E}(t-\tau)\right]^* dt$$

After performing the multiplication of the integrand, we can separate the products into three temporal averages

$$\begin{aligned} I_d &= \frac{1}{T}\int_{t_0-(T/2)}^{t_0+(T/2)} \mathcal{E}(t)\mathcal{E}^*(t)\,dt + \frac{1}{T}\int_{t_0-(T/2)}^{t_0+(T/2)} \mathcal{E}(t-\tau)\mathcal{E}^*(t-\tau)\,dt \\ &+ \frac{1}{T}\int_{t_0-(T/2)}^{t_0+(T/2)} \left[\mathcal{E}^*(t)\mathcal{E}(t-\tau) + \mathcal{E}(t)\mathcal{E}^*(t-\tau)\right] dt \end{aligned} \tag{8-22}$$

The first two integrals in (**8-22**) are individually equal to the intensity in one of the interferometer's arms; here, we will assume that they are each equal to I_0. The third integral contains all the interference effects. If we assume that the statistics are stationary, i.e., independent of the origin of time, then the intensity output of the interferometer is

$$I_d = 2I_0 + \frac{2}{T}\int_{t_0-(T/2)}^{t_0+(T/2)} \mathcal{E}(t)\mathcal{E}^*(t-\tau)\,dt = 2I_0\left[1 + \gamma(\tau)\right] \tag{8-23}$$

This is of the same form as (**8-6**). Equating like terms, we see that

$$\gamma(\tau) = \frac{\langle \mathcal{E}(t)\mathcal{E}^*(t-\tau)\rangle}{\langle \mathcal{E}(t)^2\rangle} = \frac{\int_{t_0-(T/2)}^{t_0+(T/2)} \mathcal{E}(t)\mathcal{E}^*(t-\tau)\,dt}{\int_{t_0-(T/2)}^{t_0+(T/2)} \mathcal{E}(t)\mathcal{E}^*(t)\,dt} \tag{8-24}$$

Comparing (**8-24**) to (**6-32**) and assuming that T is so large that the integral is independent of t_0, i.e., in the limit as $T \to \infty$, we find that the degree of coherence is now defined in terms of the normalized autocorrelation function. The assumption that T is very large is equivalent to the earlier assumption that the detector could not record the beat frequency.

In general, the intensities in the two arms of the interferometer would not be equal and we would define the degree of coherence in terms of the cross correlation of the two waves

$$\begin{aligned} \gamma_{12}(\tau) &= \frac{\int_{-\infty}^{\infty} \mathcal{E}_1(t)\mathcal{E}_2^*(t-\tau)\,dt}{\left[\int_{-\infty}^{\infty} \mathcal{E}_1(t)\mathcal{E}_1^*(t)\,dt\right]^{1/2}\left[\int_{-\infty}^{\infty} \mathcal{E}_2(t)\mathcal{E}_2^*(t)\,dt\right]^{1/2}} \\ &= \frac{\langle \mathcal{E}_1(t)\mathcal{E}_2^*(t-\tau)\rangle}{\sqrt{\langle \mathcal{E}_1(t)\mathcal{E}_1^*(t)\rangle}\sqrt{\langle \mathcal{E}_2(t)\mathcal{E}_2^*(t)\rangle}} \end{aligned} \tag{8-25}$$

[We have been somewhat sloppy in our notation in allowing $\gamma(\tau)$ to represent a real function sometimes and a complex function other times. We are only interested in $\mathcal{Re}\{\gamma(\tau)\}$ and, as always, expect the reader to take the necessary care in manipulating the function when complex notation is used.]

We now have two definitions for $\gamma(\tau)$

1. The Fourier transform of the power spectrum.
2. A correlation between two samples of the wave.

For the two definitions to be equivalent, we must assume that the beats are unobservable. We have made this assumption by interpreting (**8-24**) as a correlation operation; thus, we can write

$$\gamma(\tau) = \frac{\langle \mathcal{E}(t)\mathcal{E}^*(t-\tau)\rangle}{\langle \mathcal{E}(t)^2\rangle} = \frac{\int_{-\infty}^{\infty} I(\omega)e^{-i\omega\tau}\, d\omega}{\int_0^{\infty} I(\omega)\, d\omega} \tag{8-26}$$

The equivalence between the two representations stated in (**8-26**) is based on the *Wiener–Khintchine theorem*. To prove this theorem would take us too far from optics,[5] but it should be noted that the key assumption for the proof of the theorem is based on the absence of interference between different frequencies. This is equivalent to stating that T is long, $T \rightarrow \infty$, and the beats between the two frequencies cannot be observed.

We can examine (8-24) in terms of a general light field

$$\begin{aligned}\mathcal{E}(t)\mathcal{E}^*(t-\tau) &= A(t)A(t-\tau)\exp\{i\,[\omega_0(t-t+\tau) + \phi(t) - \phi(t-\tau)]\} \\ &= A(t)A(t-\tau)\exp\{i\,[\omega_0\tau + \phi(t) - \phi(t-\tau)]\}\end{aligned} \tag{8-27}$$

Only the real part of the coherence function is of interest

$$\begin{aligned}\gamma(\tau) &= \frac{\int_{-\infty}^{\infty} A(t)A(t-\tau)\cos[\omega_0\tau + \phi(t) - \phi(t-\tau)]\, dt}{\int_0^{\infty} A(t)^2\, dt} \\ &= \frac{\langle A(t)A(t-\tau)\cos[\omega_0\tau + \phi(t) - \phi(t-\tau)]\rangle}{\langle A(t)^2\rangle}\end{aligned} \tag{8-28}$$

This relationship states that the interference pattern, represented by $\gamma(\tau)$, is for small fluctuations a function of $\cos\omega_0\tau$. For signals where $A(t)$ is a constant, a condition obtainable with a laser, the result is in agreement with (**8-3**). The phase difference $\Delta\phi = \phi(t) - \phi(t-\tau)$ in this case would be interpreted as arising from random frequency variations about a mean frequency.

SPATIAL COHERENCE

Temporal coherence is associated with the wave's properties along the direction of propagation. A second type of coherence can be defined that is associated with the wave's properties transverse to the direction of propagation. If the wave is a perfect plane wave, then there is a uniform phase in a plane perpendicular to the direction of propagation. The phase may exhibit fluctuations, but all points in the plane will have identical fluctuations. The perfect plane wave is said to be *spatially coherent*. The temporal coherence is associated with the frequency distribution of the source, whereas the spatial coherence is associated with a distribution of propagation vectors $\mathbf{k}$ associated with the wave, i.e., with a departure of the wave from the ideal plane wave. We will see, in our later discussion of diffraction, that complete spatial coherence is as difficult to obtain as complete temporal coherence.

The theoretical development of this concept is identical to the development just completed for temporal coherence. By a simple redefinition of the retardation time τ and a generalization of the degree of coherence function, we will demonstrate that the same equations apply to the two classes of coherence. This is due to the fact that the two forms of coherence are really not different. We separate the two types only to increase our ability to understand coherence.

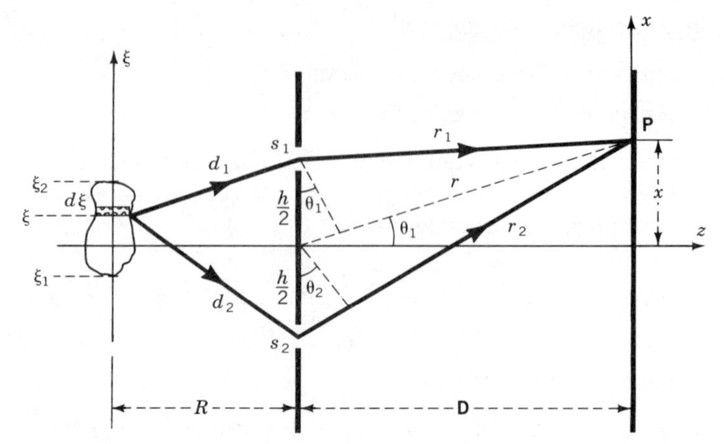

Figure 8-3. Young's two slit experiment with an extended source.

In our discussion, Young's two-slit experiment, as displayed in Figure 8-3, will play the same role as Michelson's interferometer did in the discussion of temporal coherence. The intensity at point P in Figure 8-3 is from **(4-11)**

$$\langle \mathbf{E}^2 \rangle = \left\langle \mathbf{E}_1^2 \left(t + \frac{d_1}{c} + \frac{r_1}{c} \right) \right\rangle + \left\langle \mathbf{E}_2^2 \left(t + \frac{d_2}{c} + \frac{r_2}{c} \right) \right\rangle$$

$$+ 2 \left\langle \mathbf{E}_1 \left(t + \frac{d_1}{c} + \frac{r_1}{c} \right) \mathbf{E}_2 \left(t + \frac{d_2}{c} + \frac{r_2}{c} \right) \right\rangle$$

where

$$t + \frac{d_1}{c} + \frac{r_1}{c}$$

is the time for light to travel from the source to point P by way of slit 1 and

$$t' = t + \frac{d_2}{c} + \frac{r_2}{c}$$

is the time to travel from the source to point P by way of slit 2. The phase difference associated with propagation from the source to the slits is given by

$$\Delta\phi = \frac{d_2 - d_1}{c}$$

and the difference in propagation time from the two slits to point P is

$$\tau = \frac{r_2 - r_1}{c}$$

In terms of these newly defined parameters, the intensity at point P is

$$\langle \mathbf{E}^2 \rangle = \langle \mathbf{E}_1^2(t' - \Delta\phi - \tau) \rangle + \langle \mathbf{E}_2^2(t') \rangle + 2\langle \mathbf{E}_1(t' - \Delta\phi - \tau)\, \mathbf{E}_2(t') \rangle$$

The difference in propagation time in the Young's two-slit experiment τ is equivalent to the retardation time used in temporal coherence. They both temporally define the waves that are to undergo interference. The retardation time associated with temporal coherence was defined in terms of the Michelson interferometer

$$\tau = \frac{2d}{c}$$

where d is the difference in length of the two arms of the Michelson interferometer. The retardation time is redefined for spatial coherence in terms of the parameters associated with Young's two-slit experiment

$$\tau = \frac{r_2 - r_1}{c}$$

All of the equations developed in temporal coherence can be utilized in spatial coherence theory by using this new retardation time.

A new quantity, called the *mutual coherence function*, $\Gamma_{12}(\tau)$ can now be defined

$$\Gamma_{12}(\tau) = \langle \mathbf{E}_1(t' - \Delta\phi - \tau)\, \mathbf{E}_2(t') \rangle$$

$$= \frac{1}{2T}\int_{-T}^{T} \mathbf{E}_1(t' - \Delta\phi - \tau)\, \mathbf{E}_2(t')\, dt'$$

This function is a correlation integral. The two waves that are compared are samples of the wavefront taken by slits 1 and 2. The intensities passing through each of the two slits can be expressed in terms of the mutual coherence function. They are the values of what are now the autocorrelation functions $\Gamma_{11}(0)$ and $\Gamma_{22}(0)$ at $\tau = 0$

$$\Gamma_{11}(0) = \langle \mathbf{E}_1^2 \rangle, \qquad \Gamma_{22}(0) = \langle \mathbf{E}_2^2 \rangle$$

The normalized version of $\Gamma_{12}(\tau)$ is the degree of spatial coherence

$$\gamma_{12}(\tau) = \frac{\langle \mathbf{E}_1(t)\mathbf{E}_2(t-\tau) \rangle}{\sqrt{\langle \mathbf{E}_1(t)^2 \rangle}\sqrt{\langle \mathbf{E}_2(t)^2 \rangle}} \tag{8-29}$$

or in terms of the Γ notation

$$\gamma_{12}(\tau) = \frac{\Gamma_{12}(\tau)}{\sqrt{\Gamma_{11}(0)\Gamma_{22}(0)}}$$

In general, $\gamma_{12}(\tau)$ is a complex function of the form

$$\gamma_{12}(\tau) = \left|\gamma_{12}(\tau)\right| e^{i\delta_{12}(\tau)}$$

where the real part of this relation is identical to (**8-13**).

$$\mathcal{R}\{\gamma(\tau)\} = \left|\gamma(\tau)\right| \cos \delta(\tau)$$

$$= v(\tau) \cos \omega_0 \tau$$

The intensity at point P is produced by interference between the waves from slit 1 and 2 is given by

$$I_P = I_1 + I_2 + 2\sqrt{I_1 I_2}\left|\gamma_{12}(\tau)\right| \cos \delta_{12}(\tau)$$

which is identical to our result in Chapter 4 (**4-12**), except for $|\gamma_{12}(\tau)|$. It is obvious from this expression that the coherence function $\gamma_{12}(\tau)$ reduces the fringe visibility when the degree of coherence is less than 1. The degree of coherence $\gamma_{12}(\tau)$ is a function of $r_2 - r_1$ and thus describes how the visibility of the interference fringes vary as the observation point P is moved over the observation plane.

A LINE SOURCE

Although the identical formalism of spatial coherence and temporal coherence makes it easy to develop coherence theory, it does not help in obtaining physical insight into the theory. We will expand our physical understanding of spatial coherence theory by performing a very simple experiment. We will examine Young's two-slit experiment, shown in Figure 8-3, following the discussion of Chapter 4; however, the experiment will be modified by the addition of a source of finite size. We will calculate the appearance of the Young's fringes due to the source size and shape. The result will anticipate findings to be obtained in the discussion of diffraction—that spatial coherence limits the region, in space, over which Young's interference fringes can be observed.

We will assume that the source is a line source lying along the ξ axis and extending from ξ_1 to ξ_2, where $\xi_2 - \xi_1 = s$ is the source's length. Initially, we will consider only an element of size $d\xi$ located at position ξ above the optical axis. The derivation we carried out in Chapter 4 has now an added complication; we must evaluate the phase change

$$\Delta\phi = \phi_2 - \phi_1$$

which here is

$$\Delta\phi = \frac{d_2 - d_1}{c}$$

The optical path difference from the source to P by way of the two slits is

$$\Delta = n\left[(d_2 + r_2) - (d_1 + r_1)\right] \tag{8-30}$$

(We will assume $n = 1$ in this discussion.) We derived, in Chapter 4, the relationship

$$c\tau = r_2 - r_1 = \frac{xh}{D}$$

We now wish to perform an equivalent derivation for

$$d_2 - d_1 = c\Delta\phi$$

using Figure 8-3.

The distances from the source to the two slits are given by

$$d_1^2 = R^2 + \left(\frac{h}{2} - \xi\right)^2, \qquad d_2^2 = R^2 + \left(\frac{h}{2} + \xi\right)^2$$

The difference between these two equations is

$$d_2^2 - d_1^2 = R^2 + \frac{h^2}{4} + h\xi + \xi^2 - R^2 - \frac{h^2}{4} + h\xi - \xi^2$$

$$= 2h\xi,$$

which can be written as the product between the sum and difference of the two distances

$$(d_2 - d_1)(d_2 + d_1) = 2h\xi$$

Since the distance from the source to the slit plane R is large and the slit spacing is very small compared to this distance, $R >> h$, we may write $d_1 \approx d_2 \approx R$. If we use this approximation, the path difference from the source to the two slits is

$$d_2 - d_1 \approx \frac{h\xi}{R} \tag{8-32}$$

From (**4-12**), we know that the intensity at P for a point source is

$$I_P = I_1 + I_2 + 2\sqrt{I_1 I_2}\cos\delta$$

where from (**4-16**),

$$\delta = k\Delta = \frac{kh\xi}{R} + \frac{khx}{D} \tag{8-33}$$

We will assume that the total light reaching P from each pinhole, illuminated by $d\xi$, is the same

$$dI_1 = dI_2 = I_0 P(\xi)\, d\xi$$

where $P(\xi)$ is the normalized spatial intensity distribution across the source. For this problem, the intensity distribution is that of a line of length $\xi_2 - \xi_1 = s$.

$$P(\xi) = \begin{cases} 1, & \xi_2 \geq \xi \geq \xi_1 \\ 0, & \text{all other } \xi \end{cases} \tag{8-34}$$

The intensity at point P from the source element $d\xi$ is

$$dI_P = 2(1 + \cos\delta) I_0 P(\xi)\, d\xi \tag{8-35}$$

To obtain the total intensity at point P, we must integrate **(8-35)** over the source

$$I_P = 2I_0 \int P(\xi)(1 + \cos\delta)\, d\xi$$

By analogy with (**8-9**), the degree of spatial coherence must be

$$\gamma = \int P(\xi)\cos\delta\, d\xi \tag{8-36}$$

$$= \int_{\xi_1}^{\xi_2} P(\xi)\cos\left(\frac{kh\xi}{R} + \frac{khx}{D}\right) d\xi \tag{8-37}$$

Using a trigonometric identity, we can write

$$\gamma = \cos\left(\frac{khx}{D}\right)\int P(\xi)\cos\left(\frac{kh\xi}{R}\right)d\xi - \sin\left(\frac{khx}{D}\right)\int P(\xi)\sin\left(\frac{kh\xi}{R}\right)d\xi \quad (8\text{-}38)$$

The two terms of (**8-38**) are the cosine and sine transforms of $P(\xi)$. As was the case for temporal coherence, where the degree of coherence is the Fourier transform of the spectral distribution of the light source, here the spatial coherence is the Fourier transform of the spatial distribution of the light source.

If the source is centered on the optical axis of the experiment (the z axis), then because $\sin(kh\xi/R)$ is odd, the second integral of (**8-38**) is zero and the problem simplifies to the calculation of an integral identical in form to (**6-17**). The coherence function for this source is

$$\gamma = s\frac{\sin(khs/2R)}{khs/2R}\cos(khx/D) \quad (8\text{-}39)$$

A computer simulation of the interference pattern that would be observed with the line source is shown in Figure 8-4. In Figure 8-4*a*, the source size was held constant and the pinhole spacing h was varied, as one might do in the measurement of the spatial coherence of a source. In Figure 8-4*b*, the pinhole spacing was held constant and the source size was varied, demonstrating the importance of source size on the observation of interference fringes.

The fact that the spatial coherence is the Fourier transform of the spatial distribution of the light source is a manifestation of the *van Cittert–Zernike theorem*. More precisely, the van Cittert-Zernike theorem states that

The complex degree of coherence is equal to the diffraction pattern of an aperture of the same size and shape as the source and illuminated by a spherical wave whose amplitude distribution is proportional to the intensity distribution across the source.[6]

Diffraction theory will be applied to Young's two-slit experiment in Chapter 10 to obtain a result identical to **(8-38)** [see **(10-31)**].

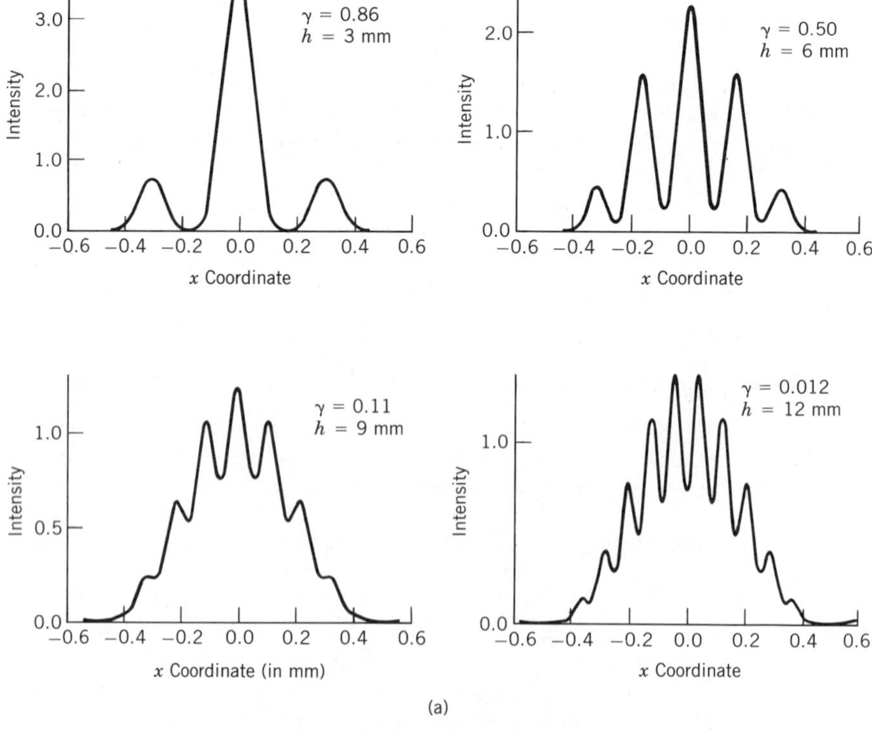

Figure 8-4*a*. Interference pattern from two pinholes with a fixed, finite source size of s = 0.1 mm. The pinhole spacing h was varied over a factor of 4 (h = 3, 6, 9, and 12 mm). The degree of coherence is described by sinc($khs/2R$) and is the cause of the fringe visibility dropping to zero and then increasing. The overall shape of each pattern is due to diffraction associated with a 1mm slit.

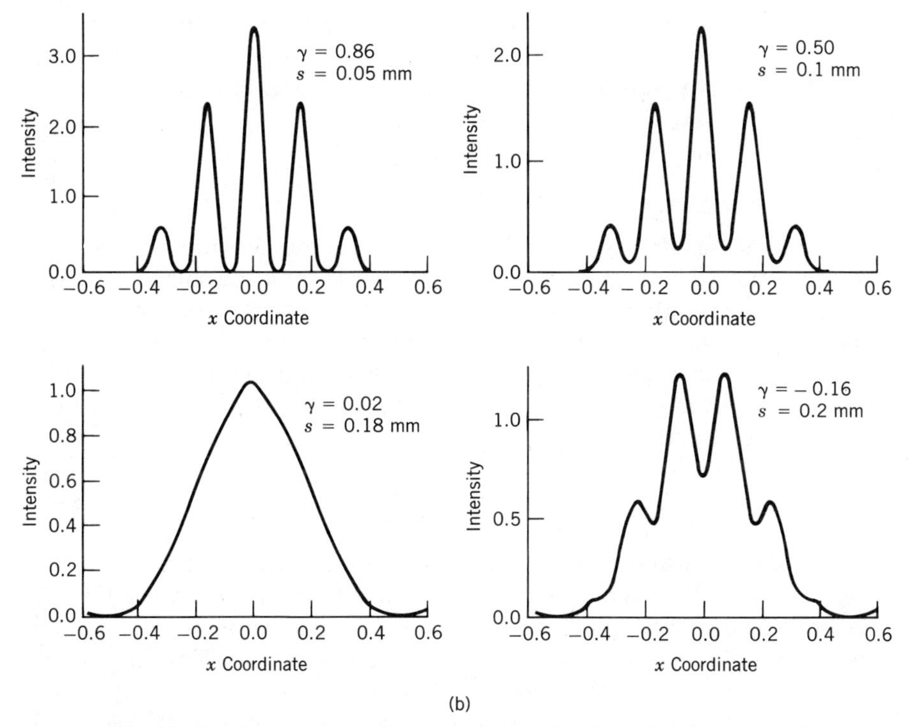

Figure 8-4*b*. The interference pattern for two pinholes with a fixed pinhole spacing of h = 6mm. The source size was varied by a factor of 4 (s = 0.05, 0.1, 0.18, and 0.2mm). All other parameters associated with these interference patterns were the same, i.e., the slit width was 1mm and $R = D =$ 2m.

SPATIAL COHERENCE LENGTH

The operational definition of the longitudinal coherence length equated it to the difference in length between the two arms of the Michelson interferometer when the fringe visibility became zero. A *transverse coherence length* ℓ_t can be defined in an analogous way, in terms of Young's two-slit experiment. It represents a distance, in a plane normal to the direction of propagation, over which the phases at two points remain correlated and equals the slit separation for which the interference fringes disappear.

The transverse coherence length can be defined in terms of the spatial extent of the light source. To derive the relationship between source size and the transverse coherence length, we use the geometry of Figure 8-3. The distance between bright bands on the plane containing point P was derived in Chapter 4 and an outline of that derivation is repeated here. The optical path change in going from one bright band to the next is

$$k(\Delta_1 - \Delta_2) = 2\pi$$

From the geometry of Figure 8-3, the path change is

$$\frac{h\xi}{R} - \frac{h\xi}{R} + \frac{hx_1}{D} - \frac{hx_2}{D} = \lambda$$

The spacing between the bright bands is therefore

$$x_1 - x_2 = \frac{\lambda D}{h}$$

which agrees with (**4-17**).

The magnitude of the fringe spacing is independent of the source position ξ. However, if the source position ξ is changed, the position of the fringes moves, as can be seen by taking the variation of (**8-33**) at constant phase δ

$$\frac{\delta\xi}{R} = -\frac{\delta x}{D} \tag{8-40}$$

where δx is the distance a band moves when the source is moved a distance $\delta\xi$ (the minus sign means that the fringes move up if the source is moved down). If there are two sets of fringes created by two points on a source located at ξ_1 and ξ_2, then the fringe visibility is destroyed when a bright band due to the light from ξ_1 falls on a dark band produced by ξ_2. This cancellation of fringe visibility occurs when $\delta x = \lambda D/2h$, which corresponds to a source displacement of

$$\delta\xi = -R\frac{\lambda}{2h} \tag{8-41}$$

Equation (**8-41**) allows the calculation of the maximum source size that can produce interference fringes at the observation point, a distance R away, when Young's slit spacing is equal to h. A source larger than this value will not produce a fringe pattern in Young's experiment, but will rather produce a uniform illumination on the observation plane. The maximum source size that can produce interference may be described as an angular width $\Delta\theta$, using the fact that $\delta\xi << R$

$$\frac{\delta\xi}{R} = \tan\frac{\Delta\theta}{2} \approx \sin\frac{\Delta\theta}{2} \approx \frac{\Delta\theta}{2}$$

The value of h, calculated using (**8-41**), is equal to the transverse coherent length ℓ_t because that slit spacing is associated with a fringe visibility of zero. The source size associated with the transverse coherence length is called the *coherent source size*

$$\Delta\theta = \frac{\lambda}{h} = \frac{\lambda}{\ell_t} \tag{8-42}$$

STELLAR INTERFEROMETER

Michelson made use of the concepts of spatial coherence and coherent source size to measure the angular dimensions associated with stellar objects. His stellar interferometer, Figure 8-5, looks a great deal like Young's two-slit experiment but the scale is a good deal larger: The two input mirrors of his device were about 6 m apart.

With the geometry of Figure 8-5, (**8-33**) becomes

$$\delta = kh_1\Delta\theta + kh_2\frac{x}{D}$$

If we assume that the light intensities gathered by the two input mirrors are equal, then at the fringe plane of Figure 8-5, the intensity pattern would be given by

$$\begin{aligned} I_P &= 2I_0(1 + \cos\delta) \\ &= 2I_0\left[1 + \cos(kh_1\Delta\theta)\cos\left(kh_2\frac{x}{D}\right) - \sin(kh_1\Delta\theta)\sin\left(kh_2\frac{x}{D}\right)\right] \end{aligned}$$

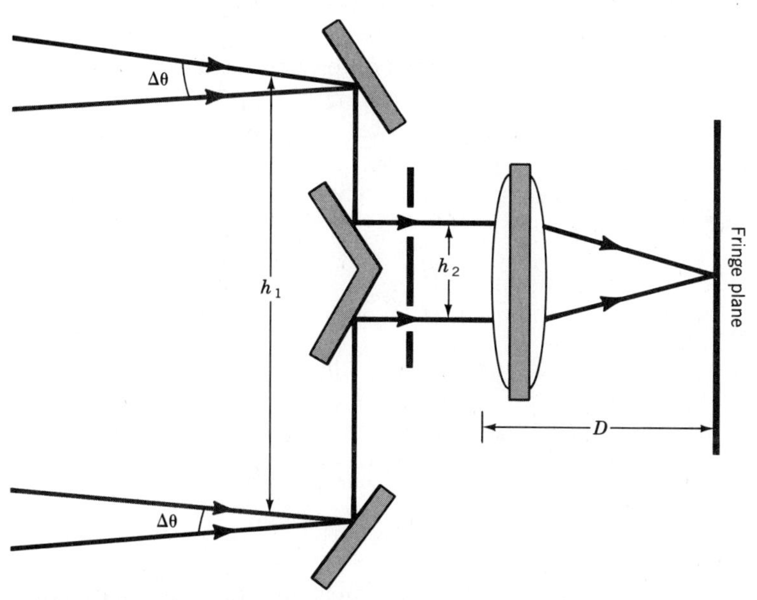

Figure 8-5. Michelson stellar interferometer. The distance h_1 is adjusted until the fringes on the fringe plane disappear. The separation h_2 of the two apertures at the front of the lens establishes the fringe spacing according to the theory developed in Chapter 4. The mirror spacing h_1 determines the fringe visibility because it determines the distance between samples of the wavefront.

To get some idea of the magnitude of ℓ_t assume the source is $\xi = 1.2$ mm wide and is $R = 2$ m from the slits; then,

$$\Delta\theta = \frac{\delta\xi}{R} = \frac{1.2}{2 \times 10^3} = 6 \times 10^{-4} \text{ rad}$$

If $\lambda = 600$ nm then the slits cannot be separated by more than 1 mm if interference is to be observed using this source, that is, $\ell_t = 1$ mm. The Sun produces an ℓ_t on the order of 1/20 mm. This coherence length is smaller than the resolution of the eye and explains why interference fringes are not normally observed.

The angular extent of the source $\Delta\theta$ is assumed to be small compared to the dimensions of the experiment, allowing the intensity distribution to be approximated by

$$I_P \approx 2I_0\left[2 - kh_1\Delta\theta \sin\left(kh_2\frac{x}{D}\right)\right]$$

The fringe spacing at the fringe plane of Figure 8-5 is determined by the separation h_2 of the two slits in the screen at the front of the lens and is calculated using (**4-17**). The two input mirrors sample the incoming wavefront at two points a distance h_1 apart and determine the contrast of the fringes. The observed fringe patterns would have an appearance similar to those in Figure 8-4*b*.

Michelson proved the interferometric technique by measuring the diameters of Jupiter's moons, which have an angular extent of 1 sec and are thus resolvable with conventional imaging techniques. Having proven the interferometric technique, Michelson applied it to the measurement of the separation of binary stars and the angular diameters of single stars. One of the largest stars measured was Betelguese with an angular diameter of 0.047 sec. The spatial coherence length ℓ_t of this star is on the order of several meters.

INTENSITY INTERFEROMETRY

The Michelson stellar interferometer is very sensitive to vibrations and fluctuations in the relative phase of the two wave samples, due to propagation through the atmosphere. These disturbances prevent the use of an interferometer with a value of h_1 larger than about 5 or 6 m. Another technique developed by **R. Hanbury Brown** and **Richard Q. Twiss**,[40] which involves measurement of intensity from two detectors spaced a distance h apart, is not sensitive to vibrations or atmospheric distortions. Dimensions associated with this type of interferometer exceed 300 m.

The theory behind the technique is based on the assumption that the fluctuations in the outputs of two detectors must be correlated if the amplitudes of the two waves are correlated. Justification for this assumption can be obtained by squaring (**8-25**)

$$\left|\gamma_{12}(\tau, h)\right|^2 = \frac{\left|\langle \mathcal{E}_1(t)\mathcal{E}_2^*(t-\tau)\rangle\right|^2}{\langle I_1\rangle\langle I_2\rangle} \geq \frac{\langle I_1(t)I_2(t-\tau)\rangle}{\langle I_1\rangle\langle I_2\rangle}$$

where the subscript $i = 1, 2$ corresponds to the detector position r_i. We see that at least a lower bound for the degree of coherence can be obtained by measuring the intensity correlation. For a classical thermal source, an exact relationship can be derived[40] through the use of statistical arguments

$$\frac{\langle I_1(t)I_2(t-\tau)\rangle}{\langle I_1\rangle\langle I_2\rangle} = 1 + \tfrac{1}{2}\left|\gamma_{12}(\tau, h)\right|^2 \qquad (8\text{-}43)$$

(A thermal source is one for which the amplitude and phase fluctuations obey Gaussian statistics, i.e., if we plot the number of occurrences of an amplitude vs the amplitude, the resultant distribution is Gaussian.)

The variance of the measured intensity is defined as

$$\delta I = I - \langle I\rangle$$

The correlation of the variances measured by two detectors located at r_1 and r_2 can be written as

$$\begin{aligned}\langle \delta I_1(t)\delta I_2(t-\tau)\rangle &= \langle [I_1(t) - \langle I_1\rangle][I_2(t-\tau) - \langle I_2\rangle]\rangle \\ &= \langle I_1(t)I_2(t-\tau)\rangle - \langle I_1\rangle\langle I_2\rangle\end{aligned} \qquad (8\text{-}44)$$

Using (**8-43**) in (**8-44**) and drawing on a basic result of coherence theory, called the *reduction property* of the degree of coherence,[5]

$$\gamma_{12}(\tau, h) = \gamma(\tau)\gamma_{12}(0, h) \qquad (8\text{-}45)$$

we can rewrite the correlation of the variances of the intensities as

$$\langle \delta I_1(t)\delta I_2(t-\tau)\rangle = \frac{\langle I_1\rangle\langle I_2\rangle}{2}\left|\gamma(\tau)\right|^2\left|\gamma_{12}(0, h)\right|^2 \qquad (8\text{-}46)$$

This result states that by measuring the fluctuations of the light intensity at two points in space, the degree of coherence can be measured.

To carry out the coherence measurement, the Michelson stellar interferometer of Figure 8-5 is replaced by two detectors spaced a distance h apart (see Figure 8-6). Amplifiers, following the detectors, remove the time-invariant components $< I_1 >$ and $< I_2 >$ by high-pass filtering. The signal from one detector is delayed electronically τ sec with respect to the other signal, and an electronic integrator performs a time average of the product of the delayed output of one detector with the prompt output of the second detector.

To prove the technique at visible wavelengths, Hanbury Brown and Twiss measured $\left|\gamma_{12}(0, h)\right|^2$ for a mercury arc lamp. Figure 8-7 displays a

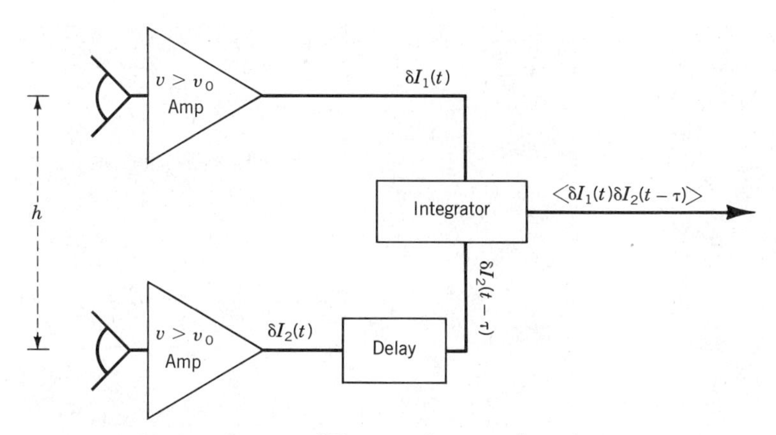

Figure 8-6. Hanbury Brown and Twiss interferometer for radio astronomy.

solid curve predicted by theory on which plotted points are obtained by experimental measurements.

The difference between the intensity interferometer and the Michelson interferometer is that the Hanbury Brown intensity interferometer measures the square of the modulus of the complex degree of coherence, whereas the Michelson interferometer also measures the phase.

The Hanbury Brown experiments introduced a great deal of controversy into the optics community during the late 1950s and early 1960s. From a quantum viewpoint, the basic assumption that the two detectors receive a correlated signal suggests that photons must arrive at the detectors in pairs. Since the stars are thermal sources, this seemed absurd to many physicists. If photons are emitted at random at the source, how could they arrive in pairs at the detectors?

The apparent conflict is a manifestation of the wave-particle duality of light. When we make certain types of measurements on light, such as interference

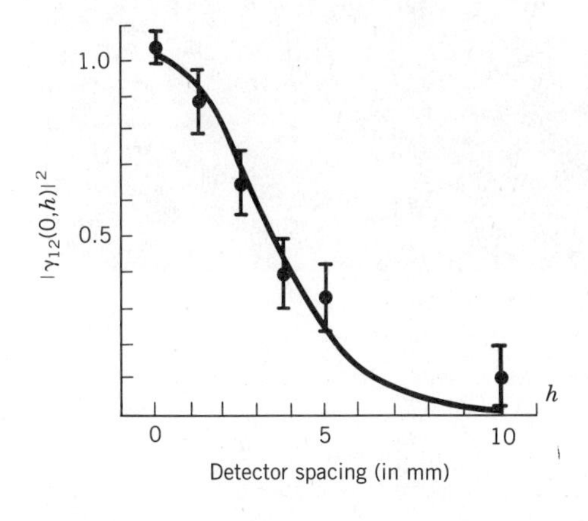

Figure 8-7. Comparison of experimental data and theoretical curve for Hanbury Brown and Twiss intensity interferometer.[41]

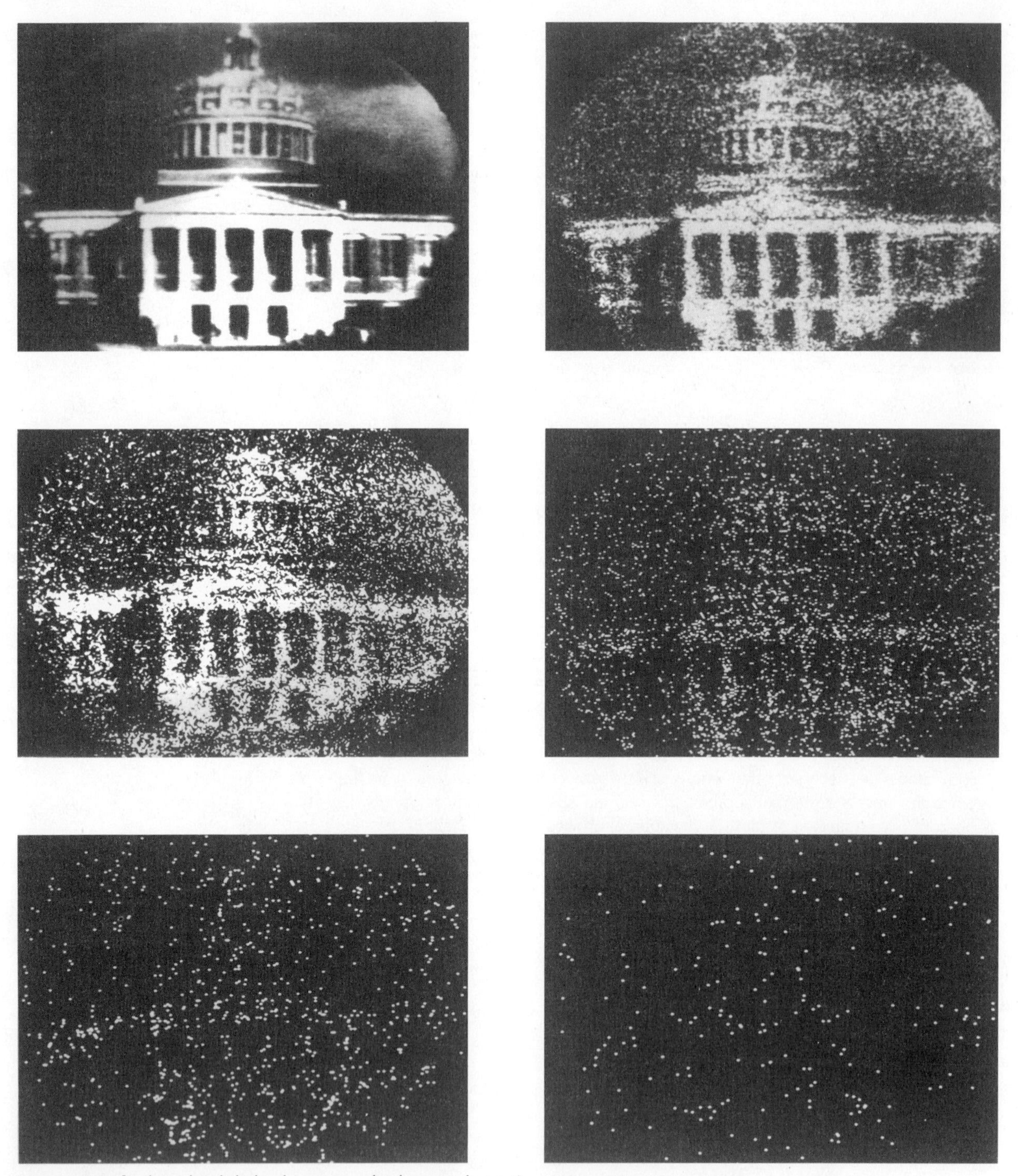

Figure 8-8. In these low-light-level images, a bright spot shows the arrival position for a single photon. The number of photons used to record the images from the upper left to the lower right were 10^7, 64×10^3, 16×10^3, 4×10^3, 10^3, and 250. (Courtesy of M. Morris, University of Rochester.)

measurements, light appears to be a wave. For other types of measurements, such as the low-light-level images of Figure 8-8, the particle nature of light is evident.

The intensity interferometry, working at low light levels, seems to emphasize the particle nature of light because it recorded the simultaneous arrival of pairs of photons. The result of the measurement, however, seems to require an interpretation based on the wave nature of light. The two detectors appear to sample portions of a wavefront associated with a wave that originated at a point on the source and diverges toward the detector. The solution to the conflict was to allow the original emitted photon to disappear and new photons to reform during propagation. The wave function associated with the photons would predict periodic increases in the probability of detection in a manner similar to the disappearance and reformation of a classical wave group (See Figure 7-3).

Many people were reluctant to accept the concept of a photon that did not retain its identity throughout the propagation of light, from emission to absorption. Purcell showed that a quantitative explanation of the observed correlation could be made based on the quantum statistical behavior of photons. Hanbury Brown and Twiss conducted a number of experiments that verified their approach and overcame the objections of other scientists. R.J. Glauber later developed the quantum theory of optical coherence that gave a firm quantum theoretical basis for the experimental results of Hanbury Brown.[39]

SUMMARY

The intensity at a point in space due to two waves of frequency ω_1 and ω_2 has been calculated. It was found that if $\Delta\omega T >> 1$, where T is the response time of the detector and $\Delta\omega = \omega_2 - \omega_1$, then the intensity of the resultant wave is obtained by adding the intensities of the individual waves. This result was generalized to N waves

$$\langle \mathbf{E}^2 \rangle = \sum_{i=1}^{N} \langle \mathbf{E}_i^2 \rangle$$

If $\Delta\omega T < 1$, then the intensity of the resultant wave is obtained by first adding the amplitudes of the individual waves and then calculating the intensity.

The Michelson interferometer can be used to interfere a wave sample with a sample of the wave delayed by a time $\tau = 2d/c$, where d is the difference in the length of the two arms of the interferometer. By varying d, the wavelength distribution of a source can be measured. The resolution that can be obtained by the Michelson interferometer is given by

$$\mathcal{R}_{\text{max}} = 2\nu_a\tau = \frac{2d_{\text{max}}}{\lambda/2}$$

where d_{max} is the maximum optical path difference that can be produced with the interferometer.

The temporal degree of coherence characterizes the ability of a wave from the source to interfere with a sample of the wave produced at an earlier time τ. The degree of coherence is a complex function of the form

$$\begin{aligned} \gamma(\tau) &= |\gamma(\tau)| e^{i\delta(\tau)} \\ &= v(\tau) e^{i\delta(\tau)} \end{aligned}$$

whose absolute value is $|\gamma(\tau)| \leq 1$. The interference intensity observed when the source is described by a coherence $\gamma(\tau)$ is

$$I_P = I_1 + I_2 + 2\sqrt{I_1 I_2}\,|\gamma(\tau)| \cos \delta(\tau)$$

A source with a continuous distribution of frequencies is characterized by its power spectrum

$$P(\omega) = \frac{I(\omega)}{\int_0^\infty I(\omega)\, d\omega} = \frac{I(\omega)}{I_0}$$

The temporal degree of coherence associated with the source is obtained by calculating the Fourier transform of its power spectrum

$$\gamma(\tau) = \int_{-\infty}^{\infty} P(\omega) e^{-i\omega\tau}\, d\omega$$

To measure the degree of coherence, the fringe visibility

$$\mathcal{V} = \frac{I_{max} - I_{min}}{I_{max} + I_{min}}$$

of the interference pattern produced by the Michelson interferometer is measured as a function of d. The temporal degree of coherence is then

$$\gamma(\tau) = \mathcal{V} \frac{I_1 + I_2}{2\sqrt{I_1 I_2}} \cos \omega_0 \tau$$

where I_1 and I_2 are the average intensities in the two arms of the interferometer; the two intensities are normally made equal.

To characterize a source's temporal coherence, a characteristic time, called the coherence time, is defined

$$\tau_c = 4 \int_0^\infty \gamma(\tau)^2\, d\tau$$

This coherence time can be used to define a longitudinal coherence length

$$\ell_\ell = c\tau_c = \frac{c}{\Delta\nu} = \frac{\lambda^2}{\Delta\lambda}$$

This length corresponds to the difference in length between the two arms of the interferometer when the fringe visibility drops to a value determined by the spectral distribution of the source. If the power spectrum was Gaussian, then the fringe visibility would have dropped to one-half of its maximum value; if the power spectrum was Lorentzian, then the fringe visibility would have dropped to $1/e$ of the maximum values.

The degree of coherence can be defined in terms of a correlation operation. For temporal coherence, the two waves used in the correlation operation are different temporal samples of the same wave; thus, the coherence function is an autocorrelation operation. This approach views coherence as a measure of the size of the amplitude and phase fluctuations occurring in the source. The resultant degree of coherence is identical to the one obtained by calculating the Fourier transform of the power spectrum,

$$\frac{\langle \mathrm{E}(t)\mathrm{E}^*(t-\tau)\rangle}{\langle \mathrm{E}(t)^2\rangle} = \frac{\int_{-\infty}^{\infty} I(\omega) e^{i\omega\tau} d\omega}{\int_{-\infty}^{\infty} I(\omega) d\omega}$$

Coherence is not only a function of time but is also a function of spatial coordinates. The spatial degree of coherence is a correlation of two samples of a wavefront, obtained at two different positions in space r_1 and r_2. The mathematical formalism used for spatial coherence is the same as that used for temporal coherence if the temporal parameter is redefined to be

$$\tau = \frac{r_2 - r_1}{c}$$

The spatial degree of coherence in terms of this newly defined temporal parameter is given by the cross correlation

$$\gamma_{12}(\tau) = \frac{\int_{-\infty}^{\infty} E_1(t)E_2^*(t-\tau)\,dt}{\left[\int_{-\infty}^{\infty} E_1(t)E_1^*(t)\,dt\right]^{1/2}\left[\int_{-\infty}^{\infty} E_2(t)E_2^*(t)\,dt\right]^{1/2}}$$

$$= \frac{\langle E_1(t)E_2^*(t-\tau)\rangle}{\sqrt{\langle E_1(t)E_1^*(t)\rangle}\sqrt{\langle E_2(t)E_2^*(t)\rangle}}$$

It is also, in general, a complex function

$$\gamma_{12}(\tau) = \left|\gamma_{12}(\tau)\right| e^{i\delta_{12}(\tau)}$$

The interference, which now would be produced using Young's two-slit experiment, can be written in terms of the degree of coherence

$$I_P = I_1 + I_2 + 2\sqrt{I_1 I_2}\left|\gamma_{12}(\tau)\right| \cos\delta_{12}(\tau)$$

where I_1 and I_2 are the intensities passing through slits 1 and 2, respectively.

The size and shape of a source determine the spatial degree of coherence. For a line source of length s, the degree of coherence is

$$\gamma = s\,\frac{\sin(khs/2R)}{khs/2R}\cos\left(\frac{khx}{D}\right)$$

As was done in temporal coherence, a spatial or transverse coherence length can be defined. This transverse coherence length is given by

$$\Delta\theta = \frac{\lambda}{\ell_t}$$

where $\Delta\theta$ is the angular extent of a coherent source. When the actual source exceeds this dimension, then $\Delta\theta$ is said to be the size of a coherent patch on the source.

The degree of coherence of a source can also be obtained by measuring the correlation of the intensity fluctuations in light from the source

$$\langle \delta I_1(t)\delta I_2(t-\tau)\rangle = \frac{\langle I_1\rangle\langle I_2\rangle}{2}\left|\gamma(\tau)\right|^2\left|\gamma_{12}(0,h)\right|^2$$

PROBLEMS

8-1. Calculate the power spectrum of a damped wave

$$f(t) = \begin{cases} Ae^{-(at-i\omega t)}, & t \geq 0 \\ 0, & t < 0 \end{cases}$$

8-2. Calculate the degree of coherence $\gamma(\tau)$ for the damped wave train in Problem 8-1.

8-3. What is the line width in nm and in Hz of light from a laser with a 10 km coherence length? The mean wavelength is 632.8 nm.

8-4. A 1 mm diameter pinhole 2 m from a double slit is illuminated by $\lambda = 589$ nm. What is the maximum slit spacing for which interference fringes are visible?

8-5. A Hg^{198} low-pressure isotope lamp emits a line at $\lambda = 546.078$ nm with a bandwidth of $\Delta\upsilon = 1000$ MHz. What is the coherence length and coherence time?

8-6. A Michelson interferometer is illuminated by red cadmium light with a mean wavelength of 643.847 nm and a linewidth of 0.0013 nm. The mirror is initially set so the two arms have equal length and then one mirror is moved until the fringes disappear (the fringe visibility goes to zero). How far must the mirror be moved?

8-7. Assume a Michelson interferometer is illuminated by two wavelengths λ_1 and λ_2 (say a Na doublet). As one mirror is moved, the fringe visibility periodically goes from maximum to minimum. Obtain a relationship between the mirror displacement Δd between two low-fringe visibility conditions and the separation of the two wavelengths $\Delta\lambda = \lambda_1 - \lambda_2$.

8-8. In a gas at high pressure, the atoms experience frequent collisions. The collisions interrupt the radiation process and the resulting line shape is Lorentzian

$$I(\omega) = \frac{A}{(\omega - \omega_0)^2 + b^2}$$

Calculate the degree of coherence.

8-9. The emission from sodium is predominantly the two Fraunhofer D lines. In a low-pressure gas discharge, these two lines, centered at 589 and 589.6 nm, are Gaussian in shape and the intensity of the second is 80% of the first

$$I(\omega) = A\left[\exp\left\{-\left[(\omega - \omega_0 + b)/a\right]^2\right\} + (0.8)\exp\left\{-\left[(\omega - \omega_0 - b)/a\right]^2\right\}\right]$$

Calculate the degree of coherence and identify the origin of the various terms.

8-10. The laser transition in a helium-neon laser has a Gaussian line shape. The Fabry–Perot cavity, which produces feedback for oscillation, will support oscillations at three different longitudinal modes under this Gaussian line; see Figure 8-9.

We can represent this system by

$$I(\omega) = A_1\delta(\omega - \omega_0) + (0.5)A_1\delta(\omega - \omega_0 - a) + (0.5)A_1\delta(\omega - \omega_0 + a)$$

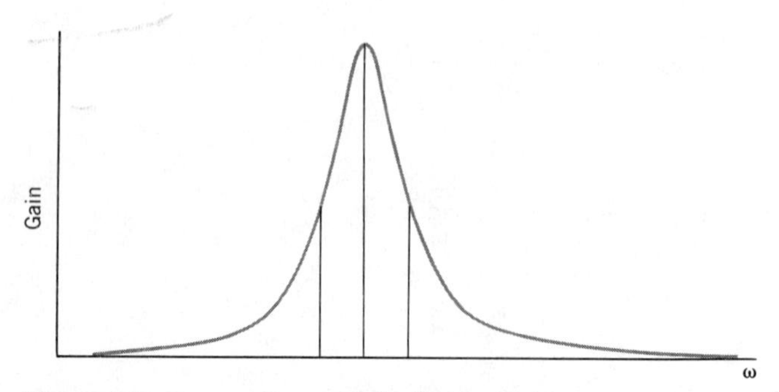

Figure 8-9. Laser cavity modes for Problem 8-10.

where we have assumed the longitudinal modes are delta functions. Calculate the degree of coherence.

8-11. Write a computer program that will reproduce Figure 8-4. Explore the effects of varying γ. Produce several curves for which the intensities from the pinholes differ.

8-12. If a monochromatic light wave with $\lambda = 488$ nm is chopped at a frequency of 40 MHz, what is the bandwidth (in nm) of the light pulses?

8-13. We stated that the transverse coherence length of the sun was 1/20 mm. Assume $\lambda = 600$ nm. What is the apparent angular diameter of the sun?

8-14. A grating monochromator with a linear dispersion of

$$\frac{\Delta x}{\Delta \lambda} = \frac{1 \text{ mm}}{2 \text{ nm}}$$

and an exit slit of 0.2 mm produces a light wave at $\lambda = 500$ nm. What is the temporal coherence length of the light exiting the slit?

8-15. Michelson's stellar interferometer had a mirror spacing of 6 m. At $\lambda = 500$ nm, what is the smallest angular diameter it could measure?

8-16. A source emits radiation at two wavelengths, 589 and 589.6 nm. If this radiation illuminates two narrow slits separated by a distance $2d$ and the interference fringes are viewed a distance of 100 cm from the slits, what is the visibility of the fringes in the region of the 20th fringe maximum?

8-17. Show that the superposition of three waves at frequencies ω, $\omega + \omega_0$, and $\omega - \omega_0$ results in an amplitude-modulated wave given by

$$E = E_0(1 + A \cos \omega_0 t) \cos \omega t$$

8-18. A multilayer bandpass filter has a center wavelength of 632.8 nm and a bandwidth of 0.1 nm. What is the coherence length of the light that is transmitted by this filter?

8-19. A local fm station operates at a frequency of 91.5 MHz and its carrier frequency has a short term stability of 5 parts in 10^8. What is the temporal coherence length of this source?

Diffraction and Gaussian Beams

Geometrical optics does a reasonable job of describing the propagation of light in free space unless that light encounters an obstacle. Geometrical optics predicts that the obstacle would cast a sharp shadow. On one side of the shadow's edge would be a bright uniform light distribution, due to the incident light, and on the other side there would be darkness. Close examination of an actual shadow edge reveals, however, dark bands in the bright region and bright bands in the dark region. This departure from the predictions of geometrical optics was given the name *diffraction* by Grimaldi.

There is no substantive difference between diffraction and interference. The separation between the two subjects is historical in origin and is retained for pedagogical reasons. Interference may be associated with the intentional formation of two or more light waves that are analyzed in their region of overlap according to the procedures outlined in previous chapters. Diffraction may be associated with the obstruction of a single wave, by either a transparent or an opaque obstacle, resulting in the obstruction casting shadows (or forming light beams) that differ from the size predicted by geometrical optics. This distinction between interference and diffraction is, however, rather fuzzy and arbitrary.

A rigorous diffraction theory is based on Maxwell's equations and the boundary conditions associated with the obstacle.[6] The boundary conditions are used to calculate a field scattered by the obstacle. The origins of this scattered field are currents induced in the obstacle by the incident field. The scattered field is allowed to interfere with the incident field to produce a resultant diffracted field. The application of the rigorous theory is very difficult and for most problems an approximate scalar theory is used.

The approximate scalar theory is based on *Huygens' principle* that states

Each point on a wavefront can be treated as a source of a spherical wavelet called a *secondary wavelet* or Huygens' *wavelet*. The envelope of these wavelets, at some later time, is constructed by finding the

tangent to the wavelets. The envelope is assumed to be the new position of the wavefront.

Rather than simply using the wavelets to construct an envelope, the scalar theory assumes that Huygens' wavelets interfere to produce a new wavefront.

Application of the scalar theory results in an integral called the *Huygens–Fresnel integral*. The traditional interpretation given to this integral is that it is the sum of a group of Huygens' wavelets. The modern interpretation of the integral is in terms of linear system theory.

Free space is viewed, by modern theory, as a linear system and the incident wave is the input to this linear system. The Huygens' wavelet is interpreted as the impulse response of free space and the Huygens-Fresnel integral is a convolution of the incident wave with the impulse response of free space.

Approximate analytic solutions of the Huygens–Fresnel integral are usually obtained through the use of two approximations. In both approximations, all dimensions are assumed to be large with respect to a wavelength. In one approximation, the viewing position is assumed to be far from the obstruction; the resulting diffraction is called the *Fraunhofer diffraction* and will be discussed in Chapter 10. The second approximation, which leads to *Fresnel diffraction,* assumes that the observation point is near the obstruction. This approximation, which is the more difficult mathematically, will be discussed in Chapter 11.

In this chapter, a simple derivation of the Huygens–Fresnel integral will be given, based on an application of Huygens' principle and the addition of waves to calculate an interference field. The derivation will be more descriptive and intuitive than formal. A more rigorous mathematical treatment of scalar diffraction theory, due to **Gustav Kirchhoff** and **Arnold Johannes Wilhelm Sommerfeld (1868–1951)**, will be given in Appendices 9-A and 9-B. Green's theorem is needed for the formal derivation and that theorem is presented in Appendix 9-C.

In this chapter, Huygens' principle will be justified by demonstrating that it can be used to derive the laws of reflection and refraction. Having justified the use of this principle, we will apply it to diffraction by assuming two pinholes act as sources for Huygens' wavelets. The interference between these two sources is a generalization of the discussion of Young's interference experiment that is used as justification for the addition of Huygens' wavelets. The addition of the Huygens' wavelets will be extended to N wavelets and then to a continuous distribution of Huygens' sources. The use of a continuous distribution leads to the Huygens–Fresnel integral for diffraction.

When using Huygens' principle, a problem arises with the spherical wavelet produced by each Huygens' source. Part of the spherical wavelet propagates backward and results in an envelope propagating toward the light source, but such a wave is not observed in nature. Huygens neglected this problem. Fresnel required the existence of an *obliquity factor* to cancel the backward wavelet but was unable to derive its functional form. When Kirchhoff placed the theory of diffraction on a firm mathematical foundation, the obliquity factor was generated quite naturally in the derivation. In the intuitive derivation of the Huygens–Fresnel integral given in this chapter, we will present an argument for treating the obliquity factor as a constant at optical wavelengths. We will then derive the constant value it must be assigned.

A quite modern approach to the subject of diffraction is to look for solutions of the differential equation describing the optical system. In this chapter, the optical system to be treated will be free space. The differential equation to be solved is the Helmholtz equation. A particular solution will be shown to be a *Gaussian wave*, i.e., a wave whose intensity distribution, transverse to the propagation direction, is a Gaussian function. We will discover that the Gaussian wave can be characterized by a *beam waist* (the diameter of the beam of light at the half-power points of the Gaussian distribution) and the *radius of curvature* of the wave's phase front.

The Gaussian beam parameters will be related to geometrical optics by showing how a Gaussian beam can be traced through an optical system. The technique will then be used to trace light through a thin lens and a Fabry–Perot cavity. We will conclude by using the connection between Gaussian waves and the *ABCD* matrix to begin the analysis of a commercial laser cavity.

The first reference to diffraction appears in the notebooks of **Leonardo Da Vinci (1452–1519)**, but a published description did not occur until the works of Grimaldi appeared posthumously. Grimaldi conducted a number of experiments involving narrow light beams grazing small obstacles. Newton repeated many of the experiments of Grimaldi, but dropped the subject when he found that he could not explain the observations in terms of his particle model. Both Grimaldi and Newton conducted their experiments in the hope of determining the nature of light but both seemed to retreat from their observations of wavelike behavior.

Young attempted to treat diffraction in terms of the interference of light reflected from an edge but was convinced by Fresnel that the approach was wrong. Scientists have returned to Young's concept of diffraction, and it is now an active research topic.[42]

Fresnel adapted a technique developed by Huygens for the explanation of birefringence to explain diffraction. Fresnel enhanced Huygens' theorem by assuming interference between wavelets and thereby was able to explain diffraction from straight edges and apertures.

Fresnel was a civil engineer who pursued optics as a hobby, after a long day of road construction. Fresnel was also active politically, supporting Napoleon at an inopportune time. His detention by the police for his political views gave Fresnel time to conduct very high-quality observations of diffraction. Fresnel submitted the theory, based on his observations, to the Paris Academy in competition for a prize in 1818. **Simèon Denis Poisson (1781–1840)**, one of the judges, presented an argument as a *reductio ad absurdum* to prove the theory was incorrect.

Poisson pointed out that a natural consequence of Fresnel's theory was the appearance of a bright spot at the center of the shadow of a circular object. **Dominique Francois Arago (1786–1853)**, also a judge, conducted an immediate experiment and verified the presence of what is now known as Poisson's spot. As you might expect, Fresnel won the prize. (Arago and Poisson had been unaware that Delisle had observed the spot about 50 years earlier).

The theory developed by Fresnel was placed on a firm mathematical basis by G. Kirchhoff in 1882. The theory as developed by Kirchhoff had problems associated with the boundary conditions. A. Sommerfeld avoided these problems by treating the case of an aperture or edge in a perfectly conducting thin screen and was the first to obtain a rigorous solution for the diffraction of light waves from a conducting obstruction (in 1894). Smythe extended this result to a vector formulation in 1947.

HUYGENS' PRINCIPLE

To discover how a wave propagates through an aperture, we must solve the vector wave equation obtained from Maxwell's equations in combination with the boundary conditions associated with the aperture. This is a very difficult mathematical problem and we will be forced to discuss only approximate solutions. The approximate solutions we will evaluate are based on the application of Huygens' principle to the propagation problem.

Huygens said that each point on a wavefront (i.e., the surface of constant phase) can be regarded as a source for a secondary wavelet in the form of a spherical wave. The envelope of the secondary wavelets gives the wavefront at some later time. We will follow the lead of Huygens and ignore the problems associated with the backward-propagating part of the wavelets.

Before applying Huygens' principle to the study of diffraction, we will demonstrate the utility of the principle and learn how to implement it. This will be accomplished by rederiving the laws of reflection and refraction through the use of Huygens' principle. Figure 9-1 shows the geometry to be used in the derivation.

We assume in Figure 9-1 that a plane wave is incident on the interface between a region of index n_1 and a region of index n_2. A wavefront of the incident wave, at a time before $t = 0$, is shown as AB. When the leading portion of this wavefront, point A, strikes the interface at point O, we set our clock to $t = 0$. Huygens' principle says that we should treat point O as a source of a wavelet. It is a point source so it will radiate a spherical wavelet that in region 1 will have a radius $v_1 t$ at time t, where $v_1 = c/n_1$ and t is selected to be equal to the time for point B to reach point P. (The wavefront of the Huygens' wavelet generated by point O is shown by the large dotted semicircle in Figure 9-1.) If the interface were not there, the wavefront at

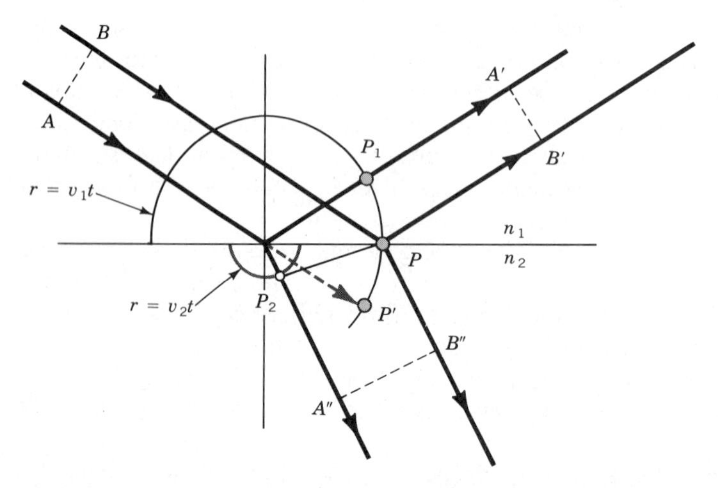

FIGURE 9-1. The geometry of reflection and refraction using Huygens' principle to construct the wavefronts.

time t would be the plane formed by the line connecting points P and P' and denoted by PP'.

The interface is present; thus, the Huygens' wavelet in the second medium is a spherical wavelet of radius v_2t, shown as the small semicircle below the interface in Figure 9-1. The wavefront of the transmitted wave in the second medium is therefore PP_2 after time t. A portion of the incident wave is reflected by the interface, and after time t, the wavefront of the reflected wave is PP_1.

The geometry of the problem is now established and can be used to prove the laws of reflection and refraction. To reduce confusion, each portion of Figure 9-1 will be redrawn in an enlarged form, as it is discussed. In Figure 9-2, we highlight the arriving wave. At $t = 0$, the incident wavefront AB is at position OO'. In time t, O' travels to P and by using Huygens' principle, a secondary wavelet forms a sphere of radius v_1t, centered on point O. From the triangles OPO',

$$\overline{O'P} = v_1 t = \overline{OP} \sin \theta_i \tag{9-1}$$

If region 2 were not present, the spherical wavelet would establish that the distance

$$\overline{OP'} = v_1 t$$

This means that the two distances $\overline{O'P}$ and $\overline{OP'}$ are identical

$$\overline{O'P} = \overline{OP'}$$

This equality states that the wavefront PP' is parallel to OO' and the application of Huygens' principle leads to rectilinear propagation.

In Figure 9-3, we highlight the reflected wave. The drawing has been distorted for illustrative purposes. Huygens' principle establishes that PP_1 is the plane of the reflected wave at time t; thus, from the triangle OP_1P,

$$\overline{OP_1} = v_1 t = \overline{OP} \sin \theta_i' \tag{9-2}$$

From **(9-1)** and **(9-2)**, we obtain

$$\sin \theta'_i = \frac{\overline{OP_1}}{\overline{OP}} = \frac{v_1 t}{\overline{OP}} = \sin \theta_i \tag{9-3}$$

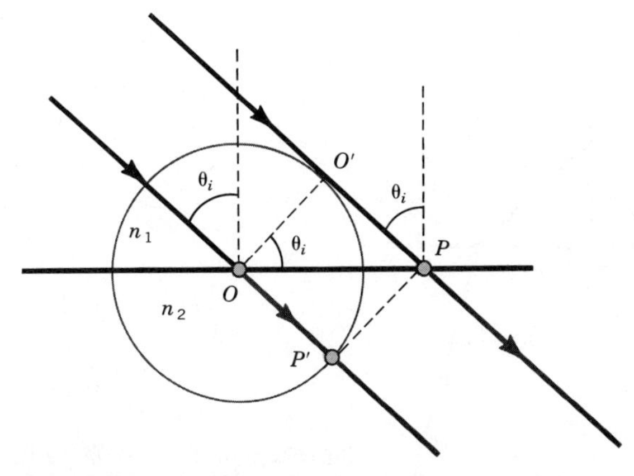

FIGURE 9-2. The incident wave from Figure 9-1. The wave with wavefront PP' does not exist. It would be present only if the interface was not there.

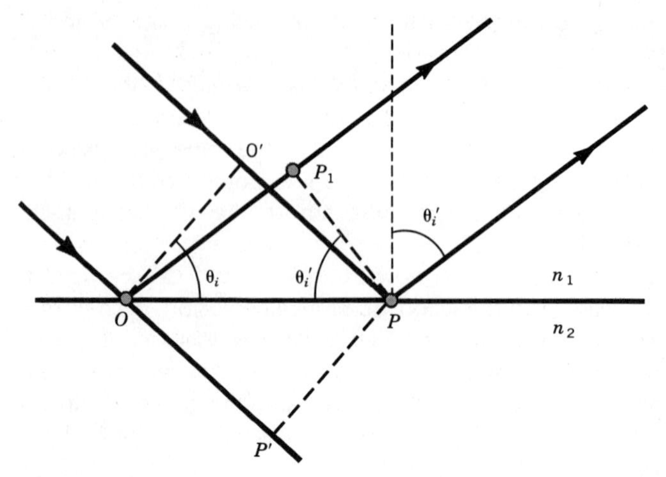

FIGURE 9-3. The reflected wave using Huygens' principle. The drawing has been distorted for illustrative purposes.

which is the law of reflection. We now see that Figure 9-3 is distorted. Distance OP_1 and $O'P$ must be the same length and O' and P_1 fall on top of one another.

The transmitted wave is highlighted in Figure 9-4. The geometry of this figure will allow the derivation of Snell's law. In medium 2, Huygens' principle states that point O will act as a source of a spherical wave of radius v_2t, where $v_2 = c/n_2$. From Figure 9-4, triangle OPP_2,

$$v_2 t = \overline{OP_2} = \overline{OP} \sin \theta_t \tag{9-4}$$

From Figure 9-2, we have

$$\overline{OP} = \frac{v_1 t}{\sin \theta_i}$$

Substituting into **(9-4)**, we obtain

$$v_1 \sin \theta_t = v_2 \sin \theta_i$$

Upon using $v_1 = c/n_1$ and $v_2 = c/n_2$, we obtain Snell's law

$$n_1 \sin \theta_i = n_2 \sin \theta_t$$

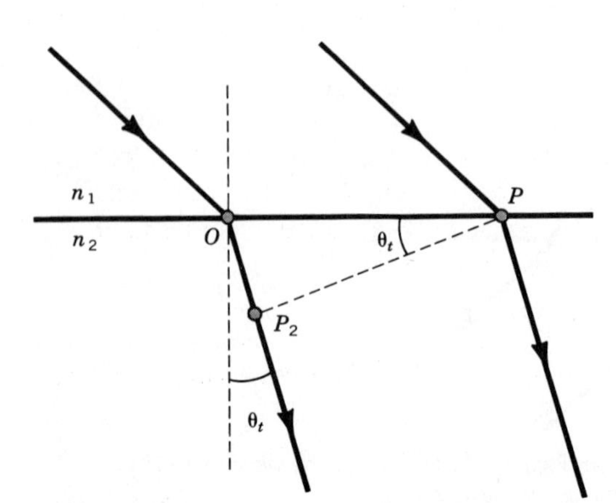

FIGURE 9-4. The refracted wave from Figure 9-1 using Huygens' principle.

We have been able to demonstrate the rectilinear propagation of light and the laws of reflection and refraction through the use of Huygens' principle. This gives some confidence in using this principle to derive the properties of diffraction.

FRESNEL FORMULATION

Fresnel used Huygens' principle as a foundation for his theoretical explanation of diffraction, but it was not until the theoretical derivations of Kirchhoff that the theory of diffraction was placed on a firm mathematical foundation. The formal derivation of the diffraction integral developed by Kirchhoff is found in Appendix 9-A. In this section, we will use a descriptive approach to obtain the Huygens–Fresnel integral of diffraction.

To apply Huygens' principle to the propagation of light through an aperture of arbitrary shape, we need to develop a mathematical description of the field from an array of Huygens' sources filling the aperture. We will begin by obtaining the field from a pinhole that is illuminated by a wave

$$E_i(\mathbf{r}, t) = E_i(\mathbf{r})e^{i\omega t}$$

The theory of interference provides the rules for combining the fields from two pinholes and allows the generalization of the result to N pinholes. Finally by allowing the areas of each pinhole to approach an infinitesimal value, we will construct an arbitrary aperture of infinitesimal pinholes. The result will be the Huygens–Fresnel integral.

We know that the wave obtained after propagation through an aperture must be a solution of the wave equation

$$\nabla^2 E = \mu\epsilon \frac{\partial^2 E}{\partial t^2}$$

We will be interested only in the spatial variation of the wave so we need only look for solutions of the Helmholtz equation

$$(\nabla^2 + k^2)E = 0$$

The problem is further simplified by replacing this vector equation with a scalar equation

$$(\nabla^2 + k^2)\,\mathcal{E}\,(x, y, z) = 0$$

This replacement is proper for those cases where $\hat{\mathbf{n}}E(x, y, z)$ (where $\hat{\mathbf{n}}$ is a unit vector) is a solution of the vector Helmholtz equation. In general, we cannot substitute $\hat{\mathbf{n}}E$ for the electric field $\mathbf{E}$ because of Maxwell's equation

$$\nabla \cdot E = 0$$

Rather than working with the magnitude of the electric field, we should use the scalar amplitude of the vector potential.[43] We will neglect this complication and assume the scalar E is a single component of the vector field $\mathbf{E}$. A complete solution would involve a scalar solution for each component of $\mathbf{E}$.

The pinhole is illuminated by a plane wave

$$E_i(\mathbf{r}, t) = E_i(\mathbf{r})e^{i\omega t}$$

The wave that leaves the pinhole will be a spherical wave that is written in complex notation as

$$\mathcal{E}(\mathbf{r})e^{i\omega t} = A\,\frac{e^{-i\delta}e^{-i\mathbf{k}\cdot\mathbf{r}}}{r}\,e^{i\omega t}$$

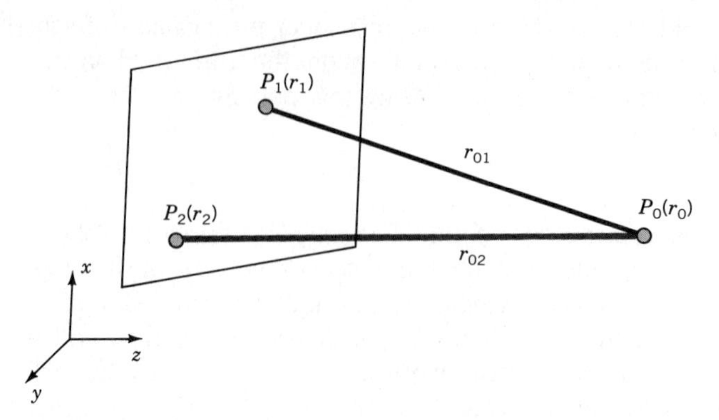

FIGURE 9-5. Geometry for application of Huygens' principle to two pinholes.

The complex amplitude

$$\mathcal{E}(\mathbf{r}) = A\frac{e^{-i\delta}e^{-i\mathbf{k}\cdot\mathbf{r}}}{r} \tag{9-5}$$

is a solution of the Helmholtz equation. The prescription for adding waves developed for interference can be applied to the problem of calculating the field at P_0 in Figure 9-5 from two pinholes: one at P_1, located a distance $r_{01} = |\mathbf{r}_0 - \mathbf{r}_1|$ from P_0, and one at P_2, located a distance $r_{02} = |\mathbf{r}_0 - \mathbf{r}_2|$ from P_0.

The two pinholes are a generalization of Young's interference experiment with the field at P_0, given by the superposition of the wavelets emitted from P_1 and P_2

$$\mathcal{E}(\mathbf{r}_0) = \frac{A_1}{r_{01}}e^{-i\mathbf{k}\cdot\mathbf{r}_{01}} + \frac{A_2}{r_{02}}e^{-i\mathbf{k}\cdot\mathbf{r}_{02}} \tag{9-6}$$

We have incorporated the phase δ_1 and δ_2 into the constants A_1 and A_2 to simplify the equations.

The light emitted from the pinholes is due to a wave E_i incident onto the screen from the left. The units of E_i are per unit area, so to obtain the amount of light passing through the pinholes, we must multiply E_i by the areas of the pinholes. If $\Delta\sigma_1$ and $\Delta\sigma_2$ are the areas of the two pinholes, respectively, then

$$A_1 \propto \mathcal{E}_i(\mathbf{r}_1)\Delta\sigma_1, \qquad A_2 \propto \mathcal{E}_i(\mathbf{r}_2)\Delta\sigma_2$$

$$\mathcal{E}(\mathbf{r}_0) = C_1\frac{\mathcal{E}_i(\mathbf{r}_1)}{r_{01}}e^{-i\mathbf{k}\cdot\mathbf{r}_{01}}\Delta\sigma_1 + C_2\frac{\mathcal{E}_i(\mathbf{r}_2)}{r_{02}}e^{-i\mathbf{k}\cdot\mathbf{r}_{02}}\Delta\sigma_2 \tag{9-7}$$

where C_i is the constant of proportionality. The constant will depend on the angle that $\mathbf{r}_{0i}$ makes with the normal to $\Delta\sigma_i$. This geometrical dependence arises from the fact that the apparent area of the pinhole decreases as the observation angle approaches 90°.

We can generalize **(9-7)** to N pinholes

$$\mathcal{E}(\mathbf{r}_0) = \sum_{j=1}^{N} C_j\frac{\mathcal{E}_i(\mathbf{r}_j)}{r_{0j}}e^{-i\mathbf{k}\cdot\mathbf{r}_{0j}}\Delta\sigma_j \tag{9-8}$$

The pinhole's diameter is assumed to be small compared to the distances to

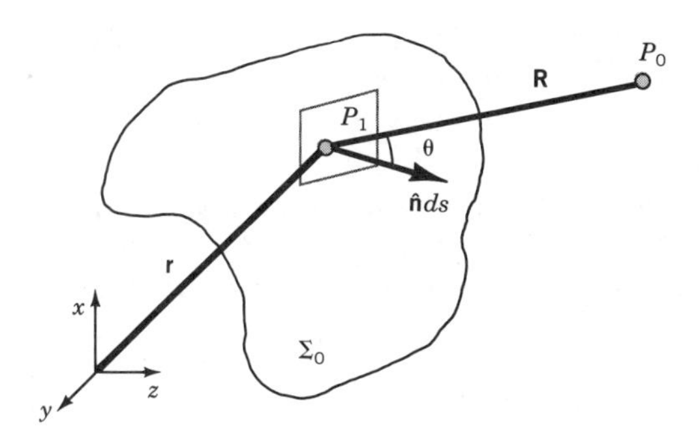

FIGURE 9-6. The geometry for calculating the field at P_0 using **(9-9)**.

the viewing position but large compared to a wavelength. In the limit as $\Delta\sigma_j$ goes to zero, the pinhole becomes a Huygens' source. By letting N become large, we can fill the aperture with these infinitesimal Huygens' sources and convert the summation to an integral.

It is in this way that we obtain the complex amplitude, at the point P_0, from a wave exiting an aperture by integrating over the area of the aperture. We replace $\mathbf{r}_{0j}$ in **(9-8)** by $\mathbf{R}$, the position of the infinitesimal Huygens' source of area ds measured with respect to the observation point P_0. We also replace $\mathbf{r}_j$ by $\mathbf{r}$, the position of the infinitesimal area ds with respect to the origin of the coordinate system. The discrete summation **(9-8)** becomes the integral

$$\mathcal{E}(\mathbf{r}_0) = \iint_A C(\mathbf{r}) \frac{\mathcal{E}_i(\mathbf{r})}{R} e^{-i\mathbf{k}\cdot\mathbf{R}} \, ds \tag{9-9}$$

This is the *Fresnel integral*. The geometry associated with **(9-9)** is shown in Figure 9-6, where the aperture is denoted as Σ_0, the observation point as P_0, and an arbitrary point in the aperture as P_1.

The constant $C(\mathbf{r})$ depends on θ, the angle between $\hat{\mathbf{n}}$, the unit vector normal to the aperture, and $\mathbf{R}$, shown in Figure 9-6. We will treat $C(\mathbf{r})$ as a constant that we may remove from the integrand. Before continuing, we must digress to justify the assumption that the angular dependence of $C(\mathbf{r})$ can be ignored.

THE OBLIQUITY FACTOR

The parameter $C(\mathbf{r})$ in **(9-9)** is called the *obliquity factor*. In Appendix 9A, the obliquity factor is shown to have an angular dependence given by

$$\frac{\cos(\hat{\mathbf{n}}, \mathbf{R}) - \cos(\hat{\mathbf{n}}, \mathbf{r}_{21})}{2}$$

which includes the geometrical effect of the incident wave arriving at the aperture, at an angle with respect to the normal to the aperture. The source is located at a position $\mathbf{r}_{21}$, with respect to the aperture, and the effective area that it illuminates is a function of the angle between $\mathbf{r}_{21}$ and the aperture normal $\hat{\mathbf{n}}$. The obliquity factor causes the amplitude per unit area of the incident and transmitted light to decrease as the viewing angle increases. In this chapter, the obliquity factor is introduced in an ad hoc fashion, but in Appendices 9-A and 9-B, the obliquity factor is generated by the formal theory. The obliquity factor increases the difficulty of working with the Fresnel integral and it is to our benefit to be able to treat it as a constant. If we neglect its angular contribution, we need to justify the approximation. Our justification will be based on the resolving power of an optical system operating at visible wavelengths.

Assume we are attempting to resolve two stars that produce plane waves at the aperture of a telescope with an aperture diameter a. The wavefronts from the two stars make an angle ψ with respect to each other; see Figure 9-7.

If the source is at infinity, then the incident wave can be treated as a plane wave, incident normal to the aperture, and we may simplify the angular dependence of the obliquity factor to

$$\frac{1 + \cos\theta}{2}$$

where θ is the angle between the normal to the aperture and the vector $\mathbf{R}$. This is the configuration shown in Figure 9-6.

The obliquity factor provides an explanation of why it is possible to ignore the backward-propagating wave that occurs in the application of Huygens' principle. For the backward wave, $\theta = \pi$, and the obliquity factor is zero.

$$\tan\psi \approx \psi = \frac{\Delta x}{a} \tag{9-10}$$

The smallest angle ψ that can be measured is determined by the smallest length Δx that can be measured. We know we can measure a fraction of wavelength with an interferometer but, without an interferometer, we can only count the crests of the waves, leading to the assumption that $\Delta x \geq \lambda$, i.e., we can measure a length no smaller than λ. This reasoning leads to the assumption that the smallest angle we can measure is

$$\psi \geq \frac{\lambda}{a} \tag{9-11}$$

The resolution limit established by the above reasoning places a limit on the minimum separation that can be produced at the back focal plane of the telescope. We will demonstrate in Chapter 10 (for now, it must be accepted on faith) that the minimum distance on the focal plane between the images of star 1 and 2 is given by

$$d = f\psi \tag{9-12}$$

From **(9-11)**,

$$d = \frac{\lambda f}{a}$$

The resolution limit of the telescope, to be derived in Appendix 10-C, can also be expressed in terms of the cone angle produced when the incident plane wave is focused on the back focal plane of the lens. From the geometry of Figure 9-8, the half-cone angle is given by

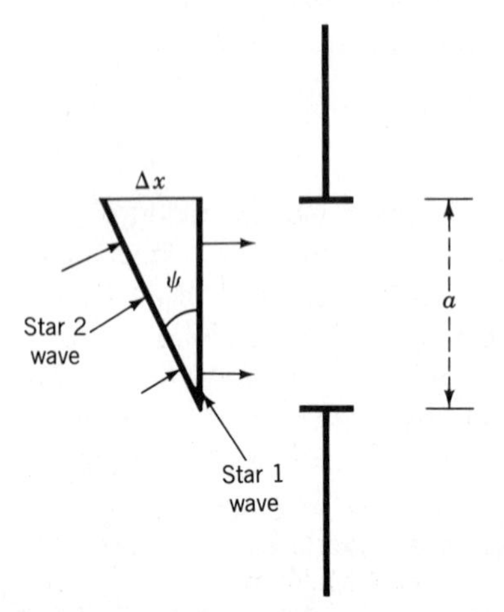

FIGURE 9-7. Light from two stars arriving at a telescope.

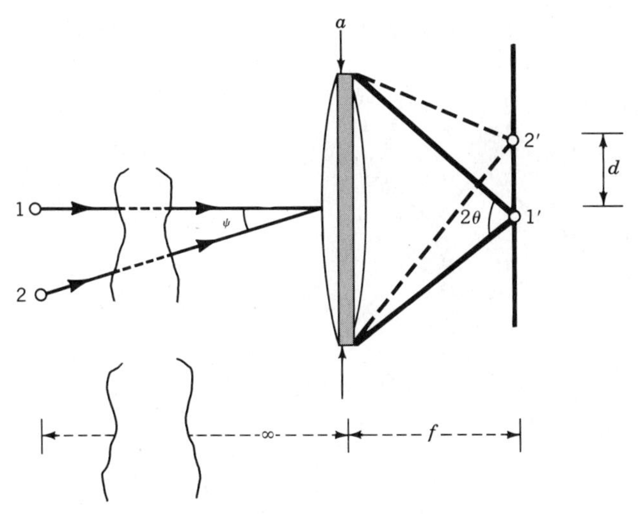

FIGURE 9-8. Telescope resolution.

$$\tan\,\theta = \frac{a}{2f}$$

The separation between the two stars at the back focal plane of the lens is thus

$$d = \frac{\lambda}{2\,\tan\,\theta} \tag{9-13}$$

This equation will establish the maximum value of θ to be encountered in a conventional optical system. We need a reasonable value for the minimum d we are capable of measuring, in order to find the maximum value of θ that will be encountered. We will assume that the minimum d is 3λ; this is four times the resolution of a typical photographic film at visible wavelengths. With this value for d in **(9-13)**, the largest obliquity angle that should be encountered in a visible imaging system is

$$\tan\,\theta = \frac{\lambda}{2d} = \frac{\lambda}{6\lambda} = \frac{1}{6} = 0.167$$

$$\theta = 9.5° \approx 10°$$

(If we had assumed that the smallest value of d was equal to λ, then $\theta = 26.6°$.) We now can make an estimate of a reasonable upper bound on the variation in the obliquity factor over the range of angles that will be encountered in a visible optical system. At $\theta = 0°$, $(1 + \cos\,\theta)/2 = 1$. At $\theta = 10°$, $\cos\,\theta = 0.985$ and $(1 + \cos\,\theta)/2 = 0.992$. The obliquity factor has only changed by 0.8% over the angular variation of 0° to 10°; thus, in most situations encountered in visible optics, we are safe in treating the factor as a constant and removing it from the integrand.

The obliquity factor undergoes a 10% change when θ varies from 0° to about 40°; therefore, the obliquity factor cannot safely be treated as a constant in any optical system that involves angles larger than 40°. This can occur in optical systems designed for wavelengths longer than visible wavelengths. (Additional comments about the variation of the obliquity factor are found in Chapter 11.)

Although we have shown that it is safe to ignore the variability of C, we still have not assigned a value to the obliquity factor. To find the proper value for C, we will compare the result obtained, using **(9-9)** with the result predicted by geometric optics.

We illuminate the aperture Σ_0 in Figure 9-9 with a plane wave of amplitude α, traveling parallel to the z axis. Geometrical optics predicts a

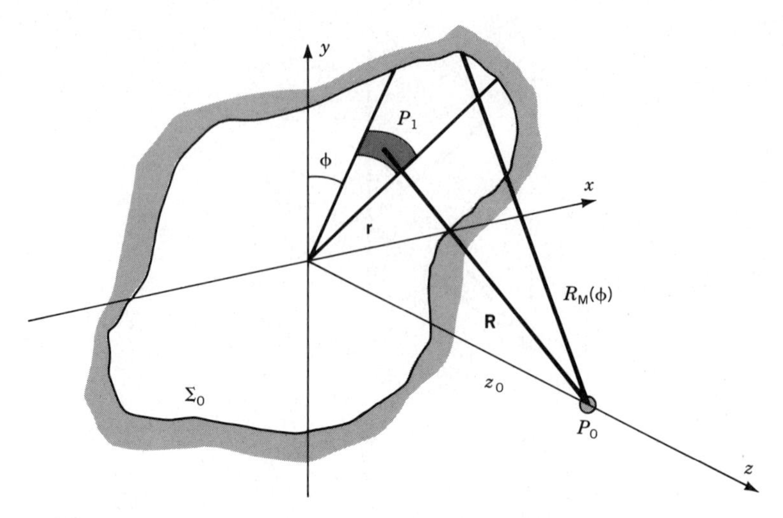

FIGURE 9-9. Geometry for evaluating the constant in the Fresnel integral.

field at P_0 on the z axis, a distance z_0 from the aperture, given by

$$\mathcal{E}_{geom} = \alpha e^{-ikz_0} \tag{9-14}$$

The area of the infinitesimal at P_1 (the Huygens' source) is

$$ds = r\,dr\,d\phi$$

The obliquity factor is assumed to be a constant C that can be removed from under the integral. The incident wave is a plane wave whose value at $z = 0$ is $E(r) = \alpha$. If we use these parameters, the Fresnel integral **(9-9)** can be written as

$$\mathcal{E}(z_0) = C\alpha \iint \frac{e^{-i\mathbf{k}\cdot\mathbf{R}}}{R}\, r\,dr\,d\phi \tag{9-15}$$

The distance from the Huygens' source to the observation point is

$$z_0^2 + r^2 = R^2$$

where z_0 is a constant equal to the distance from the observation point to the aperture plane. The variable of integration can be written in terms of R

$$r\,dr = R\,dR$$

The limits of integration over the aperture extend from $R = z_0$ to the maximum value of R, $R = R_m(\phi)$.

$$\mathcal{E}(z_0) = C\alpha \int_0^{2\pi} \int_{z_0}^{R_m(\phi)} e^{-i\mathbf{k}\cdot\mathbf{R}}\, dR\, d\phi \tag{9-16}$$

The integration over R can now be carried out to yield

$$\mathcal{E}(z_0) = \frac{C\alpha}{ik} e^{-ikz_0} \int_0^{2\pi} d\phi - \frac{C\alpha}{ik} \int_0^{2\pi} \exp[-ikR_m(\phi)]\, d\phi \tag{9-17}$$

The first integration in **(9-17)** is easy to perform. The second integration cannot be calculated because we are performing the evaluation of the Fresnel integral for a general aperture and the functional form of $R_m(\phi)$ is not known.

However, the physical significance of the second integral can be obtained. The first term in **(9-17)** is the amplitude due to geometric optics. The second term may be interpreted as the sum of the waves diffracted by the boundary of the aperture. An equivalent statement is that the second term is the interference of the waves scattered from the aperture's boundary. This is the interpretation of diffraction that was first suggested by Young.

The aperture is irregular in shape, at least on the scale of a wavelength; thus, $kR_m(\phi)$ will vary many multiples of 2π as we integrate around the aperture. For this reason, we should be able to ignore the second integral in **(9-17)**, if we confine our attention to the light distribution on the z axis. After neglecting the second term, we are left with only the geometrical optics component of **(9-17)**

$$\mathcal{E}(z_0) = \frac{2\pi C}{ik} \alpha e^{-ikz_0} \tag{9-18}$$

For **(9-18)** to agree with the prediction of geometric optics **(9-14)** the constant C must be equal to

$$C = \frac{ik}{2\pi} = \frac{i}{\lambda} \tag{9-19}$$

Using this result in **(9-17)**, we obtain

$$\mathcal{E}(z_0) = \alpha e^{-ikz_0} - \frac{1}{2\pi}\int_0^{2\pi} \alpha e^{-ikR_m(\phi)}\, d\phi$$

The Huygens–Fresnel integral can be written using the value for C just derived

$$\mathcal{E}(\mathbf{r}_0) = \frac{i}{\lambda}\iint_\Sigma \frac{\mathcal{E}_i(\mathbf{r})}{R} e^{-i\mathbf{k}\cdot\mathbf{R}}\, ds \tag{9-20}$$

The Huygens–Fresnel integral can be interpreted in two equivalent ways. The classical interpretation views

$$\frac{i}{\lambda}\frac{e^{-i\mathbf{k}\cdot\mathbf{R}}}{R}$$

as Huygens' wavelet. This is a spherical wave of unit amplitude that when multiplied by the value of the incident wave $E_i(\mathbf{r})$ at a point $\mathbf{r}$ in space, yields the Huygens' wavelet radiated by that point. The integral adds all of the interfering wavelets together to produce the resultant wave.

A more modern interpretation of the Huygens–Fresnel integral is to view it as a convolution integral. As a derived result, considering free space as a linear system, one finds that wave propagation can be calculated by convolving the incident (input) wave with the impulse response of free space

$$\frac{i}{\lambda}\frac{e^{-i\mathbf{k}\cdot\mathbf{R}}}{R}$$

The job of calculating diffraction has only begun with the derivation of the Huygens–Fresnel integral. In general, an analytic expression for the integral cannot be found because of the difficulty of performing the integration over R, (Figure 9-9). There are two approximations we can make that will allow us to obtain analytic expressions of the Huygens–Fresnel integral. The approximations, called Fraunhofer and Fresnel diffraction, will be discussed in the next two chapters.

The appendices of this chapter contain a formal derivation of the Huygens–Fresnel integral. The details contained in these derivations are not required to understand diffraction, but will provide a more formal derivation of the diffraction integral for the reader not satisfied with the intuitive approach taken above.

GAUSSIAN BEAMS

We have just introduced a view of diffraction that treats the propagation of light by an integral method, as modifications made on an input beam. A second approach to describing the propagation of light is to develop the differential equation of the optical system. The output of the system is then obtained, given the input wave from the solutions of the differential equation. For free space, the differential equation that must be satisfied to determine the spatial behavior of a wave is the Helmholtz equation.

We will find that if we make the paraxial approximation, solutions of the Helmholtz equation, with Gaussian amplitude distributions, are obtained. We will show that the waves described by these particular solutions can be characterized by two simple parameters. These parameters are the beam waist and the radius of curvature of the phasefront of the wave. The *beam waist* is defined as the half-width at an amplitude equal to $1/e$ of the maximum amplitude of the wave, transverse to the propagation direction. The *radius of curvature* describes the radius of curvature of the phasefront of the wave, as measured from the position of minimum beam waist.

The mathematical manipulations required to obtain the characteristic parameters of a Gaussian wave are long, but not complicated. The final results are worth the effort in that they allow the inclusion of diffraction in the analysis of complicated optical systems.[19] The transverse amplitude distribution of optical beams from lasers has a Gaussian amplitude distribution, as do the propagation modes of some optical fibers and the cavity modes of Fabry–Perot resonators with spherical mirrors.

To derive the characteristics of a Gaussian wave, we use the paraxial approximation. A plane wave is assumed to be propagating nearly parallel to the z direction, allowing it to be described by a scalar wave of the form

$$\mathcal{E}(\mathbf{r}) = \Psi(x, y, z)\, e^{-ikz} \tag{9-21}$$

i.e., the wave does not propagate in the x or y direction.

We will substitute this general wave into the Helmholtz equation to produce the scalar, paraxial, wave equation that must be solved

$$\left(\frac{\partial^2\Psi}{\partial x^2} + \frac{\partial^2\Psi}{\partial y^2} + \frac{\partial^2\Psi}{\partial z^2}\right) e^{-ikz} + k^2\Psi e^{-ikz} - 2ik\frac{\partial\Psi}{\partial z}e^{-ikz} - k^2\Psi e^{-ikz} = 0 \tag{9-22}$$

The assumption that Ψ changes very slowly with z (linearly with z will do it) allows $\partial^2\Psi/\partial z^2$ to be neglected. The resulting scalar wave equation is called the *paraxial wave equation*

$$\frac{\partial^2\Psi}{\partial x^2} + \frac{\partial^2\Psi}{\partial y^2} - 2ik\frac{\partial\Psi}{\partial z} = 0 \tag{9-23}$$

The paraxial wave equation and its solutions lead to a description that can be shown to be equivalent to the Fresnel description of diffraction.[43]

We assume that the solution of **(9-23)** has the form

$$\Psi = e^{-iQ(z)(x^2+y^2)}\, e^{-iP(z)} \tag{9-24}$$

In this form, we will find it easy to discover the requirements on P and Q when **(9-24)** is a solution of **(9-23)**; however, in this form, **(9-24)** does not appear to be a plane wave with a Gaussian amplitude distribution, but by proper specification of $P(z)$ and $Q(z)$, **(9-24)** will assume a Gaussian form. We will discover that $Q(z)$ must be a complex variable associated with the reciprocal of the Gaussian width and that $P(z)$ must contain information on the phase of the wave.

By the selection of **(9-24)** as the solution of **(9-23)**, we have implicitly assumed that the transverse dependence of the wave is only a function of $(x^2 + y^2)$. The assumption results in the simplest Gaussian wave solutions, i.e., waves that have circular symmetry. We will not discuss the higher-order Gaussian beam modes that arise by eliminating this assumption.[43]

We first determine the derivatives of **(9-24)**, to be substituted into **(9-23)**.

$$\frac{\partial^2\Psi}{\partial x^2} = -4x^2Q^2\Psi - 2iQ\Psi, \qquad \frac{\partial^2\Psi}{\partial y^2} = -4y^2Q^2\Psi - 2iQ\Psi$$

$$\frac{\partial\Psi}{\partial z} = -i\frac{\partial P}{\partial z}\Psi - i(x^2 + y^2)\frac{\partial Q}{\partial z}\Psi \tag{9-25}$$

Substituting **(9-25)** into **(9-23)** results in an expression that will yield the required form of $P(z)$ and $Q(z)$

$$-4(x^2 + y^2)Q^2 - 4iQ - 2k\frac{\partial P}{\partial z} - 2k(x^2 + y^2)\frac{\partial Q}{\partial z} = 0 \tag{9-26}$$

Since **(9-26)** must hold for all values of x and y, we may equate the coefficients of the different powers of x and y to zero

$$k\frac{\partial P}{\partial z} + 2iQ = 0 \tag{9-27}$$

$$2Q^2 + k\frac{\partial Q}{\partial z} = 0 \tag{9-28}$$

[Equation **(9-28)** is called Riccati's equation.]

We make a change of variables

$$q = \frac{k}{2Q} \tag{9-29}$$

In a moment, q will be identified as the desired Gaussian width of the amplitude distribution of the wave. Using the new variable, we may write

$$\frac{\partial q}{\partial z} = -\frac{k}{2Q^2}\frac{\partial Q}{\partial z}$$

We can use this new variable to rewrite **(9-27)** and **(9-28)** as

$$\frac{\partial q}{\partial z} = 1 \tag{9-30}$$

$$\frac{\partial P}{\partial z} = \frac{-i}{q} \tag{9-31}$$

Equation **(9-30)** integrates to give

$$q = q_0 + z \tag{9-32}$$

where we have chosen the constant of integration q_0 to be purely imaginary. By moving q_0 off the real axis, we remove the singularity that would occur in $\partial Q/\partial z$ if $q_0 = -z_0$. Requiring q to be complex also makes it possible to interpret **(9-24)** as a wave with a Gaussian amplitude distribution.

If q is known in one plane, we can calculate q in a plane, a distance z away, by using **(9-32)**. The solution of the paraxial wave equation provides a capability similar to that provided by Huygens' principle. The determination of new properties of a wave can be made using the old properties of the wave.

The derived form of q **(9-32)** can be substituted into **(9-31)**

$$\frac{\partial P}{\partial z} = \frac{-i}{q_0 + z}$$

to obtain the function $P(z)$.

$$P(z) = -i \ln\left(1 + \frac{z}{q_0}\right) \tag{9-33}$$

The results obtained for $P(z)$ and $1/Q(z)$ can now be substituted into **(9-24)** to produce the wave solution of the paraxial Helmholtz equation

$$\Psi = \exp\left\{-i\left[-i \ln\left(1 + \frac{z}{q_0}\right) + \frac{k}{2(q_0 + z)}(x^2 + y^2)\right]\right\} \tag{9-34}$$

Since q_0 is imaginary, it is possible to make the following substitution:

$$\ln\left(1 + \frac{z}{q_0}\right) = \ln\left(1 - \frac{iz}{q_0}\right)$$

where q_0 is now a real quantity. The identity

$$\ln(x \pm iy) = \ln\sqrt{x^2 + y^2} \pm i \tan^{-1}\left(\frac{y}{x}\right)$$

can now be used to obtain a new formulation of **(9-34)**

$$\Psi = \frac{1}{\sqrt{1 + (z/q_0)^2}} \exp\left[-\frac{kq_0(x^2 + y^2)}{2(z^2 + q_0^2)}\right] \exp\left[i \tan^{-1}\left(\frac{z}{q_0}\right) - \frac{ikz(x^2 + y^2)}{2(z^2 + q_0^2)}\right] \tag{9-35}$$

Evaluating **(9-35)** at $z = 0$ allows the identification of the physical significance of the amplitude of Ψ. At $z = 0$, **(9-35)** reduces to

$$\Psi_0 = \exp\left[-\frac{k(x^2 + y^2)}{2q_0}\right] \tag{9-36}$$

Equation **(9-36)** is a Gaussian function, as we can see by comparing it with a Gaussian spatial amplitude distribution, given by

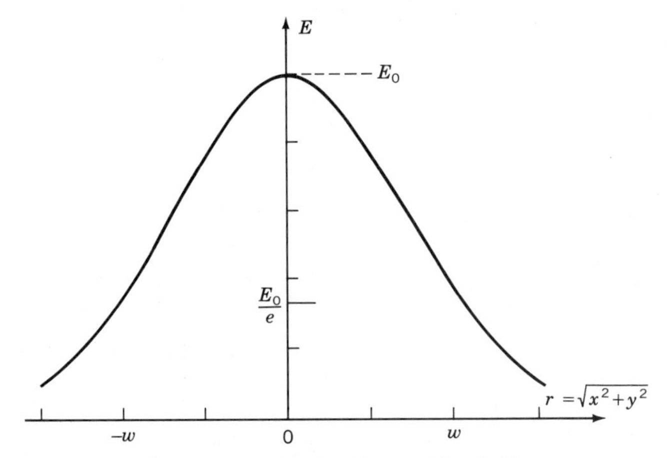

FIGURE 9-10. Gaussian wave of width w and height E_0.

$$E = E_0 \exp\left[-(x^2 + y^2/w^2)\right]$$

and plotted in Figure 9-10. The comparison demonstrates that Ψ can be interpreted as a wave, whose transverse amplitude distribution is a Gaussian.

The parameter w of the Gaussian distribution equals the half-width of the Gaussian function at the point where the amplitude is $1/e$ of its maximum value (see Figure 9-10). We define $w = w_0$ as the minimum half-width of the Gaussian function. Comparing the Gaussian function with the equation for Ψ_0 allows the definition of a minimum width for the Gaussian wave, called the *minimum beam waist,* in terms of q_0

$$w_0^2 \equiv \frac{2q_0}{k} \tag{9-37}$$

The connection between q_0 and the minimum beam waist has been established when $z = 0$; thus, the coordinate system must have its origin at the minimum beam diameter. The real valued constant

$$q_0 \equiv \frac{kw_0^2}{2} = \frac{\pi w_0^2}{\lambda} \tag{9-38}$$

is often called the *confocal parameter* (its significance will be identified in a few moments). The complex constant of integration

$$\mathfrak{q}_0 = iq_0$$

introduced in **(9-32)** can now be interpreted as the minimum value of the *complex size parameter* of a Gaussian wave

$$\mathfrak{q}_0 = \frac{i\pi w_0^2}{\lambda} \tag{9-39}$$

This analysis allows the association of the complex function $\mathfrak{q}$ with the physical variables of the beam. With this association, the physical interpretation of **(9-32)** is that as z increases, the beam variables evolve. The minimum value of the parameter is the minimum beam waist w_0 and its position is established as the origin of the coordinate system. The value of $\mathfrak{q}$ at a distance z from the beam waist is obtained from **(9-32)**

$$q = q_0 + z = z + \frac{i\pi w_0^2}{\lambda}$$

Using the definition of q_0 given in **(9-38)**, we can write **(9-35)** as

$$\Psi = \frac{1}{\sqrt{1 + (\lambda z/\pi w_0^2)^2}}$$

$$\exp\left\{-\frac{(x^2 + y^2)}{w_0^2\left[1 + (\lambda z/\pi w_0^2)^2\right]}\right\} \exp\left\{i \tan^{-1}\left(\frac{\lambda z}{\pi w_0^2}\right) - \frac{i\pi(x^2 + y^2)}{\lambda z\left[1 + (\pi w_0^2/\lambda z)^2\right]}\right\} \tag{9-40}$$

We began our analysis by assuming a paraxial wave, i.e., the wave must propagate in a direction that is nearly parallel with the z axis. Examination of that assumption will allow us to assign physical significance to the quantities contained in **(9-40)**. We allow the wave of interest to be a spherical wave, propagating away from the origin, with a wavefront of radius of curvature $R(z)$ a distance z from the origin, as is shown in Figure 9-11.

The assumption of a paraxial wave implies that this spherical wave can be approximated by a *paraxial, spherical wave* of the form

$$\frac{1}{R}e^{-ikR} = \frac{1}{R}\exp\left[-ik\sqrt{z^2 + x^2 + y^2}\right]$$

$$\approx \frac{1}{R}e^{-ikz}\exp\left[-ik(x^2 + y^2/2z)\right]$$

where were have assumed that $z^2 >> x^2 + y^2$. The paraxial assumption is that $z \approx R$, so that

$$\frac{1}{R}e^{-ikR} \approx \frac{1}{z}e^{-ikz}\exp\left[-ik(x^2 + y^2/2R)\right]$$

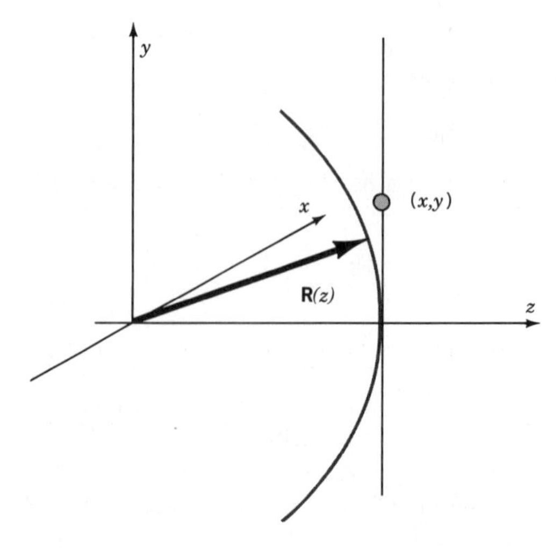

FIGURE 9-11. Propagation of a spherical wave. If point (x, y) is not too far from the z axis, we may approximate the spherical wave by a nearly plane wave called the paraxial, spherical wave.

All of the terms in the equation for a Gaussian wave **(9-40)** can now be given a physical interpretation. Upon examining **(9-40)**, we find that it can be written in the same form as the paraxial, spherical wave if we make the association

$$\frac{k}{2R} = \frac{\pi}{\lambda z\left[1 + \left(\pi w_0^2/\lambda z\right)^2\right]}$$

The radius of curvature of the phasefront of a paraxial spherical wave, described by **(9-40)**, is therefore determined by the function

$$R(z) = z\left[1 + \left(\frac{\pi w_0^2}{\lambda z}\right)^2\right] \tag{9-41}$$

At large values of z, $R \approx z$. The sign convention is such that the beam shown in Figure 9-11 has a positive radius of curvature.

This convention is consistent with the sign convention for the radius of curvature of optical surfaces, defined in Chapter 5. The radius of curvature of an optical surface is measured from the surface to the center of curvature. Here, the radius is measured, in the opposite sense, from the center of curvature to the phase front.

There is an extra phase term in **(9-40)**

$$\phi(z) = \tan^{-1}\left(\frac{\lambda z}{\pi w_0^2}\right) \tag{9-42}$$

$\phi(z)$, defined by **(9-42)**, is the phase difference between an ideal plane wave, shown by the dotted line through the point (x, y) in Figure 9-11, and the "nearly plane," spherical wave of this theory.

The first exponent in **(9-40)** describes the amplitude distribution across the wave. It has the same functional form as a Gaussian distribution with a beam width, sometimes called the beam's *spot size*, given by

$$w(z)^2 = w_0^2\left[1 + \left(\frac{\lambda z}{\pi w_0^2}\right)^2\right] \tag{9-43}$$

The curve created by connecting the $1/e$ points of the transverse amplitude of the Gaussian beam, along its propagation path, is described by **(9-43)**. The curve is a hyperbola along the wave's propagation path. At large z, the asymptotic representation of **(9-43)** is a straight line, the geometric ray

$$w(z) = \left(\frac{\lambda}{\pi w_0}\right) z$$

originating at the origin and propagating in the positive z direction. The ray is inclined, with respect to the z axis, at the *diffraction angle*

$$\theta = \frac{\lambda}{\pi w_0} \tag{9-44}$$

We will discover in the next chapter that the Fresnel formalism also yields **(9-44)**.

The diffraction angle can be used to calculate the beam diameter at a distance z from the beam waist

$$w(z)^2 = w_0^2 + \theta^2 z^2 \tag{9-45}$$

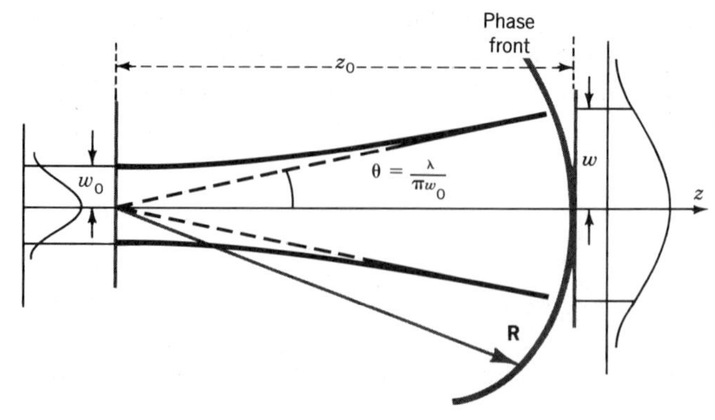

FIGURE 9-12. The propagation of a Gaussian wave from its beam waist to the far field.

The physical interpretation given to **(9-44)** is that diffraction causes a wave of diameter $2w_0$ to spread by an amount

$$\frac{2Z\lambda}{\pi w_0}$$

after propagating a distance Z. The larger the value of the beam waist w_0, the smaller the beam will spread due to diffraction. This is why a megaphone works for a cheerleader. The megaphone changes the effective aperture, producing sound from the small diameter of a human mouth, about 50 mm, to a much larger diameter, something over 30 cm. The propagation properties of laser beams are closely modeled by **(9-44)**. For this reason, lasers are often characterized by their *divergence angle,* which is equal to the diffraction angle.

Figure 9-12 schematically represents the propagation of a Gaussian beam. The dark lines are the hyperbola, given by **(9-43)**, and the dashed straight lines are asymptotes of the hyperbola, i.e., the geometric rays inclined at an angle θ calculated from **(9-44)**.

The significance of the confocal parameter can be identified by examining **(9-43)**. When the definition of the confocal parameter **(9-38)** is substituted into **(9-43)**, the resulting equation

$$w(z)^2 = w_0^2\left[1 + \left(\frac{z}{q_0}\right)^2\right]$$

demonstrates that the confocal parameter, also called the *Rayleigh range,* is the propagation distance z over which the wave's width increases from $w_0 \rightarrow \sqrt{2}w_0$. The confocal parameter, therefore, characterizes a Gaussian beam's convergent or divergent properties. To emphasize the connection with the divergence of the beam, we rewrite the Rayleigh range using **(9-38)** and **(9-44)** as

$$q_0 = \frac{w_0}{\theta}$$

Using the newly defined parameters, we can write **(9-40)** as

$$\Psi = \underbrace{\left[\frac{w_0}{w(z)}\right]\exp\left[-\frac{x^2+y^2}{w(z)^2}\right]}_{\text{amplitude}}\underbrace{\exp\left[-ik\frac{x^2+y^2}{2R(z)}\right]}_{\text{paraxial wave}}\underbrace{\exp\left[-i\phi(z)\right]}_{\text{phase}} \tag{9-46}$$

The Gaussian wave described by **(9-46)** has as its property that its cross section is everywhere given by the same function, a Gaussian. There are other solutions to **(9-23)** with this property. These additional solutions, arising when the transverse amplitude distribution is not constrained to have circular symmetry, combine, with the function we have been discussing, to form a complete orthogonal set called the *modes of propagation.*

If we divide $R(z)$ by $w(z)$, we can use the result to obtain expressions for w_0 and z in terms of R and w

$$w_0^2 = \frac{w^2}{\left[1 + (\pi w^2/\lambda R)^2\right]} \tag{9-47}$$

$$z = \frac{R}{\left[1 + (\lambda R/\pi w^2)^2\right]} \tag{9-48}$$

From **(9-47)**, we see that at the minimum beam waist, the phasefront of the Gaussian wave is a plane, i.e., $R = \infty$. These equations can be used to locate the beam waist of any Gaussian wave. To obtain the parameters of a wave at any point along its propagation path, the *ABCD* matrix can be used, as we will learn in the next section.

THE *ABCD* LAW

We will establish a relationship between the Gaussian beam parameters we have just introduced and geometrical optics that will allow the calculation of Gaussian beam parameters after the wave has passed through an optical system. We first discover how the radius of curvature of a Gaussian wavefront is changed by a thin lens. We will learn that the radius of curvature and the complex beam parameter are governed by the same propagation equations. This finding will allow the construction of an *ABCD* law for a Gaussian wave.

In Chapter 5, we found the *ABCD* matrix, which relates the input and output parameters of an optical system, using the paraxial approximation

$$\begin{pmatrix} x_2 \\ \gamma_2 \end{pmatrix} = \begin{pmatrix} A & B \\ C & D \end{pmatrix} \begin{pmatrix} x_1 \\ \gamma_1 \end{pmatrix}$$

$$x_2 = Ax_1 + B\gamma_1, \qquad \gamma_2 = Cx_1 + D\gamma_1$$

The variable x_1 is the coordinate position above the optical (z) axis of the ray entering the optical system, x_2 is the coordinate position of the ray leaving the system, and the γ's are the ray slopes. The ray slope for a Gaussian wave of radius R shown in Figure 9-13 is

$$\gamma = \frac{dx}{dz} = \tan\,\gamma \approx \frac{x}{R} \tag{9-49}$$

so that

$$R = \frac{x}{\gamma} \tag{9-50}$$

The radius of curvature of the Gaussian wave leaving the optical system described by the *ABCD* matrix is given by

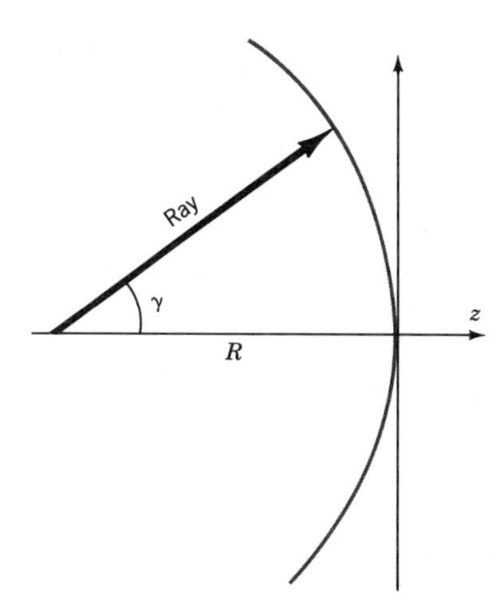

FIGURE 9-13. Geometry for a Gaussian wave in geometrical optics.

$$R_2 = \frac{x_2}{\gamma_2} = \frac{\gamma_1\left(A\frac{x_1}{\gamma_1} + B\right)}{\gamma_1\left(C\frac{x_1}{\gamma_1} + D\right)} = \frac{AR_1 + B}{CR_1 + D} \tag{9-51}$$

where R_1 is the radius of curvature of the Gaussian wave entering the optical system.

To determine the radius of curvature of the phasefront of a Gaussian wave after it has passed through a simple lens, we substitute into **(9-51)** the *ABCD* matrix from Figure 5A-4

$$R_2 = \frac{R_1}{(-R_1/f) + 1}$$

$$\frac{1}{R_2} = \frac{1 - (R_1/f)}{R_1} = \frac{1}{R_1} - \frac{1}{f} \tag{9-52}$$

When the Gaussian wave is propagating through free space, we use the *ABCD* matrix **(5-11)** to obtain the radius of curvature of the phasefront R_2 after the wave has propagated a distance d

$$R_2 = R_1 + d \tag{9-53}$$

The Gaussian wave's complex size parameter q provides a concise description of the propagation of a Gaussian beam through an optical system and it extends easily to higher-order modes. In fact, the higher order modes have the same w, R, and q as the fundamental mode we are discussing; only the phase ϕ is different. For a Gaussian wave propagating through free space, from the beam waist to a position z, the complex size parameter is

$$q_1 = q_0 + z$$

If we propagate from z to $z + d$, the q parameter becomes

$$q_2 = q_1 + d \tag{9-54}$$

The complex size parameter obeys the same rule as the radius of curvature for a wave propagating in free space.

To analyze the effects of a simple lens on a Gaussian wave, recall that the complex size parameter q can be written as

$$\frac{1}{q} = \frac{1}{(z + i\pi w_0^2/\lambda)} = \frac{z - (i\pi w_0^2/\lambda)}{z^2 + (\pi w_0^2/\lambda)^2}$$

Using **(9-41)** and **(9-43)**, we can rewrite this as

$$\frac{1}{q} = \frac{1}{R} - \frac{i\lambda}{\pi w^2} \tag{9-55}$$

For a thin lens, the spot size w is the same at the front and back surfaces of the lens (remember from Appendix 5-A that the front and back vertices of a thin lens define planes of unit magnification); thus, $w_2 = w_1$. The beam radius of curvature should change according to **(9-52)**, allowing us to write **(9-55)** as

$$\begin{aligned} \frac{1}{q_2} &= \frac{1}{R_2} - \frac{i\lambda}{\pi w_2^2} \\ &= \left(\frac{1}{R_1} - \frac{1}{f}\right) - \frac{i\lambda}{\pi w_1^2} \end{aligned} \tag{9-56}$$

Rearranging the terms yields

$$\frac{1}{q_2} = \frac{1}{q_1} - \frac{1}{f} \tag{9-57}$$

Comparing **(9-52)** and **(9-57)** leads us to the conclusion that the complex beam parameter q plays a role corresponding to the one played by the radius of curvature R of a spherical wave. This should not surprise us because **(9-55)** defines the real part of $1/q$ as being equal to $1/R$. We can rename q as the *complex curvature* of a Gaussian wave.

Because of the formal equivalence between q and R, **(9-51)** can be used to write

$$q_2 = \frac{Aq_1 + B}{Cq_1 + D} \tag{9-58}$$

Equation **(9-58)** allows a Gaussian beam to be traced through any optical system. Several examples of the application of **(9-58)** will show the usefulness of this result.

Thin Lens

As the first example of the use of **(9-58)**, a Gaussian beam will be followed through a thin lens. Assume that a plane wave uniformly illuminates a lens of diameter D. Because it is a plane wave $R_1 = \infty$ and the aperture of the lens is uniformly illuminated, $w = D/2$. The q parameter at the left surface of the lens is, therefore, given by

$$\frac{1}{q_1} = 0 - \frac{4i\lambda}{\pi D^2}$$

The *ABCD* matrix for a thin lens can be obtained from Figure 5A-4 and can be used to calculate the q parameter, after passing through the lens

$$q_2 = \frac{q_1}{1 - (q_1/f)}$$

$$\frac{1}{q_2} = -\frac{1}{f} - \frac{i\lambda}{\pi (D/2)^2}$$

After passing through the thin lens, the beam waist remains the same size, $w = D/2$, but the radius of curvature of the phase front becomes

$$R_2 = -f$$

i.e., the beam is now converging with the center of curvature of the wavefront on the right of the wavefront.

To find the location of the minimum beam waist, we use **(9-48)**

$$z = \frac{-f}{1 + (4\lambda f/\pi D^2)^2} \approx -f$$

The minus sign signifies that the current position is to the left of the minimum beam waist, i.e., the minimum beam waist lies to the right of the lens, a distance f from the lens. The size of the minimum beam waist is given by **(9-47)**

$$w_0^2 = \frac{D^2/4}{1 + (\pi D^2/4\lambda f)^2} \approx \frac{4\lambda^2 f^2}{D^2}$$

$$w_0 \approx \frac{2\lambda f}{D}$$

The conclusion of this analysis is that parallel light filling the aperture of a thin lens is brought to a focus at the back focal plane of the lens. Diffraction by the aperture of the lens prevents the beam from being focused to a spot smaller than the minimum beam waist. The focal spot size is inversely proportional to the lens aperture and linearly proportional to the focal length of the lens.

Fabry–Perot Resonator

The stability condition of a Fabry–Perot resonator, derived in Chapter 5, can be obtained by using **(9-58)**. For a mode to be stable, we require that the q parameter, at any arbitrary reference plane, must reproduce itself after a round trip in the cavity, i.e.,

$$q = \frac{Aq + B}{Cq + D}$$

where A, B, C, and D are elements of the $ABCD$ matrix for the Fabry–Perot resonator, the matrix elements for a reference plane located at mirror 1 of Figure 5-12 are given in **(5-14)**. The equation may be solved for $1/q$ to obtain

$$\frac{1}{q} = \frac{(D - A) \pm \sqrt{(D - A)^2 + 4BC}}{2B}$$

The $ABCD$ determinant for the resonator must equal 1 because the index of refraction is a constant in the resonator, i.e., $AD - BC = 1$. This fact allows the equation for $1/q$ to be simplified.

$$\frac{1}{q} = \frac{(D - A)}{2B} \pm \frac{i\sqrt{1 - (D + A)^2/4}}{B}$$

This equation is in the standard form of the q parameter

$$\frac{1}{q} = \frac{1}{R} - \frac{i\lambda}{\pi w^2}$$

The real part of the equation for the q parameter can be extracted to discover the radius of curvature of the Gaussian wave in the Fabry–Perot resonator

$$R = \frac{2B}{D - A}$$

In order to insure that the q parameter be complex, we require that

$$\left|\frac{D + A}{2}\right| < 1$$

This is the same stability condition derived using geometrical optics **(5-21)**.

In Chapter 5 **(5-14)**, we calculated the *ABCD* matrix for a ray whose round trip left it at the surface of mirror 1 in Figure 5-12. Using this *ABCD* matrix, we will find that the radius of curvature of the stable Gaussian wave is equal to the radius of curvature of mirror 1

$$R = \frac{4d\left(1 - \frac{d}{R_2}\right)}{\left(1 - \frac{2d}{R_2}\right) - 1 + \frac{4d}{R_1} + \frac{2d}{R_2}\left(1 - \frac{2d}{R_1}\right)} = R_1$$

If the *ABCD* matrix for a wave ending at mirror 2 were used to calculate R, then we would discover that at that mirror, the radius of curvature of a reproducing Gaussian wave is $R = R_2$.

We are led to the conclusion that if a stable, reproducing Gaussian wave exists in a Fabry–Perot resonator, then the radii of curvature of the mirrors making up the resonator match the wave's wavefront curvature. Figure 9-14 shows a typical Gaussian wave. This wave would be a mode of a Fabry–Perot resonator if the mirrors of the resonator were positioned so that their curvature matched the wavefront curvature, shown by the gray lines in Figure 9-14. The mirror's diameter would be selected so that it intercepted, say, 99% of the beam at that location on the optical axis.

Laser Cavity

As a final example, we will analyze a commercial HeNe laser designed to operate at $\lambda = 632.8$ nm. The optical layout of the laser cavity is shown in Figure 9-15. We will derive the *ABCD* matrix for this design but will leave the details of the calculation to the reader (see Problem 9-4).

Inside the laser cavity, the phase front curvature of the Gaussian wave

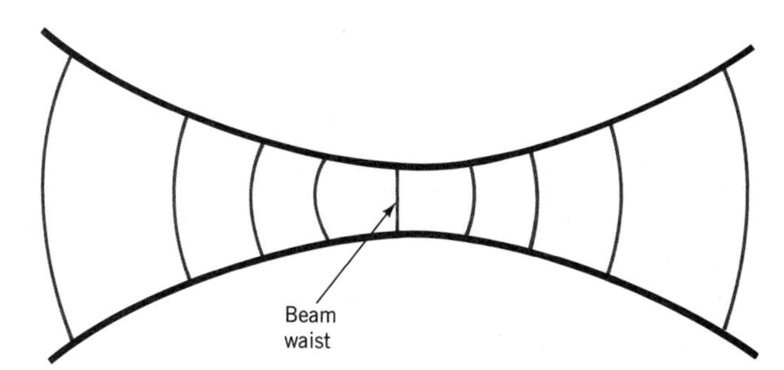

FIGURE 9-14. The black curves are the locus of the beam waists of a Gaussian wave in a Fabry–Perot cavity. The radius of curvature of the wave is shown by the gray curves. The minimum beam waist occurs at the point where the radius of curvature of the phase is infinite, i.e., a plane phasefront.

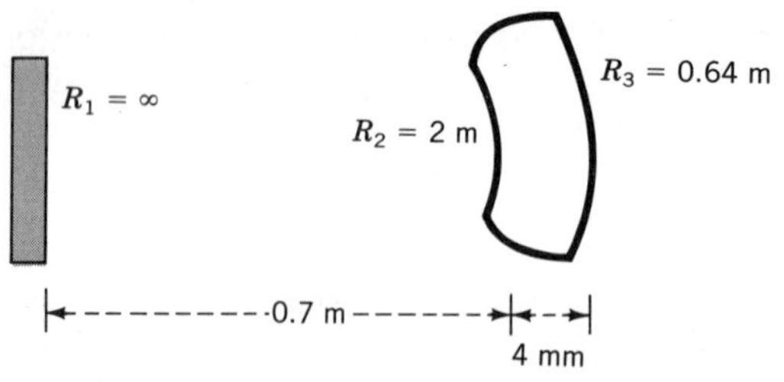

FIGURE 9-15. HeNe laser cavity design.

must match the curvature of the mirrors in the cavity. This means that at the plane mirror, on the left of Figure 9-15, the radius of curvature is infinite. From **(9-47)**, we see that the beam waist always occurs at the point where the radius of curvature is infinite (see Figure 9-14). We will make all of our measurements from the beam waist that we now know to be at the plane mirror. The second mirror is a lens whose concave surface has a reflective dielectric coating. At mirror 2,

$$z = 0.7 \text{ m}, \qquad R = 2 \text{ m}$$

We use **(9-41)** to find the size of the beam waist w_0. From w_0, we can calculate the complex beam parameter q_1.

Light leaves the laser cavity through the lens, whose concave surface serves as one of the Fabry–Perot mirrors. To locate the beam waist and find its size outside the cavity, we must calculate the *ABCD* matrix. From left to right in Figure 9-15, we have the following matrices (see Figure 5A-4 for the proper matrix formulas):

1. The propagation in the laser cavity from mirror 1 to mirror 2. We assume the index of refraction in the cavity is $n_1 = 1.0$

$$\begin{pmatrix} 1 & 0.7 \\ 0 & 1 \end{pmatrix}$$

2. Refraction at the surface of mirror 2. We assume the index of refraction of the lens, which also serves as mirror surface 2, is $n_2 = 1.5$

$$\begin{pmatrix} 1 & 0 \\ \dfrac{n_2 - n_1}{n_2 R_2} & \dfrac{n_1}{n_2} \end{pmatrix}$$

3. Propagation of light through the glass between the surfaces of mirror 2. The mirror thickness is $t = 4$ mm

$$\begin{pmatrix} 1 & \dfrac{t}{n_2} \\ 0 & 1 \end{pmatrix}$$

4. Refraction at the back surface of mirror 2. We assume the index of refraction outside the laser is $n_1 = 1$

$$\begin{pmatrix} 1 & 0 \\ \dfrac{n_1 - n_2}{n_1 R_3} & \dfrac{n_2}{n_1} \end{pmatrix}$$

5. Propagation to the beam waist outside of the laser cavity. This matrix contains the quantity of interest d

$$\begin{pmatrix} 1 & d \\ 0 & 1 \end{pmatrix}$$

The *ABCD* matrix for the system is obtained by multiplying all of the above matrices. The resultant matrix is then used in **(9-58)** to find q_2. The beam waist found outside the laser cavity is a minimum beam waist so that, at the minimum waist, $R(z) = \infty$. This means that q_2 must be completely imaginary

$$\frac{1}{q_2} = -\frac{i\lambda}{\pi w_0^2}$$

SUMMARY

Huygens' principle states that the wavefront of a propagating wave can be obtained by treating all of the points on a wavefront, at an earlier time, as sources of spherical waves called Huygens' wavelets. Fresnel was able to describe diffraction by assuming that the new wavefront should be determined by allowing Huygens' wavelets to interfere. The result of Fresnel's theory is the Huygens–Fresnel integral

$$\mathcal{E}(\mathbf{r}_0) = \frac{i}{\lambda} \iint_\Sigma \frac{\mathcal{E}_i(\mathbf{r})}{R} e^{-i\mathbf{k}\cdot\mathbf{R}}\, ds$$

where $\mathcal{E}_i(\mathbf{r})$ is the wave incident on the aperture, denoted by Σ. R is the distance from a point in the aperture to the observation point located at position $\mathbf{r}_0$. The integral is an area integral taken over the aperture Σ.

In this approach to diffraction, free space is assumed to be a linear system with an impulse response given by

$$\frac{i}{\lambda} \frac{e^{-i\mathbf{k}\cdot\mathbf{R}}}{R}$$

The Huygens–Fresnel integral is interpreted as a convolution of the impulse response with the input wave.

A second approach to diffraction is often used when the light wave's transverse amplitude distribution is Gaussian. The Gaussian wave is characterized by its width

$$w(z)^2 = w_0^2\left[1 + \left(\frac{z}{q_0}\right)^2\right]$$

and the radius of curvature of its wavefront

$$R(z) = z\left[1 + \left(\frac{q_0}{z}\right)^2\right]$$

where

$$q_0 = \frac{\pi w_0^2}{\lambda}$$

is the Rayleigh range, the parameter w_0 is the minimum beam width (or waist), and z is the distance from the minimum beam waist to the observation point.

Given the beam waist w and the radius of curvature R at any point, the minimum beam waist and its location can be found by using

$$w_0^2 = \frac{w^2}{\left[1 + \left(\pi w^2/\lambda R\right)^2\right]}$$

$$z = \frac{R}{\left[1 + \left(\lambda R/\pi w^2\right)^2\right]}$$

The *ABCD* matrix, introduced in Chapter 5, can be used to trace the behavior of a Gaussian wave as it propagates through an optical system. The Gaussian wave at any point z is characterized by a q parameter

$$\frac{1}{q(z)} = \frac{1}{R(z)} - \frac{i\lambda}{\pi w(z)^2}$$

The q parameter after propagating through an optical system is given by

$$q_2 = \frac{Aq_1 + B}{Cq_1 + D}$$

where A, B, C, and D are the elements of the *ABCD* matrix.

PROBLEMS

9-1. Calculate the *ABCD* matrix for a ray that has made a round trip in the Fabry–Perot resonator of Figure 5-12, ending at mirror 2. Find the radius of curvature of the Gaussian wave at this reference plane.

9-2. Find the location of the beam waist, relative to mirror 1, in the Fabry–Perot resonator of Figure 5-12.

9-3. A Fabry–Perot resonator has a spherical front mirror of radius 2ℓ and a plane back mirror. It is illuminated by a Gaussian beam with a waist w_0 at $z = 0$. The Fabry–Perot's front mirror is located at $z = 3\ell$, its back mirror at $z = 4\ell$. A lens of focal length f at $z = 2\ell$ couples the Gaussian beam into the resonator. What values should f and w_0 have so that the beam will match the resonator's fundamental mode?

9-4. Complete the details of the calculation for the laser cavity design started in the chapter. You will find a second beam waist, the image of the first, outside the cavity. Can you think of any reasons the optical designer placed this waist outside the cavity?

9-5. Find the value of z where the radius of curvature of the phase of a Gaussian wave is a minimum. What Gaussian parameter can be used to specify this position?

9-6. What is the beam waist of a HeNe laser ($\lambda = 632.8$ nm) with a beam divergence of 0.7 mrad? What is the Rayleigh range q_0?

9-7. Use the object image matrix to show that

$$q_2 = \frac{\left(1 - \frac{z_2}{f}\right)q_1 + \left(z_1 + z_2 - \frac{z_1 z_2}{f}\right)}{\left(1 - \frac{z_1}{f}\right) - \frac{q_1}{f}} \tag{9-59}$$

where q_1 is the beam parameter in the object plane of a positive thin lens and q_2 the beam parameter in the conjugate (image) plane. [*Hint*: Use **(9-32)** to follow the beam from the first waist to the lens and from the lens to the second waist. Use **(9-58)** to obtain the effect of the lens. Use **(9-55)** at each waist to relate the complex-size parameter to the Gaussian parameters of radius of curvature and spot size.]

9-8. Assume the beam waist w_{01} of a Gaussian beam is located a distance z_1 from a positive lens of focal length f. (a) Show that there will be a new beam waist w_{02} located at distance

$$z_2 = f + \frac{(z_1 - f)f^2}{(z_1 - f)^2 + q_0^2} \tag{9-60}$$

from the lens. (b) Also show that the beam waist will be given by

$$w_{02}^2 = \frac{f^2}{(z_1 - f)^2 + q_0^2} w_{01}^2 \tag{9-61}$$

(*Hint*: Remember that at a Gaussian beam's waist, $R = \infty$.)

9-9. Find approximate expressions for **(9-60)** and **(9-61)** when $z_1, f << q_0$.

9-10. We wish to focus the HeNe laser described in Problem 9-6 using a lens with a focal length of 100 mm. Using **(9-60)**, find the beam waist and position when $z_1 << q_0$.

9-11. Find approximate expressions for **(9-60)** and **(9-61)** when $z_1 >> q_0, f$.

9-12. Assume that in the object plane of Problem 9-7 there is a beam waist of w_{01}, and in the image plane, there is a new beam waist of w_{02}. Show

$$\frac{z_1 - f}{z_2 - f} = \frac{w_{01}^2}{w_{02}^2} \tag{9-60}$$

9-13. Find the beam waist for an argon laser cavity with two concave mirrors with radii of 3 and 1.5 m, separated by 1 m. Assume the operating wavelength is 488 nm. Prove that this is a stable configuration.

9-14. Calculate the Raleigh range and the beam divergence of the laser in Problem 9-13.

9-15. We wish to couple energy from an argon laser ($\lambda = 488$ nm), with a beam waist of 350 μm, into a single mode fiber. The beam waist of the fiber mode is 2 μm. What are the separations between the laser waist and the fiber face if as a lens, we use a microscope objective with a power of 10x? *Note*: The focal length of a microscope objective is given by

$$f = \frac{160 \text{ mm}}{\text{power}}$$

9-16. We wish to relocate the position of a beam waist but not change its size. The lens that performs this operation is called a *relay lens*. Derive a relationship for the distance between the two waist locations $z_1 + z_2$ and the focal length of the lens. Remember $w_{01} = w_{02}$.

9-17. We would like to move the beam waist of a HeNe laser ($\lambda = 632.8$ nm, with a beam divergence of 0.7 mrad) to a new location with a relay lens of focal length 1.65 m. How far apart are the two beam waists? How does this agree with the predictions of the thin lens equation? What happens if we try to use a shorter-focal-length lens?

9-18. We wish to use a 1 m relay lens to move the beam waist of a dye laser to a new position. The dye laser is operating at a wavelength of 600 nm with a beam divergence of 1 mrad. How far apart are the two waists?

Appendix 9-A

FRESNEL-KIRCHHOFF DIFFRACTION

Kirchhoff, in 1887, developed a rigorous theory of diffraction that demonstrated that the results of Fresnel and Huygens could be obtained from the wave equation. His scalar theory was based on the elastic theory of light but can be reformulated into a vector theory.[3] We will be using Green's theorem (see Appendix 9-C) to formally develop the Kirchhoff theory of diffraction. Given an incident wave φ, we wish to calculate the optical wave at point P_0 in Figure 9A-1, located at $\mathbf{r}_0$, in terms of the wave's value on a surface S that we construct about the observation point.

To obtain a solution to the wave equation at the point P_0, we select as a Green's function a unit amplitude spherical wave, expanding about the point at $\mathbf{r}_0$, denoted by Ψ, and then apply Green's theorem from Appendix 9-C. At an observation point $\mathbf{r}_1$, Green's function is given by

$$\Psi(\mathbf{r}_1) = \frac{e^{-i\mathbf{k}\cdot\mathbf{r}_{01}}}{r_{01}} \tag{9A-1}$$

where

$$r_{01} = |\mathbf{r}_{01}| = |\mathbf{r}_1 - \mathbf{r}_0|$$

is the distance from the source of the Green's function at $\mathbf{r}_0$ to the observation position at $\mathbf{r}_1$.

Green's theorem requires that there be no sources (singularities) inside the surface S. This requirement is met by constructing a small spherical surface S_ϵ of radius ϵ about $\mathbf{r}_0$, excluding the singularity at $\mathbf{r}_0$ from the volume of interest, shown in gray in Figure 9A-1. The surface integration that must be performed in Green's theorem is over the surface $S' = S + S_\epsilon$.

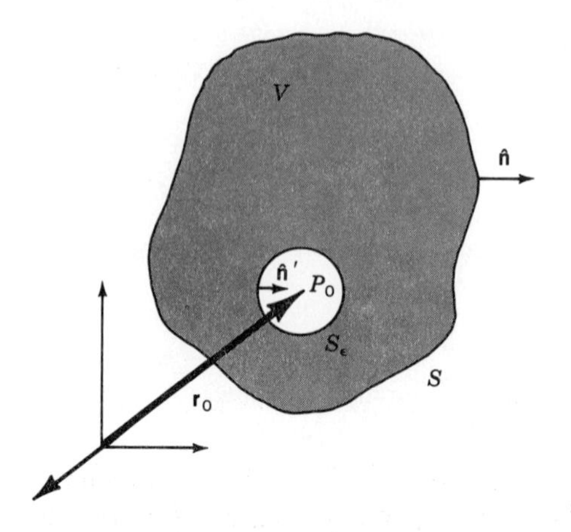

FIGURE 9A-1. Region of integration for solution of Kirchhoff diffraction integral.

Within the volume enclosed by S', the Green's function Ψ and the incident wave φ satisfy the scalar Helmholtz equation so that the volume integral can be written as

$$\iiint_V (\Psi\nabla^2\varphi - \varphi\nabla^2\Psi)\, dv = -\iiint_V (\Psi\varphi k^2 - \varphi\Psi k^2)\, dv \tag{9A-2}$$

The right side of **(9A-2)** is identically equal to zero. This fact allows us to use **(9C-5)** to produce a simplified statement of Green's theorem

$$\iint_{S'} \left(\Psi\frac{\partial\varphi}{\partial n} - \varphi\frac{\partial\Psi}{\partial n}\right) ds = 0 \tag{9A-3}$$

Because the integral over the two surfaces S and S_ϵ is equal to zero, the integral over the surface S must equal the negative of the integral over the surface S_ϵ

$$-\iint_{S_\epsilon} \left(\Psi\frac{\partial\varphi}{\partial n} - \varphi\frac{\partial\Psi}{\partial n}\right) ds = \iint_S \left(\Psi\frac{\partial\varphi}{\partial n} - \varphi\frac{\partial\Psi}{\partial n}\right) ds \tag{9A-4}$$

To perform the surface integrals, we must evaluate the Green's function on the surface S'. We, therefore, select $\mathbf{r}_1$, the vector defining the observation point, to be on either of the two surfaces that make up S'. The derivative $\partial\Psi/\partial n$ to be evaluated is

$$\frac{\partial\Psi}{\partial n} = \frac{\partial\Psi}{\partial r}\frac{\partial r}{\partial n}$$

$$= \left(-ik\frac{e^{-i\mathbf{k}\cdot\mathbf{r}_{01}}}{r_{01}} - \frac{e^{-i\mathbf{k}\cdot\mathbf{r}_{01}}}{r_{01}^2}\right)\cos(\hat{\mathbf{n}}, \mathbf{r}_{01}) \tag{9A-5}$$

where $\cos(\hat{\mathbf{n}}, \mathbf{r}_{01})$ is the cosine of the angle between the outward normal $\hat{\mathbf{n}}$ and the vector $\mathbf{r}_{01}$, the vector between points P_0 and P_1. (Note from Figure 9A-1 that the outward normal on S_ϵ is inward, toward P_0, whereas the normal on S is outward, away from P_0.) In particular, if $\mathbf{r}_1$ were on S_ϵ, then

$$\cos(\hat{\mathbf{n}}, \mathbf{r}_{01}) = -1$$

$$\Psi(\mathbf{r}_1) = \frac{e^{-ik\epsilon}}{\epsilon}$$

$$\frac{\partial\Psi(\mathbf{r}_1)}{\partial n} = \left(\frac{1}{\epsilon} + ik\right)\frac{e^{-ik\epsilon}}{\epsilon}$$

where ϵ is the radius of the small sphere around the point at P_0.

We now use these results to rewrite the two integrals in **(9A-4)**. The integral over the surface S_ϵ is

$$\iint_{S_\epsilon} \left(\Psi\frac{\partial\varphi}{\partial n} - \varphi\frac{\partial\Psi}{\partial n}\right) ds = \iint_{S_\epsilon} \left[\frac{\partial\varphi}{\partial n}\frac{e^{-ik\epsilon}}{\epsilon} - \varphi\frac{e^{-ik\epsilon}}{\epsilon}\left(\frac{1}{\epsilon} + ik\right)\right]\epsilon^2 \sin\theta\, d\theta\, d\phi \tag{9A-6}$$

The integral over the surface S is

$$\iint_S \left[\frac{e^{-i\mathbf{k}\cdot\mathbf{r}_{01}}}{r_{01}}\frac{\partial\varphi}{\partial n} - \varphi\frac{\partial}{\partial n}\left(\frac{e^{-i\mathbf{k}\cdot\mathbf{r}_{01}}}{r_{01}}\right)\right] r^2 \sin\theta\, d\theta\, d\phi \tag{9A-7}$$

The omitted volume contained within the surface S_ϵ is allowed to shrink to zero by taking the limit as $\epsilon \to 0$. Equation **(9A-6)** but not **(9A-7)** will be affected by taking the limit. The first and last terms of the right side of **(9A-6)** go to zero as $\epsilon \to 0$ because they contain $\epsilon e^{-ik\epsilon}$. The second term in **(9A-6)** contains $e^{-ik\epsilon}$, which goes to 1 as $\epsilon \to 0$. Therefore, in the limit as $\epsilon \to 0$, **(9A-4)** becomes

$$\varphi(\mathbf{r}_0) = \frac{1}{4\pi}\iint_S \left[\frac{e^{-i\mathbf{k}\cdot\mathbf{r}_{01}}}{r_{01}}\frac{\partial\varphi}{\partial n} - \varphi\frac{\partial}{\partial n}\left(\frac{e^{-i\mathbf{k}\cdot\mathbf{r}_{01}}}{r_{01}}\right)\right] ds \tag{9A-8}$$

which is sometimes called the *integral theorem of Kirchhoff*. We can apply this result to the problem of diffraction by an aperture in an infinite opaque screen, if the surface of integration is carefully selected.

Assume that a source at P_2 in Figure 9A-2 produces a spherical wave that is incident on an infinite, opaque screen from the left. To find the field at P_0 in Figure 9A-2, we apply **(9A-8)** to the surface $S_1 + S_2 + \Sigma$, where S_1 is a plane surface adjacent to the screen, Σ is that portion of S_1 in the aperture, and S_2 is a large spherical surface of radius R centered on P_0. See Figure 9A-2. As R increases, Ψ and φ will decrease as $1/R$. However, the area of integration increases as R^2 so the $1/R$ fall-off is not a sufficient reason for neglecting the contribution of the integration over S_2.

On the surface S_2, the Green's function and its derivative are

$$\Psi = \frac{e^{-i\mathbf{k}\cdot\mathbf{R}}}{R} \tag{9A-9}$$

$$\frac{\partial\Psi}{\partial n} = \left(-ik - \frac{1}{R}\right)\frac{e^{-i\mathbf{k}\cdot\mathbf{R}}}{R} \approx -ik\Psi \tag{9A-10}$$

where the approximate equality is for large R. Thus, the integral over S_2 for large R is

$$\iint_{S_2}\left[\Psi\frac{\partial\varphi}{\partial n} + \varphi(ik\Psi)\right] ds = \iint_{S_2}\Psi\left(\frac{\partial\varphi}{\partial n} + ik\varphi\right) R^2 \sin\theta\, d\theta\, d\phi \tag{9A-11}$$

The quantity

$$R\Psi = e^{-i\mathbf{k}\cdot\mathbf{R}}$$

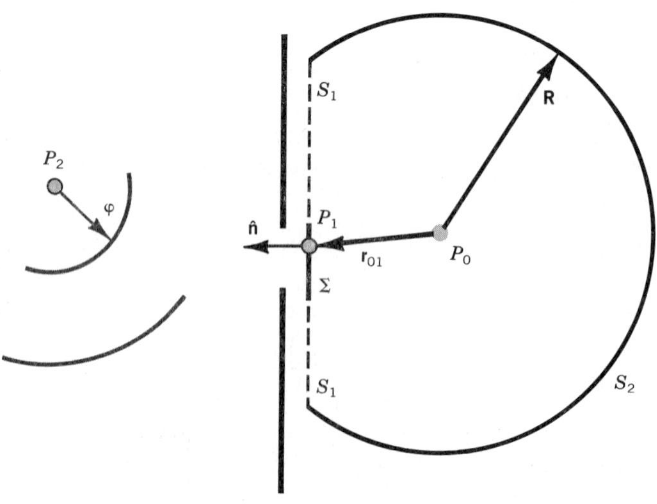

FIGURE 9A-2. Integration surface for diffraction calculation. P_2 is the source and P_0 is the point where the field is to be calculated.

is finite as $R \to \infty$ so for the integral to vanish, we must have

$$\lim_{R\to\infty} R\left(\frac{\partial \varphi}{\partial n} + ik\varphi\right) = 0 \qquad (9A\text{-}12)$$

This requirement is called the *Sommerfeld radiation condition* and is satisfied if $\varphi \to 0$ as fast as $1/R$ (for example, a spherical wave). Since the illumination of the screen will be a spherical wave, or at least a linear combination of spherical waves, we should expect the integral over S_2 to make no contribution (it is exactly zero).

We should expect the major contribution in the integral over S_1 to come from the portion of the surface in the aperture Σ. We make the following assumptions about the incident wave φ, (these assumptions are known as *St. Venaut's Hypothesis* or *Kirchhoff's boundary conditions*):

1. We assume that in the aperture φ and $\partial\varphi/\partial n$ have the values they would have if the screen was not in place.

2. On the portion of S_1 not in the aperture, φ and $\partial\varphi/\partial n$ are identically zero.

Another way to insure that surface S_2 makes no contribution is to assume that the light turns on at time t_0. At a later time t, when we desire to know the field at P_0, the radius of S_2 is $R > c(t-t_0)$, which means physically that the light has not had time to reach S_2. In this situation, there can be no contribution from S_2. This is not a perfect solution because the wave can no longer be considered monochromatic.

The Kirchhoff boundary conditions allow the screen to be neglected, greatly simplifying the problem and yielding very accurate results (as long as polarization is not important). However, mathematically the assumptions are incorrect.[44] The two boundary conditions imply that the field is zero everywhere behind the screen, except in the aperture that makes Ψ and $\partial\Psi/\partial n$ discontinuous on the boundary of the aperture. Another problem with the boundary conditions is that Ψ and $\partial\Psi/\partial n$ are known only if the problem has already been solved. We must make assumptions about the values of Ψ and $\partial\Psi/\partial n$, resulting in only approximate solutions of the diffraction problem.

Even with the problems concerning the boundary conditions, if the aperture is large with respect to a wavelength, and we do not get too close to the aperture, the theory does work. Sommerfeld removed some of the problems created by the boundary conditions; see Appendix 9B.

As a result of the above discussion, the surface integral, **(9A-8)**, is only performed over the surface Σ in the aperture. The evaluation of Green's function on this surface can be simplified by noting that normally, $r_{01} >> \lambda$, that is, $k >> 1/r_{01}$. Thus on Σ,

$$\frac{\partial \Psi}{\partial n} = \cos(\hat{\mathbf{n}}, \mathbf{r}_{01})\left(-ik - \frac{1}{r_{01}}\right)\frac{e^{-i\mathbf{k}\cdot\mathbf{r}_{01}}}{r_{01}}$$

$$\frac{\partial \Psi}{\partial n} \approx -ik\cos(\hat{\mathbf{n}}, \mathbf{r}_{01})\frac{e^{-i\mathbf{k}\cdot\mathbf{r}_{01}}}{r_{01}} \qquad (9A\text{-}13)$$

Substituting this approximate evaluation of the derivative into **(9A-8)** yields

$$\varphi(\mathbf{r}_0) = \frac{1}{4\pi}\iint_\Sigma \frac{e^{-i\mathbf{k}\cdot\mathbf{r}_{01}}}{r_{01}}\left[\frac{\partial \varphi}{\partial n} + ik\varphi\cos(\hat{\mathbf{n}}, \mathbf{r}_{01})\right] ds \qquad (9A\text{-}14)$$

The source of the incident wave is a point source located at P_2 with a position vector $\mathbf{r}_2$, measured with respect to the coordinate system and a distance $|\mathbf{r}_{21}|$ away from P_1, a point in the aperture (see Figure 9A-3). The incident wave is therefore a spherical wave

$$\varphi(\mathbf{r}_{21}) = A\frac{e^{-i\mathbf{k}\cdot\mathbf{r}_{21}}}{r_{21}} \qquad (9A\text{-}15)$$

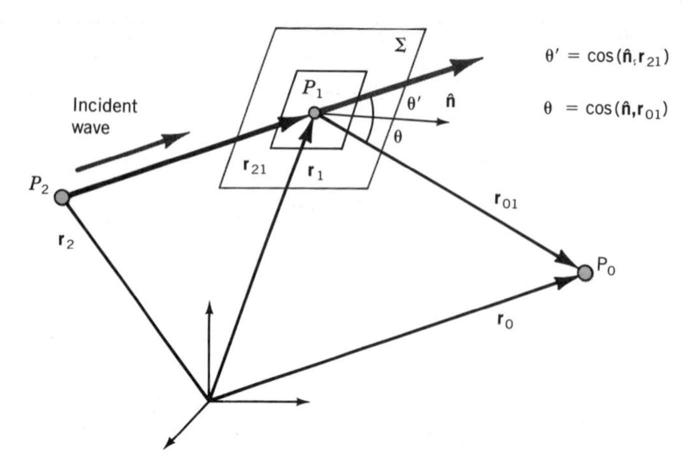

FIGURE 9A-3. Geometry for the Fresnel–Kirchhoff diffraction formula.

which fills the aperture. Here also, we will assume that $\mathbf{r}_{21} >> \lambda$, so that the derivative of the incident wave assumes the same form as **(9A-13)**. Equation **(9A-14)** can now be written

$$\varphi(\mathbf{r}_0) = \frac{iA}{\lambda} \iint_\Sigma \frac{e^{-i\mathbf{k}\bullet(\mathbf{r}_{21}+\mathbf{r}_{01})}}{\mathbf{r}_{21}\mathbf{r}_{01}} \left[\frac{\cos(\hat{\mathbf{n}},\ \mathbf{r}_{01}) - \cos(\hat{\mathbf{n}},\ \mathbf{r}_{21})}{2}\right] ds \tag{9A-16}$$

This relationship is called the *Fresnel–Kirchhoff diffraction formula*. It is symmetric with respect to r_{01} and r_{21}, making the problem identical when the source and measurement point are interchanged.

A physical understanding of this equation can be obtained if it is rewritten as

$$\varphi(\mathbf{r}_0) = \iint_\Sigma \Phi(\mathbf{r}_{21}) \frac{e^{-i\mathbf{k}\bullet\mathbf{r}_{01}}}{r_{01}} ds \tag{9A-17}$$

where the integrand contains

$$\Phi(\mathbf{r}_{21}) = \frac{i}{\lambda}\left(A\frac{e^{-i\mathbf{k}\bullet\mathbf{r}_{21}}}{r_{21}}\right)\left[\frac{\cos(\hat{\mathbf{n}},\ \mathbf{r}_{01}) - \cos(\hat{\mathbf{n}},\ \mathbf{r}_{21})}{2}\right] \tag{9A-18}$$

The field at P_0 is due to the sum of an infinite number of secondary Huygens' sources in the aperture Σ. The secondary sources are point sources radiating spherical waves of the form

$$\Phi(\mathbf{r}_{21})\frac{e^{-i\mathbf{k}\bullet\mathbf{r}_{01}}}{r_{01}}$$

with amplitude $\Phi(\mathbf{r}_{21})$, defined in **(9A-18)**. The obliquity factor

$$\tfrac{1}{2}\left[\cos(\hat{\mathbf{n}},\ \mathbf{r}_{01}) - \cos(\hat{\mathbf{n}},\ \mathbf{r}_{21})\right]$$

causes the secondary sources to have a forward-directed radiation pattern. The imaginary constant i causes the wavelets from each of these secondary sources to be phase-shifted with respect to the incident wave. (This notation agrees with the notation used in Chapter 9 if the substitution $r_{01} = R$ is made.)

Appendix 9-B

RAYLEIGH-SOMMERFELD FORMULA

A number of problems are associated with the Kirchhoff formulation of the scalar diffraction theory.[44] Sommerfeld corrected the problem associated with the boundary conditions by selecting a different Green's function for the solution of the wave equation. Here, we will introduce two possible Green's functions and show the consequence of selecting either of the two functions.

The modification that must be made to the Kirchhoff theorem **(9A-8)** concerns the assumption about the Green's function Ψ. Instead of assuming that the wave Ψ is due to a single point source **(9A-1)**, we assume that there are two point sources, one at P_0 and one at P_0' (see Figure 9B-1), where the source P_0' is positioned to be the mirror image of the source at P_0 on the opposite side of the screen. Thus,

$$r_{01} = r_{01}', \qquad \cos(\hat{\mathbf{n}}, \mathbf{r}_{01}) = -\cos(\hat{\mathbf{n}}, \mathbf{r}_{01}')$$

Evaluating the new wave, denoted by Ψ', and its derivative in the aperture yields

$$\Psi' = \frac{e^{-i\mathbf{k}\cdot\mathbf{r}_{01}}}{r_{01}} - \frac{e^{-i\mathbf{k}\cdot\mathbf{r}_{01}'}}{r_{01}'} \tag{9B-1}$$

$$\frac{\partial\Psi'}{\partial n} = \cos(\hat{\mathbf{n}}, \mathbf{r}_{01})\left(-ik - \frac{1}{r_{01}}\right)\frac{e^{-i\mathbf{k}\cdot\mathbf{r}_{01}}}{r_{01}}$$

$$- \cos(\hat{\mathbf{n}}, \mathbf{r}_{01}')\left(-ik - \frac{1}{r_{01}'}\right)\frac{e^{-i\mathbf{k}\cdot\mathbf{r}_{01}'}}{r_{01}'} \tag{9B-2}$$

Because the point P_0' is the mirror image of P_0, the value of **(9B-1)** and **(9B-2)** at any point P_1 in the aperture is

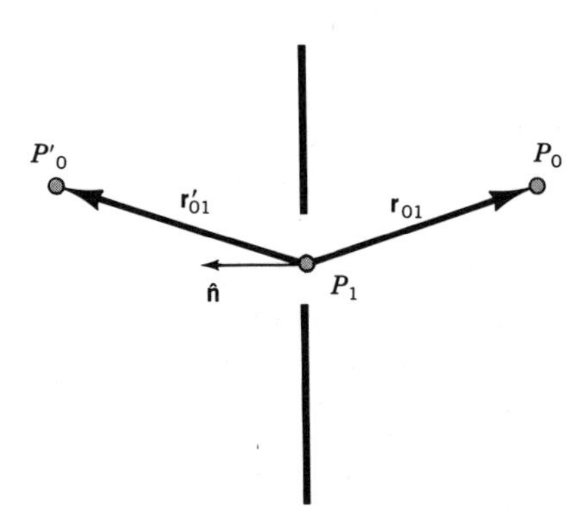

FIGURE 9B-1. Geometry for the Sommerfeld derivation.

$$\Psi'\big|_{P_1} = 0$$

$$\left.\frac{\partial \Psi'}{\partial n}\right|_{P_1} = 2\cos(\hat{\mathbf{n}}, \mathbf{r}_{01})\left(-ik - \frac{1}{r_{01}}\right)\frac{e^{-i\mathbf{k}\cdot\mathbf{r}_{01}}}{r_{01}}$$

Another possible Green's function that could be used is

$$\Psi' = \frac{e^{-i\mathbf{k}\cdot\mathbf{r}_{01}}}{r_{01}} + \frac{e^{-i\mathbf{k}\cdot\mathbf{r}'_{01}}}{r'_{01}} \tag{9B-3}$$

With this Green's function, we have in the aperture

$$\left.\frac{\partial \Psi'}{\partial n}\right|_{P_1} = 0$$

If we replace the simple point source wave function that was used for a Green's function in Appendix 9-A with either **(9B-1)** or **(9B-3)**, i.e., a double point source, we obtain the *Sommerfeld integral*

$$\varphi(\mathbf{r}_0) = \frac{i}{\lambda}\iint_{\Sigma} \varphi(\mathbf{r}_1)\frac{e^{-i\mathbf{k}\cdot\mathbf{r}_{01}}}{r_{01}}\cos(\hat{\mathbf{n}}, \mathbf{r}_{01})\, ds \tag{9B-4}$$

where the assumption has been made that $r_{01} >> \lambda$. The only difference between the Sommerfeld integral, **(9B-4)** and **(9A-16)** is in the obliquity factor.

Kirchhoff's boundary conditions are now applied to either φ for **(9B-1)** or $\partial\varphi/\partial n$ for **(9B-3)**. Since we no longer must specify boundary conditions for both φ and $\partial\varphi/\partial n$, the inconsistencies of the Kirchhoff theory are removed. The assumption that the value of φ in the aperture is the same as would be obtained in the absences of the screen is still a questionable assumption.

Appendix 9-C

GREEN'S THEOREM

To calculate the complex amplitude $\mathcal{E}(\mathbf{r})$ of a wave at an observation point defined by the vector $\mathbf{r}$, we need to use a mathematical relation known as Green's theorem. Green's theorem states that if φ and Ψ are two scalar functions that are well behaved, then

$$\iint_S (\varphi \nabla \Psi \cdot \hat{\mathbf{n}} - \Psi \nabla \varphi \cdot \hat{\mathbf{n}})\, ds = \iiint_V (\varphi \nabla^2 \Psi - \Psi \nabla^2 \varphi)\, dv \qquad \text{(9C-1)}$$

This theorem comes from a basic theorem of vector analysis known as Gauss' theorem. It is somewhat similar in principle to the theorem regarding integration by parts in calculus.

Gauss' Theorem

Gauss' theorem says if $\mathbf{F}$ is a vector function of position, then

$$\iint_S \mathbf{F} \cdot \hat{\mathbf{n}}\, ds = \iiint_V \nabla \cdot \mathbf{F}\, dv \qquad \text{(9C-2)}$$

$\mathbf{F} \cdot \hat{\mathbf{n}}$ is the normal component of $\mathbf{F}$ on the surface S (pointing outward). The surface and volume differentials to be used in **(9C-2)** are presented in Table 9C.1 for three of the most common coordinate systems.

We can obtain Green's theorem from Gauss' theorem by the substitution

$$\mathbf{F} = \varphi \nabla \Psi$$

where φ and Ψ are *scalars*.

$$\iint_S (\varphi \nabla \Psi \cdot \hat{\mathbf{n}})\, ds = \iiint_V (\varphi \nabla^2 \Psi + \nabla \varphi \cdot \nabla \Psi)\, dv \qquad \text{(9C-3)}$$

If we interchange φ and Ψ, we produce another form similar to **(9C-3)** that we subtract from **(9C-3)** to generate Green's theorem, i.e., **(9C-1)**.

TABLE 9C.1 Surface and Volume Differentials

	Coordinate System		
Differential	Rectangular	Cylindrical	Spherical
ds	$dx\, dy$	$r d\theta\, dz$	$r^2 \sin\theta\, d\theta\, d\phi$
dv	$dx\, dy\, dz$	$r\, dr\, d\theta\, dz$	$r^2 \sin\theta dr\, d\theta\, d\phi$

The vector identities

$$\nabla\Psi \cdot \hat{\mathbf{n}} = \frac{\partial \Psi}{\partial n}, \qquad \nabla\varphi \cdot \mathbf{n} = \frac{\partial \varphi}{\partial n} \tag{9C-4}$$

allow Green's theorem to be written as

$$\iint_S \left\{ \varphi \frac{\partial \Psi}{\partial n} - \Psi \frac{\partial \varphi}{\partial n} \right\} ds = \iiint_V (\varphi \nabla^2 \Psi - \Psi \nabla^2 \varphi)\, dv \tag{9C-5}$$

This equation is the prime foundation of scalar diffraction theory, but only the proper choice of Ψ, φ, and the surface S allows direct application to the diffraction problem.

Fraunhofer Diffraction

The job of calculating diffraction only began with the derivation of the Huygens–Fresnel integral **(9-20)**. Rigorous solutions exist for only a few idealized obstructions. To allow discussion of the general properties of diffraction, it is necessary to use approximate solutions. The problem of obtaining a general solution lies in the evaluation of the diffraction component of the Huygens–Fresnel integral **(9-20)**, which we rewrite here in one dimension

$$\int \frac{\mathcal{E}_i(\mathbf{r})}{R} e^{-i\mathbf{k}\cdot\mathbf{R}}\, dx$$

For this diffraction component to contribute to the field at Point P, in Figure 10-1, the phase of the exponent must not vary over 2π when the integration is performed over the aperture shown in Figure 10-1, that is,

$$\boldsymbol{\Delta} = \mathbf{k}\cdot(\mathbf{R}_1 - \mathbf{R}_1' + \mathbf{R}_2 - \mathbf{R}_2') < 2\pi$$

From the geometry of Figure 10-1, the two paths from the source S to the observation point P_0 are

$$R_1 + R_2 = \sqrt{z_1^2 + (x_1 + b)^2} + \sqrt{z_2^2 + (x_2 + b)^2}$$

$$R_1' + R_2' = \sqrt{z_1^2 + x_1^2} + \sqrt{z_2^2 + x_2^2}$$

By assuming that the aperture of width b is small compared to the distances z_1 and z_2, the difference between these distances Δ can be rewritten as an expansion in terms of b (details of the expansion will be given in the next section)

$$\boldsymbol{\Delta} \approx k\left(\frac{x_1}{z_1} + \frac{x_2}{z_2}\right) b + \frac{k}{2}\left(\frac{1}{z_1} + \frac{1}{z_2}\right) b^2 + \cdots \tag{10-1}$$

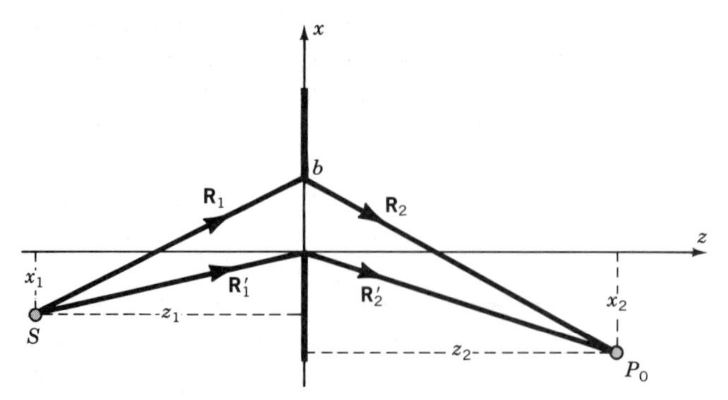

FIGURE 10-1. One-dimensional slit of width b used to establish Fraunhofer and Fresnel approximations.

If we assume that the second term of this expansion is small, i.e.,

$$\frac{k}{2}\left(\frac{1}{z_1}+\frac{1}{z_2}\right)b^2 << 2\pi$$

$$\frac{b^2}{2}\left(\frac{1}{z_1}+\frac{1}{z_2}\right) < \lambda$$

then we may treat the variation of the phase across the aperture as if it were linear. Physically, this means that all waves are assumed to be plane waves. The diffraction predicted by the theory based on this assumption is called *Fraunhofer diffraction* and is the subject of this chapter. [Retention of the first two terms of **(10-1)** leads to Fresnel diffraction, the subject of the next chapter.]

Since diffraction and interference are actually the same physical process, we expect the observation of diffraction to be a strong function of the coherence of the illumination. With incoherent sources, geometrical optics is usually all that is needed to predict the performance of an optical system and diffraction can be ignored. With light sources having a large degree of spatial coherence, it is impossible to neglect diffraction in any analysis. Even though diffraction produced by most common light sources and objects is a small effect, it is possible to observe diffraction without special equipment. By viewing a light source through a slit formed by two fingers that are nearly touching, fringes can be observed.

Fraunhofer diffraction is used by photographers and television cameramen to provide artistic highlights in an image containing small light sources. They accomplish this by placing a screen in front of the camera lens. Light from point sources in the field of view is diffracted by the screen and produces "stars" in the image. The stars are the Fraunhofer diffraction patterns of the screen, produced by each point source. By observing a mercury or sodium streetlight, a few hundred meters away, through a sheer curtain, a Fraunhofer diffraction pattern of the fabric weave can easily be observed.

Joseph Fraunhofer (1787–1826) developed in 1823 the theory to be discussed in this chapter. (It is interesting that Fraunhofer started out as a lens grinder and before he died at the age of 39 had become an owner of an optical company and a professor at the University of Munich.) The approach to Fraunhofer diffraction to be taken here is based on a novel interpretation given to the theory in 1873 by **Ernst Abbe (1840–1905)**. In not too clear a fashion, Abbe described Fraunhofer diffraction as a decomposition, of a spatial distribution of light, into a series of plane waves. Abbe's objective was the explanation of resolution in high-power microscopes. His theory described image resolution in terms of the spatial frequencies contained in the plane wave decomposition. The first application based on Abbe's interpretation of Fraunhofer diffraction was the phase contrast microscope invented by **F. Zernike** in 1935.

The view of the Fraunhofer theory to be taken in this chapter is a modern extension of Abbe's theory, based on linear system theory. Linear system theory uses an impulse response function (called the Green's function in mathematics and the point spread function in optics), convolved with an input function, to generate the output function of the linear system. The initial stimulus for linear system theory's use in optics began with the publication of a book in French by **P.M. Duffieux**.[22] Several books in English have been written to introduce this approach to optics.[45]

By applying the approximation discussed above to the Huygens–Fresnel integral, the fundamental equation of Fraunhofer diffraction will be derived. We will discover that the integral, used to calculate Fraunhofer diffraction, is mathematically identical to the Fourier transform integral. Because we have already developed the tools needed to evaluate Fourier transforms, we will find it quite easy to obtain Fraunhofer diffraction from rectangular and circular apertures. We will derive a theorem from Fourier transform theory, called the array theorem, that will allow the diffraction associated with Young's two-slit experiment to be calculated. The array theorem will also allow the analysis of an important optical component, the diffraction grating.

We will demonstrate, using an intuitive argument, that a lens generates a Fraunhofer diffraction pattern in its back focal plane. A more formal derivation of this fact will be given in Appendix 10-A. The Fourier transform relationship between the amplitude distribution function in the front focal plane of a lens and the Fraunhofer diffraction pattern in the back focal plane is the basis of Abbe's theory of a microscope. The theory, discussed in Appendix 10-B, is used as the basis of understanding the phase contrast microscope, aperture apodization, and optical pattern recognition.

The theory of coherent and incoherent imaging systems will be discussed in Appendix 10-C, where resolution limits of imaging systems will be established. A unique noise that limits the resolving capability of coherent optical systems will be introduced.

FRAUNHOFER DIFFRACTION

In this section, we will derive the equation describing Fraunhofer diffraction. We will assume the source of light is at infinity, $z_1 = \infty$ in Figure 10-1, so that the aperture is illuminated by a plane wave traveling parallel to the z axis. We then derive an approximate expression, analogous to **(10-1)** for the position vector **R** in Figure 10-2. We obtain an approximate expression for the position vector of the observation point P_0, relative to the aperture, by assuming that the aperture size is small, relative to the distance to the observation point. Using this approximate expression and the paraxial approximation, we are able to formulate the Huygens–Fresnel integral as a two-dimensional Fourier transform.

In Fraunhofer diffraction, we require that the source of light and the observation point P_0 be far from the aperture so that the incident and diffracted wave can be approximated by plane waves. A consequence of this requirement is that the entire waveform passing through the aperture contributes to the observed diffraction. We will see in the next chapter that for Fresnel diffraction this is not the case.

The geometry to be used in this derivation is shown in Figure 10-2. The wave incident on the aperture Σ is a plane wave and the objective of the calculation is to find the departure of the transmitted wave from its geometrical optic path. The calculation will provide the light distribution, transmitted by the aperture, as a function of the angle the light is deflected from the incident direction. We assume that diffraction makes only a small perturbation on the predictions of geometrical optics. The deflection angles encountered in this derivation are, therefore, small and we will be able to use the paraxial approximation.

The distance from a point P in the aperture to the observation point P_0 of Figure 10-2 is

$$R^2 = (x - \xi)^2 + (y - \eta)^2 + Z^2 \tag{10-2}$$

From Figure 10-2, we see R_0 is the distance from the center of the screen to the observation point P_0

$$R_0^2 = \xi^2 + \eta^2 + Z^2 \tag{10-3}$$

The difference between these two vectors is

$$\begin{aligned} R_0^2 - R^2 &= \xi^2 + \eta^2 + Z^2 - Z^2 - (x^2 - 2x\xi + \xi^2) - (y^2 - 2y\eta + \eta^2) \\ &= 2(x\xi + y\eta) - (x^2 + y^2) \end{aligned} \tag{10-4}$$

We can rewrite the difference between the two vectors as

$$R_0^2 - R^2 = (R_0 - R)(R_0 + R) \tag{10-5}$$

Using **(10-4)**, we can write an equation for the position of point P in the aperture in terms of **(10-5)**

$$\begin{aligned} r = R_0 - R &= \frac{R_0^2 - R^2}{R_0 + R} \\ &= [2(x\xi + y\eta) - (x^2 + y^2)]\frac{1}{R_0 + R} \end{aligned}$$

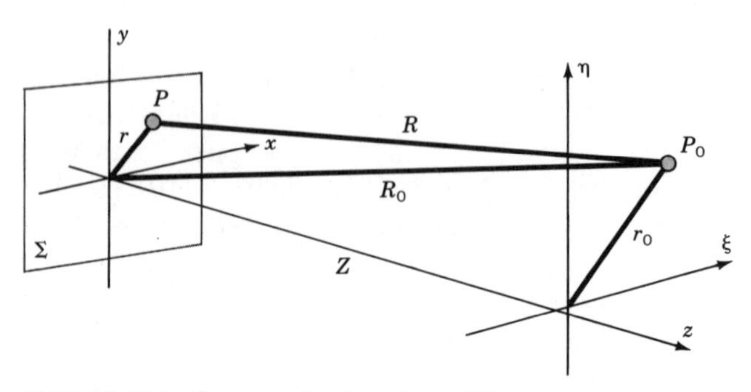

FIGURE 10-2. Geometry for Fraunhofer diffraction.

The reciprocal of $(R_0 + R)$ can be written as

$$\frac{1}{R_0 + R} = \frac{1}{2R_0 + R - R_0}$$
$$= \frac{1}{2R_0}\left(1 + \frac{R - R_0}{2R_0}\right)^{-1} \qquad (10\text{-}6)$$

Now using **(10-6)** yields

$$R_0 - R = \left(\frac{x\xi + y\eta}{R_0} - \frac{x^2 + y^2}{2R_0}\right)\left(1 - \frac{R_0 - R}{2R_0}\right)^{-1}$$

If the diffraction integral is to have a finite (nonzero) value, then

$$k|R_0 - R| << kR_0$$

This requirement insures that all the Huygens' wavelets, produced over the half of the aperture from the center out to the position r will have similar phases and will interfere to produce a nonzero amplitude at P_0. The requirement that the phase changes are small can be written as

$$\frac{1}{1 - (R_0 - R)/2R_0} = \frac{1}{1 - (r/2R_0)} \approx 1$$

The approximation can be seen to be equivalent to assuming that the aperture is small.

By making the approximation, we obtain the diffraction integral

$$\mathcal{E}_P = \frac{iAe^{-ik\cdot R_0}}{\lambda R_0}\iint_{\Sigma} f(x, y)\exp\left[+ik\left(\frac{x\xi + y\eta}{R_0} - \frac{x^2 + y^2}{2R_0}\right)\right]dx\,dy \qquad (10\text{-}7)$$

where A is the amplitude of the plane wave illuminating the aperture. The change in the amplitude of the wave due to the change in R as we move across the aperture is neglected, allowing R in the denominator of the Huygens–Fresnel integral to be replaced by R_0 and moved outside of the integral. We have introduced the complex transmission function $f(x, y)$ of the aperture to allow a very general aperture to be treated. If the aperture function described the variation in absorption of the aperture as a function of position, as would be produced by a photographic negative, then $f(x, y)$ would be a real function. If the aperture function described the variation in transmission of a biological sample, it might be entirely imaginary.

The argument of the exponent in **(10-7)** is

$$ik\left(\frac{x\xi + y\eta}{R_0} - \frac{x^2 + y^2}{2R_0}\right) = i2\pi\left(\frac{x\xi + y\eta}{\lambda R_0} - \frac{x^2 + y^2}{2\lambda R_0}\right) \qquad (10\text{-}8)$$

If the observation point P_0 is far from the screen, we can neglect the second term and treat the phase variation across the aperture as a linear function of position. This is equivalent to assuming that the diffracted wave is a collection of plane waves. Mathematically, the second term in **(10-8)** can be neglected if

$$\frac{x^2 + y^2}{2\lambda R_0} << 2\pi \qquad (10\text{-}9)$$

This is called the *far-field approximation*.

The first term in **(10-8)** contains the direction cosines

$$L = \frac{\xi}{R_0}, \qquad M = \frac{\eta}{R_0} \tag{10-10}$$

These cosines can remain finite even in the far field. As the unit of measure at the aperture screen, we use the illuminating wavelength that allows the aperture coordinates to be defined as

$$X = \frac{x}{\lambda}, \qquad Y = \frac{y}{\lambda} \tag{10-11}$$

We can rewrite **(10-7)** by using the approximation allowed by **(10-9)** and the definitions of **(10-10)** and **(10-11)**

$$\mathcal{E}_P(L, M) = \frac{iAe^{-i\mathbf{k}\cdot\mathbf{R}_0}}{\lambda R_0} \iint_\Sigma f(X, Y) \exp[+i2\pi(LX + MY)]\, dX\, dY \tag{10-12}$$

We define the spatial frequencies with a negative sign to allow equations involving spatial frequencies to have the same form as those involving temporal frequencies. The negative sign is required because ωt and $\mathbf{k}\cdot\mathbf{r}$ appear in the phase of the wave with opposite signs. Rather than use a negative spatial frequency, we could have redefined the Fourier transform using a positive exponential for spatial transforms.

In this form, the integral may not be recognizable but, by defining two spatial frequencies, we can obtain a familiar form. The spatial frequencies in the x and y direction are defined as

$$\omega_x = kL = -\frac{2\pi\xi}{\lambda R_0}, \qquad \omega_y = kM = -\frac{2\pi\eta}{\lambda R_0} \tag{10-13}$$

We equate the spatial frequencies to coordinate positions in the observation plane by applying the paraxial approximation; this makes the mapping from the aperture to the observation plane easy to interpret. With the variables defined in **(10-13)**, the integral becomes

$$\mathcal{E}_P(\omega_x, \omega_y) = \frac{iAe^{-i\mathbf{k}\cdot\mathbf{R}_0}}{\lambda R_0} \iint_\Sigma f(x, y) \exp[-i(\omega_x x + \omega_y y)]\, dx\, dy \tag{10-14}$$

The Fraunhofer diffraction field $\mathcal{E}_P$ equals the two-dimensional Fourier transform of the aperture's transmission function.

To summarize, by assuming that the illumination is a plane wave and that the observation position is in the far field, the diffraction of an aperture is found to be given by the Fourier transform of the function describing the amplitude transmission of the aperture. The resulting Fraunhofer diffraction spectrum has an angular distribution

$$\mathcal{E}_P(L, M)$$

that equals the spatial frequency spectrum of the diffracting screen. The amplitude transmission of the aperture $f(x, y)$ may thus be interpreted as the superposition of mutually coherent plane waves leaving the diffracting screen in directions given by (L, M).

FOURIER TRANSFORMS VIA A LENS

If we place a point source in the focal plane of a positive lens, the image of the point source occurs at infinity. The inverse is also true: The image of a point source at infinity is found in the focal plane of the lens. Because the focal plane and a plane at infinity are conjugate planes, it seems plausible that a lens would form an image of the far-field diffraction pattern in its focal plane. To support our supposition that the lens would create the Fraunhofer diffraction pattern in its focal plane, we will examine how a lens transforms a plane wave. We will demonstrate that the transformation is equivalent to a Fourier transform.

A formal demonstration of the Fourier transform properties of a lens is found in Appendix 10-A, but because the mathematical manipulations make it difficult to follow the basic physics, we will here use a more intuitive presentation to justify our expectation.

It is possible to demonstrate that the angle of arrival of a plane wave is mapped onto a unique spatial position in the back focal plane of a lens. This demonstration can be accomplished by the use of geometrical optics. We will be able to show that a lens with an off-axis point source in its front focal plane will produce a plane wave traveling at an angle to the optical axis of the lens.

In Appendix 5-A, we defined the back and front focal planes [**(5A-4)** and Figure 5A-2] as the image planes of objects at infinity. This means that a lens will produce a plane wave if a source is located in one of the focal planes. The *ABCD* matrix can be used to demonstrate this fact. We use Figure 5A-4 to create the *ABCD* matrix for a light ray propagating from the front to back focal plane of a simple lens

$$\begin{pmatrix} x' \\ \gamma' \end{pmatrix} = \begin{pmatrix} 1 & f \\ 0 & 1 \end{pmatrix} \begin{pmatrix} 1 & 0 \\ -\frac{1}{f} & 0 \end{pmatrix} \begin{pmatrix} 1 & f \\ 0 & 1 \end{pmatrix} \begin{pmatrix} x_0 \\ \gamma \end{pmatrix}$$

$$\begin{pmatrix} x' \\ \gamma' \end{pmatrix} = \begin{pmatrix} 0 & f \\ -\frac{1}{f} & 0 \end{pmatrix} \begin{pmatrix} x_0 \\ \gamma \end{pmatrix}$$

From this, we discover that

$$\gamma' = -\frac{x_0}{f}$$

Thus, a source located in the front focal plane of the lens, at a position x_0 above the optical axis, will generate a plane wave, which makes an angle

$$\gamma' = -\frac{x_0}{f}$$

with the optical axis in the back focal plane. In Figure 10-3, we show that the plane wave will be traveling downward at an angle γ', as predicted by the negative sign.

The result for a single point source, shown in Figure 10-3, agrees with the Fourier transform calculated in Chapter 6 **(6-24)** for a single delta function. The Fourier transform of a delta function located at the origin $\delta(x)$ is a constant. In the optical analog, the delta function represents a point source, located on the optical axis. The resulting Fourier transform produced by the lens corresponds to a plane wave propagating parallel to the optical axis. When the delta function is moved off axis, the shifting property of a Fourier transform **(6A-5)** can be applied to yield a constant with a varying phase $e^{-i\omega_x x}$. The shifted delta function $\delta(x - x_0)$ corresponds to the point source located off axis at position x_0 in Figure 10-3, and the constant with a linearly varying phase corresponds to the plane wave propagating at an angle $\gamma' = \omega_x$ to the optical axis.

A numerical estimate of the far-field distance can be obtained by first restating **(10-9)** in one dimension

$$\frac{x^2}{4\pi\lambda} << R_0$$

For an illumination wavelength of

$$\lambda = 6 \times 10^{-7}\text{m}$$

and an aperture width of 2.5 cm, the location of the far-field observation point must exceed $R_0 >> 1600$ m. This demonstrates why this requirement is called the far-field condition. It is not necessary to go to such distances to observe Fraunhofer diffraction, as we will show in this section.

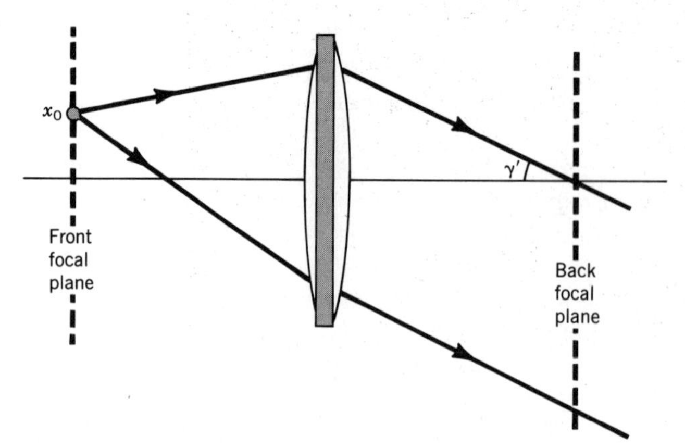

FIGURE 10-3. Connection between direction of plane wave and position of spot in focal plane.

If a lens is added to Young's two-slit experiment to collect the light from the two slits, then we will discover that the light distribution in the back focal plane of the lens is the Fourier transform of the light distribution at the front focal plane of the lens.

Two slits are placed in the front focal plane of a lens. The wave illuminating the slit is polarized along the y axis and propagating in the x-z plane. The light from the two slits that overlaps in the back focal plane will produce interference. The light distribution should be given by **(4-12)** with the phase angle δ given by **(4-16)**. To see that this is the case, consider that the two slits will act, in one dimension, as two point sources. The lens transforms light from the two slits into two plane waves propagating at angles γ_1 and γ_2 with respect to the z axis. We can write the phase difference between the two plane waves as

$$g_2 - h_2 = kz(\cos\gamma_1 - \cos\gamma_2) + kx(\sin\gamma_1 - \sin\gamma_2)$$

From Figure 4-6, we have

$$\gamma_1 = \frac{h}{2f} \quad \text{and} \quad \gamma_2 = -\frac{h}{2f}$$

so that

$$g_2 - h_2 = 2kx \sin\left(\frac{h}{2f}\right) \approx \frac{kxh}{f} \tag{10-15}$$

which agrees with **(4-16)** if we replace the distance between the slit and observation plane D by the focal plane of the lens. The conclusion we can draw from this analysis is that the intensity distribution in the back focal plane of the lens, with two equally spaced slits in the front focal plane, is proportional to

$$\cos^2\left(\frac{kxh}{2f}\right)$$

The two slits can be viewed as an experimental representation of two delta functions

$$\delta\left(x - \frac{h}{2}\right) \quad \text{and} \quad \delta\left(x + \frac{h}{2}\right)$$

FIGURE 2-11. Propagation of light in an absorbing medium. Light from a HeCd laser propagates through first a layer of xylene and then water containing the dye, Rhodamine 6G, in solution. The blue laser light is absorbed by the Rhodamine as the wave propagates through the water. The Rhodamine reemits the energy in the form of a red fluorescence. The reemitted light is radiated in all directions, making the beam appear diffuse. Some of the absorped energy is not reemitted but instead heats the Rhodamine.

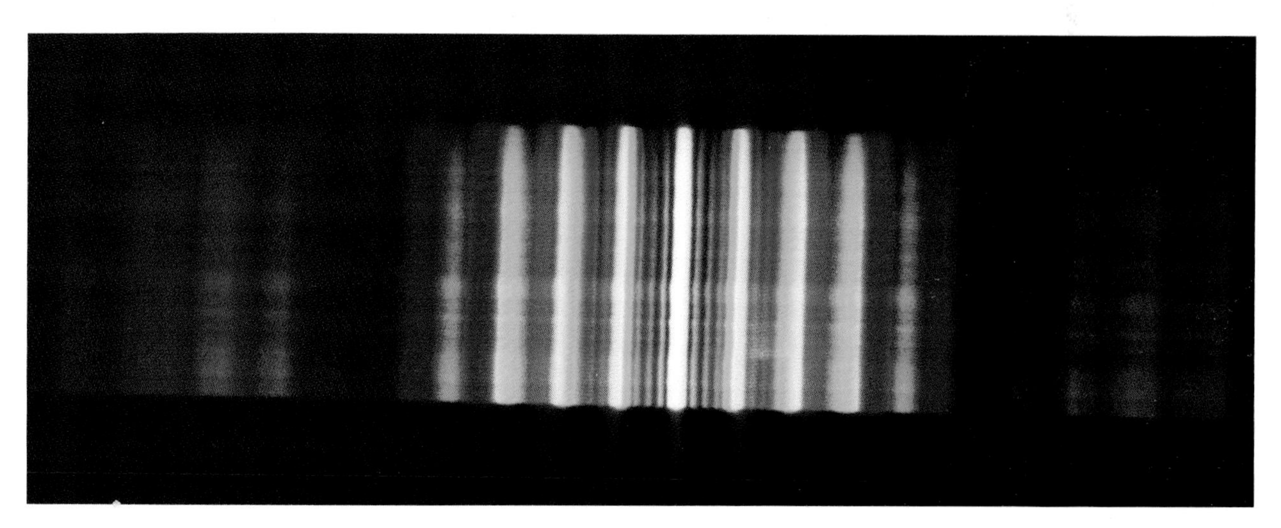

FIGURE 4.7. Fringes from a Young's two-slit experiment.

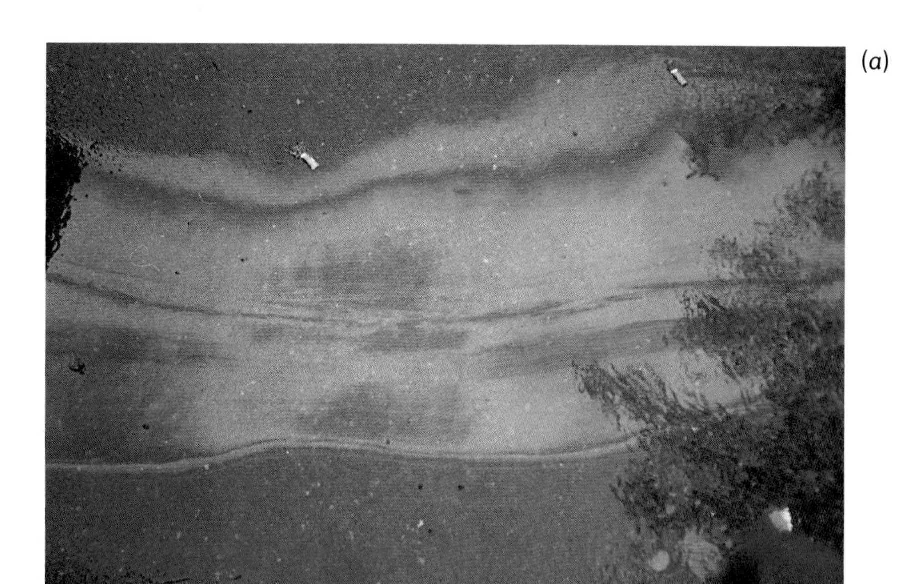

(a)

(b)

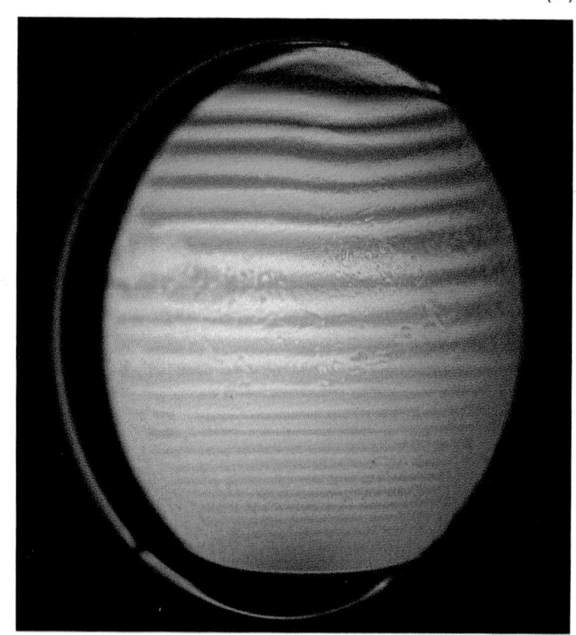

FIGURE 4-11. (a) An oil film on a wet street. (b) A soap film similar to the one shown in 10b, but here the soap film is illuminated by white light. The color pattern repeats when the thickness changes by one wavelength. Thus, each cycle in color corresponds to one order, m, in the interference pattern.

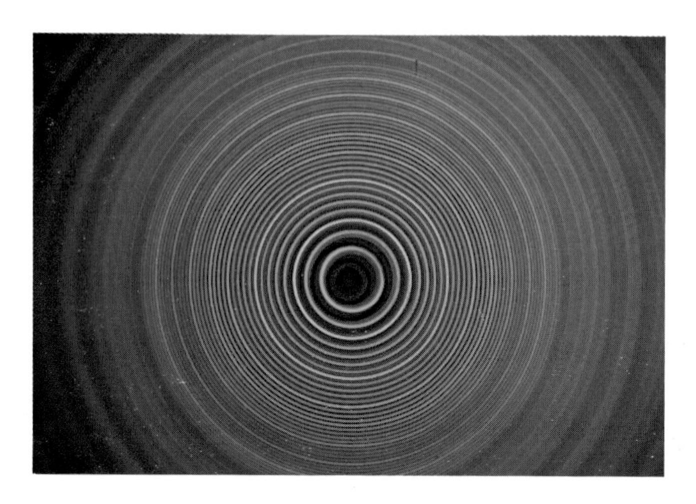

FIGURE 4-19. The output fringes from a Fabry-Perot interferometer. Two sets of fringes due to two different colors are present in this photograph. (Courtesy of Fredrick L. Roesler, University of Wisconsin.)

(*a*)

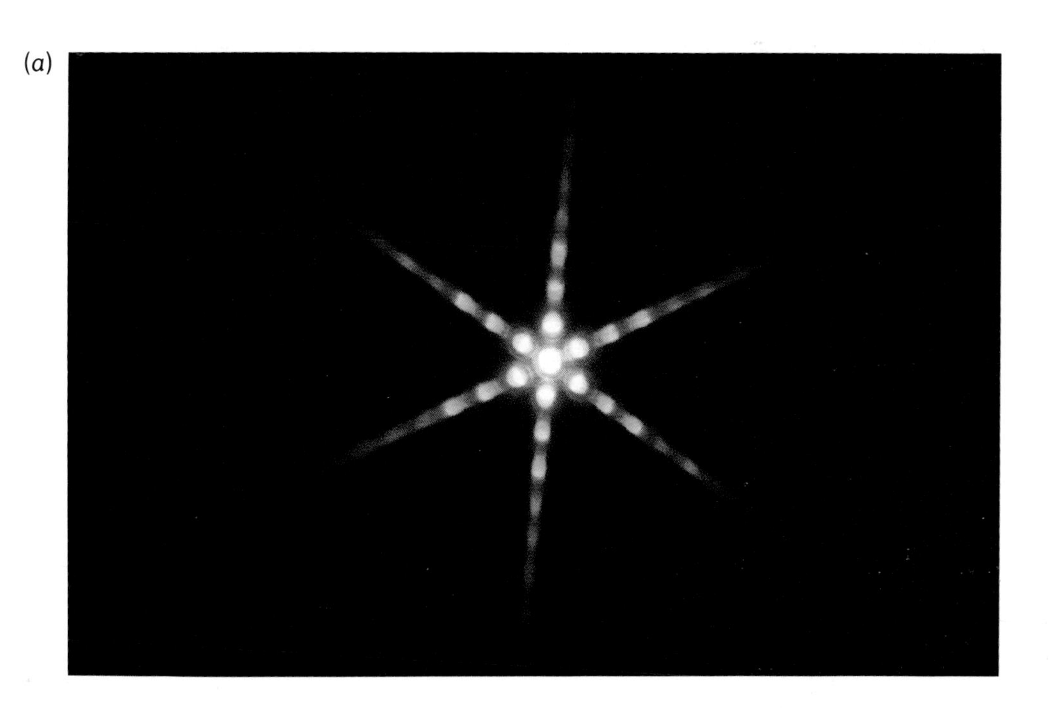

(*b*)

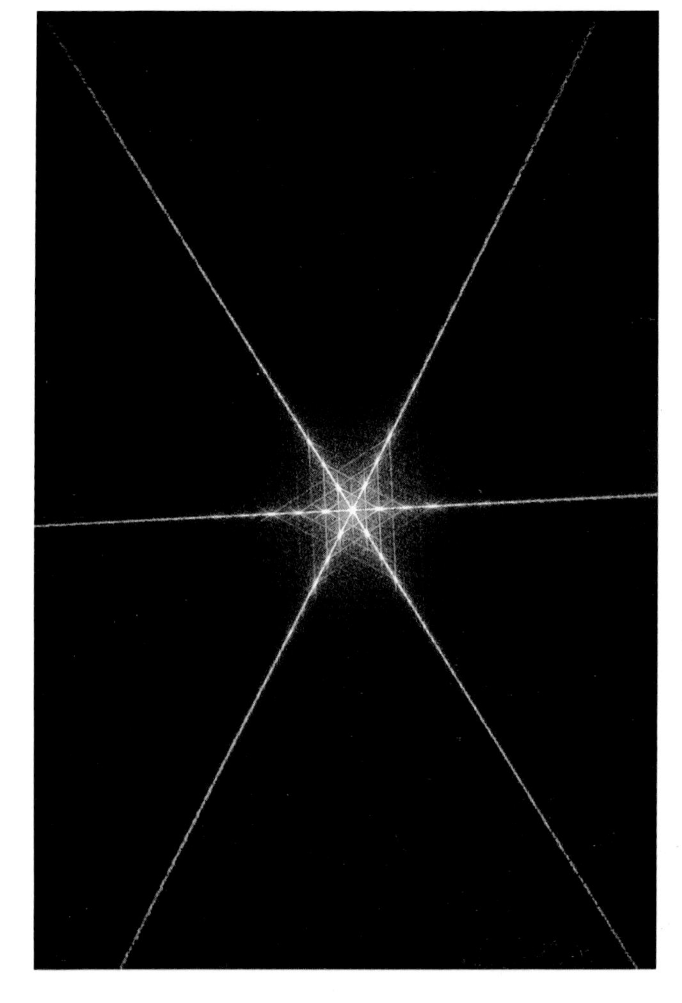

FIGURE 10B-5. The Fraunhofer diffraction pattern produced by a nested set of triangles that form three sets of gratings oriented at 120° with respect to each other. (*a*) The illuminating light was spatially coherent white light. The spots are very broad due to the spread in wavelengths in the source. In a colored photograph, blue light can be seen at one side of each spot and red light can be seen at the other. (*b*) The illuminating light producing this diffraction pattern contained only a single wavelength, 488 nm from an Argon laser. (Nested triangles courtesy of N. George, University of Rochester.)

(*a*)

(*b*)

FIGURE 12-31. A hologram of a ceramic statue reconstructed at (*a*) the recording wavelength; (*b*) three discrete wavelengths. (Hologram courtesy of N. George, University of Rochester.)

FIGURE 14-12. In the manufacture of plastic French curves, components of stress are locked in the plastic. Here is shown a French curve, illuminated with polarized light and viewed through a polarizer oriented with its axis normal to the axis of polarization of the illumination. Sunlight Rayleigh scattered through 90° by the atmosphere is polarized and was used to illuminate the French curve.

In Chapter 6, the Fourier transform of N delta functions was calculated in **(6-26)** and was shown to equal a cosine function for $N = 2$. For the two slits, the experimental observation is equivalent to the formal mathematical treatment. We have again demonstrated that a Fourier transform relationship exists between the amplitude distribution in the front focal plane and the amplitude distribution in the back focal plane of the lens. It seems reasonable to extend the results for two point sources to N sources. A larger leap in reasoning is needed to extend this result to a continuous distribution, but it seems to be a reasonable extension and in Appendix 10-A it will be formally demonstrated.

We therefore conclude that a lens will generate, in its back focal plane, the Fraunhofer diffraction pattern (or equivalently, the Fourier transform) of the amplitude distribution in its front focal plane. (It should be noted that a second lens will take the positive mapping of spatial frequency onto coordinate position in the focal plane and convert it into negative-going plane waves. Therefore, two lenses perform two Fourier transforms, yielding the original function. The recovered function, however, is displayed in a coordinate system that is the negative of the original coordinate system, i.e., $x \Rightarrow -x$ and $y \Rightarrow -y$.)

PLANE WAVE REPRESENTATION

We have demonstrated that the Fraunhofer diffraction pattern of an aperture is calculated by performing a spatial Fourier transform of the aperture's amplitude transmission function. We proved that the point at x_0, in the front focal plane of the lens in Figure 10-3, produces a plane wave propagating at an angle γ' to the optical axis. We have extended this concept to a large number of points, allowing us to conclude that the Fraunhofer diffraction pattern is equal to the interference pattern produced by a collection of plane waves.

Because of the reciprocal relationship between a function and its Fourier transform, we should be able to treat the plane waves that form the diffraction pattern as a unique representation of the aperture. This view is equivalent to the one taken of a temporal wave packet as a frequency distribution of sine and cosine functions (Figures 6-6 and 6-7).

We can define an aperture as a collection of plane waves distributed over a group of propagating directions. In this section, we will demonstrate that this representation is possible by showing that the spatial distribution of a wave is obtained by integrating over a distribution of plane waves. The integral will have the form of the inverse Fourier transform.

We saw in our discussion of Fourier transforms, Figure 6-7, that a wave of finite temporal duration could not be represented by a single frequency but instead required a distribution of frequencies. We now conclude that a beam of light that has a finite width cannot be represented by a single plane wave but requires a distribution of plane waves, distributed over propagation directions. Consider a nearly plane wave traveling in the x, z plane with

$$\mathbf{k} \cdot \mathbf{r} = k_x x + k_z z$$

The wave propagating with a wave vector between

$$k_x - \frac{dk_x}{2} < k_x < k_x + \frac{dk_x}{2}$$

is

$$d\mathcal{E} = A(k_x) \exp\left[i(\omega t - \mathbf{k}\cdot\mathbf{r})\right] dk_x \qquad (10\text{-}16)$$

The amplitude A is allowed to be a function of k_x, which goes to zero for $|k_x| > \Delta$.

The resultant wave (if we use the principle of superposition and treat plane waves as coherent waves) is given by the integral of **(10-16)**. The mean direction of the resultant wave is along the z direction, but it is spread over angles corresponding to range of wave vectors defined by the inequality

$$-\Delta \le k_x \le \Delta$$

We will consider the wave at $t = 0$ to remove all temporal dependence of the wave

$$\mathcal{E}(x) = \int_{-\Delta}^{\Delta} A(k_x) e^{-ik_x x}\, dk_x \qquad (10\text{-}17)$$

Since $A(k_x) = 0$ for values of kx outside the range $\pm\Delta$, we can extend the limits to $\pm\infty$. If we multiply the integral by $1/2\pi$, we have the inverse Fourier transform **(6-14)** with x corresponding to t and $-k_x$ to ω. We have demonstrated that a distribution of plane waves yields a resultant wave with a bounded spatial extent given by the inverse Fourier transform **(10-17)**. The result supports the premise that an aperture can be described in terms of a distribution of plane waves.

DIFFRACTION BY A RECTANGULAR APERTURE

We will now use Fourier transform theory to calculate the Fraunhofer diffraction pattern from a rectangular slit and will point out the reciprocal relationship between the size of the diffraction pattern and the size of the aperture.

Consider a rectangular aperture with a transmission function given by

$$f(x, y) = \begin{cases} 1, & |x| \le x_0 \\ & |y| \le y_0 \\ 0, & \text{all other } x \text{ and } y \end{cases}$$

Because the aperture is two-dimensional, we need to apply a two-dimensional Fourier transform **(6-41)**. The amplitude transmission function is separable, in x and y, so we may use **(6-42)** and write the diffraction amplitude distribution from a rectangular slit as

$$\mathcal{E}_P = \frac{iA}{\lambda R_0} e^{-i\mathbf{k}\cdot\mathbf{R}_0} \int_{-x_0}^{x_0} f(x) e^{-i\omega_x x}\, dx \int_{-y_0}^{y_0} f(y) e^{-i\omega_y y}\, dy \qquad (10\text{-}18)$$

Since both $f(x)$ and $f(y)$ are defined as symmetric functions, we need only calculate the cosine transforms **(6-15a)**

$$\mathcal{E}_P = i\frac{4x_0 y_0 A}{\lambda R_0} e^{-i\mathbf{k}\cdot\mathbf{R}_0} \frac{\sin \omega_x x_0}{\omega_x x_0} \frac{\sin \omega_y y_0}{\omega_y y_0} \qquad (10\text{-}19)$$

The intensity distribution of the Fraunhofer diffraction produced by the rectangular aperture is

$$I_P = I_0 \frac{\sin^2 \omega_x x_0}{(\omega_x x_0)^2} \frac{\sin^2 \omega_y y_0}{(\omega_y y_0)^2} \qquad (10\text{-}20)$$

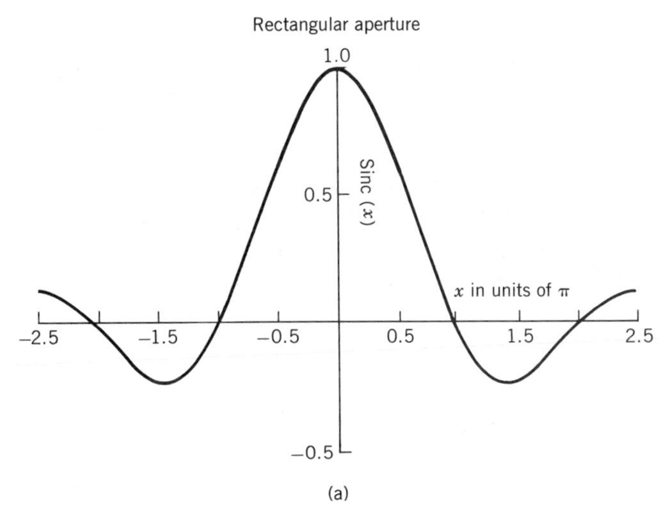

FIGURE 10-4a. Diffraction from a infinite slit. This sinc function describes the light wave's amplitude that would exist in the x direction.

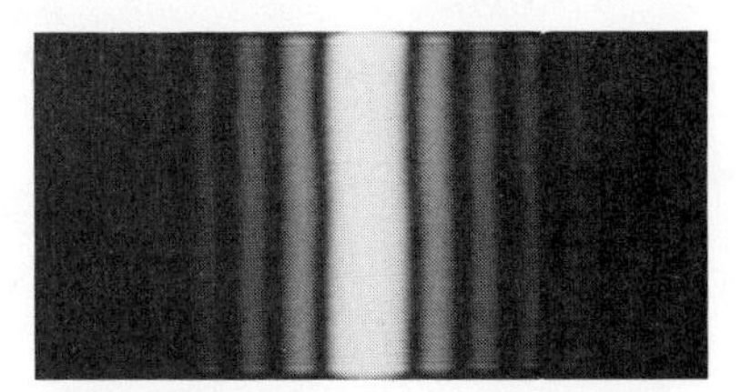

FIGURE 10-4b. Experimentally generated Fraunhofer diffraction pattern of a rectangular aperture.

where the spatial frequencies are defined as

$$\omega_x = -\frac{2\pi \sin\theta_x}{\lambda} = -\frac{2\pi\xi}{\lambda R_0}, \qquad \omega_y = -\frac{2\pi \sin\theta_y}{\lambda} = -\frac{2\pi\eta}{\lambda R_0}$$

The maximum intensities in the x and y directions occur at $\omega_x x_0 = \omega_y y_0 = 0$. Remembering that the area of this rectangular aperture is defined as $\mathcal{A} = 4x_0y_0$, we may write the maximum intensity as

$$I_0 = \left(\frac{i4x_0y_0Ae^{-i\mathbf{k}\cdot\mathbf{R}_0}}{\lambda R_0}\right)^2 = \frac{\mathcal{A}^2A^2}{\lambda^2R_0^2}$$

The minima of this function occur when $\omega_x x_0 = n\pi$ or $\omega_y y_0 = m\pi$. The location of the zeroes can be specified as a dimension in the observation plane or, if we use the paraxial approximation, in terms of an angle

$$\sin\theta_x \approx \theta_x \approx \frac{\xi}{R_0} = \frac{n\lambda}{2x_0}, \qquad \sin\theta_y \approx \theta_y \approx \frac{\eta}{R_0} = \frac{m\lambda}{2y_0}$$

The dimensions of the diffraction pattern are characterized by the location of the first zero, i.e., when $n = m = 1$ and are given by the observation plane coordinates ξ and η. The dimensions of the diffraction pattern are inversely proportional to the dimensions of the aperture. As the aperture dimension expands, the width of the diffraction pattern decreases until, in the limit of an infinitely wide aperture, the diffraction pattern becomes a delta function.

Figure 10-4*a* is a theoretical plot of the diffraction pattern from a rectangular slit; Figure 10-4*b* is an experimentally obtained Fraunhofer diffraction pattern.

If a lens is used to display the diffraction pattern, then R_0 is replaced by the focal length of the lens f in all of the equations. If the aperture of the lens is very large, we can consider a plane wave incident on it as infinitely wide. For this plane wave, the lens would produce a focal spot in its back focal plane that approximated a delta function, the diffraction pattern of an infinite plane wave.

DIFFRACTION FROM A CIRCULAR APERTURE

From the discussion of two-dimensional transforms in Chapter 6, we can immediately obtain the diffraction pattern from a circular aperture of diameter **a** by using the transform **(6-44)**. A second procedure will be used here to obtain the diffraction pattern. The cylindrical geometry shown in Figure 10-5 is used to convert the Huygens–Fresnel integral from rectangular to

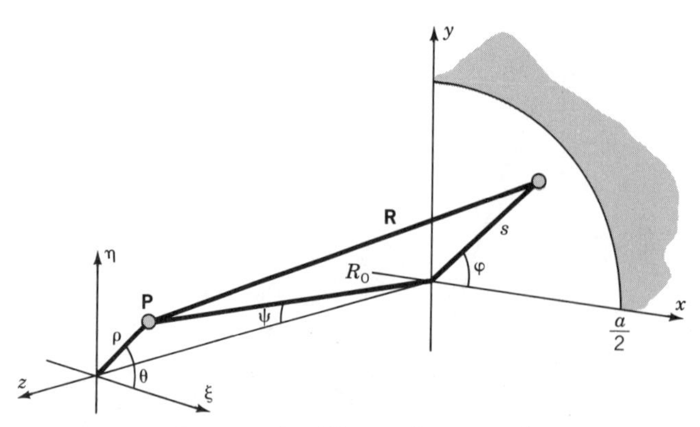

FIGURE 10-5. Geometry for diffraction from a circular aperture.

cylindrical coordinates. To transform to the new coordinate system, we make use of the following equations. At the aperture plane,

$$x = s \cdot \cos\varphi, \qquad y = s \cdot \sin\varphi$$
$$f(x, y) = f(s, \varphi), \qquad dx\,dy = s\,ds\,d\varphi \tag{10-21}$$

At the observation plane,

$$\xi = \rho\cos\theta, \qquad \eta = \rho\sin\theta \tag{10-22}$$

In the new, cylindrical coordinate system at the observation plane, the spatial frequencies are written as

$$\begin{aligned} \omega_x &= -\frac{k\xi}{R_0} = -\frac{k\rho}{R_0}\cos\theta \\ \omega_y &= -\frac{k\eta}{R_0} = -\frac{k\rho}{R_0}\sin\theta \end{aligned} \tag{10-23}$$

Using **(10-21)** and **(10-23)**, we may write

$$\begin{aligned} \omega_x x + \omega_y y &= -\frac{ks\rho}{R_0}(\cos\theta\cos\varphi + \sin\theta\sin\varphi) \\ &= -\frac{ks\rho}{R_0}\cos(\theta - \varphi) \end{aligned} \tag{10-24}$$

From Figure 10-5 we see that the observation point P can be defined in terms of the angle ψ, where

$$\sin\psi = \frac{\rho}{R_0}$$

This allows an angular representation for the size of the diffraction pattern if it is desired.

The Huygens–Fresnel integral can now be written in terms of cylindrical coordinates as

$$\mathcal{E}_P = \frac{iA}{\lambda R_0} e^{-i\mathbf{k}\cdot\mathbf{R}_0} \int_0^{a/2}\int_0^{2\pi} f(s, \varphi)\exp\left[-ik\frac{s\rho}{R_0}\cos(\theta - \varphi)\right] s\,ds\,d\varphi \tag{10-25}$$

We can demonstrate the use of **(10-25)** by using it to calculate the diffraction amplitude from a clear aperture of diameter **a**, defined by the equation

$$f(s, \varphi) = \begin{cases} 1, & s \leq \frac{a}{2}, \text{ all } \varphi \\ 0, & s > \frac{a}{2} \end{cases}$$

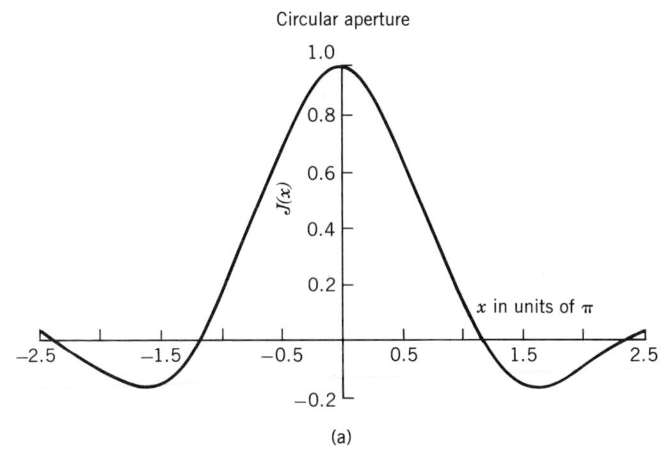

FIGURE 10-6a. Diffraction amplitude from a circular aperture. The observed light distribution is constructed by rotating the square of the Bessel function around the optical axis.

The symmetry of this problem is such that $f(s, \varphi) = f(s)$, which makes **(10-25)** identical to **(6-44)**. We may therefore use **(6-45)** to write the diffraction pattern amplitude

$$\mathcal{E}_P = \frac{iA}{\lambda} e^{-i\mathbf{k}\cdot\mathbf{R}_0} \left[\frac{\pi a}{k\rho} \mathbf{J}_1\left(\frac{ka\rho}{2R_0} \right) \right] \tag{10-26}$$

A plot of the function in the bracket is given in Figure 10-6*a*.
If we define

$$u = \frac{ka\rho}{2R_0}$$

then the spatial distribution of intensity in the diffraction pattern can be written in a form known as the *Airy formula*

$$I = I_0 \left[\frac{2\mathbf{J}_1(u)}{u} \right]^2 \tag{10-27}$$

where we have defined

$$I_0 = \left(\frac{A\mathcal{A}}{\lambda R_0} \right)^2$$

$\mathcal{A}$ is the area of the aperture

$$\mathcal{A} = \pi \left(\frac{a}{2} \right)^2$$

The intensity pattern described by **(10-27)** and shown in Figure 10-6*b* is called the Airy pattern. The intensity at $u = 0$ in **(10-27)** is the same as was obtained for a rectangular aperture of the same area **(10-20)** because in the limit

$$\lim_{u \to 0} \left[\frac{2\mathbf{J}_1(u)}{u} \right] = 1$$

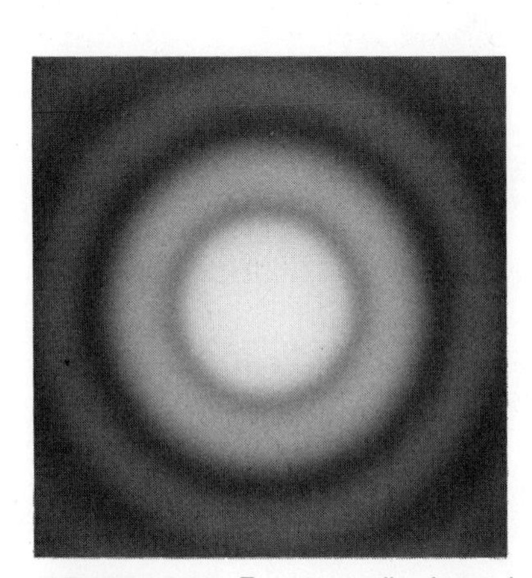

FIGURE 10-6b. Experimentally obtained Fraunhofer diffraction pattern from circular aperture.

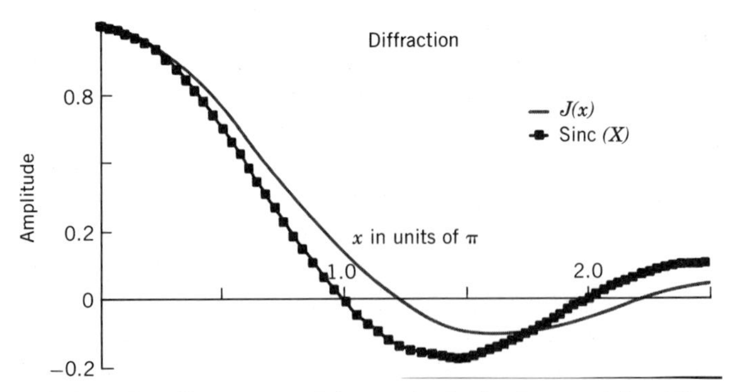

FIGURE 10-7. Comparison of the amplitude distribution of the diffraction pattern for a rectangular aperture and circular aperture.

At first glance, Figures 10-5 and 10-6 seem to have the same functional form, but a close examination of the location of the zeroes of the two functions reveals the difference. Figure 10-7 displays the two functions on the same graph. The first few zeroes of the sinc function occur at π, 2π, 3π, 4π, . . . , and the first few zeroes of the Airy function are 1.22π, 2.233π, 3.238π, ...

The location of the first zero is used to characterize the size of the diffraction pattern. The circular area defined by the first zero of **(10-27)** is called the *Airy disk*. The Airy disk formed by a lens of diameter a uniformly illuminated by a plane wave is obtained by solving

$$u = \frac{ka\rho}{2f} = \frac{\pi a\rho}{\lambda f} = 1.22\pi$$

for ρ. The diameter of the Airy disk 2ρ that we will define as w is given by

$$2\rho = w = \frac{2.44\lambda f}{a}$$

This is in agreement with the result from Chapter 9, obtained by using the Gaussian wave formalism of diffraction. In Chapter 9, we found that the minimum beam waist produced, in the back focal plane of a lens, uniformly illuminated by a plane wave is

$$w_0 \approx \frac{2\lambda f}{a}$$

The mathematical description of the amplitude distribution in the two spots is different. Thus, the constants in the two formula differ but the dependence on the experimental parameters is the same. Previously, we stated that a uniformly illuminated lens produced a delta function in the focal plane. Now we learn that this is only approximately true. Diffraction by the lens aperture produces a focal spot of finite extent.

For the Airy pattern, 84% of the total area is contained in the Airy disk, and 91% of the light is contained within the circle bounded by the second minimum at 2.233π. The intensities in the secondary maxima of the diffraction pattern of a rectangular aperture are much larger than those in the Airy pattern of a circular aperture. The peak intensities, relative to the central maximum, of the first three secondary maxima of a rectangular aperture are 4.7, 1.6, and 0.8%, respectively. For a circular aperture, the same quantities are 1.70, 0.04, and 0.02%, respectively.

ARRAY THEOREM

There is an elegant mathematical technique for handling multiple apertures called the array theorem. The theorem is based on the convolution integral discussed in Chapter 6 **(6-35)** and makes use of the fact that the Fourier transform of a convolution of two functions is the product of the Fourier transforms of the individual functions, **(6-38)**. We will prove the theorem for one dimension in which the functions represent slit apertures. The results can be extended to two dimensions in a straightforward way.

Assume that we have a collection of identical apertures, shown on the right of Figure 10-8. If one of the apertures is located at the origin of the aperture plane, its transmission function is $\psi(x)$. The transmission function of an aperture located at a point x_n can be written in terms of a generalized aperture function $\psi(x-\alpha)$ by the use of the sifting property of the delta function

$$\psi(x - x_n) = \int \psi(x-\alpha)\delta(\alpha - x_n)\, d\alpha \tag{10-28}$$

This convolution integral will allow the application of the convolution theorem to complete the derivation of the array theorem.

Equation **(10-28)** can be recognized as the convolution integral if we redefine the terms

$$t = \alpha - x_n, \qquad dt = d\alpha$$

$$\psi(x - x_n) = \int \psi(x - x_n - t)\delta(t)\, dt$$

By defining $x - x_n = \tau$, we may rewrite the integral in the same form as the convolution integral **(6-35)**.

The aperture transmission function representing an array of apertures will be the sum of the distributions of the individual apertures, represented graphically in Figure 10-8 and mathematically by the summation

$$\boldsymbol{\psi}(x) = \sum_{n=1}^{N} \psi(x - x_n)$$

The Fraunhofer diffraction from this array is the Fourier transform of $\Psi(x)$

$$\boldsymbol{\Phi}(\omega_x) = \int_{-\infty}^{\infty} \boldsymbol{\Psi}(x) e^{-i\omega_x x}\, dx$$

From **(6A-3)**, we can write

$$\boldsymbol{\Phi}(\omega_x) = \sum_{n=1}^{N} \int_{-\infty}^{\infty} \psi(x - x_n) e^{-i\omega_x x}\, dx$$

We now make use of the fact that $\psi(x - x_n)$ can be expressed in terms of a convolution integral. The Fourier transform of $\psi(x - x_n)$ is from the convolution theorem **(6A-8)** the product of the Fourier transforms of the individual functions that make up the convolution

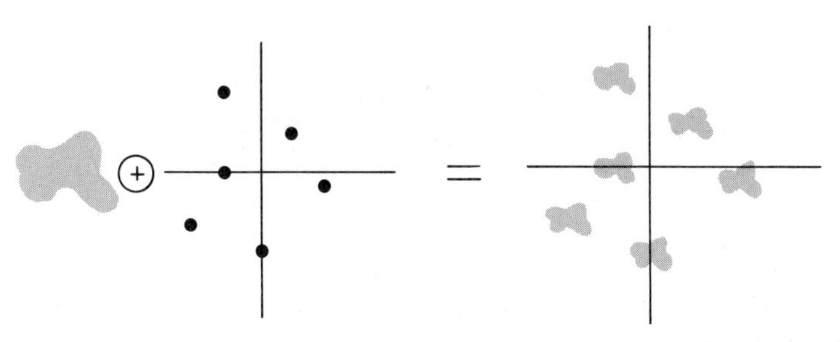

FIGURE 10-8. The convolution of an aperture with an array of delta functions will produce an array of identical apertures, each located at the position of one of the delta functions.

$$\Phi(\omega_x) = \sum_{n=1}^{N} \mathcal{F}\{\psi(x-\alpha)\}\,\mathcal{F}\{\delta(x-x_n)\}$$

$$= \mathcal{F}\{\psi(x-\alpha)\}\sum_{n=1}^{N} \mathcal{F}\{\delta(x-x_n)\}$$

Again applying **(6A-3)** yields

$$\Phi(\omega_x) = \mathcal{F}\{\psi(x-\alpha)\}\,\mathcal{F}\left\{\sum_{n=1}^{N}\delta(x-x_n)\right\} \tag{10-29}$$

The first transform in **(10-29)** is the diffraction pattern of an individual aperture and the second transform is the diffraction pattern produced by a set of point sources with the same spatial distribution as the array of identical apertures. We will call this second transform the array function. In one dimension, the array function is the comb function whose Fourier transform has been calculated **(6-28)**.

To summarize, the array theorem states that the diffraction pattern of an array of similar apertures is given by the product of the diffraction pattern from a single aperture and the diffraction (or interference) pattern of an identically distributed array of point sources.

Figure 10-9 is a physical realization of the array theorem. Displayed in Figure 10-9*b* is the diffraction pattern of a random array of circular apertures shown in Figure 10-9*a*. The overall diffraction pattern is an Airy pattern due to the diffraction of an individual circular aperture. The intensity distribution in the Airy disk is a random distribution of intensity maxima and minima. This "speckled" distribution is called speckle noise and is due to the interference between the waves from the random array of circular apertures. We will mention the speckle noise again in Appendix 10-C.

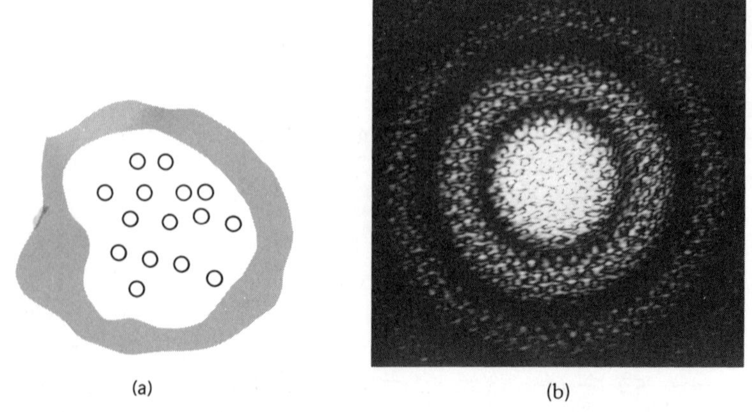

FIGURE 10-9. A random array of circular apertures, as shown in (a) produces the Fraunhofer diffraction pattern shown in (b). As predicted by the array theorem, the overall diffraction pattern (the shape factor) is the Airy pattern of a circular aperture, whereas the intensity distribution in the Airy disk (the grating factor) is due to the interference between the waves from the random array of apertures. The speckled intensity pattern produced by the interference is called speckle noise. We will make some additional comments about this type of noise in Appendix 10-C.

N RECTANGULAR SLITS

An array of N identical apertures is called a *diffraction grating* in optics. The Fraunhofer diffraction patterns produced by such an array have two important properties: A number of very narrow beams are produced by the array and the beam positions are a function of the wavelength of illumination of the apertures and the relative phase of the waves radiated by each of the apertures.

Because of these properties, arrays of diffracting apertures have been used in a number of applications. At radio frequencies, arrays of dipole antennas are used to both radiate and receive signals in radar and radio astronomy systems. One advantage offered by diffracting arrays at radio frequencies is that the beam produced by the array can be electrically steered by adjusting the relative phase of the individual dipoles.

An optical realization of a two-element array of radiators is Young's two-slit experiment and a realization of a two-element array of receivers is Michelson's stellar interferometer. Two optical implementations of arrays, containing more diffracting elements, are diffraction grating and holograms. In nature, periodic arrays of diffracting elements are the origin of the colors observed on some invertebrates.

Many solids are naturally arranged in three-dimensional arrays of atoms or molecules that act as diffraction gratings when illuminated by x-ray wavelengths. The resulting Fraunhofer diffraction patterns are used to analyze the ordered structure of the solids.

The array theorem can be used to calculate the diffraction pattern from N slits, each of width a and separation d. The aperture function of a single slit is equal to

$$\psi(x, y) = \begin{cases} 1, & |x| \leq (a/2) \\ & |y| \leq y_0 \\ 0, & \text{all other } x \text{ and } y \end{cases}$$

The Fraunhofer diffraction pattern of this aperture has already been calculated and is given by **(10-19)**.

$$\mathcal{F}\{\psi(x, y)\} = K\frac{\sin\alpha}{\alpha}$$

where the constant **K** and the variable α are

$$K = \frac{i2ay_0A}{\lambda R_0}e^{-i\mathbf{k}\cdot\mathbf{R}_0}\frac{\sin\omega_y y_0}{\omega_y y_0}$$

$$\alpha = \frac{ka}{2}\sin\theta_x$$

The array function is the comb function **(6-25)**

$$A(x) = \sum_{n=1}^{N}\delta(x - x_n)$$

where $x_n = (n-1)d$. The comb function's Fourier transform is given by **(6-28)**

$$\mathcal{F}\{A(x)\} = \frac{\sin N\beta}{\sin\beta}$$

$$\beta = \frac{kd}{2}\sin\theta_x$$

The Fraunhofer diffraction pattern's intensity distribution in the x direction is thus given by

$$I_\theta = I_0 \frac{\sin^2 \alpha}{\alpha^2} \frac{\sin^2 N\beta}{\sin^2 \beta} \tag{10-30}$$

We have combined the variation in intensity in the y direction into the constant I_0 because we assume that the intensity variation in the x direction will be measured at a constant value of y.

A physical interpretation of **(10-30)** views the first factor as arising from the diffraction associated with a single slit; it is called the *shape factor*. In Figure 10-9, the shape factor is the Airy pattern. The second factor arises from the interference between light from different slits; it is called the *grating factor*. In Figure 10-9, the grating factor is a random function. The fine detail in the spatial light distribution of the diffraction pattern is described by the grating factor and arises from the coarse detail in the diffracting object. The coarse, overall light distribution in the diffraction pattern is described by the shape factor and arises from the fine detail in the diffracting object.

Young's Double Slit

The array theorem makes the analysis of Young's two-slit experiment a trivial exercise. The application of the array theorem will combine the effects of diffraction with the interference effects discussed in Chapter 4 and will demonstrate that the interference between the two slits arises naturally from an application of diffraction theory. The result of this analysis will support a previous assertion that interference describes the same physical process as diffraction and the division of the two subjects is an arbitrary one.

In Chapter 4, we discussed interference from Young's two-slit experiment without considering diffraction from the slits. We did, however, assume that diffraction would spread the light from each slit, causing the two waves to overlap. By using the results from the array theorem, **(10-30)** with $N = 2$, the effects of diffraction can be included explicitly in the theoretical explanation of Young's experiment.

The intensity of the diffraction pattern from two slits is obtained from **(10-30)** by setting $N = 2$

$$I_\theta = I_0 \frac{\sin^2 \alpha}{\alpha^2} \cos^2 \beta \tag{10-31}$$

The sinc function describes the energy distribution of the overall diffraction pattern, whereas the cosine function describes the energy distribution created by interference between the light waves from the two slits. Physically, α is a measure of the phase difference between points in one slit and β is a measure of the phase difference between similar points in the two slits. Zeroes in the diffraction intensity occur whenever $\alpha = n\pi$ or $\beta = (2n + 1)\pi/2$. Figure 10-10 shows the interference maxima from the grating factor under the central maximum described by the shape factor. The number of interference maxima appearing within the central diffraction maximum can be shown to be given by (see Problem 10-1)

$$\frac{2d}{a} - 1$$

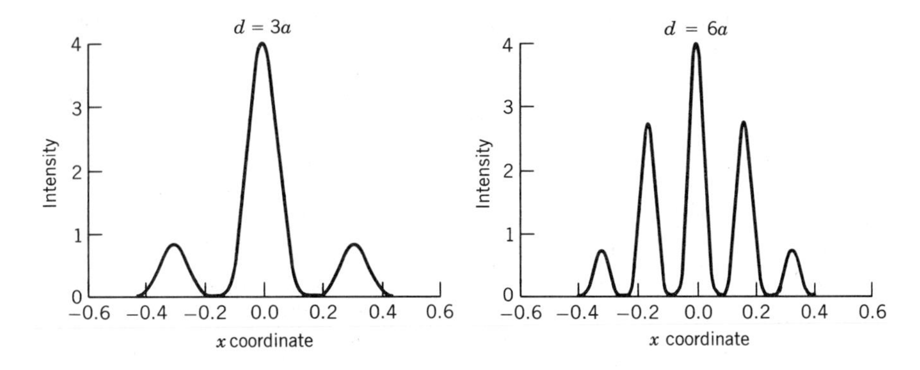

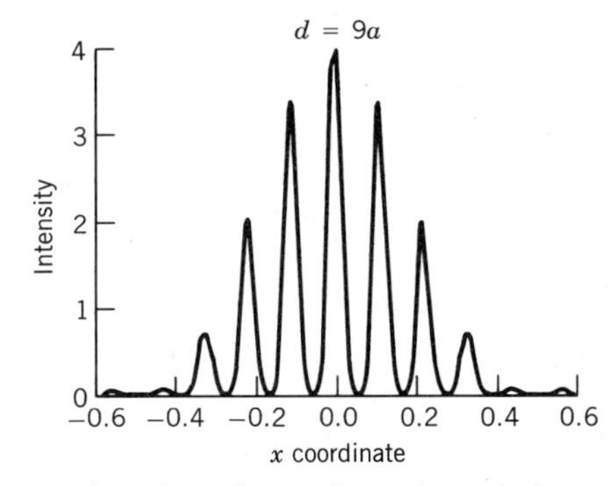

FIGURE 10-10. The number of interference fringes beneath the main diffraction peak of a rectangular aperture in a Young's two slit experiment.

In Figure 10-10, the diffraction pattern for three values of the ratio d/a is shown.

The Diffraction Grating

In this section, we will use the array theorem to derive the diffraction intensity distribution of a large number of identical apertures. We will discover that the positions of the principal maxima are a function of the illuminating wavelength. This functional relationship has led to the application of a diffraction grating to wavelength measurements.

The diffraction grating normally used for wavelength measurements is not a large number of diffracting apertures, but rather a large number of reflecting grooves cut in a surface such as gold or aluminum. The theory to be derived also applies to these reflection gratings, but a modification must be made to the theory because the shape of the grooves in the reflection grating controls the fraction of light diffracted into a principal maximum. A grating whose groove shape has been controlled to enhance the energy contained in a particular principal maximum is called a *blazed grating*. The use of special groove shapes is equivalent to the modification of the phase in an antenna array at radio frequencies.

The construction of a large number of diffracting elements into an optical device for measuring wavelength was first suggested by **David Rittenhouse**, an American astronomer, in 1785, but the idea was ignored until Fraunhofer reinvented the concept in 1819. Fraunhofer's first gratings were fine wires spaced by wrapping the wires in the threads of two parallel screws. He later made gratings by cutting (*ruling*) grooves in gold films deposited on the surface of glass. **H.A. Rowland** made a number of well-designed ruling machines that made possible the production of large area gratings. Following a suggestion by Lord Rayleigh, **Robert Williams Wood (1868–1955)** developed the capability to control the shape of the grooves.

If N is allowed to assume values much larger than 2, the appearance of the interference fringes, predicted by the grating factor, changes from a simple sinusoidal variation to a set of narrow maxima, called principal maxima, surrounded by much smaller, secondary maxima, as is shown in Figure 10-11 for $N = 2, 3, 4, 5$, and 6.

To understand the images in Figure 10-11, we must interpret the predictions of **(10-30)** when N is a large number. Whenever $N\beta = m\pi$, where $m = 0, 1, 2, \ldots$, the numerator of the second factor in **(10-30)** will be zero, leading to an intensity that is zero, $I_\theta = 0$. The denominator of the second factor in **(10-30)** is zero when $\beta = \ell\pi$, $\ell = 0, 1, 2, \ldots$ If both conditions occur at the same time, the ratio of m and N is equal to an integer, $m/N = \ell$, and an indeterminate value for the intensity $I_\theta = 0/0$ is obtained. For this situation, we must apply L'Hospital's rule

$$\lim_{\beta\to\ell\pi} \frac{\sin N\beta}{\sin\beta} = \lim_{\beta\to\ell\pi} \frac{N\cos N\beta}{\cos\beta} = N$$

L'Hospital's rule predicts that whenever

$$\beta = \frac{m}{N}\pi$$

where m/N is an integer, a principal maximum in the intensity will occur with a value given by

$$I_{\theta P} = N^2 I_0 \frac{\sin^2\alpha}{\alpha} \tag{10-32}$$

Secondary maxima, much weaker than the principal maxima, occur when

$$\beta = \left(\frac{2m+1}{2N}\right)\pi, \qquad m = 1, 2, \ldots \tag{10-33}$$

[When $m = 0$, m/N is an integer; thus, the first value that m can have in **(10-33)** is $m = 1$.] The intensity of each secondary maximum is given by

$$I_{\theta S} = I_0 \frac{\sin^2\alpha}{\alpha^2}\frac{\sin^2 N\beta}{\sin^2\beta}$$

$$= I_0 \frac{\sin^2\alpha}{\alpha^2}\left[\frac{\sin^2\left(\frac{2m+1}{2}\right)\pi}{\sin^2\left(\frac{2m+1}{2N}\right)\pi}\right]$$

$$= I_0 \frac{\sin^2\alpha}{\alpha^2}\left[\frac{1}{\sin\left(\frac{2m+1}{2N}\right)\pi}\right]^2$$

The quantity $(2m + 1)/2N$ is a small number due to the size of N, allowing the small angle approximation to be made

(a)

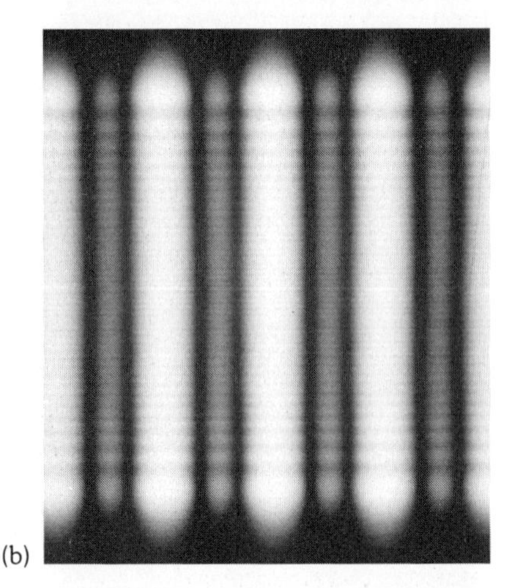

(b)

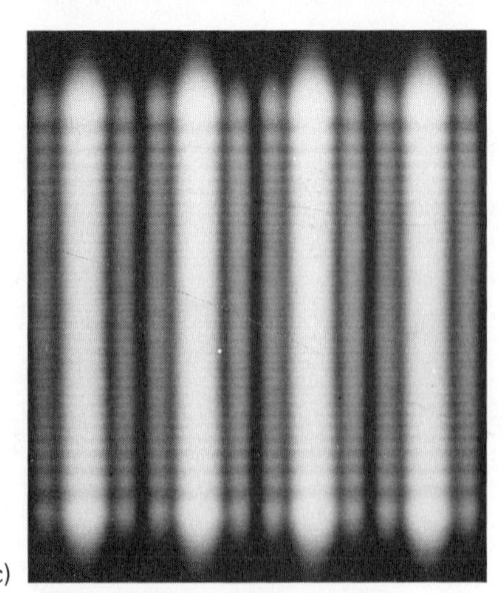

(c)

FIGURE 10-11. Diffraction patterns from N slits, each identical to the one used to produce Figure 10-4*b*. (a) $N=2$, (b) $N=3$, (c) $N=4$, (d) $N=5$, (e) $N=6$.

$$I_{\theta S} \approx I_0 \frac{\sin^2 \alpha}{\alpha^2} \left[\frac{2N}{\pi(2m + 1)} \right]^2 \tag{10-34}$$

The ratio of the intensity of a secondary maximum and a principal maximum is given by

$$\frac{I_{\theta S}}{I_{\theta P}} = \left[\frac{2}{\pi(2m + 1)} \right]^2$$

The strongest secondary maximum occurs for $m = 1$ and, for large N, has an intensity that is about 4.5% of the intensity of the neighboring principal maximum.

The position of principal maxima occurs at angles specified by the *grating formula.*

$$\begin{aligned} \beta &= \ell\pi = \frac{m}{N}\pi \\ &= \frac{kd \sin \theta}{2} \\ &= \frac{\pi d \sin \theta}{\lambda} \end{aligned} \tag{10-35}$$

The angular position of the principal maxima is thus

$$\sin \theta_x = \frac{m\lambda}{Nd}$$

where m is called the *interference order.* [We will encounter this same equation later in the analysis of holography **(12-13)** and the acoustooptic effect **(14-62)**. It is called the *Bragg equation.*]

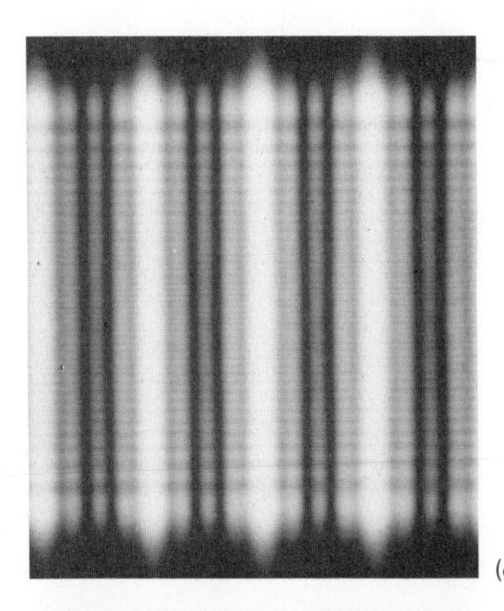

(d)

(e)

FIGURE 10-11. *(continued)*

The model we have used to obtain the Bragg equation is based on a periodic array of identical apertures. The transmission function of this array would be a periodic square wave. If we, for the moment, treat the grating as infinite in size, we discover that the principal maxima in the diffraction pattern correspond to the terms of the Fourier series describing the square wave transmission function; see Figure 6-3.

The zero order, $m = 0$, corresponds to the a_0 term in the Fourier series and has an intensity proportional to the spatially averaged transmission of the grating. Because of its equivalence to the temporal average of a time-varying signal, the zero-order principal maximum is often called the *dc term.*

The first-order, principal maximum corresponds to the fundamental spatial frequency of the grating and the higher orders correspond to the harmonics of this frequency.

The dc term provides no information about the wavelength of the illumination. Information about the wavelength of the illuminating light can only be obtained by measuring the angular position of the first- or higher-order principal maximum.

The fact that the grating is finite in size causes each of the orders to have an angular width that limits the resolution with which the illuminating wavelength can be measured. To calculate the resolving power of the grating, we first determine the angular width of a principle maximum. This is accomplished by measuring the angular change, of the principal maximum's position, when β changes from $\beta = \ell\pi = m\pi/N$ to $\beta = (m+1)\pi/N$, i.e., $\Delta\beta = \pi/N$. Using the definition of β yields

$$\beta = \frac{kd \sin\theta}{2}$$

$$\Delta\beta = \frac{\pi d \cos\theta\Delta\theta}{\lambda}$$

The angular width is then

$$\Delta\theta = \frac{\lambda}{Nd \cos\theta} \tag{10-36}$$

The derivative of the grating formula gives

$$\Delta\lambda = \frac{d}{\ell}\cos\theta\,\Delta\theta$$

$$\Delta\theta = \frac{\ell\Delta\lambda}{d\cos\theta}$$

Equating this result to **(10-36)** yields

$$\frac{\ell\Delta\lambda}{d\cos\theta} = \frac{\lambda}{Nd\cos\theta}$$

The resolving power of a grating is therefore

$$\frac{\lambda}{\Delta\lambda} = N\ell \tag{10-37}$$

The dependence of the principal maxima on N and d can be seen by comparing the curves shown in Figure 10-12. The separation of the principal maxima is displayed as a function of sin θ. The angular separation of principal maxima can be converted to a linear dimension by assuming a distance R from the grating to the observation plane. In the lower right-hand curve of Figure 10-12, a distance of 2 m was assumed. Grating spectrometers are classified by the distance R used in their design. The larger the R, the easier it is to resolve wavelength differences. For example, a 1 m spectrometer is a higher-resolution instrument than a 1/4 m spectrometer.

The improvement of resolving power with N can be seen in Figure 10-12. A grating 2 in. wide and containing 15,000 grooves per inch would have a resolving power in second order ($\ell = 2$) of 6×10^4. At a wavelength of 600 nm, this grating could resolve two waves, differing in wavelength by 0.01 nm.

The diffraction grating is limited by overlapping orders, as was the Fabry–Perot interferometer discussed in Chapter 4. If two wavelengths λ and $\lambda + \Delta\lambda$ have successive orders that are coincident, then

$$(m+1)\lambda = m(\lambda + \Delta\lambda)$$

The minimum wavelength difference for which this occurs is defined as the free spectral range of the diffraction grating

$$(\Delta\lambda)_{SR} = \frac{\lambda}{m}$$

This result is identical to the result obtained for the Fabry–Perot interferometer **(4-54)**.

We have been discussing amplitude transmission gratings. Amplitude transmission gratings have little practical use because they waste light. The light loss is from a number of sources.

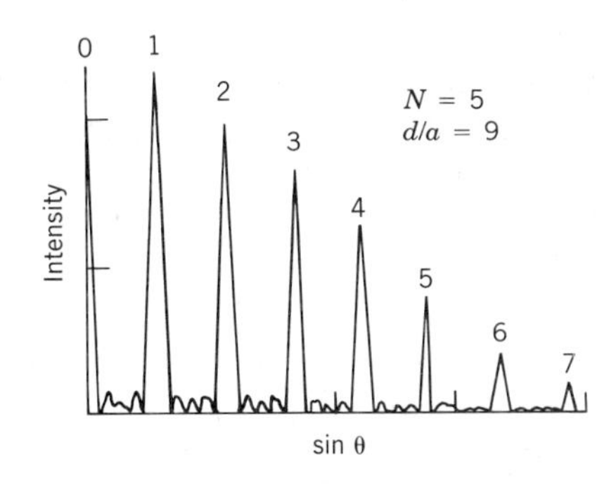

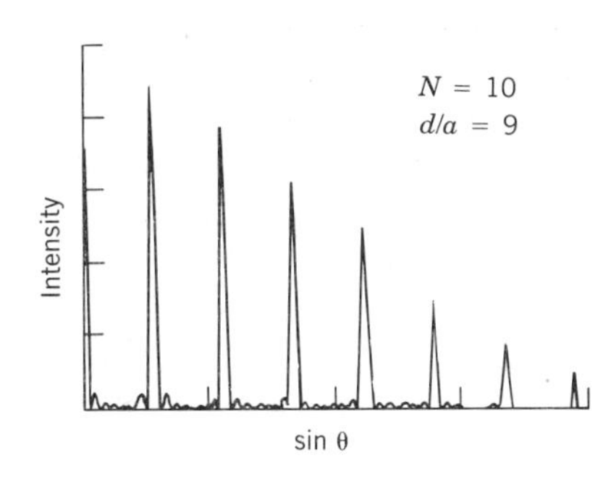

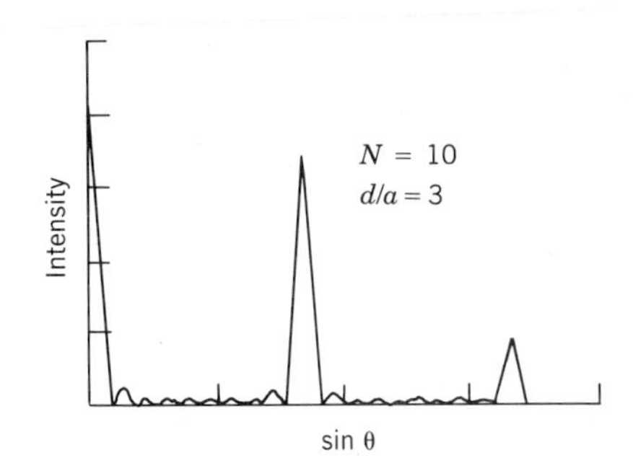

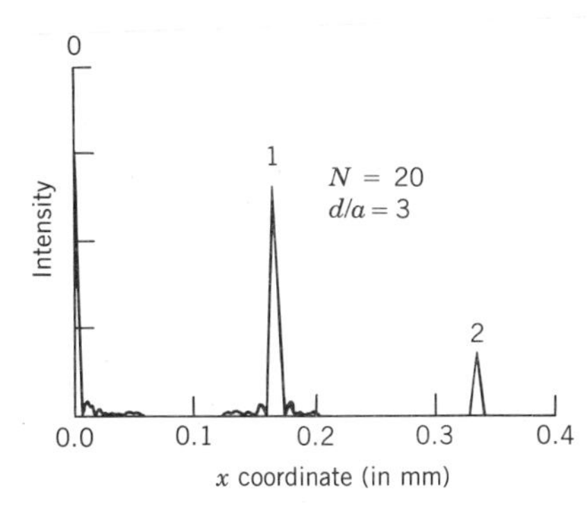

FIGURE 10-12. Decrease in the width of the principal maxima of a transmission grating with an increasing number of slits. The various principal maxima are called orders, numbering from zero at the origin to a value as large as 7 in this example. Also shown is the effect of different ratios, d/a, on the number of visible orders.

1. Light is diffracted simultaneously into both positive and negative orders (the positive and negative frequencies of the Fourier transform). The negative diffraction orders contain redundant information and waste light.
2. In an amplitude transmission grating, light is thrown away because of the opaque portions of the slit array.
3. The width of an aperture leads to a shape factor

$$\text{sinc}^2 \alpha = \frac{\sin^2 \alpha}{\alpha^2}$$

   for a rectangular aperture, which modulates the grating factor and causes the amplitude of the orders to rapidly decrease. This can be observed in Figure 10-12, where the second order is very weak (for $d/a = 3$). Because of the loss in intensity at higher orders, only the first few orders ($\ell = 1$, 2, or 3) are useful. The shape factor also causes a decrease in diffracted light intensity with increasing wavelength for higher-order principal maxima.
4. The location of the maximum in the diffracted light, i.e., the angular position for which the shape factor is a maximum, coincides with the location of the principal maximum due to the zero-order interference. This zero-order maximum is independent of wavelength and not of much use.

One solution to the problems created by transmission gratings would be the use of a grating that modified only the phase of the transmitted wave. Such gratings would operate using the same physical processes as a microwave phased array antenna, where the location of the shape factor's maximum is controlled by adding a constant phase shift to each antenna element. The construction of an optical transmission phase grating with a uniform phase variation, across the aperture of the grating, is very difficult. For this reason a second approach, based on the use of reflection gratings is the practical solution to the problems listed above.

By tilting the reflecting surface of each groove of a reflection grating, (Figure 10-13), the position of the shape factor's maximum can be controlled. Problems 1, 3, and 4 are eliminated because the shape factor maximum is moved from the optical axis out to some angle with respect to the axis. The use of reflection gratings also removes Problem 2 because all of the incident light is reflected by the grating.

Robert Wood in 1910 developed the technique of producing grooves of a desired shape in a reflective grating. Gratings, with grooves shaped to enhance their performance at a particular wavelength, are said to be *blazed* for that wavelength. The physical properties on which blazed gratings are based can be understood by using Figure 10-13. The shape of the grooves is controlled so that they can be treated as an array of mirror surfaces. The normal to each of the groove faces makes an angle θ_B with the normal to the grating surface. We can measure the angle of incidence and the angle of diffraction with respect to the grating normal or the groove normal, as shown in Figure 10-13.

From Figure 10-13, we can write a relationship between the angles

$$\theta_i = \varphi_i - \theta_B, \qquad -\theta_d = -\varphi_d + \theta_B$$

(The sign convention is in accord with the one used in geometric optics; positive angles are those measured in a counterclockwise rotation from the normal to the surface. Therefore, θ_B is a negative angle.) The blaze angle provides an extra degree of freedom that will allow independent adjustment of the angular location of the principal maxima of the grating factor and the zero-order, single-aperture diffraction maximum. To see how this is accomplished, we must determine first the effect of off-axis illumination of a diffraction grating.

In the discussion of spatial coherence, we mentioned briefly the effect of an off-axis source on the interference pattern produced by two slits. We saw

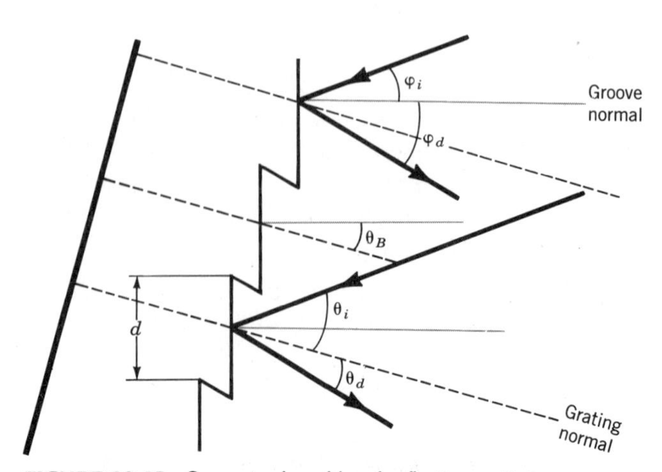

FIGURE 10-13. Geometry for a blazed reflection grating.

in **(8-33)** that the interference pattern is shifted by off-axis illumination. Off-axis illumination is easy to incorporate into the equation for the diffraction intensity from an array. To include the effect of an off-axis source, the phase of the illuminating wave is modified by changing the incident illumination from a plane wave of amplitude $\mathcal{E}$, traveling parallel to the optical axis, to a plane wave with the same amplitude, traveling at an angle θ_i to the optical axis

$$\mathcal{E}e^{-ikx \sin \theta_i}$$

(Because we are interested only in the effects in a plane normal to the direction of propagation, we ignore the phase associated with propagation along the z direction, $kz \cos \theta_i$.) The off-axis illumination results in a modification of the parameter for single-aperture diffraction from

$$\alpha = \frac{ka}{2} \sin \theta_d$$

to

$$\alpha = \frac{ka}{2} (\sin \theta_i + \sin \theta_d) \tag{10-38}$$

and for the multiple-aperture interference from

$$\beta = \frac{kd}{2} \sin \theta_d$$

to

$$\beta = \frac{kd}{2} (\sin \theta_i + \sin \theta_d) \tag{10-39}$$

where we have relabeled θ_x in the equations for α and β as θ_d, the angle the diffracted light makes with the grating normal.

The zero-order, single-aperture diffraction peak occurs when $\alpha = 0$. If we measure the angles with respect to the groove face, this occurs when

$$\alpha = \frac{ka}{2} (\sin \varphi_i + \sin \varphi_d) = 0$$

The angles are therefore related by

$$\sin \varphi_i = -\sin \varphi_d$$

$$\varphi_i = -\varphi_d$$

We see that the single-aperture diffraction maximum (the shape factor's maximum) occurs at the same angle that reflection from the groove faces occurs. We can write this result in terms of the angles measured with respect to the grating normal

$$\theta_i = -(\theta_d + 2\theta_B) \tag{10-40}$$

The blaze condition requires the single-aperture-diffraction maximum to occur at the ℓth principal maximum for wavelength λ_B. At that position,

$$\ell \pi = \frac{2\pi d}{\lambda_B} (\sin \theta_i + \sin \theta_d)$$

$$\ell \lambda_B = 2d \sin \frac{1}{2}(\theta_i + \theta_d) \cos \frac{1}{2}(\theta_i - \theta_d)$$

For the special geometrical configuration called the *Littrow condition*, where $\theta_i = \theta_d$, we find that **(10-40)**, leads to the equation

$$\ell \lambda_B = 2d \sin \theta_B$$

A moment's thought will reveal that the physical significance of the blaze condition, in the Littrow configuration, is that the groove face must be normal to the incident wave so that $\theta_d = \theta_B$. By adjusting the blaze angle, the single-aperture diffraction peak can be positioned on any order of the interference pattern. Typical blaze angles are between 15 and 30°, but gratings are made with larger blaze angles.

SUMMARY

An approximate formulation of the Huygens–Fresnel diffraction integral was obtained by assuming that the distance from the diffracting obstruction to the observation point R_0 was large enough to fulfill the following far-field condition:

$$R_0 >> \frac{r^2}{4\pi\lambda}$$

where r is the dimension of the obstruction. Diffraction meeting this condition is called Fraunhofer diffraction and is described by the integral

$$\mathcal{E}_P(\omega_x, \omega_y) = \frac{iAe^{-i\mathbf{k}\cdot\mathbf{R}_0}}{\lambda R_0} \iint_\Sigma f(x, y) \exp[-i(\omega_x x + \omega_y y)]\, dx\, dy$$

This is a two-dimensional spatial Fourier transform of the aperture transmission function $f(x, y)$ with spatial frequencies defined by

$$\omega_x = -\frac{2\pi\xi}{\lambda R_0}, \qquad \omega_y = -\frac{2\pi\eta}{\lambda R_0}$$

We discovered that a lens would produce in its back focal plane the Fraunhofer diffraction pattern of an object in its front focal plane. Thus, a lens could be thought of as a Fourier transform operator. The above equations applied to a lens if we replaced R_0 by the focal length of the lens. The negative signs used in these definitions allow an identical mathematical formalism for spatial frequencies and temporal frequencies. The Fraunhofer diffraction patterns for two simple but important apertures were calculated.

Rectangular Slit

The amplitude transmission function was assumed to be

$$f(x, y) = \begin{cases} 1, & |x| \le x_0 \\ & |y| \le y_0 \\ 0, & \text{all other } x \text{ and } y \end{cases}$$

The intensity of the diffracted light was given by

$$I_P = I_0 \frac{\sin^2 \omega_x x_0}{(\omega_x x_0)^2} \cdot \frac{\sin^2 \omega_y y_0}{(\omega_y y_0)^2}$$

where the spatial frequencies can also be defined in terms of angles

$$\omega_x = -\frac{2\pi\xi}{\lambda R_0} = \frac{2\pi \sin \theta_x}{\lambda}, \qquad \omega_y = -\frac{2\pi\eta}{\lambda R_0} = \frac{2\pi \sin \theta_y}{\lambda}$$

Circular Aperture

$$f(s, \varphi) = \begin{cases} 1, & s \le \dfrac{a}{2} \\ & \text{all } \varphi \\ 0, & s > \dfrac{a}{2} \end{cases}$$

The intensity of the diffracted wave is

$$I_P = \left(\frac{A\mathcal{A}}{\lambda R_0}\right)^2 \left[\frac{2J_1(u)}{u}\right]^2$$

where $\mathcal{A}$ is the area of the aperture

$$\mathcal{A} = \pi\left(\frac{a}{2}\right)^2$$

and the argument of the Bessel function is

$$u = \frac{ka\rho}{2R_0}$$

The array theorem was used to find the diffraction patterns from an array of N identical apertures. The intensity distribution of the diffraction pattern was found to be

$$I_\theta = I_0 \frac{\sin^2 \alpha}{\alpha^2} \frac{\sin^2 N\beta}{\sin^2 \beta}$$

where the variable α is defined as

$$\alpha = \frac{ka}{2} \sin \theta_x$$

and a is the x dimension of a single aperture. The variable β is defined as

$$\beta = \frac{kd}{2} \sin \theta_x$$

where d is the separation in the x direction of the individual apertures.

Diffraction Grating

The model of an array of N identical apertures was used to describe the operation of a diffraction grating. The diffraction pattern of a grating contains a set of principal maxima with intensity given by

$$I_{\theta P} = N^2 I_0 \frac{\sin^2 \alpha}{\alpha^2}$$

The grating equation

$$\beta = \frac{\pi d \sin \theta_x}{\lambda}$$

was solved for the case of $\beta = m\pi/N$ yielding the Bragg equation

$$\sin \theta_x = \frac{m\lambda}{Nd}$$

The Bragg equation specifies the angular position of each of the principal maxima produced in the diffraction pattern. The resolution of the diffraction grating was found to depend on the size of the grating as shown by the equation for the resolving power

$$\mathcal{R} = \frac{\lambda}{\Delta\lambda} = \ell N$$

PROBLEMS

10-1. In a double-slit Fraunhofer diffraction experiment, missing orders occur at those values of sin θ which satisfy, at the same time, the condition for interference maxima and the condition for diffraction minima. Show that this leads to the condition (d/a) = integer, where a is the slit width and d is the distance between slits. Show the approximate relation $d \sin \theta = m\lambda$ as the condition for interference maxima.

10-2. Using the results of Problem 10-1, show that the number of interference maxima under the central diffraction maximum of the double-slit diffraction pattern is given by $(2d/a) - 1$, where a is the slit width and d is the distance between slits. What is the physical significance of values of d/a that are noninteger?

10-3. The Fraunhofer pattern of a double slit under $\lambda = 650$ nm illumination appears, at the back focal plane of an 80 cm focal-length lens. Using measurements taken from Figure 10-14, calculate the width and spacing of the slits.

10-4. How many lines must be ruled on a transmission grating so that it will just resolve the sodium doublet (589.592 and 588.995 nm) in the first-order spectrum?

10-5. Is there any limit to the number of principal maxima in the far-field pattern of an N-slit array? Explain the answer.

10-6. A grating is ruled at 1000 lines/mm. How wide must it be in order to resolve the spectral content of an HeNe laser ($\lambda = 632.8$ nm)? The frequency content of the HeNe laser consists of three very narrow emission lines: one at 632.8 nm with two others spaced 450 MHz on either side of the 632.8 nm line. Would there be a better instrument to resolve this spectrum?

10-7. Monochromatic radiation is incident normal to a grating surface from a narrow slit so that the principal maxima appear as thin bright bands. Show that the width $\Delta\theta$ of each band is inversely proportional to the width of the grating. See Figure 10-15.

10-8. Formulate a rough estimate of the extent of the Airy disk at visible wavelengths for a lens in terms of its f-number (the ratio of its focal length to its diameter).

FIGURE 10-14. The Fraunhofer diffraction pattern from a double slit. Use the information gathered from measurements of their pattern in Problem 10-3 to determine the width and spacing of the slits. The image has been enlarged by a factor of 10.

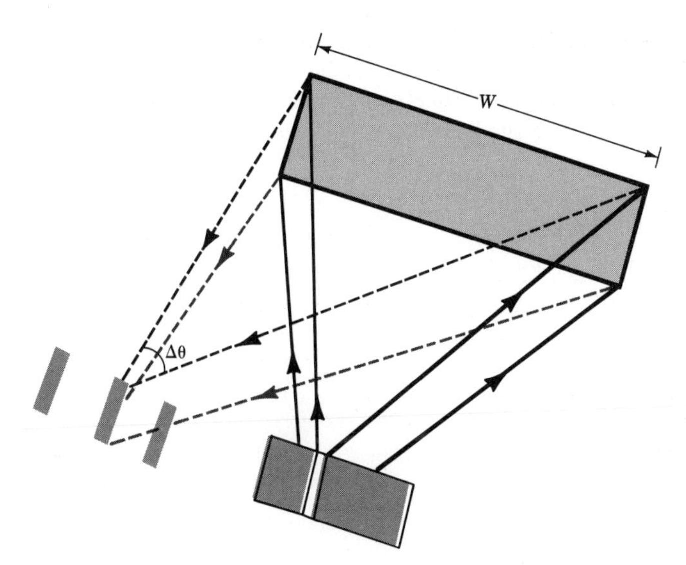

FIGURE 10-15. Geometry for Problem 10-7.

10-9. A point source of yellow light ($\lambda = 589$ nm) is viewed through a piece of light, closely woven material 30 m from the observer. A square array of bright points individually separated by 30 cm is observed. What are the number of threads per meter in the material?

10-10. What does an increase in wavelength do to a Fraunhofer diffraction pattern?

10-11. What happens when we increase the focal length of the lens used to produce Fraunhofer diffraction?

10-12. The transmission function of a slit is given by

$$f(x) = \begin{cases} \dfrac{a - |x|}{a}, & -a < x < a \\ 0, & \text{all other } x \end{cases}$$

Calculate the diffraction pattern. Calculate the peak intensity in the sidebands and compare it to the intensity observed in the sidebands produced by a uniformly illuminated slit. Intentional modification of an aperture's transmission function is called apodization (see Appendix 10-B).

10-13. Calculate the diffraction pattern of an apodized slit for which the transmission function is

$$f(x) = \frac{1}{a}\sqrt{\frac{2}{\pi}}\exp\left[-2\left(\frac{x}{a}\right)^2\right]$$

Comment on the sidebands of this diffraction pattern.

10-14. Prove that the secondary maxima of a single-slit diffraction pattern occur when $\alpha = \tan\alpha$.

10-15. What diameter telescope aperture is needed to resolve a double star with a separation of 10^8 km if the star is located 10 light years from the earth? (Assume $\lambda = 500$ nm.)

10-16. A single-slit Fraunhofer diffraction pattern is formed using white light. For what wavelength does the second minimum coincide with the third minimum produced by a wavelength of $\lambda = 400$ nm?

10-17. A rectangular, horizontal hole 0.25 mm high and 1.0 mm wide is illuminated by a plane wave from an argon laser ($\lambda = 488$ nm). A lens of focal length

2.5 m forms the diffraction pattern in its back focal plane. What is the size of the central maximum?

10-18. A collimated beam ($\lambda = 600$ nm) is incident normally onto a 1.2 cm diameter, 50 cm focal-length converging lens. What is the size of the Airy disk?

10-19. From Rayleigh's criterion (see Appendix 10-C), what would be the smallest angular separation between two equally bright stars that could be resolved by the 200 in. Hale telescope? Assume $\lambda = 550$ nm.

10-20. Explain why the first minimum of

$$\mathcal{E}_a(\xi) = \frac{2\pi}{b}\left[\frac{\cos(\xi b/2)}{(\pi/b)^2 - \xi^2}\right]$$

does not occur when $\xi = \pi/b$. What is the value of E_a at $\xi = \pi/b$?

10-21. Under what conditions does the integral

$$\mathcal{E}_P = \frac{iA}{\lambda R_0} e^{-i\mathbf{k}\cdot\mathbf{R}_0} \int_{-x_0}^{x_0} f(x) e^{i\omega_x x}\, dx$$

yield the same result as the integral

$$\mathcal{E}_P = \frac{iA}{\lambda R_0} e^{-i\mathbf{k}\cdot\mathbf{R}_0} \int_{-x_0}^{x_0} f(x) e^{-i\omega_x x}\, dx$$

Appendix 10-A

FRAUNHOFER DIFFRACTION BY A LENS

We have proven that the Fraunhofer diffraction pattern of an aperture is equivalent to the Fourier transform of the aperture's amplitude transmission function **(10-14)**. We will now show that, within certain limits, the Fourier transform of the amplitude transmission function in the front focal plane of a lens appears in the lens' back focal plane. To obtain this result, we will assume that a plane wave of amplitude α is incident on the source plane (x_S, y_S) in Figure 10A-1. By using a plane wave, we reduce the number of integrations we must perform and the number of variables we must follow, but this assumption is not required to demonstrate the Fourier transform property of a lens.

If the transmission function of the source plane (x_S, y_S) is $f(x_S, y_S)$, then we can apply the Huygens–Fresnel integral to find the field distribution just to the left of the lens

$$\mathcal{E}(x_L, y_L) = \left(\frac{i\alpha}{\lambda}\right)\iint_A f(x_S, y_S)\frac{e^{-i\mathbf{k}\cdot\mathbf{R}'}}{R'}\,dx_S\,dy_S \qquad \text{(10A-1)}$$

To further simplify the mathematical bookkeeping, we will assume $f(x, y) = f(x)f(y)$ in all planes, reducing the analysis to one dimension. The distance R' from a point S in the source plane to an arbitrary point in the aperture of the lens can be approximated by

$$\begin{aligned} R' &= \sqrt{(x_S - x_L)^2 + Z'^2} \\ &\approx Z' + \frac{(x_S - x_L)^2}{2Z'} \end{aligned} \qquad \text{(10A-2)}$$

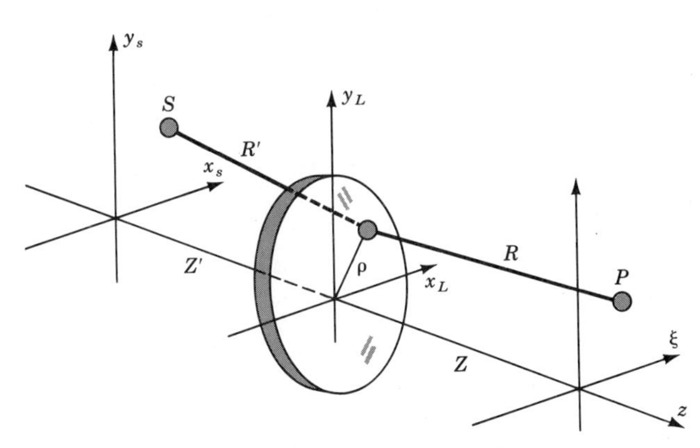

FIGURE 10A-1. Geometry for the observation of Fraunhofer diffraction by a lens.

If we use this approximate expression for R', the field at the left of the lens surface is

$$\mathcal{E}(x_L) = \sqrt{\frac{i\alpha}{\lambda Z'}} \int_{-\infty}^{\infty} f(x_s) \exp\left[-ikZ'\right] \exp\left[-ik\frac{(x_s - x_L)^2}{2Z'}\right] dx_s \qquad \text{(10A-3)}$$

Here, we have assumed that $R' \approx Z'$ in the denominator of the integrand. In the exponent, we must retain higher-order terms because of the sensitivity of the integrand to small changes in the exponent. The factor

$$\sqrt{\frac{i\alpha}{\lambda Z'}}$$

appears because we have reduced the problem to one dimension and are evaluating $E(x_L)$ instead of $E(x_L)E(y_L)$.

The lens introduces a phase shift of the wave. To obtain the phase shift, we use the geometry shown in Figure 10A-2. A plane wave is incident on the lens. After passing through the lens, the wave is a spherical wave, converging to the focal point I of the lens, a distance f (the focal length) behind the lens. From the pythagorean theorem,

$$(f - z_0)^2 + \rho^2 = f^2$$

$$2z_0 f = \rho^2 - z_0^2$$

We limit the evaluation to paraxial optics for which z_0 is small, allowing z_0^2 to be neglected (this is called the *sagittal approximation* and z_0 is the sagitta, or sag, that we first saw in Appendix 5-B).

$$z_0 \approx \frac{\rho^2}{2f}$$

The phase change introduced by the lens is then

$$\phi = kz_0 = \frac{k\rho^2}{2f}$$

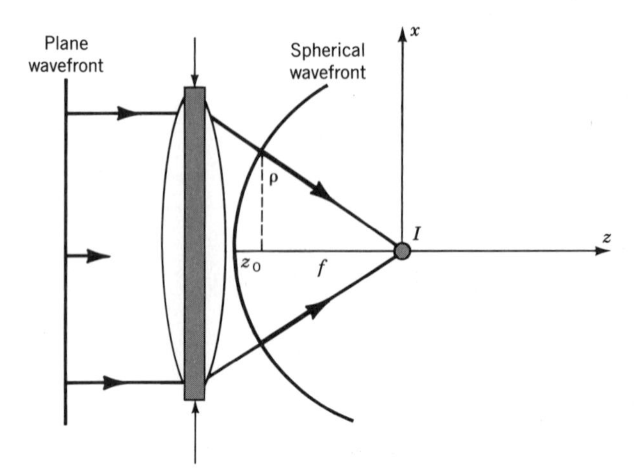

FIGURE 10A-2. Geometry for deriving the phase shift produced by a simple lens.

Looking at Figure 10A-1, we see that

$$\rho^2 = x_L^2 + y_L^2$$

where x_L and y_L are the coordinates measured in the lens plane. Because the analysis is limited to one dimension, the y_L dependence will be ignored.

The electric field, after passing through the lens, is

$$\boldsymbol{\Psi}(x_L) = \mathcal{E}(x_L) \exp\left(ik\frac{x_L^2}{2f}\right) \tag{10A-4}$$

The field at the plane (η, ξ), a distance Z from the lens, is obtained by again using the Huygens–Fresnel integral. An approximation for R, equivalent to **(10A-2)**, is used in the Huygens–Fresnel integral

$$\mathcal{P}(\xi) = \sqrt{\frac{i\alpha}{\lambda Z}} \int_{-\infty}^{\infty} \boldsymbol{\Psi}(x_L) e^{-ikZ} \exp\left[-ik\frac{(x_L - \xi)^2}{2Z}\right] dx_L \tag{10A-5}$$

Using **(10A-3)** and **(10A-4)**, we can write

$$\mathcal{P}(\xi) = \frac{i\alpha}{\lambda\sqrt{ZZ'}} e^{ik(Z+Z')} e^{-(ik\xi^2)/(2Z)}$$

$$\times \int\int_{-\infty}^{\infty} f(x_s) \exp\left[-ik\frac{x_L^2}{2}\left(\frac{1}{Z'} + \frac{1}{Z} - \frac{1}{f}\right)\right] \tag{10A-6}$$

$$\exp\left(-ik\frac{x_s^2}{2Z'}\right) \exp\left[ikx_L\left(\frac{x_s}{Z'} + \frac{\xi}{Z}\right)\right] dx_s\, dx_L$$

The integration over x_L, the lens aperture, is performed first. We will neglect the finite extent of the aperture and assume that the aperture's transmission function is a constant equal to 1. With these assumptions, the integral can be calculated by completing the square

$$\int_{-\infty}^{\infty} \exp\left[-ik\frac{x_L^2}{2}\left(\frac{1}{Z'} + \frac{1}{Z} - \frac{1}{f}\right)\right] \exp\left[ikx_L\left(\frac{x_s}{Z'} + \frac{\xi}{Z}\right)\right] dx_L$$

$$= \int_{-\infty}^{\infty} \exp\left[-ik\frac{a^2}{2}\left(x_L^2 - \frac{2b}{a^2}x_L + \frac{b^2}{a^4} - \frac{b^2}{a^4}\right)\right] dx_L$$

where

$$b = \frac{x_s}{Z'} + \frac{\xi}{Z} \quad \text{and} \quad a^2 = \frac{1}{Z'} + \frac{1}{Z} - \frac{1}{f}$$

From a table of definite integrals,

$$\int_0^{\infty} e^{-\alpha^2 u^2} du = \frac{\sqrt{\pi}}{2\alpha}$$

which allows the integral to be written as

$$2\exp\left(ik\frac{b^2}{2a^2}\right)\int_0^\infty \exp\left[-ik\frac{a^2}{2}\left(x_L-\frac{b}{a^2}\right)^2\right]dx_L=\sqrt{\frac{-i\lambda}{a^2}}\exp\left(ik\frac{b^2}{2a^2}\right)\quad(10A\text{-}7)$$

The parameters in this equation are

$$\sqrt{\frac{-i\lambda}{a^2}}=\frac{-i\lambda Z'Zf}{f(Z+Z')-ZZ'}\quad(10A\text{-}8)$$

$$(b/a)^2=\left[\frac{f}{f(Z+Z')-ZZ'}\right]\left[x_s^2(Z/Z')+\xi^2(Z'/Z)+2x_s\xi\right]\quad(10A\text{-}9)$$

Using **(10A-7)** through **(10A-9)** in **(10A-6)**, we obtain

$$\mathcal{P}(\xi)=\alpha\sqrt{\frac{if}{\lambda[f(Z+Z')-ZZ']}}\exp[-ik(Z+Z')]\exp\left[-ik(\xi^2/2Z)\right]$$
$$\exp\left\{ik\frac{f\xi^2(Z'/Z)}{2[f(Z+Z')-ZZ']}\right\}\quad(10A\text{-}10)$$
$$\times\int_{-\infty}^{\infty}f(x_s)\exp\left[-ik\,(x_s^2/2Z')\right]\exp\left\{ik\frac{fx_s^2(Z/Z')}{2[f(Z+Z')-ZZ']}\right\}$$
$$\exp\left[ik\frac{fx_s\xi}{f(Z+Z')-ZZ'}\right]dx_s$$

The exponentials involving x_s^2 under the integral in **(10A-10)**, will cancel whenever

$$\frac{1}{Z'}=\frac{(Z/Z')f}{f(Z+Z')-ZZ'}\quad(10A\text{-}11)$$

or

$$Zf=f(Z+Z')-ZZ'$$

When the quadratic dependence is removed, **(10A-10)** is within a phase factor of being equal to the Fourier transform of the source plane's amplitude transmission function $f(x_s)$. The quadratic dependence in the exponents that multiply the integral is the offending phase factor, but it can be eliminated if

$$\frac{1}{Z}=\frac{(Z'/Z)f}{f(Z+Z')-ZZ'}\quad(10A\text{-}12)$$

or

$$Z'f=f(Z+Z')-ZZ'$$

Combining **(10A-11)** and **(10A-12)** provides the requirement that $Z=Z'$ if **(10A-10)** is to reduce to a Fourier transform. The input function and its Fourier transform will be located at equal distances on either side of the lens. Using this result in either **(10A-11)** or **(10A-12)** yields

$$Zf=2Zf-Z^2$$
$$Z=f$$

With the input function located in the front focal plane of the lens, the

Fourier transform will appear in the back focal plane of the lens. To verify this statement, we will substitute $Z = Z' = f$ into **(10A-10)**

$$\mathcal{P}(\xi) = \alpha\sqrt{\frac{i}{\lambda f}}\exp[-2ikf]\int_{-\infty}^{\infty} f(x_s)\exp[\,i(k/f)x_s\xi\,]\,dx_s \tag{10A-13}$$

Defining the spatial frequency as

$$\omega_x = -\frac{k\xi}{f}$$

allows us to write

$$\mathcal{P}\left(\frac{\omega_x}{k}f\right) = \sqrt{\frac{i}{\lambda f}}\exp[-2ikf]\int_{-\infty}^{\infty} f(x_s)\exp(-i\omega_x x_s)\,dx_s$$

This is the Fourier transform relationship defined in terms of the angular frequency ω_x. Figure 10B-4 shows the formation of a Fourier transform of a periodic array of slits (called a Ronchi ruling) by a single lens, supporting experimentally the results of this derivation.

It is of interest to determine what conditions must be met for a lens to generate the Fourier transform of an aperture function when the source is a finite distance away from the lens. To simplify the calculation, we will assume the source is in plane (x_s, y_s), and the aperture function $f(x, y)$ is in plane (x_L, y_L) of the lens. With these assumptions, only minor modifications need be made to the original equation **(10A-6)**

$$\mathcal{P}(\xi) = \frac{i\alpha e^{-ik(Z+Z')}}{\lambda\sqrt{ZZ'}}\exp\left(-ik\frac{\xi^2}{2Z}\right)\iint_{-\infty}^{\infty} f(x)f(x_s)$$

$$\exp\left[-ik\frac{x^2}{2}\left(\frac{1}{Z'}+\frac{1}{Z}-\frac{1}{f}\right)\right]\exp\left(-ik\frac{x_s^2}{2Z'}\right)\exp\left[ikx\left(\frac{x_s}{Z'}+\frac{\xi}{Z}\right)\right]dx_s\,dx$$

where $f(x_s)$ is the source distribution function. As a further simplification, we assume that we have a point source at the origin of plane (x_s, y_s)

$$f(x_s) = \delta(x_s)$$

The integration over the source plane can now be accomplished using the sifting property of the delta function

$$\mathcal{P}(\xi) = \frac{i\alpha}{\lambda\sqrt{ZZ'}}\exp[-ik(Z+Z')]\exp\left(-ik\frac{\xi^2}{2Z}\right)\int_{-\infty}^{\infty} f(x)$$

$$\times\exp\left[-ik\frac{x^2}{2}\left(\frac{1}{Z'}+\frac{1}{Z}-\frac{1}{f}\right)\right]\exp\left(ik\frac{x\xi}{Z}\right)dx \tag{10A-14}$$

To cancel the quadratic component of the exponent in the integrand, we require

$$\frac{1}{Z'}+\frac{1}{Z}=\frac{1}{f}$$

This relationship can be recognized as the lens equation **(5A-1)** if we apply the sign convention of Chapter 5, $Z' = -S$ and $Z = S'$. The source and Fourier transform planes are conjugate planes, related as object and image planes with respect to the lens.

Another interpretation can be given to **(10A-14)**, based on linear system theory. An imaging system is assumed to be a linear system. The impulse response of the linear system is equal to the output of the system when a delta function is the input function. Here, a point source is used as the input to the imaging system, of focal length f, and the impulse response of the imaging system is shown by **(10A-14)** to be the Fourier transform of the aperture function of the imaging system.

If we use the linear system viewpoint, the image produced by an optical system will be given by the convolution of the aperture function of the lens with the object. Equivalently, the Fourier transform of the image could be described in terms of the product of the Fourier transforms of the aperture function and object. This interpretation makes the importance of diffraction on the imaging process apparent. The linear system formulation of the imaging process will be discussed in Appendix 10-C.

Appendix 10-B

In 1871, while developing a theory of the microscope for Karl Zeiss, Abbe introduced a view of coherent imaging that has led to the field of optics called *optical signal processing*. Abbe's theory assumes the object to be imaged can be decomposed into a number of elemental gratings; each grating diffracts light at an angle that is a function of the grating's period and orientation. The diffracted light beams are plane waves that are focused by the imaging lens to diffraction patterns of light in the lens' back focal plane. The shapes of the diffraction patterns are determined by the aperture of the imaging lens. The diffraction patterns act as sources of waves that propagate, from the focal plane, to the image plane where they interfere to produce the image.

In 1896 Rayleigh developed a similar approach to coherent imaging that assumes that the object can be treated as a large number of point sources. The imaging lens images each of the points into a diffraction pattern, whose shape is determined by the lens aperture, in the image plane. The diffraction patterns interfere at the image plane to produce the image. In the following discussion, only Abbe's theory will be addressed.

Once Abbe's theory of imaging has been developed, a few applications to the field of signal processing will be discussed; all are based on the premise of Abbe's theory that the Fourier transform of the object is present in the back focal plane of an imaging lens. The applications to be discussed modify the Fourier transform of an object by

1. Changing its amplitude distribution.
2. Changing its phase distribution.
3. Operating on both the phase and amplitude of the transform.

The objective of two of the applications is to modify the image produced by the optical system; thus, they could be called image processing applications. The objective of the third application is to recognize and locate a particular object in the field of view of the optical system; thus, it could be called a pattern recognition application.

We will assume the light is both temporally and spatially coherent. The demands on temporal coherence made by the theory are much less stringent than the demands on spatial coherence. We will mention the different coherence requirements during the development of the theory.

ABBE'S THEORY OF IMAGING

A simple lens, shown in Figure 10B-1, is used to image an extended object. The extended object is located a distance $-S$ to the left of the lens and the image of the object is located a distance S' to the right of the lens.

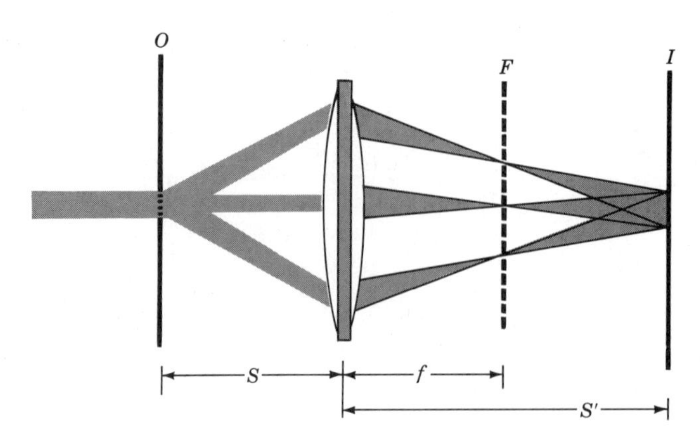

FIGURE 10B-1. Abbe's theory of imaging.

We have already developed in Appendix 10-A the necessary Huygens–Fresnel integral **(10A-6)** for the electric field distribution at the back focal plane of the lens. The field distribution is obtained by setting $Z' = -S$ and $Z = f$ in **(10A-10)**, yielding the field at the back focal plane of the lens

$$\mathcal{P}(\xi) = \alpha\sqrt{\frac{i}{\lambda f}}\exp\left[-ik(f-S)\right]\exp\left[-ik\left(\frac{\xi}{f}\right)^2\left(\frac{f+S}{2}\right)\right]$$

$$\int_{-a/2}^{a/2} f(x_s)\exp\left[ikx_s(\xi/f)\right]dx_s \quad (10B\text{-}1)$$

where the limits of integration are established by the size of the object. We will normally let those limits go to infinity by assuming that no light is produced outside the boundary of the object. (The analysis we will conduct assumes that the limiting aperture in the experiment is the object's spatial extent, a condition quite easy to obtain experimentally. The result of this assumption is that the lens' transmission function will be neglected in the analysis.) Using the definition **(10-13)**, we can write

$$\mathcal{P}(\omega_x) = \alpha\sqrt{\frac{i}{\lambda f}}e^{-ik(f-S)}\exp\left(-i\omega_x^2\frac{f+S}{2}\right)\int_{-\infty}^{\infty} f(x_s)e^{-i\omega_x x_s}dx_s \quad (10B\text{-}2)$$

Equation **(10B-2)** demonstrates that the back focal plane of the lens contains the Fourier transform of the object. This is one of the two basic ideas contained in Abbe's theory. We decompose the light from the object into a set of plane waves that are focused to Airy disks (we will assume the limiting aperture is circular) in the back focal plane of the lens. The location of an Airy disk in the focal plane, as we saw in Chapter 10, is a function of the angle the plane wave makes with the optical axis.

An equivalent view of this first step in the imaging process is that the object is made up of a number of sinusoidal gratings, each with a different period and amplitude. Each grating diffracts light into a plane wave, at an angle that is a function of the period and orientation of the grating. The plane waves produced by each grating are then focused to a set of Airy disks by the lens. Figure 10B-1 illustrates the imaging process when the object is a simple sinusoidal grating.

The second fundamental idea contained in Abbe's theory is that each Airy disk in the back focal plane of the lens acts as a source, producing

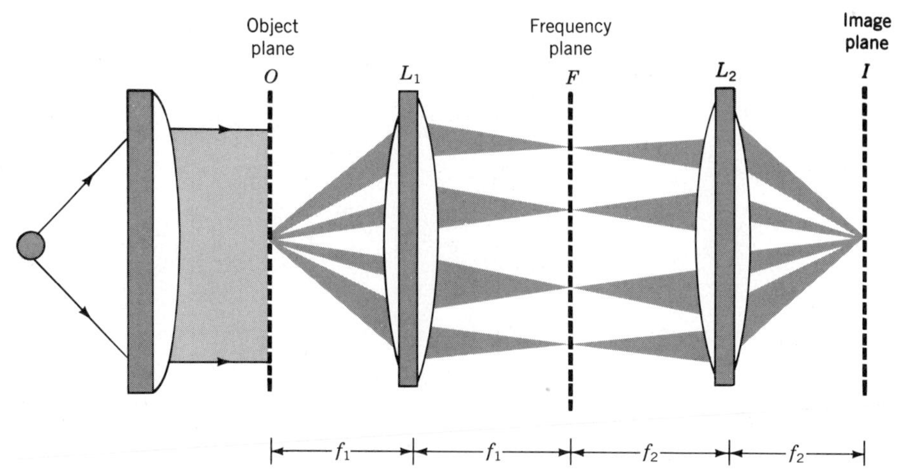

FIGURE 10B-2. Two-lens imaging system. The first lens is to produce a plane wave that illuminates the object. Here, the object is assumed to be a transparency.

a spherical wave. These spherical waves interfere at the image plane to produce the image of the object. We will not write out the Huygens–Fresnel integral for this second step, but instead refer the interested reader to the book by Klein.[5]

To understand the ideas contained in Abbe's theory and see how they might be applied, it is useful to replace the simple lens of Figure 10B-1 with a pair of lenses (see Figure 10B-2). There are three lenses in Figure 10B-2 but the first lens is present only to produce a plane wave to illuminate the object. We will ignore this lens in our discussion.

Assume for this discussion that both lenses in the two-lens system, shown in Figure 10B-2, have the same focal length, making the magnification of the system unity, $\beta = 1$ (see Problem 5-12). The object plane for the two-lens system is the front focal plane of lens L_1 and the image plane, which contains an inverted replica of the object, is the back focal plane of lens L_2.

We will follow in a stepwise fashion the light as it propagates from the object to the image plane in Figure 10B-2. The lens L_1 produces a Fourier transform of the object in plane F, the back focal plane of L_1, designated as the *frequency plane*. The light distribution in the frequency plane acts as the source for lens L_2, which produces the Fourier transform of the frequency plane at plane I. Thus, the field distribution in plane I is the Fourier transform of the Fourier transform of the object. This light distribution is the image of the object.

A logical question to ask is what would happen to the image of the object if we modified the light distribution in the frequency plane F. To answer this question, we will examine a simple sinusoidal grating with amplitude $d \leq 1$ and of finite extent in the object plane O of Figure 10B-2.

$$f(x, y) = \begin{cases} \dfrac{1}{2} + \dfrac{d}{2} \cos \omega_0 x, & |x| \leq \dfrac{\ell}{2} \quad |y| \leq \dfrac{\ell}{2} \\ 0, & \text{all other } x \text{ and } y \end{cases} \tag{10B-3}$$

The multiplicative factor d is known as the *amplitude modulation* of the grating. Because in optics we measure intensity rather than amplitude, it is more useful to define object *contrast*

$$C = \frac{I_{\text{max}} - I_{\text{min}}}{I_{\text{max}} + I_{\text{min}}}$$

For the object **(10B-3)**, the contrast is

$$C_0 = \frac{2d}{1 + d^2} \approx 2d$$

We will have more to say about contrast in Appendix 10-C.

The source must have a spatial coherence length larger than the width and height of the aperture, defined by the grating. If the source meets the spatial coherence requirement, the amplitude distribution in plane F of Figure 10B-2 will be

$$\mathcal{E}(x_0, y_0) = \frac{i\ell^2}{2\lambda f} e^{-2ikf} \operatorname{sinc}\left(\frac{\ell y_0}{\lambda f}\right) \times \left\{ \operatorname{sinc}\left(\frac{\ell x_0}{\lambda f}\right) + \frac{d}{2}\left[\operatorname{sinc}\frac{\ell}{\lambda f}\left(x_0 + \frac{\omega_0 f}{k}\right) + \sin\frac{\ell}{\lambda f}\left(x_0 - \frac{\omega_0 f}{k}\right)\right]\right\} \tag{10B-4}$$

The intensity distribution in plane F is

$$\begin{aligned} I(x_0, y_0) = &\left(\frac{\ell^2}{2\lambda f}\right)^2 \operatorname{sinc}^2\left(\frac{\ell y_0}{\lambda f}\right) \\ &\times \left\{ \operatorname{sinc}^2\left(\frac{\ell x_0}{\lambda f}\right) \right. \\ &+ \frac{d^2}{4}\operatorname{sinc}^2\frac{\ell}{\lambda f}\left(x_0 + \frac{\omega_0 f}{k}\right) \\ &\left. + \frac{d^2}{4}\operatorname{sinc}^2\frac{\ell}{\lambda f}\left(x_0 - \frac{\omega_0 f}{k}\right)\right\} \end{aligned} \tag{10B-5}$$

A schematic representation of the process described by **(10B-5)** is shown in Figure 10B-3. The distribution described by **(10B-5)** is also the origin of the graphic display in Figure 10B-1, where an object diffraction pattern, located in plane F of the lens, is shown to consist of a light spot on the optical axis and a spot above and below the optical axis.

The multiplier of the quantity in braces on the right side of **(10B-5)**

$$\operatorname{sinc}^2\left(\frac{\ell y_0}{\lambda f}\right)$$

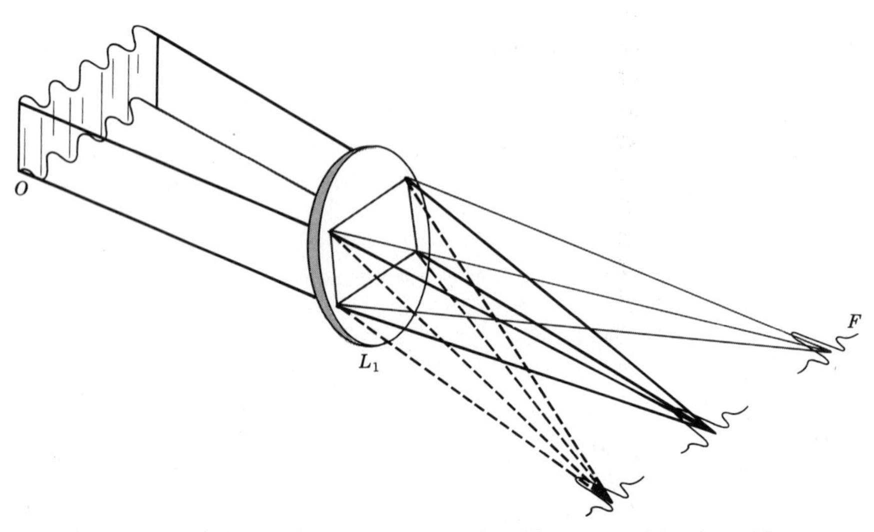

FIGURE 10B-3. Fraunhofer diffraction pattern produced by a sinusoidal grating of finite size.

gives the diffraction intensity distribution due to the aperture dimension of the grating in the y direction. The first term within the braces of **(10B-5)** involving

$$\operatorname{sinc}^2\left(\frac{\ell x_0}{\lambda f}\right)$$

is the diffraction intensity due to the aperture dimension of the grating in the x direction. The product of these two terms gives the intensity distribution on the optical axis. This intensity distribution is often called the *dc component* or the *zero order*. It corresponds to the zero-order principal maximum of a grating and equals the spatial average of the transmission function of the grating.

The sinusoidal transmission variation of the grating produces two additional side patterns (called the *first-order components*) or spatial sidebands, located a distance $\omega_0 f/k$ on either side of the dc component

$$\operatorname{sinc}^2\left[\frac{\ell}{\lambda f}\left(x_0 \pm \frac{\omega_0 f}{k}\right)\right]$$

The intensity distribution of each of these spots is identical to the distribution of the dc spot and is determined by the shape of the grating aperture.

If the grating were not sinusoidal but instead represented by a periodic square wave, then the resulting diffraction pattern would consist of a large number of sinc functions as shown in Figure 10B-4. This behavior is consistent with the interpretation of the diffraction pattern as a Fourier transform, or for an infinite grating, a Fourier series. The Fourier series of the sinusoidal grating has only one term, but the Fourier series of a square wave would be an infinite series of terms involving harmonics of the fundamental frequency of the square wave; see **(6-9)**.

The maximum amplitude transmission in the object plane occurs when the amplitude transmission function, $\cos \omega_0 x$ is a maximum; the separation between each maxima is equal to the period of the cosine function

$$\Delta x = \frac{2\pi}{\omega_0}$$

We will denote Δx as the *resolution* of the object. The periodic variations in transmission associated with the grating structure result in large-scale variations in the frequency plane, that is, the three sinc functions in **(10B-5)**. The large-scale variations in the object plane, dictated by the rectangular aperture that limits the grating, result in the small-scale variations in the frequency plane, that is, the sinc distribution at each spatial frequency.

The size of a diffraction pattern will change with the wavelength. If the temporal coherence of the source is small and a frequency spectrum of light is present, then the diffraction pattern formed by light on the blue end of the spectrum is smaller than the pattern formed by light on the red end, causing the sidebands for each wavelength to occur at different positions. Figure (10B-5*a*) shows the Fourier transform of a nested set of triangles, illuminated by a temporally incoherent light source. The spread in wavelengths present in the white light used to create Figure 10B-5*a* causes the spots of light in the photo to appear much larger than would be predicted by the size of the set of triangles. (In a color reproduction of Figure 10B-5*a*, shown in the insert,

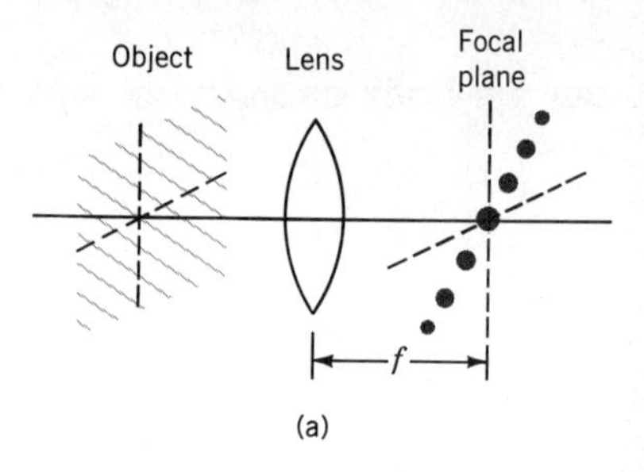

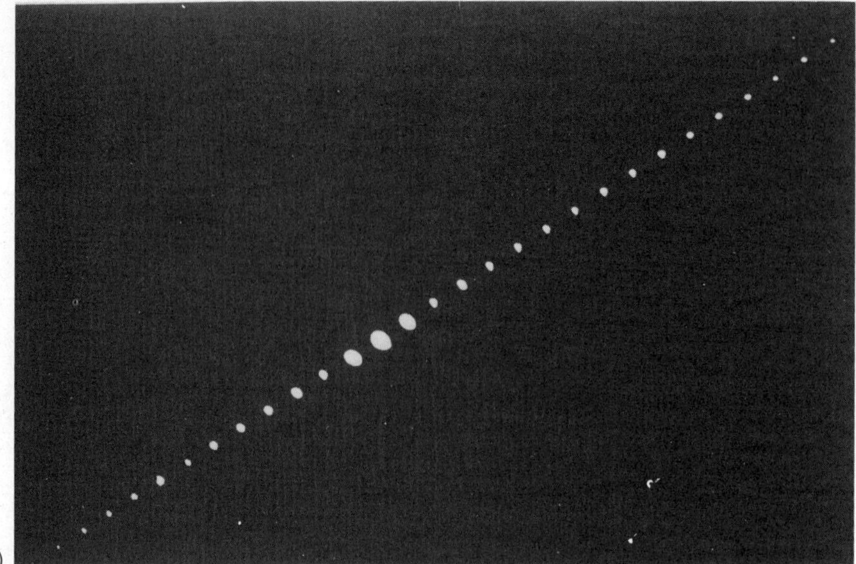

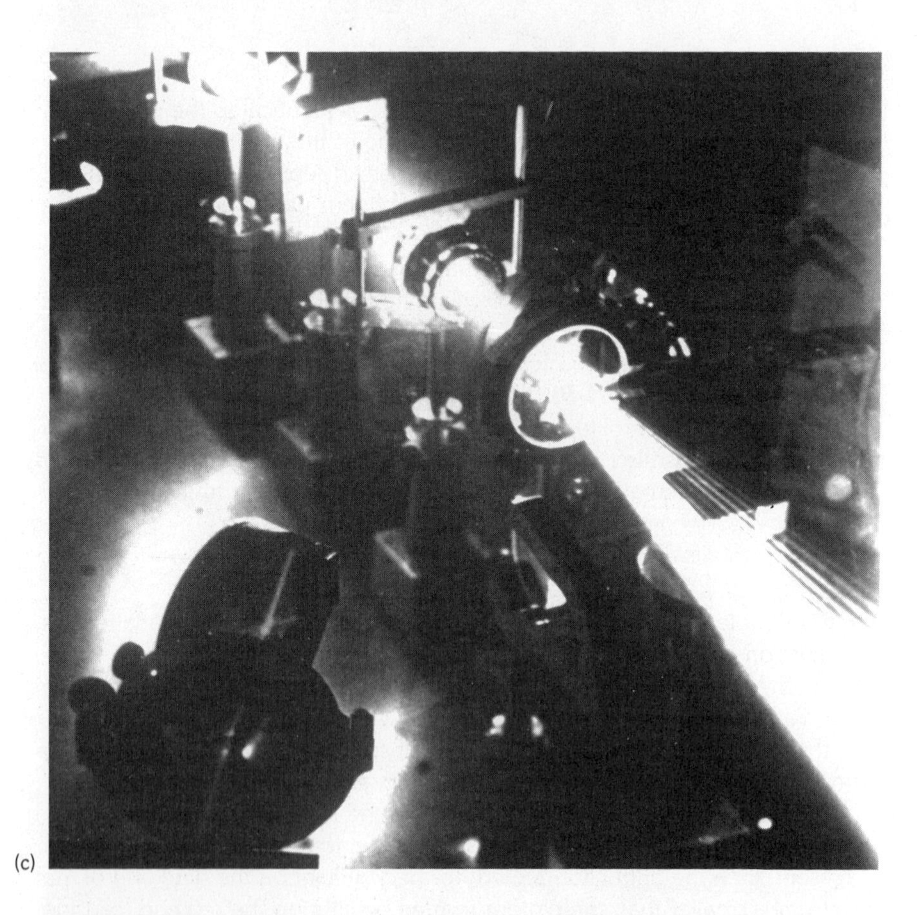

FIGURE 10B-4. A Ronchi ruling is a periodic array of opaque lines separated by clear regions of the same width. The transmission function of the Ronchi ruling is a square wave. The Fourier transform of a square wave is a set of harmonic functions. The diffraction pattern of the Ronchi ruling thus should be a set of bright spots whose intensity corresponds to the coefficients of the harmonic functions of the Fourier transform. If the Ronchi ruling is placed in the front focal plane of a lens (a), then the intensity distribution shown in (b) is observed in the back focal plane. In (c), we see the formation of the spots of light by the lens. If a screen is placed so it intercepts the beams of light seen in (c), then the image of (b) is observed. (Figure 10B-4c is courtesy of Robert Leighty, U.S. Army Engineering Topographic Laboratory.)

(a)

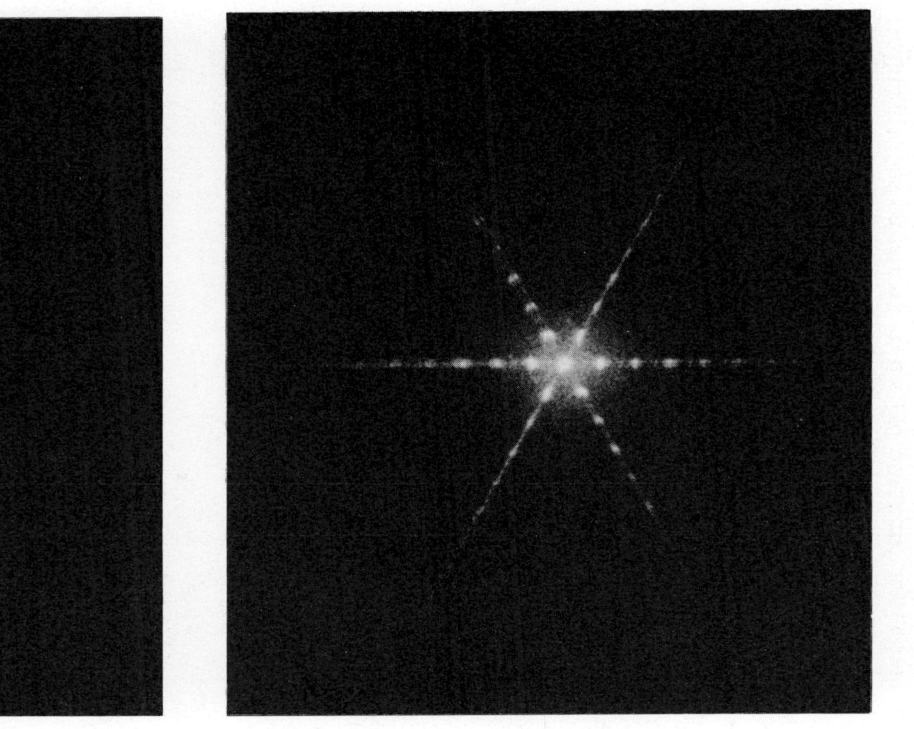

(b)

FIGURE 10B-5. The Fraunhofer diffraction pattern produced by a nested set of triangles that form three sets of gratings oriented at 120° with respect to each other. (a) The illuminating light was spatially coherent white light. The spots are very broad due to the spread in wavelengths in the source. In the colored photo in the insert, blue light can be seen at one side of each spot and red light at the other. (b) The illuminating light producing this diffraction pattern contained only a single wavelength, 488 nm from an Argon laser. (Figure *a* courtesy N. George, University of Rochester.)

the inner edges of the sideband spots are blue and the outer edges are red.) Figure 10B-5*b* (and its color representation in the insert) shows an identical Fraunhofer diffraction pattern of the nested set of triangles obtained using a single wavelength of light. A comparison of the spots of light in Figures 10B-5*a* and 10B-5*b* shows the effect of temporal coherence on the spots that make up the diffraction pattern.

We can establish the wavelength spread that would destroy the diffraction pattern observed in Figure 10B-4*a* by assuming that the sinc functions for the two extremes in wavelength, λ_1 and λ_2, are separated by the width of the sinc function of λ_1. With this assumption, the allowable spread in wavelength is

$$\frac{\Delta\lambda}{\lambda} = \frac{\lambda_1 - \lambda_2}{\lambda_1} = \frac{4\pi^2}{\ell\omega_0} = \frac{2\pi\Delta x}{\ell} \tag{10B-6}$$

This result states that as the resolution in the object increases (i.e., as the value of Δx decreases) or the size of the object ℓ increases, the temporal coherence of the source must increase if the diffraction pattern is still to be observed.

Suppose an aperture is placed in plane F of Figure 10B-2 with its symmetry axis aligned with the optical axis of the system. If the size of the aperture, denoted by D, is too small to allow the two sidebands of **(10B-5)** to pass on to plane I, we would not see the sinusoidal grating at the image plane. Stated mathematically, if

$$D < \frac{2\omega_0 f}{k} + \frac{2\pi\lambda f}{\ell} = \frac{2f\lambda}{\Delta x} + \frac{2\pi\lambda f}{\ell}$$

then the optical system cannot resolve the grating. If D is less than this dimension, the light distribution in the image plane is a constant, equal to the Fourier transform of

$$\operatorname{sinc}\left(\frac{\ell y_0}{\lambda f}\right)\operatorname{sinc}\left(\frac{\ell x_0}{\lambda f}\right)$$

Any aperture less than the above dimension, but larger than a minimum dimension of

$$D = \frac{2\pi\lambda f}{\ell}$$

yields an image that is a uniform light distribution over a rectangle the size of the grating. If a small aperture, equal to the minimum dimension, could be moved up or down in the frequency plane so that it allowed either of the sidebands to pass, we would find minimal change in the appearance of the image plane. Because each of the three components is a simple sinc function, the light distribution in plane I, when any one of the three components is allowed to pass is the image of the rectangular aperture with uniform illumination, i.e., a plane wave with limited extent. If the zero order is passed through, the plane wave will propagate parallel to the z axis, whereas if a sideband is passed through, the plane wave will propagate at an angle with respect to the z axis.

Rayleigh postulated a criterion for resolution, otherwise known as *the Rayleigh criterion*, which states that for an imaging system with a circular aperture

Two Airy functions in the image plane can be just resolved if the central maximum of one function falls on the first zero of the other function.

This postulated criterion is in agreement with the relationship between resolution and beam waist established by the theories developed here. (The Rayleigh criterion is discussed in more detail in Appendix 10-C.)

By neglecting the effect of a finite-sized grating, the minimum diameter required to resolve a grating with resolution Δx can be simplified to

$$D = \frac{2f\lambda}{\Delta x} \tag{10B-7}$$

This result should be compared to the diameter of an optical system needed to produce a Gaussian beam waist of w, calculated in Chapter 9,

$$D = \frac{2\lambda f}{w}$$

The minimum resolution of an optical system is associated with the minimum spot size that the optical system can produce.

We now set the size of the aperture in plane F equal to the value

$$D = \frac{f\lambda}{\Delta x} + \frac{2\pi\lambda f}{\ell}$$

and move its symmetry axis above the optical axis, so that only the zero order and one sideband are allowed to pass through to the image plane. This configuration reduces the number of diffraction patterns in the frequency plane to two. The two diffraction patterns, sinc functions in the example we are considering, act as point sources. The transition from the frequency plane to the image plane is equivalent to Young's experiment.

In the image plane, light from the two sources will interfere to produce a sinusoidal interference pattern. From **(4-17)**, the period of this interference pattern is

$$T = \frac{\lambda D}{h}$$

yielding a spatial frequency of the sinusoidal interference pattern equal to

$$\omega = \frac{2\pi}{T} = \frac{kh}{D}$$

In the example we are considering, the distance to the observation plane is given by $D = f$ and the point source separation is

$$h = \frac{\omega_0 f}{k}$$

Thus, the spatial frequency of the interference pattern is ω_0, the frequency of the input object.

The peak intensities of the two sources in plane F are

$$I_1 = \left(\frac{\ell^2}{2\lambda f}\right)^2, \qquad I_2 = \left(\frac{\ell^2}{2\lambda f}\right)^2 \frac{d^2}{4}$$

From **(4-13)**, we can write the intensity distribution in the image plane I as

$$I = \left(\frac{\ell^2}{2\lambda f}\right)^2 \left[\left(1 + \frac{d^2}{4}\right) + d \cos \omega_0 x\right] \tag{10B-7}$$

Previously, the contrast of the object was found to be $C_0 = 2d$, and the contrast of the image is given by

$$C_I = \frac{d}{1 + (d/2)^2} \approx d$$

The image of the object retains the object's spatial frequency content, but at a reduced contrast.

If we neglect the term associated with the size of the grating, we find that the aperture diameter required to resolve the grating is half the size of the aperture needed to pass both sidebands. (This is equivalent to single sideband transmission in communication theory.) If the noise associated with the object is small enough to allow detection of the image at the reduced contrast, then the aperture (or equivalently, the spatial bandwidth) of the imaging system can be reduced by one-half with no loss in resolution.

Instead of displacing the aperture in the frequency plane above the optical axis, we could illuminate the object with a plane wave incident from below the optical axis at an angle γ; see Figure 10B-6.

$$\gamma = \frac{x_0}{f} = \frac{\omega_0}{2k} = \frac{\lambda}{2\Delta x}$$

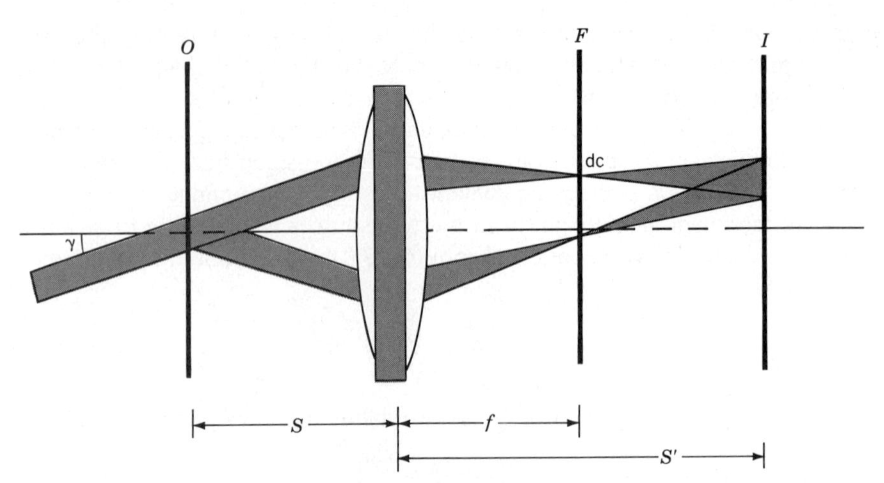

FIGURE 10B-6. Off-axis illumination allows an aperture to provide the same resolution as an aperture twice its size operating with illumination parallel to the optical axis.

The off-axis illumination causes the diffraction pattern in the frequency plane to move upward, half the distance between the dc and first-order sideband

$$x_0 = \frac{\omega_0 f}{2k}$$

For the off-axis illumination, shown in Figure 10B-6, an aperture of size

$$D_\theta = \frac{f\lambda}{\Delta x}$$

centered on the optical axis will pass the dc component and first-order sideband. We therefore find that by illuminating the object at an angle with respect to the optical axis, the imaging system is able to resolve the object grating with an aperture only one-half the size needed to resolve the same grating with illumination incident normal to the grating.

If the dc is prevented from passing through the system by placing a ring aperture in plane F, then we again simulate Young's experiment. Now, however, the spacing of the two interfering sources is

$$h = \frac{2\omega_0 f}{k}$$

The spatial frequency of the interference pattern for this case is

$$\omega = \frac{kh}{f} = 2\omega_0$$

The spatial frequency is twice the object's spatial frequency, and the contrast of the image is independent of the object's contrast. This is the appearance of false detail described first by Abbe.

The origin of the false detail we have discovered is quite easy to understand mathematically. The original object has an amplitude transmission that is a sinusoidal amplitude variation with a bias term added, as is shown on the left

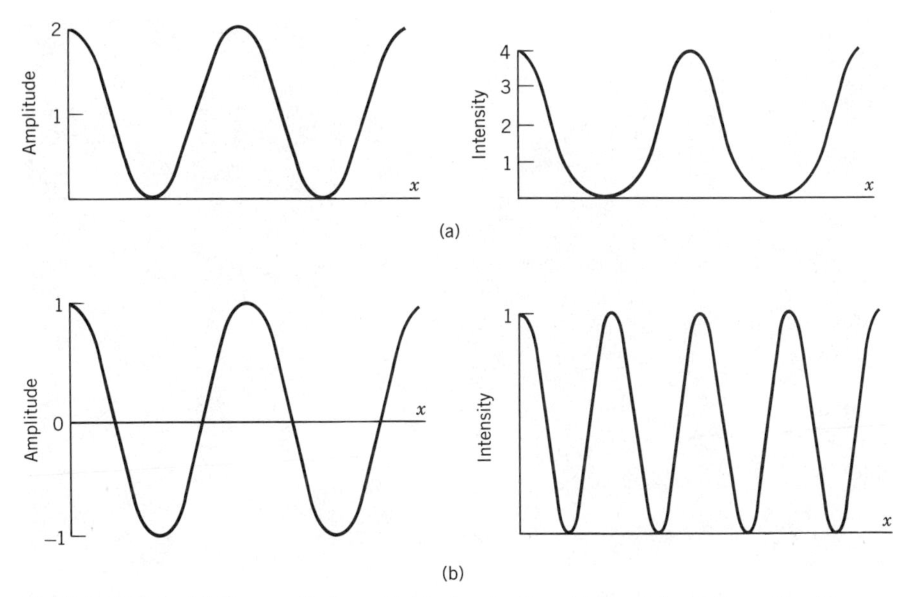

FIGURE 10B-7. (a) The amplitude and intensity of a sinusoidal spatial variation with a bias term added. (b) The amplitude distribution of a sinusoidal grating with no bias term and the observed intensity. The false resolution can be seen in the plot of the intensity shown on the right of the figure.

of Figure 10B-7*a*. When the dc and sidebands are retained, then the image, which is detected as the square of the amplitude at plane *I*, is a function with the same period as the object, as is shown on the right of Figure 10B-7*a*. The detected image has the same spatial period as the original object. If the dc component is removed in the frequency plane, then at the image plane the spatial variation of the amplitude of the image is the waveform shown on the left of Figure 10B-7*b*. The intensity distribution of the image, shown on the right of Figure 10B-7*b*, has twice the spatial period as the original object.

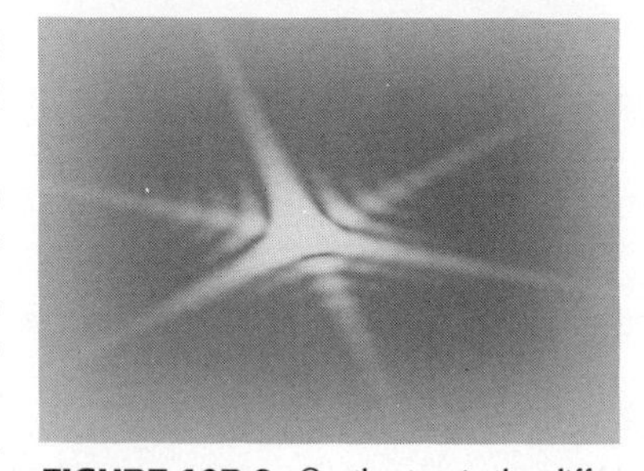

FIGURE 10B-8. On the top is the diffraction pattern produced by an equilateral triangle and on the bottom is the diffraction pattern produced by a scalene triangle. Both diffraction patterns have the threefold symmetry of a triangle, but note the reduction in the symmetry of the diffraction pattern from the scalene triangle. (Courtesy T. Stone, University of Rochester.)

The ideas of Abbe were not well received, particularly the statement that false detail could appear in an image. A.B. Porter was able to construct experiments, without the objectionable components of Abbe's experiments, that demonstrated Abbe's conclusions. Today, a number of applications utilize Abbe's theory and demonstrate its accuracy. As we will see in our discussion, Abbe's concepts allow the methods of signal processing, used in communication theory, for one-dimensional signals to be applied to two-dimensional signals, such as images.

The intensity distribution found in the frequency plane reflects the symmetry of the object, as can be seen in Figure 10B-8. In Figure 10B-8*a* the diffraction pattern of an equilateral triangle is shown. This pattern should be compared to the diffraction pattern of Figure 10B-8*b*, produced by a scalene triangle. It is possible to use the diffraction patterns produced by different areas of a complex scene to rapidly sort aerial images, as is illustrated by the images contained in Figure 10B-9. Of particular interest is the diffraction patterns produced by surface waves on the ocean. From the orientation and scale of this diffraction pattern, the wave pattern can be analyzed.

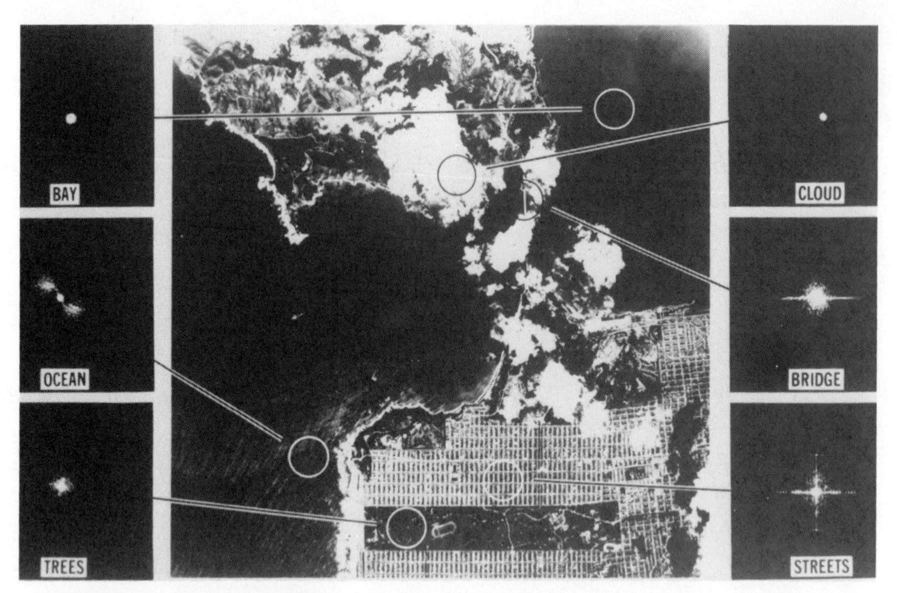

FIGURE 10B-9. An aerial photograph is surrounded by Fraunhofer diffraction patterns taken of small areas in the photo. The symmetry and extent of the diffraction pattern can be used to recognize geographical classes in the photo. For example, the diffraction pattern of the ocean can be used to analyze the wave structure on the ocean surface. (Courtesy of Robert Leighty, U.S. Army Engineering Topographic Laboratory.)

AMPLITUDE SPATIAL FILTERING

A more complicated object than the sinusoidal one we have considered would contain many more frequencies but the analysis, just completed, would still apply. An aperture in the frequency plane would restrict the maximum frequency that could pass through the imaging system, limiting the resolution of the resulting image. If we use the terminology of communication theory, the aperture would act as a *low-pass filter*. An example of low-pass spatial filtering of an image is shown in Figure 10B-10*a*. In the figure, a resolution test target and an enlargement of a small section of the target are shown at the top. The imaging system, as constructed, had no problem resolving 10 lines/mm. In fact, the spatial bandpass of the imaging system was 63 lines/mm. The lower pictures in Figure 10B-9 are identical to the upper ones, but now the imaging system has a circular aperture, centered on the optical axis, in the frequency plane that reduces the spatial bandwidth of the imaging system from 63 to 6 lines/mm by blocking the passage of the high spatial frequencies. After the reduction in spatial bandwidth, the imaging system cannot resolve the 8 lines/mm pattern.

A disk-shaped obstruction placed on the optical axis would act as a *high-pass filter*, preventing low spatial frequencies from contributing to the image. High-pass filtering with such an obstruction is shown in Figure 10B-10*b*. The imaging system, with no obstructions in the frequency plane, had a spatial bandwidth of 63 lines/mm, as shown in the upper left-hand image of Figure 10B-10*b*. With a disk centered on the optical axis in the frequency plane, the spatial bandwidth of the imaging system is reduced by restricting its low spatial frequency response. Three different spatial bandwidths are shown in Figure 10B-10. The low frequency bars are eliminated by the high-pass filter, but their edges are still observable. In image process-

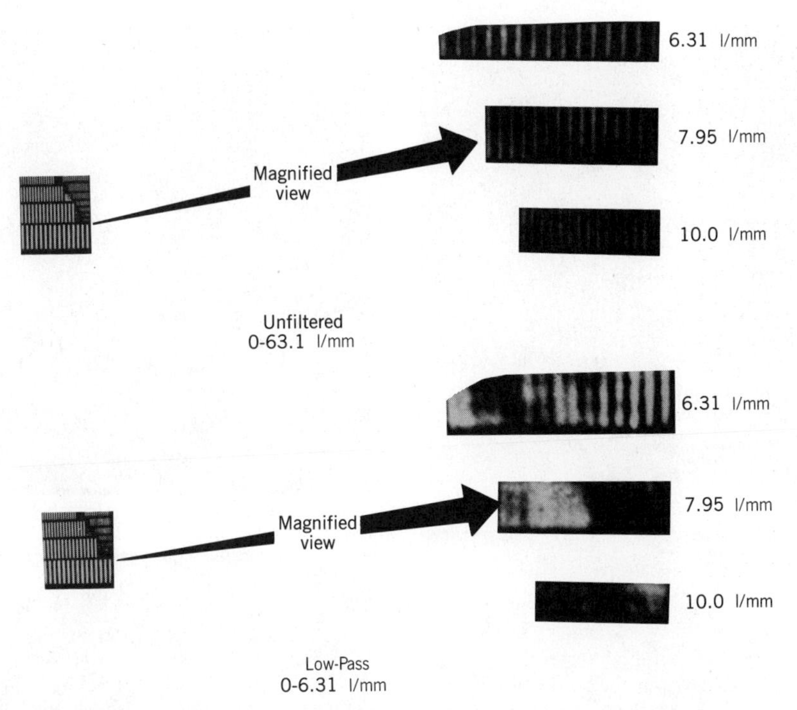

FIGURE 10B-10*a*. Optical low-pass filtering performed by introducing a circular aperture in the frequency plane of an imaging system similar to the one shown in Figure 10B-2.

ing, this is called *edge enhancement*. The high spatial frequencies, needed to describe the rapidly changing transmission, at the edge of an object are allowed to pass through the system; the slowly varying transmission from the bulk of the object, described by low spatial frequencies, is stopped by the disk.

A frequent application of amplitude spatial filtering is the use of a simple circular aperture to smooth the output beam of a laser. The fundamental (TEM_{00}) mode of a laser, using a conventional Fabry–Perot resonator as the cavity, is a Gaussian wave with an amplitude

$$E = A \exp\left(-\frac{x^2 + y^2}{\rho^2}\right)$$

where A and ρ are the wave constants that were defined in Chapter 9; see **(9-46)**. The Fourier transform of the Gaussian wave is

$$\mathcal{F}\{E\} = A\pi\rho^2 \exp[-\pi^2\rho^2(\xi^2 + \eta^2)]$$

This result assumes an ideal beam; however, the laser beam departs from this simple theoretical function due to imperfections in the laser. We can treat the departures from the ideal beam by adding a correction $E_r = E + \Delta E$ to the laser's light field. The fluctuations described by ΔE will create spatial frequencies at much higher values than $1/\rho$. If we place in the back focal plane of a lens of focal length f an aperture with a diameter of approximately $\lambda f/\rho$, then we can eliminate the fluctuations at high spatial frequencies and generate an ideal beam.

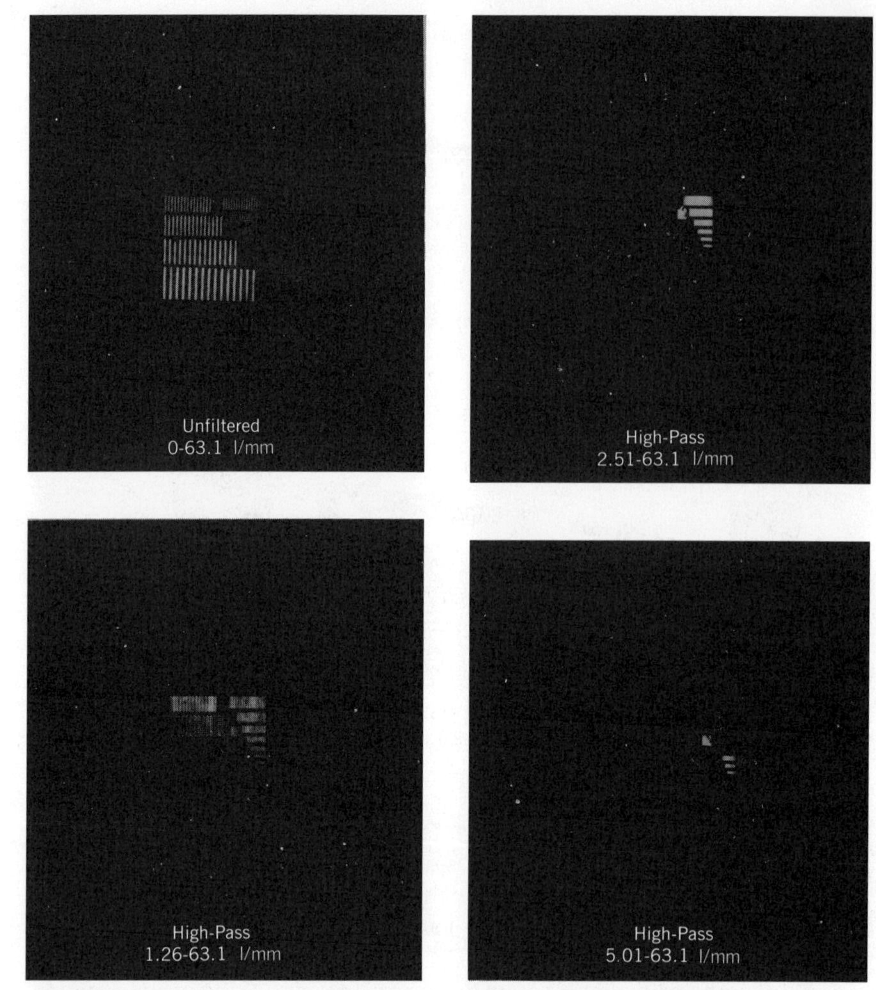

FIGURE 10B-10*b*. Optical high-pass spacial filtering accomplished by placing a disk in the frequency plane of the imaging system shown in Figure 10B-2. The upper left-hand image is identical to the upper, left-hand image of (a) and displays the maximum performance of the imaging system. The other three pictures show images produced for the three disks in the frequency plane whose diameters were in the ratio of 1:2:4.

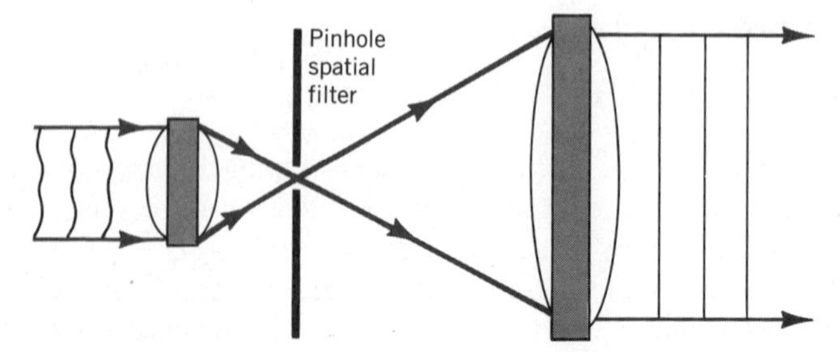

FIGURE 10B-11. Spatial filter to produce a plane wave. A laser beam strikes a microscope objective from the left. The beam's amplitude departs from a uniform amplitude wave as shown in the figure. The wave is focused down to a spot by the microscope objective and a pinhole is positioned to allow only the dc component to pass on to a second lens. Often, the focal length of the second lens is selected to produce a larger-diameter beam of light.

A typical experimental arrangement used to produce a plane wave is shown in Figure 10B-11. A laser beam strikes a microscope objective from the left. The wave is focused down to a spot by the microscope objective and a pinhole is positioned so that it allows only the dc component to pass on to a second lens. As is shown schematically in Figure 10B-11, the beam's wavefront departs from a plane wave, but the pinhole removes the higher spatial frequencies contained in the initial beam to produce a final beam with a plane wavefront. Figure 10B-12 shows an image of a beam before (on the top) and after spatial filtering. The focal length of the second lens is usually much larger than the microscope objective in order to magnify the diameter of the beam.

APODIZATION

The amplitude distribution of a wave, after passing through an aperture, has a central maximum but also has a number of secondary maxima, called sidelobes in radio frequency antenna design. It is desirable, when operating an imaging system near its diffraction-limiting resolution, to reduce the sidelobes. This can be done by a process known as *apodization* (apodization is derived from the Greek word meaning foot, and has as its objective the removal of the secondary diffraction maxima produced by an aperture with sharp boundaries).

We have just discussed modifying a beam by removing spatial frequencies with an opaque screen in the spatial frequency plane. By modifying the diffracting aperture's edge, the secondary maxima contained in the Fraunhofer diffraction pattern can also be modified. A simple example will make this application of spatial filtering clear.

Assume the aperture of the optical system is a simple slit and ignore the y dependency

$$f(x) = \begin{cases} 1, & -\dfrac{b}{2} \le x \le \dfrac{b}{2} \\ 0, & \text{all other } x \end{cases}$$

The Fraunhofer diffraction pattern produced by this aperture is equal to the Fourier transform of the aperture

$$F(\omega_x) = \int_{-b/2}^{b/2} e^{-i\omega_x x}\, dx = \frac{b \sin \omega_x b/2}{\omega_x b/2}$$

a result identical to that obtained in Chapter 10; see **(10-19)**. We now apodize the aperture by changing its transmission function to a cosine function

$$f(x) = \begin{cases} \cos \dfrac{\pi x}{b}, & -\dfrac{b}{2} \le x \le \dfrac{b}{2} \\ 0, & \text{all other } x \end{cases}$$

The Fraunhofer diffraction pattern produced by this aperture is

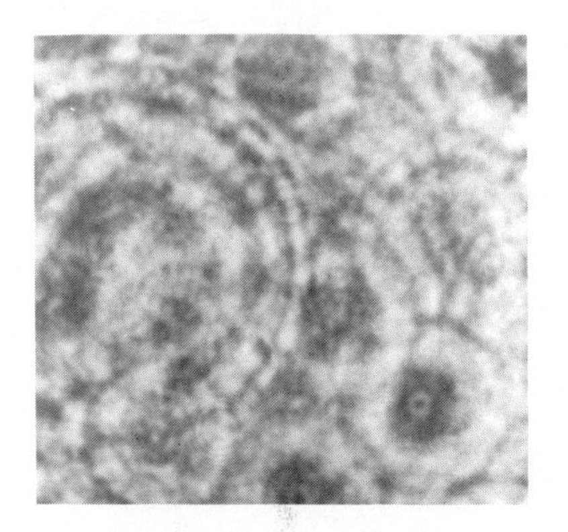

FIGURE 10B-12. A laser beam before spatial filtering is shown on the top and after spatial filtering on the bottom.

$$F_a(\omega_x) = \int_{-b/2}^{b/2} \cos\frac{\pi x}{b}\, e^{-i\omega_x x}\, dx$$

$$= \cos\frac{\omega_x b}{2}\left[\frac{1}{\omega_x + \pi/b} - \frac{1}{\omega_x - \pi/b}\right]$$

The apodization smoothes the aperture's edge from a discontinuous change in transmission to a continuous change. The result of apodization is a diffraction amplitude given by

$$F_a(\omega_x) = \frac{2\pi}{b}\left[\frac{\cos \omega_x b/2}{(\pi/b)^2 - \omega_x}\right]$$

Comparing the amplitudes of the diffraction patterns for these two slit functions, we find that the amplitude on the optical axis is given by $E_a(0) = 2b/\pi$ and $E(0) = b$. Apodization has reduced the intensity, on the optical axis of the apodized aperture, to 41% of the intensity of the slit aperture.

If the widths of the central peaks of the diffraction patterns are compared, we find that the apodized central maximum is broader. The first minimum in the diffraction from a slit occurs when $\omega_x = 2\pi/b$. The first minimum in the apodized aperture's diffraction pattern occurs when $\omega_x = 3\pi/b$. The central spot in the diffraction pattern from the apodized aperture is wider than the central spot produced by the slit.

From the analysis so far, the advantage provided by apodization is not clear. The advantage provided by apodization becomes apparent when the intensity in the sidelobes of the diffraction patterns is compared. The first secondary maximum in the diffraction pattern for the apodized aperture occurs when $\omega_x b/2 = 2\pi$. The amplitude of the diffraction wave at this secondary maximum is $2b/15\pi$. The ratio of the secondary maximum intensity to the on-axis intensity for the apodized pattern is

$$\left[\frac{F_a(4\pi/b)}{F_a(0)}\right]^2 = \left[\frac{2b/15\pi}{2b/\pi}\right]^2 = 0.0044$$

When we calculate the same intensity ratio for the diffraction pattern from a rectangular aperture, we obtain 0.047.

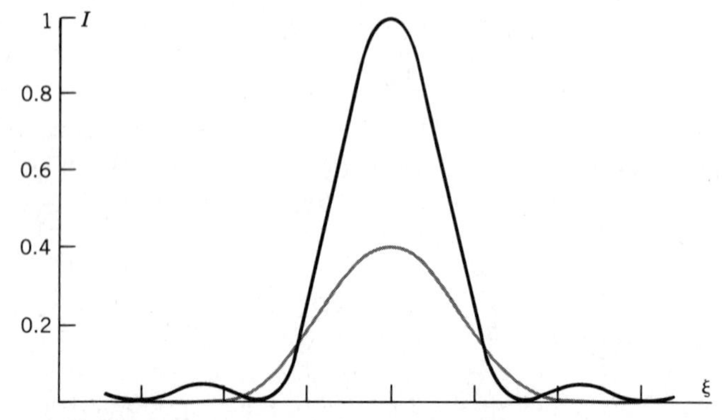

FIGURE 10B-13. The solid curve is the diffraction pattern for a slit and the gray curve is the diffraction pattern for a slit apodized by a cosine function.

The advantage of apodization is now clear. The relative intensity of the first secondary maximum in the apodized aperture's diffraction pattern is a factor of 10 smaller than that found in the diffraction pattern of the rectangular aperture. A comparison of the two diffraction patterns is shown in Figure 10B-13. The improvement in sidelobe intensity has come at the expense of on-axis intensity and central beam width.

PHASE FILTERING

With a coherent optical system, both the phase and amplitude of the light wave are available. Several applications in the field of optical processing utilize spatial filtering operations to observe the phase changes experienced by a light wave.

The *dark-ground illumination* technique uses a circular obstruction to prevent the dc component from propagating past the frequency plane. (The circular obstruction is sometimes called a *dc stop* because it is used to remove the dc component of the diffraction field.) The image generated is an edge-enhanced image. The large bars in Figure 10B-8 are examples of edge-enhanced images. This technique permits the observation of objects that are transparent but scatter a small amount of light. The scattered light is at high spatial frequencies and misses the dc stop; the majority of the light, which is unscattered, is removed from the image by the obstruction.

A second phase-filtering technique, called the *Schlieren method*, replaces the circular obstruction of the dark-ground technique by a knife edge that interrupts all spatial frequencies on one side of the dc component. This technique produces an intensity distribution that is proportional to the phase change produced by the object. The Schlieren method is used in aerodynamics to measure density changes associated with airflow.

A special case of the Schlieren method, *the Foucault knife-edge test*, is used to evaluate lens and mirrors. This technique brings a knife edge into the dc spot in the back focal plane of an optical component under test; see Figure 10B-14.

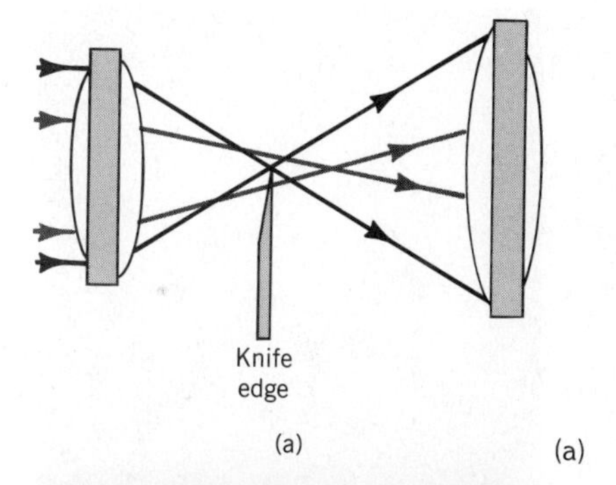

(a)

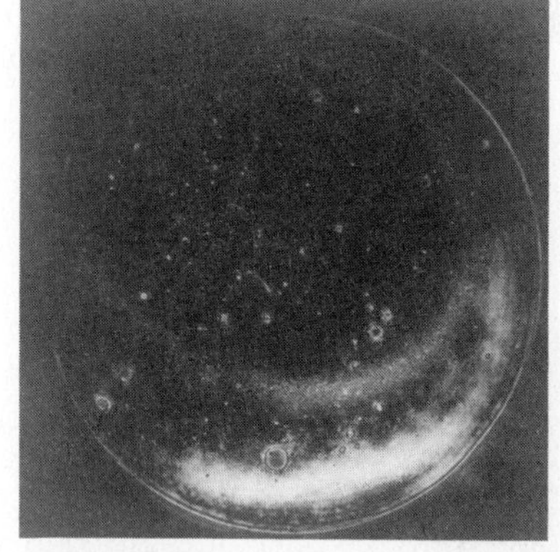

(b)

FIGURE 10B-14. (a) Foucault knife-edge test. In the example shown, a knife edge is brought into the focus of the marginal rays of a lens with spherical aberration. Some of the paraxial rays are not interrupted, resulting in the observation of light over the lower portion of the inner segment of the lens image. (b) A knife-edge test of the mirror whose interferogram is shown in Figure 4-14*b*.

Aberrations in the optical system will cause the light from different portions of the optical element to be focused at different positions along the optical axis; see Appendix 5-B. The portion of the lens producing these rays can be identified by interrupting light at different positions along the optical axis.

The above two processing schemes are special cases of amplitude filtering. There is also a processing technique that operates only on the phase of the light wave. This technique was developed by **Fritz Zernike (1888–1966)** to generate images of microscopic organisms. Zernike received the Nobel Prize for this invention, called the *phase-contrast microscope*. This microscope is now used extensively in the study of biological specimens.

Rather than removing the dc spot in the frequency plane, Zernike shifted the phase of this portion of the spatial frequency spectrum by $\pi/2$ (a quarter-wave) relative to the phase of the rest of the spatial spectrum. The technique allows the observation of transparent objects that absorb very little light by converting phase changes introduced by the transparent object into intensity changes. The conversion from spatial phase variations to equivalent variations in the intensity is accomplished by placing a transparent plate in the frequency plane of the optical system with a quarter-wave coating placed over the low-frequency portion of the spatial spectrum.

To understand the Zernike phase-contrast microscope, assume that the object absorbs no light but rather has a transmission function that is a spatial modulation of only the phase

$$f(x) = e^{i\phi(x)}$$

We will also assume that the index of refraction is very nearly that of its surroundings. For a living organism in water, this approximation is very good and allows the exponent to be approximated by the first two terms of its power series

$$f(x) \approx 1 + i\phi(x)$$

The Fourier transform of this transmission function is of the form

$$E(\omega_x) = \int_{-\infty}^{\infty} [1 + i\phi(x)]\, e^{-i\omega_x x}\, dx = \mathcal{F}\{1\} + i\mathcal{F}\{\phi(x)\} = A(\omega_x) + i\Phi(\omega_x)$$

The functions $A(\omega_x)$ and $\Phi(\omega_x)$ have a relative phase difference of 90°, as indicated by the multiplicative factor i (see Appendix 1-B). If the relative phase of the two functions can be retarded or advanced, we will demonstrate that it is possible to produce an image of the phase object.

We construct a spatial filter that modifies only the phase of the light by the deposition of a quarter-wave dielectric layer on an optically flat plate. By using a flat plate of transparent material, which is made thicker near the optical axis, we can shift the phase of the low spatial frequencies by more than the high spatial frequencies. This phase plate is placed in the F plane of Figure 10B-2, which represents a microscope when the proper values for f_1 and f_2 are selected.

$A(\omega_x)$ is the diffraction pattern of the entire aperture and because there is no amplitude modulation in the aperture, it is nonzero only near $\omega_x = 0$, that is, $A(\omega)$ contains only low spatial frequencies. The phase plate will only modify the phase of $A(\omega)$ because of the limited extent of the dielectric layer. The diffraction pattern modified by the phase plate is

$$E'(\omega_x) = A(\omega_x)e^{\pm i(\pi/2)} + i\Phi(\omega_x) = \pm i[A(\omega_x) \pm \Phi(\omega_x)]$$

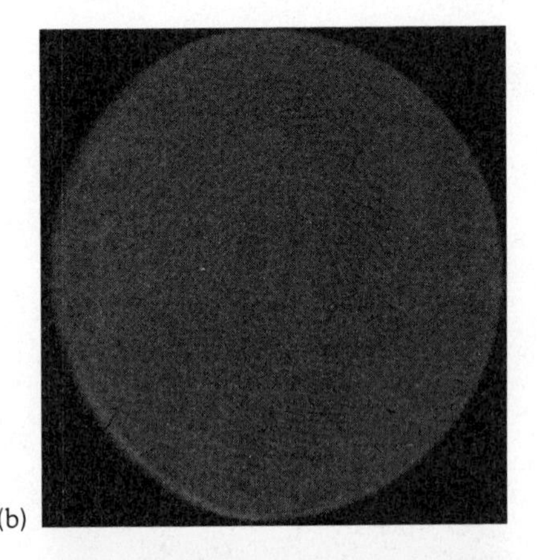
(b)

(c)

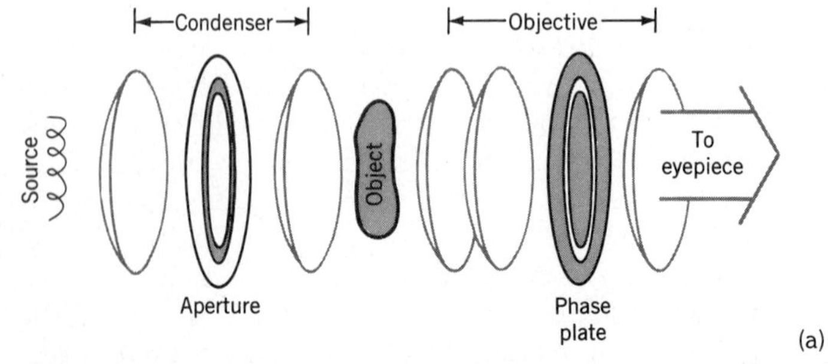

(a)

FIGURE 10B-15. (a) Phase-contrast microscope. The aperture is focused on the phase plate. Low spatial frequencies are given a quarter-wave phase shift relative to higher spatial frequencies. (b) A conventional image of a fingerprint produced by skin oils. (c) The fingerprint of (b) viewed through a phase-contrast microscope. (Figures *b* and *c* courtesy N. George, University of Rochester.)

[The positive and negative signs allow the use of a quarter-wave layer in the dc region of the spectrum (a positive phase plate) or a quarter-wave layer in the high-frequency region of the spatial spectrum (a negative phase plate) to obtain the relative phase shift.]

The image at plane I of Figure 10B-2 is the inverse transform of the phase-modified signal. The inverse transform is given by $f' = \pm i(1 \pm \phi)$ and has an intensity distribution (if we neglect ϕ^2) given by $1 \pm \phi(x)$. To the degree that the approximation of small spatial phase variation is correct, the phase filter provides an image whose variation in intensity is directly proportional to the variation in phase in the object.

The actual optical design is slightly more complicated than implied by the above discussion. To increase the amount of available light and increase the depth of field, which will eliminate the confusion of out-of-focus images, a source in the form of an annular ring is used. With an annular source, the phase plate must have a quarter-wave layer in the form of a ring (see Figure 10B-15).

PHASE AND AMPLITUDE FILTER

We have repeatedly used the fact that a lens can generate the complex representation of the Fourier transform $G(\omega_x, \omega_y)$ of a function $g(x, y)$ to perform simple signal processing operations on $g(x, y)$. The operations that have been discussed are limited by the complexity of the filter placed in the frequency plane. To perform more complicated operations, filters that operate on both the amplitude and phase of the input function are needed. One useful complex operation that can be performed, with a complex filter, is the convolution of two functions. By storing the complex representation $G(\omega_x, \omega_y)$ in the frequency plane of the lens, a convolution operation can be performed between $g(x, y)$ and any other function, say, $f(x, y)$. The calculation of the convolution is accomplished by multiplying Fourier transforms, according to **(6A-10)**, and generating the Fourier transform of the product. Optically, a lens configuration similar to the one shown in Figure 10B-2 will perform the convolution operation.

Initially, the function $g(x, y)$ is placed in the front focal plane of lens L_1, labeled plane O, and its Fourier transform is stored in the frequency plane F. See Figure 10B-16*a* where the labels identify the optical components of Figure 10B-2. The function $g(x, y)$ in plane O is then replaced by a new function $f(x, y)$ and its Fourier transform is projected onto the stored representation of $G(\omega_x, \omega_y)$ by L_1, as is shown in Figure 10B-16*b*. The light transmitted through the complex filter in the frequency plane F is the product of $G(\omega_x, \omega_y)$ and $F(\omega_x, \omega_y)$. Lens L_2 behind the frequency plane generates the Fourier transform of the product and in plane I is a light distribution representing the convolution of f and g.

The technique used for storing the complex filter $G(\omega_x, \omega_y)$ shown in Figure 10B-16*a*, was developed by **Anthony VanderLugt**. (The first successful experiments were conducted on December 7, 1962.) The phase and amplitude of the complex transform are stored as an amplitude transmittance variation, produced by an interferometric technique to be discussed in Appendix 12-B. The technique allows the storage of not only the complex function $G(\omega_x, \omega_y)$, but also its complex conjugate $G^*(\omega_x, \omega_y)$. The presence of the complex conjugate means that the correlation function of $f(x, y)$ and

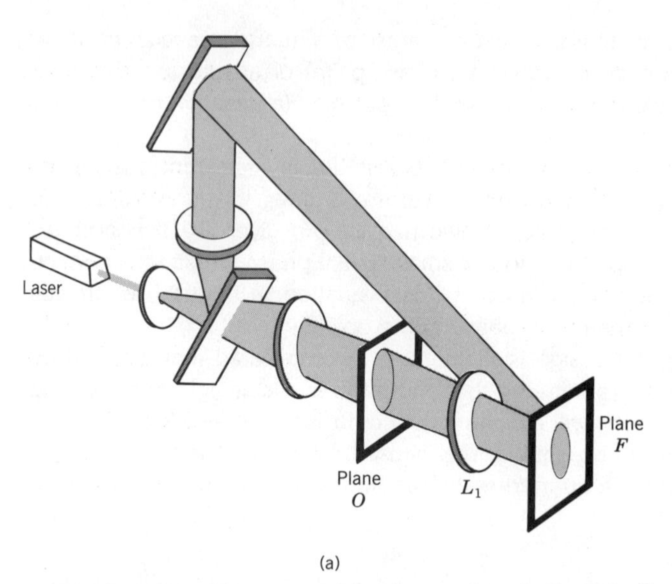

(a)

FIGURE 10B-16*a*. The storage of the phase and amplitude of the Fourier transform of the input function $g(x, y)$ is accomplished by interfering the Fourier transform with a reference beam and recording the interference pattern.

$g(x, y)$ may also be generated, using the configuration displayed in Figure 10B-16*b*.

The demonstration of the calculation of the correlation between g and f can be accomplished both mathematically and experimentally by the proper choice of the functions f and g. The family of Gaussian functions introduced

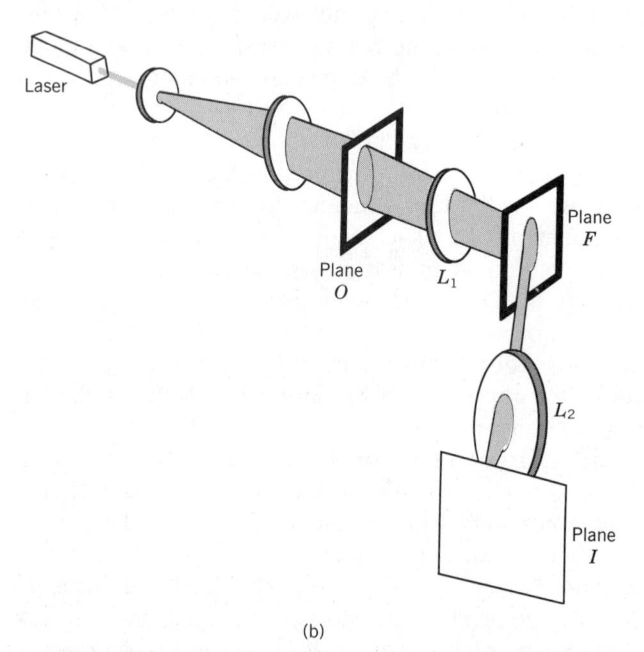

(b)

FIGURE 10B-16*b*. The optical configuration used to calculate the correlation and convolution of pairs of two-dimensional functions.

in Chapter 9 is a good choice for the demonstration of optical correlation. The Fourier transform of any of the Gaussian functions yields the function itself; thus, the mathematical calculation of the correlation operation is very simple.

Experimentally, this class of functions is easy to produce because it forms the resonant modes of a Fabry–Perot resonator. By properly adjusting a Fabry–Perot resonator, a function $g(x, y)$ can be produced and made to fall on plane O of Figure 10B-16*a*. In the example to be shown here, the simplest Gaussian wave, a Gaussian function, was used as $g(x, y)$.

After storing the reference $G^*(\omega_x, \omega_y)$, a generalized output of the Fabry–Perot was made to fall on the plane O of Figure 10B-16*b*. The two-dimensional intensity distributions of the various modes produced by the Fabry–Perot are shown in Figures 10B-17*b–e*. The top of Figure 10B-17*a* displays the transmission of the Fabry-Perot on axis as the mirror separation is changed. The various modes are identified with the peaks in transmitted intensity. The transmitted intensity on the optical axis at plane I is shown on the bottom of Figure 10B-17*a*.

Of more interest are the two-dimensional intensity distributions recorded in plane I. These are the correlation functions generated by the optical system and can be compared to the calculated correlation functions shown in Figure 10B-18. The calculated functions show the intensity distribution measured in plane I outward from the optical axis. Below each recording of the intensity distribution in plane I in Figure 10B-17f-g is an intensity scan across plane I. The agreement between these scan lines and the calculated correlation functions, labeled 0, 1, and 4, in Figure 10B-18 are excellent.

The correlation operation compares two signals and generates a measure of the similarity of the two functions. This is why the correlation operation can be used for the complex pattern recognition task of identifying and tracking a car. The Fourier transform of the car, shown in Figure 10B-19, is stored in plane F of Figure 10B-16*b* as $G(\omega_x, \omega_y)$, and the images shown

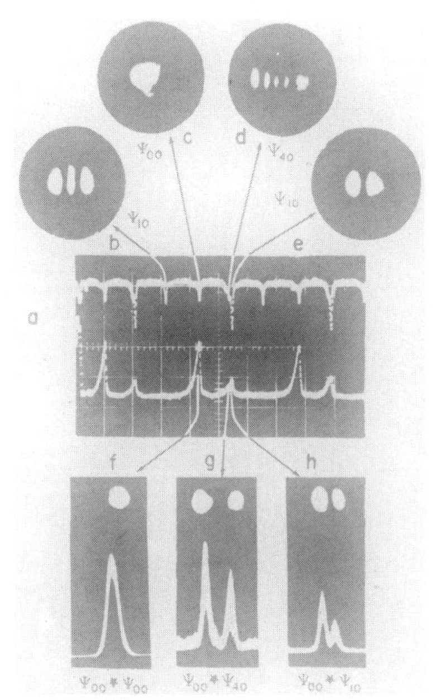

FIGURE 10B-17. Pattern recognition using an optical processor. Here is demonstrated the ability to store a recording of the Fourier transform of a function and then to correlate that function with a number of similar functions. (a) The transmission of a Fabry–Perot interferometer on axis is recorded as the mirror separation is changed. The Fabry-Perot was aligned so that a number of transverse modes were present. The spatial intensity distribution of these modes is identified in (b–e). A filter was recorded of the fundamental Gaussian mode (c). (This is the mode discussed in Chapter 9.) The autocorrelation of this function is shown in (f) and the cross-correlation with two other spatial modes is shown in (g) and (h). Figure 10B-18 displays the theoretical calculation of the correlation functions. [C.C. Aleksoff and B.D. Guenther, "Cross Correlation Discrimination for Optical Cavity Modes," *Applied Optics* **15**:206 (1976).]

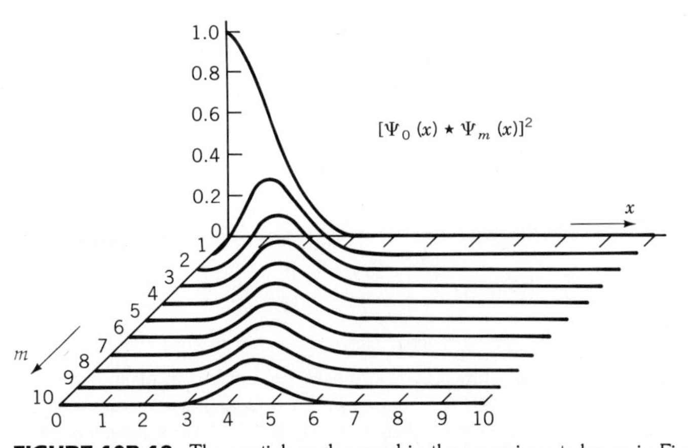

FIGURE 10B-18. The spatial modes used in the experiment shown in Figure 10B-17 can be represented by a set of analytic functions and those functions used to calculate the correlation functions. The theoretical correlation functions associated with the experiment of Figure 10B-17 are shown here. The calculation represents the intensity distribution from the center of plane I along the x axis. The curves labeled 0, 4, and 1 should be compared with the images in Figures 10B-17 *f, g,* and *h*, respectively.

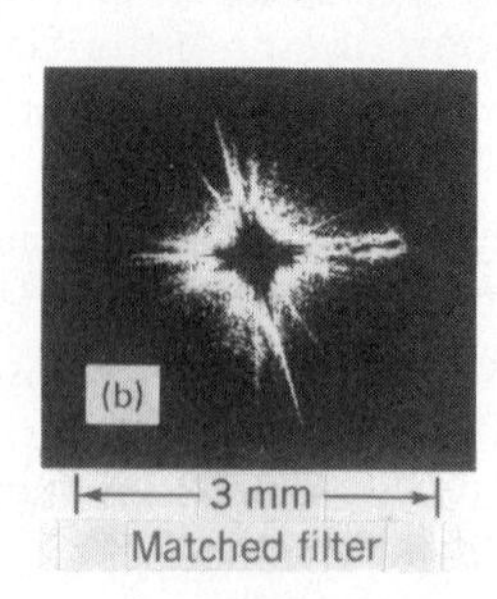

FIGURE 10B-19. (a) This photograph shows the image used in the optical system of Figure 10-16. (b) This photograph shows a magnified image of the Fourier transform recorded in Plane F by the optical system of Figure 10-16*a*.

in the left column of Figure 10B-20 are the functions $f(x, y)$ displayed in plane O. The light distribution in plane I is shown in the right-hand column of Figure 10B-20. A bright spot of light occurs at the same location in plane I as the car does in plane O, and as the car moves through the object field, the spot of light moves through the correlation field in plane I.

FIGURE 10B-20. The reference filter shown in Figure 10B-19 is placed in plane F of the optical processor shown in Figure 10B-16*b*. A TV tape of a car driving down a road is played into plane O and the spatial intensity distribution seen in plane I is recorded. In the left column are several frames from the TV tape. In the middle column is the equivalent frame recorded at plane I. Note the bright spot of light that identifies the automobile and provides its location in the input scene. Structureless correlation functions are often obtained when complicated images with low symmetry are used. [B.D. Guenther, C.R. Christensen, and J. Upatnieks, "Coherent Optical Processing: Another Approach," *IEEE J. Quant Elect QE-15*: 1348–1362 (1979).]

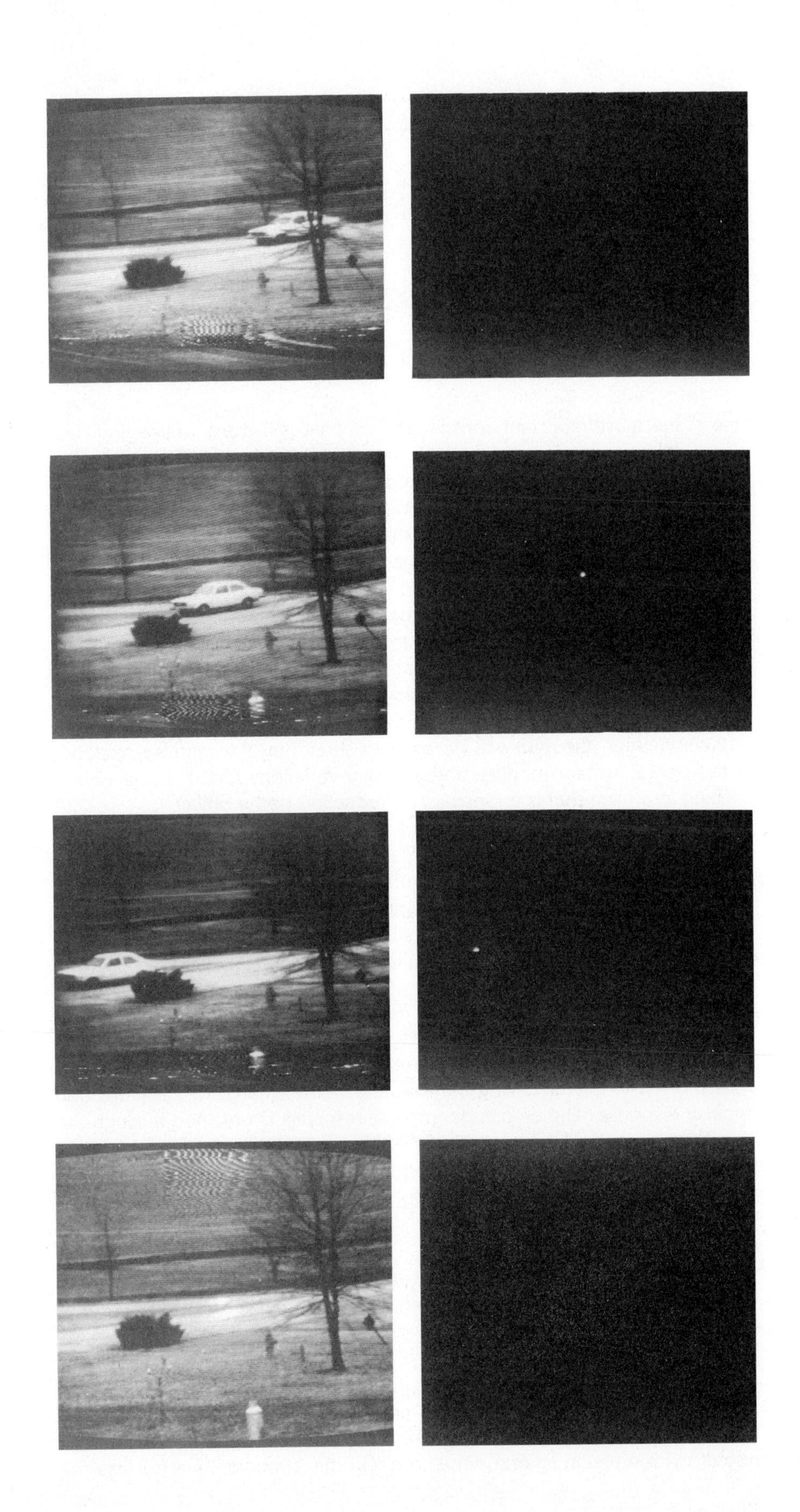

Appendix 10-C

One of the most important applications of optics is the task of imaging. The theory of image formation developed independently from diffraction, but as the capability to image improved, it became necessary to consider the limitations imposed on imaging by diffraction. Most imaging tasks encountered in our daily experience involve incoherent images so we will examine incoherent illumination first. The approach we will take is based on communication theory.[46]

We will theoretically develop a one-dimensional model of an incoherent imaging system and use the model to discuss two criteria that have been used to characterize the imaging capability of an optical system.

The optical system will be treated as a linear system, characterized by either its impulse response or the Fourier transform of the impulse response. The convolution theorem will be used to show that the impulse response of the optical system modifies the intensity variations found in the object, creating an image that is an inexact reproduction of the object.

We will show that the incoherent imaging system is linear in intensity, whereas a coherent imaging system is linear in field amplitude. We will discover that the criterion for resolution must be applied with care to a coherent system. When properly applied, we will find that the resolution criterion requires a coherent optical system to have a larger aperture than an incoherent system to produce equivalent imaging capability.

INCOHERENT IMAGING

An optical system utilizing incoherent illumination can be considered to be linear in intensity. The impulse response used to describe this linear system is found by calculating the absolute value squared of the amplitude impulse response. The amplitude impulse response is obtained through the use of the Huygens–Fresnel integral **(9-20)** to follow the propagation of light from a point source, through an optical system, to the image plane. This procedure used in Appendix 10-A to obtain **(10A-14)** demonstrated that the impulse response of an imaging system is the Fourier transform of its aperture function.

Linear system theory states that an optical system with an amplitude impulse response $\mathcal{R}(x)$ and an input wave $\mathcal{E}_O(x, t)$ has a scalar wave output of

$$\mathcal{E}_I(x, t) = \int_{-\infty}^{\infty} \mathcal{E}_O(\xi, t)\mathcal{R}(x - \xi)\, d\xi$$

where the subscripts O and I represent object and image waves. We will restrict the discussion to one dimension. [In this appendix, we will use the spatial coordinate ξ rather than the spatial frequency ω_x, in our discussions as the coordinate is directly related to the image. The results we derive can easily be extended to two dimensions. For example, if the aperture of the optical system had a rectangular shape, then $\mathcal{R}(x, y)$ would be given by **(10-19)**, and if the aperture had a circular shape, then $\mathcal{R}(r)$ would be given by **(10-26)**.]

The time average of the square of the amplitude, of the output wave, is the measured quantity. This intensity function is proportional to

$$\langle \mathcal{E}_I(x, t)\mathcal{E}_I^*(x, t)\rangle = \left\langle \left(\int_{-\infty}^{\infty} \mathcal{E}_O(\xi, t)\mathcal{R}(x-\xi)\, d\xi \right) \left(\int_{-\infty}^{\infty} \mathcal{E}_O^*(\xi', t)\mathcal{R}^*(x-\xi')\, d\xi' \right) \right\rangle$$

The impulse response of the linear system is independent of time; thus, we may remove it and its complex conjugate from the time averages

$$\langle \mathcal{E}_I(x, t)\mathcal{E}_I^*(x, t)\rangle = \iint_{-\infty}^{\infty} \langle \mathcal{E}_O(\xi, t)\mathcal{E}_O^*(\xi', t)\rangle \mathcal{R}(x-\xi)\,\mathcal{R}^*(x-\xi')\, d\xi\, d\xi'$$

The time average in the integral can be identified as the unnormalized mutual coherence function introduced in Chapter 8 **(8-29)**. (In making the connection with the mutual coherence function, we implicitly assume that the light is temporally coherent.)

For spatially incoherent light, it can be shown using coherence theory that

$$\langle \mathcal{E}_O(\xi, t)\mathcal{E}_O^*(\xi', t)\rangle = I_O(\xi)\delta(\xi - \xi')$$

which is a statement of the fact that all points on a spatially incoherent source are uncorrelated. (The correlation function is a delta function and is zero as soon as we move away from the source point.) If we use this identity relation, the integration over ξ' can quickly be performed. The result of the integration states that the output image of the optical system is the convolution of the input intensity with the square of the system's amplitude impulse response. The square of the amplitude impulse response is physically equal to the Fraunhofer diffraction, intensity pattern of the system's aperture.

Some sort of criterion is needed to specify how well an incoherent imaging system performs. Historically, the first criterion was used to specify how well an imaging system could resolve two point objects. We will derive the criterion for an optical system with a rectangular aperture, allowing the use of a one-dimensional theory. The result for a two-dimensional system with a circular aperture will be stated, but not derived.

Assume that the aperture of the one-dimensional system, shown in Figure 10C-1, is a slit of width $2a$. The system's impulse response is from **(10-20)**

$$\begin{aligned} \mathcal{R}(\xi)\mathcal{R}^*(\xi) &= s(\xi) = \frac{I_d(\xi)}{I_0} \\ &= \frac{\sin^2(k\xi a/S')}{(k\xi a/S')^2} = \operatorname{sinc}^2 \frac{k\xi a}{S'} \end{aligned} \tag{10C-1}$$

From Figure 10C-1, the source distribution consists of two incoherent point sources A and B, located equally spaced about the optical axis a distance $2b$ apart

$$I(x) = I_O[\delta(x-b) + \delta(x+b)] \tag{10C-2a}$$

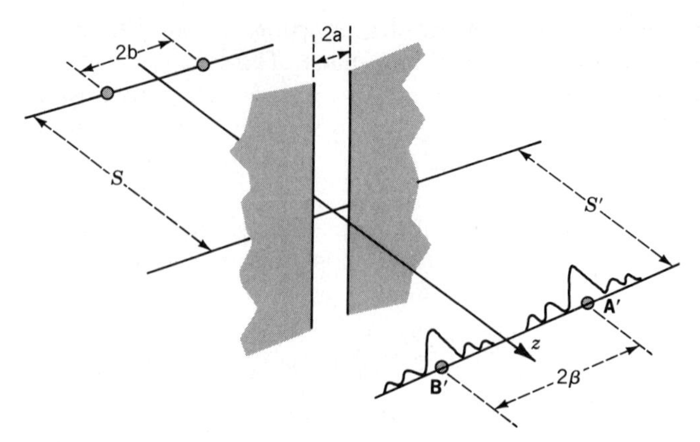

FIGURE 10C-1. Geometry for imaging points A and B onto the positions A' and B' using an imaging system with a rectangular aperture.

This distribution is imaged onto the image points A' and B' whose locations at positions

$$\beta = \pm\frac{bS'}{S}$$

are found by using geometrical optics **(5A-3)**. In the image plane, geometrical optics predicts a light distribution of

$$I(x') = I_O[\delta(x' - \beta) + \delta(x' + \beta)] \qquad (10C\text{-}2b)$$

The image intensity is obtained by convolving **(10C-1)** with **(10C-2b)**

$$I(\xi) = \int_{-\infty}^{\infty} I_O[\delta(x' + \beta) + \delta(x' - \beta)]\,\text{sinc}^2\frac{ka}{S'}(x' - \xi)\,dx'$$

$$= I_O\left[\text{sinc}^2\frac{ka}{S'}(\xi + \beta) + \text{sinc}^2\frac{ka}{S'}(\xi - \beta)\right]$$

We wish to discover the minimum value of b for which we can just resolve image points A' and B'. One criterion that can be applied is Rayleigh's criterion, restated here from Appendix 10-B as

> **Two image points A′ and B′ in the image plane can be just resolved if the central diffraction maximum of one image falls on the first diffraction pattern zero of the second image.**

The location of the central maximum and the first diffraction zero are obtained from

$$\frac{ka}{S'}(\xi - \beta) = 0, \qquad \frac{ka}{S'}(\xi + \beta) = \pi$$

Arranging the central maximum of image A' to fall on the first minimum of image B' results in

$$2\beta = \frac{\lambda S'}{2a}$$

This resolution criterion is equivalent to the resolution condition derived in Appendix 10-B, using Abbe's theory **(10B-7)**

$$\Delta x = \frac{2\lambda f}{D}$$

and is also equivalent to the result obtained in Chapter 9 using Gaussian wave theory for an optical system's minimum beam waist

$$w = \frac{2\lambda f}{D}$$

From the geometrical optics relationship between b and β, the minimum object separation that can be resolved in the image plane is given by

$$2b = \frac{\lambda S}{2a} \tag{10C-3a}$$

The resolution criterion is usually stated in terms of the angular extent of the object

$$\sin \Psi \approx \tan \Psi = \frac{2b}{S}$$

Thus, the minimum angular extent of an object that can just be resolved by the optical system is

$$\sin \Psi \approx \frac{\lambda}{2a} \tag{10C-3b}$$

If we used a circular aperture of radius a, then the resolution criterion would be

$$\sin \Psi = 1.22\frac{\lambda}{2a}$$

The Rayleigh criterion is not a law, but rather a rule of thumb. For that reason, other criteria can be used; one of the most popular is *Sparrow's criterion* which states that

Two image points A′ and B′ in the image plane can be just resolved if the intensity is a constant as we move from image point A′ to point B′ in the image plane.

Mathematically, Sparrow's criterion is stated in terms of the second spatial derivative of the intensity

$$\left[\frac{\partial^2 I(\xi)}{\partial \xi^2}\right]_{\xi=0} = 0 \tag{10C-4}$$

We have evaluated the second derivative at the origin because we assume that, as in Figure 10C-1, the two objects and their images are separated by equal distances on either side of the optical axis. The second derivative is obtained by using the relations

$$\frac{\partial}{\partial x}\left(\frac{\sin^2 x}{x^2}\right) = \frac{x^2 \sin 2x - 2x \sin^2 x}{x^4}$$

$$\frac{\partial^2}{\partial x^2}\left(\frac{\sin^2 x}{x^2}\right) = \frac{2x^2 \cos 2x - 2x \sin 2x}{x^4} - \frac{2x^3 \sin 2x - 6x^2 \sin^2 x}{x^6}$$

$$= \frac{2 \cos 2x}{x^2} - \frac{4 \sin 2x}{x^3} + \frac{6 \sin^2 x}{x^4}$$

We also make use of the fact that the sinc function is an even function, $\text{sinc}(x) = \text{sinc}(-x)$

$$\frac{\partial^2 I_I(\xi)}{\partial \xi^2} = \cos\left(\frac{2ka\beta}{S'}\right) - \left(\frac{S'}{2ka\beta}\right)\sin\left(\frac{2ka\beta}{S'}\right) + 3\left(\frac{S'}{2ka\beta}\right)^2 \sin^2\left(\frac{ka\beta}{S'}\right) = 0$$

This equation can be solved to yield

$$\sin \Psi = \frac{2.606\lambda}{\pi 2a} = 0.83\frac{\lambda}{2a} \tag{10C-5}$$

For a circular aperture, the Sparrow criterion becomes

$$\sin \Psi = \frac{2.976\lambda}{\pi 2a} = 0.95\frac{\lambda}{2a} \tag{10C-6}$$

In Figure 10C-2, the image intensity distributions are shown for a one-dimensional imaging system. One object was placed at the origin in the object plane and the second object was located at the coordinate position equal to the minimum resolvable distance.

The Rayleigh and Sparrow criteria are too restrictive for general imagery and attempting to treat a general image as a continuous distribution of point sources is too cumbersome, and often intractable, because of the convolution integral. Instead, another approach based on linear system theory is used. We move to spatial frequency space, via the Fourier transform, and describe the linear optical system in terms of its spatial frequency response. The imaging process is described in terms of the convolution integral. The most straightforward implementation of this would describe the imaging process by convolving the object with the impulse response of the optical system. It is usually much easier, however, to utilize the Convolution theorem **(6A-8)** and write the convolution as the Fourier transform of the product of the Fourier transform of the object function with the Fourier transform of the impulse response.

Optical Transfer Function

The Fourier transform of the impulse response of an optical system is called the *optical transfer function* **(OTF)**. It draws its name from the analogy that can be formed between optical and electronic systems. In a linear electronic system, a sinusoidal input wave will result in a sinusoidal output wave, with an amplitude and phase modified by the transfer function of the electronic system. We will demonstrate, by using the one-dimensional model

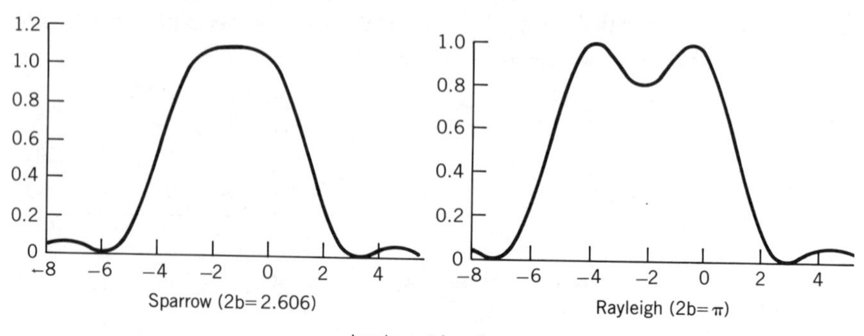

FIGURE 10C-2. The Rayleigh and Sparrow criteria for a slit aperture and incoherent illumination. Here one of the objects was placed at the origin.

of an imaging system, that the **OTF** of an optical system similarly modifies the intensity of a simple sinusoidal object. It will become apparent that an imaging system can be characterized by the modification it produces on a distribution of single spatial frequency objects.

For the imaging system in Figure 10C-1, the Fourier transform of the impulse response is

$$S(\omega_x) = \mathcal{F}\{s(x)\} = \int_{-\infty}^{\infty} \operatorname{sinc}^2\left(\frac{kax}{S'}\right) e^{-i\omega_x x}\, dx \tag{10C-7}$$

This unnormalized transfer function can be evaluated by using a mathematical trick, which will also provide another interpretation of the **OTF**. The integrand of **(10C-7)** is the product of two sinc functions

$$\left[\operatorname{sinc}\left(\frac{kax}{S'}\right)\right]\left[\operatorname{sinc}\left(\frac{kax}{S'}\right)\right]^*$$

From **(6-17)**, we find that a sinc function is the Fourier transform of a rectangular function

$$\mathcal{F}\left\{\operatorname{sinc}\left(\frac{kax}{S'}\right)\right\} = \operatorname{rect}(x)$$

Because of the relationship between the Fourier transform of a correlation and the product of the Fourier transforms of the correlated functions **(6A-10)**, the integral **(10C-7)** can now be viewed as the correlation of two identical rectangular functions. The rectangular function is equal to the aperture function of the imaging system. Thus, the **OTF** of the one-dimensional model of an imaging system is equal to the autocorrelation of the aperture function of the system.

Using the result displayed in Figure 6-15 for the autocorrelation of a rectangular function, we can write **(10C-7)** as a triangular function

$$S(\omega_x) = \begin{cases} 2a\left(1 - \dfrac{S'}{2ak}|\omega_x|\right), & |\omega_x| \leq \dfrac{2ak}{S'} \\ 0, & \text{all other } \omega \end{cases} \tag{10C-8}$$

The normalized transfer function is defined as

$$\tau(\omega_x) = \frac{S(\omega_x)}{S(0)} = \begin{cases} 1 - \dfrac{S'}{2ak}|\omega_x|, & |\omega_x| \leq \dfrac{2ak}{S'} \\ 0, & \text{all other } \omega \end{cases} \tag{10C-9}$$

and is shown in Figure 10C-3.

The result obtained for the simple, one-dimensional model can be extended to two dimensions, allowing the optical transfer function for any imaging system to be generated by calculating the autocorrelation of the aperture function of the imaging system. The behavior of the transfer function shown in Figure 10C-3 exhibits the general behavior of all imaging systems. We can obtain the properties of a typical imaging system by assuming a two-dimensional, circularly symmetric aperture. We replace S' by the focal length of the optical system and use the definition that the f-number f/# of the optical system is [see **(5A-13)**]

$$f/\# = \frac{f}{2a} \tag{10C-10}$$

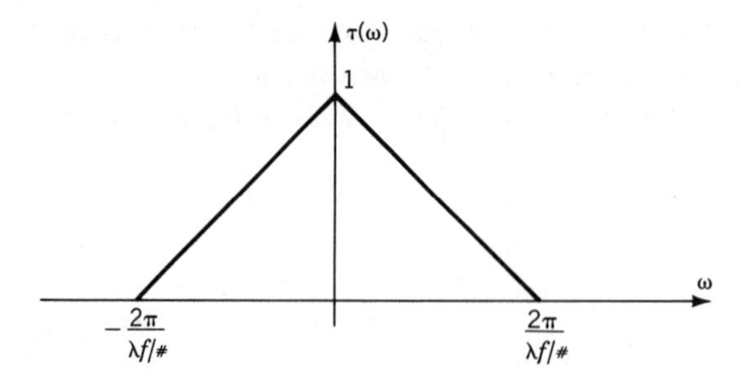

FIGURE 10C-3. Incoherent transfer function for a rectangular aperture.

where a is now the radius of the aperture of the system. With these replacements, Figure 10C-3 demonstrates that spatial frequencies beyond $1/\lambda f/\#$ are not imaged by an incoherent imaging system.

The **OTF** of an optical system is consistent with other views of the imaging process. The **OTF** of an optical system is zero for frequencies greater than

$$\omega_{\text{max}} = \frac{D}{\lambda f}$$

where D is the system's aperture diameter and f its focal length, and therefore, no frequencies higher than ω_{max} can pass through the system. From **(10B-7)** we can rewrite this cutoff frequency as

$$\omega_{\text{max}} = \frac{2}{\Delta x}$$

The maximum spatial frequency that an optical system will pass is inversely proportional to the resolution of the system. The factor of 2 in the relationship is consistent with the sampling theorem that requires two samples at the highest frequency to accurately represent a signal.

To find the physical interpretation that should be associated with the **OTF** in Figure 10C-3, we evaluate the response of the optical system to a simple sinusoidal intensity distribution

$$I_O(x) = 1 + d \cos \omega_0 x \qquad \text{(10C-11)}$$

In electronics, the constant d is called the modulation of the signal; in optics we define the *contrast* of the object; see Appendix 10-B.

$$C_O(\omega_0) = \frac{I_{\text{max}} - I_{\text{min}}}{I_{\text{max}} + I_{\text{min}}} = d \qquad \text{(10C-12)}$$

We expect the contrast to exhibit a dependence on the spatial frequency and we emphasize that functional dependence in **(10C-12)** by writing $C_O(\omega_0)$.

The intensity distribution of the image is

$$I_I(x) = \int_{-\infty}^{\infty} (1 + d \cos \omega_0 \xi) s(x - \xi) d\xi = \mathcal{F}\{\mathcal{F}\{I_O(x)\}\mathcal{F}\{s(x)\}\}$$

Using the results from **(6-22)** and **(6-26)**, we can write the spatial frequency spectrum of the object as

$$\mathcal{F}\{I_O(x)\} = \delta(\omega) + \frac{d}{2}[\delta(\omega - \omega_0) + \delta(\omega + \omega_0)]$$

If we multiply this function by the transfer function shown in Figure 10C-3 and perform the inverse transform, we obtain the image intensity

$$I_I(x) = 1 + \tau(\omega_0) d \cos \omega_0 x$$

The final image has its contrast reduced by

$$\tau(\omega_0) = 1 - \frac{S'|\omega_0|}{2ka}$$

This result is generally true, as we will demonstrate by using a general transfer function with the sinusoidal object's frequency spectrum.

Modulation Transfer Function

We assume that the impulse response of the imaging system is **s(x)** and the optical transfer function is $S(\omega)$. In general, it is a complex function of the form

$$S(\omega_0) = |S(\omega_0)| e^{i\phi(\omega_0)}$$

where $|S(\omega_0)|$ is the modulus of the transfer function and $\phi(\omega_o)$ the phase. The image intensity is then

$$I_I(x) = \int \left\{ \delta(\omega) + \frac{d}{2}[\delta(\omega - \omega_0) + \delta(\omega + \omega_0)] \right\} S(\omega) e^{-i\omega x}\, d\omega$$

$$= S(0) + \frac{d}{2}[S(\omega_0) e^{i\omega_0 x} + S(-\omega_0) e^{-i\omega_0 x}]$$

The impulse response of an incoherent system $s(x)$ is always a real function because it equals the square of the amplitude impulse response. From the discussions in Chapter 6, the Fourier transform of a real function is

$$S(\omega) = K[\cos \omega - i \sin \omega]$$

We can therefore write

$$S(-\omega) = S^*(\omega)$$

and

$$I_I(x) = S(0) + d \cdot \mathcal{Re}\{S(\omega_0) e^{i\omega_0 x}\}$$

$$= S(0) \left[1 + d \frac{|S(\omega_0)|}{S(0)} \cos(\omega_0 x + \phi) \right] \qquad \text{(10C-13)}$$

The use of the **OTF** to obtain an image of a simple sinusoidal function demonstrates that the image contrast is equal to the object contrast, modified by the modulus of the transfer function. Because d is called the modulation in electronics, the modifier of d is called *the modulation transfer function* **(MTF)** in optics. The image contrast of an optical system is thus

$$C_I(\omega_0) = C_O(\omega_0) \frac{|S(\omega_0)|}{S(0)} \qquad \text{(10C-14)}$$

$$= C_O(\omega_0) \tau(\omega_0)$$

where $\tau(\omega_0)$ is the **MTF**.

This relationship suggests a method for measuring the **MTF**. The contrast in the image is measured using different spatial frequencies as object

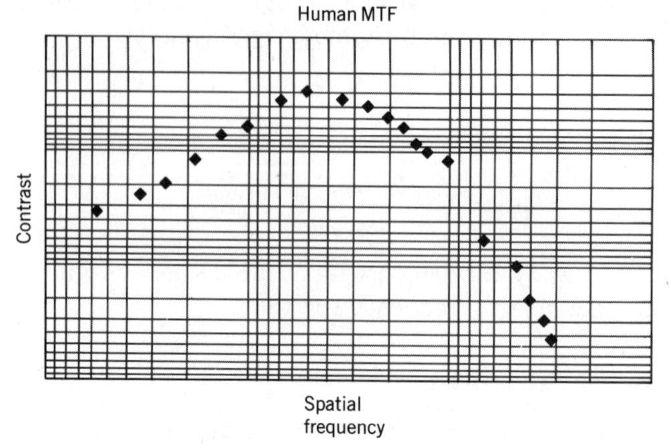

FIGURE 10C-4. The modulation transfer function for the human eye.

functions. An example of an **MTF** is shown in Figure 10C-4. This **MTF** for the human eye differs from manmade imaging systems in that it has a maximum response frequency that is not at dc. Manmade imaging systems usually have an **MTF** that is a monotonic decreasing function of spatial frequency as shown in Figure 10C-3.

COHERENT IMAGING

The same one-dimensional model of an imaging system, Figure 10C-1, will be used to evaluate coherent imaging. An added complication exists for coherent imaging systems. We not only have the intensity images of A and B, as we did in the incoherent case, but we now also have a light distribution due to interference between sources A and B. In Chapter 8, we discovered during the discussion of spatial coherence that the intensity at some observation point, due to two point sources, is given by

$$I_P = I_1 + I_2 + 2\sqrt{I_1 I_2}|\gamma_{12}| \cos \delta_{12}$$

Using this result, we can write, as we did for the incoherent case, the intensity distribution due to the imaging system in Figure 10C-1

$$\begin{aligned} I_I(\xi) = 4a^2 I_0 \Big[& \operatorname{sinc}^2 \frac{ka}{S'}(\xi - \beta) + \operatorname{sinc}^2 \frac{ka}{S'}(\xi + \beta) \\ & + 2\gamma_{AB} \operatorname{sinc} \frac{ka}{S'}(\xi - \beta) \operatorname{sinc} \frac{ka}{S'}(\xi + \beta) \Big] \end{aligned} \tag{10C-15}$$

For complete coherence, the spatial coherence function is $\gamma_{AB} = 1$ and the image intensity is

$$I_I(\xi) = 4a^2 I_o \left[\operatorname{sinc} \frac{ka}{S'}(\xi - \beta) + \operatorname{sinc} \frac{ka}{S'}(\xi + \beta) \right]^2$$

This implies that for coherent imaging

$$I_I(x) = \left[\int_{-\infty}^{\infty} E(\xi, t)\mathcal{R}(x - \xi)\, d\xi \right]^2 \tag{10C-16}$$

The coherent system is linear in amplitude rather than intensity, as was the case for the incoherent system.

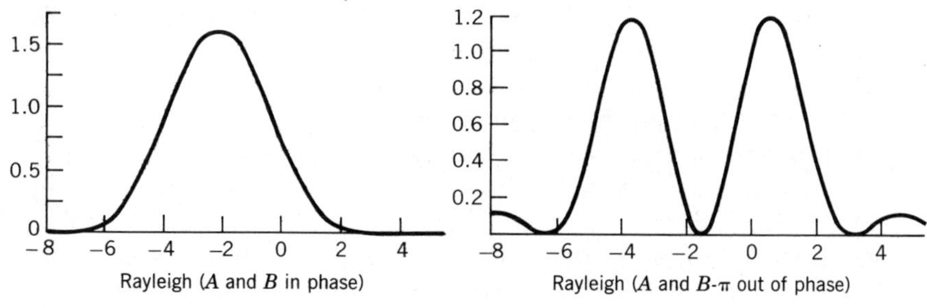

FIGURE 10C-5. Application of the Rayleigh criterion with coherent sources A and B in Figure 10C-1. The relative phase of the two sources determines if resolution is possible, making this criterion useless for a coherent system.

Rayleigh's criterion is inappropriate for coherent imaging. Figure 10C-5 displays the image of A and B separated by a distance equal to Rayleigh's criterion. The images A' and B' are not resolved when objects A and B are in phase, but A' and B' are easily resolved when objects A and B are out of phase.

Unlike Rayleigh's criterion, Sparrow's criterion can be used to obtain a minimum resolution for coherent imaging. We simply set the second derivative of **(10C-15)** with respect to ξ equal to zero to obtain the minimum angular resolution

$$\sin \Psi = \frac{4.164}{ka} = 1.33\frac{\lambda}{2a} \tag{10C-17}$$

for a slit of width $2a$ and

$$\sin \Psi = \frac{4.60}{ka} = 1.46\frac{\lambda}{2a} \tag{10C-18}$$

for a circular aperture of radius a. Figure 10C-6 shows the image of two in-phase, coherent point sources, separated by a distance equal to Sparrow's criterion.

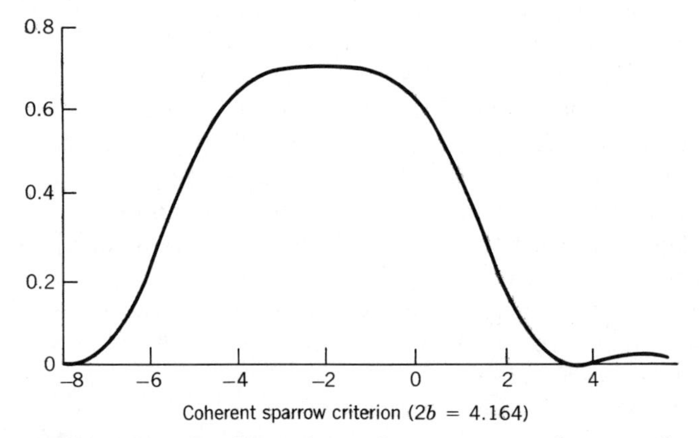

FIGURE 10C-6. Resolution of two coherent sources in phase using Sparrow's criterion. If the sources are out of phase, the discrimination between the two sources is better as is shown in Figure 10C-5.

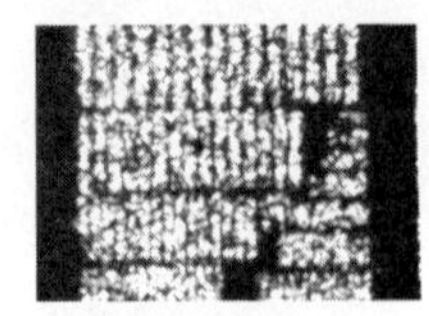

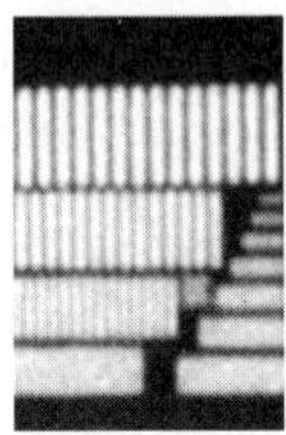

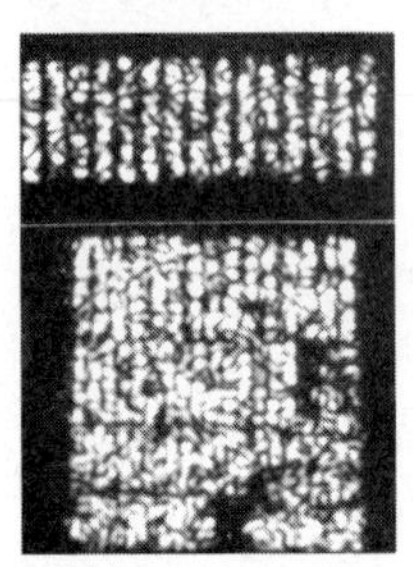

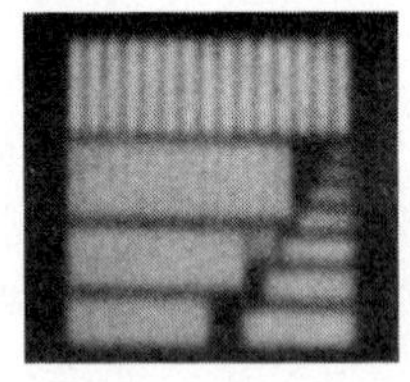

FIGURE 10C-7. Images of a standard resolution chart containing rows of high-contrast bars at different spatial frequencies are shown. The top images of each pair were made using coherent light at apertures of 12 and 8 mm. The lower images were made with incoherent light at apertures of half the diameter used in the upper images, i.e., 6 and 4 mm. The smallest spatial frequency that can be detected in the coherent images is about the same as in the incoherent images. To produce the same resolution, the coherent system required an aperture twice the size required by the incoherent optical system. (Courtesy C.R. Christensen, U.S. Army Missile Command.)

Equations **(10C-17)** and **(10C-18)** state that the incoherent resolution for point objects is better than the coherent resolution. Figure 10C-7 displays images of a standard resolution chart with the spatial frequencies associated with the various bars identified in the figure. Coherent images are shown for imaging systems with two different apertures. Below each coherent image is shown an incoherent image with the same resolution. The incoherent image could be obtained with an aperture of about one-half of the size of the aperture needed to produce the coherent image with the same resolution. These images of a standard resolution chart support the statement that the resolution of a coherent imaging system is less than the resolution of an incoherent system.

Coherent Noise

Another effect that reduces the resolution of a coherent system is speckle noise. Coherent light that travels different optical path lengths will interfere and produce an image with intensity variations called speckle. The intensity variations depend on the atmospheric path and/or the surface roughness of the object. The speckle noise is often completely random and can be smoothed only by averaging statistically independent samples of the image.

Figure 10C-8 shows a test pattern consisting of a matrix of disks. The diameter of the disks in each row is identical, but the diameters decrease by a factor of 2 from row to row. The disks in each column have the same contrast but the contrast decreases from left to right. (This test pattern was developed by A. Rose for the evaluation of television systems.) The image at the lower-right hand portion of the figure, labeled $N = 1$, is immersed in speckle noise, whereas the other images are created by recording multiple exposures (N denotes the number of exposures) of statistically independent images. The images contributing to the multiple exposed recordings differed only in their speckle noise. The intensity averaging produced by the multiple exposures increases the signal to noise in the images, allowing small disks of lower contrast to be detected when N is large.

Amplitude Transfer Function

We introduced a transfer function for the incoherent imaging case. By analogy, we could define an amplitude transfer function, for a coherent imaging system, by applying the convolution theorem to **(10C-16)**. If we followed through the mathematical development, we would discover that the amplitude transfer function is the Fourier transform of the amplitude impulse response. From **(10A-14)**, the amplitude transfer function is the aperture function. The maximum spatial frequency that can be imaged by a coherent system is, therefore, one-half of the maximum spatial frequency imaged by the incoherent system. This result agrees with the images displayed in Figure 10C-7. The contrast for a coherent imaging system does not decrease monotonically, as it does for an incoherent imaging system, but instead remains equal to 1 up to the frequency cut-off. The coherent transfer function of a rectangular aperture is shown in Figure 10C-9.

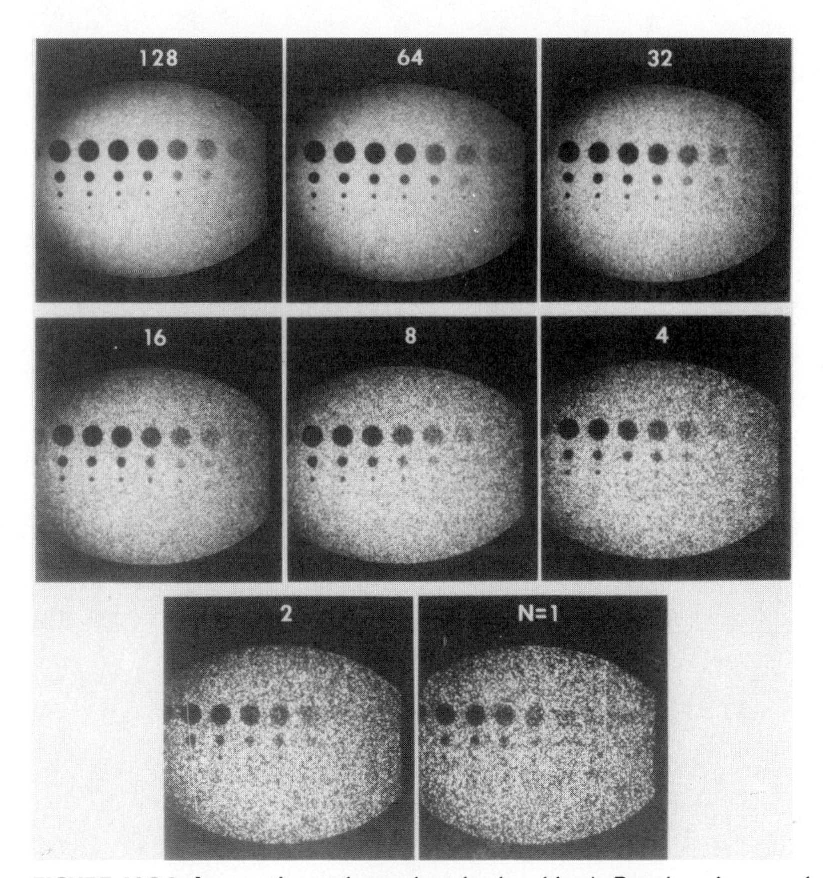

FIGURE 10C-8. Images of a resolution chart developed by A. Rose have been produced using coherent light. The contrast decreases from left to right along each row and the size of the disks decreases from top to bottom along each column. The value of N shown on the images is equal to the number of exposures of the object made to produce the image. The noise, most apparent in the images with low values of N, is called speckle and is a coherent noise that limits the imaging performance of a coherent optical system. The noise is due to interference between waves that have traveled over slightly different optical paths from the object to image. The same speckle noise is observable in Figures 10C-7 and 10-9. The speckle can be reduced by statistically averaging a number of independent samples of the noise field, as was done in the multiple exposures shown here.

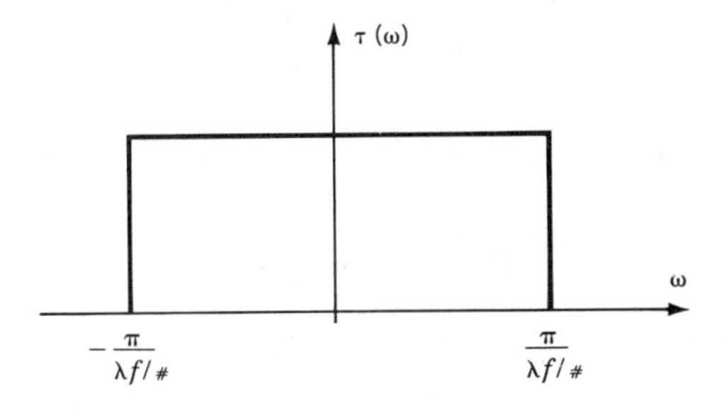

FIGURE 10C-9. Coherent transfer function for a rectangular aperture.

Fresnel Diffraction

In the first approximate solution to the Huygens–Fresnel integral, presented in Chapter 10, it was assumed that the phase of the wavefront in the aperture was a linear function of the aperture's coordinates. This assumption led to Fraunhofer diffraction theory. We now return to the Huygens–Fresnel integral to discuss another approximate solution of the integral. The approximate solution, called *Fresnel diffraction,* assumes that the phase of the wavefront in the aperture has a quadratic dependence on aperture coordinates. The curvature of the wavefront increases the mathematical difficulty of the diffraction problem over that of Fraunhofer diffraction. We will discover that, contrary to Fraunhofer diffraction, only a portion of the wave in the aperture contributes to Fresnel diffraction. That portion of the aperture that contributes to the diffraction amplitude lies near the line connecting the source and observation point, which means that, as with Fraunhofer diffraction, we are limiting our considerations to small departures from geometrical optics.

We will reformulate the Huygens–Fresnel integral

$$\int \frac{\mathcal{E}_i}{R} e^{-i\mathbf{k}\cdot\mathbf{R}}\, dx$$

using the approximation derived in the previous chapter **(10-1)**

$$\Delta \approx k\left(\frac{x_1}{z_1} + \frac{x_2}{z_2}\right)b + \frac{k}{2}\left(\frac{1}{z_1} + \frac{1}{z_2}\right)b^2 + \cdots$$

In this chapter, we will retain both terms of the expansion. For those apertures whose transmission functions can be separated into two independent one-dimensional functions of the aperture coordinates, i.e., $h(x, y) = f(x)g(y)$, the integral with the quadratic approximation can be separated into two integrals, called Fresnel integrals.

The Fresnel integrals have been evaluated numerically and a table of the integrals (found in Appendix 11-B) will be used to calculate the Fresnel diffraction from a straight edge. A plot of the Fresnel integrals called the Cornu spiral (found in Appendix 11-B) can be used to obtain a graphical solution of Fresnel diffraction. The graphical technique will be demonstrated by obtaining diffraction from a straight edge and comparing the results to the solution obtained using the tabulated data.

Fresnel developed a technique for the solution of the Huygens–Fresnel integral based on a geometrical construction called Fresnel zones. The Fresnel zone approach provides a unique insight into the diffraction process and can provide a partial description of Fresnel diffraction from circularly symmetric objects. The concept of Fresnel zones will be used to discuss diffraction from a circular aperture and a disk and to obtain additional insight into Fermat's principle.

In Appendix 11-A, a mathematical theorem called Babinet's principle will be discussed. The theorem connects the diffraction patterns from similar diffracting objects such as a circular aperture and a disk. This theorem provides a useful tool for calculating diffraction and will be demonstrated by recalculating diffraction by a circular disk using the Fresnel diffraction from a circular aperture.

The fact that Fraunhofer diffraction is an integral part of the theory of imaging makes Fraunhofer diffraction a more important subject of optics than Fresnel diffraction. Because there are fewer applications of Fresnel diffraction, only two applications based on the Fresnel zone approach will be discussed.

FRESNEL APPROXIMATION

In Fraunhofer diffraction, the phase of the wave in the aperture is assumed to vary linearly across the aperture. This would occur if, for example, a plane wave were incident on the aperture at an angle with respect to the optical axis. In Fresnel diffraction, we replace the assumption of a linear phase variation with a quadratic phase variation. This is equivalent to assuming that a spherical wave of amplitude α from a point source at position (x_S, y_S, Z') illuminates the aperture; see Figure 11-1.

We found the relation that is needed for the solution of this diffraction problem in Appendix 9-A. Here, we will continue to treat the obliquity factor as a constant and rewrite **(9A-14)** as

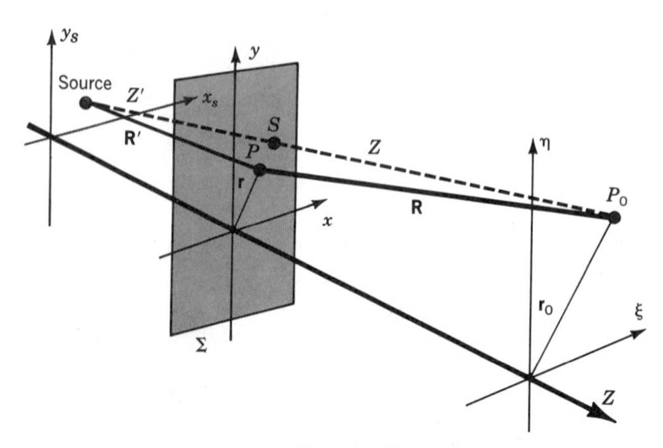

FIGURE 11-1. Geometry for Fresnel diffraction.

$$\mathcal{E}_P = \frac{i\alpha}{\lambda}\int\!\!\int_{\Sigma} f(x, y)\frac{e^{-i\mathbf{k}\cdot(\mathbf{R}+\mathbf{R}')}}{RR'}\,dx\,dy \tag{11-1}$$

In our discussion of the Huygens–Fresnel integral in Chapter 9, we pointed out that the integral is nonzero only when the phase of the integrand is stationary. For Fresnel diffraction, we can insure that the phase is nearly constant if R does not differ appreciably from Z, or R' from Z'. This is equivalent to stating that only the wave in the aperture around a point **S**, called the *stationary point,* will contribute to $\mathcal{E}_P$. Physically, only light propagating over paths nearly equal to the path predicted by geometrical optics (obtained from Fermat's principle) will contribute to $\mathcal{E}_P$. The stationary point **S** is the point in the aperture plane where the line connecting the source and observation positions intersects the plane.

The geometry of Figure 11-1 allows the distances R and R' to be written as

$$\begin{aligned} R^2 &= (x-\xi)^2 + (y-\eta)^2 + Z^2 \\ R'^2 &= (x_s - x)^2 + (y_s - y)^2 + Z'^2 \end{aligned} \tag{11-2}$$

The same procedure used in Chapter 10 to obtain **(10-6)** could be used here to find an approximate expression for R and R'. We will, instead, apply the binomial expansion of a square root

$$\sqrt{1+b} = 1 + \frac{b}{2} - \frac{b^2}{8} + \cdots \tag{11-3}$$

to obtain R and R'. In the Fresnel approximation, the first two terms of the binomial expansion **(11-3)** are retained

$$R \approx Z + \frac{(x-\xi)^2 + (y-\eta)^2}{2Z} \tag{11-4a}$$

$$R' \approx Z' + \frac{(x_s - x)^2 + (y_s - y)^2}{2Z'} \tag{11-4b}$$

In the denominator of **(11-1)**, R is replaced by Z and R' by Z'. (By making this replacement, we are implicitly assuming that the amplitudes of the spherical waves are not modified by the differences in propagation distances over the area of integration. This is a reasonable approximation because the two propagation distances are within a few hundred wavelengths of each other.) In the exponent, we must use **(11-4)** because the phase changes by a large fraction of a wavelength as we move from point to point in the aperture. This procedure differs from the one used in Chapter 10 for Fraunhofer diffraction, only in the fact that here we will retain the quadratic terms of the binomial expansion

$$R + R' \approx Z + Z' + \left[\frac{(x-\xi)^2}{2Z} + \frac{(x_s - x)^2}{2Z'}\right] + \left[\frac{(y-\eta)^2}{2Z} + \frac{(y_s - y)^2}{2Z'}\right]$$

The integral obtained using these terms is quite complicated because it assumes that the source wave and the diffracted wave are both spherical.

A less complicated expression of the Huygens–Fresnel integral would be obtained if the wave incident on the aperture were a plane wave. The Huygens–Fresnel integral

$$E_P = \frac{i\alpha}{\lambda}\int\!\!\int_{\Sigma} f(x, y)\frac{e^{-i\mathbf{k}\cdot\mathbf{R}}}{R}\,dx\,dy$$

would be rewritten, using the approximate expression for R given by **(11-4)**

$$E_P = \frac{i\alpha}{\lambda}\frac{e^{-ikZ}}{Z}\int\int_{\Sigma} f(x, y)\exp\left\{\frac{-ik}{2Z}\left[(x-\xi)^2 + (y-\eta)^2\right]\right\} dx\, dy \quad (11\text{-}5)$$

The physical interpretation of this equation states that when a plane wave illuminates the obstruction, the field at point P is a spherical wave originating at the aperture, a distance Z away from P

$$\frac{e^{-ikZ}}{Z}$$

The amplitude and phase of this spherical wave are modified by an integral, with a quadratic phase dependent on the obstruction's spatial coordinates.

By defining three new parameters,

$$\rho = \frac{ZZ'}{Z + Z'} \quad \text{or} \quad \frac{1}{\rho} = \frac{1}{Z} + \frac{1}{Z'} \quad (11\text{-}6a)$$

$$x_0 = \frac{Z'\xi + Zx_s}{Z + Z'} \quad (11\text{-}6b)$$

$$y_0 = \frac{Z'\eta + Zy_s}{Z + Z'} \quad (11\text{-}6c)$$

the more general expression for Fresnel diffraction of a spherical wave can be placed in the same format as the expression obtained for an incident plane wave.

The parameters x_0 and y_0 are the coordinates in the aperture plane of the stationary point **S**, which lies on a line joining the source and observation point. To illustrate the derivation of the stationary point's coordinates, the x coordinate will be obtained using the geometry in Figure 11-2.

$$\tan\theta = \frac{\xi - x_s}{Z + Z'} \quad \text{and} \quad \tan\theta = \frac{x_0 - x_s}{Z'}$$

Equating these two relationships and solving for x_0 yield **(11-6b)**.

The parameters in **(11-6)** can be used to express, after some manipulation, the spatial dependence of the phase in the integrand of **(11-1)**

$$R + R' = Z + Z' + \frac{(\xi - x_s)^2 + (\eta - y_s)^2}{2(Z + Z')} + \left[\frac{(x - x_0)^2 + (y - y_0)^2}{2\rho}\right]$$

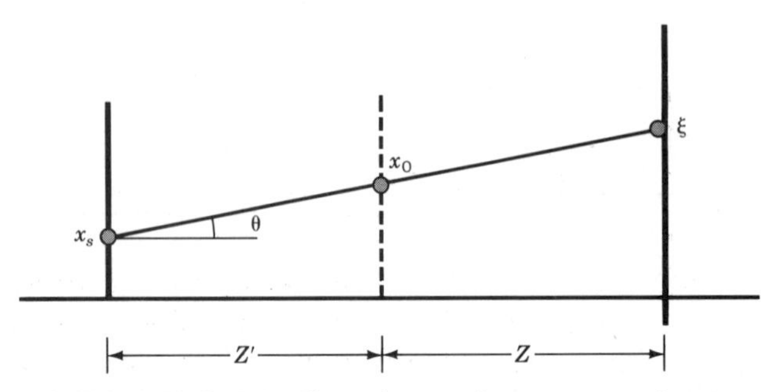

FIGURE 11-2. Geometry for evaluation of parameters used in equation for Fresnel diffraction.

A further simplification to the Fresnel diffraction integral can be made by obtaining an expression for the distance D between the source and observation point. We will demonstrate the derivation in one dimension by using Figure 11-2. The total distance from x_s, the source position, to ξ, the observation position, in Figure 11-2 is

$$\sqrt{(Z+Z')^2+(\xi-x_s)^2} \approx Z+Z'+\frac{(\xi-x_s)^2}{2(Z+Z')}$$

If this one-dimensional calculation is extended to two dimensions, the distance between the source and observation point can be defined as

$$D = Z+Z'+\frac{(\xi-x_s)^2+(\eta-y_s)^2}{2(Z+Z')} \tag{11-7}$$

We use this definition of D to also write

$$\frac{1}{ZZ'}=\frac{Z+Z'}{ZZ'}\frac{1}{Z+Z'}\approx\frac{1}{\rho D}$$

Using the parameters we have just defined, we may rewrite **(11-1)** as

$$E_P=\frac{i\alpha}{\lambda\rho D}e^{-ikD}\iint f(x,y)\exp\left\{-\frac{ik}{2\rho}\left[(x-x_0)^2+(y-y_0)^2\right]\right\}dx\,dy \tag{11-8}$$

By introducing the variables x_0, y_0, ρ, and D, the physical significance of the general expression of the Huygens–Fresnel integral can be understood. At point P, a spherical wave

$$\frac{e^{-ikD}}{D}$$

originating at the source, a distance D away, is observed. This wave would be observed if no obstruction were present. Because of the obstruction, the amplitude and phase of the spherical wave are modified by the integral in **(11-8)**. The modification of the spherical wave made by the integral is called Fresnel diffraction.

Equation **(11-8)** can be shown to be equivalent to **(11-5)** by taking the limit as the source is moved to ∞

$$\lim_{Z'\to\infty}\rho=Z,\qquad \lim_{Z'\to\infty}x_0=\xi,\qquad \lim_{Z'\to\infty}y_0=\eta$$

The spherical wave containing the variable Z' becomes a plane wave as Z' approaches infinity

$$\lim_{Z'\to\infty}\frac{\alpha e^{-ikD}}{\rho D}=\lim_{Z'\to\infty}\left(\frac{e^{-ikZ}}{Z}\frac{\alpha e^{-ikZ'}}{Z'}\right)=\frac{\alpha e^{-ikZ}}{Z}$$

Mathematically, **(11-8)** is an application of the method of *stationary phase,* a technique developed in 1887 by Lord Kelvin to calculate the form of a boat's wake. The integration is nonzero only in the region of the critical point we have labeled **S**. Physically, the light distribution at the observation point is due to wavelets from the region around **S**. The phase variations of light coming from other regions in the aperture are so rapid that the value of the integral over those spatial coordinates is zero.

The calculation of the integral for Fresnel diffraction is more complicated than Fraunhofer diffraction because when the observation point is moved, we must perform the integration over a new region in the aperture.

RECTANGULAR APERTURES

If the aperture function $f(x, y)$ is separable in the spatial coordinates of the aperture, then we can rewrite **(11-8)** as

$$E_P = \frac{i\alpha}{\lambda\rho D} e^{-ikD} \int_{-\infty}^{\infty} f(x) e^{-ig(x)}\, dx \int_{-\infty}^{\infty} f(y) e^{-ig(y)}\, dy$$

$$\mathcal{E}_P = A\left[C(x) - i\,S(x)\right]\left[C(y) - i\,S(y)\right]$$

The spherical wave from the source is represented by

$$A = \frac{i\alpha}{2D} e^{-ikD}$$

If we treat the aperture function as a simple constant, C and S are integrals of the form

$$C(x) = \int_{x_1}^{x_2} \cos\left[g(x)\right] dx \tag{11-9a}$$

$$S(x) = \int_{x_1}^{x_2} \sin\left[g(x)\right] dx \tag{11-9b}$$

The integrals $C(x)$ and $S(x)$ have been evaluated numerically and are found in the table of *Fresnel integrals* in Appendix 11-B. More complete listings can be found in collections of mathematical tables.[48] To use the tabulated values for the integrals, **(11-9)** must be written in a general form

$$C(w) = \int_0^w \cos\left(\frac{\pi u^2}{2}\right) du \tag{11-10}$$

$$S(w) = \int_0^w \sin\left(\frac{\pi u^2}{2}\right) du \tag{11-11}$$

The variable u is an aperture coordinate, measured relative to the stationary point **S** in units of $\sqrt{\lambda\rho/2}$

$$u = \sqrt{\frac{2}{\lambda\rho}}\,(x - x_0) \qquad \text{or} \qquad u = \sqrt{\frac{2}{\lambda\rho}}\,(y - y_0) \tag{11-12}$$

The parameter w in **(11-10)** and **(11-11)** specifies the location of the aperture edge relative to the stationary point **S**. The parameter w is calculated through the use of **(11-12)**.

An example will clarify the method of calculating the parameter w for use in evaluating the Fresnel integrals. Assume the diffracting aperture is a rectangular aperture, defined as

$$f(x, y) = \begin{cases} 1, & x_1 \le x \le x_2 \\ & y_1 \le y \le y_2 \\ 0, & \text{all other } x \text{ and } y \end{cases}$$

The upper edge of the aperture is

$$w_{x_2} = \sqrt{\frac{2}{\lambda \rho}}\,(x_2 - x_0)$$

and the right edge of the aperture is

$$w_{y_2} = \sqrt{\frac{2}{\lambda \rho}}\,(y_2 - y_0)$$

When the observation point is moved, the coordinates x_0 and y_0 of **S** change and the origin of the aperture's coordinate system moves. New values for the w's must therefore be calculated for each observation point.

The values from the Fresnel integral table are used to calculate the light wave's amplitude $\mathcal{E}_P$ and intensity I_P

$$I_P = \frac{A^2}{2}\left\{\left[C(w_{x_2}) - C(w_{x_1})\right]^2 + \left[S(w_{x_2}) - S(w_{x_1})\right]^2\right\}$$
$$\times \left\{\left[C(w_{y_2}) - C(w_{y_1})\right]^2 + \left[S(w_{y_2}) - S(w_{y_1})\right]^2\right\} \qquad \text{(11-13)}$$

The calculation of the two integrals is simplified because the integrals are odd functions

$$C(-w) = -C(w), \qquad S(-w) = -S(w) \qquad \text{(11-14)}$$

This is evident in the plot of $C(w)$ vs $S(w)$ in Figure 11-3.

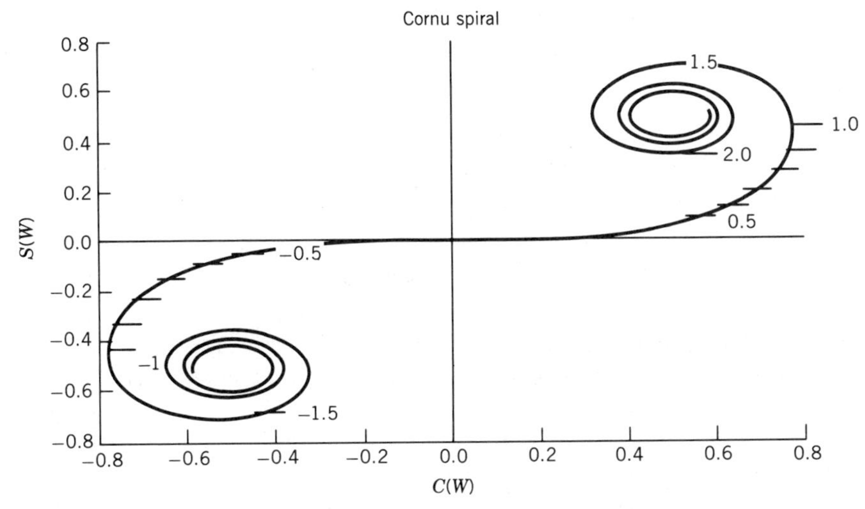

FIGURE 11-3. Cornu spiral obtained by plotting the values for $C(w)$ vs $S(w)$ from the **table of Fresnel integrals**, Appendix 11-B. The numbers indicated along the arc of the spiral are values of w. A more accurate rendition is found in Figure 11B-1.

The plot of $S(w)$ vs $C(w)$, shown in Figure 11-3, is called the Cornu spiral in honor of **M. Alfred Cornu (1841–1902)** who was the first to use this plot for the graphical evaluation of the Fresnel integrals. [The vertical axis of Figure 11-3 must be multiplied by $-i$ if the plot is to represent the complex amplitude of **(11-8)**. A more accurate plot of the Cornu spiral can be found in Appendix 11-B, Figure 11B-1.]

To use the Cornu spiral, the limits w_1 and w_2 of the aperture are located along the arc of the spiral. The length of the straight-line segment drawn from w_1 to w_2 gives the magnitude of the integral. For example, if there were no aperture present, then for the x dimension, $w_1 = -\infty$ and $w_2 = \infty$. The length of the line segment from the point $(-1/2, -1/2)$ to $(1/2, 1/2)$ would be the value of $\mathcal{E}_P$, i.e., $\sqrt{2}$. An identical value is obtained for the y dimension, so that

$$\mathcal{E}_P = 2A = \frac{\alpha e^{-ikD}}{D}$$

This is the spherical wave that is expected when no aperture is present.

An example will clarify the use of the Cornu spiral and the table of Fresnel integrals. Assume that the obstruction is an infinitely long straight edge, reducing the problem to one dimension. The straight edge is assumed to block the negative half-plane with its edge located at $x_1 = 0$. The other limit to be used in **(11-9)** is $x_2 = \infty$, i.e., the straight edge is treated as an infinitely wide slit with one edge located at infinity. Figure 11-4 shows the geometry of the problem. A set of observation points P_1—P_5 on the observation screen is selected. The origin of the coordinate system in the aperture plane is relocated to a new position (given by the position of **S**) when a new observation point is selected. The value of w_1, the position of the edge with respect to **S**, must be recalculated using **(11-12)** for each observation point. The distance from the origin to the straight edge w_1 is positive for P_1 and P_2, zero for P_3, and negative for P_4 and P_5.

Figure 11-5 shows the geometrical method used to calculate the intensity values at each of the observation points in Figure 11-4 using the Cornu

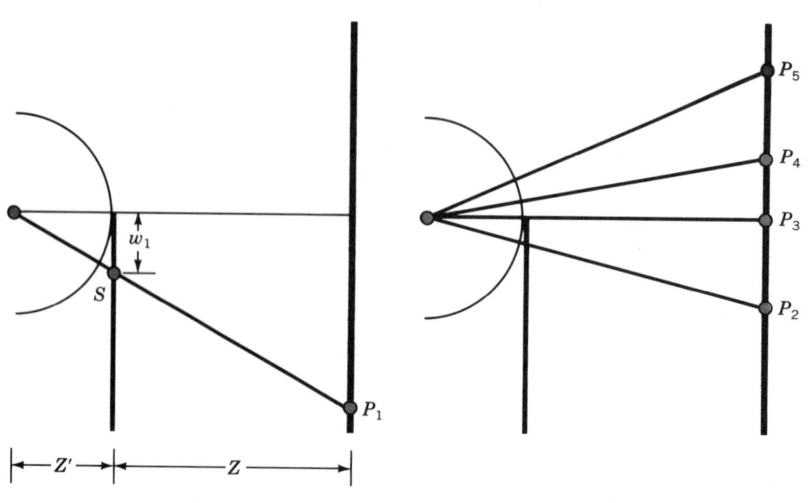

FIGURE 11-4. Geometry of diffraction from a straight edge.

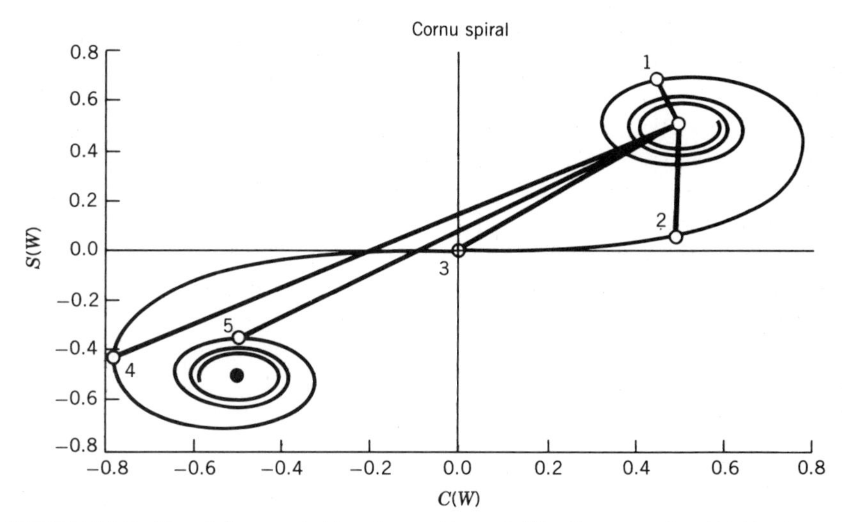

FIGURE 11-5. Use of Cornu spiral to solve problem of diffraction around a straight edge.

spiral. The numbers labeling the straight-line segments in Figure 11-5 are associated with the labels of the observation points in Figure 11-4.

To obtain an accurate calculation of Fresnel diffraction from a straight edge, the table of Fresnel integrals in Appendix 11-B must be used. Before the calculation can proceed, **(11-13)** must be modified to apply to the geometry of this problem

$$I_P = \frac{I_0}{4}\left\{\left[\tfrac{1}{2} - C(w_1)\right]^2 + \left[\tfrac{1}{2} - S(w_1)\right]^2\right\}$$

where $I_0 = 2A^2$. Table 11.1 shows the values extracted from the table of Fresnel integrals and used in the modified version of **(11-13)** to find the relative intensity at various observation points.

The result obtained by using either method for calculating the light distribution in the observation plane, due to the straight edge in Figure 11-4, is plotted in Figure 11-6. The relative intensities at the observation points depicted in Figure 11-4 are labeled on the diffraction curve of Figure 11-6.

TABLE 11-1 Fresnel Integrals for Straight Edge

w_1	$C(w_1)$	$S(w_1)$	I_P/I_0	
∞	0.5	0.5	0	
2.0	0.4882	0.3434	0.01	
1.5	0.4453	0.6975	0.021	1
1.0	0.7799	0.4383	0.04	
0.5	0.4923	0.0647	0.09	2
0	0	0	0.25	3
−0.5	−0.4923	−0.0647	0.65	
−1.0	−0.7799	−0.4383	1.26	4
−1.2	−0.7154	−0.6234	1.37	
−1.5	−0.4453	−0.6975	1.16	
−2.0	−0.4882	−0.3434	0.84	5
−2.5	−0.4574	−0.6192	1.08	
$-\infty$	−0.5	−0.5	1.0	

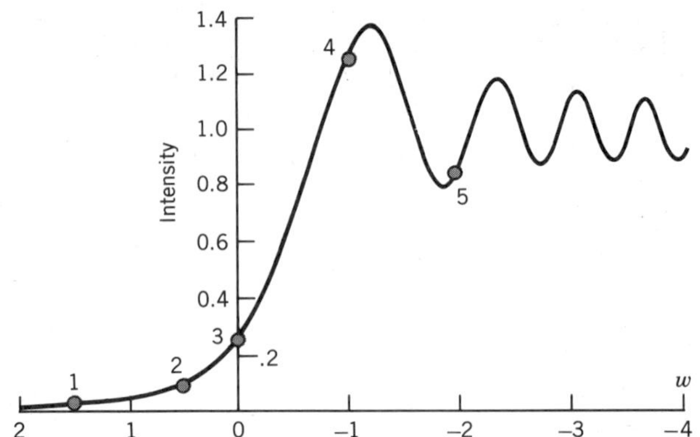

FIGURE 11-6. Light distribution around a straight edge with the edge located at $w = 0$.

FIGURE 11-7. Fresnel diffraction pattern from a large rectangular aperture produced by illuminating the aperture with a plane wave from a HeNe laser. Each edge of the aperture produces the fringe pattern calculated in Figures 11-5 and 11-6. The fringes produced by the second edge extend into the shadow region of the first edge, as can be seen by close examination of this image.

An experimentally obtained Fresnel diffraction pattern from a straight edge along one side of a large rectangular aperture is shown in Figure 11-7. The pattern is not as simple as the one shown in Figure 11-6 because fringes from one side of the aperture extend into the shadow region of the second side.

The same procedure can be used to calculate the light distribution of a slit of finite width. The only modification that must be made to the straight-edge calculation in order to treat a slit is the recalculation of w_2 for each observation point. The calculation of w_2 is accomplished by using the same procedure used to calculate w_1 in the straight-edge problem. For Fresnel diffraction from a slit, the arc length on the Cornu spiral $s = (w_2 - w_1)$ is a constant proportional to the slit width $(x_2 - x_1)$

$$\begin{aligned} s &= w_2 - w_1 \\ &= \sqrt{\frac{2}{\lambda\rho}}\left[(x_2 - x_0) - (x_1 - x_0)\right] \\ &= \sqrt{\frac{2}{\lambda\rho}}(x_2 - x_1) \end{aligned}$$

The position of the center of this arc segment moves because the location of the center of the aperture relative to **S** changes with observation position. The center of the arc's position is given by

$$\begin{aligned} \frac{w_2 + w_1}{2} &= \frac{1}{2}\sqrt{\frac{2}{\lambda\rho}}\left[(x_2 - x_0) + (x_1 - x_0)\right] \\ &= \sqrt{\frac{2}{\lambda\rho}}\left[\left(\frac{x_2 + x_1}{2}\right) - x_0\right] \end{aligned}$$

The square of the length of the chord spanning the arc gives the intensity of light at the observation point. Fresnel diffraction from a rectangular aperture is shown in Figure 11-8.

We are forced to stop the development of general solutions to Fresnel diffraction at this point. The calculations of apertures that are more complex than simple rectangular apertures, with uniform amplitude transmission functions, are very demanding and are best left to the specialist. The calculations

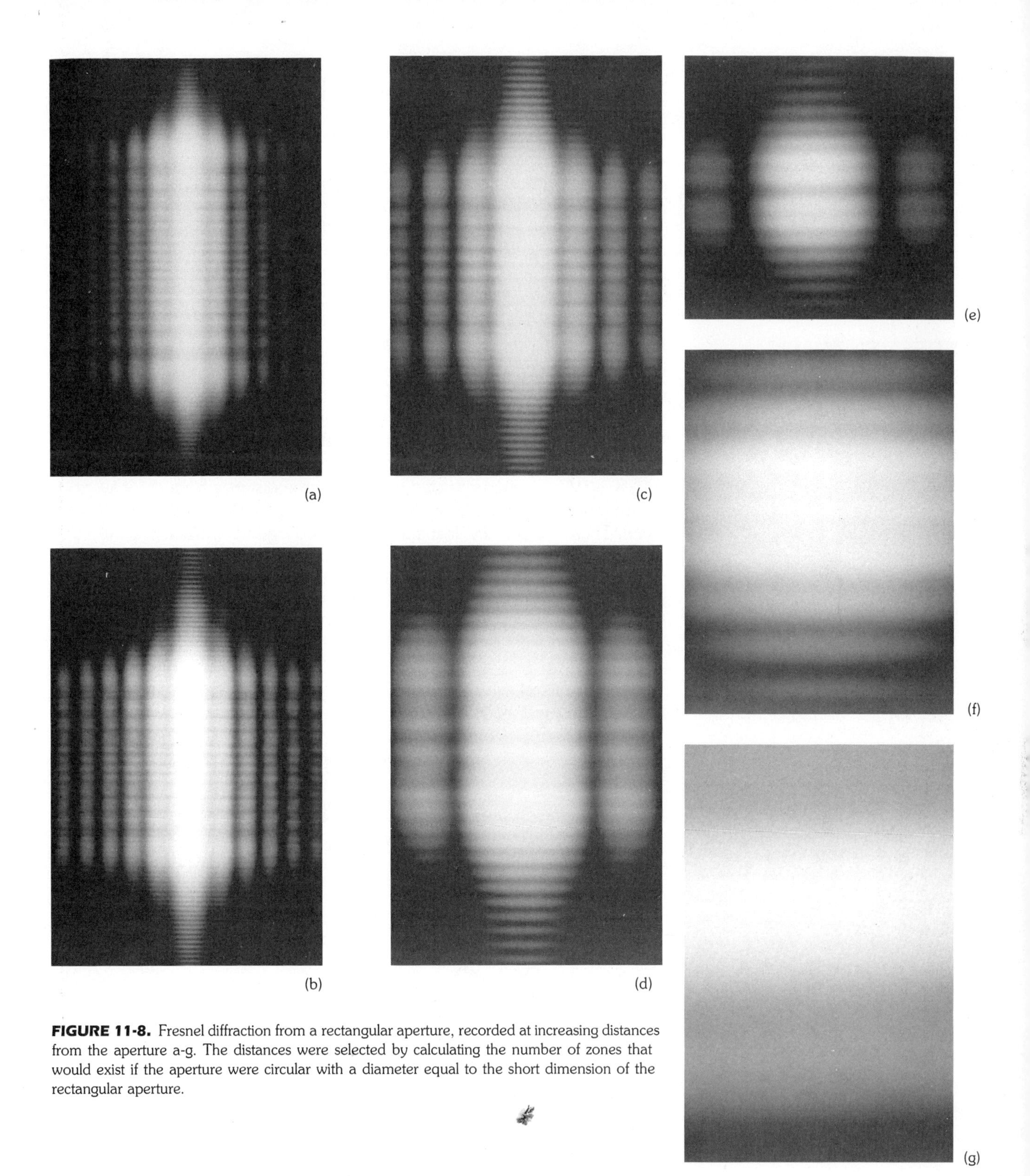

FIGURE 11-8. Fresnel diffraction from a rectangular aperture, recorded at increasing distances from the aperture a-g. The distances were selected by calculating the number of zones that would exist if the aperture were circular with a diameter equal to the short dimension of the rectangular aperture.

usually require numerical evaluation of **(11-8)**. The reader can get an idea of the next degree of difficulty in diffraction problems by referring to Born and Wolf,[6] where examples are given of the use of Fresnel diffraction to calculate the optical fields near the focal plane of optical systems, including systems with aberrations.

We will now develop a semiquantitative method for visualizing Fresnel diffraction that is particularly useful for predicting diffraction by a circular aperture. The method uses the concept of Fresnel zones.

FRESNEL ZONES

Fresnel used the geometrical construction, now called the Fresnel zone, to evaluate the Huygens–Fresnel integral. The Fresnel zone is a mathematical construct serving the role of a Huygens' source in the description of wave propagation. Assume that at time t, a spherical wavefront from a source at P_1 has a radius of R'. To determine the field at the observation point P_0 due to this wavefront, a set of concentric spheres of radii Z, $Z + \lambda/2$, $Z + 2\lambda/2$, $\ldots, Z + j\lambda/2, \ldots$ is constructed, where Z is the distance from the wavefront to the observation point on the line connecting P_1 and P_0 (see Figure 11-9). These spheres divide the wavefront into a number of zones $\varsigma_1, \varsigma_2, \ldots,$ $\varsigma_j, \ldots.$ called *Fresnel zones*, or *half-period zones*.

We treat each zone as a circular aperture illuminated from the left by a spherical wave of the form

$$\mathcal{E}_j(\mathbf{R}') = \frac{Ae^{-i\mathbf{k}\cdot\mathbf{R}'}}{R'} = \frac{Ae^{-ikR'}}{R'}$$

R' is the radius of the spherical wave. The field at P_0 due to the jth zone is obtained by using **(9-9)**

$$\mathcal{E}_j(P_0) = \frac{A}{R'}e^{-i\mathbf{k}\cdot\mathbf{R}'}\iint_{\varsigma_j} C(\varphi)\frac{e^{-i\mathbf{k}\cdot\mathbf{R}}}{R}\,ds \tag{11-15}$$

For the integration over the jth zone, the surface element is

$$ds = R'^2 \sin\theta \, d\theta \, d\phi \tag{11-16}$$

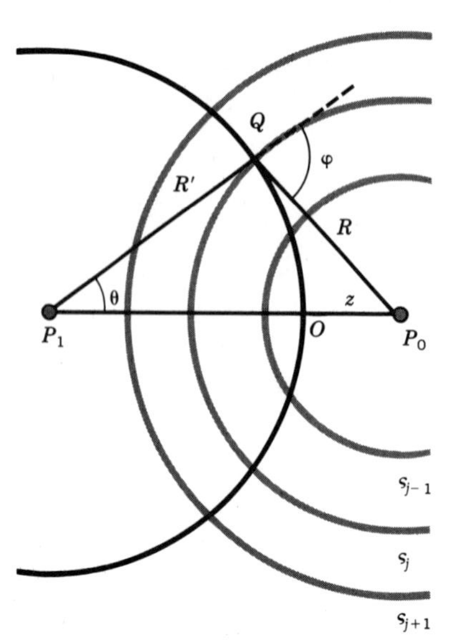

FIGURE 11-9. Construction of Fresnel zones for a spherical wave.

The limits of integration extend over the range

$$Z + (j - 1)\lambda/2 \leq R \leq Z + j\lambda/2$$

The variable of integration is R; thus, a relationship between R' and R must be found. This is accomplished by using the geometrical construction shown in Figure 11-10.

A perpendicular is drawn from the point Q on the spherical wave to the line connecting P_1 and P_0; see Figure 11-10. In the notation of this chapter, the distance from the source to the observation point is $Z' + Z$ and the distance from the source to the plane of the zone is the radius of the incident spherical wave, $Z' = R'$. The distance from P_1 to P_0 can be written

$$x_1 + x_2 = R' + Z$$

The distance from the observation point to the zone is

$$R^2 = x_2^2 + y^2 = \left[(R' + Z) - x_1\right]^2 + (R' \sin\theta)^2$$

$$= R'^2 + (R' + Z)^2 - 2R'(R' + Z)\cos\theta$$

The derivative of this expression yields

$$R\,dR = R'(R' + Z)\sin\theta\,d\theta \tag{11-17}$$

Substituting **(11-17)** into **(11-16)** gives

$$ds = \frac{R'}{R' + Z} R\,dR\,d\phi \tag{11-18}$$

The integration over ϕ is accomplished by rotating the surface element about the P_1P_0 axis. After integrating over ϕ between the limits of 0 and 2π, we obtain

$$\mathcal{E}_j(P_0) = \frac{2\pi A}{R' + Z} e^{-i\mathbf{k}\cdot\mathbf{R}'} \int_{Z+(j-1)(\lambda/2)}^{Z+(j\lambda/2)} e^{-i\mathbf{k}\cdot\mathbf{R}} C(\varphi)\,dR \tag{11-19}$$

We assume that $R', Z >> \lambda$ so that the obliquity factor is a constant over a single zone, i.e., $C(\varphi) = C_j$. To demonstrate that this assumption is reasonable, a list of obliquity factors for the first 12 zones is shown in Table 11.2. The parameters used to calculate these factors were $Z = 1$ m and $\lambda = 500$ nm.

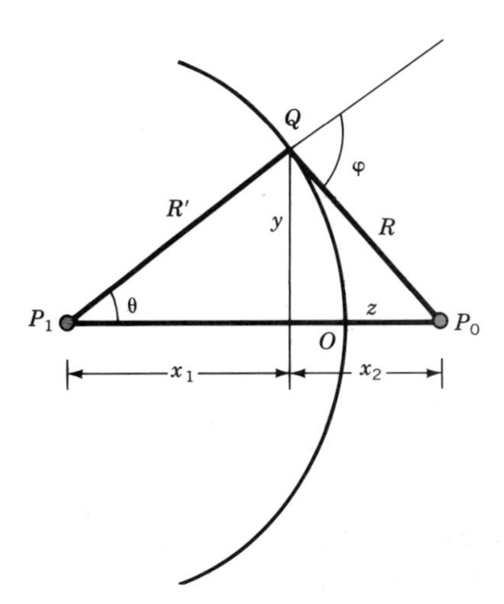

FIGURE 11-10. Geometry for finding R.

TABLE 11-2 Obliquity Factor

Zone	Angle	Obliquity
0	0	1.0
1	0.057295768	0.99999975
2	0.081028435	0.9999995
3	0.099239139	0.99999925
4	0.114591464	0.999999
5	0.128117124	0.99999875
6	0.140345249	0.9999985
7	0.151590163	0.99999825
8	0.162056667	0.999998
9	0.171887016	0.99999775
10	0.181184786	0.9999975
11	0.190028167	0.99999725
12	0.198477906	0.999997

As can be seen in Table 11.2, the obliquity factor only changes two parts in 10^{-7} across one zone. Applying the assumption that the obliquity factor is a constant over a zone allows the integral in **(11-19)** to be calculated

$$\mathcal{E}_j(P_0) = \frac{2\pi i C_j A}{k(R' + Z)} e^{-ik(R'+Z)} e^{-ikj\lambda/2}\left(1 - e^{-ik(\lambda/2)}\right) \tag{11-20}$$

If we use the identity $k\lambda = 2\pi$ and the definition for the distance between the source and observation point $D = R' + Z$, **(11-20)** can be simplified

$$\mathcal{E}_j(P_0) = 2i\lambda(-1)^j \frac{C_j A}{D} e^{-ikD} \tag{11-21}$$

Not only is the obliquity factor C_j a constant over a single zone, but it is a very slowly varying parameter of j as is shown in Figure 11-11, where the obliquity factor can be seen to change by less than two parts in 10^4 when j changes by 500. We are therefore justified in treating the absolute value of **(11-21)** as a constant, as j varies over a large range of values.

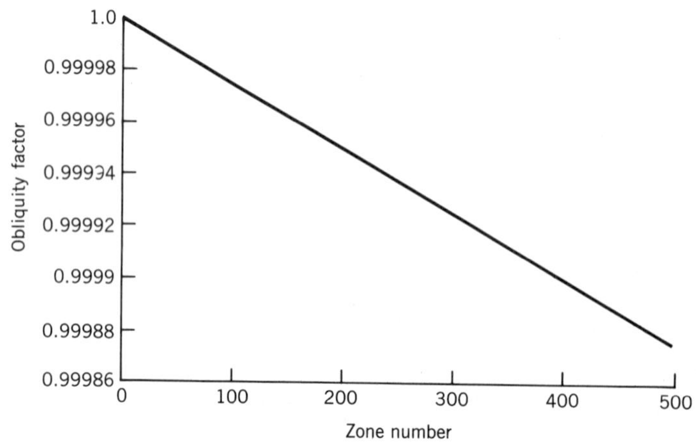

FIGURE 11-11. Obliquity factor as a function of the zone number. For this example $Z = 1$ m and the wavelength was 500 nm.

The physical reasons for the behavior predicted by **(11-21)** are quite easy to understand. The distance from P_0 to a zone changes by only $\lambda/2$ as we move from zone to zone, and the area of a zone is almost a constant independent of the zone number; thus, the amplitudes of Huygens' wavelets from each zone should be approximately equal. The alternation in sign from zone to zone is due to the phase change of the light wave from adjacent zones because the propagation path for adjacent zones differs by $\lambda/2$.

To find the total field strength at P_0 due to N zones, the collection of Huygens' wavelets is added

$$\begin{aligned} E(P_0) &= \sum_{j=1}^{N} \mathcal{E}_j(P_0) \\ &= \frac{2i\lambda A}{D} e^{-ikD} \sum_{j=1}^{N} (-1)^j C_j \end{aligned} \qquad (11\text{-}22)$$

To evaluate the sum, the elements of the sum are regrouped and rewritten as

$$-\sum_{j=1}^{N} (-1)^j C_j = \frac{C_1}{2} + \left(\frac{C_1}{2} - C_2 + \frac{C_3}{2}\right) + \left(\frac{C_3}{2} - C_4 + \frac{C_5}{2}\right) + \cdots$$

Because the C's are very slowly varying functions of j even as shown in Figure 11-11 out to 500 zones, we are justified in setting the quantities in parentheses equal to zero. With this approximation, the summation can be set equal to one of two values, depending on whether there are an even or odd number of terms in the summation

$$-\sum_{j=1}^{N} (-1)^j C_j = \begin{cases} \frac{1}{2}(C_1 + C_N), & N \text{ odd} \\ \frac{1}{2}(C_1 - C_N), & N \text{ even} \end{cases}$$

For very large N, the obliquity factor approaches zero, $C_N \to 0$, as is demonstrated in Figure 11-12. Thus, the theory has led to the conclusion that the total field produced by an unobstructed wave is equal to one-half the contribution from the first Fresnel zone, i.e., $E = E_1/2$. Stating this result in a slightly different way, we see that a surprising result has been obtained: The contribution from the first Fresnel zone is twice the amplitude of the unobstructed wave!

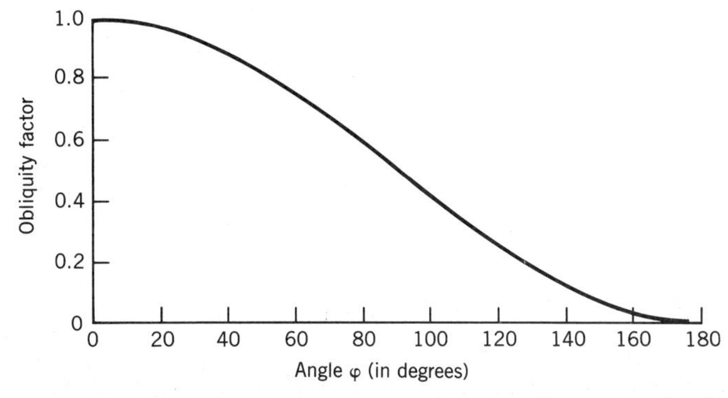

FIGURE 11-12. The obliquity factor as a function of the angle defined in Figure 11-8.

The zone construction can be used to analyze the effect of the obstruction of all or part of a zone. For example, by constructing a circular aperture with a diameter equal to the diameter of the first Fresnel zone, we have just demonstrated that it is possible to produce an intensity at the point P_0 equal to four times the intensity that would be observed if no aperture was present. To analyze the effects of a smaller aperture, we subdivide a half-period zone into a number of subzones such that there is a constant phase difference between each subzone. To add the waves from each of the subzones, the vector approach introduced in Chapter 4 (see Figure 4-2) will be used. The individual vectors form a curve know as the *vibrational curve*. Figure 11-13*a* shows the vibrational curve produced by the addition of waves from nine subzones of the first Fresnel zone. The vibrational curve in Figure 11-13*a* is an arc with the appearance of a half-circle. The radius of curvature of the arc is defined as

$$\frac{\Delta s}{\Delta \alpha} = \rho$$

where s and α are defined as the arc length between two subzones and the phase difference between two subzones, respectively. (See Figure 11-15*b* for a geometrical representation of these parameters.) If the radius of curvature of the arc were calculated, we would discover that it is a constant, except for the contribution of the obliquity factor

$$\rho = \frac{\lambda R'}{R' + Z} C(\varphi)$$

Because the obliquity factor for a single zone is assumed to be a constant, the radius of curvature of the vibration curve is a constant over a single zone. If we let the number of subzones approach infinity (Figure 11-13*b*), the vibrational curve becomes a semicircle whose chord is equal to the wavelet produced by zone one, i.e., E_1. If we subdivide additional zones and add the subzone contributions, we create other half-circles whose radii decrease

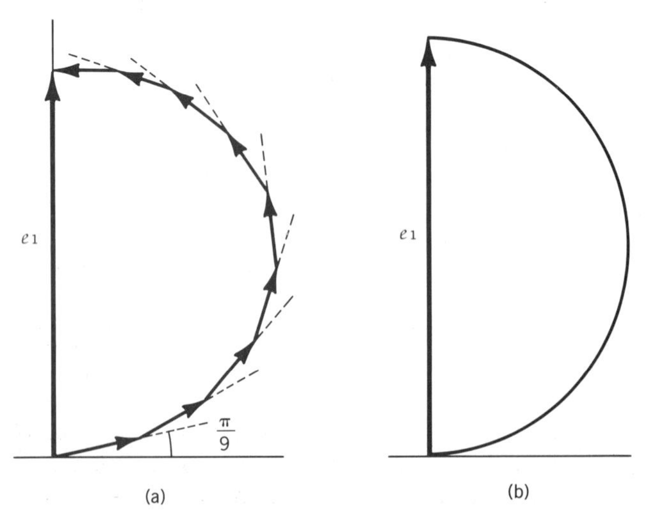

FIGURE 11-13. (a) Vector addition of waves from nine subzones from the first Fresnel zone. (b) Vector addition of waves from an infinite number of subzones from the first Fresnel zone.

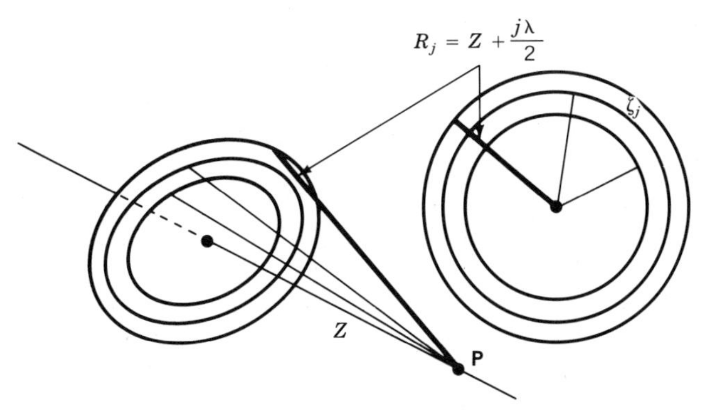

FIGURE 11-14. Construction of Fresnel zones for a plane wave.

at the same rate as the obliquity factor. The vibrational curve for the total wave is a spiral, constructed of semicircles that converge to a point halfway between the first half-circle. The length of the vector from the start to the end of the spiral is $E = E_1/2$, as we derived above.

To appreciate the utility of the vibrational curve, we will use the spiral just described to evaluate the Fresnel diffraction due to a circular aperture and a circular disk.

The Fresnel zones have been defined as spherical rings constructed on a spherical wave. By replacing the incident spherical wave with a plane wave, the problem can be greatly simplified. With an incident plane wave, the zone is defined on a plane, removing the need for solid geometry. The use of an incident plane wave does not, however, reduce the physical insight provided by the zone construction.

Plane wave, Fresnel zones consist of a set of concentric rings, constructed by drawing circles of radius r_j; see Figure 11-14. The nth plane wave, Fresnel zone shown in Figure 11-14 has a radius of

$$r_n^2 = R_n^2 - Z^2$$

$$r_n^2 = \left(Z + \frac{n\lambda}{2}\right)^2 - Z^2 = nZ\lambda + \left(\frac{n\lambda}{2}\right)^2 \tag{11-23}$$

For small values of n, the *nth* zone has a radius of

$$r_n \approx \sqrt{nZ\lambda} \tag{11-24}$$

In the discussion of Fresnel zones, when it is necessary to calculate the area of a zone, it will be assumed that the incident wave is a plane wave. This assumption will reduce the mathematical complexity of the examples.

CIRCULAR APERTURE

The vector addition technique, described in Figure 11-13, can be used to evaluate Fresnel diffraction at a point P_0 from a circular aperture and yields the intensity distribution along the axis of symmetry of the circular aperture. The zone concept will also allow a qualitative description of the light distribution normal to this axis.

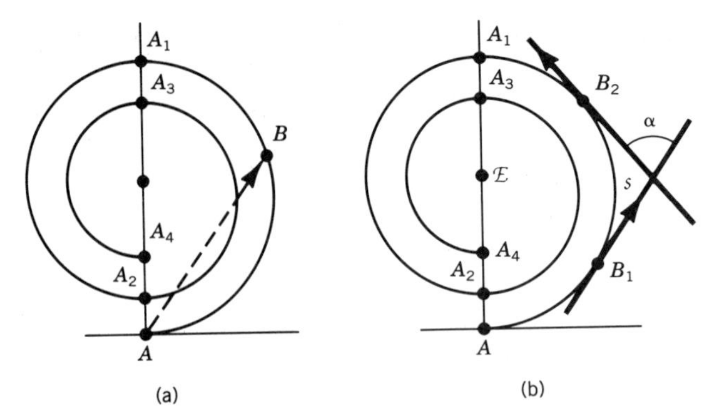

FIGURE 11-15. (a) Vibration curve for determining the Fresnel diffraction from a circular aperture. The change of the diameter of the half-circles making up the spiral has been exaggerated for easy visualization. The actual changes are those displayed in Figure 11-11. (b) B_1 and B_2 are two points on the incident wave surface a distance R_1 and R_2 from the observation point. The arc length s is the arithmetic sum of the amplitude of the wavelets from B_1 and B_2. The phase difference between the wavelets from B_1 and B_2 is α.

To develop a quantitative estimate of the intensity at an observation point on the axis of symmetry of the circular aperture, we construct a spiral (Figure 11-15) to represent the Fresnel zones of a spherical wave incident on the aperture.

The point B on the spiral shown in Figure 11-15*a* corresponds to the portion of the spherical wave unobstructed by the screen. The length of the chord AB represents the amplitude of the light wave at the observation point P_0. As the diameter of the aperture increases, B moves along the spiral in a counterclockwise fashion away from A. AB first reaches a maximum when B reaches the point labeled A_1 in Figure 11-15; the aperture has then uncovered the first Fresnel zone. At this point, the amplitude is twice what it would be with no obstruction! Four times the intensity!

If the aperture's diameter continues to increase, B reaches the point labeled A_2 in Figure 11-15 and the amplitude is very nearly zero; two zones are now exposed in the aperture. Further maxima occur when an odd number of zones are in the aperture and further minima when an even number of zones are exposed. Figure 11-16 shows an aperture containing

FIGURE 11-16. Aperture with four Fresnel zones exposed.

four exposed Fresnel zones. The amplitude at the observation point would correspond to the chord drawn from A to A_4 in Figure 11-15.

The aperture diameter can be fixed and the observation point P_0 can move along, or perpendicular to, the axis of symmetry of the circular aperture. As P_0 is moved away from the aperture along the symmetry axis, i.e., as Z increases, the radius of the Fresnel zones increases without limit. [See **(11-24)** to calculate the zone radius for an incident plane wave.] If Z is large enough, the aperture radius a will be smaller than the radius of the first zone

$$Z_{\text{max}} = \frac{a^2}{\lambda} \tag{11-25}$$

For values of Z that exceed **(11-25)**, Fraunhofer diffraction will be observed. At Z_{max}, the light intensity is a maximum, given by the chord length from A to A_1 in Figure 11-15. If $\lambda = 500$ nm and $a = 0.5$ mm, then this maximum occurs when $Z = 0.5$ m.

If we start at Z_{max} and move toward the aperture along the axis, as Z decreases in value, a point will be reached when the intensity on the axis becomes a minimum. The value of Z where the first minimum in intensity is observed is equal to

$$Z_{\text{min}} = \frac{a^2}{2\lambda}$$

In Figure 11-15, the chord would extend from A to A_2.

As the observation point P_0 is moved along the axis toward the aperture and Z assumes values less than Z_{min}, the point B in Figure 11-15 spirals inward, toward the center of the spiral and the intensity cycles through maximum and minimum values. The cycling of the intensity, as the observation point moves toward the aperture, will not continue indefinitely. At some point, the field on the axis must approach the field observed without the aperture. It is easy to demonstrate that the theory predicts this behavior by calculating the distance between maximum and minimum values of intensity.

Intensity near the Aperture

We can derive an expression for the distance between the maximum and minimum intensity along the symmetry axis of the circular aperture. We will demonstrate that this distance decreases, as the observation point approaches the aperture, until the distance between maxima is equal to a wavelength. Once that position has been reached, the on-axis intensity can be treated as a constant.

Using the geometry of Figure 11-14 yields

$$R_n^2 = Z^2 + a^2$$

where a is the aperture radius. We define the distance

$$q_n = R_n - Z$$

as the increase in distance from the wavefront to the observation point, if we move out on the wavefront to the nth zone. The change in q, as we move between two adjacent zones, is

$$\Delta q = \frac{\lambda}{2}$$

Since

$$R_n^2 = (q_n + Z)^2$$

we may write

$$(q_n + Z)^2 = Z^2 + a^2$$

Differentiating this equation yields

$$\Delta Z = -\frac{q_n + Z}{q_n}\Delta q$$

$$= -\frac{R_n}{R_n - Z}\Delta q \qquad (11\text{-}26)$$

$$= -\frac{R_n}{R_n - Z}\frac{\lambda}{2}$$

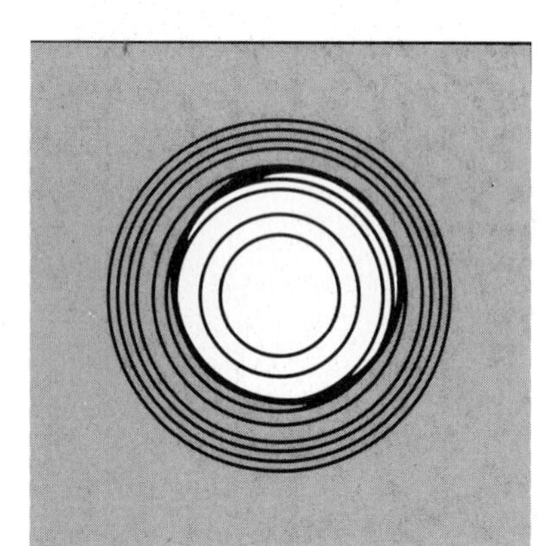

Figure 11-17. The Fresnel zones in the aperture of Figure 11-16 as the observation point moves to the left.

When the observation point is very near the aperture and Z is not very large compared to the aperture radius a, then $Z \approx R_n \approx a$. We have already stated that

$$R_n^2 - Z^2 = (R_n - Z)(R_n + Z) = a^2$$

so that

$$R_n - Z \approx \frac{R_n}{2}$$

Substituting this result into **(11-26)** demonstrates that ΔZ is on the order of a wavelength. Thus, when the observation point gets close to the aperture, the cycles between intensity maxima occur over a distance equal to a wavelength. The intensity changes are then unobservable and the on-axis intensity is a constant.

FIGURE 11-18. The Fresnel zones exposed for the aperture of Figure 11-16 just as we move into the geometrical shadow region.

Off-Axis Intensity

We can also determine in a qualitative way the light distribution perpendicular to the axis of a circular aperture through the use of Fresnel zones. Assume that a plane wave is incident on a circular aperture of radius a and that the observation point is located a distance

$$Z = \frac{a^2}{4\lambda}$$

away from the aperture. There are four Fresnel zones in the aperture at this observation point and the light intensity on axis is very nearly zero; see Figure 11-16. As the observation point is moved off axis to the left, the fourth zone becomes partially obstructed and the fifth zone begins to appear; see Figure 11-17.

Since the negative contribution to the amplitude is decreasing and the positive contribution is increasing, the light intensity increases as the obser-

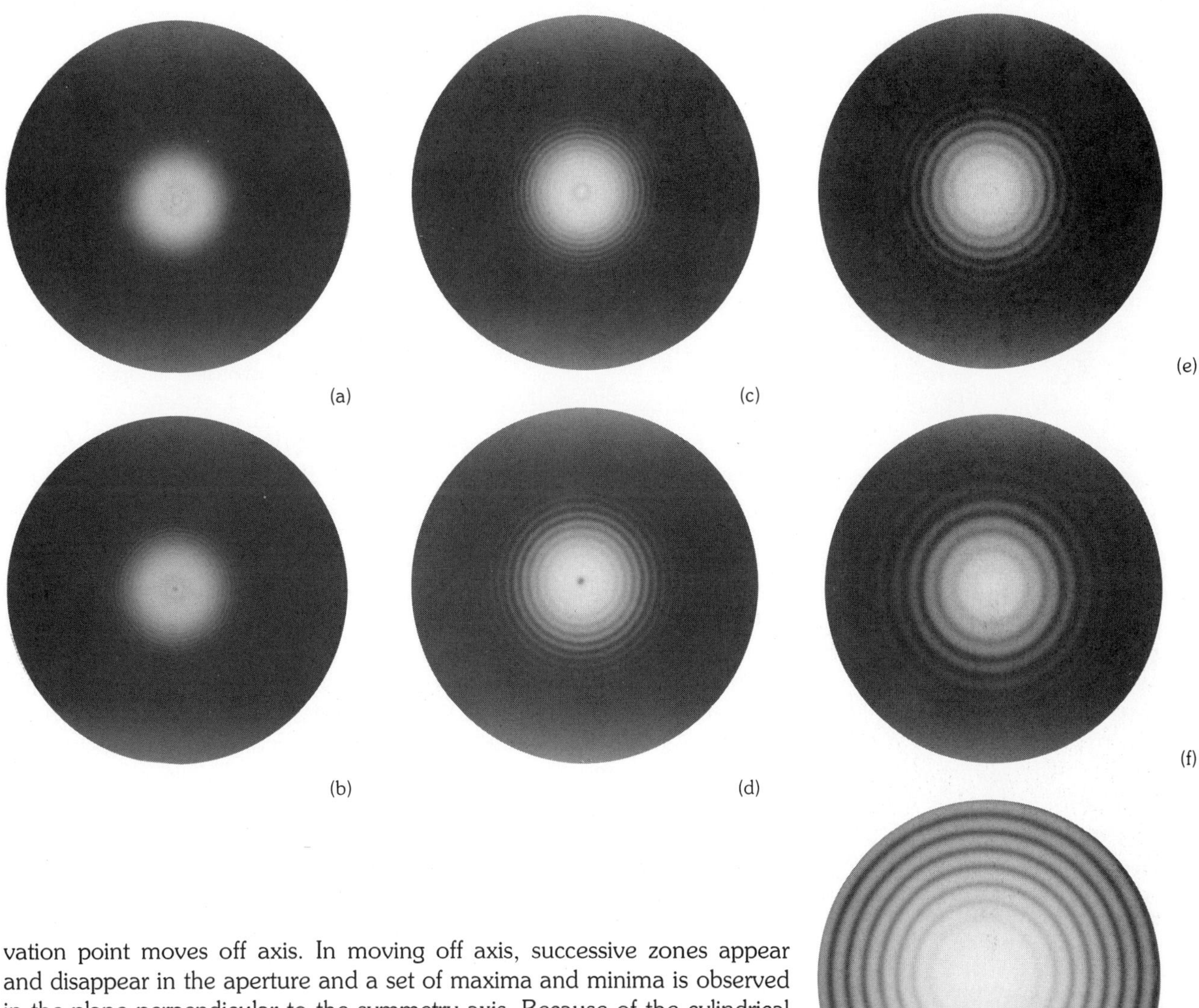

FIGURE 11-19. Fresnel diffraction from a circular aperture measured at positions from the aperture corresponding to (a) 5, (b) 4, (c) 3, (d) 2, (e) 1, (f) 1/2, and (g) 1/3 zones present in the aperture.

vation point moves off axis. In moving off axis, successive zones appear and disappear in the aperture and a set of maxima and minima is observed in the plane perpendicular to the symmetry axis. Because of the cylindrical symmetry of the aperture, lines of constant intensity will be circles about the axis.

If the observation point is moved far off axis (Figure 11-18), the central zone becomes obscured, the contributions from the many exposed partial zones cancel one another, and the intensity falls to zero. Experimental verification of these theoretical predictions is given in Figure 11-19. The images in Figure 11-19 were recorded at various distances from a 1 mm aperture illuminated by a plane wave of wavelength 632.8 nm.

OPAQUE SCREEN

If the screen containing a circular aperture of radius a is replaced by an opaque disk of radius a, the intensity distribution on the symmetry axis behind the disk is found to be equal to the value that would be observed with no disk present. This prediction was first derived by Poisson to demonstrate that wave theory was incorrect; however, experimental observation supported the prediction and verified the theory. We will develop a qualitative derivation of this prediction here, and in Appendix 12-B, a quantitative description will be obtained using Babinet's principle.

(e)

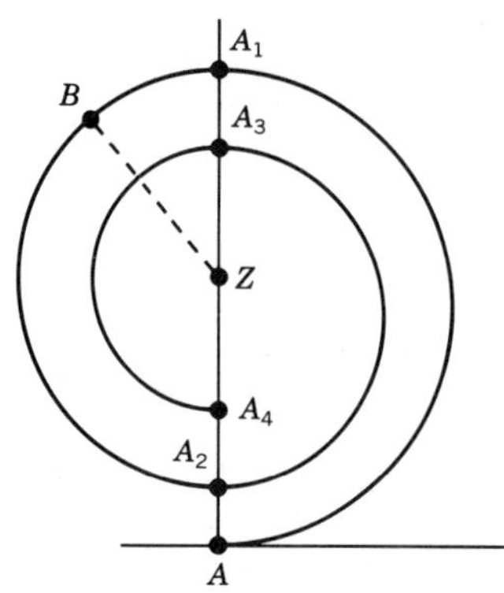

FIGURE 11-20. Vibration curve for opaque disk.

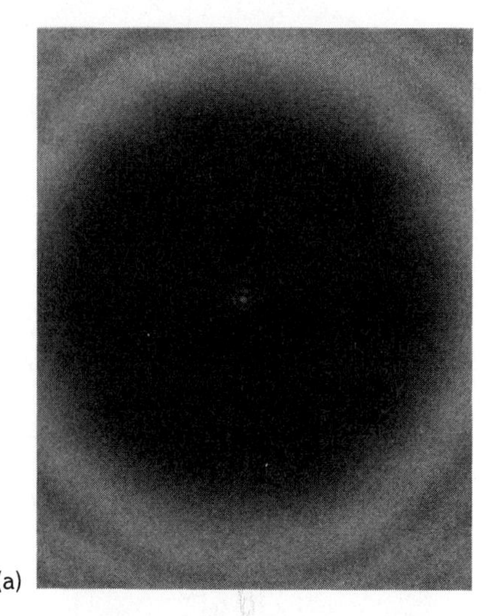
(a)

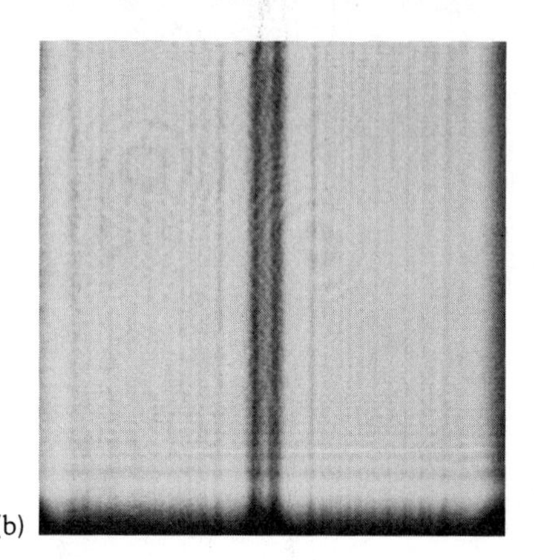
(b)

FIGURE 11-21. Poisson's spot observed in the shadow of (a) a ball bearing and (b) a thin wire illuminated by a plane wave from a HeNe laser.

We construct a spiral, shown in Figure 11-20, similar to the one for the circular aperture in Figure 11-15. The point B on the spiral represents the edge of the disk. The portion of the spiral from A to B does not contribute because that portion of the wave is covered by the disk, and the zones associated with that portion of the wave cannot be seen from the observation point. The amplitude at P_0 is the length of the chord from B to Z shown in Figure 11-20. If the observation point moves toward the disk, then B moves along the spiral toward Z. There is always intensity on axis for this configuration, although it slowly decreases until it reaches zero when the observation point reaches the disk; this corresponds to point B reaching point Z on the spiral. Physically, zero intensity occurs when the disk blocks the entire light wave. There are no maxima or minima observed as the disk diameter a increases or the observation distance changes. If the observation point is moved perpendicular to the symmetry axis, a set of concentric bright rings is observed. The origin of these bright rings can be explained using Fresnel zones, in a manner similar to the one used to explain the bright rings observed in a Fresnel diffraction pattern from a circular aperture. A photo of Poisson's spot is shown in Figure 11-21*a*. A similar bright area is found in the shadow of a rectangular obstruction; see Figure 11-21*b*. The bright area exhibits the same symmetry as the obstruction.

ZONE PLATE

In the construction of Fresnel zones, each zone was assumed to produce a Huygens' wavelet out of phase with the wavelets produced by its nearest neighbors. If every other zone were blocked, then there would be no negative contributions to **(11-22)**. The intensity on axis would be equal to the square of the sum of the amplitudes produced by the in-phase zones, exceeding the intensity of the incident wave. An optical component made by the obstruction of either the odd or even zones could therefore be used to concentrate the energy in a light wave.

We will use the geometry shown in Figure 11-22 to determine how to construct an aperture that would block every other Fresnel zone. We will neglect the curvature of the wavefront, shown by the gray line in Figure 11-22. By ignoring the wavefront curvature, we introduce aberrations into any optical component designed with this theory.

The radius of an arbitrary zone is given by y and can be found by using the pythagorean theorem

$$y^2 = R'^2 - Z'^2, \qquad y^2 = R^2 - Z^2$$

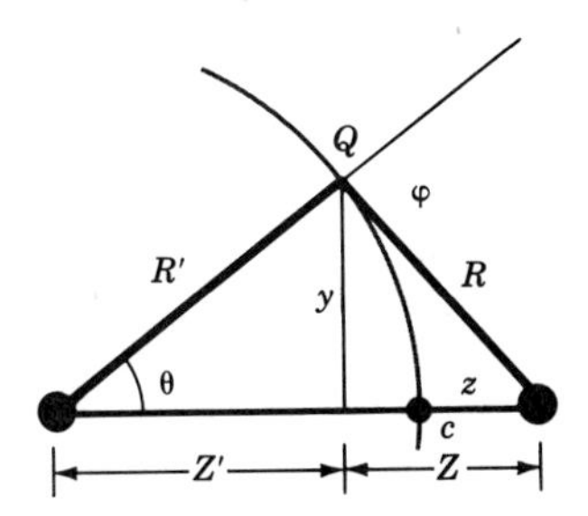

FIGURE 11-22. Geometry for the design of a zone Plate.

The difference in the path length from the center of the wavefront to the point Q, a height y above the optical axis, can be obtained from these two equations

$$R' - Z' = \frac{y^2}{R' + Z'}, \qquad R - Z = \frac{y^2}{R + Z}$$

We will assume that $R' \approx Z'$ and $R \approx Z$, allowing these equations to be rewritten as

$$R' - Z' \approx \frac{y^2}{2Z'}, \qquad R - Z \approx \frac{y^2}{2Z}$$

When $y = r_n$, the radius of the nth Fresnel zone, then the difference in propagation paths for a light ray, propagating from the source to the opening's edge and then to point P_0 and a light ray propagating from the source P_1 to P_0 along the axis, is given by

$$(R' + R) - (Z' + Z) = \pm \frac{n\lambda}{2}$$

By constructing an obstruction in the shape of a ring with a minimum radius given by $y = r_1$, we prevent the out-of-phase light wave of zone 1 from reaching the observation point. The opaque ring will block the out-of-phase light rays as long as its radius satisfies the equation

$$\pm \frac{\lambda}{r_1^2} = \frac{1}{Z'} + \frac{1}{Z} \tag{11-27}$$

Equation **(11-27)** can be generalized to allow the radius of a ring to be calculated that will block a preselected zone, given the source and observation positions

$$\pm \frac{n\lambda}{r_n^2} = \frac{1}{Z'} + \frac{1}{Z}$$

The boundaries of the opaque zones used to block out-of-phase wavelets are seen to increase as the square root of the integers. An array of opaque rings constructed according to this prescription is called a *zone plate;* see Figure 11-23.

By comparing **(11-27)** to **(5A-1)** and **(5A-4)**, we see that a zone plate with zone radii constructed according to the rule just derived will perform like a lens with focal length

$$f = \pm \frac{r_1^2}{\lambda} \tag{11-28}$$

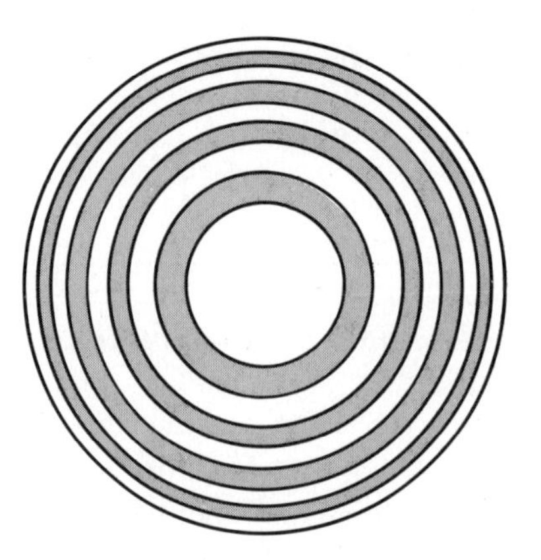

FIGURE 11-23. Zone plate.

The zone plate's operation is based on diffraction, whereas a lens' operation is based on the law of refraction. For this reason, the performance of a zone plate is different than the performance of a conventional lens.

The zone plate shown in Figure 11-24 will act as both a positive and negative lens. What we originally called the source can now be labeled the object point **O**, and what we called the observation point can now be labeled the image point **I**. The light passing through the zone plate is diffracted into two paths, labeled **C** and **D** in Figure 11-24. The light waves, labeled **C**, converge to a real image point **I**. For these waves, the zone plate performs the role of a positive lens with a focal length given by the positive value of **(11-28)**. The light waves, labeled **D** in Figure 11-24, appear to originate from the virtual image point labeled ▯. For these waves, the zone plate performs the role of a negative lens with a focal length given by the negative value of **(11-28)**.

The zone plate will not have a single focus, as is the case for a refractive optical element, but rather it will have multiple foci. As we move toward the zone plate from the first focus, given by **(11-28)**, the effective Fresnel zones will decrease in diameter. The zone plate will no longer obstruct out-of-

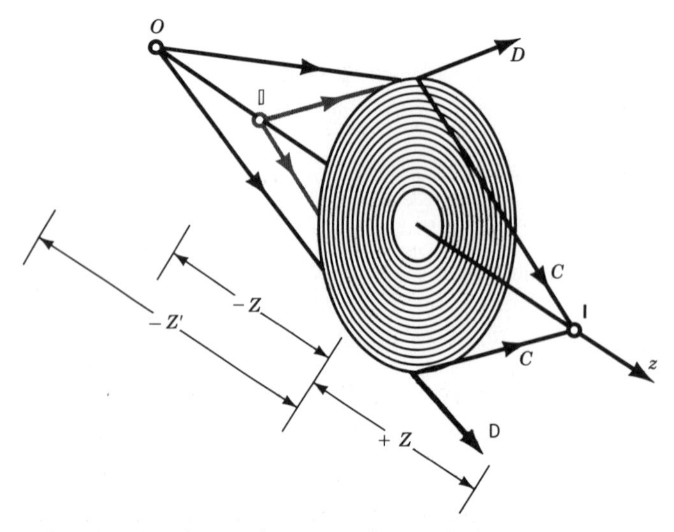

FIGURE 11-24. A zone plate acts as if it were both a positive and negative lens.

phase Fresnel zones and the light intensity on the axis will decrease. However, additional maxima of the on-axis intensity will be observed at values of Z for which the first zone plate opening contains an odd number of zones. These positions can also be labeled as foci of the zone plate; however, the intensity at each of these foci will be less than the intensity at the primary focus.

Lord Rayleigh suggested that an improvement of the zone plate design would result if, instead of blocking every other zone, we shifted the phase of alternate zones by 180°. The resulting zone plate, called a *phase-reversal zone plate*, would more efficiently utilize the incident light. R.W. Wood was the first to make such a zone plate. The difficulty of construction, however, limits practical application of this type of zone plate to the microwave and millimeter wave regions of the spectrum. For example, polyethylene has an index of refraction of 2.25 at 33 GHz. A 9 mm thick sheet of this plastic is about one wavelength thick. By cutting circular slots of radii given by **(11-28)** into a 9 mm sheet of polyethylene, we can produce an efficient millimeter wave, Fresnel zone plate. (Holography provides an optical method of constructing the phase-reversal zone plates in the visible region of the spectrum. Holography also allows the control of the foci produced by the zone plate; see Chapter 12.)

The resolving power of a zone plate is a function of the number of zones contained in the plate. When the number of zones exceeds 200, the zone plate's resolution approaches that of a lens.[49]

Pinhole Camera

A special case of the Fresnel zone plate is an aperture of radius r_1. Only the central zone would exist in this aperture and an image of an object will be generated in a plane whose position is determined by **(11-27)**. An aperture designed using **(11-27)** produces images for a single object or image distance and a single wavelength, but the performance of the device is tolerant of variations of these parameters. The single Fresnel zone pinhole is very small compared to the other dimensions of the camera. If we use geometrical optics to analyze the camera's performance, we find that the pinhole's size prevents more than one ray from an object point from reaching the image plane. For this reason, we expect to observe little image blurring. Figure 11-25 shows an image recorded using such an aperture. The aperture generates acceptable but blurry images over a wide range of wavelengths and object distances.

The aperture containing a single Fresnel zone is called a *pinhole camera* and is quite easy to construct. The camera used to produce Figure 11-25 was made of cardboard and had nominal dimensions of 35–35–45 mm. It was sized to use with 126 film cartridges. The pinhole was about 500 μm in diameter, but the size is not critical. The pinhole can be produced in aluminum foil using a straight pin.

From Figure 11-25, we can see several properties of a pinhole camera

FIGURE 11-25. Photo produced with a pinhole camera.

1. No linear distortion: Observe the tree trunks in all parts of the field.
2. Large depth of field: Near and far objects are in focus.
3. Long exposure times are required: The flowers moved in the breeze during exposure, resulting in a blurred image.

The geometrical optics view of a pinhole camera is not adequate. The modulation transfer function approach discussed in Appendix 10-C is a more correct approach for the analysis of the imaging properties of the pinhole. When this approach is taken, one finds that the pinhole diameter given by **(11-28)** is too large.[50] The proper focal length suggested by a detailed analysis is

$$f = \frac{r_1^2}{3.8\lambda}$$

The pinhole constructed according to this formula covers about 9/10 of the first Fresnel zone. The modification of **(11-28)** is a result of balancing the aberrations, due to the neglect of wavefront curvature, against the resolving power, due to the aperture size.

FERMAT'S PRINCIPLE

The Fresnel zone construction provides physical insight into the interpretation given to Fermat's principle in Chapter 5. Fermat's principle states that if the optical path length of a light ray is varied in a neighborhood about the "true" path, there is no change in path length. By constructing a set of Fresnel zones about the optical path of a light ray, we can discover the size of the neighborhood.

The rules for constructing a Fresnel zone require that all rays passing through a given Fresnel zone have the same optical path length. The true path will pass through the center of the first Fresnel zone constructed on the actual wavefront. A neighborhood must be the area of the first Fresnel zone, for it is over this area that the optical path length does not change. Figure 11-26*a* shows the neighborhood, about the true optical path, as a cross-hatched region equal to the first Fresnel zone.

In Chapter 5, we stated that light waves did not travel over "wrong" paths because "The phase differences for those waves that travel over the 'wrong' paths are such that they destructively interfere. . . ." By moving the neighborhood, defined in the previous paragraph, to a region displaced from the "true" optical path, we can use the zone construction to see that this statement is correct. In Figure 11-26*b*, the neighborhood is constructed about a ray that is improper according to Fermat's principle. We see that this region of space would contribute no energy at the observation point because of the destructive interference between the large number of partial zones contained in the neighborhood.

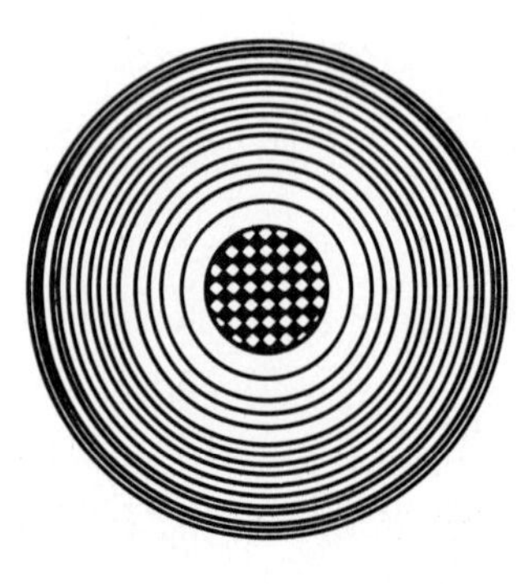

FIGURE 11-26*a*. A set of Fresnel zones has been constructed about the optical path taken by some light ray. If the optical path of the ray is varied over the cross-hatched area shown in the figure, then the optical path length does not change. This cross-hatched area is equal to the first Fresnel zone and is described as the neighborhood of the light ray.

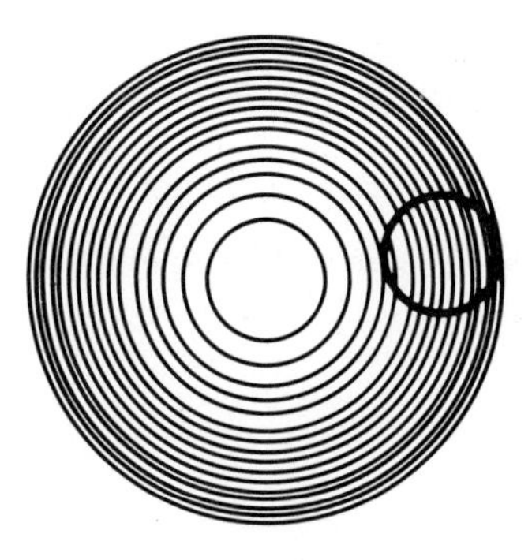

FIGURE 11-26*b*. The neighborhood defined in Figure 11-26*a* is moved so that it surrounds an incorrect optical path for a light ray. We see that this region of space would contribute no wave amplitude at the observation point because of the destructive interference between the large number of partial zones contained in the neighborhood.

The technique of zones introduced in this chapter to obtain information about Fresnel diffraction from a circular aperture can also be applied to rectangular apertures. The vibrational curve, produced using rectangular zones, has a radius of curvature α that is a quadratic function of the arc length s. The equation for the vibrational curve is

$$\alpha = Ks^2$$

where K is a constant of proportionality.[51] This is the equation for a Cornu spiral. Thus, the vibrational curve generated by applying the Fresnel zone construction to apertures of rectangular symmetry is the same Cornu spiral that was obtained by a direct solution of the Huygens–Fresnel integral. It is comforting to see that the two approaches lead to the same result.

A vibrational curve, equivalent to the Cornu spiral, could be constructed for Fraunhofer diffraction. The equation describing such a vibrational curve is

$$\alpha = \frac{\lambda}{2\pi \sin \theta}$$

where

$$\sin \theta = \frac{\xi}{R_0}$$

The radius is a constant, i.e., the vibrational curve is a circle. Such a construction is not nearly as useful for understanding Fraunhofer diffraction as the equivalent construction is for understanding Fresnel diffraction.

All of the techniques introduced in the study of diffraction could be applied generally but are not, because each technique provides some unique physical insight to aid in understanding a particular diffraction problem.

SUMMARY

The Huygens–Fresnel integral is reformulated in this chapter by approximating the distance from a point in the aperture to the observation point and source, respectively

$$R \approx Z + \frac{(x - \xi)^2 + (y - \eta)^2}{2Z}$$

$$R' \approx Z' + \frac{(x_s - x)^2 + (y_s - y)^2}{2Z'}$$

The physical interpretation of this approximation is that the wavefront of the diffracted light is spherical. The integral resulting from the approximation describes Fresnel diffraction

$$\mathcal{E}_P = \frac{i\alpha}{\lambda\rho D} e^{-ikD} \iint f(x, y) \exp\left\{-\frac{ik}{2\rho}\left[(x - x_0)^2 + (y - y_0)^2\right]\right\} dx\, dy$$

The variables in the integral are defined as follows. D, the distance from the source to the observation point, is

$$D = Z + Z' + \frac{(\xi - x_s)^2 + (\eta - y_s)^2}{2(Z + Z')}$$

An effective radius of curvature for the diffracted wave ρ is

$$\rho = \frac{ZZ'}{Z + Z'} \qquad \text{or} \qquad \frac{1}{\rho} = \frac{1}{Z} + \frac{1}{Z'}$$

The diffracted wave is assumed to be only a slightly modified version of the original spherical wave. Finally, the coordinates of the stationary point, about which the integration is performed, are

$$x_0 = \frac{Z'\xi + Zx_s}{Z + Z'}$$

$$y_0 = \frac{Z'\eta + Zy_s}{Z + Z'}$$

Fresnel Integrals

In general, the Fresnel diffraction integral can only be evaluated numerically. For a simple aperture with rectangular symmetry, the integral can be written as

$$\mathcal{E}_P = \frac{i\alpha}{2D} e^{-ikD}\left[C(x) - i\,S(x)\right]\left[C(y) - i\,S(y)\right]$$

where the Fresnel integrals C and S are defined as

$$C(w) = \int_0^w \cos\frac{\pi u^2}{2}\, du$$

$$S(w) = \int_0^w \sin\frac{\pi u^2}{2}\, du$$

The variable w is the coordinate of the aperture edge and is calculated using the equation for the dummy variable u

$$u = \sqrt{\frac{2}{\lambda\rho}}\,(x - x_0) \qquad \text{or} \qquad u = \sqrt{\frac{2}{\lambda\rho}}\,(y - y_0)$$

By plotting $C(w)$ vs $S(w)$, a graphic representation called the Cornu spiral (see Appendix 11-B) is obtained that can be used to calculate the light distribution due to Fresnel diffraction. If this graphic solution is not of sufficient accuracy, then tables of the two Fresnel integrals are used (see Appendix 11-B).

Fresnel Zones

The most useful concept introduced in this chapter is the geometrical construction, called the Fresnel zone. For incident plane waves, the radius of the nth Fresnel zone is given approximately by

$$r_n = \sqrt{nZ\lambda}$$

When the diameter of the zone for $n = 1$ exceeds the diameter of the aperture, then

Fraunhofer diffraction is observed. When the converse is true, Fresnel diffraction is observed. The distance where Fresnel diffraction gives way to Fraunhofer diffraction is given by the Rayleigh range introduced in Chapter 9.

If we use the concept of Fresnel zones, a semiquantitative evaluation of Fresnel diffraction from obstructions with circular symmetry can be obtained. However, a more important benefit gained from using the concept of Fresnel zones is the intuitive understanding of light propagation that can be developed.

PROBLEMS

11-1. Use the table of Fresnel integrals to calculate the points $w = +1.5$, $w = -1.7$ and $w = -1.3$ in the diffraction pattern of a straight edge.

11-2. A HeNe laser ($\lambda = 632.8$ nm) produces a plane wave, which passes through a lens of focal length 25 cm. A 0.5 mm diameter straight wire is oriented vertically in the beam 225 cm from the lens, and the Fresnel pattern is observed 3 m from the wire. Use the Cornu spiral to find intensity (in terms of incident intensity I_0) on axis.

11-3. A point source ($\lambda = 500$ nm) is placed 1 m from an aperture consisting of a 1 mm radius hole with an opaque disk 0.5 mm radius at its center. The observation point is 1 m away. What is the intensity with the aperture present, compared to the intensity without the aperture?

11-4. A plane wave from a HeNe laser ($\lambda = 632.8$ nm) impinges normally on a circular aperture of radius 0.795 mm. What is the intensity 2 m from the hole in terms of the incident intensity?

11-5. A plane wave ($\lambda = 500$ nm) is normally incident on a 1 cm diameter hole. How many Fresnel zones are visible when the aperture is viewed from a position on axis 0.5 m away?

11-6. The innermost zone of a zone plate has a diameter of 0.4 mm. What is the focal length and the first subsidiary focal length when it is illuminated by a plane wave with $\lambda = 441.6$ nm?

11-7. A 3 mm circular aperture is illuminated by $\lambda = 500$ nm plane waves. What are the locations of the first three maxima and minima along the optical axis?

11-8. A plane wave with a wavelength of $\lambda = 500$ nm is incident normal to a square hole 4 mm on a side. The square hole is oriented with its sides parallel to the x and y axis. What is the intensity, relative to the incident intensity, 0.1 mm from the optical axis along the positive y axis, 4 m from the hole?

11-9. A plane monochromatic wave ($\lambda = 488$ nm) is incident normal to an opaque screen containing an aperture. The aperture is an annulus of radii 1 mm and 1.414 mm. What is the amplitude of the electric field on axis 2.05 m away from the screen in terms of the incident field E_0?

11-10. A plane wave ($\lambda = 500$ nm) of intensity I_0 is normally incident on a square aperture 2 mm on a side. What is the intensity 4 m from the screen and 0.1 mm below the optical axis ($-x$ direction) and 0.1 mm to the right ($-y$ direction)?

11-11. Calculate the distance between intensity maxima on the axis of symmetry of a circular aperture using **(11A-6)**. Compare the result to the distance predicted using Fresnel zones **(11-26)**.

11-12. A plane monochromatic wave ($\lambda = 488$ nm) is incident normal to an opaque ring with an outer radius of 1.414 mm and an inner radius of 1 mm. What is the amplitude of the electric field on axis 2.05 m away from the screen in terms of the incident field E_0?

11-13. Using **(10B-7)**, calculate an expression for the resolution of a zone plate in terms of the number of zones and the maximum zone radius.

11-14. What is the longitudinal chromatic aberration of a pinhole camera with a pinhole diameter of 100 μm? Use C and F light in making the calculation.

11-15. A HeNe laser ($\lambda = 632.8$ nm) produces a plane wave that is incident on a 0.5 mm diameter straight wire oriented vertically in the beam. Use the Cornu spiral to find the Fresnel pattern, on axis 3 m from the wire, in terms of incident intensity I_0. Make a plot similar to the one in Figure 11-6.

11-16. An argon laser of wavelength $\lambda = 488$ nm is focused to a point 50 cm in front of a pair of slits. The slits are each 0.1 mm wide and are spaced 0.05 mm apart. Use the Cornu spiral to generate a plot of the Fresnel diffraction pattern a distance 50 cm behind the slits. Plot a graph of the relative intensity vs w over the range $-2.0 \le w \le 2.0$.

Appendix 11-A

BABINET'S PRINCIPLE

In the development of the theory of diffraction, we saw that the diffraction field is due to a surface integral, but no restrictions were imposed on the choice of the surface over which the integration must be performed. This fact leads to a very useful property called *Babinet's principle*, first stated by **Jacques Babinet (1794–1872)** in 1837 for scalar waves. We will discuss only the scalar Babinet's principle; discussion of the rigorous vector formulation can be found in a number of books on electromagnetic theory.[3]

To introduce Babinet's principle, we label a plane separating the source S and observation point P as Σ. If no obstructions are present, a surface integration over Σ yields the light distribution at P. If we place an opaque screen in this plane with a clear aperture Σ_1, then the field at P is given by integrating over only Σ_1; contributions from Σ outside of Σ_1 are zero.

We may define a screen Σ_2 as *complementary* to Σ_1 if the screen is constructed by replacing the transparent regions of Σ, i.e., Σ_1 by opaque surfaces and the opaque regions of Σ by clear apertures. Figure 11A-1 shows two complementary screens where the shaded region indicates an opaque surface.

The surface integral over Σ generates the field E in the absence of any screen. If screen Σ_1 is present, then the diffraction field is E_1, obtained by integrating over Σ_1. According to Babinet's principle, the diffraction field due to screen Σ_2 must be

$$\mathcal{E}_2 = \mathcal{E} - \mathcal{E}_1$$

We will look at examples of the application of the principle for both Fraunhofer and Fresnel diffraction.

Fraunhofer Diffraction

Assume that $a(x)$ is the amplitude transmission of an aperture in a screen, $b(x)$ the amplitude transmission function of the complementary screen, and g the amplitude of the wave when no screen is present. The Fourier transforms of the amplitude transmission functions of the two apertures are equal to the Fraunhofer diffraction patterns that will be generated by the two apertures

$$A(k) = \int_{-\infty}^{\infty} a(x)e^{-ikx}\,dx$$

$$B(k) = \int_{-\infty}^{\infty} b(x)e^{-ikx}\,dx$$

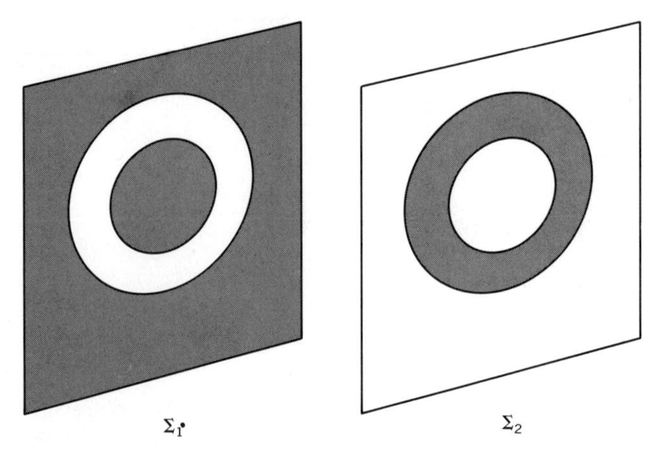

FIGURE 11A-1. An aperture Σ_1 shown by the unshaded region and its complement Σ_2.

With no aperture present, the far-field amplitude is

$$G(k) = g\int_{-\infty}^{\infty} e^{-ikx}\,dx$$

Babinet's principle states that

$$B(x) = G - A(x)$$

must be the diffraction field of the complementary aperture. We may rewrite this equation for the diffraction field as

$$B(k) = g\delta(k) + A(k)e^{i\pi} \tag{11A-1}$$

The first term in the Fraunhofer diffraction field **(11A-1)** is located at the origin of the observation plane and is proportional to the amplitude of the unobstructed wave. The second term in the equation for the Fraunhofer diffraction pattern **(11A-1)** is identical to the Fraunhofer diffraction pattern of the original aperture, except for a constant phase. Thus, the Fraunhofer diffraction from the two complementary apertures is equal, except for a constant phase and the bias term at the origin. Physically, this means that the diffraction intensity distributions of complementary apertures will be identical but their brightness will differ!

Fresnel Diffraction

We can calculate the Fresnel diffraction from an opaque disk by applying Babinet's principle to the diffraction pattern calculated for a circular aperture. The discussion of Fresnel diffraction from a circular aperture in Chapter 12 was only qualitative so we will first derive a quantitative expression for Fresnel diffraction on axis from a circular aperture.

We assume that a circular aperture of radius a is illuminated by a point source a distance Z' from the aperture. We observe the transmitted wave at the point P, located a distance Z from the aperture; see Figure 11A-2.

The Fresnel diffraction integral **(11-8)** must be expressed in cylindrical coordinates. Before doing so, the parameters of the integral are simplified to fit the geometry of Figure 11A-2. Both the observation point and source are

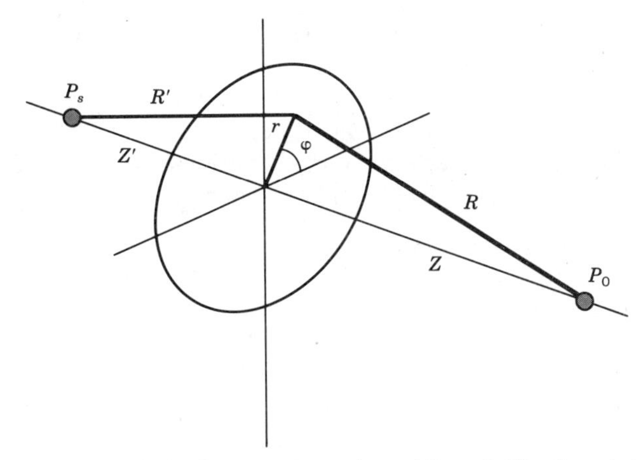

FIGURE 11A-2. Geometry for analysis of Fresnel diffraction of circular aperture.

located on the axis of symmetry of the aperture that lies along the z axis of a cylindrical coordinate system; thus,

$$x_0 = y_0 = \xi = \eta = 0 \qquad \text{and} \qquad D = Z' + Z$$

The approximation for the distance from the source and observation point to the aperture yields

$$R + R' \approx Z + Z' + \frac{r^2}{2\rho} = D + \frac{r^2}{2\rho} \tag{11A-2}$$

where $r^2 = x^2 + y^2$. The cylindrical coordinate version of **(11-8)** is then

$$\mathcal{E}_P = \frac{i\alpha}{\lambda\rho D} e^{-ikD} \int_0^a \int_0^{2\pi} f(r, \varphi) \exp\left[-i\left(kr^2/2\rho\right)\right] r\,dr\,d\varphi \tag{11A-3}$$

We assume that the transmission function of the aperture is a constant, $f(r, \varphi) = 1$, to make the integral as simple as possible. After performing the integration over the angle φ, **(11A-3)** can be rewritten as

$$\mathcal{E}_{Pa} = \frac{i\alpha\pi}{D} e^{-ikD} \int_0^{a^2/\rho\lambda} \exp\left[-i\pi\left(r^2/\rho\lambda\right)\right] d\left(r^2/\rho\lambda\right) \tag{11A-4}$$

Performing the integration of **(11A-4)** results in the Fresnel diffraction amplitude

$$\mathcal{E}_{Pa} = \frac{\alpha}{D} e^{-ikD}\left[1 - \exp\left(-i\pi\frac{a^2}{\rho\lambda}\right)\right] \tag{11A-5}$$

The intensity of the diffraction field is

$$\begin{aligned} I_{Pa} &= 2I_0\left(1 - \cos\frac{\pi a^2}{\rho\lambda}\right) \\ &= 4I_0 \sin^2\frac{\pi a^2}{2\rho\lambda} \end{aligned} \tag{11A-6}$$

Equation **(11A-6)** predicts a sinusoidal variation in intensity as the observation point moves toward the aperture. The separation between maxima in intensity predicted by this result is equivalent to the separation in intensity maxima predicted by **(11-26)**; see Problem 11-11.

Babinet's principle can now be used to evaluate Fresnel diffraction by an opaque disk the same size as the circular aperture, i.e., of radius a. The field for the disk is obtained by subtracting the field, diffracted by a circular aperture, from the field of an unobstructed spherical wave E_P.

$$\mathcal{E}_{Pd} = \mathcal{E}_P - \mathcal{E}_{Pa}$$

$$\mathcal{E}_{Pd} = \frac{\alpha}{D} e^{-ikD} - \mathcal{E}_{Pa} \qquad (11A\text{-}7)$$

$$= \frac{\alpha}{D} e^{-ikD} e^{-i\pi(a^2/\rho\lambda)}$$

The intensity is

$$I_{Pd} = I_0$$

The result of the application of Babinet's principle is the conclusion that at the center of geometrical shadow cast by the disk, there is a bright spot with the same intensity as would exist if no disk were present. This is Poisson's spot. The intensity of this spot is independent of the choice of the observation point. This result agrees with the findings in Chapter 11, obtained by using the Fresnel zone construction.

Appendix 11-B

TABLE 11B.1 Table of Fresnel Integrals

w	$C(w)$	$S(w)$	w	$C(w)$	$S(w)$	w	$C(w)$	$S(w)$
0	0	0	3.6	0.588	0.4923	6.2	0.4676	0.5398
0.1	0.1	0.0005	3.7	0.542	0.575	6.25	0.4493	0.4954
0.2	0.1999	0.0042	3.8	0.4481	0.5656	6.3	0.476	0.4555
0.3	0.2994	0.0141	3.9	0.4223	0.4752	6.35	0.524	0.456
0.4	0.3975	0.0334	4	0.4984	0.4204	6.4	0.5496	0.4965
0.5	0.4923	0.0647	4.1	0.5738	0.4758	6.45	0.5292	0.5398
0.6	0.5811	0.1105	4.2	0.5418	0.5633	6.5	0.4816	0.5454
0.7	0.6597	0.1721	4.3	0.4494	0.554	6.55	0.452	0.5078
0.8	0.723	0.2493	4.4	0.4383	0.4622	6.6	0.469	0.4631
0.9	0.7648	0.3398	4.5	0.5261	0.4342	6.65	0.5161	0.4549
1	0.7799	0.4383	4.6	0.5673	0.5162	6.7	0.5467	0.4915
1.1	0.7638	0.5365	4.7	0.4914	0.5672	6.75	0.5302	0.5362
1.2	0.7154	0.6234	4.8	0.4338	0.4968	6.8	0.4831	0.5436
1.3	0.6386	0.6863	4.9	0.5002	0.435	6.85	0.4539	0.506
1.4	0.5431	0.7135	5	0.5637	0.4992	6.9	0.4732	0.4624
1.5	0.4453	0.6975	5.1	0.4998	0.5624	6.95	0.5207	0.4591
1.6	0.3655	0.6389	5.2	0.4389	0.4969	7	0.5455	0.4997
1.7	0.3238	0.5492	5.25	0.461	0.4536	7.05	0.4733	0.536
1.8	0.3336	0.4508	5.3	0.5078	0.4405	7.1	0.4887	0.4572
1.9	0.3944	0.3734	5.35	0.549	0.4662	7.15	0.5393	0.5199
2	0.4882	0.3434	5.4	0.5573	0.514	7.25	0.4601	0.5161
2.1	0.5815	0.3743	5.45	0.5269	0.5519	7.35	0.516	0.4607
2.2	0.6363	0.4557	5.5	0.4784	0.5537	7.45	0.5156	0.5389
2.3	0.6266	0.5531	5.55	0.4456	0.5181	7.55	0.4628	0.482
2.4	0.555	0.6197	5.6	0.4517	0.47	7.65	0.5395	0.4896
2.5	0.4574	0.6192	5.65	0.4926	0.4441	7.75	0.476	0.5323
2.6	0.389	0.55	5.7	0.5385	0.4595	7.85	0.4998	0.4602
2.7	0.3925	0.4529	5.75	0.5551	0.5049	7.95	0.5228	0.532
2.8	0.4675	0.3915	5.8	0.5298	0.5461	8.05	0.4638	0.4859
2.9	0.5624	0.4101	5.85	0.4819	0.5513	8.15	0.5378	0.4932
3	0.6058	0.4963	5.9	0.4486	0.5163	8.25	0.4709	0.5243
3.1	0.5616	0.5818	5.95	0.4566	0.4688	8.35	0.5142	0.4653
3.2	0.4664	0.5933	6	0.4995	0.447	∞	0.5	0.5
3.3	0.4058	0.5192	6.05	0.5424	0.4689			
3.4	0.4385	0.4296	6.1	0.5495	0.5165			
3.5	0.5326	0.4152	6.15	0.5146	0.5496			

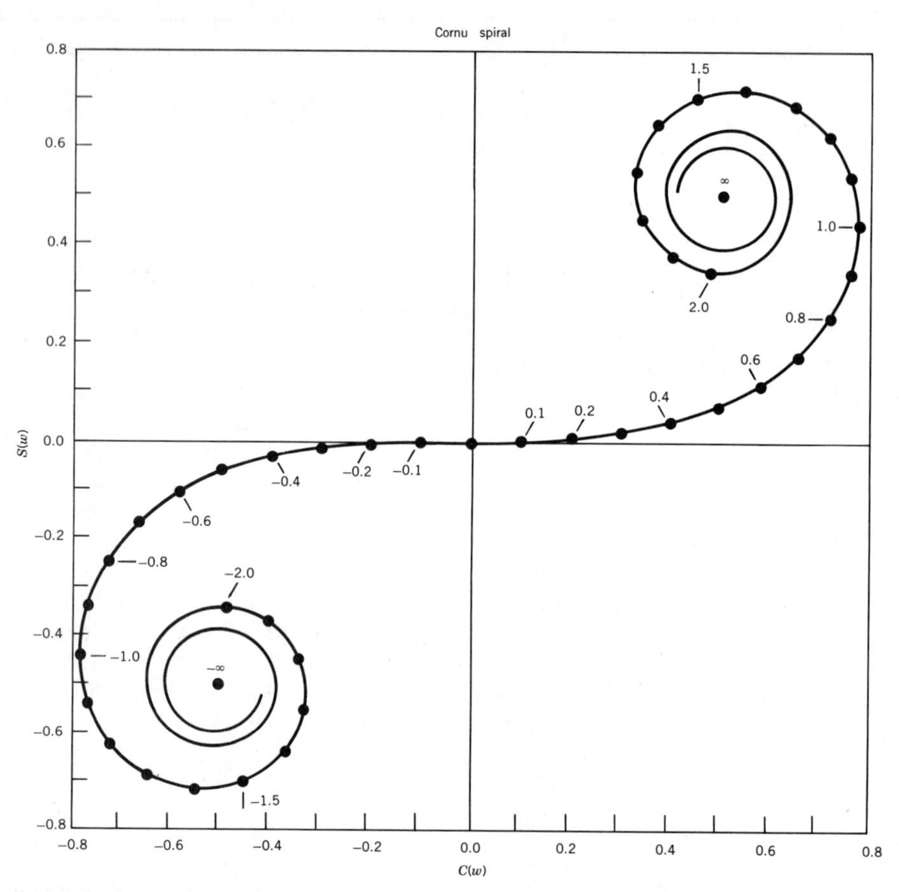

FIGURE 11B-1. Cornu spiral.

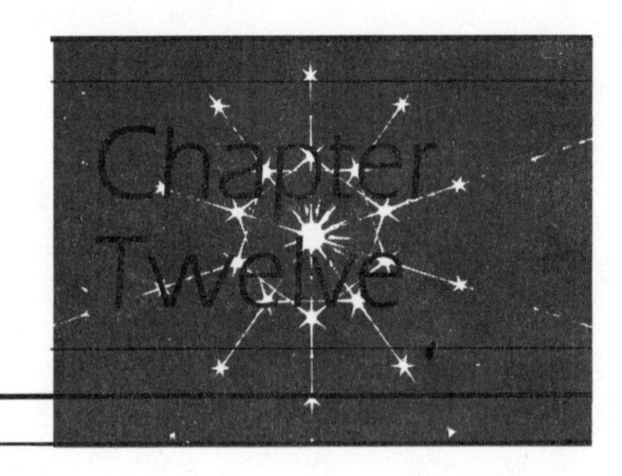

Holography

All detectors of optical radiation respond to the intensity of the incident radiation, but the measurement of intensity does not uniquely determine the optical field. For example, assume two light signals exist, with the time average of the spatial amplitude distribution in the x direction given by $\cos x$ and $|\cos x|$, as shown on the left of Figure 12-1*a*. The measured intensity distributions of these two signals are identical, as shown on the right of Figure 12-1*a*. The inability of a detector to characterize the optical field occurs in all spectral regions and complicates the measurement process. For example, if both the amplitude and phase of a diffracted x-ray pattern could be measured, then a crystal structure could be determined by a simple Fourier transformation of the scattered field.

An interferometric technique can be used to overcome the shortcoming of optical detectors. By the coherent addition of a uniform background (a bias) to the signal that we wish to record, the difference between $\cos x$ and $|\cos x|$ can be detected, as shown in Figure 12-1*b*. The coherent addition of two light fields produces an interference pattern whose recording is called a *hologram*. *Holography*, the study of the production and application of holograms, differs from the study of interference only in the complexity of the waves that are allowed to interfere and the geometry of the interference experiments.

The word hologram was coined from the Greek word *holos*, meaning whole, by **Dennis Gabor**, because the recording of the interference pattern contains all of the information of an image field. Holography was invented by Dr. Gabor in 1948 with the objective of improving the image quality of the electron microscope. (The Nobel Lecture of Dr. Gabor[52] contains a very readable review of the early history of holography.)

Holography did not prove to be useful to electron microscopy and did not attract much interest until the discovery of the laser in 1960. With the development of the off-axis recording technique by **Emmett N. Leith and Juris Upatnieks**,[53] it became possible to produce bright, sharp, three-

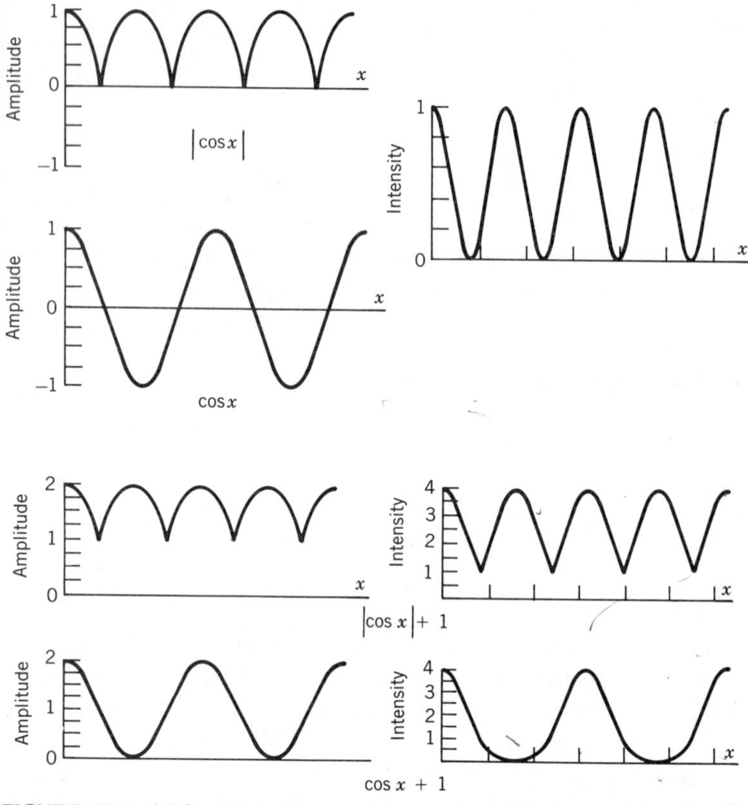

FIGURE 12-1. (a) Spatial distribution of two possible optical fields and the detected spatial intensity distribution. (b) The two spatial field distributions from (a) with a bias added are shown on the left. On the right are the detected intensity distributions.

dimensional images. Both the artistic and scientific communities were stimulated to explore this technology.

A hologram is capable of producing a three-dimensional image of any object. Three-dimensional images produced by holograms have been used for advertising, educational, technical, and artistic displays. Two examples of such applications are shown in Figure 12-2. A typical arrangement used to make a hologram is shown in Figure 12-3. A light wave is divided into two mutually coherent waves. One wave, called the *reference wave*, illuminates the recording medium; the reference wave provides the bias needed for recording both the amplitude and phase of the signal wave. The second coherent wave illuminates the object of interest and the light, scattered from the object, creates the *signal wave*. The signal and reference wave interfere at the hologram plane and a light-sensitive medium records the spatial intensity distribution of the interference.

At some later time, the recorded fringes can be illuminated by a reproduction of the reference wave, called the *reconstruction wave*. The recorded interference pattern diffracts the reconstruction wave into an *image wave*. If an observer looks through the recording, called a *transmission hologram*, he will see a three-dimensional image of the original object created by the image wave. In Figure 12-4, of a rickshaw and the letters "U of M," the basic properties of a hologram are displayed. Comparing Figures 12-4*c* and 12-4*d* demonstrates that the image produced by the hologram has a large

(a)

(b)

FIGURE 12-2. (a) A hologram used to help sell a box of cereal. (b) An artistic display hologram.

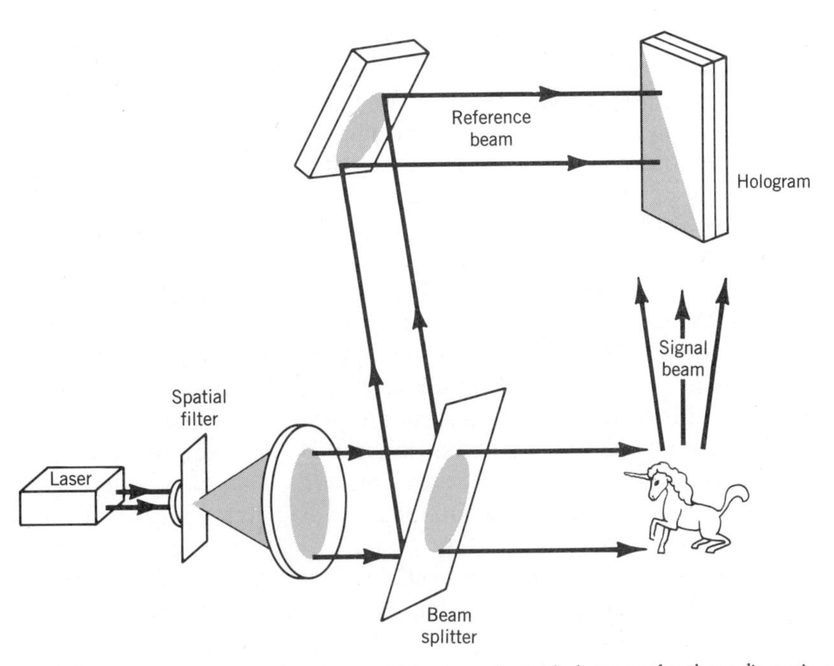

FIGURE 12-3. Experimental arrangement for recording a hologram of a three-dimensional object.

(a)

(b)

FIGURE 12-4. These photos of an early hologram display the three-dimensional properties of the holographic image. The parallax associated with a three-dimensional image can be seen by comparing images (a) and (b). The depth contained in the images is demonstrated by focusing the camera on first the foreground image (c) and then the background image (d). (Courtesy J. Upatnieks, Environmental Research Institute of Michigan.)

depth-of-field. Figure 12-4*b* demonstrates, by the obstruction of the word "of," that the holographic image displays parallax, that is, we can look around near-field objects. The three-dimensional image is so striking that it is easy to forget that a hologram has applications other than display. A number of the applications of holography[54] are actually more important to technology than displays.

Holograms of large volumes of air containing microscopic particles can be recorded and later examined, in detail, to measure the particle size distributions. These distributions are of interest in the study of such diverse fields as meteorology, the study of fog formation; biology, the study of marine plankton; and elementary particle physics, the analysis of bubble chamber tracks.

The recorded interference pattern, created by the reference and signal waves, forms a diffraction grating. If the two interfering waves are simple plane waves, the holographic grating has the same appearance as a ruled grating. If photoresist is used as the recording medium, exposure and development will produce a relief image of the interference pattern; by proper alignment, the relief pattern produced is triangular in shape, producing a blazed grating. If the relief image is coated with a metal film, a reflection grating is produced. Holographic gratings offer several advantages over ruled gratings. Holography is a low-cost method of producing not only plane gratings but also gratings on curved shapes and gratings with curved grooves. Gratings produced holographically are free from both periodic and random errors in grating period, defects normally found in ruled gratings. The holographic gratings are not only used for wavelength measurement; specially designed gratings are now used as optical scanners in supermarkets, copy machines, and printers.

The hologram can be viewed as an optical component that transforms the reconstruction wave into the signal wave. If the reconstruction wave were a plane wave and the signal wave were a spherical wave, then the hologram would perform the same transformation performed by a simple lens. This type of hologram, called a *holographic optical element* (HOE), has been used in a number of specialized optical applications. One example is the "heads-up" display that projects important instrument readings into the forward field of view of an aircraft pilot.

Because holography is an extension of interference, it is not surprising to find a large number of applications of holography in interferometry. Holography allows the study of nontransparent objects, a capability not available to conventional interferometry. Holography also provides interferometry the capability of comparing a stored wavefront with a wave produced at a different location or at a different time. For example, in nondestructive testing, a hologram is made of a component and later, that same component is modified by the application of stress, induced by, for example, pressure or temperature. By making a double exposure or viewing the component through a previously exposed and developed hologram, the result of the applied stress can be measured.

Other interferometric applications utilizing holography can be found in medical and dental research, aerodynamics, plasma diagnostics, and the measurement of the amplitude and phase of vibrating objects.

A number of applications of holography have been suggested, but have yet to be extensively implemented. One such application is holographic information storage, either as an alternate implementation of a conventional memory or an associative memory. (An associative memory allows the recovery of data without complete knowledge of its location.) A second suggested application is the use of holography to record the amplitude and phase of Fourier transforms, for use in optical correlators and convolvers, as discussed in Chapter 10. A brief discussion of the theory of the latter type of hologram is contained in Appendix 12-B. In the future, the use of four-wave mixing, discussed in Chapter 15, will allow the use of a dynamic holographic device.

In this chapter, we will attempt to develop an intuitive understanding of holography by developing a theory based on point objects. The first examination of holography will treat the recording and reconstruction of interference patterns, using collinearly propagating reference and signal waves. The recording will be assumed to be an amplitude recording; the use of phase recording will be briefly discussed in Appendix 12-A. In the discussion of holography, we will ignore effects of hologram thickness. A more detailed analysis of holography can be found in a book by Collier, Burckhardt, and Lin.[55]

We will demonstrate that the hologram of a collinearly propagating reference and signal wave can produce the amplitude and phase of the signal wave when illuminated by the proper reconstruction wave. We will find that a number of unwanted waves are also present when this configuration is used. By allowing the reference wave to propagate at an angle with respect to the signal wave, we will find that the unwanted waves can be spatially separated from the desired signal wave. Initially, Fourier transforms are used to analyze the *off-axis holograms*, recorded using noncollinearly propagating waves.

An intuitive understanding of off-axis holograms can be developed by viewing the hologram as a large number of Fresnel zone lenses. We will develop a theory around this view, based on a hologram recording of the interference between a spherical signal wave and a plane reference wave.

The simple model developed allows specifications to be placed on the recording material. We will discover that the properties of a hologram make unique demands on the recording material. These demands exceed the requirements of ordinary photography.

By using a point source for both signal and reference waves, a geometrical model of holographic imaging is developed. The very simple model allows the introduction of a number of the unique properties of holograms. It also allows the discussion of a number of special recording geometries that are experimentally important.

The coherency requirements must be addressed in any process based on interference. Holography is not an exception. The last topic of this chapter will be the coherence requirements that are placed on the light waves used in holography. One of the most interesting results of the brief analysis is that the reconstruction wave's coherence affects the resolution of images produced by a hologram.

(c)

(d)

FIGURE 12-4. *(Continued)*

HOLOGRAPHIC RECORDING

We will demonstrate that the holographic technique can be used to record, using a conventional intensity recording, the amplitude and phase of a wave. More important, we will demonstrate that it is possible to use the recording to regenerate either of the recorded waves. For this demonstration, we will examine the recording and playback of a hologram, without regard to the wave's propagation direction. We will do this by assuming that all of the waves propagate in a collinear fashion. In the following section, we will consider the more general, and more useful, situation where the waves propagate at an angle with respect to each other.

Assume that the signal wave can be represented by the complex amplitude

$$a(x, y) = a_0(x, y)\, e^{-i\phi(x, y)} \tag{12-1}$$

and the reference wave by the complex amplitude

$$A(x, y) = A_0(x, y)\, e^{-i\Psi(x, y)} \tag{12-2}$$

The theory of interference from Chapter 4 (**4-12**) gives the resultant intensity

$$\begin{aligned} I(x, y) &= AA^* + aa^* + A^*a + Aa^* \\ &= |A(x, y)|^2 + |a(x, y)|^2 + 2A_0a_0 \cos \delta \qquad (12\text{-}3) \\ \delta(x, y) &= \Psi(x, y) - \phi(x, y) \end{aligned}$$

If the two waves were spherical waves produced by point sources, then the interference fringes described by (**12-3**) would appear as the dark lines shown in Figure 12-5. We will assume that the recording medium is located in a plane normal to the line, parallel to the z axis, connecting the two sources. In this situation, all of the waves in the problem are collinearly propagating.

A number of methods exist for recording the intensity in (**12-3**). Conventional photographic film records the fringes as variations in the absorption of the light transmitted through the film. For obvious reasons, this is called an *absorption hologram*. The silver in the developed absorption hologram can be replaced, through a process called *bleaching*, by a

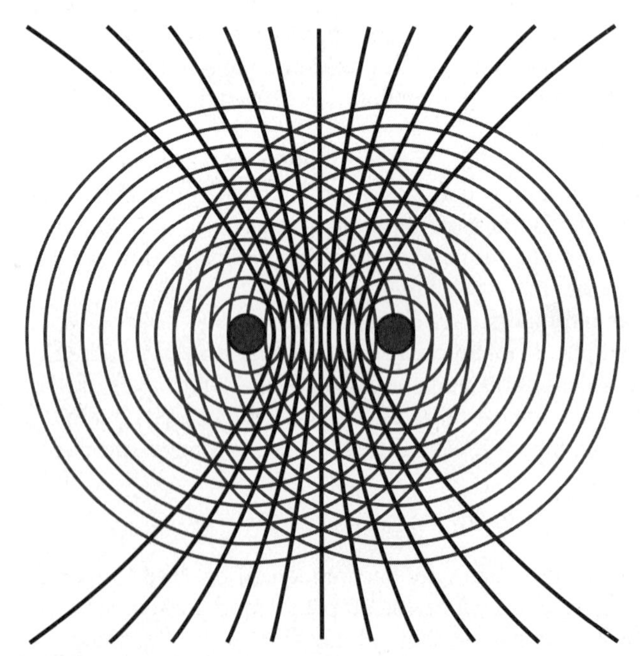

FIGURE 12-5. Interference fringe produced by two point sources is displayed by the dark lines in the drawing. The light lines represent the wavefronts produced by the two point sources. The point sources are the two dark spots shown in the drawing.

transparent compound with an index of refraction different than the film's gelatin base; the result is a *phase hologram*. In this type of hologram, the reconstruction wave's phase is modified, at each spatial position on the hologram, by an amount proportional to the intensity of the interference pattern at that position. It is also possible to produce phase holograms directly in certain photopolymers or gelatin sensitized with ammonium dichromate. Finally, the thickness of the recording medium can be varied to produce a phase hologram. Thermoplastic films and photoresist have been used to accomplish the required surface deformation. We will limit our discussion to amplitude recording with only a brief discussion of phase recording in Appendix 12-A.

Whatever medium is used, the recording medium must linearly map the intensity, of the interference pattern incident during the recording, into a complex amplitude transmitted by the medium after development. This requirement results in a proper exposure that differs from the exposure selected for a linear intensity recording in incoherent light. An example will clarify this requirement.

We will select photographic film as the recording medium because we are familiar with the proper exposure of this medium in incoherent light. The recording characteristics of a photographic film are given by the H&D curve (first used by Hurter and Driffield in 1890); see Figure 12-6. The H&D curve, characterizing a photographic emulsion, displays the variation in the density of the developed emulsion as a function of the logarithm of the energy used to expose the undeveloped emulsion E. The exposure energy is defined as

$$\mathbf{E} = I_e \tau_e$$

where I_e is the exposure illumination received by the emulsion and τ_e is the exposure time. The optical density of the developed emulsion is defined as

$$\mathbf{D} = \log\left(\frac{I}{I_0}\right) = \log\left(\frac{1}{\mathbf{T}}\right) = \log\left(\frac{1}{\mathbf{t}^2}\right)$$

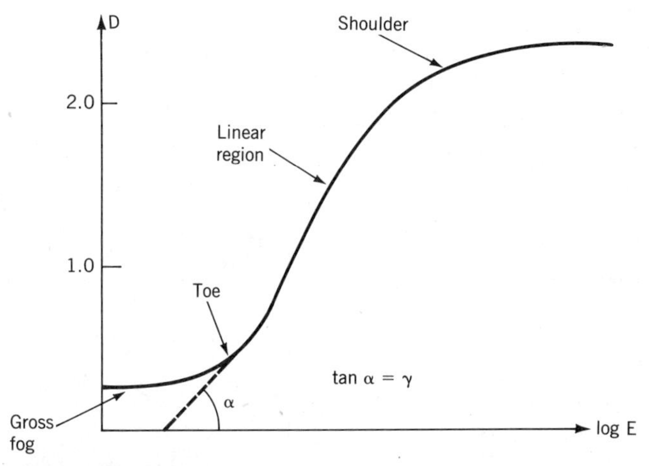

FIGURE 12-6. A typical H&D curve. The horizontal axis is the log of the exposure energy (intensity • time) and the vertical axis is the exposed film's density, $D = \log(I/I_0)$, where I is the transmitted intensity and I_0 the incident intensity. The slope of the linear region of the curve is called the film's γ.

where **T** is the transmissivity and **t** the amplitude transmission coefficient of the emulsion [the transmission coefficient has the same form as (**3-45**) with the real indices of refraction replaced by complex indices].

The process of generating a conventional linear photographic recording is illustrated in Figure 12-7. On the right of Figure 12-7 is displayed the linear recording of a spatial intensity distribution, labeled the signal. To obtain the linear intensity recording, an exposure time τ_e is selected to insure, if possible, that the maximum and minimum illumination intensity fall on the linear portion of the H&D curve.

With the proper exposure energy, the density of the developed film is

$$\mathbf{D} = \gamma_N \log \mathbf{E} - \mathbf{D}_0$$
$$= \gamma_N \log \left(I_e \tau_e\right) - \mathbf{D}_0$$

where γ is the slope of the linear portion of the H&D curve, called the film's *gamma*. The resulting photo is a negative of the original signal and is indicated by the subscript N on the gamma of the film. If the negative, produced by developing, is illuminated by a reconstruction intensity I_0, the transmitted intensity is given by

$$I_t = I_0 \mathbf{t}^2 = I_0 \mathbf{T}$$
$$= I_0 (10)^{-\mathbf{D}}$$

This relationship can be rewritten in terms of the original signal intensity

$$I_t = I_0 (10)^{\mathbf{D}_0} \tau_e^{-\gamma_N} I_e^{-\gamma_N}$$
$$= K I_e^{-\gamma_N} I_0$$

The transmitted intensity is labeled "playback" in Figure 12-7. The resulting spatial intensity distribution is not a linear reproduction of the scene intensity

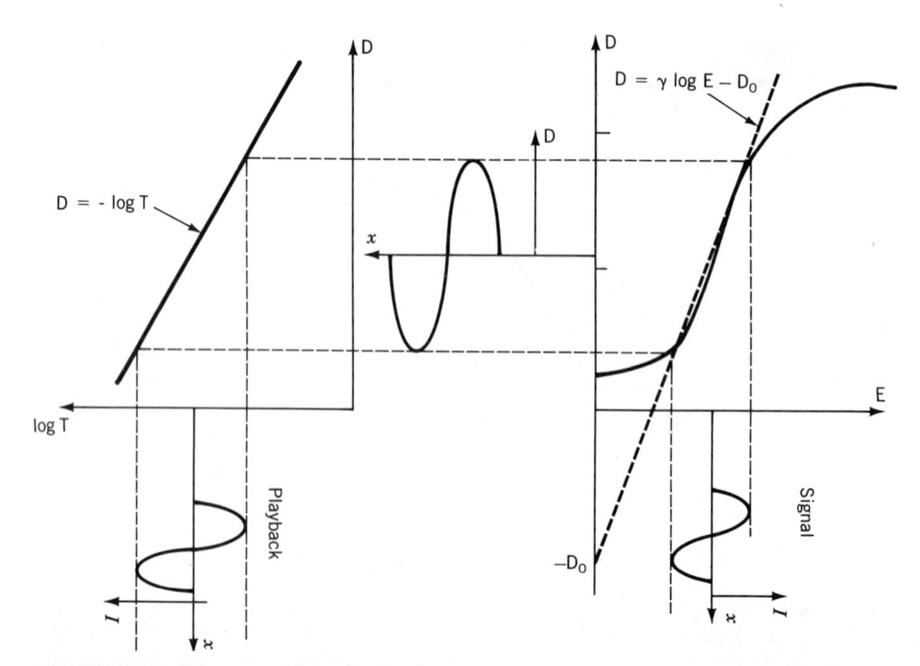

FIGURE 12-7. Linear recording of a signal.

I_e we wished to record. As is normally done in photography, a print of the negative is made, yielding a positive representation of the original signal. The transmitted intensity of the positive copy is

$$I_{tP} = K_P I_e^{\gamma_P \gamma_N} I_0.$$

The relationship between this playback intensity I_{tP} and the signal intensity I_e is linear if the product $\gamma_N \gamma_p$ is equal to 1, where γ_p is the gamma associated with the recording of the positive copy.

In holography, the recording process should provide, after development, a transmitted wave whose complex amplitude is a linear reproduction of the interference pattern, produced by the signal and reference waves. The intensities of the reconstruction wave I_0 and the transmitted wave I_t are obtained from the square of the amplitudes of these waves

$$I_t = E_t E_t^*, \qquad I_0 = E_0 E_0^*$$

The playback amplitude is therefore

$$E_{tP} = E_0 K_P I_e^{\frac{\gamma_P \gamma_N}{2}}$$

where I_e is given by (**12-3**). This leads to the conclusion that $\gamma_n \gamma_p = 2$ is required for a hologram to produce a linear reproduction of the signal wave. The use of off-axis holograms, to be discussed a bit later, removes the need for a positive copy of the hologram, but we will still find it more convenient to use the amplitude transmission coefficient curve, shown in Figure 12-8, to characterize the recording medium.

The correct exposure for a linear holographic recording is selected to lie not on the linear portion of the H&D curve, but rather on the linear portion of the **t** vs **E** curve of Figure 12-8. We also require that the reference wave amplitude A be much larger than the signal wave's amplitude a so that the variation in the transmission coefficient remains on the linear portion of the

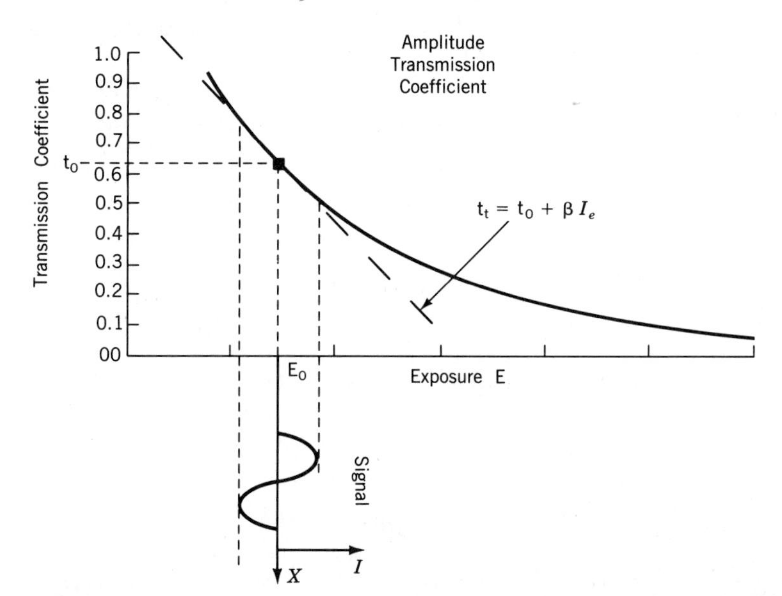

FIGURE 12-8. Amplitude transmission coefficient as a function of exposure for Kodak high-resolution plates developed in Kodak D19 developer. The point on the curve occurs at an exposure $|A|^2\tau_e = E_0$ and a transmission coefficient t_0.

t-E curve in Figure 12-8. Restricting the recording parameters in this fashion allows the recorded transmission coefficient to be approximated by the first two terms in a Taylor expansion. (In practice, this restriction on the size of a may be ignored, as we will discuss later.)

The optimum exposure for a linear amplitude recording selected in Figure 12-8 occurs near the toe of the H&D curve as shown in Figure 12-9. Thus, the optimally recorded hologram would appear underexposed by conventional photographic standards.

Assume that an exposure time τ_e has been selected that places the spatial average of the transmission coefficient in the linear region of the amplitude transmittance curve; the exposure energy

$$E_0 = \left(|A|^2 + |\ll a \gg|^2\right)\tau_e$$

where $\ll a \gg$ is the spatial average of the signal beam, yields a transmission coefficient t_0 in Figure 12-8. After the film is exposed and developed, we obtain a transmission coefficient described by the equation

$$t_h(x, y) = t_0 + \beta(A^*a + Aa^*) \tag{12-4}$$

To simplify the notation, we have defined

$$\beta = \beta'\tau_e$$

where β' is the slope of the exposure vs amplitude transmittance curve.

The developed film is illuminated by a reconstruction (or playback) wave $B(x, y)$, resulting in a transmitted wave of amplitude

$$B(x, y)\, t_h(x, y) = t_0 B + \beta A^* B a + \beta A B a^* \tag{12-5}$$

If the reconstruction wave is identical to the reference wave $B = A$, the second term of (**12-5**) is $\beta|A|^2 a(x, y)$, and the original signal wave is reproduced. If the reconstruction wave is the conjugate of the reference wave $B = A^*$, the third term is $\beta|A|^2 a^*(x, y)$, and the conjugate of the original signal beam is produced. Either of these terms produces a usable duplication of the original signal wave's amplitude and phase.

To understand the meaning of the conjugate of a wave, consider a diverging spherical wave generated by a point source, displayed graphically

There is a change in the thickness of the developed film due to the replacement of silver halide by silver in the exposed regions and the removal of silver halide in the unexposed region. This thickness change introduces a phase modulation of the transmitted wave. The phase modulation due to the developing process can be removed experimentally so we will ignore it.

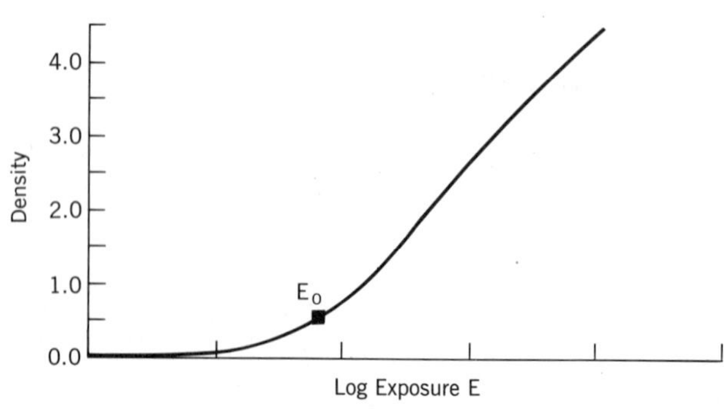

FIGURE 12-9. The density vs exposure curve for the amplitude transmittance curve shown in Figure 12-4. The same operating point E_0 shown in Figure 12-8 is indicated by the square.

on the left of Figure 12-10*a*, by the wave labeled a. The conjugate of a, that is, a^*, is identical to a but is propagating backward in time (the time reversal of a) and backward in space. The conjugate wave is shown graphically on the right of Figure 12-10*a* as a wave converging toward a point.

The terms in **(12-5)** can now be interpreted by first assuming that a is a diverging spherical wave. The first term in **(12-5)** is equal to the average amplitude transmission of the developed film. The second term of **(12-5)** can be interpreted as containing the original diverging spherical wave a, when B is identical to A. The transmitted wave appears to originate from a nonexistent point source; thus, a virtual image of the original point source has been reconstructed. The third term of **(12-5)** can be interpreted as a wave, converging toward a point, when B is equal to A^*. For this case, a real image of the original point source has been reconstructed.

In holography, the complex amplitude of the signal wave is stored in the hologram in the form of an interference pattern; all information concerning the temporal character of the wave is lost. If the recording medium is thin, then the information about the phase of the wave along the propagation direction is also lost. To first order then, a hologram only records information about the phase of the signal wave transverse to the propagation direction. Because the propagation information is lost upon reconstruction, the transmitted wave assumes the propagation properties of the reconstruction wave. When it is appropriate to ignore the hologram thickness, a second type of conjugate wave, the space conjugate wave, can be defined in terms of the complex amplitude of the wave. On the left of Figure 12-10*b*, the wave vector of a plane wave propagating in the x, z plane is shown in black.

We assume the hologram is positioned normal to the z axis. The wave stored in the hologram is given by

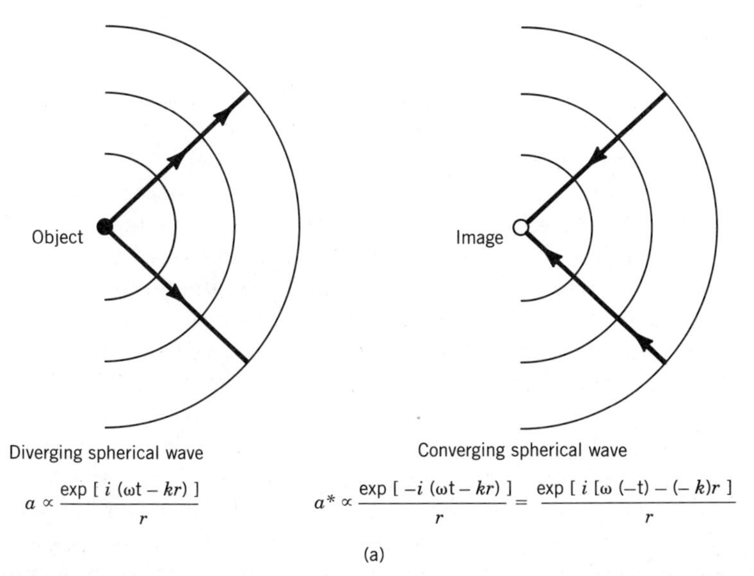

FIGURE 12-10*a*. A spherical wave and its conjugate. The conjugate wave is identical to the original wave but it propagates backward in time $(-t)$ and in the opposite direction $(-k)$.

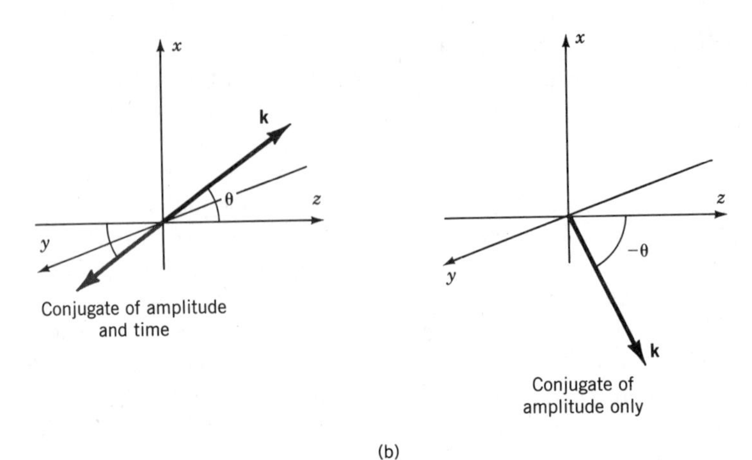

FIGURE 12-10*b*. The wave vector of a plane wave is shown in black on the left. The conjugate wave when both temporal and spatial characteristics of a wave must be considered is shown on the left as a shaded wave vector. The conjugate wave when only the complex amplitude of a wave must be considered is shown on the right as a black wave vector.

$$a = a_0 e^{-i\mathbf{k}\cdot\mathbf{r}} \approx a_0 e^{-ikx \sin\theta}$$

When both the temporal and spatial characters of the wave are considered, the previous interpretation of the conjugate wave yields a wave with the wave vector shown, as a shaded vector, on the left of Figure 12-10*b*. If only the stored amplitude of the wave is considered, the space conjugate wave would have the wave vector shown on the right of Figure 12-10*b*. The mathematical representation of the space conjugate wave is

$$\begin{aligned} a^* &= a_0^* e^{+ikx \sin\theta} \\ &= a_0^* e^{-ik(-x)\sin\theta} \end{aligned}$$

OFF-AXIS HOLOGRAPHY

We have just demonstrated, through **(12-5)**, that by recording the interference pattern produced by a signal and reference waveform, the signal waveform can be reconstructed at a later time. The reconstruction is accomplished by using a "reconstruction" waveform identical to the original reference waveform. There are, however, two unwanted signals that must be separated from the desired signal waveform. The method used to separate the unwanted and desired signals is to use noncollinear waveforms. This off-axis technique seems to be a trivial modification of the collinear holographic method introduced by Gabor; however, it was over 10 years after the original papers by Gabor before Leith and Upatnieks introduced the off-axis technique.

To understand the off-axis technique and obtain information about how the off-axis angle affects the properties of the hologram, we will use Young's two-slit experiment as a model. The necessary equations have been derived in previous chapters. Here, we need only make the proper physical interpretation of the mathematics to understand the operation of an off-axis hologram.

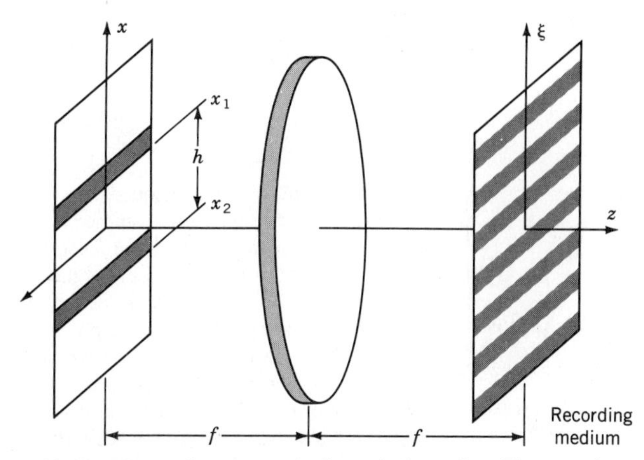

FIGURE 12-11. Geometry of off-axis holography. The two slits and the lens produce two plane waves arriving at the recording medium at angles of θ_1 and θ_2 with respect to the normal of the recording medium where $\theta_i = x_i/f$.

Recording

The experimental arrangement that corresponds to the mathematical model is shown in Figure 12-11. Two slits of infinitesimal width in the x direction and infinite width in the y dimension are located at the front focal plane of the lens: slit 1 at position x_1 and slit 2 at position $-x_2$, measured from the optical axis of the lens, the z axis. These slits act as sources for two spherical waves, the signal and reference waves, that arrive at the recording medium, placed in the back focal plane of the lens, as plane waves propagating at the angles

$$\theta_1 = -\frac{x_1}{f}, \qquad \theta_2 = \frac{x_2}{f}$$

with respect to the optical axis. A magnified view of the recording medium is shown in Figure 12-12. The interference pattern that will be recorded is seen in Figure 12-12 to extend throughout the recording medium. The fringes are parallel to the bisector of the angle $\theta_1 + \theta_2$ between the signal and reference waves.

The mathematical description of the interference pattern produced in the back focal plane of the lens in Figure 12-11 is equal to the Fraunhofer diffraction pattern produced by the two slits and is given by **(10-31)**

$$I_D = I_0 \frac{\sin^2 \alpha}{\alpha^2} \cos^2 \beta_h$$

Because we have assumed that the widths of the slits are infinitesimal, we may assume that sinc $(\alpha) \approx 1$. The fringe pattern to be recorded in the hologram positioned in the back focal plane of the lens is, therefore, a simple sinusoidal intensity distribution given by

$$\begin{aligned} I_D &= I_0 \cos^2 \beta_h \\ &= \frac{I_0}{2}(1 + \cos 2\beta_h) \end{aligned} \tag{12-6}$$

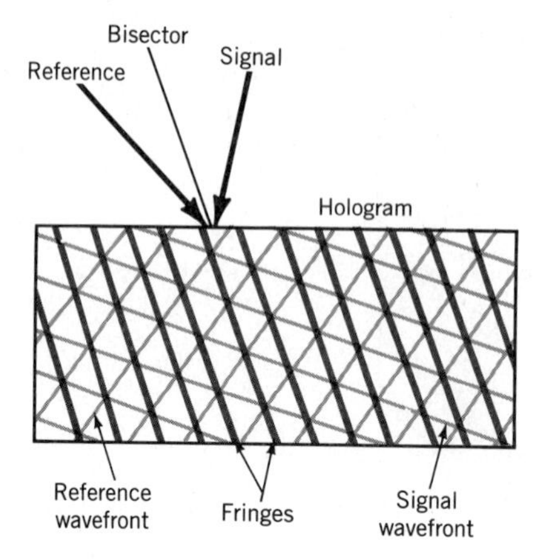

FIGURE 12-12. Fringes, produced by the interference of the signal and reference waves, are shown by the gray areas in the holographic recording medium. The wavefronts of the two waves are shown by the two types of light gray lines. The fringes are parallel to the bisector of the angle between the two waves. If the bisector of the angle is parallel to the normal to the surface of the recording medium, then the fringes will also be normal to the surface.

where

$$\beta_h = \frac{k\,(x_1 + x_2)}{2}\frac{\xi}{f} = \frac{\pi h \xi}{\lambda f}$$

Development

After recording the interference fringes in the x, y plane and developing the recorded intensity distribution, the transmission coefficient of the hologram is given by

$$\begin{aligned} t_H &= t_0(1 + d_H \cos 2\beta_h) \\ &= t_0 + \frac{t_0 d_H}{2}(e^{-i2\beta_h} + e^{i2\beta_h}) \end{aligned} \qquad (12\text{-}7)$$

where

$$t_0 = \frac{\beta' I_0 \tau_e}{2}$$

An additional factor d_H, the amplitude contrast, has been added to this equation. The contrast allows the inclusion of the visibility of the interference fringes into the equation for the recording of the interference pattern from the two slits. (It should be noted, however, that a contrast less than 1 may result not only from a fringe visibility that is less than 1, but also from a nonlinear recording process.)

Reconstruction

The developed hologram is placed in the front focal plane of the lens and illuminated with a plane wave of amplitude B, as shown in Figure 12-13. The light transmitted by the hologram is diffracted into three plane waves that are focused onto the back focal plane of the lens. The mathematical description of the diffraction pattern is given by **(10B-4)**.

$$\mathcal{E}_H(\xi) = B t_0 \left[\mathrm{sinc}(\xi) + d_H \mathrm{sinc}(\xi + x_2) + d_H\, \mathrm{sinc}(\xi - x_1)\right]$$

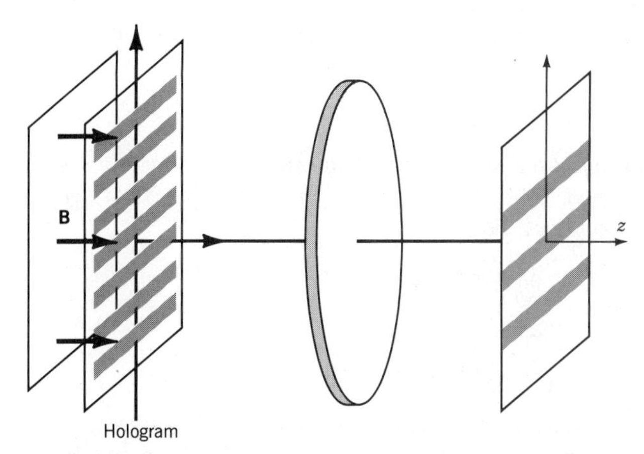

FIGURE 12-13. Reconstruction of the hologram. By looking at the Fraunhofer diffraction produced by the hologram, we can obtain the requirements for recording a signal.

The hologram produces a sinc function on axis that is the diffraction pattern due to the finite size of the hologram. The hologram also produces a real image of slit 2 at a distance x_2 above the optical axis and a real, inverted image of slit 2 at a distance $-x_1$ below the optical axis, as shown in Figure 12-13. Each image is a sinc function because of the finite size of the hologram.

Two real images are generated because the reconstruction wave B does not reproduce the reference wave. The origin of these two images can be discovered by decomposing the complex amplitude of the reconstruction wave B into two components: B_1 parallel to the propagation direction of the plane wave associated with slit 1 and B_2, the space conjugate of B_1, i.e., $B_2 = B_1^*$. The wave reconstructed by B_1 creates the real image of slit 2 and the wave reconstructed by B_2 produces an inverted real image of slit 2, the conjugate image.

SPATIAL SPECTRUM OF OFF-AXIS HOLOGRAMS

The result just obtained can be generalized to describe a hologram of an arbitrary signal by assuming that the signal wave is produced by a distribution of point sources located about the position $-x_2$. The hologram would contain a distribution of sinusoidal gratings oriented at different angles with respect to the hologram's surface normal. The distribution of gratings would diffract the reconstruction wave into a large number of plane waves, which would be focused by the lens in Figure 12-13 to an array of real image points, duplicating the distribution of point sources in the original object.

To simplify the discussion, assume that the location of the slit producing the reference wave is located at a distance $x_1 = |x_2|$ above the optical axis and the object is located the distance x_2 below the optical axis. The geometry of the slit experiment, except for this slight modification in distances, is shown in Figure 12-11. The reference wave at the recording plane (which we here define as $z = 0$ to suppress the z dependence) is

$$A = A_0 e^{-ik_x x}$$

and the signal wave is

$$a = a_0 e^{ik_x x}$$

where the component of the propagation vector along the x direction is

$$k_x = \frac{2\pi \sin \theta/2}{\lambda}$$

The complex variable a_0 contains the spatial information about the object that creates the signal wave. After exposure and development, **(12-4)** yields a film with a transmission coefficient of

$$t_H(x, y) = \beta\,(A_0A_0^* + a_0a_0^* + A_0^*a_0e^{i2k_x x} + A_0a_0^*e^{-i2k_x x}) \tag{12-8}$$

where the elements of the bias term

$$t_0 = \beta\,(A_0A_0^* + a_0a_0^*)$$

have been shown for this discussion. The hologram with this transmission coefficient is reconstructed using the geometry of Figure 12-13. The Fourier transform of jth term of **(12-8)** ($j = 1, \ldots, 4$) is defined as

$$G_j(\omega_x) = \mathcal{F}\{t_{Hj}\} = \int_{-\infty}^{\infty} t_{Hj}(x)e^{-i\omega_x x}\, dx$$

where $t_{Hj}(x)$ represents the jth term of **(12-8)**.

The first term is

$$G_1 = \mathcal{F}\{\beta A_0^2\} = \beta A_0^2\delta(\omega_x)$$

This term is associated with the diffraction pattern created by the limiting aperture of the hologram. Here, it is a delta function because we are assuming that the hologram is infinite in size.

The second term is found by applying the correlation theorem **(6A-10)**

$$G_2 = \beta\mathcal{F}\{|a|^2\} = \beta\int_{-\infty}^{\infty} G_a(\xi)G_a^*(\xi - \omega_x)\, d\xi \tag{12-9}$$

where G_a is the Fourier transform of the signal beam. The wave associated with **(12-9)** is created by the interference between the various point scatterers that make up the object. In the hologram of two slits, this term was not present because we assumed that the slits had infinitesimal widths.

From the shift theorem **(6A-5)**,

$$G_3 = \beta\mathcal{F}\{A_0^*a_0e^{i2k_x x}\} = \beta A_0^*G_a(\omega_x - 2k_x) \tag{12-10}$$

$$G_4 = \beta\mathcal{F}\{A_0a_0^*e^{-i2k_x x}\} = \beta A_0G_a(\omega_x + 2k_x) \tag{12-11}$$

For the purpose of illustration, we will assume that G_a, the Fourier transform of the signal, is a rectangle of width $2k_W$, i.e., the signal is a sinc function. With this assumption, **(12-9)** becomes a triangular function. The total spatial spectrum of the hologram will appear as shown in Figure 12-14.

The frequency spectrum displayed in Figure 12-14 provides a method for determining the minimum allowable angle between the reference and signal waves needed to separate the various diffracted waves produced by the hologram. We insure this separation by making the angle between the signal and reference beam greater than

$$\theta_{\min} = \sin^{-1}\left(\frac{3k_W\lambda}{2\pi}\right) \tag{12-12}$$

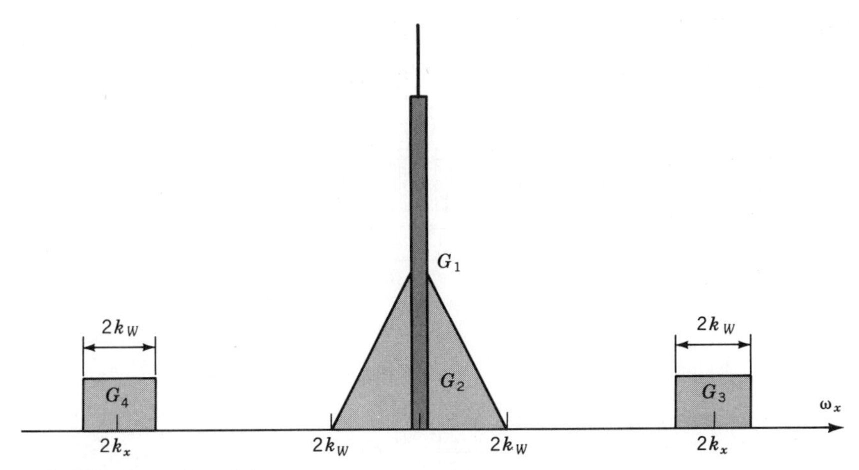

FIGURE 12-14. Spatial frequency spectrum of the holographic recording described by the transmission function in (**12-8**).

This angle insures that $2k_x \geq 3k_W$ and allows the signal wave to be separated from the other terms by simply propagating some distance.

If the smallest object spacing of interest in the signal is 0.1 mm (this is about the minimum resolution of the unaided eye when viewing an object 0.1 m away and corresponds to a spatial frequency of 10 lines/mm), we have an object spatial frequency bandwidth of

$$k_W = 2\pi \times 10^4 \text{ lines/m}$$

For an object with this spatial frequency content, $\theta_{\min}$ must be greater than 1.1° if the illumination is at a wavelength of $\lambda = 633$ nm. Of course, this angle neglects the constraint created by the finite size of the hologram. The reference and signal beams would have to propagate 260 cm before they did not overlap if this offset angle was used to record a 5 cm diameter hologram.

An arbitrarily large offset angle may not be possible because the offset angle determines the resolving capability needed by the recording medium. If we record a hologram using the minimum offset angle, then the maximum spatial frequency that the hologram must record is $4k_W$, four times the maximum spatial frequency in the signal.

Nonlinear recording will produce additional terms that diffract at larger angles and will generally not overlap the desired signal. For this reason, we can use a signal wave with an amplitude that is comparable to the amplitude of the reference wave and still recover the signal wave.

CLASSIFICATION OF HOLOGRAMS

The angle between the reference and signal beam affects the properties of the hologram and can be used to classify holograms. The geometry associated with each of the classifications is shown in Figure 12-15. The classification of holograms according to the angle between the reference and signal waves is as follows:

1. θ is small: For most photographic emulsions, which are 5 to 16 μm thick, this is an angle less than about 10°. The fringe spacing, produced by the interference between the signal and reference waves, is about the

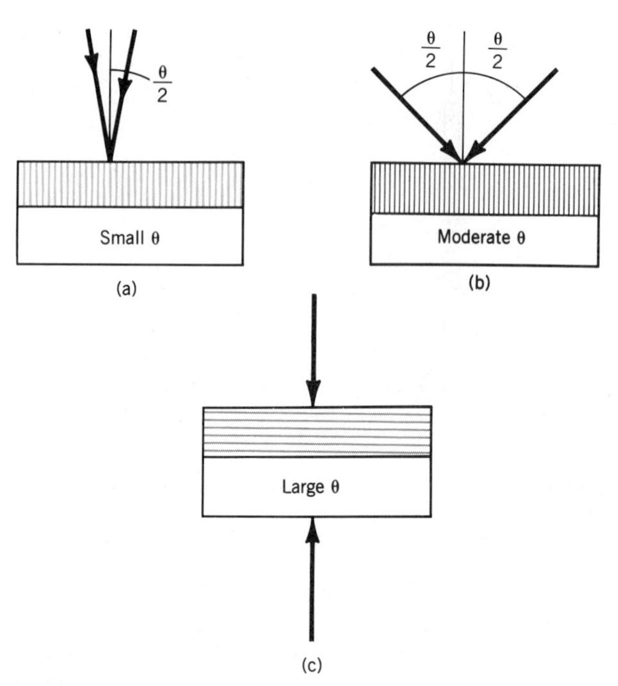

FIGURE 12-15. The geometry associated with three classes of holograms. (a) A thin hologram, where the fringe spacing is large relative to the thickness of the recording medium. (b) A thick hologram, where the fringe spacing is small relative to the thickness of the recording medium. (c) A reflection hologram, where the fringes are parallel to the surface of the recording medium.

same size as the recording medium thickness and the recording medium behaves as if it were two-dimensional. This type of hologram is called a *thin* hologram. The diffraction by the thin hologram's fringes obeys the grating equation **(10-39)**. Figure 12-16*a* displays a cross section of a thin absorption hologram where the fringe maxima can be seen as bright lines. The fringes are normal to the hologram surface because the reference and signal waves, which were plane waves, were incident at equal angles

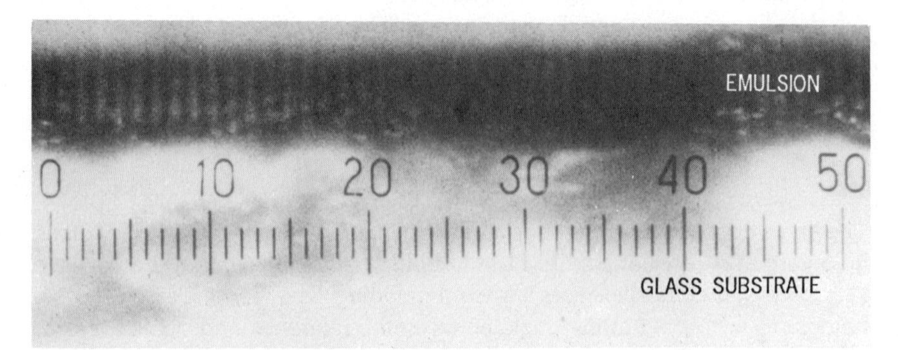

FIGURE 12-16. (a) A cross section of the developed gelatin of an absorption hologram exposed to two interfering plane waves. The bright lines are layers of silver produced by bright interference bands. [C.R. Bendall and B.D. Guenther, "Technique for Obtaining a Photomicrograph of a Cross Section of Photographic Emultion," *Appl. Opt.* **11**:1653 (1972).] (b) A cross section of a thick-phase hologram. The fringes in this photopolymer are voids separated by polymer.

FIGURE 12-16. *(Continued)*

with respect to the normal of the hologram surface. The theory describing diffraction from a hologram is identical to the theory used to describe the diffraction of light by acoustic waves (see Chapter 14 and Appendix 14-B), and the Raman–Nath approximation of that theory generates solutions that apply to the thin hologram.

2. θ is moderately large: For photographic emulsions, this is an angle in the range of 10 to 120°. Holograms of this class are called *thick* holograms because the fringe spacing is small compared to the recording medium thickness; see Figure 12-16*b*. The theory used to describe light diffracted by the gratings in a thick hologram is identical to the theory based on the Bragg approximation of acoustic wave diffraction of light discussed in Chapter 14 and Appendix 14-B. The result of the Bragg approximation is an equation called *Bragg's law* (derived by **W.L. Bragg** in 1912 for x-ray diffraction by a crystal).

A simple derivation of Bragg's law can be made by using Figure 12-17. The horizontal lines in Figure 12-17 represent the location of the fringe maxima in a thick hologram. In photographic film, these lines would be deposits of silver.

A parallel beam of light with wavefront AA' is incident on the fringe maxima in the hologram at an angle θ. If the light reflected from the silver atoms in the two adjacent fringe maxima is to be in phase, so that a plane wave with wavefront CC' exits the hologram, then

$$\overline{DB'} + \overline{B'E} = n\lambda$$

From the geometry of Figure 12-17

$$\overline{DB'} = \overline{B'E} = d \sin \theta$$

Reflections of waves propagating parallel to the plane wave AA' by other equally spaced, parallel fringe planes create plane waves parallel to the plane

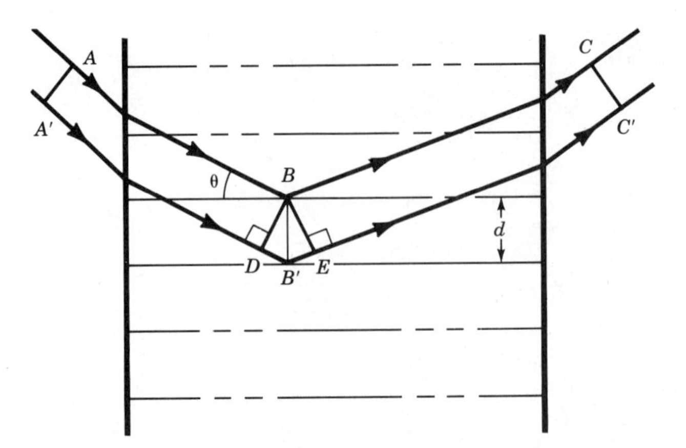

FIGURE 12-17. Geometry for Bragg diffraction.

wave CC'. These waves constructively interfere, producing a maximum in the light diffracted by the fringe system. The angle θ for which this maximum in diffracted light occurs is given by Bragg's law

$$2d \sin \theta = n\lambda \tag{12-13}$$

The reconstruction of the thick hologram is very sensitive to the angle that the reconstruction wave makes with the developed hologram. A small departure from the angle obtained from **(12-13)** will result in a loss of the signal wave. Because of the angular sensitivity, a number of different holograms can be stored in a single piece of recording material by using reference beams at different angles. Each stored hologram can generate its signal wave without generating the other stored signal waves. Signal to noise ratios below 20 db in stored images prevent storage of more than about 20 complex images with this technique. The diffracted intensity is also a sensitive parameter of the reconstruction wavelength, one of the variables in **(12-13)**. The reconstruction wavelength must be within tens of nanometers of the recording wavelength to obtain a signal wave.

3. θ is large: The reference and signal beams enter from opposite sides of the recording medium and the recorded fringes are nearly parallel to the surface of the medium with a period of about $\lambda'/2$ (where λ' is the wavelength in the recording medium). The experimental arrangement for constructing a hologram meeting this condition can be seen in Figure 12-18. This type of hologram is called a *reflection* or *white light* hologram. It can be illuminated by white light and will act as a wavelength filter, reconstructing the signal wave for a narrow band of wavelengths determined by the recording wavelength. It is relatively insensitive to a change in direction of the illuminating wave but most sensitive to wavelength.

Figure 12-18 shows a recording medium positioned in the fringe field produced by two point sources (a redrawn version of Figure 12-5). The upper rectangle in Figure 12-18 is in position to record either a type-1 or type-2 hologram, and the lower rectangle in Figure 12-18 is in the position to record a reflection (type-3) hologram.

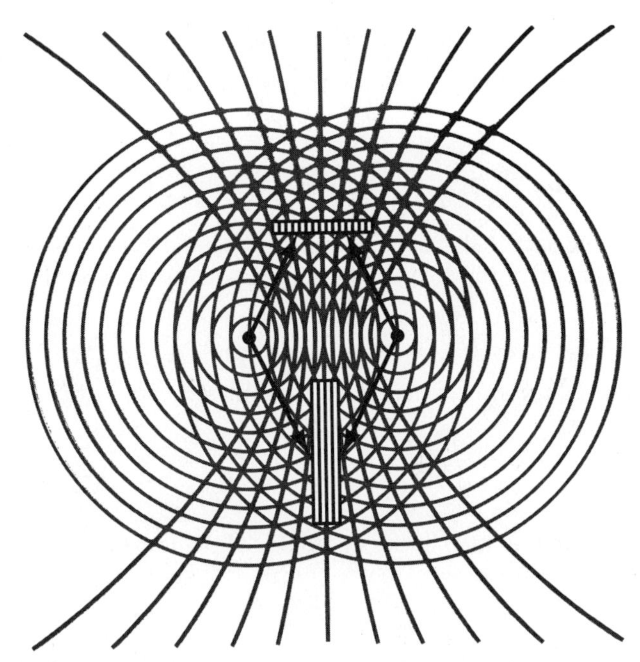

FIGURE 12-18. The rectangles represent holograms produced by recording the interference fringes from two point sources. (The lightly shaded circles are the wavefronts from the two sources and the shaded curves are the interference fringes generated, see Figure 12-5.) The upper hologram has both sources on one side of the recording medium. This experimental configuration produces either thin or thick holograms, depending on the angle between the reference and signal waves. The lower hologram has a source on each side of the recording medium, producing a reflection hologram.

DIFFRACTION EFFICIENCY

A hologram is recorded so that at a later time, the signal wave can be reproduced. An important characteristic of a hologram is the fraction of energy diffracted from the reconstruction wave into the signal wave. The parameter used to characterize this energy fraction is called the *diffraction efficiency* of the hologram.

$$\eta = \frac{(\beta B A^* a)^2}{|B|^2} \tag{12-14}$$

where B is the light amplitude of the reconstruction wave. The most useful hologram is normally one with a large diffraction efficiency. (A hologram's diffraction efficiency may be designed to be less than the maximum possible because of the desired application. For example, a hologram used to produce copies should have a diffraction efficiency around 50%.)

To measure the diffraction efficiency of a particular recording material, two plane waves A_0 and a_0 are used as the reference and signal waves. The measurement of η is then not confused by the spatial variation in the signal beam. In this section, we will relate the diffraction efficiency to the interference pattern produced by two plane waves. The resulting equation can be used to establish a maximum diffraction efficiency for thin holograms.

The diffraction efficiency η depends on the visibility of the fringes produced during the recording process. To display the dependence explicitly, we use the definition of the fringe visibility

$$V = \frac{I_{\max} - I_{\min}}{I_{\max} + I_{\min}}$$

The maximum and minimum intensities during interference are

$$I_{max} = A_0^2 + a_0^2 + 2A_0a_0, \qquad I_{min} = A_0^2 + a_0^2 - 2A_0a_0$$

and the visibility becomes

$$V = \frac{2A_0a_0}{A_0^2 + a_0^2}$$

We also make use of the fact that the hologram's amplitude transmission coefficient is linearly related to the exposure energy used to generate the hologram and given by

$$\text{t}_0 = \beta'\text{E}_0 = \beta'\tau_e(A_0^2 + a_0^2) = \beta(A_0^2 + a_0^2)$$

We can now write the diffraction efficiency as an explicit function of the fringe visibility

$$\eta = \left(\frac{\beta'\text{E}_0V}{2}\right)^2 \tag{12-15}$$

In the holographic literature, the ratio of the intensities of the two beams A_0^2 and a_0^2 is usually called the *K ratio*

$$\text{K} = \frac{A_0^2}{a_0^2}$$

We see that the maximum efficiency is obtained when $V = 1$, that is, when $A_0 = a_0$ and the K ratio is 1.

It would appear that any diffraction efficiency desired could be obtained by the proper selection of the exposure energy. Real materials, however, depart from the simple relationship of **(12-7)** because the recording medium departs from a linear relationship between t_0 and E_0. This departure is demonstrated for a common holographic film in Figure 12-8. In particular, the recording material saturates above some exposure energy, preventing the hologram from exceeding a maximum diffraction efficiency.

The maximum diffraction efficiency η that can be exhibited by a thin hologram can be estimated by assuming that the hologram is produced by the interference of two plane waves. The hologram's transmission coefficient is given by **(12-7)** that can be used to calculate the diffraction efficiency. If the hologram is illuminated by a reconstruction wave of amplitude B, then the maximum light amplitude transmitted by the hologram is $\text{t}_0B(1 + d_H)$ and the light diffracted into one order is

$$\text{t}_0B\frac{d_H}{2}$$

yielding a diffraction efficiency of

$$\eta = \left[\frac{\text{t}_0Bd_H}{2\text{t}_0B(1 + d_H)}\right]^2$$

$$= \left[\frac{1}{2\left(1 + \dfrac{1}{d_H}\right)}\right]^2$$

The maximum fringe contrast that can be produced occurs when $d_H = 1$ and is equal to $\eta \leq 6.25\%$. Other types of holograms exhibit higher diffraction efficiencies (see Appendix 12-A).

HOLOGRAPHY AND ZONE PLATES

An intuitive feel for the performance of a hologram can be obtained by developing a theoretical description of a hologram produced by a point source, which generates both the reference wave, and in an indirect fashion, the signal wave. One portion of the wave from this point source is scattered by a single point object and then interferes with the primary portion of the wave to generate the hologram. Figure 12-19 displays the geometry for this simple configuration. Upon calculating the resultant interference pattern, we will find that the hologram produced by this interference pattern acts like a simple zone plate.

The concept of a hologram as a zone plate can be extended to an arbitrary object, viewed as a set of point scatters. The hologram of the object, a recording of the fringes resulting from the interference between each scattering point and the reference beam, can be viewed as a set of overlapping zone plates. The image produced by the hologram is a set of image points, conjugate to the set of object points, created by the zone plates.

Consider a single point source at position B, a distance Z_r from the recording plane in Figure 12-19, and one point scatterer at position O, a distance Z_s from the recording plane. Light reaches point P in the recording plane from the source at B, a distance t away, and from the scattering object, point O, a distance s from P. The light scattered by O is produced by the source at B; therefore, the phase difference between light waves that have traversed the two different paths along t and along Z_0 plus s is given by

$$\begin{aligned}\delta &= kt - k(Z_0 + s)\\ &= k(t + Z_s - Z_r - s)\end{aligned}$$

The geometric optics sign convention we selected in Chapter 6 will be applied here. The coordinate system is positioned so that the x plane, containing P, is at the origin. An approximate expression for the phase difference δ is obtained by using the binomial expansion

$$\begin{aligned}\delta &\approx k\left(Z_s - Z_r + Z_r + \frac{x_1^2}{2Z_r} Z_s - \frac{x_1^2}{2Z_s}\right)\\ &\approx \frac{kx_1^2}{2}\left(\frac{1}{Z_r} - \frac{1}{Z_s}\right)\end{aligned}$$

The bright bands of the interference fringes are separated by a distance Δx associated with a phase change of $\Delta\delta = 2\pi n$. The first bright band, at

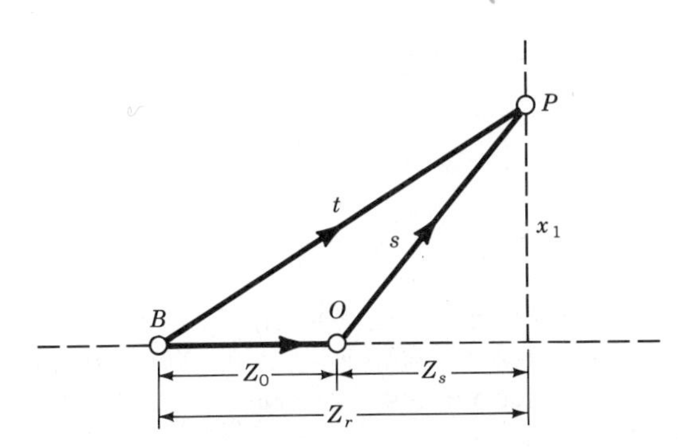

FIGURE 12-19. Geometry for the interference of two in-line point sources. The interference pattern can be recorded to produce a zone plate.

$n = 0$, is on the z axis and the second bright band, $n = 1$, is located at $\Delta x = x_1$, leading to the result

$$2\pi = \frac{2\pi}{\lambda}\frac{x_1^2}{2}\left(\frac{1}{Z_r} - \frac{1}{Z_s}\right)$$

$$\frac{\lambda}{x_1^2} = \frac{1}{Z_r} - \frac{1}{Z_s} \tag{12-16}$$

This equation is identical to **(11-27)**, derived for a zone plate. [Note that in **(11-27)** we must use $Z = -Z_s$ to meet the requirements of the sign convention.]

If the field from the source has an amplitude A and the scattered field an amplitude a, then the intensity at P is

$$I = |A|^2 + |a|^2 + 2Aa\cos\delta$$

$$I = (|A|^2 + |a|^2)\left\{1 + 2V\cos\left[\frac{kx_1^2}{2}\left(\frac{1}{Z_r} - \frac{1}{Z_s}\right)\right]\right\} \tag{12-17}$$

We define the focal length f of the zone plate described by **(12-17)** as

$$\frac{1}{f} = \frac{1}{Z_r} - \frac{1}{Z_s}$$

Using this definition in the expression for the transmission function of the recorded intensity distribution, we obtain

$$t(x) = t_0 + 2\beta V\cos\frac{\pi x_1^2}{\lambda f}$$

This is the transmission function of a sinusoidal zone plate.

The field distribution produced by any zone plate along the optical axis of the zone plate (here the z axis) can be calculated by using the transmission function $t(r, \varphi)$ of the zone plate in the cylindrical coordinate version of the Fresnel diffraction integral **(11A-3)**. Because the zone plate is circularly symmetric, **(11A-3)** becomes

$$\mathcal{E}_P = \frac{ik\alpha}{\rho D}e^{-ikD}\int_0^{x_{max}} t(x)e^{-i(kx^2/2\rho)}\,x\,dx$$

By a change of variables, we may write

$$\mathcal{E}_P = \frac{ik\alpha}{D}e^{-ikD}\int_0^{u_{max}} t(u)e^{-iku}\,du$$

where

$$u_{max} = \frac{x_{max}^2}{2\rho}, \qquad \frac{1}{\rho} = \frac{1}{Z_s} + \frac{1}{Z_r}, \qquad D = Z_s + Z_r$$

This integral can be identified as a Fourier transform **(6-13)**. Thus, the light distribution along the optical axis of the zone plate is equal to the Fourier transform of the radial transmission function of the zone plate.

The sinusoidal zone plate would have a transmission function in the current notation given by

$$t(u) = t_0 + 2\beta V \cos\frac{\pi u}{\lambda f}$$

which is mathematically identical to **(12-7)**. The Fourier transform of this amplitude transmission function leads to two intensity maxima, and thus two foci. If we allowed the recording process to be nonlinear, additional higher-order foci would appear. The binary zone plate discussed in Chapter 11 has a radial transmission function that is a square wave of varying frequency (called a chirped grating because the frequency increases with increasing x). The Fourier transform of this radial transmission function would have an infinite number of components. The binary zone plate thus produces an infinite number of foci.

The two foci of the sinusoidal zone plate can be obtained by calculating the Fourier transform of its transmission function, which is a cosine

$$\frac{1}{f_\pm} = \mp\frac{\lambda_2}{x_1^2} \qquad (12\text{-}18)$$

We have introduced a second wavelength λ_2, the reconstruction wavelength, which we allow to be different from the recording wavelength λ_1. The radius of the first zone is given by **(12-16)**. Substituting the expression for the radius of the first zone x_1 into **(12-18)** yields

$$\frac{1}{f_\pm} = \mp\frac{\lambda_2}{\lambda_1}\left(\frac{1}{Z_r} - \frac{1}{Z_s}\right) \qquad (12\text{-}19)$$

where f_+ represents the focal length of a positive lens and f_- the focal length of a negative lens. Equation **(12-19)** describes the imaging property of the hologram and suggests that the magnification of the hologram could be modified by reconstructing the image at a longer wavelength than the one used to record the hologram.

The hologram we record, using the configuration shown in Figure 12-19, behaves simultaneously as a positive and negative lens; see Figure 11-24. From this Fresnel zone viewpoint of holography, we find, as we did in **(12-5)**, that the hologram produces two images, a real image at Z_s and a virtual image at $-Z_s$, that must be separated.

A more complicated object would contain a collection of point scatterers and we assume the hologram would consist of a summation of zone plates, one for each scatterer. Each of these zone plates acts independent of the others to produce a field; all of the fields add to produce the image. The use of a zone plate to interpret the behavior of a hologram is useful, but is flawed because it neglects the interference between the waves from each point scatterer and considers only the interference between the individual scattered waves and reference wave.

When we move our recording medium off axis, we obtain an interference pattern, as shown in Figure 12-20, where we assume for illustrative purposes that the reference beam is a plane wave. The shaded rectangle in Figure 12-20 is assumed to be the recording medium. We see that the

The spatial frequency of the fringes producing the zone plate can be defined as

$$\frac{1}{2\pi}\frac{\partial\delta}{\partial x} = \frac{kx}{2\pi}\left(\frac{1}{Z_r} - \frac{1}{Z_s}\right) = \frac{x}{\lambda f}$$

As with a normal zone plate, the fringe frequency increases linearly with increasing x. At some point x_{max}, the photosensitive material can no longer record the fringes—we will exceed the resolving capability of the medium. This x_{max} defines the limiting aperture of the hologram. The effects of this limiting aperture on the resolving capability of a hologram will be explored in a moment.

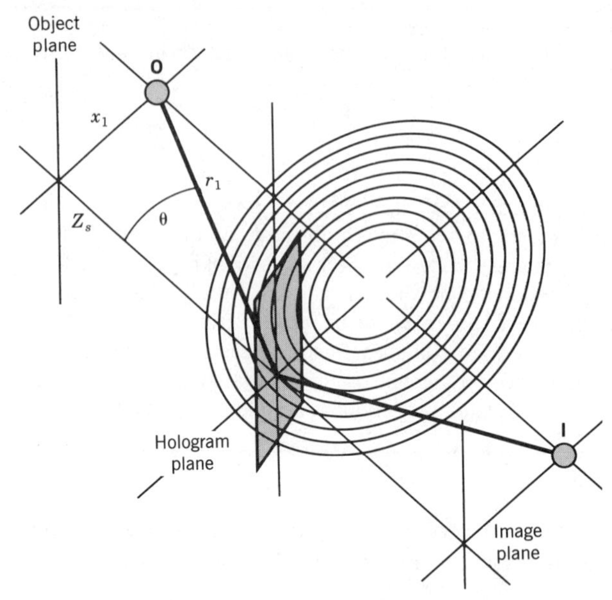

FIGURE 12-20. Fresnel zone picture of an off-axis hologram. Point *O* is the object during recording and the virtual image during playback. Point *I* is the real image during playback. The reference and reconstruction beam are assumed to be plane waves.

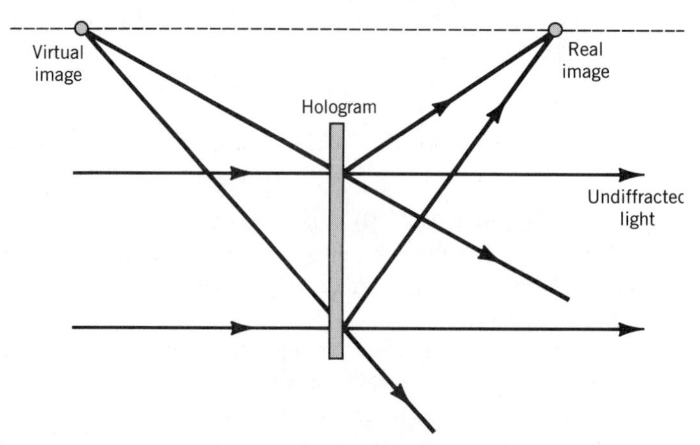

FIGURE 12-21. The real and virtual image of an off-axis hologram.

object *O* and the real image *I* would lie on the optical axis of the zone plate. The various waves produced by the developed hologram are shown schematically in Figure 12-21. Simply put, we can separate the two images of the hologram, if we offset the reference, so that the central Fresnel zone is not recorded.

RESOLUTION REQUIREMENTS

To find the resolution requirements imposed on a holographic recording medium, we need to define three angles, shown in Figure 12-22. The object subtends an angle θ_O at the recording surface, and the recording surface subtends an angle θ_R at the object. These angles are defined so that they

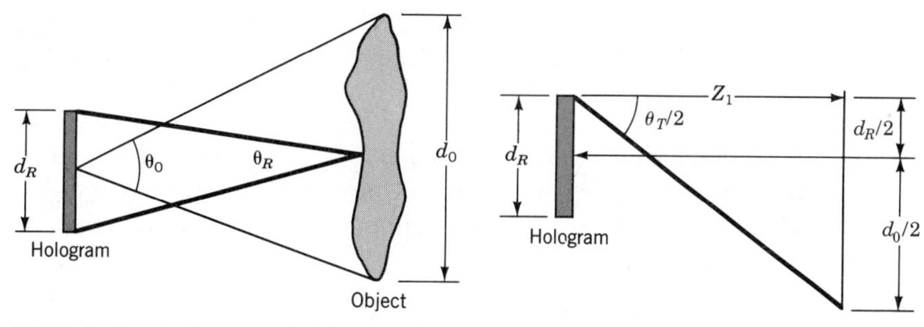

FIGURE 12-22. Geometrical definition of angular bandwidth of holographic recording.

are the largest angles that will be encountered. In this simple analysis, we will assume the reference wave is a plane wave propagating in a direction normal to the hologram. The largest angle encountered at the recording plane between the reference wave and a wave from the object is θ_T. The half-angle $\theta_T/2$ is shown on the right of Figure 12-22. Interference between the wave propagating at θ_T and the reference wave will create the highest spatial frequency that must be recorded; this maximum spatial frequency is given by

$$k^T = \frac{2\pi \sin(\theta_T/2)}{\lambda}$$

The spatial frequency bandwidth is defined as

$$k_W = 2k^T = \frac{4\pi}{\lambda} \sin \frac{\theta_T}{2} \tag{12-20}$$

The geometry of Figure 12-22 can be used to relate the three angles just defined

$$\tan \frac{\theta_T}{2} = \frac{1}{Z}\left(\frac{d_O}{2} + \frac{d_R}{2}\right) = \tan \frac{\theta_O}{2} + \tan \frac{\theta_R}{2} \tag{12-21}$$

By using the object angle θ_O we can calculate the maximum value of the object's spatial bandwidth

$$k_W^O = \frac{4\pi}{\lambda} \sin \frac{\theta_O}{2}$$

Similarly, the recording's maximum spatial bandwidth can be calculated using the hologram angle θ_R

$$k_W^R = \frac{4\pi}{\lambda} \sin \frac{\theta_R}{2}$$

Using these limiting definitions of spatial bandwidth, we may write **(12-21)** as

$$\frac{k_W^T}{\cos(\theta_T/2)} = \frac{k_W^O}{\cos(\theta_O/2)} + \frac{k_W^R}{\cos(\theta_R/2)}$$

From the definitions of the three angles, we know that $\theta_T > \theta_R$ or θ_O, leading to the inequality

$$k_W^T \leq k_W^O + k_W^R \tag{12-22}$$

If the object to be recorded is much larger than the hologram, we may neglect the hologram's spatial bandwidth. For example, suppose the vibrations of a guitar face are to be studied. The standard photographic plate is 4 in. × 5 in. At a distance of 0.9 m from the guitar, the hologram subtends an angle of approximately 16° at the guitar. The guitar, which is 0.5 m long, subtends an angle of 60°. The sine of the half-angle of the object is $\sin 30° = 1/2$ and $k_W^T \leq 2\pi/\lambda$. For green light, $\lambda = 500$ nm and the required spatial bandwidth becomes

$$\frac{k_W^T}{2\pi} \leq 2000 \frac{\text{lines}}{\text{mm}}$$

This spatial bandwidth requirement means that the recording medium must record detail beyond the resolution capability of conventional photographic film by a factor of 20. High-resolution spectroscopic plates are capable of recording the required resolution of 5×10^{-4} mm.

The resolution requirement is due to the wide field of view recorded by the hologram. If the field of view of the hologram were limited to that covered by the fovea region of the human eye, 2°, then the spatial bandwidth the hologram must record would only be 70 lines/mm.

IMAGING PROPERTIES OF OFF-AXIS HOLOGRAMS

We saw in **(12-19)** that the image produced by a hologram can be magnified by moving the position of the reference or changing the reconstruction wavelength. We will use a geometry similar to the one shown in Figure 12-19 to examine the issue of magnification in more detail. We will modify the geometry of Figure 12-19 by introducing a second source to produce the reference beam. The recorded geometry is shown in Figure 12-23.

The object is illuminated by a spherical wave of wavelength λ_1 from point A; when this wave reaches the object at O, it has a phase $k_1 r_0$. The wave from A is scattered by the object (without phase change), arriving at P with a phase of $k_1(r_0 + r_1)$. The reference wave, also of wavelength λ_1, travels from a source, in phase with A but located at B, to the recording plane at P, arriving with a phase of $k_1 r_2$. The coordinate system is positioned so that the hologram, both before and after recording, is located at $Z = 0$.

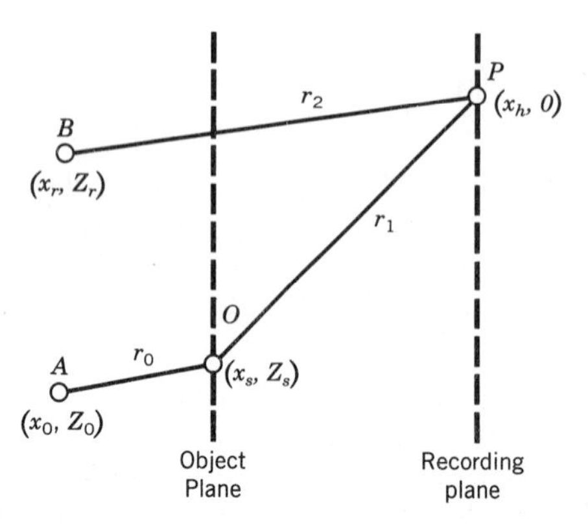

FIGURE 12-23. Geometry for recording a hologram.

The interference produced at P is due to the phase difference between these two waves

$$\delta = k_1(r_2 - r_1 - r_0) \tag{12-23}$$

We use the binomial expansion to obtain a first-order approximation for the phase in terms of the x and z coordinates

$$\delta \approx k_1\left[Z_r + \frac{(x_h - x_r)^2}{2Z_r} - Z_s - \frac{(x_h - x_s)^2}{2Z_s} - Z_0 - \frac{(x_s - x_0)^2}{2Z_0}\right] \tag{12-24}$$

(Any study of aberrations in holograms would involve an evaluation of higher-order terms in this expansion.)

A parameter p is used to modify the hologram's x coordinate from x_h to x_h/p. This parameter takes into account any change in the dimension of the hologram produced by the recording and processing of the recording. Such dimensional changes could be intentional but most often are due to an unwanted, and unavoidable, shrinkage or swelling of the emulsion during development. After development, the recorded fringes are described by

$$\delta_R \approx k_1(Z_r - Z_s - Z_0) + k_1\left[\frac{\left(\dfrac{x_h}{p} - x_r\right)^2}{2Z_r} - \frac{\left(\dfrac{x_h}{p} - x_s\right)^2}{2Z_s} - \frac{(x_s - x_0)^2}{2Z_0}\right] \tag{12-25}$$

To observe the image wave reconstructed by diffraction in the hologram, a source of wavelength λ_2 illuminates the hologram. The geometry of the reconstruction process is shown in Figure 12-24 with the reconstruction source located at C. The transmitted wave will have a phase modulation given by

$$\delta_t = k_2 r_3 \pm \delta_R \tag{12-26}$$

If an image exists at position I, then the wave transmitted by the hologram must be a spherical wave with a phase of the form

$$\delta_I = C + k_2 r_4$$

where C contains the phases of all waves except the wave associated with the image. The sign convention is the same one used in the discussion of geometrical optics. For example, if the image I is located at the position

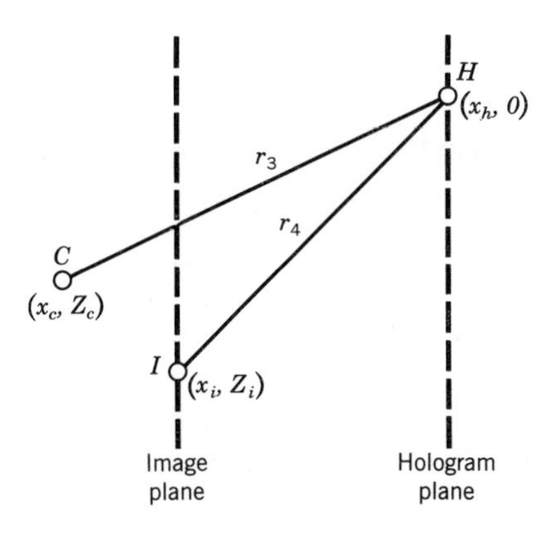

FIGURE 12-24. Geometry to play back a hologram recorded in the layout shown in Figure 12-23.

shown in Figure 12-24, then r_4 is negative because Z_i is negative [see **(12-27)**] and the image is a virtual image. If the image I is located to the right of the hologram, then Z_i is positive and the image is a real image, produced by a converging wave.

We assume that an image does exist by equating the phase of the wave transmitted by the hologram to the assumed image phase $\delta_t = \delta_I$. Writing the terms of δ_t and δ_I explicitly, we have

$$k_2 Z_c \pm k_1 (Z_r - Z_s - Z_0) \pm k_1 \left[\frac{\left(\frac{x_h}{p} - x_r\right)^2}{2Z_r} - \frac{\left(\frac{x_h}{p} - x_s\right)^2}{2Z_s} - \frac{(x_s - x_0)^2}{2Z_0} \right] \tag{12-27}$$

$$+ \frac{k_2 (x_h - x_c)^2}{2Z_c} = C + k_2 Z_i + \frac{k_2}{2Z_i}(x_h^2 - 2x_i x_h + x_i^2)$$

Since we are evaluating the phases at the point on the hologram $(x_h, 0)$, we will equate the coefficients of powers of x_h. Equating the coefficients of x_h^2 yields

$$\frac{k_2}{Z_i} = \frac{k_2}{Z_c} \pm \frac{k_1}{p^2}\left(\frac{1}{Z_r} - \frac{1}{Z_s}\right)$$

This relationship allows the derivation of an equation for the z component of the image. The distance from the hologram to the image is given by

$$Z_i = \frac{p^2 Z_s Z_r Z_c}{p^2 Z_s Z_r \pm (\lambda_2/\lambda_1) Z_c (Z_s - Z_r)}$$

which can be simplified to read

$$\frac{1}{Z_i} = \frac{1}{Z_c} \pm \frac{1}{p^2}\left(\frac{\lambda_2}{\lambda_1}\right)\left(\frac{1}{Z_r} - \frac{1}{Z_s}\right) \tag{12-28}$$

Equating the coefficients of x_h yields an equation for the x coordinate of the image

$$x_i = \frac{p^2 x_c Z_s Z_r \pm (\lambda_2/\lambda_1) p Z_c (x_r Z_s - x_s Z_r)}{p^2 Z_s Z_r \pm (\lambda_2/\lambda_1) Z_c (Z_s - Z_r)} \tag{12-29}$$

It is easy to become confused by the large number of coordinates. Table 12.1 identifies the coordinates of the various objects, sources, and images in this geometrical model of holography.

Lateral Magnification

The magnification of the dimension, transverse to the propagation direction, is defined as the *lateral magnification*.

TABLE 12.1 Coordinates for Holography Model

Reference	x_r, Z_r
Signal	x_s, Z_s
Reconstruction	x_c, Z_c
Image	x_i, Z_i
Hologram	$x_h, 0$

$$\beta = \frac{dx_i}{dx_s} = \frac{\pm(\lambda_2/\lambda_1)pZ_cZ_r}{\pm(\lambda_2/\lambda_1)Z_c(Z_r - Z_s) - p^2Z_sZ_r} \tag{12-30}$$

If we simplify **(12-30)**, the lateral magnification becomes

$$\beta = \frac{1}{\dfrac{1}{p} \mp p\dfrac{Z_s}{Z_c}\left(\dfrac{\lambda_1}{\lambda_2}\right) - \dfrac{1}{p}\dfrac{Z_s}{Z_r}} \tag{12-31}$$

The hologram that is recorded should generate an undistorted image if it is to be useful. To determine if, in fact, an undistorted image is produced, the lateral magnification β must be compared to the magnification in the direction of propagation.

Longitudinal Magnification

The magnification of an image along the direction of propagation is called the *longitudinal magnification* and is defined as

$$\beta_z = \frac{\partial Z_i}{\partial Z_s} = \pm\frac{\lambda_1}{\lambda_2}\beta^2 \tag{12-32}$$

To obtain an undistorted image, the longitudinal and lateral magnification must be identical, $\beta_z = \beta$. For no distortion, therefore, we must have

$$\beta = \frac{\lambda_2}{\lambda_1}$$

FRESNEL HOLOGRAM

A hologram constructed with a reference wave located at an arbitrary position but with the recording medium in the near field of the object is called a *Fresnel hologram*. The real and virtual images of a Fresnel hologram are located symmetrically about the recording, as shown in Figure 12-21. Two special configurations of this type of hologram are easy to evaluate and will allow the demonstration of the basic properties of the Fresnel hologram.

To analyze the special configuration, we rewrite **(12-28)**

$$Z_i = \frac{p^2Z_s}{p^2(Z_s/Z_c) \pm (\lambda_2/\lambda_1)[(Z_s/Z_r) - 1]}$$

and **(12-29)**

$$x_i = \frac{p^2\,(x_c/Z_c)\,Z_s \pm p(\lambda_2/\lambda_1)\left[(x_r/Z_r)Z_s - x_s\right]}{p^2(Z_s/Z_c) \pm (\lambda_2/\lambda_1)[(Z_s/Z_r) - 1]}$$

$$= \frac{p^2Z_s \tan\,\alpha_c \pm p\,(\lambda_2/\lambda_1)(Z_s \tan\,\alpha_r - x_s)}{p^2(Z_s/Z_c) \pm (\lambda_2/\lambda_1)[(Z_s/Z_r) - 1]}$$

where

$$\tan\alpha_c = \frac{x_c}{Z_c}, \qquad \tan\alpha_r = \frac{x_r}{Z_r}$$

The new equations allow the easy evaluation of the image coordinates in the limit as Z_r, $Z_c \to -\infty$. In these limits, the resulting reference and reconstruction waves are incident on the hologram at the angles α_r and α_c, respectively.

Reconstruction With a Plane Wave

If the reconstruction wave were a plane wave, then $Z_c = -\infty$, and the lateral magnification is

$$\beta = \frac{pZ_r}{Z_r - Z_s} \tag{12-33}$$

The magnification is independent of the recording and reconstruction wavelengths. The coordinate of the image in the x position is given by

$$x_i = \frac{p(x_s - Z_s \tan \alpha_r) \mp (\lambda_1/\lambda_2)p^2 Z_s \tan \alpha_c}{1 - (Z_s/Z_r)} \tag{12-34}$$

If the reconstruction wave is propagating parallel to the optical axis, then

$$x_i = \frac{p(x_s - Z_s \tan \alpha_r)}{1 - (Z_s/Z_r)}$$

The coordinate of the image in the z direction is given by

$$Z_i = \frac{p^2}{\pm\left(\frac{\lambda_2}{\lambda_1}\right)\left(\frac{1}{Z_s} - \frac{1}{Z_r}\right)} = \mp p^2 f_{\pm} \tag{12-35}$$

where we use **(12-19)** to write the image's location in terms of the focal length of the zone plate produced by the interference of the two point sources. The two images of this Fresnel hologram are located symmetric about the hologram plane, as is demonstrated by **(12-35)**. The images are located along a line, parallel to, and displaced a distance x_i from the z axis; see Figure 12-20.

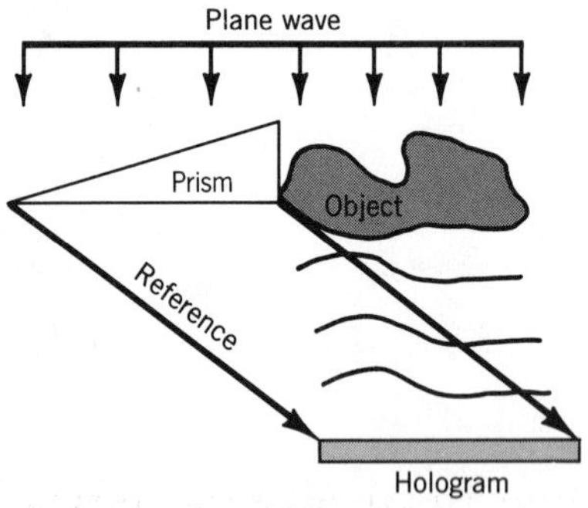

Plane Reference Wave

If the reference wave and the reconstruction wave are both plane waves, then $Z_r = Z_c = -\infty$. This configuration is a special case of the preceding hologram. Its lateral magnification is given by

$$\beta = p \tag{12-36}$$

A possible geometry for constructing a hologram with a plane wave as the reference wave is shown in Figure 12-25.

FIGURE 12-25. A possible geometry for recording an off-axis hologram with a reference wave that is a plane wave.

Spherical reference and reconstruction waves can produce the lateral magnification given by **(12-36)** if the location of the point sources of these waves satisfies the equation

$$\frac{Z_r}{Z_c} = \mp \frac{\lambda_2}{\lambda_1 p^2} \tag{12-37}$$

For spherical waves satisfying **(12-37)** or for reference and reconstruction plane waves, propagating parallel to the optical axis, the image's x coordinate is

$$x_i = px_s \tag{12-38}$$

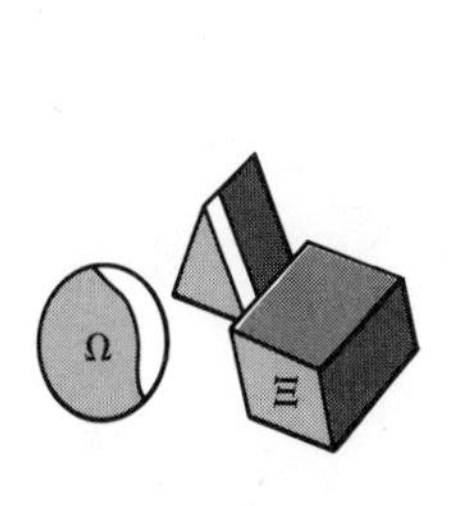

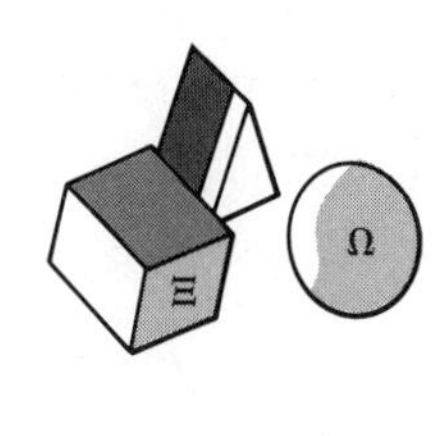

FIGURE 12-26. A virtual and pseudoscopic image are shown in their positions relative to a hologram. The virtual image is on the left and the pseudoscopic image is on the right. When the pseudoscopic image is viewed from the right, the nearest object is obstructed by an object that is farther away.

and its z coordinate is

$$Z_i = \pm p^2(\lambda_1/\lambda_2)Z_s \tag{12-39}$$

From **(12-39)**, we see that the real and virtual images are located at equal distances from, and on either side of, the hologram plane, as is indicated in Figure 12-21. If there is no change in the recording medium's dimension, $p = 1$, and both the recording and reconstruction wavelengths are the same, $\lambda_1 = \lambda_2$, then the virtual image is located at the same position occupied by the object during recording.

The real image is a little unusual as we can see by considering two objects Ω and Ξ, located such that $Z_\Omega > Z_\Xi$. The object Ξ obstructs a portion of Ω as we view the objects from a position on the far right; see Figure 12-26. The images seen in the hologram are located such that $Z'_\Omega > Z'_\Xi$ measured with respect to the hologram. When the real image is observed from the far right, Ξ still obstructs Ω even though Ω is closer to the observer than Ξ. This type of image is called a *pseudoscopic image*. A virtual and pseudoscopic image are shown positioned about a Fresnel hologram in Figure 12-26.

A pseudoscopic image of your face would have your nose receding into your head. *Orthoscopic* real images, that is, images with normal three-dimensional relationships between their parts, can be produced if the hologram is generated using the image produced by a lens or if the pseudoscopic image produced by another hologram is used as the object.

FOURIER TRANSFORM HOLOGRAM

If the location of the reference source is in the plane of the object, $Z_s = Z_r$, the hologram is called a *lensless Fourier transform hologram*. For this recording geometry, the bandwidth requirements placed on the recording medium are minimized. A possible recording geometry for this type of hologram is shown in Figure 12-27.

The magnification produced by a Fourier transform hologram is

$$\beta = \mp \frac{1}{p}\frac{\lambda_2}{\lambda_1}\frac{Z_c}{Z_s} \tag{12-40}$$

Equation **(12-19)** implied that the magnification of a hologram was deter-

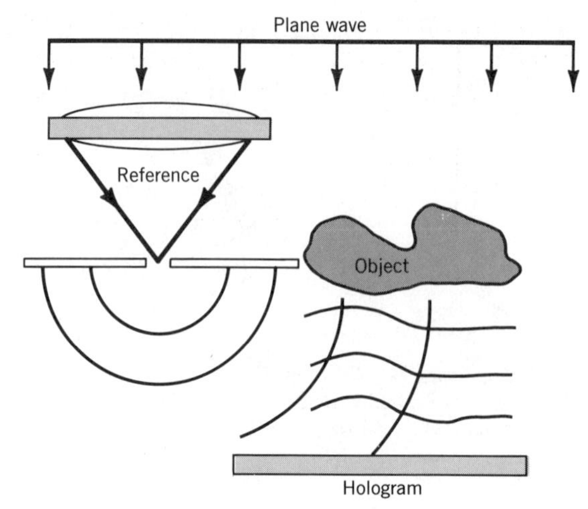

FIGURE 12-27. Geometry for recording a lensless Fourier transform hologram.

mined by the ratio λ_2/λ_1. We have discovered from the geometrical analysis that only for the special case of a Fourier transform hologram **(12-40)** is the magnification simply proportional to the wavelength.

If the source used to reconstruct a Fourier transform hologram is located in the identical plane of the reference source, $Z_r = Z_c$, there are two virtual images formed. These images are located in the same plane as the reconstruction source and symmetrically displaced on either side of the reconstruction source.

If the reference and reconstruction sources are not only in the same plane but are located in identical positions, $Z_r = Z_c$ and $x_r = x_c$, the virtual images will be found in the same z plane, $Z_i = Z_s$, as the object. The images' x coordinates in the object plane are given by

$$x_i = x_r \pm \frac{1}{p}(\lambda_2/\lambda_1)(x_r - x_s) \tag{12-41}$$

If either of the following two conditions hold, there is no lateral magnification in a Fourier transform hologram:

1. If the ratio of the wavelengths is equal to the recording medium's dimension change, $\lambda_2/\lambda_1 = p$.
2. If the wavelengths of the reconstruction and reference waves are equal, $\lambda_1 = \lambda_2$, and there is no dimension change, $p = 1$.

When there is no lateral magnification, the two images are located at the positions

$$x_i = \begin{cases} 2x_r - x_s, & \text{image 1} \\ x_s, & \text{image 2} \end{cases}$$

If the reference source is located on the z axis, these two virtual images would be displayed symmetrically about the z axis.

If the reconstruction wave is a plane wave, then the virtual images are located at infinity. For this configuration, two real images can be generated in the back focal plane of a lens. This is the geometry shown in Figure 12-13 and described by **(12-7)**.

COHERENCY REQUIREMENTS

After the discussion of interference and coherence in Chapter 8, it is our hope that no one is surprised that the source needed to produce a hologram must be both spatially and temporally coherent. It may be a surprise, however, to learn that the reconstruction wave must also be coherent if maximum resolution of the image is to be obtained. After briefly examining the coherency requirements placed on the recording light by the special geometries associated with holography, we will examine the coherency requirements for reconstruction of a holographic image.

Temporal Coherence

Recording an off-axis hologram places demands on the temporal coherence of the recording source. Consider the typical layout of Figure 12-28. The reference ray at the top of the hologram has traveled

$$\ell = a \sin \alpha_r$$

farther than the reference ray at the bottom of the hologram. If we adjust the paths of the signal and reference beams, so that they are in phase at the center of the hologram, the recording source must have sufficient temporal coherence so that its longitudinal coherence length exceeds

$$\ell_R = \frac{\ell}{2}$$

If this condition is met, fringes will be produced across the hologram.

The information content of the signal places additional demands on the temporal coherence of the source used to record the hologram. For ease of visualization, consider the case in which the object is a transparency (Figure 12-29.) The highest spatial frequency in the object diffracts the signal beam at an angle θ_M. This portion of the signal travels a distance

$$\ell_s = \frac{d}{\cos \theta_M} - d$$

farther than the undiffracted signal beam. Thus, to record the spatial information in the signal beam requires a longitudinal coherence length in excess of ℓ_s. The requirements on coherence made by the off-axis reference beam

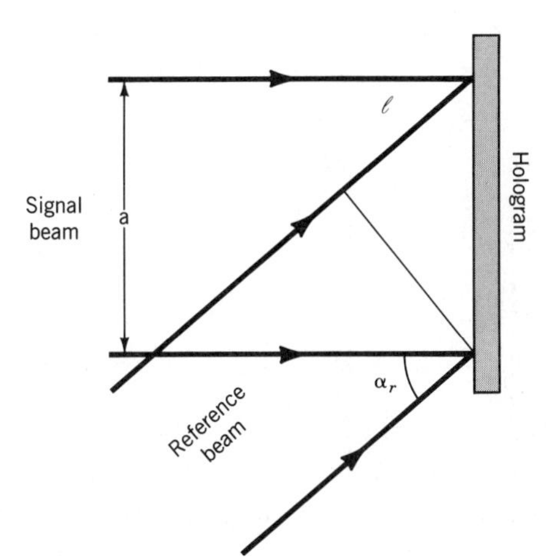

FIGURE 12-28. Recording geometry for off-axis holograms.

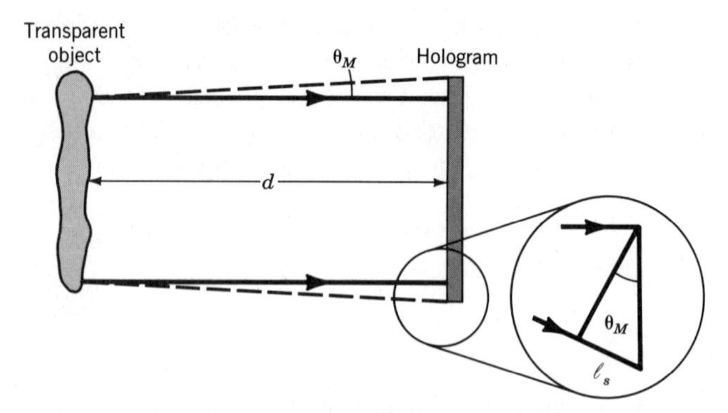

FIGURE 12-29. Effects on signal beam path by the information content of the object.

and the information content of the signal beam demand a recording source with a longitudinal coherence length of

$$\ell_\ell > \ell_R + \ell_s$$

Spatial Coherence

The requirements on the spatial coherence of the reference and reconstruction waves can be calculated by obtaining the point spread function of a hologram. In this approach, the hologram is treated as a linear system. The impulse response of the linear system is found by calculating the image produced by a hologram, generated by a point source. The point source acts as the impulse and the point spread function is the impulse response function. This general treatment cannot be carried to completion so we will use a geometrical analysis of a source of finite dimension and one emitting several wavelengths, to obtain a statement of the coherency requirements.

The hologram to be evaluated is a Fourier transform hologram. Fourier transform holograms are discussed in Appendix 12-B; here, we will give the briefest of details so that we may immediately see the effects of spatial coherence on image resolution.

The object light distribution at the hologram plane, shown in Figure 12-30, is

$$O(\xi) = \mathcal{F}\{a(x)\} \tag{12-42}$$

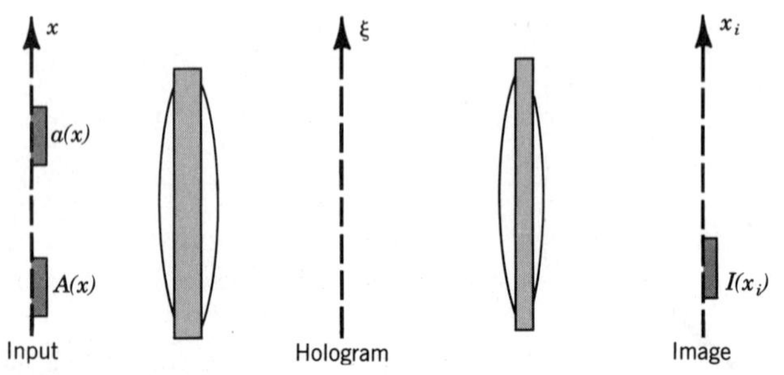

FIGURE 12-30. Geometry for producing a Fourier transform hologram. See Appendix 12-B for further details.

The reference light distribution at the hologram plane is

$$R(\xi) = \mathcal{F}\{A(x)\} \tag{12-43}$$

After exposure and development, we obtain a hologram that we illuminate with a reconstruction wave. The light distribution of the reconstruction, or playback, wave at the hologram is

$$P(\xi) = \mathcal{F}\{B(x)\} \tag{12-44}$$

The light transmitted by the hologram is imaged at the back focal plane of a second lens, labeled the image plane in Figure 12-30. We will only follow the image wave to this plane; its amplitude distribution in the image plane is

$$I(x_i) = \mathcal{F}\{POR^*\} \tag{12-45}$$

The point spread function of the hologram can be calculated by assuming the object is a point source located at position x_0 in the front focal plane of the first lens, that is,

$$a(x) = \delta(x - x_0)$$

The Fourier transform of a delta function is from **(6-24)**

$$O(\xi) = \mathcal{F}\{\delta(x - x_0)\} = \exp\left(-i\frac{k}{f}\xi x_0\right)$$

We can now use the rule for conjugation of a Fourier transform **(6A-6)** and the shifting property of a Fourier transform **(6A-5)** to write

$$I(x_i) = \mathcal{F}\{\mathcal{F}\{B(x)\}\mathcal{F}\{A^*(x - x_0)\}\} \tag{12-46}$$

From the correlation theorem **(6A-10)**, we see that the amplitude point spread function of the hologram is the correlation of the spatial distribution of the reference and reconstruction sources.

Image Blur and Spatial Coherence

A less formal approach, based on a geometrical analysis, will provide a more intuitive description of the effects of coherence on the hologram's resolution. Assume that the hologram was made with a perfect point source, i.e., the source distribution is a delta function and the spatial coherence function is 1. The reconstruction source has a finite, but small, size given by

$$\frac{\Delta r}{2} = x_c - x_r$$

that is, the reconstruction source distribution function is a rectangular function and the spatial coherence function is a sinc function. We will also assume that the reconstruction and reference sources have the same wavelength, i.e., $\lambda_1 = \lambda_2$. The effect of the finite size of the reconstruction source can be found by using the equations for the image coordinates, **(12-28)** and **(12-29)**.

We will set $p = 1$ to reduce the clutter in the equations and, for clarity, we will limit our attention to the image coordinates of the virtual image. The image coordinates for light coming from the center of the reconstruction source, located at the original reference position (x_r, Z_r), are given by the equations

$$x_i = \frac{x_r Z_s Z_r \pm x_r Z_s Z_r \mp x_s Z_r^2}{Z_s Z_r \pm Z_s Z_r \mp Z_r^2} \tag{12-47}$$

$$Z_i = \frac{Z_s Z_r}{Z_s \pm Z_s \mp Z_r} \tag{12-48}$$

The virtual image coordinates, obtained from **(12-47)** and **(12-48)**, are (x_s, Z_s), identical to the object's coordinates. The light from the upper extreme of the reconstruction source has coordinates (x_c, Z_r). The virtual image coordinates due to this wave are

$$x_i = x_s + (Z_s/Z_r)(x_c - x_r) \tag{12-49}$$

$$Z_i = Z_s \tag{12-50}$$

We define a quantity called the *image blur* as

$$\Delta x = 2(x_i - x_s)$$

From **(12-49)**, we obtain the image blur produced by the finite size of the reconstruction source

$$\Delta x = \frac{Z_s}{Z_r} \Delta r \tag{12-51}$$

We can rewrite **(12-51)** in terms of the transverse coherence length ℓ_t, defined in Chapter 8 as a function of the source size **(8-42)**. The source size can be defined in terms of its angular extent

$$\frac{\Delta\theta}{2} = \frac{\Delta r}{Z_r}$$

allowing the image blur to be written as

$$\Delta x = (\lambda/2\ell_t)\, Z_s \tag{12-52}$$

We see that the transverse coherence length of the reconstruction source establishes the minimum resolvable image dimension. Another important consequence of **(12-52)** is that the closer the image is to the hologram (i.e., the smaller Z_s), the lower the requirements on spatial coherence for a given image resolution.

Image Blur and Temporal Coherence

We can use the same geometrical analysis to learn the effects of temporal coherence of the reconstruction source on the resolution of the holographic image. We will assume that the reference and reconstruction sources are point sources, with identical coordinates. We will also assume that these sources are far from the hologram so that at the hologram, they are nearly plane waves. The signal and reference waves have the same wavelength λ_1, but the reconstruction source has a spread

$$\frac{\Delta\lambda}{2} = \lambda_1 - \lambda_2$$

in wavelength. The virtual image coordinates produced by that part of the reconstruction wave at λ_1 are

$$x_i = x_s, \qquad Z_i = Z_s$$

For the portion of the reconstruction source at wavelength λ_2, the image coordinates are

$$x_i = \frac{x_r Z_s \pm (\lambda_2/\lambda_1)(x_r Z_s - x_s Z_r)}{Z_s \pm (\lambda_2/\lambda_1)(Z_s - Z_r)} \tag{12-53}$$

$$Z_i = \frac{Z_s Z_r}{Z_s \pm (\lambda_2/\lambda_1)(Z_s - Z_r)} \tag{12-54}$$

To simplify these equations, we take the limit as the reference wave position is moved to infinity, $Z_r \to \infty$. The coordinates of the image are then

$$x_i = \frac{x_r}{Z_r} Z_s \left(\frac{\lambda_1 - \lambda_2}{\lambda_2} \right) + x_s \tag{12-55}$$

$$Z_i = \frac{\lambda_1 Z_s}{\lambda_2} \tag{12-56}$$

The angle that the reference beam makes with the hologram, the offset angle, is given by

$$\tan \alpha_r = \frac{x_r}{Z_r}$$

In the paraxial approximation, the offset angle can be used in **(12-55)** to write the resolution limits due to the temporal coherence of the reconstruction source as

$$\Delta x = Z_s \alpha_r \frac{\Delta\lambda}{\lambda_2} \tag{12-57}$$

$$\Delta Z = Z_s \frac{\Delta\lambda}{\lambda_2} \tag{12-58}$$

or in terms of the longitudinal coherence length

$$\Delta x = Z_s \alpha_r \frac{\lambda}{\ell_\ell} \tag{12-59}$$

$$\Delta Z = Z_s \frac{\lambda}{\ell_\ell} \tag{12-60}$$

Figure 12-31 displays a hologram reconstructed by three discrete wavelengths. The color image on the color insert can be used to observe the dependence of the image coordinates on wavelength.

Since the image position, relative to the hologram, controls the coherency requirement placed on the reconstruction source, we may reduce the coherency requirements by using the proper geometry. For example, by using a lens to image the object we wish to record onto the hologram plane, we can make $Z_s \approx 0$.

Equation **(12-59)** reinforces our previous discussion of the effects of offset angle on coherence requirements. As we see, large offset angles must be accompanied by a large coherence length if we wish high-resolution (small Δx) images.

(a)

(b)

FIGURE 12-31. A hologram of a ceramic statue reconstructed at (a) the recording wavelength and (b) three discrete wavelengths. (Hologram courtesy of N. George, University of Rochester.)

SUMMARY

By recording the fringe pattern produced by two coherent waves, a hologram can be produced with an amplitude transmission function given by

$$\mathrm{t}_h(x, y) = \mathrm{t}_0 + \beta(A^*a + Aa^*)$$

If the hologram is illuminated by a wave of amplitude B, then the wave transmitted by the hologram will have an amplitude given by

$$B(x, y)\,\mathrm{t}_h(x, y) = \mathrm{t}_0 B + \beta A^* B a + \beta A B a^*$$

The transmitted wave contains a reproduction of the complex amplitude of both waves. The wave we wish to reproduce is identified as the signal wave and its complex amplitude is denoted by a.

The wave transmitted by the hologram contains not only the desired signal wave, but also a number of unwanted waves. To separate the signal wave from these unwanted waves, we produce the hologram by illuminating the recording medium with a signal wave and a reference wave, which propagate at an angle with respect to each other. The minimum angle between the reference and signal wave that will allow an uncluttered reproduction of the signal wave is given by

$$\theta_{\min} = \sin^{-1}\left(\frac{3k_W\lambda}{2\pi}\right)$$

where k_W is the spatial frequency bandwidth of the signal wave.

Holograms can be classified by the size of the angle between the reference and signal waves. If the angle is small, then the hologram's thickness can be neglected and the grating equation can be used to describe the performance of the hologram. If the angle is large, then the hologram is classified as thick and its performance is described by Bragg's law

$$2d \sin\theta = n\lambda$$

The fraction of the reconstruction wave that is diffracted into the signal wave by the hologram is called the diffraction efficiency of the hologram

$$\eta = \frac{(\beta B A^* a)^2}{|B|^2}$$

The diffraction efficiency is a function of the exposure energy E_0 and the visibility V of the interference fringes

$$\eta = \left(\frac{\beta' \mathrm{E}_0 V}{2}\right)^2$$

The fringe visibility is determined by the relative amplitudes of the signal and reference waves called the K ratio

$$\mathrm{K} = \frac{A_0^2}{a_0^2}$$

A hologram can be viewed as a collection of Fresnel zone plates. The focal length of a zone plate created by interfering the waves from two point sources would be given by

$$\frac{1}{f_\pm} = \mp\left(\frac{1}{Z_r} - \frac{1}{Z_s}\right)$$

The reconstruction of the zone plate would be described by the equation

$$\frac{1}{f_\pm} = \mp\frac{\lambda_2}{\lambda_1}\left(\frac{1}{Z_c} - \frac{1}{Z_i}\right)$$

where λ_1 is the recording wavelength, λ_2 the reconstruction wavelength, Z_c the position of the reconstruction source, and Z_i the position of the image of the signal wave source.

A simple two-dimensional geometric model of a hologram can be created using the associations between coordinate positions and components of the holographic experiment shown in Table 12.1. The first-order theory yields an equation relating the coordinates of the image to the coordinates of the other objects. The distance from the hologram to the image is given by

$$\frac{1}{Z_i} = \frac{1}{Z_c} \pm \frac{1}{p^2}(\lambda_2/\lambda_1)\left(\frac{1}{Z_r} - \frac{1}{Z_s}\right)$$

The x component of the image is given by

$$x_i = \frac{x_c \pm \frac{1}{p}(\lambda_2/\lambda_1)\, Z_c\left(\frac{x_r}{Z_r} - \frac{x_s}{Z_s}\right)}{1 \pm \frac{1}{p^2}(\lambda_2/\lambda_1)\, Z_c\left(\frac{1}{Z_r} - \frac{1}{Z_s}\right)}$$

The lateral magnification of the image produced by the hologram is given by

$$\beta = \frac{p}{1 - (Z_s/Z_r) \mp p^2(\lambda_1/\lambda_2)(Z_s/Z_c)}$$

and the longitudinal magnification

$$\beta_z = \frac{\partial Z_i}{\partial Z_s} = \pm\frac{\lambda_1}{\lambda_2}\beta^2$$

The coherence of the reconstruction wave affects the resolution of the image produced by the hologram. The spatial coherence of the reconstruction wave is inversely proportional to the lateral image blur

$$\Delta x = \frac{\lambda}{2\ell_t} Z_s$$

The temporal coherence is inversely proportional to both the lateral and longitudinal image blur

$$\Delta x = \frac{\alpha_r \lambda}{\ell_\ell} Z_s$$

$$\Delta Z = \frac{\lambda}{\ell_\ell} Z_s$$

PROBLEMS

12-1. Assume that the reference and signal positions are 100 mm apart and are both 100 mm from the recording medium. If the wavelength is 632.8 nm, what is the minimum resolution the recording medium must have? What type of hologram will be created?

12-2. Assume a hologram is made with a plane wave of wavelength $\lambda = 632.8$ nm and a signal of the same wavelength located 1 m from the recording plane. If the hologram is reconstructed with a plane wave of wavelength $\lambda = 488$ nm, where is the virtual image located?

12-3. A holographic diffraction grating with 1000 lines/mm is to be constructed using plane waves from a HeNe laser of wavelength $\lambda = 632.8$ nm. What is the angle between the waves? Draw the experimental configuration.

12-4. Kodak Tri-X photographic film has a maximum resolution of about 50 lines/mm. What is the maximum angle between the reference and signal waves for this film if we use $\lambda = 441.6$ nm? What if we use $\lambda = 632.8$ nm?

12-5. With a wavelength of 514.5 nm, a hologram is recorded. The reference point source is located at (−0.5 m, −1 m) and the signal at (0.5 m, −0.5 m). After development, the hologram is illuminated by a reconstruction wave of $\lambda = 488$ nm located at (−0.5, −0.5 m). What are the lateral and longitudinal magnification? Where are the real and virtual images?

12-6. Where are the images in Problem 12-5 if the developed hologram is enlarged during processing by 10%?

12-7. We wish to record a hologram 10 cm on a side with a resolution of 0.01 mm using an argon laser with a wavelength of $\lambda = 488$ nm. What is the minimum angle between the signal and reference beam if the recording and viewing distance must be 20 cm?

12-8. A zone plate is created on a photographic film using two point sources: one 20 cm and the other 50 cm from the recording film. If the recording and playback wavelengths are both 632.8 nm, what is the focal length of the zone plate?

12-9. If the recording film used in Problem 12-8 can record 2000 lines/mm, what size zone plate can be recorded?

12-10. We wish to record a hologram of an automobile that is 18 ft × 4 ft. The hologram will be recorded on an 8 in. × 10 in. glass plate using 488 nm radiation from an argon laser. How far from the car should we locate the film plate if the film can record 2500 lines/mm?

12-11. Calculate the magnification and find the location of the images for a hologram made with a reference beam located at (15, −200) and a signal at (0, −100). [In these problems, the coordinates are presented as (x, z).] The reconstruction wave is a plane wave arriving at the hologram at an angle of −30° to the hologram normal. Assume that there is no change in the dimensions of the hologram during processing and that the wavelengths used for construction and reconstruction are the same.

12-12. Calculate the magnification and find the location of the images for a hologram made at a wavelength of 514.5 nm with a reference beam located at (0, −200) and a signal at (5, −100). The reconstruction wave of wavelength 488 nm has coordinates of (0, −190). Assume that the hologram swelled by 5% during processing.

12-13. Calculate the magnification and find the location of the images for a hologram made at a wavelength of 514.5 nm with a reference beam located at (−20, −200) and a signal at (15, −200). The reconstruction wave of wavelength 488 nm has coordinates of (−20, −200). Assume that the hologram swelled by 5% during processing.

12-14. Calculate the magnification and find the location of the images for a hologram made at a wavelength of 488 nm with a reference beam located at (0, −200) and a signal at (15, −200). The reconstruction wave of wavelength 488 nm has coordinates of (0, −200). Assume that no dimension changes occurred during processing.

12-15. Calculate the magnification and find the location of the images for a hologram made at a wavelength of 488 nm with a reference beam located at (0, −200) and a signal at (15, −200). The reconstruction wave of wavelength 488 nm has coordinates of (0, −∞). Assume that no dimension changes occurred during processing. How should this hologram be viewed?

12-16. A recording medium is placed at the origin of a coordinate system in the x, y plane and normal to the z axis. The reference wave's propagation vector is

$$k(\sin\,\theta_r\hat{\mathbf{i}} + \cos\,\theta_r\hat{\mathbf{k}})$$

The signal wave's propagation vector is

$$k(-\sin\,\theta_s\hat{\mathbf{i}} + \cos\,\theta_s\hat{\mathbf{k}})$$

Derive an expression for the fringe spacing along the x and z directions. *Hint:* This is a generalization of Problem 4-17.

12-17. Using the arrangement described in Problem 12-16, find the fringe spacing along the bisector of the angle between the signal and reference wave $\theta_r + \theta_s$. Assume the bisector makes an angle φ with respect to the z axis.

12-18. Using the arrangement described in Problem 12-16, find the fringe spacing along the normal to the bisector of the angle between the signal and reference wave.

12-19. We have constructed a hologram of an object 10 cm long. We used a lens to image the object at the hologram plate so the reconstruction appears to be centered at the hologram. What is the transverse coherence length needed to reconstruct the hologram with an image blur less than 100 μm?

12-20. We wish to use sunlight and a filter centered at 500 nm with a spectral width of 50 nm to display a hologram. The hologram was made with a reference angle of 30°. What is the maximum depth we can expect to see in the hologram if we require an image blur of 100 μm?

Appendix 12-A

PHASE HOLOGRAMS

In the discussion of holography in Chapter 12, the holographic recording was always assumed to be an amplitude recording. We now wish to indicate briefly the properties of a hologram produced by phase recording. In this type of recording, the amplitude transmission coefficient of the developed hologram will be assumed to be constant, independent of spatial position, and the phase change produced by the developed hologram will depend linearly on the exposure energy.

The signal wave will be represented by

$$a(x) = a_0 e^{i\varphi(x)} \tag{12A-1}$$

and the reference wave will be assumed to be a plane wave

$$A(x) = A_0 e^{ikx} \tag{12A-2}$$

where

$$k = \frac{2\pi}{\lambda} \sin \theta_r$$

We assume that the recording material will linearly record the spatial variation of the exposure energy $E(x)$ as a spatial phase variation given by

$$\Phi(x) = \beta E(x)$$

After exposure and development, the transmission coefficient of the hologram will be

$$t_h(x) = t_0 e^{i\beta a_0^2} e^{i\beta A_0^2} \exp\left[2i\beta a_0 A_0 \cos(\varphi - kx)\right] \tag{12A-3}$$

To simplify the equation, we define an average complex transmission coefficient of the developed hologram as

$$\tilde{t}_c = t_0 e^{i\beta a_0^2} e^{i\beta A_0^2}$$

We also define the phase change due to the amplitude variation of the interference pattern as

$$\Gamma = 2\boldsymbol{\beta} a_0 A_0$$

and the spatial variation of the phase due to the phase variation of the interference pattern as

$$\psi = \varphi - kx$$

If we use these newly defined parameters, the transmission coefficient of the hologram becomes

$$\begin{aligned} t_h(x) &= \tilde{t}_c e^{i\Gamma \cos \psi} \\ &= \tilde{t}_c[\cos(\Gamma \cos \psi) + i \sin(\Gamma \cos \psi)] \end{aligned} \tag{12A-4}$$

The sine and cosine functions of **(12A-4)** can be expressed as a series of Bessel functions[48] (see relation 9.1.44 and 9.1.45 in the reference)

$$\cos(\Gamma\cos\psi) = J_0(\Gamma) + 2\sum_{n=1}^{\infty}(-1)^n J_{2n}(\Gamma)\cos(2n\psi) \tag{12A-5}$$

$$\sin(\Gamma\cos\psi) = 2\sum_{n=0}^{\infty}(-1)^{n+2} J_{2n+1}(\Gamma)\cos[(2n+1)\psi] \tag{12A-6}$$

Equation **(12A-4)** can be expressed in terms of **(12A-5)** and **(12A-6)**,

$$t_h(x) = \tilde{t}_c\left\{J_0(\Gamma) + 2\sum_{n=1}^{\infty}(-1)^n J_{2n}(\Gamma)\cos(2n\psi) + 2i\sum_{n=0}^{\infty}(-1)^{n+2}J_{2n+1}(\Gamma)\cos[(2n+1)\psi]\right\}$$

We see that a large number of diffraction orders, one for each value of n, will be reconstructed by the thin phase hologram. This is in contrast to only two diffraction orders in a linearly recorded amplitude hologram; see **(12-5)**. The amplitude of the nth diffracted order is proportional to the nth order Bessel function J_n. Figure 12A-1 displays the amplitude variation of the first five Bessel functions.

We will limit our attention to the $n = 0$ term corresponding to the first diffracted order. The hologram's transmission coefficient associated with the first diffracted order is labeled as t_{h0}.

$$\begin{aligned} t_{h0}(x) &= 2i\,\tilde{t}_c J_1(\Gamma)\cos\psi \\ &= i\,\tilde{t}_c J_1(\Gamma)[e^{i(\varphi-kx)} + e^{-i(\varphi-kx)}] \end{aligned} \tag{12A-7}$$

If we illuminate the exposed hologram with a duplicate of the reference wave, then the first diffracted order of the reconstruction wave is

$$\begin{aligned} Bt_{h0}(x) &= A_0 e^{ikx}\,\tilde{t}_c e^{i(\pi/2)} J_1(\Gamma)\left[e^{i(\varphi-kx)} + e^{-i(\varphi-kx)}\right] \\ &= A_0 t_0 e^{i\left(\beta a_0^2 + \beta A_0^2 + \pi/2\right)} J_1(\Gamma)\left[e^{i\varphi} + e^{-i(\varphi-2kx)}\right] \end{aligned} \tag{12A-8}$$

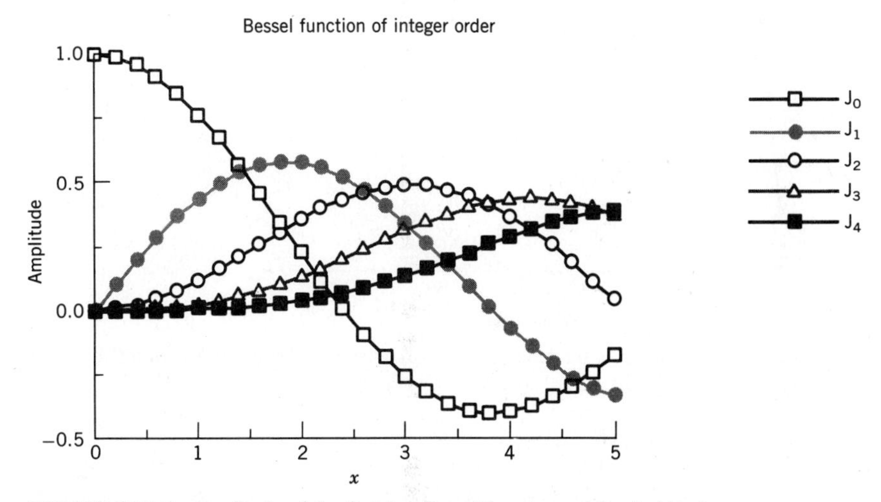

FIGURE 12A-1. Amplitude of the first five Bessel functions of the first kind.

This wave does not appear to be a reasonable reproduction of the original signal beam. However, if $a_0 << A_0$, then we can replace $J_1(\Gamma)$ with the first term of its series expansion

$$J_1(2x) = x - \frac{x^3}{2} + \frac{x^5}{2^2 \cdot 3} - \cdots$$

Using the definition of Γ, we can approximate the first-order Bessel function by

$$J_1(\Gamma) \approx \beta a_0 A_0$$

whenever the signal beam is small relative to the reference beam. This enables us to write an expression for the virtual image wave

$$A_0^2 \beta t_0 \exp\left[i\left(\beta a_0^2 + \beta A_0^2 + \frac{\pi}{2}\right)\right]\{a_0 e^{i\varphi}\} \tag{12A-9}$$

The term in braces is the original signal wave. The intensity distribution of the diffracted wave reproduces the signal wave's intensity distribution and the diffracted wave is within a constant phase factor of the original signal wave.

The maximum transmissivity of the phase recording is $(A_0 t_0)^2$ and the maximum intensity diffracted into the first order is given by the maximum value of $(J_1)^2$ multiplied times $(A_0 t_0)^2$, which is $(0.582)^2 (A_0 t_0)^2 = (0.339)(A_0 t_0)^2$. The maximum diffraction efficiency of a thin-phase hologram is therefore 33.9%.

The amplitude holograms we discussed in Chapter 12 and the thin-phase holograms we have just discussed have marginally useful diffraction efficiencies. If we could do no better, holography would be of limited utility. However, thick-phase holograms have a maximum diffraction efficiency of 100% and simple holographic optical elements have approached this theoretical diffraction efficiency; see Figure 12A-2.

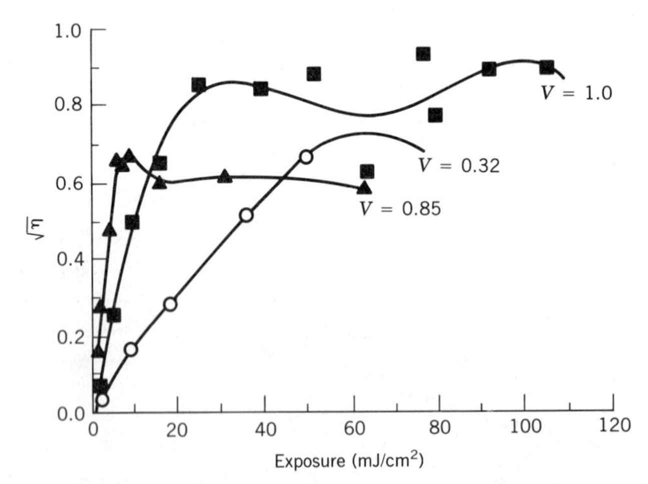

FIGURE 12A-2. Diffraction efficiency vs exposure energy for several values of fringe visibility V. The hologram is a phase recording using a photopolymer.

Appendix 12-B

VANDERLUGT FILTER

As discussed in Appendix 10-B, the calculation of a correlation or a convolution integral can be accomplished by recording the amplitude and phase of a Fourier transform. From the theoretical development made in Chapter 12, it is apparent that this type of recording is a Fourier transform hologram. A. VanderLugt in 1963 was the first to introduce the use of a Fourier transform hologram for use in pattern recognition operations, and the hologram is often called a *VanderLugt filter*. The technique works because both the phase and amplitude of the Fourier transform are recorded in the hologram.

To construct a VanderLugt filter, we place a functions $g(x, y)$, which we will call the reference function, in the front focal plane of lens L_1, a distance f_1 from L_1, as shown in Figure 12B-1.

The Fourier transform $G(p, q)$ of $g(x, y)$ appears in the back focal plane of L_1 (the transform plane), a distance f_1 behind L_1. We add a reference plane wave $R(p, q)$ interferometrically to $G(p, q)$ at the transform plane, the recording plane of the hologram. For simplicity, we will limit the analysis to one dimension with the dimension in the signal plane represented by x and in the recording plane by p. The reference wave of amplitude R_0 is incident at an angle θ to the transform plane; this angle is the hologram's offset angle

$$R(p) = R_0 e^{ikp \sin\theta} \tag{12B-1}$$

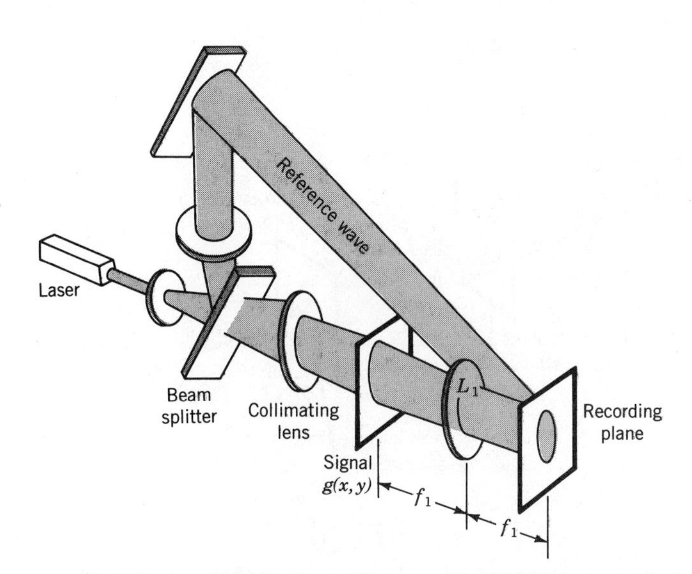

FIGURE 12B-1. Geometry for recording a Fourier transform hologram.

The amplitude distribution of the interference pattern in the transform plane

$$\mathcal{E}(p) = R_0 e^{ikp\sin\theta} + \sqrt{(i/\lambda f_1)} \int_{-\infty}^{\infty} g(x) \exp\left[i\,(k/f_1)\,px\right] dx \tag{12B-2}$$

$$= R(p) + G(p)$$

is recorded on photosensitive material. The exposure is made using an exposure time τ_e such that the average exposure energy $|R_0|^2\tau_e$ creates an amplitude transmission coefficient of t_0, centered in the linear portion of the amplitude transmission coefficient curve of the developed hologram (Figure 12-8).

The amplitude transmission coefficient of the developed hologram is

$$t_h = t_0 + \beta|G(p)|^2 + \beta R_0 G(p) e^{-ikp\sin\theta} + \beta R_0 G^*(p) e^{ikp\sin\theta} \tag{12B-3}$$

No provision has been made in **(12B-3)** for dimensional changes in the hologram due to processing. Also, it is implicitly assumed here that the entire Fourier transform can be linearly recorded by the recording medium.

The recording is illuminated with a reconstruction wave obtained by replacing $g(x, y)$ by a new function $f(x, y)$ in the front focal plane of L_1. After passing through L_1, the reconstruction field at the hologram is described by the Fourier transform of $f(x, y)$

$$F(p) = \sqrt{(i/\lambda f_1)} \int_{-\infty}^{\infty} f(x') \exp\left[i\,(k/f_1)\,px'\right] dx' \tag{12B-4}$$

The hologram diffracts $F(p)$ into four component waves

$$\begin{aligned}\Psi(p) &= t_0 F(p) + \beta|G(p)|^2 F(p) \\ &\quad + \beta R_0 G(p) F(p) e^{-ikp\sin\theta} \\ &\quad + \beta R_0 G^*(p) F(p) e^{ikp\sin\theta}\end{aligned} \tag{12B-5}$$

The various terms of this diffracted field can be associated with the wave directions shown in Figure 12B-2. The first two terms,

$$t_0 F(p) + \beta|G(p)|^2 F(p)$$

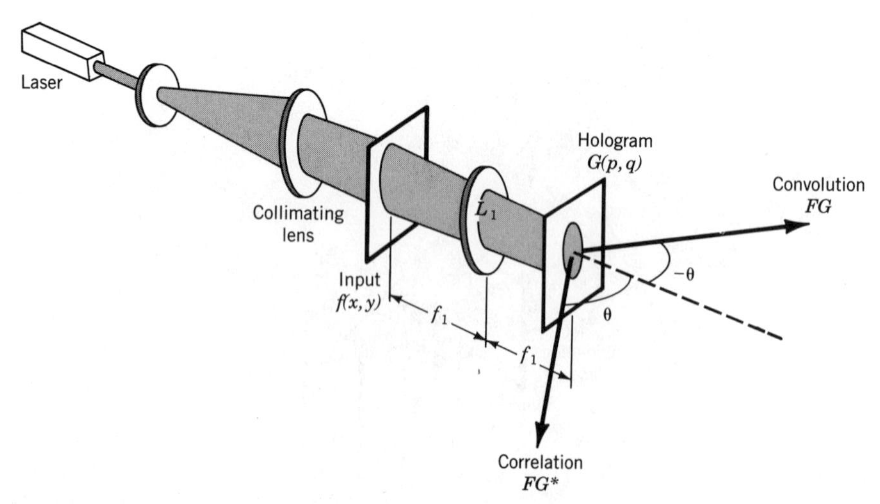

FIGURE 12B-2. The waves reconstructed by the Fourier transform hologram when illuminated by the reconstruction wave $f(x, y)$.

propagate along the optical axis of lens L_1 and are not of interest here. The third term, propagating at an angle $-\theta$ to the optical axis, is labeled "convolution" in Figure 12B-2. The fourth term, propagating at an angle θ to the optical axis, is labeled "correlation" in Figure 12B-2. We will restrict our attention to the fourth term of **(12B-5)**.

A lens L_2 of focal length f_2 is positioned with its optical axis at an angle θ with respect to the optical axis of L_1 so that it collects all of the fourth term of **(12B-5)**, as is shown in Figure 12B-3. At the front focal plane of L_2, the fourth term of **(12-5)** can be written

$$\Psi_4(p) = \frac{\beta R_0}{\lambda f_1} e^{ikp\sin\theta} \int_{-\infty}^{\infty} f(x') e^{i(k/f_1)px'} dx' \int_{-\infty}^{\infty} g^*(x) \exp\left[-i(k/f_1)px\right] dx \tag{12B-6}$$

$$\Psi_4(p) = \frac{\beta R_0}{\lambda f_1} e^{ikp\sin\theta} \int\int_{-\infty}^{\infty} f(x') g^*(x) \exp\left[i(k/f_1)p(x'-x)\right] dx'\,dx \tag{12B-7}$$

The second lens L_2 forms the Fourier transform of **(12B-7)** in its back focal plane a distance f_2 from L_2. The coordinate in the back focal plane of L_2 is represented by ξ.

$$\mathrm{R}(\xi) = \sqrt{\frac{i}{\lambda f_2}} \int_{-\infty}^{\infty} \Psi_4(p) \exp\left[i(k/f_2)p\xi\right] dp \tag{12B-8}$$

To simplify the mathematical manipulation, we assume that the focal length of L_1 and L_2 is the same, $f_1 = f_2 = f$. The field in the back focal plane of L_2, labeled correlation plane in Figure 12B-3, is obtained by substituting **(12B-7)** into **(12B-8)**

$$R(\xi) = \frac{\beta R_0}{\lambda f} \sqrt{\frac{i}{\lambda f}} \int\int\int_{-\infty}^{\infty} g^*(x) f(x') \exp\left[i(k/f)p(x'-x+\xi+f\sin\theta)\right] dx\,dx'\,dp \tag{12B-9}$$

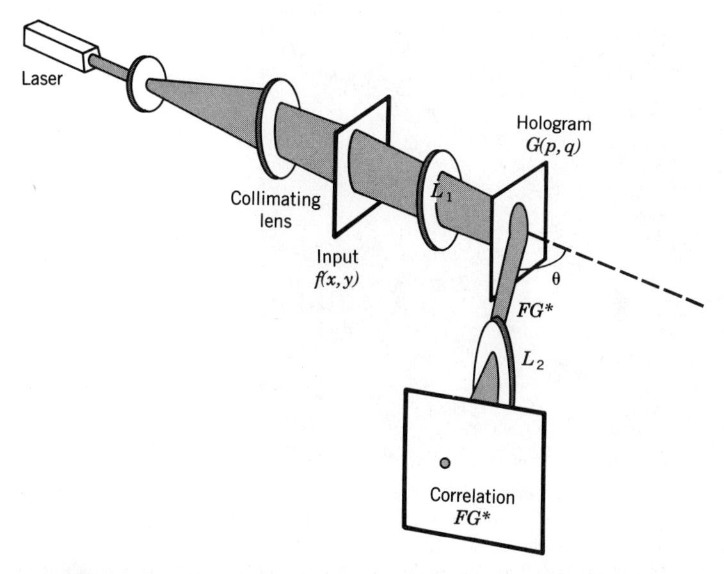

FIGURE 12B-3. Geometry for reconstruction of a Fourier transform hologram in a correlation configuration.

The sifting property of the delta function **(6-23)** simplifies **(12B-9)** to

$$R(\xi) = \frac{\beta R_0}{\lambda f}\sqrt{\frac{i}{\lambda f}}\, 2\pi \int\int_{-\infty}^{\infty} g^*(x) f(x')\, \delta\left(\frac{k}{f}\left[x' - x + \xi + f \sin\theta\right]\right) dx\, dx'$$

$$= \beta R_0 \sqrt{\frac{i}{\lambda f}} \int_{-\infty}^{\infty} g^*(x) f(x - \xi - f \sin\theta)\, dx \qquad \text{(12B-10)}$$

Equation **(12B-10)** can be identified as the correlation of the input function $f(x, y)$ and the function $g(x, y)$.

An operational interpretation of **(12B-10)** can be obtained by assuming that $f(x, y)$ is identical to $g(x, y)$. The Fourier transform of f, which is now $G(p, q)$, is used as the reconstruction wave for the hologram stored in the back focal plane of L_1. From our knowledge of holograms, we know that if one of the two waves used to record a hologram is used as the reconstruction wave, then the second recording wave will be reproduced. Here, $G(p, q)$ is to be used as the reconstruction wave; thus, the hologram will form through diffraction a wave identical to the reference wave $R(p)$ when it is illuminated with $G(p, q)$. The reference wave, a plane wave, is focused by L_2 to a spot in the correlation plane. The occurrence of the spot of light in the correlation plane will thus indicate that $g(x, y)$ is at the input of the optical system shown in Figure 12B-3.

An argument identical to the one leading to **(12B-10)** can be used to demonstrate that the third term of **(12B-5)**

$$\Psi_3(p) = \beta R_0 G(p) e^{-ikp \sin\theta}$$

can be used to generate the convolution of $f(x)$ and $g(x)$. To do so, we need only position a lens with its optical axis aligned along the direction $-\theta$ shown in Figure 12B-2. The convolution will then be formed in the back focal plane of the lens by the diffracted wave described by $\Psi_3(p)$.

The range of amplitudes in the Fourier transform of most signals exceeds

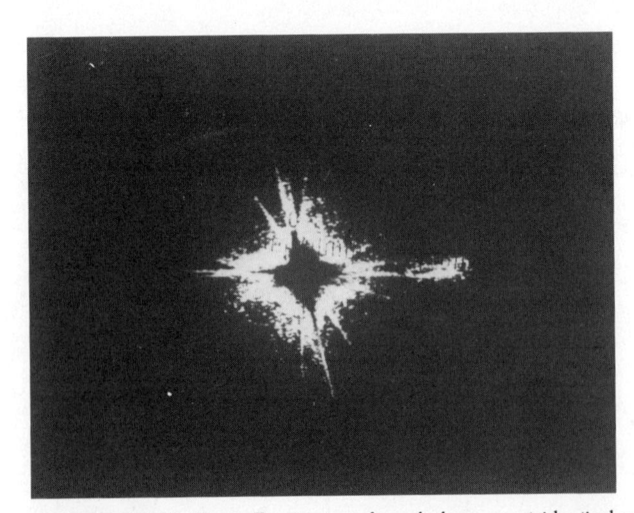

FIGURE 12B-4. Three Fourier transform holograms at identical magnifications are shown here. The bright region in the three images is due to light diffracted by the holograms and indicates the active region of the recording. The exposure time and reference wave intensity were adjusted to produce band-limited Fourier transform holograms that would operate over different spatial frequency regions (Courtesy C.R. Christensen, U.S. Army Missile Command.)

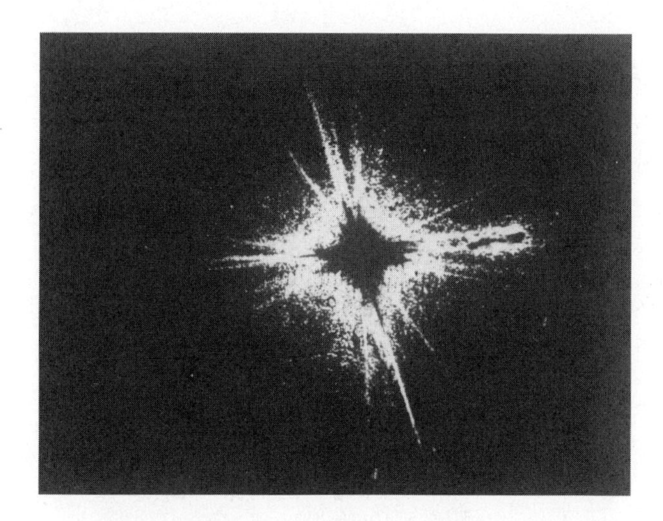

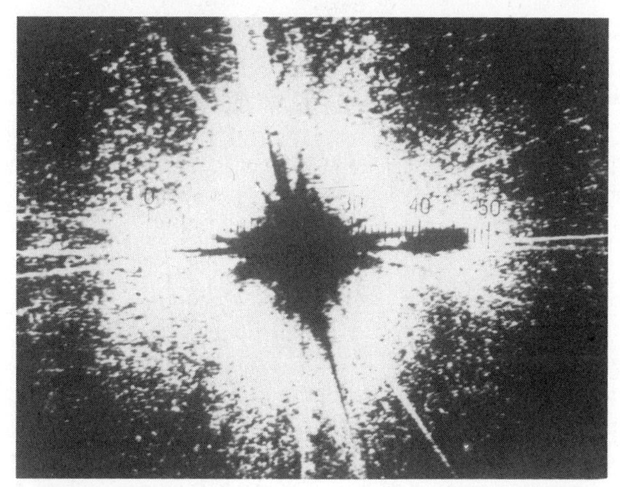

FIGURE 12B-4. *(Continued)*

the dynamic range of holographic recording mediums. For this reason, the hologram recorded in the configuration shown in Figure 12B-1 does not reproduce the entire Fourier transform of the reference object. Instead, a "band-limited" version of the Fourier transform is recorded, i.e., the recorded hologram only reproduces a limited number of spatial frequencies of the object. Very low spatial frequencies have a very large amplitude and overexpose the recording medium, whereas very high spatial frequencies have very small amplitudes and underexpose the recording medium. If we look at the light diffracted by a typical Fourier transform hologram (Figure 12B-4), we find that only a limited range of spatial frequencies contributes to the diffracted light. Control of the exposure time and the reference wave amplitude can be used to select the spatial frequency bandwidth of the Fourier transform recorded in the hologram. This technique was used to generate the Fourier transform holograms of different bandwidths shown in Figure 12B-4.

If the Fourier transform hologram of limited bandwidth is reconstructed using a duplicate of the original reference wave, the original signal is not

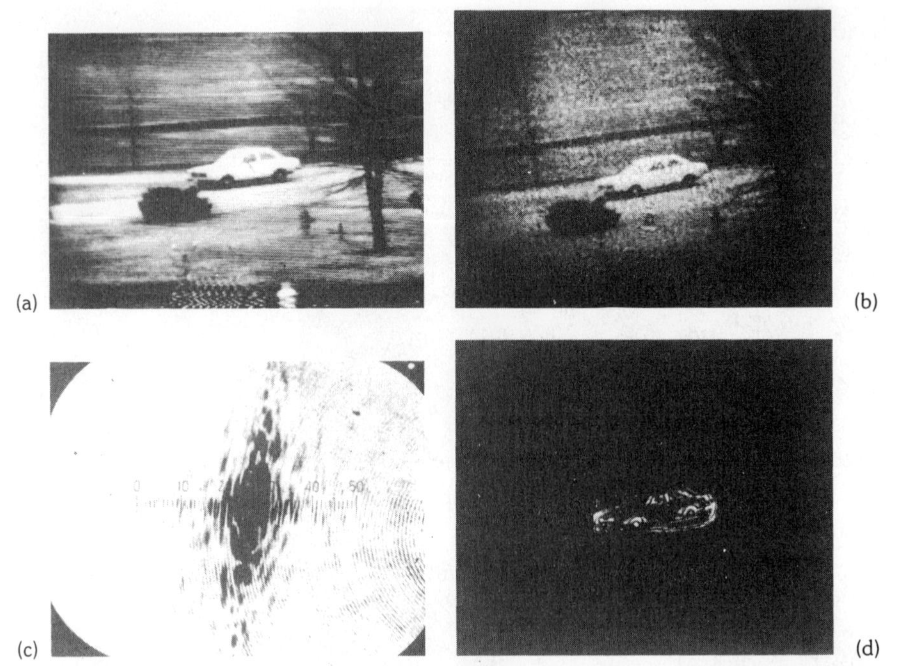

FIGURE 12B-5. (a) The image used in the optical system of Figure 12B-1. (b) The image at the signal plane of Figure 12B-1. (c) A magnified image of the Fourier transform hologram. (d) The reconstructed image generated by the Fourier transform hologram when exposed to a plane wave.

reproduced, but instead a rather fuzzy, edge-enhanced version of the original object is obtained, as can be seen in Figure 12B-5. In Figure 12B-5*a*, an automobile is used as the original object. The appearance of the object at the signal plane of Figure 12B-1 is shown in Figure 12B-5*b*. A photo of the Fourier transform hologram is shown in Figure 12B-5*c*. When the matched filter is placed in the optical system of Figure 12B-1 and illuminated by only the reference wave, the image shown in Figure 12B-5*d* is observed when a lens is used to collect the light propagating along the optical axis of lens L_1.

Anisotropy

Up to now, we have assumed that light's propagation medium had isotropic optical properties, that is, the index of refraction did not depend on the direction of propagation in the medium. The study of most physical theories begins with a similar assumption, that the medium has uniform physical properties; however, in nature, it is more common to observe nonuniform physical properties. When a physical property of a material varies with direction, the material is said to be *anisotropic*. A nonoptical example of an anisotropic property is the mechanical strength of wood. Along the grain direction, the strength of wood is much lower than it is across the grain.

Optical anisotropic materials have an index of refraction that varies with the propagation direction in the material. These materials are said to be *birefringent*. When an object is viewed through a birefringent material, two images of the object are observed that move relative to one another as the birefringent material is rotated. The first recorded observation of this behavior was made in 1669 by **Erasmus Bartholinus (1625–1692)** with calcite, a natural occurring mineral (calcite, the mineral $CaCO_3$, is called Iceland spar in older textbooks). Bartholinus believed that the observed behavior was a special case of refraction and thus called the effect *double refraction*. (Refringence is an archaic synonym of refraction. The derivative, birefringence, remains the common descriptor for double refraction.)

Bartholinus' observation led Huygens in 1690 to describe birefringence using his construction technique. Huygens assumed that a point on the object produced two wavelets in the calcite: a spherical wave, called the *ordinary wave*, that obeyed Snell's law, and an ellipsoidal wave, called the *extraordinary wave*, that did not obey Snell's Law. Huygens found that the light that passed through calcite could be extinguished by passing it through a second, properly oriented, calcite crystal but he was unable to explain the origin of his observations.

Newton developed an explanation of the observations of birefringence based on his particle theory of light. He believed that birefringence and rectilinear propagation were the important obstacles to the acceptance of a wave theory of light. One reason for Newton's failure to associate birefringence with wave behavior was the inability to envision a transverse, polarized wave. During Newton's time, all waves were assumed to be longitudinal waves.

In 1808 **Ètienne Louis Malus (1775–1812)** accidentally observed, while watching a sunset reflected from the windows across the street from his apartment, that the two images of the sun produced by a piece of calcite would alternately dim and brighten as the calcite was rotated. He coined the word *polarization* for the observed property of light, relating the orientation of the preferred direction of the light to the Earth's poles. Malus defined the plane of polarization of the reflected light as the plane of incidence. This was an arbitrary definition made with no knowledge of electromagnetic theory; in fact, Malus' view of light was based on Newton's particle theory. The definition of the plane of polarization used by Malus is orthogonal to the current definition of polarization in terms of the electric field. Many authors use the term *plane of vibration* for the plane containing the electric field to differentiate the modern definition from the one made by Malus.

The present concept of polarization had to await the suggestion by Young in 1817 that light propagated as a transverse wave. Fresnel claimed to have made the association of light and transverse waves prior to Young, but even if he did not, he did construct a formal theory of the concept in 1824. The final step in understanding the polarization properties of light came with the introduction of Maxwell's theory.

We introduced the concept and description of polarized light in Chapter 2. Here, we wish to explore methods for the production and control of polarized light. There are materials, called *dichroic*, whose absorption constant varies with polarization. The anisotropic absorption properties can be used to detect the polarization of a wave or produce a wave of a desired polarization. The use of the selective reflection of a single polarization at Brewster's angle can also be used to manipulate the polarization of waves.

Of the most interest is the application of the observation of Malus, that the two images produced by a birefringent material are orthogonally polarized. In fact, the prime objective in this presentation of polarization is to introduce birefringent materials and discuss their properties. We will find, in later chapters, that birefringence is useful, not only for the manipulation of polarized light, but also in the design of optical modulators and use of nonlinear optics for harmonic generation.

The mathematics needed to describe birefringence relies on the use of tensors. Some of the properties of tensors and a geometrical construct used to aid in the interpretation of symmetric, second-rank tensors are presented in Appendix 13-A. The nonzero components of the tensors are determined by the symmetry of the birefringent crystals. The nonzero elements of second-order tensors are presented for each of the crystal symmetry classes in Appendix 13-A.

The first encounter with a tensor is in relating the **E** and **D** fields in an anisotropic medium through the constitutive relationship. We discover that if the anisotropic material has no loss mechanism, then a second-order, symmetric tensor, represented by a graphical construct in the shape of an ellipsoid, describes the relationship between **E** and **D**.

The need for a tensor relationship between **E** and **D** is because in an anisotropic material, the two vectors are no longer parallel to one another. In fact, we will demonstrate in the chapter that the electric field is no longer perpendicular to the propagation vector **k**, and the Poynting vector is no longer parallel to the propagation vector. However, Maxwell's equations require that **D** and **k** be perpendicular to each other and the definition of the Poynting vector requires that **E** and **S** be perpendicular to each other.

Because the two pairs of vectors $\mathbf{E} \perp \mathbf{S}$ and $\mathbf{D} \perp \mathbf{k}$ are orthogonal, it is possible to formulate two descriptions of propagation in an anisotropic material. The descriptions are based on a coordinate system established by either **D** and **k** or by **E** and **S**. We will use the former formulation in this chapter but provide a discussion of the latter in Appendix 13-B. (The multiple descriptions of anisotropy and the geometrical constructions associated with the descriptions cause a great deal of confusion. To help avoid confusion, only one geometrical construction will be presented in this chapter. All other constructions are discussed in separate appendices.)

In this chapter, the ellipsoid called the optical indicatrix, describing **E** in terms of **D**, will be introduced and its use in determining **E**, **k**, and **S**, given **D**, will be demonstrated. The reciprocal formulation based on the tensor describing **D** in terms of **E** is presented in Appendix 13-B.

Maxwell's equations are used to discuss the propagation of plane waves in an anisotropic material. The nonzero scalar product, $\mathbf{E} \cdot \mathbf{k} \neq 0$, leads to an equation, called Fresnel's equation. The solution of the Fresnel's equation yields two indices of refraction for each possible propagation vector. The two indices are associated with two mutually orthogonal displacement vectors, **D**, each orthogonal to the propagation vector **k**. It is this association that allows the construction of the optical component called a retarder. A retarder can be used to rotate the direction of linear polarization or convert linearly polarized light into elliptically polarized light.

The operation of a polarizer or retarder can be described through the use of the Mueller or Jones matrices. The matrices for the most common components are given and examples of their use presented.

To complete the discussion of anisotropic materials, a discussion of optically active materials is presented. The effects of an optically active material on light are described by assigning a different index of refraction for right-circularly and left-circularly polarized light.

Two additional constructions are used in the discussion of wave propagation in an anisotropic material. They have been placed in the appendices to reduce the confusion these additional geometrical surfaces create. A construction based on the two indices of refraction associated with each propagation vector, called a *normal surface*, is discussed in Appendix 13-C. (The Fermi surface used in the study of electron transport in solids is a normal surface for electron motion when electrons are treated as waves.)

A construction similar to the normal surface, called the *ray surface*, is discussed in Appendix 13-D. It is constructed using the two ray velocities associated with each possible direction of the Poynting vector. The normal surface is of use if quantitative information about the propagation of a wave is desired and the ray surface is of use in the application of Huygens' principal for a qualitative picture of light propagation.

DICHROIC POLARIZERS

In a *dichroic polarizer*, one component of polarization is absorbed to a greater extent than its transverse partner. Anisotropic absorption occurs in nature and can be engineered by man, both on the molecular and laboratory scale.

Crystals

When the optical absorption in a crystal depends on the direction of propagation in the crystal and the state of polarization, the crystal is said to exhibit *dichroism*. The dichroic property is due to a complex index of refraction that depends on crystallographic direction. The effect was first observed by **Jean Baptiste Biot (1774–1862)** in 1815. Examples of naturally occurring dichroic materials are tourmaline and herapathite (iodosulfate of quinine) in the visible and pyrolytic graphite in the infrared. By present-day standards, dichroic materials make poor polarizers but some dichroic crystals have been studied as possible optical storage materials and laser hosts. These crystals also make interesting gem stones because the color of the crystals can vary with the viewing angle.

The term dichroic means two-color and is also used to describe the change in color with concentration of dye solutions, to label a filter that transmits and reflects different colors, or to label a filter that transmits two colors. In mineralogy, the term *pleochroism*, meaning many colors, is used to describe crystals that absorb light differently in different crystallographic directions.

Wire Grids

We can construct a dichroic polarizer by stringing a set of wires as shown in Figure 13-1. Heinrich Hertz invented this type of polarizer in 1888 to study the properties of microwaves.

Our first thought would be that Figure 13-1 is drawn improperly because the transmitted polarization is perpendicular to the openings between the wires. Our intuition is wrong: The grid attenuates $\mathbf{E}_v$ because $\mathbf{E}_v$ induces currents in the wires and the resistance of the wires dissipates the energy provided by the wave. The grid performs badly if the wires are thick ($\mathbf{E}_h$ is then attenuated), the spaces between the wires are large ($\mathbf{E}_v$ can leak through the grid), or the wires have a high resistivity.

The wire grid polarizer has been shown to operate over a large range of angles of incidence, from 0 to 45°, and over a large wavelength region. The long wavelength limit of this polarizer is set by absorption in the supporting substrate, if any, and the short wavelength limit is established by the grid spacing. The minimum operating wavelength of a wire grid polarizer is equal to twice the grid spacing.

Polaroid sheet

"H-sheet" Polaroid is the most popular and familiar dichroic polarizer. This type of polarizer was invented in 1928 by **Edwin Herbert Land** when he was a 19-year-old undergraduate student. Long-chain polymeric molecules that make up polyvinyl alcohol are aligned by stretching sheets of the plastic. After stretching, the polyvinyl plastic is cemented to a rigid base such as cellulose acetate to prevent relaxation to the original alignment. Iodine is then

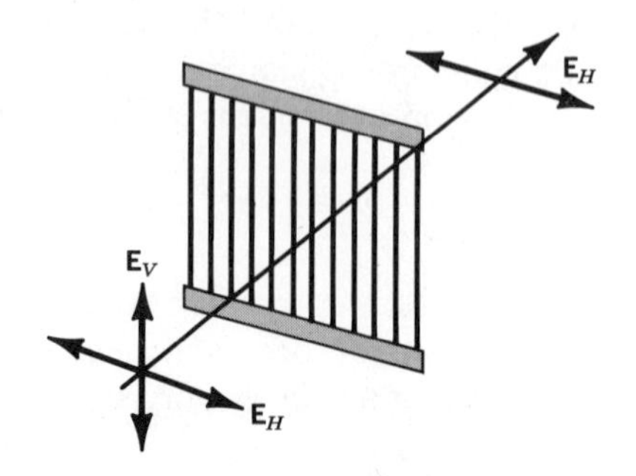

FIGURE 13-1. Wire grid polarizer.

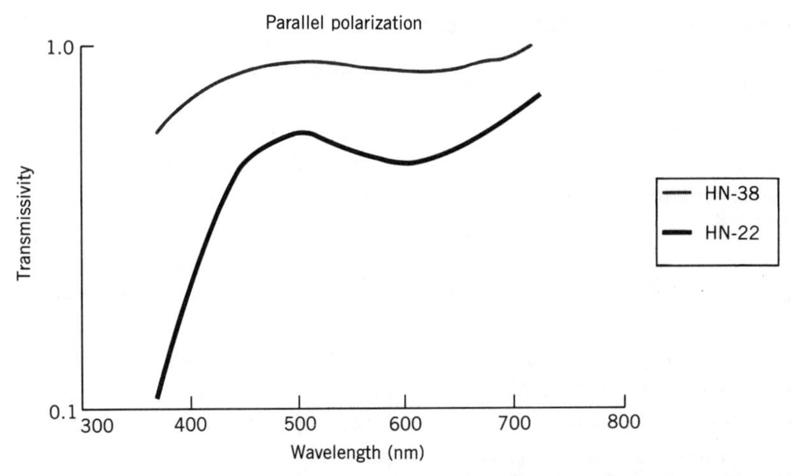

FIGURE 13-2. The transmission of light polarized parallel to the Polaroid axis. Two different grades of H-sheet are shown. These curves should be compared to those shown in Figure 13-3.

diffused into the polyvinyl alcohol layer with the iodine atoms distributing themselves along the polymer molecule. The electrons from the iodine are free to move along the polymer chain. A moment of thought reveals that the Polaroid film is a molecular analog of the wire grid. The first version of Polaroid film (called J-sheet) were sheets of nitrocellulose with microscopic herapathite crystals imbedded in the plastic. The J-sheet was later replaced by the more achromatic H-sheet. A number of different grades of H-sheet are manufactured. The performance of two types, with visible transmission of unpolarized light of about 38% and 22%, respectively, is shown in Figures 13-2 and 13-3. In these figures, the transmission of the polarizer is displayed as a function of wavelength. Linearly polarized light is incident on the polarizer.

Figure 13-2 shows the spectral transmission properties of Polaroid H-sheet when the axis of the polarizer is parallel to the incident wave's direction of polarization. Figure 13-3 is similar to the curves in Figure 13-2,

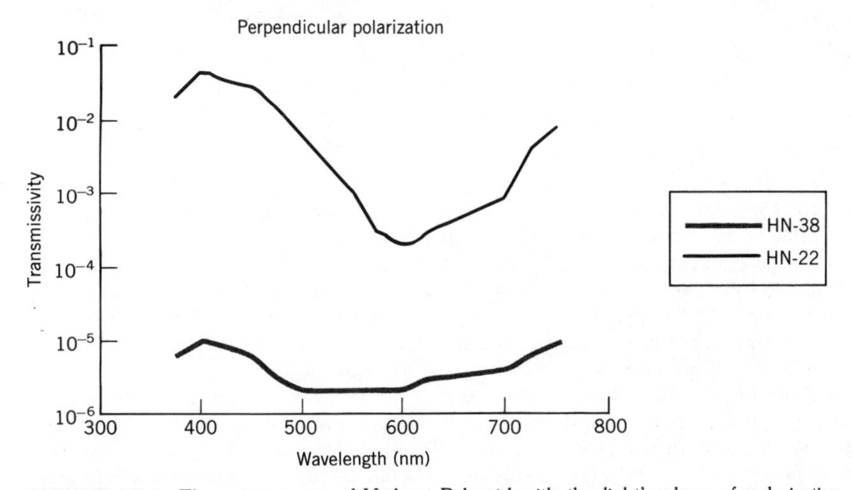

FIGURE 13-3. The transmission of H-sheet Polaroid with the light's plane of polarization normal to the axis of the Polaroid film.

The quality of a polarizer is measured by the ratio, the *extinction ratio*, of the maximum transmission of the polarizer, as shown in Figure 13-2, to the minimum transmission, as shown in Figure 13-3. Examples of the extinction ratio for a few types of polarizers are listed in Table 13.1.

TABLE 13.1 Extinction Ratio of Polarizers

Material	Extinction Ratio
Polaroid	10^{-3}–10^{-5}
Tourmaline	10^{-2}
Wire grid	10^{-2}–10^{-3}
Four-plate reflection	
($n = 2.46$)	10^{-2}
($n = 4.0$)	10^{-4}
Calcite	10^{-7}

with the exception that the axis of the polarizer is transverse to the direction of polarization of the incident wave. (The polarization axis defines the direction, parallel to the electric field of an incident linearly polarized wave, when the transmission through the polarizer is a maximum.)

REFLECTION POLARIZERS

The Fresnel equations describing the reflection of light from an interface (see Chapter 3), depend on the polarization of the incident light. Polarizers can be designed using this polarization-dependent reflectivity.

Brewster's Angle Polarizers

During our discussion of the Fresnel equations for reflection (Chapter 3), we found that at the Brewster's angle, the reflected wave is polarized perpendicular to the plane of incidence. A device, based on the polarization of a wave reflected from a stack of dielectric plates, was invented by Arago in 1812. The polarization of light by a stack of thin glass plates is shown in Figure 13-4.

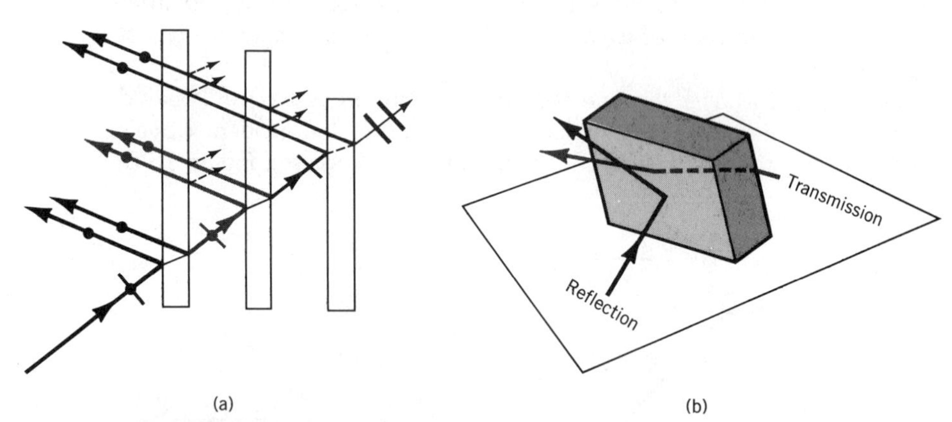

FIGURE 13-4. (a) Polarization of light by a stack of glass plates. This is a schematic display of the operation of a stack of glass plates that use Brewster's angle to polarize light. The experimental demonstration of this effect is shown in (b) through (e). (b) The word "transmission" is positioned behind a stack of very thin glass plates and the word "reflection" is in the foreground at the front of the stack. In (c), we see a mirror image of the word reflection reflected off the front surfaces of the glass plates and the word transmission through the stack of glass plates. Without the use of a polarizer to analyze the light, there is no indication of the polarization properties of the two images. An H-sheet Polaroid polarizer with its axis of polarization oriented along its long dimension is placed in the foreground in (d) and (e). In (d), the polarizer has its axis oriented normal to the plane of incidence. The word transmission cannot be seen, but the word reflection is visible. In (e), the polarizer is oriented with its axis in the plane of incidence. Here, the word transmission can be seen through the plates, but the reflection of the word reflection cannot be seen.

Polarizers that use reflected light have two drawbacks. The light beam is deviated through a large angle and the polarizer becomes quite long when other reflectors are added to redirect the deviated beam. To overcome the drawbacks of the reflective device, a transmissive Brewster's angle polarizer is used. The transmitted beam is partially polarized by virtue of the fact that some of the light polarized perpendicular to the plane of incidence has been removed through reflection. This is indicated schematically in Figure 13-4*a*. By assembling a stack of glass plates and transmitting light through the stack, the degree of polarization of the transmitted beam can be increased.

To demonstrate polarization by a stack of glass plates, the experimental arrangement shown in Figure 13-4*b* was assembled. The word "Transmission" is positioned behind a stack of very thin glass plates and the word "Reflection" is placed in the foreground of the stack. In this configuration, we see a mirror image of the word Reflection from the front surfaces of the glass plates and the word Transmission through the stack of plates. A photo of the actual experiment is shown in Figure 13-4*c*. The two images can be seen, but there is no indication of the polarization properties of the two images. An H-sheet Polaroid polarizer with its axis of polarization oriented along its long dimension is placed in the foreground in Figures 13-4*d* and 13-4*e*. In Figure 13-4*d* the polarizer has its axis oriented normal to the plane of incidence. The word transmission cannot be see but the word reflection is visible. In 13-4*e*, the polarizer is oriented with its axis in the plane of incidence. Here, the word Transmission can be seen through the plates but no reflection of the word Reflection seen.

Any nonmetallic, smooth object can act as a Brewster's angle polarizer, but the carefully designed reflection polarizers (see Figure 13-5) tend to be bulky and expensive. The fraction of polarized light produced is often too low and there is too much leakage of the unwanted polarization. The utilization of the reflection polarizer, with the exception of lasers, is limited to the infrared and ultraviolet where it is often the only option.

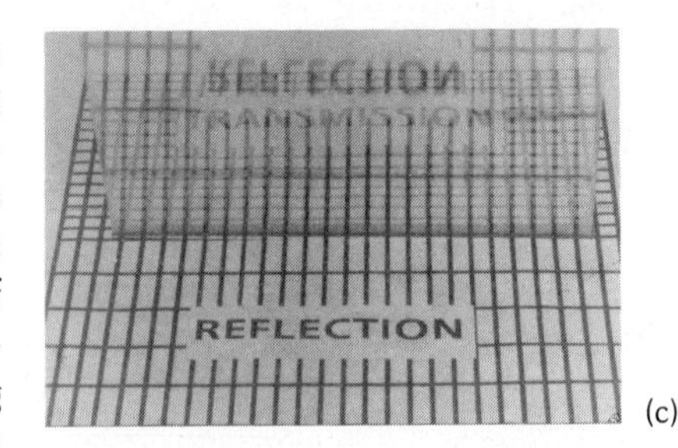

(c)

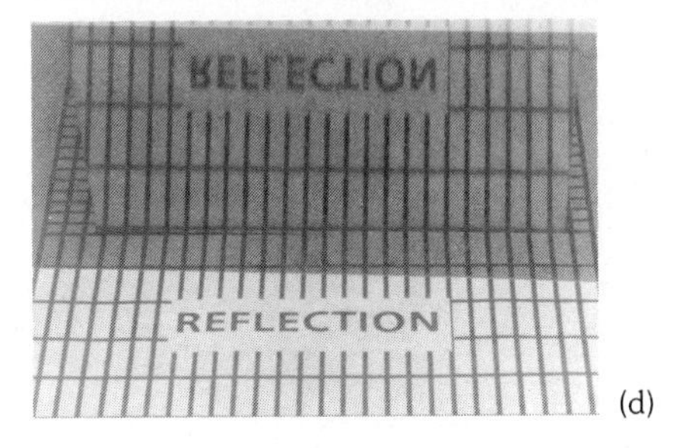

(d)

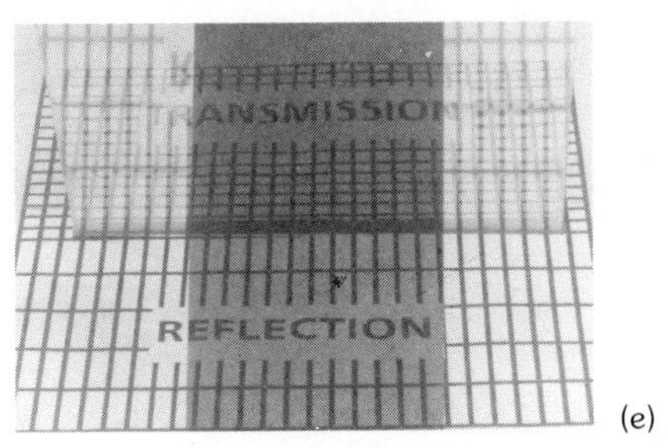

(e)

FIGURE 13-4. *(Continued)*

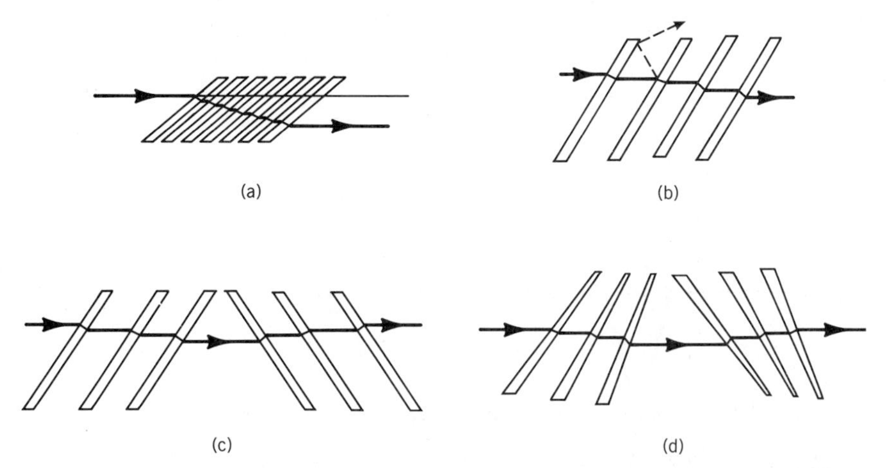

FIGURE 13-5. In Figure 13-4, we implied that a reflection polarizer would look much like (a). However, a number of modifications must be made if the polarizer is to function properly. (b) is a design that allows the escape of unwanted reflected light by increasing the spacing of the plates. (c) is a design to keep the transmitted light on the optical axis. (d) displays a carefully designed polarizer with wedged plates so that multiple reflections can be removed with apertures and the transmitted beam remains on the optical axis.

The polarized output of lasers is a result of surfaces oriented at the Brewster's angle in the laser cavity to reduce reflective loss due to the surfaces. The polarization that is not reflected, because of Brewster's law, has very low loss and thus is preferentially amplified by the gain medium in the laser.

Interference Polarizers

If a transmission polarizer is needed over a limited wavelength range, then it can be constructed using multilayer dielectric coatings. The approach makes use of the theory of multilayer reflector design introduced in Appendix 4-A. Here we will examine the design philosophy using a single dielectric layer, as displayed in Figure 13-6.

The reflection coefficients for the electric field parallel to the plane of incidence from the two interfaces are equated, and the thickness of the dielectric layer is adjusted to insure destructive interference between the waves reflected from the two interfaces. The result of this design procedure is a dielectric mirror that reflects only the normal component of polarization.

The reflectivity of the dielectric layer in Figure 13-6 can be calculated using **(4-40)**

$$R = 1 - T = \frac{r_1^2 + r_2^2 + 2r_1r_2 \cos 2\delta}{1 + r_1^2r_2^2 + 2r_1r_2 \cos 2\delta}$$

where we have replaced the single reflection coefficient of **(4-40)** with the two reflection coefficients needed in Figure 13-6, that is, $r^2 = r_1r_2$. Following the prescription just outlined, the phase from **(4-34)** is set equal to an odd multiple of π

$$\delta = (2m + 1)\pi = \frac{4\pi n_1 d \cos \theta_{t1}}{\lambda}$$

establishing the required film thickness as

$$d = \frac{(2m + 1)\lambda}{4n_1 \cos \theta_{t1}}$$

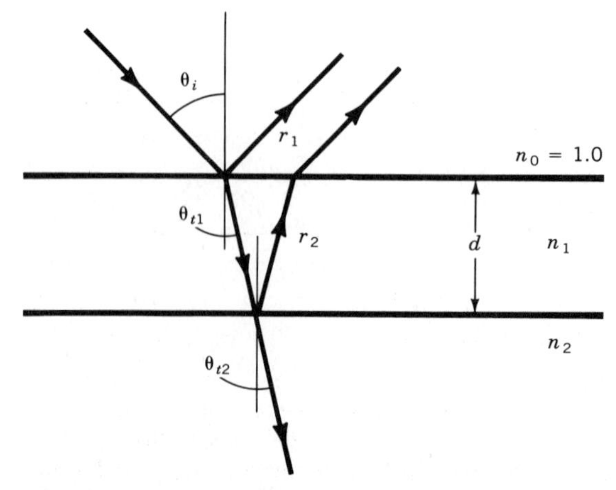

FIGURE 13-6. Use of interference from a dielectric layer to produce a reflective polarizer.

The reflection coefficients for the parallel polarization components from the two interfaces are set equal to each other

$$\frac{\tan(\theta_i - \theta_{t1})}{\tan(\theta_i + \theta_{t1})} = \frac{\tan(\theta_{t1} - \theta_{t2})}{\tan(\theta_{t1} + \theta_{t2})}$$

This equality insures complete cancellation of the parallel component of polarization. After applying these design rules, the reflectivity of the normal component of polarization from the dielectric layer is

$$R = \frac{(n_1^2 - 1)\sin^2\theta_i - \cos\theta_i\sqrt{n_2^2 - \sin^2\theta_i}}{(n_1^2 - 1)\sin^2\theta_i + \cos\theta_i\sqrt{n_2^2 - \sin^2\theta_i}}$$

Interference polarizers are used in laser systems when the incident radiation will strike the dielectric layer at an angle. The polarizing beam splitter is a popular example of this design. The radiation is incident on the dielectric layer at an angle of 45°. The wave transmitted by the layer is polarized with its electric vector in the plane of incidence, whereas the reflected wave has its electric vector normal to the plane of incidence.

POLARIZATION BY BIREFRINGENCE

In crystals with cubic symmetry, the propagation properties of light are isotropic, there is a single index of refraction, and the crystal behaves optically like a noncrystalline material such as glass (at least to first order). All other classes of crystals are optically anisotropic. The index of refraction (and naturally, the dielectric constant) depends on the direction of propagation of the light, relative to the crystal axes. We must use tensor calculus (see Appendix 13-A) to discuss the propagation of light in these materials. In general, to analyze the propagation of light in an anisotropic medium, the wave is divided into two waves with orthogonal polarizations (the polarization direction in the medium will be denoted by the direction of the displacement **D**). The two waves, with orthogonal polarizations $\mathbf{D}_1$ and $\mathbf{D}_2$, will not have the same propagation vector and will exhibit double refraction. There are, however, special directions in an anisotropic crystal for which both polarizations have the same propagation velocity. The special direction is called an *optical axis*, and if two such directions exist, the crystals are called *biaxial*. If there is only one direction for which the two orthogonal polarizations have the same propagation velocity, the crystal is called *uniaxial*.

The physical origins of birefringence can be understood by modifying the classical model of dispersion introduced in Chapter 7. The single spring constant used in the model **(7-24)**, can be replaced by three spring constants, one for each of the coordinate directions. The application of the procedure used in Chapter 7 to this modified model leads to three indices of refraction for the material, that is, a biaxial crystal; examples are listed in Table 13.2. If the spring constants are equal in two coordinate directions, then the model predicts two indices of refraction, that is a uniaxial crystal, (see Table 13.3). If a different damping term for each coordinate direction is included in the model **(7-26)**, the result is a complex index of refraction that has directional dependence. This model would describe pleochroism.

To help understand the physical origin of the different coupling constants, consider the unit cell of the uniaxial crystal calcite shown in Figure 13-7. A crystal of calcite is made up of a three-dimensional array of these

TABLE 13.2 Biaxial Crystals

Mineral	Index of Refraction n_α	n_β	n_γ
Tridymite	1.469	1.47	1.473
Mica(muscovite)	1.5601	1.5936	1.5977
Turquoise	1.61	1.62	1.65
Topaz	1.619	1.62	1.627
Sulfur	1.95	2.043	2.240
Borax	1.447	1.47	1.472
Lanthanite	1.52	1.587	1.613
Stibnite(Sb_2S_3)	3.194	4.303	4.46

TABLE 13.3 Uniaxial Crystals

Mineral	Index of Refraction (Na-D) n_O	n_e
Ice(H_2O)	1.309	1.313
Sellaite(MgF_2)	1.378	1.390
Quartz	1.54424	1.55335
Wurtzite(ZnS)	2.356	2.378
Rutile(TiO_2)	2.616	2.903
Cinnabar(HgS)	2.854	3.201
Calcite($CaO \cdot CO_2$)	1.658	1.486
Tourmaline	1.669	1.638
Sapphire	1.7681	1.7599

unit cells. Each carbon atom in the crystal lies at the center of an imaginary equilateral triangle with oxygen atoms at each corner. The direction, normal to the planes containing these triangles, is the optical axis of calcite. From the arrangement of atoms, it seems reasonable to assume that the binding energy in the planes containing the oxygen atoms is different than the binding energy normal to these planes. When light is incident on a crystal of calcite parallel to the optical axis, shown in Figure 13-7, the electric displacement of the light wave lies in the same plane as the oxygen atoms. The electric polarization induced by the electric field is the same for all polarization directions in this case and the resulting index of diffraction would be independent of polarization direction.

When a light wave is incident on the calcite crystal normal to the optical axis, the electric polarization induced by the light wave's electric field in the plane of the oxygen atoms is different than the polarization induced by the electric field normal to that plane, resulting in two indices of refraction.

A light wave propagating in an arbitrary direction through this crystal can be decomposed into two waves, with orthogonal polarizations, propagating at different velocities. Any wave whose electric displacement lies in the plane of the oxygen atoms, and thus is perpendicular to the optical axis, obeys Snell's law. For this polarization orientation, the propagation velocity is isotropic and the wave is called the *ordinary wave*. Any other wave with its electric displacement at an angle to the plane containing the oxygen atoms has an anisotropic propagation velocity and does not obey Snell's law. This wave is called the *extraordinary wave* (see Figure 13-8).

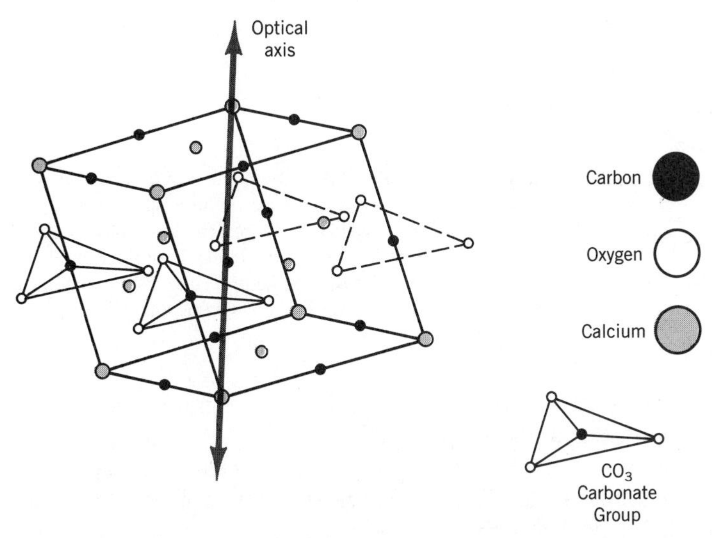

FIGURE 13-7. The unit cell of calcite. Only four of the carbonate groups that make up the unit cell are shown in their entirety; the other groups are represented by the carbon atoms at their center.

The electrons can move quite easily in the planes containing the oxygen atoms so the spring constant for these atoms is small relative to the direction normal to the planes. In the direction normal to the oxygen planes, the electrons experience strong binding forces. For this reason, the propagation velocity in the direction of the optical axis, the ordinary wave's velocity, is less than the propagation velocity normal to that direction. The index of refraction associated with the ordinary wave is thus greater than the index associated with the extraordinary wave, $n_O > n_e$. Crystals with this property are called *negative* uniaxial crystals $(n_e - n_O < 0)$. The polarization direction for the ordinary wave of a negative, uniaxial crystal is called the *slow axis*. The polarization direction (the direction of **D**) for the wave that has the

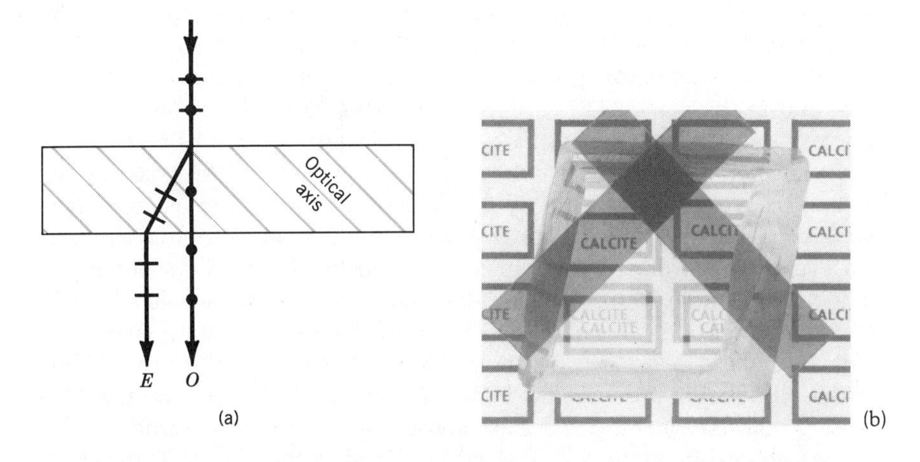

FIGURE 13-8. (a) Uniaxial crystal showing the failure of the extraordinary wave to obey Snell's law. The lines in the crystal indicate the direction called the optical axis. (b) The birefringence shown in (a) results in a double image. The polarizations of the two images are orthogonal, as indicated by the images transmitted by the two Polaroid strips. The long axis of each strip is parallel to the polarization axis.

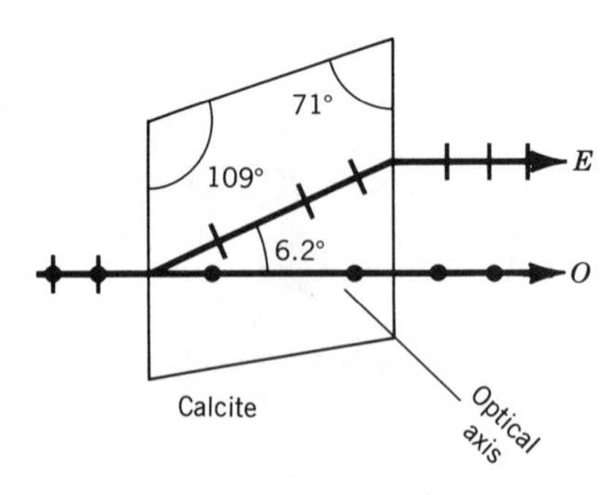

FIGURE 13-9. Separation of the extraordinary (the upper beam) and ordinary (lower beam) waves in calcite. The line labeled optical axis indicates the direction of the optical axis. The rhombohedron shown is the naturally occurring form of a single crystal of calcite.

higher propagation velocity is called the *fast axis*, and for a *positive* uniaxial crystal ($n_e - n_O > 0$), this wave is the ordinary wave.

In general, when an unpolarized beam of light is incident normal to the surface of a plane parallel plate of an uniaxial crystal, there will be two beams emerging from the back side of the crystal, as is shown in Figure 13-8. The ordinary wave, labeled O in Figure 13-8, is polarized with its displacement vector **D** normal to the plane containing the optical axis. The extraordinary wave labeled E in Figure 13-8 is polarized with its displacement vector in the plane containing the optical axis. The two waves indicated in Figure 13-8*a* produce two images with perpendicular polarizations, as shown in Figure 13-8*b*. Polarizers are designed to utilize the spatial separation of the two beams shown in Figure 13-8 to select a desired polarization direction.

Polarizers are constructed using the uniaxial material; the simplest design utilizes the naturally occurring birefringent crystal calcite and a set of "stops" to remove either of the two beams. See Figure 13-9. Because the separation of the two waves is small (only 6.2° in calcite), this technique can only be used with very narrow beams.

The Wollaston and Rochon polarizers increase the separation of the two beams over that obtainable with a single crystal by using two single crystals, usually quartz, cut and polished into two prisms (see Figure 13-10). The Rochon polarizer has an entrance prism with its optical axis oriented perpendicular to the incident face of the polarizer in the plane of Figure 13-10 and a second exit prism, glued to the first with its optical axis perpendicular to the first prism's axis, perpendicular to the plane of Figure 13-10.

The light wave whose polarization is perpendicular to the second prism's axis, and in the plane of Figure 13-10, sees no index discontinuity at the interface between the two prisms because the wave also has its polarization perpendicular to the optical axis of the first prism. For this reason, the ordinary wave is not deviated by the Rochon polarizer. The extraordinary wave of the second prism with its polarization parallel to the optical axis, however, is an ordinary wave in the first prism. It sees a discontinuous change in the index of ($n_e - n_O$) as it crossed the boundary between the two prisms. The prism angle θ and the difference between the extraordinary and ordinary indices determine the angle φ between the E and O beams. The Wollaston polarizer produces twice the deviation as the Rochon polarizer by causing both the ordinary and extraordinary wave to see the same discontinuous change in the index. This occurs because the extraordinary wave and ordinary waves interchange roles in the two prisms. The advantage of the Rochon prism is that it keeps the ordinary wave on the optical axis.

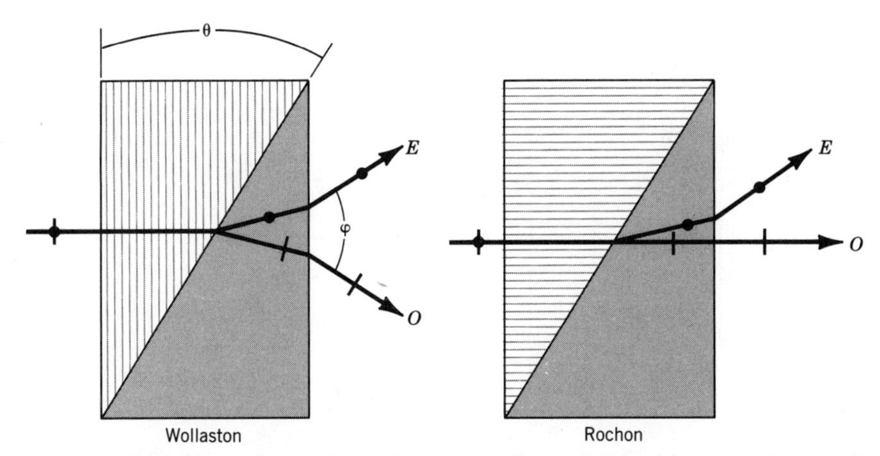

FIGURE 13-10. Two polarizers designed to increase the separation of the extraordinary and ordinary waves. The entrance prism on both polarizers has its ordinary wave polarized out of the paper. The second prism has its ordinary wave polarized in the plane of the paper. The lines and dots used to shade the two prisms indicate the direction of the optical axis.

Another method of separating the two orthogonal polarizations in a birefringent crystal is the use of total reflection to deflect one beam toward an absorber; see Figure 13-11. The Nicol prism named after **William Nicol (1768–1857)** was the first of this type of polarizer. A calcite crystal is cut into two prisms and polished to obtain the angles shown in Figure 13-11. The two prisms are then glued together with a cement called Canada balsam. The cement used to assemble the two prisms is selected so that the ordinary wave experiences total reflection at the glue joint, but the extraordinary wave passes through the polarizer. (Just as is the case with an optical fiber, which depends on total reflection, the Nicol polarizer will operate only with rays arriving within a cone whose angle, called the acceptance angle, is 24°. The acceptance angle of this type of polarizer is determined by the difference between the critical angle of the ordinary and extraordinary rays.) The cement restricts the wavelength range of the polarizer so designs were developed that eliminated the cement.

The Glan–Foucault is a more modern polarizer design (see Figure 13-12) based on the same principal as the Nicol. Its advantages are that it replaces glue with air at the interface of total reflection and the incident and exit faces are perpendicular to the light wave. The acceptance angle of this polarizer is only 7°, but it will operate from 0.23 to 5.0 μm. If the

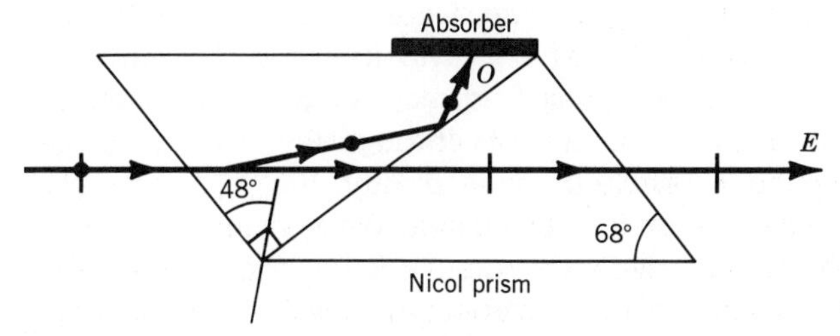

FIGURE 13-11. The geometry of a Nicol prism polarizer. The direction of the optical axis is indicated by the line in the lower left corner.

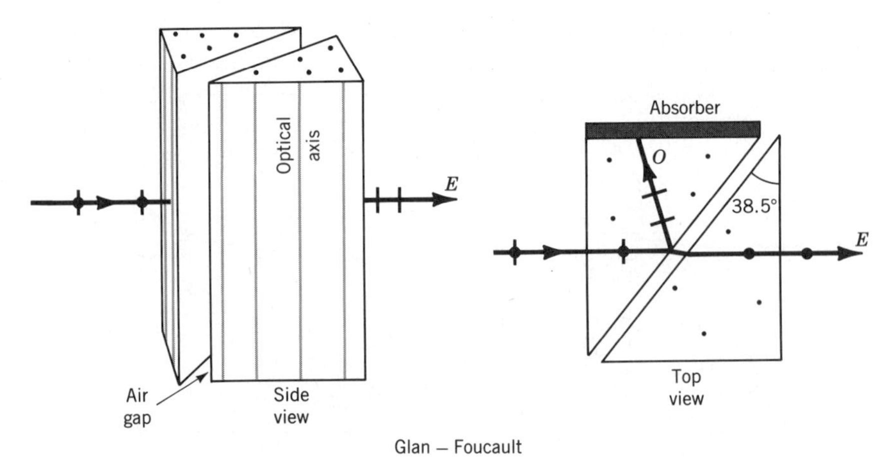

FIGURE 13-12. Two views of the Glan-Foucault polarizer showing its operation. The parallel lines indicate the direction of the optical axis.

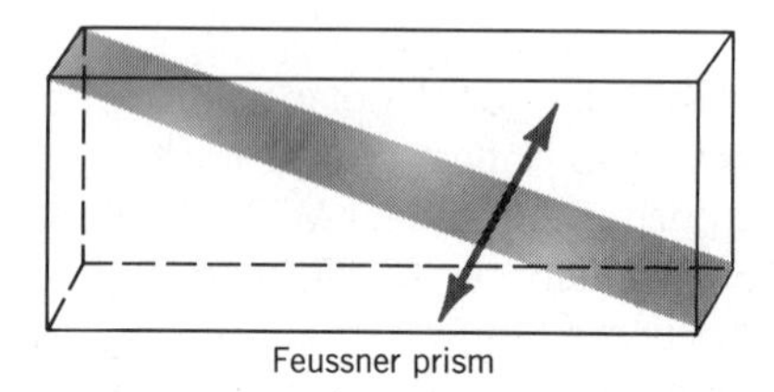

FIGURE 13-13. A Feussner polarizer. The optical axis of the birefringent material is normal to the slab of material as shown by the arrows.

two prisms of the Glan–Foucault polarizer are cemented together, the device is called a Glan–Thompson polarizer. This design modification increases the field of view to 30°, but it cannot be used in the uv.

One of the problems in constructing polarizers using birefringent material is finding or growing a birefringent crystal of a size and optical quality to be useful. One polarizer design uses isotropic prisms with a thin plate of birefringent material sandwiched between the prisms; see Figure 13-13. The two prisms are made of a glass with an index equal to the higher index of the birefringent material. If the birefringent material is calcite, then the ordinary ray is transmitted. The optical axis can be oriented normal to the slab (Feussner polarizer) or parallel to the entrance face (Bertrand type). (Polarizers in optics are like beakers in chemistry; even a slight modification of a design results in a new device bearing the designer's name.)

OPTICAL INDICATRIX

In the previous discussions about light waves, we assumed that the medium through which the light propagated was homogeneous and isotropic. In those cases, we could relate the electromagnetic field vectors **E** and **B** to the electric displacement **D** and the magnetic vector **H** through the scalar quantities ϵ (dielectric constant) and μ (magnetic permeability). We will continue to assume that the material is isotropic with regard to the magnetic field, so that μ remains a scalar, but we will now assume that the dielectric properties can vary with direction. The impact of discarding the assumption of isotropic electrical properties is the need to use tensor relationships between **E** and **D** because the electric field **E** will, in general, not be parallel to **D**.

The complexity of the tensor can be reduced by assuming that there are no losses in the medium. We will show that this assumption also allows the use of a geometrical construction to aid in understanding the properties of the anisotropic medium. The geometrical construction, called the optical indicatrix, demonstrates the origin of two propagation velocities in the anisotropic material and can be used to determine **E**, **k**, and **S** from the specification of **D**.

The simple relationship between **D** and **E** that we have used up to now must be replaced by the more general equations

$$\begin{aligned} D_x &= \epsilon_{xx} E_x + \epsilon_{xy} E_y + \epsilon_{xz} E_z \\ D_y &= \epsilon_{yx} E_x + \epsilon_{yy} E_y + \epsilon_{yz} E_z \\ D_z &= \epsilon_{zx} E_x + \epsilon_{zy} E_y + \epsilon_{zz} E_z \end{aligned} \tag{13-1}$$

or in more compact notation,

$$D_i = \sum_{j=1}^{3} \epsilon_{ij} E_j$$

In many books, the Einstein notation is used to represent tensor equations such as **(13-1)**. In this notation,

$$D_i = \epsilon_{ij} E_j$$

The summation sign is supressed and the summation is indicated by the repeating index j. We will not use this notation.

The ϵ_{ij} are the components of a tensor of second rank. In general, there would be nine such components, but energy considerations reduce the number of independent elements in the dielectric tensor to six or less. To prove this statement, we must make several assumptions

1. The energy of the electromagnetic field that we used in Chapter 2 **(2-21)** for an isotropic medium is valid for an anisotropic medium.
2. The energy flux (Poynting vector) that we used in Chapter 2 **(2-22)** for an isotropic medium is valid for an anisotropic medium.
3. There is no energy loss in the medium.

The Poynting vector **(2-22)** is used to obtain the unit energy flow into a unit volume

$$\nabla \cdot \mathbf{S} = \nabla \cdot (\mathbf{E} \times \mathbf{H}) \tag{13-2}$$

We use the identity **(2A-13)** to rewrite **(13-2)** as

$$\nabla \cdot \mathbf{S} = \mathbf{E} \cdot \left(\frac{\partial \mathbf{D}}{\partial t} \right) + \mathbf{H} \cdot \left(\frac{\partial \mathbf{H}}{\partial t} \right) \tag{13-3}$$

$$\nabla \cdot \mathbf{S} = \left(\sum_{k=1}^{3} \sum_{j=1}^{3} E_k \epsilon_{kj} \frac{\partial E_j}{\partial t} \right) + \mathbf{H} \cdot \frac{\partial \mathbf{H}}{\partial t}$$

where we assume that the components of the dielectric tensor are independent of time.

We are only interested in the electric energy flux that is the first term of **(13-3)**. The steady-state value of the electric energy flux is obtained by applying **(2-26)** to write

$$\langle \nabla \cdot \mathbf{S} \rangle_E = \frac{1}{2} \left(\sum_{j=1}^{3} E_j \sum_{k=1}^{3} \epsilon_{jk}^{*} \frac{\partial E_k^{*}}{\partial t} + \sum_{j=1}^{3} E_j^{*} \sum_{k=1}^{3} \epsilon_{jk} \frac{\partial E_k}{\partial t} \right) \tag{13-4}$$

Using **(2-14)**, we can rewrite **(13-4)** as

$$\langle \nabla \cdot \mathbf{S} \rangle_E = \frac{i\omega}{2} \sum_{j=1}^{3} E_j \sum_{k=1}^{3} E_k^{*} (\epsilon_{kj} - \epsilon_{jk}^{*}) \tag{13-5}$$

The study of anisotropy is confusing because different names are used to indicate the optical indicatrix. Some of these are listed in Table 13.4.

TABLE 13.4 Names for Optical Indicatrix

Index ellipsoid
Optical indicatrix
Reciprocal ellipsoid
Poinsot ellipsoid
Ellipsoid of wave normals

A second confusing fact is that the optical indicatrix is not the only surface used in the description of optical anisotropy. The first step in reading the optical anisotropy literature is to determine what surface the author is discussing. Other surfaces used in the discussion of optical anisotropy are introduced in the appendices.

We now use the third assumption that the medium is a lossless one, which means that **(13-5)** will be equal to zero, i.e., there is no change in energy flow through the medium. This results in the requirement that

$$\epsilon_{jk}^{*} = \epsilon_{kj}$$

Thus the dielectric tensor $\tilde{\epsilon}$ must be Hermitian.

We will treat the dielectric tensor as real, resulting in the requirement that the tensor $\tilde{\epsilon}$ must be symmetric. The assumption of a lossless medium reduces the maximum number of tensor components that must be considered to six. A further reduction in the number of components to three can be made by the proper choice of the coordinate system. In the properly oriented coordinate system (with coordinate axes parallel to the *principal dielectric axes*), the nonzero tensor components are called the *principal dielectric constants*. As discussed in Appendix 13-A, the assumption of a lossless medium allows the use of a geometrical surface to aid in understanding the properties of the tensor. The geometrical surface used to represent a second-order, symmetric, dielectric tensor is called the *optical indicatrix*.

Starting with the equation for the electric energy density

$$2U_E = \mathbf{D}\cdot\mathbf{E}$$

(2-21), we select as the coordinate system the principal dielectric axes, allowing us to write

$$D_x = \epsilon_x E_x, \qquad D_y = \epsilon_y E_y, \qquad D_z = \epsilon_z E_z$$

In this coordinate system, we may rewrite the energy density equation as

$$\frac{D_x^2}{2U_E\epsilon_x} + \frac{D_y^2}{2U_E\epsilon_y} + \frac{D_z^2}{2U_E\epsilon_z} = 1 \tag{13-6}$$

Equation **(13-6)** is the equation of an ellipsoid **(13A-5)**, whose semiaxes are equal to the square roots of the principal dielectric constants, or equivalently, the principal indices of refraction. This ellipsoid (see Figure 13A-1) is called the optical indicatrix.

The ellipsoid defined by **(13-6)** is not a representative of the dielectric tensor but its reciprocal called the *impermeability tensor*, defined by the relation

$$E_i = \sum_{j=1}^{3} \ni_{ij} D_j$$

The dielectric tensor is discussed in Appendix 13-B; see **(13B-4)**.

The optical indicatrix can be used to determine **E**, **k**, and **S**, given **D**. The geometrical construction shown in Figure 13-14 will demonstrate the method used to determine these parameters. Assume that **D** is known and represented by the vector from *O* to *P*, where *P* is a point on the ellipsoid in Figure 13-14. Construct first a surface, tangent to the ellipsoid at *P*, with a normal $\hat{\mathbf{n}}$; then construct the line *OQ* to be parallel to the surface normal $\hat{\mathbf{n}}$ at *P*. The point *Q* is in the plane tangent to *P*; see Figure 13-14. The magnitude of **E** is given by 1/*OQ* where *OQ* is the distance from the origin to the plane tangent to *P*. The direction of **E** is parallel to $\hat{\mathbf{n}}$ and the direction of the Poynting vector **S** is parallel to the line *PQ* in Figure 13-14.

If the propagation vector **k** of a wave is known, we can determine **D** and **E**. This is accomplished by constructing a plane through the origin normal to

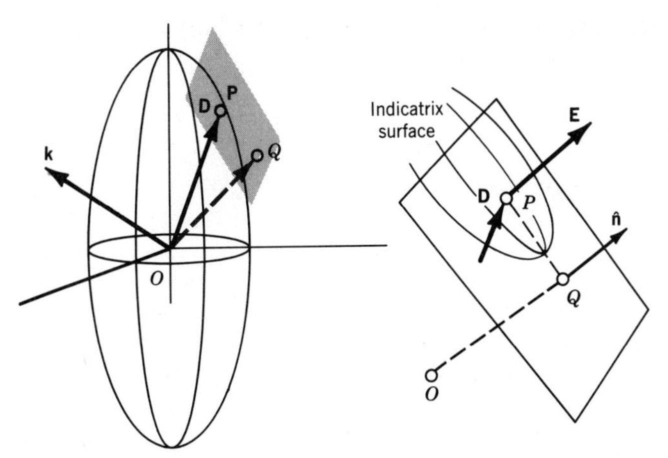

FIGURE 13-14. Use of the optical indicatrix to find the electric field given **D**.

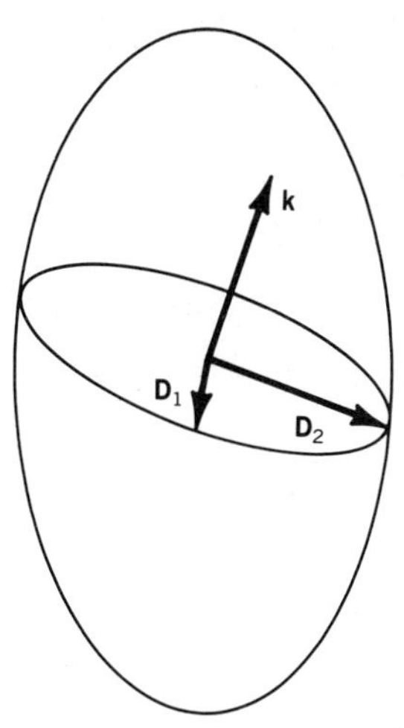

FIGURE 13-15. The determination of **D** given the propagation vector **k**.

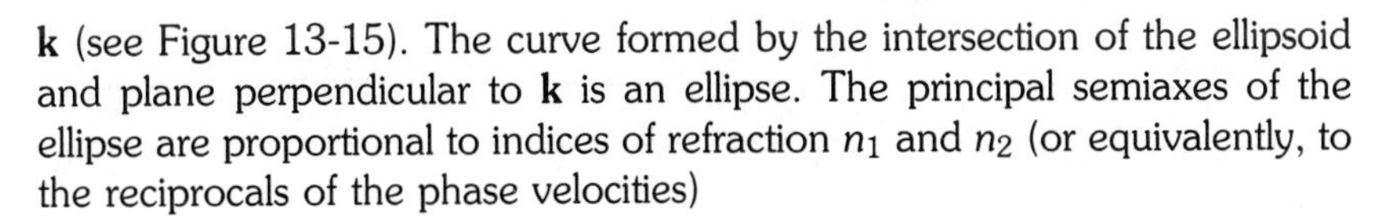

k (see Figure 13-15). The curve formed by the intersection of the ellipsoid and plane perpendicular to **k** is an ellipse. The principal semiaxes of the ellipse are proportional to indices of refraction n_1 and n_2 (or equivalently, to the reciprocals of the phase velocities)

$$n_1 = \frac{c}{v_{p1}}, \qquad n_2 = \frac{c}{v_{p2}}$$

The directions of the principal semiaxes coincide with $\mathbf{D}_1$ and $\mathbf{D}_2$ that are the two orthogonal polarizations for the wave, with wave vector **k**.

For certain directions of **k**, the plane normal to **k** will cut the ellipsoid so as to form an intersecting curve that is a circle. These special directions are called the *optic axes* (of the wave normals) of the crystal. If there is only one such direction, the crystal is uniaxial. The maximum number of directions that can be found is two. Crystals with two optical axes are biaxial.

FRESNEL'S EQUATION

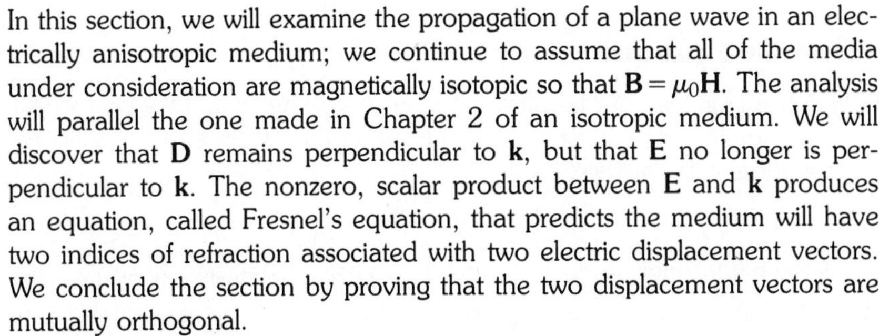

In this section, we will examine the propagation of a plane wave in an electrically anisotropic medium; we continue to assume that all of the media under consideration are magnetically isotopic so that $\mathbf{B} = \mu_0 \mathbf{H}$. The analysis will parallel the one made in Chapter 2 of an isotropic medium. We will discover that **D** remains perpendicular to **k**, but that **E** no longer is perpendicular to **k**. The nonzero, scalar product between **E** and **k** produces an equation, called Fresnel's equation, that predicts the medium will have two indices of refraction associated with two electric displacement vectors. We conclude the section by proving that the two displacement vectors are mutually orthogonal.

We will continue to use the assumption that the medium is nonconducting and that there are no currents or charges present. If we allowed a conductive medium, the analysis would also apply to dichroic materials. Maxwell's equations, as given by **(2-6)**, no longer apply because the constitutive relation **(2-6f)**, $\mathbf{D} = \epsilon \mathbf{E}$, no longer holds. The modified equations are

$$\nabla \cdot \mathbf{B} = 0, \qquad \nabla \cdot \mathbf{D} = 0$$
$$\nabla \times \mathbf{H} = \frac{\partial \mathbf{D}}{\partial t}, \qquad \nabla \times \mathbf{E} = -\frac{\partial \mathbf{B}}{\partial t} \tag{13-7}$$

We will interpret Maxwell's equations, as we did in Chapter 2, by the use of plane waves

$$\mathbf{E} = \mathbf{E}_0 \exp[i(\omega t - \mathbf{k}\cdot\mathbf{r} + \phi)], \qquad \mathbf{D} = \mathbf{D}_0 \exp[i(\omega t - \mathbf{k}\cdot\mathbf{r} + \phi)]$$

$$\mathbf{B} = \mathbf{B}_0 \exp[i(\omega t - \mathbf{k}\cdot\mathbf{r} + \phi)], \qquad \mathbf{H} = \mu\mathbf{B}$$

Transverse Waves

Substituting the plane wave solutions into the first pair of Maxwell's equations yields, for the electric field

$$\nabla\cdot\mathbf{D} = -i\mathbf{k}\cdot\mathbf{D} = 0$$

D remains perpendicular to **k**, but this fact no longer implies that **E** is perpendicular to **k**. For the magnetic field,

$$\nabla\cdot\mathbf{B} = \mu\nabla\cdot\mathbf{H} = -i\,\mathbf{k}\cdot\mathbf{H} = 0$$

Since we continue to assume that the material is magnetically isotropic, we still find that both **B** and **H** are perpendicular to **k**.

Interdependence of D and H

Continuing to examine Maxwell's equations, we substitute the plane wave equations into the second pair of equations

$$\nabla\times\mathbf{H} = \frac{\partial\mathbf{D}}{\partial t}, \qquad \nabla\times\mathbf{E} = -\frac{\partial\mathbf{B}}{\partial t}$$

$$i\,\mathbf{k}\times\mathbf{H} = -i\omega\,\mathbf{D}, \qquad i\,\mathbf{k}\times\mathbf{E} = i\omega\,\mathbf{B} = i\omega\mu_0\mathbf{H}$$

The plane wave solutions of Maxwell's equations require that **H** be perpendicular to both **k** and **D**, and also **H** must be perpendicular to both **k** and **E**.

Fresnel's Equation

The fact that **E** may not be perpendicular to **k** requires that we now depart from the analysis that was followed in Chapter 2 to discover the significance of the nonzero scalar product

$$\mathbf{k}\cdot\mathbf{E} \neq 0$$

To find what this dot product is equal to, we use Maxwell's equation to write

$$\mathbf{k}\times\mathbf{E} = \omega\mu_0\mathbf{H}$$

and then the cross product of this equation with **k** yields

$$\mathbf{k}\times(\mathbf{k}\times\mathbf{E}) = \omega\mu_0\mathbf{k}\times\mathbf{H} = \omega\mu_0(-\omega\mathbf{D})$$

We can rewrite this equation by using the vector identity **(2A-3)**

$$\mathbf{k}(\mathbf{k}\cdot\mathbf{E}) - \mathbf{E}(\mathbf{k}\cdot\mathbf{k}) = -\omega^2\mu_0\mathbf{D} \tag{13-8}$$

The propagation constant is given by

$$k^2 = \left(\frac{n\omega}{c}\right)^2$$

This allows **(13-8)** to be rewritten to yield an expression for the dot product **k•E**

$$\frac{\mathbf{k}}{k^2}(\mathbf{k}\cdot\mathbf{E}) = \mathbf{E} - \frac{c^2\mu_0}{n^2}\mathbf{D} \tag{13-9}$$

D and **E** must satisfy **(13-9)** if they are to satisfy Maxwell's equations.

To allow a physical interpretation of **(13-9)**, we evaluate the components of this vector equation along the three principal directions. The *jth* component of the vector equation is

$$\frac{k_j}{k^2}(\mathbf{k}\cdot\mathbf{E}) = \left(1 - \frac{c^2\mu_0\epsilon_j}{n^2}\right)E_j \tag{13-10}$$

where the ϵ_j's are the principal dielectric constants. By multiplying both sides of the jth component by k_j, we obtain an eigenvalue equation. We have simplified the equation by making the substitution

$$c^2 = \frac{1}{\mu_0\epsilon_0}$$

$$\frac{k_j^2}{k^2\left(1 - \dfrac{\epsilon_j}{n^2\epsilon_0}\right)}(\mathbf{k}\cdot\mathbf{E}) = k_j E_j \tag{13-11}$$

Adding the three component equations of the form **(13-11)** together yields a new expression of the vector equation **(13-9)**

$$(\mathbf{k}\cdot\mathbf{E})\sum_{j=1}^{3}\frac{k_j^2}{k^2\left(1 - \dfrac{\epsilon_j}{n^2\epsilon_0}\right)} = \mathbf{k}\cdot\mathbf{E}$$

If $\mathbf{k}\cdot\mathbf{E} \neq 0$, as we implied might be the case, then we may divide both sides of the equation by the scalar product. We also remove n^2 from the summation to obtain

$$\sum_{j=1}^{3}\frac{k_j^2}{k^2\left(n^2 - \dfrac{\epsilon_j}{\epsilon_0}\right)} = \frac{1}{n^2} \tag{13-12}$$

A new parameter, which we will call *the principal refractive index*, is definite as

$$n_j^2 = \frac{\epsilon_j}{\epsilon_0} \tag{13-13}$$

Using the new definition, we can write **(13-12)** in the form called *Fresnel's equation*

$$\sum_{j=1}^{3}\frac{k_j^2}{k^2(n^2 - n_j^2)} = \frac{1}{n^2} \tag{13-14}$$

This equation allows the calculation of the index of diffraction for an arbitrary propagation direction. It appears to be cubic in n^2, but the equation is actually only quadratic in n^2 with two positive roots: n_1^2 and n_2^2. Thus, Fresnel's equation **(13-14)** states that for any propagation direction **k**, there are, in general, two values of the refractive index (n_1 and n_2). If the solutions of **(13-14)** are substituted into **(13-10)**, two values of the electric field ($\mathbf{E}_1$

and $\mathbf{E}_2$) are obtained for the selected propagation direction. Finally, the two solutions for the electric field corresponding to n_1 and n_2 can be substituted into **(13-9)** to yield two values for the electric displacement ($\mathbf{D}_1$ and $\mathbf{D}_2$).

The two electric displacement vectors $\mathbf{D}_1$ and $\mathbf{D}_2$ associated with a selected propagation vector are perpendicular to one another, as will be demonstrated below. These vectors specify the polarizations of two light waves propagating at two different velocities, given by n_1 and n_2, in the anisotropic material. From Maxwell's equations, we have generated a theoretical explanation of the birefringent observations in any optically anisotropic material.

We can demonstrate that the two electric displacement vectors are orthogonal by first decomposing $\mathbf{E}$ into components parallel ($\mathbf{E}_\parallel$) and perpendicular ($\mathbf{E}_\perp$) to $\mathbf{k}$

$$\mathbf{E}_\parallel = \frac{\mathbf{k}}{k^2}(\mathbf{k}\cdot\mathbf{E}), \qquad \mathbf{E}_\perp = \mathbf{E} - \frac{\mathbf{k}}{k^2}(\mathbf{k}\cdot\mathbf{E}) \tag{13-15}$$

By rewriting **(13-9)**, we discover a relationship between $\mathbf{D}$ and $\mathbf{E}_\perp$

$$\mathbf{D} = \frac{k^2}{\mu_0\omega^2}\left[\mathbf{E} - \frac{\mathbf{k}}{k^2}(\mathbf{k}\cdot\mathbf{E})\right] = \frac{n^2}{\mu_0 c^2}\mathbf{E}_\perp \tag{13-16}$$

Since we have two indices of refraction, **(13-16)** produces two equations

$$\mathbf{D}_1 = \frac{n_1^2}{\mu_0 c^2}\mathbf{E}_{1\perp}, \qquad \mathbf{D}_2 = \frac{n_2^2}{\mu_0 c^2}\mathbf{E}_{2\perp} \tag{13-17}$$

To evaluate the scalar product $\mathbf{D}_1\cdot\mathbf{D}_2$, we first evaluate the scalar product

$$\mathbf{E}_2\cdot\mathbf{D}_1 = \sum_{i=1}^{3} E_{2i} D_{1i} = \sum_{i=1}^{3}\sum_{j=1}^{3} E_{2i}\epsilon_{ij} E_{1j}$$

Since the dielectric tensor must be symmetric, the scalar product between $\mathbf{E}$ and $\mathbf{D}$ is identical for either index of refraction

$$\sum_{i=1}^{3}\sum_{j=1}^{3} E_{2i}\epsilon_{ij} E_{1j} = \sum_{j=1}^{3}\sum_{i=1}^{3} E_{1j}\epsilon_{ji} E_{2i} = \sum_{j=1}^{3} E_{1j} D_{2j}$$

and we may write

$$\mathbf{E}_2\cdot\mathbf{D}_1 = \mathbf{E}_1\cdot\mathbf{D}_2 \tag{13-18}$$

The electric field can be written as the sum of the components parallel and perpendicular to $\mathbf{k}$

$$(\mathbf{E}_{2\perp} + \mathbf{E}_{2\parallel})\cdot\mathbf{D}_1 = (\mathbf{E}_{1\perp} + E_{1\parallel})\cdot\mathbf{D}_2$$

From the definition in **(13-15)**, we know that $\mathbf{E}_\parallel\cdot\mathbf{E}_\perp = 0$ for all i and j. We combine this fact with **(13-17)** to write

$$(n_1^2 - n_2^2)(\mathbf{E}_{2\perp}\cdot\mathbf{E}_{1\perp}) = 0$$

If $n_1^2 \neq n_2^2$, then

$$\mathbf{E}_{2\perp}\cdot\mathbf{E}_{1\perp} = 0$$

and **(13-17)** allows us to write

$$\mathbf{D}_1\cdot\mathbf{D}_2 = \frac{n_1^2 n_2^2}{(\mu_0 c^2)^2}(\mathbf{E}_{1\perp}\cdot\mathbf{E}_{2\perp}) = 0 \tag{13-19}$$

When $n_1^2 = n_2^2$, the directions of $\mathbf{D}_1$ and $\mathbf{D}_2$ are arbitrary, but it is convenient to choose them to also be perpendicular.

RETARDER

Retarders modify the polarization of an incident wave by changing the relative phase of two orthogonally polarized waves that form the incident wave. Either birefringence or reflection can be used to produce the desired phase change.

To understand how birefringent retarders operate and learn how they are constructed, we will examine a uniaxial, birefringent plate, cut with the optic axis parallel to the plate's face. In Figure 13-16, we show a linearly polarized wave incident on such a crystal with the electric displacement **D** making an angle θ with the optic axis. Any polarized wave can be decomposed into two linearly polarized waves with orthogonal polarization directions. These waves are the ordinary and extraordinary waves in the birefringent crystal. In this geometry, the ordinary and extraordinary waves propagate parallel to one another but at different velocities. The different propagation velocities cause the component of polarization parallel to the slow axis (vertical in Figure 13-16) to lag the component parallel to the fast axis.

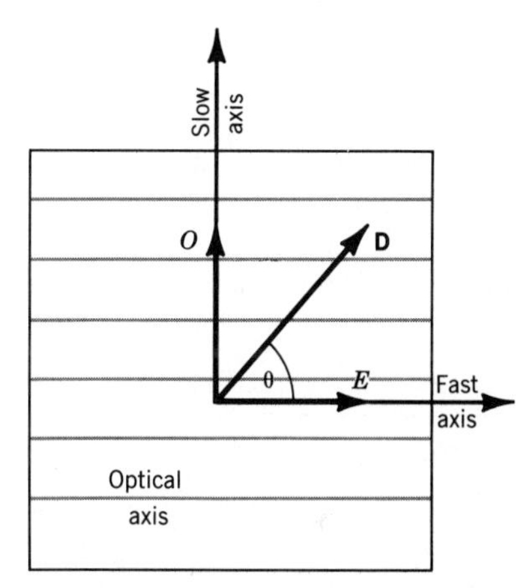

FIGURE 13-16. A linearly polarized wave is incident on a negative uniaxial crystal cut so that its face is parallel to the optical axis.

The birefringent plate modifies the polarization of the incident light because of the different propagation velocities, characterized by the two indices n_O and n_e. If the retarder plate has a thickness d, the optical path difference between the two orthogonally polarized waves is

$$N\lambda = \pm(n_e - n_O)d \tag{13-20}$$

N is called the *retardation* and is expressed in fractions of a wavelength. For example, $N = 1/4$ corresponds to a quarter-wave retardation.

The phase difference between the two orthogonally polarized waves generated by propagating through the retarder plate is simply 2π times the retardation

$$\delta = 2\pi N = \pm\frac{2\pi d(n_e - n_O)}{\lambda} \tag{13-21}$$

This phase is used to determine the polarization after propagating through the retarder plate through the use of **(2-35)**.

If the displacement **D** of the linearly polarized light is incident on the birefringent plate, with the direction of polarization parallel to either the fast or slow axis of the plate, the plate will not modify the polarization. In all other orientations, the plate will modify the polarization of the incident light because of the phase difference between the polarization components parallel to the fast and slow axes.

Quarter-Wave Plate

If the plate's thickness results in **(13-21)** having a value of $\delta = \pi/2$, a phase retardation equivalent to a shift of one-quarter of a sinusoidal wave is generated between the ordinary and extraordinary waves and the plate is called a *quarter-wave plate*, $N = 1/4$. Mica or quartz are normally used for retarders because it is easy to produce plates of the desired thickness. In quartz, a plate 13.7 μm thick produces a quarter-wave phase retardation, whereas in mica, the quarter-wave thickness is 22.3 μm at $\lambda = 500$ nm.

Mica, a naturally occuring mineral, is actually biaxial with the angle between the two optic axes ranging from 0° to 42°. The mica selected for retarders has a biaxial angle near zero degrees. Mica can be cleaved into very thin plates, by virtue of its crystal structure, and this ease overrides any complication introduced by the small departure from uniaxial behavior. The cleavage planes that allow mica to be formed into thin plates are not parallel to either optic axes. For the crystal orientation with the faces of the retarder determined by the cleavage planes, the ordinary and extraordinary indices are $n_0 = n_\gamma$ and $n_e = n_\beta$; see Table 13.3.

If linearly polarized light is incident on a quarter-wave plate such that $\theta = 45°$ in Figure 13-16, then the linear polarization is converted to circularly polarized light (see Chapter 2). If elliptically polarized light is incident on a quarter-wave plate, the elliptically polarized light will be converted to linearly polarized light when the slow and fast axes of the quarter-wave plate coincide with the major and minor axes of the elliptically polarized light.

Half-Wave Plate

If the plate's thickness results in **(13-21)** having a value of $\delta = \pi$, then the plate is called a *half-wave plate*. The half-wave plate changes the orientation of the polarization, but does not change its form. For example, suppose in Figure 13-16 that **D** represents the polarization of an incident plane wave or the major axis of an incident elliptically polarized wave; the emerging wave will have **D** oriented at an angle of $-\theta$ with respect to the optical axis. The half-wave plate will have rotated the polarization through an angle of 2θ.

Compensator

Because the desired phase retardation produced by a wave plate occurs only at the design wavelength, the plate will be too thick for shorter wavelengths and too thin for longer wavelengths. It is useful to have a component with a continuously variable phase retardation for use at any wavelength. Such a device, called a *compensator*, would also allow the measurement of an unknown retardation by comparison with the known retardation of the compensator.

One way to produce a continuously varying phase retardation is to construct a wedge using a birefringent material—the upper wedge shown in Figure 13-17 would meet this requirement. The simple wedge is not too useful because it cannot produce zero phase retardation. If two wedges are used, with the slow axis of one normal to the slow axis of the other, then by moving one wedge relative to the other, a continuous variation of phase retardation can be produced. This design is called a *Babinet compensator*.

The *Soleil–Babinet compensator*, shown in Figure 13-17, produces the same phase retardation as the Babinet compensator but over a wide aperture. The aperture of the compensator is determined by the small wedge and plate at the bottom of Figure 13-17. When the small wedge is positioned as shown at the left of Figure 13-17, the phase shift produced by the rectangular plate is exactly canceled by the two overlapping wedges, producing zero retardation. When the small wedge is positioned to the extreme right of the large wedge, as shown on the right of Figure 13-17, the two

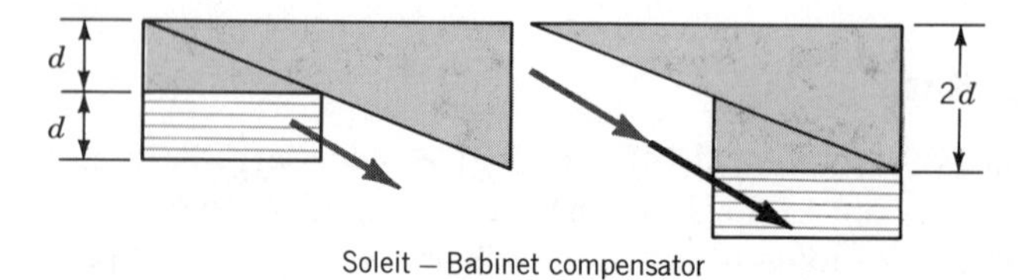

FIGURE 13-17. Soleit–Babinet compensator. On the left is the configuration of the compensator that produces no retardation. On the right, the maximum retardation is produced. The compensator is designed to produce a uniform retardation across the aperture of the device, which is the width of the small sliding component.

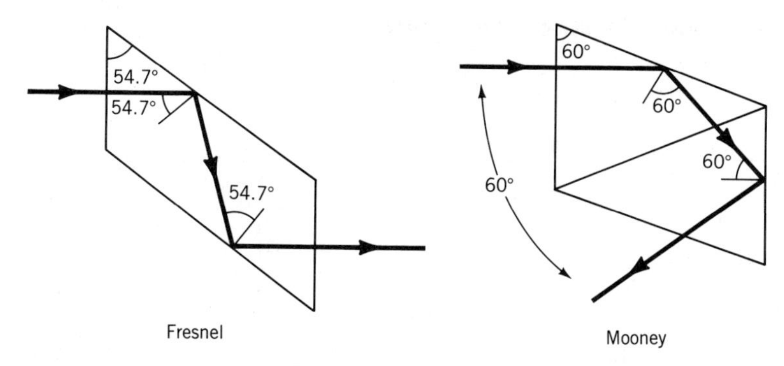

FIGURE 13-18. Two types of rhombs.

wedges produce a phase retardation equal to twice the negative of the phase retardation of the rectangular plate. This design can produce as much as two waves of retardation.

Rhomb

The rhomb-type retarder is a popular substitute when the strong wavelength dependence of the birefringent, phase-shifter design is a problem. The rhombs derive their phase retardation from the phase change introduced by total reflection [see **(3-55)**, **(3-56)**, and Figure 3-13]. Figure 13-18 shows two typical examples from a large number of rhomb designs. All of the designs require that the rhomb be constructed of a homogeneous, isotropic material.

MUELLER CALCULUS

Soleillet discovered in 1929 that an optical device performed a linear transformation on the input wave, and in 1942 Perrin put this fact into a matrix formalism

$$\boldsymbol{S}_1 = \boldsymbol{M} \cdot \boldsymbol{S}_0 \tag{13-22}$$

Mueller used experimentally derived 4–4 matrices, the $\boldsymbol{M}$ in **(13-22)**, to describe the effect of an optical device on a light wave's polarization. The matrices are based on an assumed linear relationship between the incident and transmitted beams, which is the usual case. (We will examine nonlinear behavior later.)

The analysis of the effect of a number of polarizers and retarders is made easier by the use of the Mueller–Stokes matrix calculus coupled with the use of the Stokes vectors **(2-44)**.

The S in **(13-22)** are column matrices whose elements are the Stokes parameters **(13-23)**

$$\boldsymbol{S} = \begin{pmatrix} s_0 \\ s_1 \\ s_2 \\ s_3 \end{pmatrix} \tag{13-23}$$

A few optical devices and their Mueller matrix are listed in Tables 13.5 and 13.6.

The Mueller matrix contains 16 parameters, but there are only 7 inde-

TABLE 13.5 Mueller Matrices for Polarizers

Linear

Transmission Axis	
Horizontal	Vertical
$\frac{1}{2}\begin{bmatrix} 1 & 1 & 0 & 0 \\ 1 & 1 & 0 & 0 \\ 0 & 0 & 0 & 0 \\ 0 & 0 & 0 & 0 \end{bmatrix}$	$\frac{1}{2}\begin{bmatrix} 1 & -1 & 0 & 0 \\ -1 & 1 & 0 & 0 \\ 0 & 0 & 0 & 0 \\ 0 & 0 & 0 & 0 \end{bmatrix}$
45°	−45°
$\frac{1}{2}\begin{bmatrix} 1 & 0 & 1 & 0 \\ 0 & 0 & 0 & 0 \\ 1 & 0 & 1 & 0 \\ 0 & 0 & 0 & 0 \end{bmatrix}$	$\frac{1}{2}\begin{bmatrix} 1 & 0 & -1 & 0 \\ 0 & 0 & 0 & 0 \\ -1 & 0 & 1 & 0 \\ 0 & 0 & 0 & 0 \end{bmatrix}$

Circular

Right	Left
$\frac{1}{2}\begin{bmatrix} 1 & 0 & 0 & 1 \\ 0 & 0 & 0 & 0 \\ 0 & 0 & 0 & 0 \\ 1 & 0 & 0 & 1 \end{bmatrix}$	$\frac{1}{2}\begin{bmatrix} 1 & 0 & 0 & -1 \\ 0 & 0 & 0 & 0 \\ 0 & 0 & 0 & 0 \\ -1 & 0 & 0 & 1 \end{bmatrix}$

TABLE 13.6 Mueller Matrices for Retarders

Quarter-Wave Plate

Horizontal	Vertical
$\begin{bmatrix} 1 & 0 & 0 & 0 \\ 0 & 1 & 0 & 0 \\ 0 & 0 & 0 & 1 \\ 0 & 0 & -1 & 0 \end{bmatrix}$	$\begin{bmatrix} 1 & 0 & 0 & 0 \\ 0 & 1 & 0 & 0 \\ 0 & 0 & 0 & -1 \\ 0 & 0 & 1 & 0 \end{bmatrix}$
45°	−45°
$\begin{bmatrix} 1 & 0 & 0 & 0 \\ 0 & 0 & 0 & -1 \\ 0 & 0 & 1 & 0 \\ 0 & 1 & 0 & 0 \end{bmatrix}$	$\begin{bmatrix} 1 & 0 & 0 & 0 \\ 0 & 0 & 0 & 1 \\ 0 & 0 & 1 & 0 \\ 0 & -1 & 0 & 0 \end{bmatrix}$

Half-Wave Plate

0 or 90°	± 45°
$\begin{bmatrix} 1 & 0 & 0 & 0 \\ 0 & 1 & 0 & 0 \\ 0 & 0 & -1 & 0 \\ 0 & 0 & 0 & -1 \end{bmatrix}$	$\begin{bmatrix} 1 & 0 & 0 & 0 \\ 0 & -1 & 0 & 0 \\ 0 & 0 & 1 & 0 \\ 0 & 0 & 0 & -1 \end{bmatrix}$

pendent parameters. The matrix contains no information about absolute phase, but it handles partially polarized and unpolarized light without modification.

JONES CALCULUS

Jones calculus[56] is complementary to Mueller calculus and operates on the Jones vector **(2-48)** in a manner similar to the way the Mueller matrix operates on the Stokes vector. The Jones matrix contains eight independent

TABLE 13.7 Jones Matrices for Polarizers

Linear

Transmission Axis			
Horizontal	Vertical	45°	−45°
$\begin{bmatrix}1 & 0\\0 & 0\end{bmatrix}$	$\begin{bmatrix}0 & 0\\0 & 1\end{bmatrix}$	$\frac{1}{2}\begin{bmatrix}1 & 1\\1 & 1\end{bmatrix}$	$\frac{1}{2}\begin{bmatrix}1 & -1\\-1 & 1\end{bmatrix}$

Circular

Right	Left
$\frac{1}{2}\begin{bmatrix}1 & -i\\i & 1\end{bmatrix}$	$\frac{1}{2}\begin{bmatrix}1 & i\\-i & 1\end{bmatrix}$

TABLE 13.8 Jones Matrices for Retarders

Quarter-Wave Plate

Horizontal	Vertical
$\begin{bmatrix}e^{i\pi/4} & 0\\0 & e^{-i\pi/4}\end{bmatrix}$	$\begin{bmatrix}e^{-i\pi/4} & 0\\0 & e^{i\pi/4}\end{bmatrix}$
45°	−45°
$\frac{1}{2}\begin{bmatrix}1 & i\\i & 1\end{bmatrix}$	$\frac{1}{2}\begin{bmatrix}1 & -i\\-i & 1\end{bmatrix}$

Half-Wave Plate

0 or 90°	± 45°
$\begin{bmatrix}1 & 0\\0 & -1\end{bmatrix}$	$\begin{bmatrix}0 & 1\\1 & 0\end{bmatrix}$

parameters with no redundancy, making it simpler than Mueller calculus. However, Jones calculus only applies to polarized light. Jones calculus can be extended, using the density matrix formalism mentioned in Chapter 2, to allow manipulation of unpolarized light, but with a loss of simplicity. The matrix equation for Jones calculus is

$$\boldsymbol{E}_{\text{out}} = \boldsymbol{\Omega}\boldsymbol{E}_{\text{in}} \tag{13-24}$$

A few optical devices and their Jones matrices, $\boldsymbol{\Omega}$, are listed in Tables 13.7 and 13.8.

For every matrix in Jones calculus, there is a matrix in Mueller calculus, but the converse is not true. For example, it is possible to construct a depolarizer by using a thick piece of opal glass in the visible or gold-covered sandpaper in the infrared. Such a device can be described in Mueller calculus, but there is no matrix for such a device in Jones calculus.

OPTICAL ACTIVITY

Arago observed in 1811 that the direction of polarization was rotated when plane polarized light was propagated through quartz, parallel to the optical axis of quartz. **Jean Baptiste Biot** in 1815 extended Arago's observations and discovered that the continuous rotation of the plane of polarization occurred in gases and liquids, as well as crystalline material. (During the experiment that led to the first observation of rotation in a gas, Biot set a church on fire.) Materials that rotate the plane of polarization are said to be *optically active*. It is sometimes known as *natural rotation* to distinguish it from magnetic rotation (the Faraday effect), which we will discuss later.

Biot observed that if one looked at the light source through optically active materials, the materials could be grouped into two classes: those that rotated the polarization to the left, *levorotatory*, and those that rotated the polarization to the right, *dextrorotatory*. (In chemistry, anything that can exhibit a "handedness" is said to be *chiral*, and atoms that differ only in the direction they rotate polarized light are said to differ in their *chirality*.) **Louis Pasteur** demonstrated that molecules which differ only in their chirality are mirror images of each other.

Fresnel developed the first theoretical description of optical activity by first demonstrating experimentally that linearly polarized light can be decomposed into right- and left-circularly polarized waves. He then developed an explanation of optical rotation by assuming that the two circularly polarized components, of a linearly polarized wave, propagated at different velocities. Fresnel suggested that the observed optical activity might be due to "a helical arrangement of molecules of the medium."

Later investigations have confirmed Fresnel's postulated cause of optical activity for crystalline quartz. The right-hand form of quartz is shown in Figure 13-19. By following along the **Si-O** bond from silicon atom to silicon atom ($A \Rightarrow B \Rightarrow C \Rightarrow A$ in Figure 13-19), a right-hand spiral is traced out. This path is shown in Figure 13-19 by the dark arrows.

If the silicon atoms are projected onto a plane, they form the vertices of an equilateral triangle, shown on the right of Figure 13-19. To aid in visualizing the helical path, a shaded version of the triangle is shown at each of four planes in the quartz unit cell.

Two classical models[57] have been developed to explain optical activity. One model is based on the interaction of the electromagnetic wave with charges confined to a helical path. This model approximates the behavior of charge carriers in quartz and in many optically active molecules.

A second classical model treats the interaction of two coupled harmonic oscillators that lie in separate planes with an electromagnetic wave. The two-oscillator model approximates the behavior of tetrahedral bonded atoms, such as the molecule shown in Figure 13-20. The coupled oscillator model

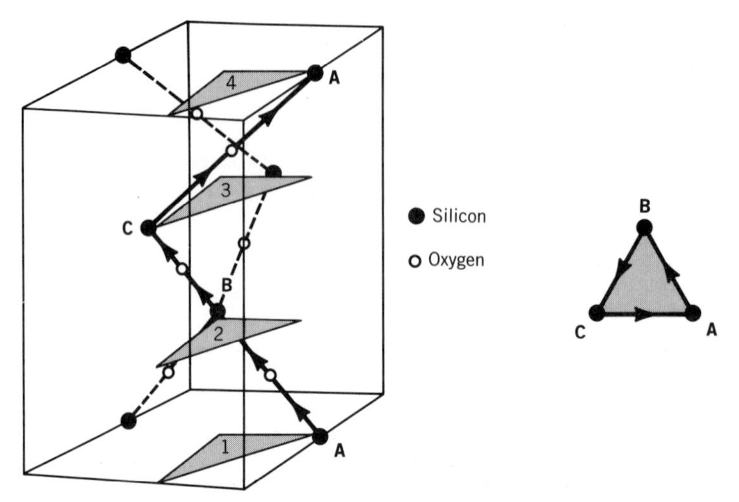

FIGURE 13-19. Crystal structure of quartz.

is equivalent to the helical model, but proof of that fact will not be given here.

The unique property possessed by optically active molecules is that a mirror image of the molecule exists. For example, the molecule and its mirror image, both shown in Figure 13-20, will lose their separate identity and optical activity if any two of the bonded atoms are identical. For example, if the fluorine atom is replaced by a hydrogen, then the mirror image does not exist because the same image can be generated by a simple rotation about the normal to the plane of the figure.

Classical and quantum mechanical models predict that there will be both electric (μ) and magnetic ($\mathbf{m}$) dipole moments induced in an optically active molecule by the incident radiation. The rotation strength is given by

$$R = Im\{\boldsymbol{\mu} \cdot \mathbf{m}\}$$

where complex notation is required to allow inclusion of the difference in phase between the two dipoles.

We have previously ignored any possible space dependence of **D** on **E**, but now we must include, at least to first order, the dependence of the space derivatives of **E** to explain the magnetic contribution; see **(2-6c)**. Drude demonstrated that the inclusion of the spatial variation of the electric field led to the prediction of optical activity by Maxwell's equations. The result of Drude's theory yielded a prediction of the wavelength dependence of optical activity given by

$$\theta \propto R \frac{\lambda_0^2}{\lambda^2 - \lambda_0^2}$$

where θ is the angle through which the plane of polarization is rotated.

Adding the dependence of **D** on $\nabla \cdot \mathbf{E}$ requires that we replace the real dielectric tensor we have been using by a complex tensor

$$\tilde{\epsilon}^0 = \tilde{\epsilon} + i\tilde{\gamma}$$

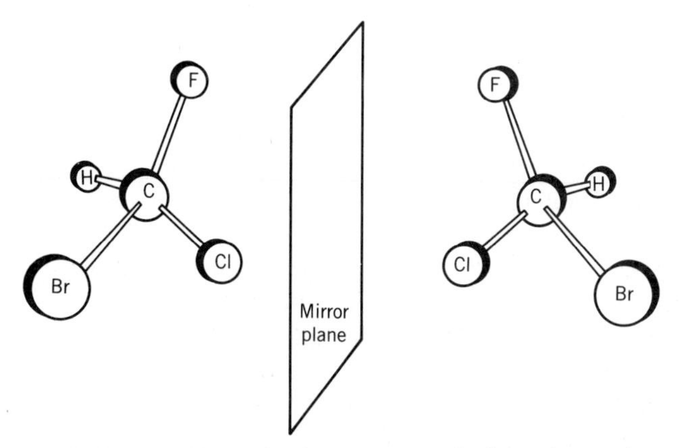

FIGURE 13-20. The molecular structure on the left is the simplest optically active molecule. The mirror image of this tetrahedral molecule, shown on the right, is a unique molecule that cannot be reproduced by rotation of the molecule on the left. If any two of the atoms on this class of tetrahedral bonded molecule are the same, then the molecule is no longer dissymmetric and it is not optically active.

We have to add an imaginary term because the space derivatives of

$$\mathbf{E} = \mathbf{E}_0 e^{i(\omega t - kx)}$$

yield terms involving i. We will continue to assume that the medium is lossless so the requirement placed on the dielectric tensor by **(13-6)** is

$$\epsilon_{jk} + i\gamma_{jk} = \epsilon_{kj} - i\gamma_{kj} \tag{13-25}$$

The real part of the dielectric tensor continues to be a symmetric tensor; however, the imaginary part must be an antisymmetric tensor. It is interesting to note that a lossless, anisotropic medium can have a complex dielectric constant; whereas a lossless, isotropic medium must have a real dielectric constant. (If loss were included, this theory would also describe *circular dichroism*: the observation of a difference in absorption of the right- and left-circularly polarized light waves.)

We may always replace an antisymmetric tensor by a vector $\boldsymbol{\gamma}$ with components

$$\gamma_1 = \gamma_{23} = -\gamma_{32}, \qquad \gamma_2 = \gamma_{31} = -\gamma_{13}, \qquad \gamma_3 = \gamma_{12} = -\gamma_{21} \tag{13-26}$$

The x component of the product $\tilde{\boldsymbol{\gamma}}\mathbf{E}$ is written as

$$\sum_{k=1}^{3} \gamma_{1k} E_k = \gamma_3 E_2 - \gamma_2 E_3$$

We may use **(2A-2)** to identify this as the x component of the vector product $\boldsymbol{\gamma}\times\mathbf{E}$. This means that **(13-1)** must be rewritten as

$$D_j = \sum_{k=1}^{3} \left[\epsilon_{jk} E_k - i(\boldsymbol{\gamma} \times \mathbf{E})_j \right] \tag{13-27}$$

If we perform an inversion operation on the medium, the components of **D**, **E** and $\boldsymbol{\gamma}\times\mathbf{E}$ change sign along with i. However, if a material has inversion symmetry, the components of **D** must not be affected by the inversion operation. This would be true for **(13-27)** only if $(\boldsymbol{\gamma}\times\mathbf{E}) = 0$. We can conclude from this discussion that if a material has inversion symmetry, $\tilde{\boldsymbol{\gamma}} = 0$ and no optical activity will be observed.

By using **(13-27)** in place of **(13-1)**, we will be able to describe optical activity. A formal derivation of the theory would demonstrate that except for differences in second order in γ, the normal surfaces and ray surfaces of an optically active crystal agree with those of a nonactive, birefringent crystal (Appendices 13-C and 13-D). Although the surfaces have the same appearance, their interpretation must be modified.

If we have a positive uniaxial crystal that does not exhibit optical activity, the index surfaces would consist of a sphere and an inscribed prolate spheroid [see **(13C-6)** and **(13C-7)**]. The radius from the center to any point on either sheet gives the index of refraction for one of the two orthogonal, linearly polarized waves. The two sheets touch at two points, and the line between those two points defines the direction of the optical axis.

Consider now the positive uniaxial crystal, quartz, which is also optically active. The second-order effects of optical activity result in two propagation velocities in the direction of the optical axis, i.e., the two normal surfaces no longer touch; see Figure 13-21. (In Figure 13-21, we have exaggerated the separation that is actually only 0.6% of the sphere's radius normal to the optical axis and 4.5×10^{-5} of the sphere's radius along the optical axis.) In addition to the appearance of two indices of refractions along the optical axis, the polarizations associated with those indices are no longer linear, but are now right- and left-circular.

Thus, along the optical axis of quartz, the normal surfaces give the index of refraction for right- and left-circular polarized light. Normal to the optical axis, the normal surfaces are as they are in a nonactive material, that is, they give the index of refraction for the two orthogonal, linearly polarized light waves. In between these two extreme conditions, the two polarizations that are represented by the two normal surfaces are right- and left-elliptically polarized light.

To follow a plane-polarized wave as it propagates through an optically active material, we assume that at the surface of the optically active material, $z = 0$, we have a vertically polarized beam. Using the Jones vector notation **(2-44)**

$$\frac{E_0}{2}\begin{pmatrix} 0 \\ \sqrt{2} \end{pmatrix} e^{i\omega t}$$

we decompose the plane-polarized wave in the optically active material into a right- and left-circularly polarized component, **(2-38)** and **(2-39)**. The wave traveling in the positive z direction is

$$\frac{E_0}{2\sqrt{2}}\begin{pmatrix} -i \\ 1 \end{pmatrix} \exp\left[i(\omega t - kz)\right] + \frac{E_0}{2\sqrt{2}}\begin{pmatrix} i \\ 1 \end{pmatrix} \exp\left[i(\omega t - kz)\right] \tag{13-28}$$

After traveling through a thickness d of a crystal of index of refraction n, the phase of light, relative to the initial wave, at the surface $z = 0$ is

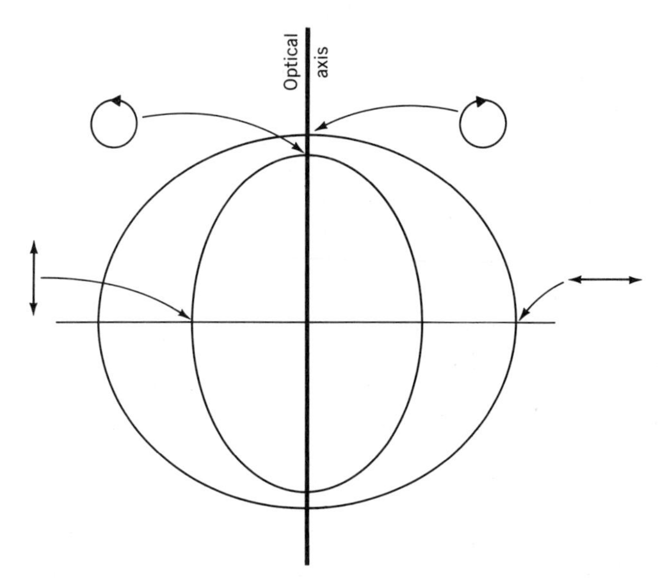

FIGURE 13-21. A very schematic representation of the normal surface of quartz. Also shown are the states of polarization that propagate in two directions in the crystal.

$$nkd = \frac{2\pi}{\lambda}nd$$

We assume that the right-circularly polarized wave sees an index of refraction n_+ and the left-circularly polarized waves sees n_-. Upon leaving the optically active material, the wave is

$$\frac{E_0}{2\sqrt{2}}\begin{pmatrix}-i\\1\end{pmatrix}\exp\left[i(\omega t - n_+kz)\right] + \frac{E_0}{2\sqrt{2}}\begin{pmatrix}i\\1\end{pmatrix}\exp\left[i(\omega t - n_-kz)\right] \tag{13-29}$$

This can be rewritten as

$$\frac{E_0}{2\sqrt{2}}\exp\left\{i\left[\omega t - (n_+ + n_-)\frac{kd}{2}\right]\right\}$$

$$\left\{\begin{pmatrix}-i\\1\end{pmatrix}\exp\left[i(n \;\; - n_+)\frac{kd}{2}\right] + \begin{pmatrix}i\\1\end{pmatrix}\exp\left[i(n_+ - n_-)\frac{kd}{2}\right]\right\} \tag{13-30}$$

To interpret the physical meaning of **(13-30)**, we simplify the equation by defining

$$\psi = \frac{kd}{2}(n_+ + n_-), \qquad \theta = \frac{kd}{2}(n_- - n_+)$$

With these definitions, we can write **(13-30)** as

$$\frac{E_0}{\sqrt{2}}e^{\,i(\omega t-\psi)}\left[\begin{pmatrix}0\\1\end{pmatrix}\left(\frac{e^{i\theta}+e^{-i\theta}}{2}\right) + \begin{pmatrix}1\\0\end{pmatrix}\left(\frac{e^{i\theta}-e^{-i\theta}}{2i}\right)\right]$$

$$\frac{E_0}{\sqrt{2}}e^{i(\omega t-\psi)}\left[\begin{pmatrix}0\\1\end{pmatrix}\cos\theta + \begin{pmatrix}1\\0\end{pmatrix}\sin\theta\right] \tag{13-31}$$

Equation **(13-31)** describes a plane polarized wave with the polarization direction at an angle

$$\theta = \frac{\pi d}{\lambda}(n_- - n_+) \tag{13-32}$$

The incident, plane-polarized wave has had its plane of polarization rotated through an angle θ proportional to the length of the path traversed in the substance. If $n_- > n_+$, then the right-circularly polarized wave travels faster than the left and the material is called dextrorotatory. For the opposite case, the material is called levorotatory.

For most crystals, $(n_- - n_+)$ is on the order of 10^{-4} or less. For quartz, the index difference is 7.1×10^{-5} at 589.3 nm; thus, a 1 mm thick plate of quartz, cut normal to the optical axis, rotates a linearly polarized wave through an angle of 21.7°. The angle through which the plane of polarization is rotated for a 1 mm thick crystal is defined as the *specific rotation* (or *rotatory power*)

$$\beta = \frac{\theta}{1\text{ mm}}$$

The specific rotation for a few solids can be found in Table 13.9.

TABLE 13.9 Specific Rotation of Solids

Material	Rotation β, deg/mm
HgS	+32.5
Lead hyposulfate	+5.5
Potassium hyposulfate	+8.4
Quartz	+21.684
$NaBrO_3$	+2.8
$NaClO_3$	+3.13

We see from **(13-32)** that the specific rotation of any material has an explicit wavelength dependence of $1/\lambda$. For quartz, the index difference $(n_- - n_+)$ also varies inversely with wavelength, so that the specific rotation of quartz should vary as $1/\lambda^2$. Throughout most of the visible region of the spectrum, the specific rotation of quartz can be described by an equation of the same form as Cauchy's equation (see Figure 13-22), which is in agreement with the prediction of Drude's theory.

As Biot discovered, optical activity is also exhibited by fluids. The specific rotation for a few liquids are listed in Table 13.10. A fluid is isotropic in the sense that the molecules are randomly oriented throughout. The observed optical activity is attributed to molecules with no center of symmetry and no plane of symmetry. The molecules are randomly oriented in a liquid or gas, but still they produce optical activity because the preferred direction of rotation associated with each molecule is independent of orientation. To aid in visualizing this property, think of a box of screws; they remain right-handed no matter what orientation they have in the box.

The vector $\boldsymbol{\gamma}$ for a liquid or gas reduces to a scalar and the two indices have the form

$$n_+ = \sqrt{\epsilon} + \frac{\gamma}{\sqrt{\epsilon}} \tag{13-33}$$

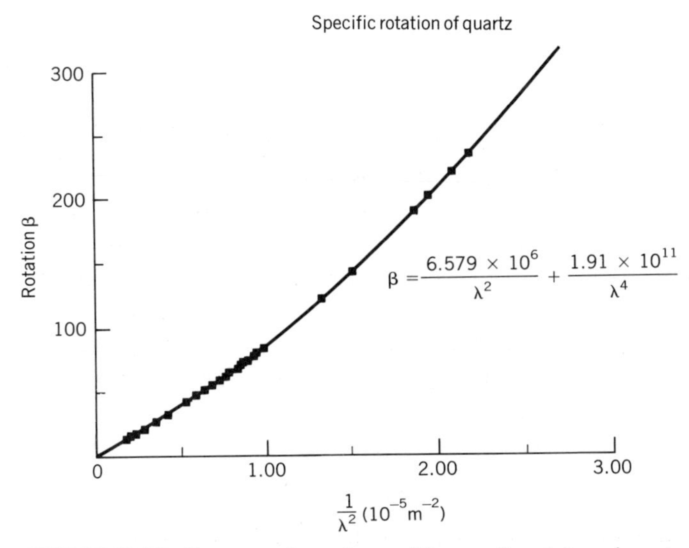

FIGURE 13-22. Frequency dependence of the specific rotation of quartz.

TABLE 13.10 Specific Rotation of Liquids

Material	Rotation β, deg/dm
Amyl alcohol	−5.7
Camphor	+70.33
Menthol	−49.7
Nicotine	−162
Turpentine	−37

TABLE 13.11 Specific Rotation of Solutions

Substance	Solvent	Rotation β, deg/dm
Camphor	Alcohol	+54.4
Camphor	Benzene	+56
Camphor	Ether	+57
Galactose	Water	+83.9
d-Glucose (dextrose)	Water	+52.5
l-Glucose	Water	−51.4
Lactose	Water	+52.4
Maltose	Water	+138.48
Nicotine	Water	−77
Nicotine	Benzene	−164
Sucrose	Water	+66.412

$$n_- = \sqrt{\epsilon} - \frac{\gamma}{\sqrt{\epsilon}} \qquad (13\text{-}34)$$

If an optically active compound is dissolved in an inactive solvent, then the rotation is nearly proportional to the amount of the compound dissolved; for this reason, the specific rotation is defined for 1 g of solute in 1 cm of solution. The specific rotation of a liquid is smaller than for a crystal, so it is usually defined for a 10 cm path length rather than a 1 mm path length. If the concentration of the solution is mg/cc, then the rotation of the plane of polarization θ produced by a solution of an optically active material is

$$\theta = \beta \frac{md}{10}$$

For a pure fluid, m is replaced by the density of the fluid. The specific rotation for a few typical solutions are listed in Table 13.11.

SUMMARY

A number of methods for the control of the polarization of light have been presented. They include

1. **Absorption**. Dichroic polarizers absorb one polarization to a greater extent than the orthogonal polarization. When the two orthogonal polarizations are right- and left-circular, then the phenomena is called circular dichroism.
2. **Reflection**. The Fresnel equations for reflection are different for light polarized in and normal to the plane of incidence. At Brewster's angle

$$\theta_B = \tan^{-1}\frac{n_t}{n_i}$$

only the polarization normal to the plane of incidence is reflected. This fact can be used to produce a linearly polarized wave.

3. **Interference**. By using the design rules for a multilayer filter, a stack of dielectric layers can be produced that will selectively reflect a desired polarization. The unwanted polarization is reflected from the interfaces of the multilayer stack and destructively interferes.
4. **Birefringence**. In an anisotropic crystal, one polarization has an isotropic propagation velocity and obeys Snell's law; the wave associated with this polarization is called the ordinary wave. The orthogonal polarization has an anisotropic propagation velocity and does not obey Snell's law. The departure from Snell's law can be used to separate the two polarized waves. Once separated, one of the two polarizations can be removed by using a stop or total reflection.
5. **Optical activity**. Birefringence of the two orthogonal circularly polarized waves leads to optical activity. An optically active material rotates the plane of polarization of a linearly polarized wave through an angle

$$\theta = \frac{\pi d}{\lambda}(n_- - n_+)$$

where d is the thickness of the optically active medium, n_+ the index of refraction for a right-circularly polarized wave, and n_- the index of refraction for a left-circularly polarized wave.

These physical processes can be used to construct a number of optical devices that can be used to manipulate a light wave's polarization. The operation of the devices can be described by Mueller matrices that multiply a wave's Stokes vector to produce the Stokes vector of the final wave. A similar set of matrices, called Jones matrices, exist for use with Jones vectors. The polarization devices discussed in this chapter were

1. **Polarizer**. This device will only transmit one polarization. If it is a linear polarizer, then it will transmit the component of the polarization parallel to its axis. A circular polarizer will transmit either a right- or a left-circularly polarized wave.
2. **Retarder**. This device converts one form of polarization into another. It does so because the two orthogonal polarizations travel at different velocities. After passing through the retarder, there is a phase difference between the two polarizations given by

$$\delta = \pm\frac{2\pi d(n_e - n_0)}{\lambda}$$

n_e is the index of refraction of the extraordinary wave and n_0 is the index of refraction of the ordinary wave. If the thickness d of the retarder produces a phase shift of $\pi/2$, then the retarder is called a quarter-wave plate and linearly polarized light is converted to circularly polarized light. If d is such that a phase shift of $\delta = \pi$ is produced, then the retarder is called a half-wave plate and linearly polarized light has its plane of polarization rotated by an angle of 2θ, where θ is the angle between the plane of polarization of the wave and the optical axis of the retarder.
3. **Compensator**. The compensator is a retarder that will produce a variable phase shift.

To describe the propagation of electromagnetic waves in an anisotropic medium, we must use a dielectric tensor. We demonstrated that if the medium is lossless, then the tensor is symmetric. The properties of the tensor can be understood by using a geometrical construct defined by the equation

$$\frac{D_x^2}{2U_E\epsilon_x} + \frac{D_y^2}{2U_E\epsilon_y} + \frac{D_z^2}{2U_E\epsilon_z} = 1$$

This equation describes an ellipsoid called the optical indicatrix. The indicatrix can be used to determine **E**, **k**, and **S**, given **D**.

There are two indices of refraction in a birefringent material. Their values, in the propagation direction specified by **k**, can be obtained using Fresnel's equation

$$\sum_{j=1}^{3} \frac{k_j^2}{k^2(n^2 - n_j^2)} = \frac{1}{n^2}$$

PROBLEMS

13-1. Discuss the reasoning behind the statement that wire polarizers should have a low resistivity for optimum performance.

13-2. Prove that Fresnel's equation

$$\sum_{j=1}^{3} \frac{k_j^2}{k^2(n^2 - n_j^2)} = \frac{1}{n^2}$$

is a quadratic equation.

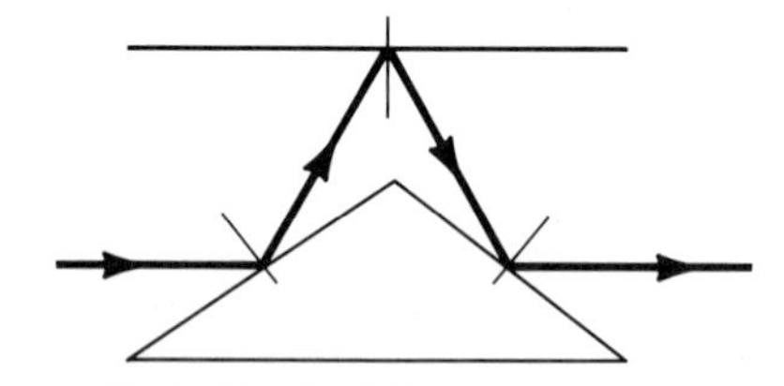

FIGURE 13-23. A K-mirror.

13-3. Calculate the maximum index of refraction that the cement in a Nicol prism can have.

13-4. Derive Equation **(13C-1)**.

13-5. The mirrors shown in Figure 13-23 can be used to rotate the polarization of an incident linearly polarized wave through 2θ when the mirrors are rotated through an angle θ about an axis parallel to the incident wave and passing through the centers of the two faces of the prism.

Describe how this rotator works. Find an equation that relates the maximum aperture of this device to the prism's angle and the distance from the tip of the prism to the top mirror. What does the result say about the angle of the prism?

13-6. What is the effect of stacking two linear half-wave retarders in series with the fast axis of one at 0° and the second at $\theta = 45°$? Use Mueller calculus to obtain your result.

13-7. Explain why the dielectric tensor of a loss-free, isotropic medium must be a real function.

13-8. Assume a uniaxial crystal with $n_1 = n_2 = n_0$, $n_3 = n_e$, and a wave propagating in the x, z plane at an angle θ with respect to the z axis (optical axis). Write the vector expressions for a unit vector in the direction of wave propagation and for **D** and **E**.

13-9. Show that for a uniaxial crystal the Poynting vector **S** for the extraordinary wave makes an angle ϕ with respect to the wave vector **k**; see Figure 13B-1, where ϕ is given by

$$\cos\phi = \frac{n_0^2 \cos^2\theta + n_e^2 \sin^2\theta}{\sqrt{n_0^4 \cos^2\theta + n_e^4 \sin^2\theta}} \tag{13-35}$$

This is called *Poynting vector walkoff.*

13-10. Show that the Poynting vector walkoff angle is also given by

$$\tan\phi = \frac{\sin\theta\cos\theta\,(n_0^2 - n_e^2)}{n_e^2\cos^2\theta + n_0^2\sin^2\theta} \tag{13-36}$$

13-11. A 1 mm diameter beam of light of wavelength $\lambda = 589.3$ nm travels through a 1 cm thick calcite crystal, at an angle of 32° with respect to the optical axis.

How far apart will the extraordinary and ordinary beam be upon exit from the crystal?

13-12. A 10 cm long cylinder contains a 20% solution of maltose. If the observed rotation of linearly polarized light after passing through the solution is 28.8°, what is the specific rotation of maltose?

13-13. A solution of camphor in alcohol in a 20 cm long cylinder rotates linearly polarized light 33°. The specific rotation of camphor is 54°/dm. What is the concentration of camphor in the solution?

13-14. Elliptically polarized light with a Jones vector

$$\begin{pmatrix} 3 \\ i \end{pmatrix}$$

passes through a quarter-wave plate oriented with its fast axis vertical. What is the Jones vector for the light after passing through the quarter-wave plate?

13-15. Use Jones calculus to describe the effect of a right-circular polarizer on vertically polarized light.

13-16. Describe the polarization of a wave, initially unpolarized, after passing through a linear polarizer and then a half-wave plate. What happens if we reverse the order of the two components?

13-17. Light is incident on eight glass plates of index $n = 1.75$ at Brewster's angle. What is the percent of polarization of the transmitted light?

13-18. A plate of quartz 0.5 mm thick has its faces parallel to the optical axis. What is the phase difference between the ordinary and extraordinary waves after passing through the plate for $\lambda = 589$ nm? For $\lambda = 1\ \mu$m?

13-19. Using Mueller calculus, describe the effect of a linear polarizer and a quarter-wave plate on an unpolarized wave. What happens if the order of the two components is reversed?

13-20. Why do most polarizers using birefringent materials contain two prisms?

Appendix 13-A

TENSORS

In this appendix, we will define a tensor. A geometrical construction of use in determining the properties of a symmetric, second-order tensor will be discussed. Second-order symmetric tensors for each of the crystal symmetry classes will also be presented.

A tensor is an abstract mathematical object that has specific components in every coordinate system and undergoes a well-defined transformation when the coordinate system undergoes a transformation, say, a rotation. We will define below several tensors that are of use in optics.

Scalar

Scalars are quantities that are not connected with direction, for example, charge. We can completely specify the value of a scalar by assigning it a single number. A scalar is a tensor of rank zero.

Vector

Vectors are quantities that must be defined with reference to a direction, for example, the electric field. To specify a vector's value at a point, we must give both its magnitude and direction or, equivalently, we may specify the vector's components along three mutually perpendicular directions

$$\mathbf{E} = \{E_1,\ E_2,\ E_3\}$$

A vector is called a tensor of rank one.

Second Rank Tensor

Tensors of second rank relate two general vectors. If we wish to relate displacement **D** and electric field **E**, we introduce the dielectric tensor $\tilde{\boldsymbol{\epsilon}}$ such that

$$D_i = \sum_{j=1}^{3} \boldsymbol{\epsilon}_{ij} E_j \tag{13A-1}$$

The constants ϵ_{ij} are the components of the tensor $\tilde{\epsilon}$ and can be given the following physical meaning. Suppose the electric field is in the x direction

$$\mathbf{E} = \{E_1,\ 0,\ 0\}$$

Then, **D** would have components, according to **(13A-1)**, along all three axes

$$\mathbf{D} = \{\epsilon_{11}E_1,\ \epsilon_{21}E_1,\ \epsilon_{31}E_1\}$$

The component ϵ_{21} gives the component of **D** in the y direction due to the electric field in the x direction.

In matrix notation, **(13A-1)** would be written

$$(D_1 \quad D_2 \quad D_3) = \begin{pmatrix} \epsilon_{11} & \epsilon_{12} & \epsilon_{13} \\ \epsilon_{21} & \epsilon_{22} & \epsilon_{23} \\ \epsilon_{31} & \epsilon_{32} & \epsilon_{33} \end{pmatrix} \begin{pmatrix} E_1 \\ E_2 \\ E_3 \end{pmatrix} \qquad \text{(13A-2)}$$

The 3×3 matrix of ϵ's is the tensor of the second rank, denoted by $\tilde{\boldsymbol{\epsilon}}$. In general, there are nine components in a second-rank tensor.

Higher-Rank Tensor

A tensor of third rank relates a vector **A** to two vectors **B** and **C**. In general, each component of **A** will depend on all the components of **B** and **C**.

$$A_i = \sum_{j=1}^{3} \sum_{k=1}^{3} S_{ijk} B_j C_k$$

The S_{ijk} are components of a third-rank tensor. In general, an Nth-rank tensor relates one vector to $(N-1)$ other vectors.

Coordinate Transformation

The components of the tensor will change if a coordinate transformation is made. In fact, the way in which a tensor changes, under a coordinate transformation, is part of its definition. The transformation from one rectangular coordinate system to another is defined by the equation

$$x_i' = \sum_{j=1}^{3} \gamma_{ij} x_j$$

To transform a second-rank tensor between the two coordinate systems defined by this transformation, we use the equation

$$\epsilon_{ij}' = \sum_{k=1}^{3} \sum_{\ell=1}^{3} \gamma_{ik} \gamma_{j\ell} \epsilon_{k\ell} \qquad \text{(13A-3)}$$

Third-rank tensors would transform according to the equation

$$S_{ijk}' = \sum_{\ell=1}^{3} \sum_{m=1}^{3} \sum_{n=1}^{3} \gamma_{i\ell} \gamma_{jm} \gamma_{kn} S_{\ell mn}$$

The physical property that the tensor represents (in the example of a second-rank tensor, the dielectric constant) should be uniquely defined by the tensor, independent of the coordinate system. The tensor selected to represent the physical property is the simplest matrix that can be constructed.

If the tensor is symmetric, that is, if $\epsilon_{kj} = \epsilon_{jk}$, then by proper choice of the coordinate system, we can write the second-rank tensor as a diagonal matrix, reducing the number of constants from six to three. The three elements of the diagonal tensor are called the principal elements of the tensor and are the quantities used to characterize the physical property described by the tensor.

Geometrical Representation

Our understanding of symmetrical tensors of the second rank can be aided by the use of a geometric construction. Equation **(13A-1)**, in its most general form, is a second-degree equation that, for an arbitrary set of unit vectors, is written

$$\sum_{i=1}^{3}\sum_{j=1}^{3} S_{ij}x_i x_j = 1$$

If the tensor is symmetric, $S_{ij} = S_{ji}$, then the second-degree equation can be written in expanded form as

$$S_{11}x_1^2 + S_{22}x_2^2 + S_{33}x_3^2 + 2(S_{23}x_2x_3 + S_{31}x_3x_1 + S_{12}x_1x_2) = 1 \quad \text{(13A-4)}$$

This is the general equation of a quadric (a second-degree surface or conicoid). The surface described by this equation may be an ellipsoid or a hyperboloid.

The coefficients S_{ij} of **(13A-4)** transform according to the same equation as the components of a symmetrical tensor of second rank **(13A-3)**, and thus, we can say that S is a tensor. The surfaces in Figure 13A-1 are aligned with the principal axes; therefore, the equation for the surfaces shown is the simplified form

$$S_1x_1^2 + S_2x_2^2 + S_3x_3^2 = 1 \quad \text{(13A-5)}$$

The standard equation of the quadric is

$$\frac{x^2}{a^2} + \frac{y^2}{b^2} + \frac{z^2}{c^2} = 1 \quad \text{(13A-6)}$$

Comparing **(13A-5)** with **(13A-6)**, we see that the semiaxes of the ellipsoid in Figure 13A-1 are equal to the principal elements of the tensor **S**.

$$a = \frac{1}{\sqrt{S_1}}, \qquad b = \frac{1}{\sqrt{S_2}}, \qquad c = \frac{1}{\sqrt{S_3}}$$

The major use of the geometrical representation of Figure 13A-1 is in providing an understanding of how symmetric tensors of second rank change with coordinate transformations.

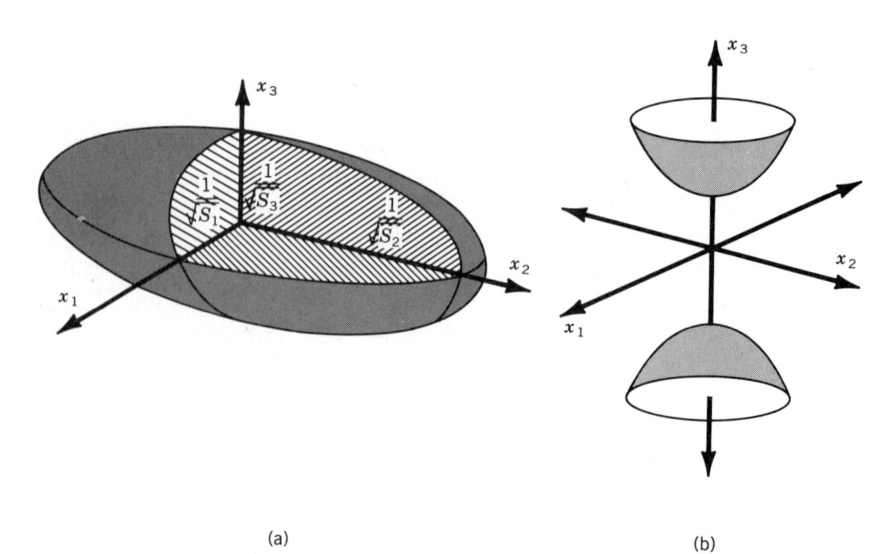

FIGURE 13A-1. The quadric surface described by **(13A-4)**. (a) An ellipsoid, (b) a hyperboloid of two sheets.

If the symmetrical tensor S is referred to an arbitrary set of axes, six independent components (S_{11}, S_{22}, S_{33}, S_{23}, S_{31}, and S_{12}) are needed to describe the surface. Naturally, we would try, when possible, to refer the tensor to its principal axes, the surface's symmetry axes, so that only the three principal components are needed.

Crystal Symmetry

Another way that the number of required components of a tensor is reduced is when the environment requires that the tensor obey certain symmetry operations. For example, the dielectric constant of a crystal must mirror the symmetry exhibited by the crystal. The crystallographic symmetry classes can be divided into three groups according to the number of symmetry axes present in the crystal. The groups, and the crystal classes that are members of the groups, are listed in Table 13A.1.

TABLE 13A.1 Crystal Classes and Their Second-Rank Tensor

Group	Quadric Surface	Orientation of Quadric	Crystal Class	Tensor	Number of Coefficients	Optical Type
Three equivalent orthogonal directions	Sphere	Unimportant	Cubic, three 4-fold axes	$\begin{bmatrix} S & 0 & 0 \\ 0 & S & 0 \\ 0 & 0 & S \end{bmatrix}$	1	Isotropic
Two equivalent orthogonal directions in a plane normal to the crystal's symmetry axis	Spheroid	One principal axis parallel to the crystal symmetry axis	Trigonal, one 3-fold symmetry axis	$\begin{bmatrix} S_1 & 0 & 0 \\ 0 & S_1 & 0 \\ 0 & 0 & S_2 \end{bmatrix}$	2	Uniaxial
			Tetragonal, one 4-fold symmetry axis			
			Hexagonal, one 6-fold symmetry axis			
No equivalent directions	General Quadric	Not specified	Triclinic, no symmetry (at most, a center of symmetry)	$\begin{bmatrix} S_{11} & S_{12} & S_{31} \\ S_{12} & S_{22} & S_{23} \\ S_{31} & S_{23} & S_{33} \end{bmatrix}$	6	Biaxial
		Principal axis parallel to symmetry axis	Monoclinic one 2-fold symmetry axis	$\begin{bmatrix} S_1 & 0 & S_{31} \\ 0 & S_2 & 0 \\ S_{31} & 0 & S_3 \end{bmatrix}$	4	
		Principal axes aligned with symmetry axes	Orthorombic, three perpendicular 2-fold axes of symmetry	$\begin{bmatrix} S_1 & 0 & 0 \\ 0 & S_2 & 0 \\ 0 & 0 & S_3 \end{bmatrix}$	3	

Appendix 13-B

POYNTING VECTOR IN ANISOTROPIC DIELECTRIC

The ray of geometrical optics was defined in Chapter 5 **(5-4)** as a unit vector $\hat{\mathbf{s}}$ normal to the wavefront and, (see Problem 5-17) parallel to the Poynting vector $\mathbf{S}$, i.e., parallel to the direction of energy flow. The theory of Chapter 13 is not suitable for use with geometrical optics or Huygens' principal because it does not follow the flow of energy in the crystal, but rather tracks the progress of $\mathbf{k}$ and the transverse displacement $\mathbf{D}$. A suitable theory of propagation for use with geometric optics must be built around the vectors $\mathbf{E}$ and $\mathbf{S}$ rather than $\mathbf{D}$ and $\mathbf{k}$. The theory follows a development parallel to the theory presented in Chapter 13; therefore, the results will be presented here in summary form.

We can determine $\mathbf{E}$ in terms of $\mathbf{D}$ in a manner analogous to **(13-16)** by using the following definitions. The unit vector in the direction of energy flow is

$$\hat{\mathbf{s}} = \frac{\mathbf{S}}{|\mathbf{S}|}$$

and the ray, or energy, index of refraction is defined as

$$n_r = \frac{c}{v_r} \tag{13B-1a}$$

where v_r is called the *ray velocity*

$$\mathbf{v}_r = \frac{\mathbf{S}}{U} \tag{13B-1b}$$

This velocity, parallel to the Poynting vector, is related to the velocity of the phase front of a wave through the following relation:

$$v_p = v_r \frac{\hat{\mathbf{s}} \cdot \mathbf{k}}{k} = v_r \cos \phi$$

The relationship that yields $\mathbf{E}$ in terms of $\mathbf{D}$ is

$$\mathbf{E} = \frac{\mu_0 c^2}{n_r^2} [\mathbf{D} - \hat{\mathbf{s}}\,(\hat{\mathbf{s}} \cdot \mathbf{D})] \tag{13B-2}$$

In Figure 13B-1, the various vectors associated with an electromagnetic wave propagating in an anisotropic medium are displayed. The Poynting vector **(2-22)** is perpendicular to both $\mathbf{E}$ and $\mathbf{H}$ and because $\mathbf{H}$ is normal to the plane formed by $\mathbf{D}$, $\mathbf{E}$, and $\mathbf{k}$, the Poynting vector $\mathbf{S}$ will also be in that plane. However, unlike the case for propagation in an isotropic medium, **(2-23)**, $\mathbf{S}$ is not parallel to $\mathbf{k}$.

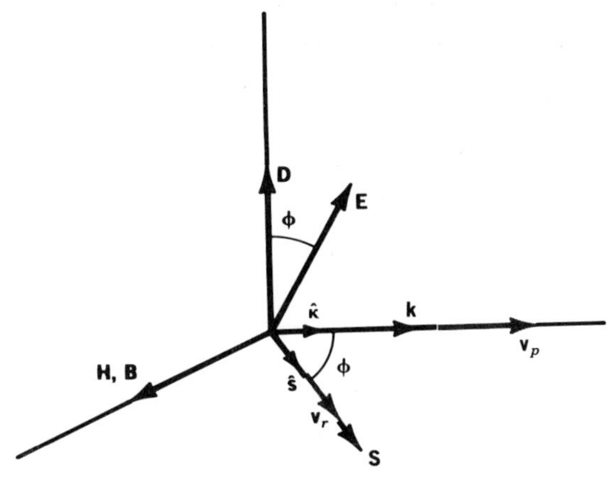

FIGURE 13B-1. Relationship between the various vectors used in the discussion of the propagation of light in an anisotropic medium.

A Fresnel equation equivalent to **(13-14)**, but based on the electric field representation of a wave, can be derived,[58] using an approach similar to the one used to obtain **(13-14)**. The resulting equation is called the *ray equation*.

$$\sum_{j=1}^{3} \frac{s_j^2}{v_r^2 - v_j^2} = \frac{1}{v_r^2} \tag{13B-3}$$

Because **(13B-3)** is quadratic in v_r^2, there are two roots v_{r1}^2 and v_{r2}^2 corresponding to two ray velocities. The *principal velocities* used in the ray equation

$$v_j^2 = \frac{\epsilon_0}{\epsilon_j}$$

are not components of a wave velocity but rather material constants.

RAY ELLIPSOID

If in the equation for the electric energy density

$$2U_E = \mathbf{D}\cdot\mathbf{E}$$

the electric field representation is used instead of the electric displacement, then **(13-6)** is replaced by

$$\epsilon_x \frac{E_x^2}{2U_E} + \epsilon_y \frac{E_y^2}{2U_E} + \epsilon_z \frac{E_z^2}{2U_E} = 1 \tag{13B-4}$$

This equation describes an ellipsoid representing the dielectric tensor and called the *ray* or *Fresnel ellipsoid*. It has semiaxes that are proportional to the ray velocities **(13B-1b)** rather than the index, as was the case for the optical indicatrix used in Chapter 13.

A plane cutting through this ellipsoid and passing through the origin, normal to **ŝ** (the unit vector parallel to the Poynting vector **S**), forms an ellipse at the intersection with the ellipsoid. The semiaxes of the ellipse have lengths proportional to the two ray velocities and directions parallel to $\mathbf{E}_1$ and $\mathbf{E}_2$.

Given the direction of **E**, we can find the direction of **D** using a construction equivalent to the one shown in Figure 13-14. In this construction, the normals to the planes that cut the ellipsoid at circular cross sections define the directions of the *optic ray axes* (or *biradials*). The optical ray axes are not in the same direction as the optical axes introduced in Chapter 13.

Appendix 13-C

NORMAL SURFACES

A geometric construction called the *normal surface* or *index surfaces* is quite helpful in understanding wave propagation in an anisotropic medium. The normal surfaces are constructed by plotting in the direction of **k** two vectors of length n_1 and n_2. The loci of the tips of all such vectors form the normal surfaces; see Figure 13C-1.

To determine the form of the normal surfaces, both sides of Fresnel's equation **(13-14)** are multiplied by n^2/k^2. The unit vector components $\kappa_i = k_i/k$ associated with the propagation vector are then defined, yielding the equation

$$\kappa_1^2 n_1^2 (n^2 - n_2^2)(n^2 - n_3^2) + \kappa_2^2 n_2^2 (n^2 - n_1^2)(n^2 - n_3^2) + \kappa_3^2 n_3^2 (n^2 - n_1^2)(n^2 - n_2^2) = 0 \quad \text{(13C-1)}$$

Biaxial Crystal

For a biaxial crystal, we retain the three indices of refraction in **(13C-1)**. To visualize the shape of the surfaces described by **(13C-1)** for a biaxial crystal, we look at a cross section of the volume enclosed by **(13C-1)** by assuming that $\hat{\boldsymbol{\kappa}} = (0, \kappa_y, \kappa_z)$

$$(n^2 - n_1^2)[\kappa_2^2 n_2^2 (n^2 - n_3^2) + \kappa_3^2 n_3^2 (n^2 - n_2^2)] = 0 \quad \text{(13C-2)}$$

The solutions of **(13C-2)** are

$$n^2 = n_1^2 \quad \text{(13C-3)}$$

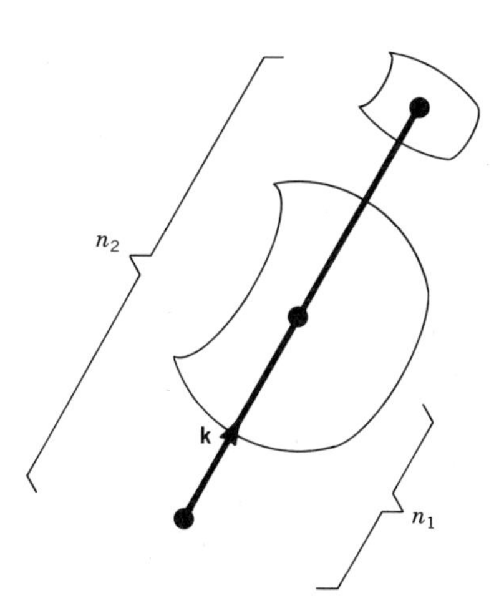

FIGURE 13C-1. Construction of normal surfaces.

$$\frac{1}{n^2} = \frac{\kappa_2^2}{n_3^2} + \frac{\kappa_3^2}{n_2^2} \tag{13C-4}$$

As can be seen from **(13C-3)**, the index n_1 is independent of the direction of propagation in the yz plane. The normal surface associated with this index is a circle in this plane. Equation **(13C-4)** is the equation of an ellipse, and if we assume that $n_1 < n_2 < n_3$, the ellipse encloses the circle of radius n_1.

For a biaxial crystal, the cross sections of the normal surfaces in all three coordinate planes can be constructed using the procedure just outlined for the yz plane. With some artistic license, the cross sections of the normal surfaces are displayed in Figure 13C-2; the differences between the indices have been exaggerated. The points of intersection of the two surfaces, labeled **A** in Figure 13C-2*b*, define the optical axes for the normal surfaces. There are two optical axes for a biaxial crystal. The two polarizations have the same refractive index when **k** is parallel to the direction defined by a line passing from **A** through the origin.

Uniaxial Crystal

To calculate the surfaces for a uniaxial crystal, we set $n_1 = n_2 = n_0$ and $n_3 = n_e$ in **(13C-1)**, yielding

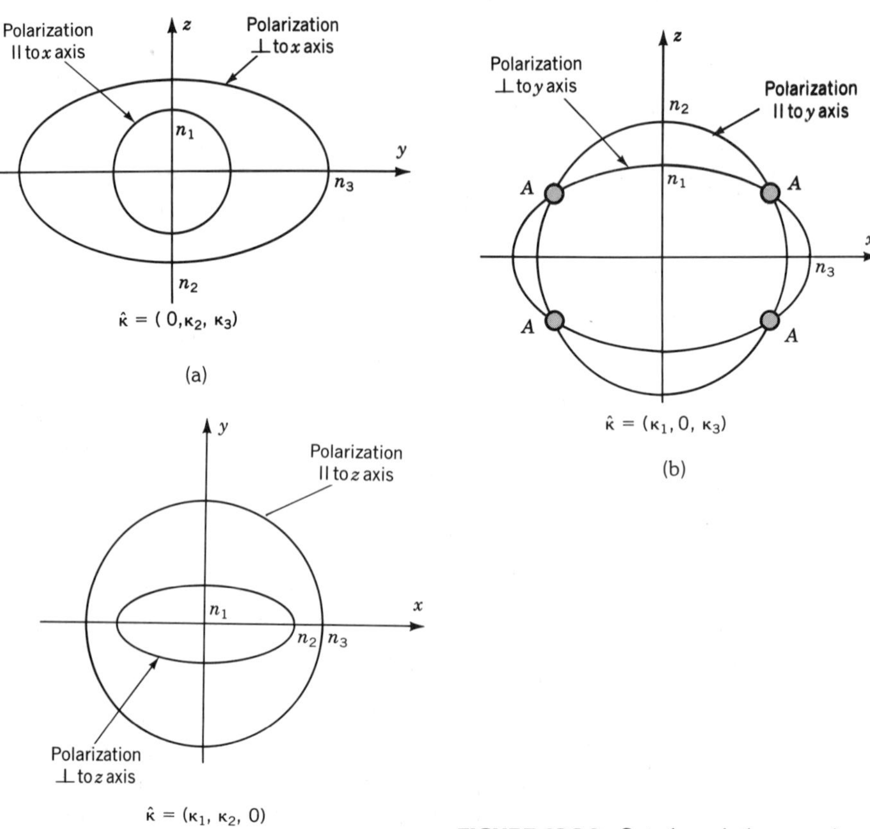

FIGURE 13C-2. Cuts through the normal surfaces for a biaxial crystal are shown in (a) the y, z plane, (b) x, z plane, and (c) x, y plane.

$$\kappa_1^2 n_0^2(n^2 - n_0^2)(n^2 - n_e^2) + \kappa_2^2 n_0^2(n^2 - n_0^2)(n^2 - n_e^2) + \kappa_3^2 n_e^2(n^2 - n_0^2)^2 = 0 \tag{13C-5}$$

Because the propagation in the x, y plane is independent of direction ($n_1 = n_2 = n_0$), we need only determine the two-dimensional normal surfaces in the x, z plane. Assume that a wave is propagating in the x, z plane at an angle θ with respect to the z axis

$$\hat{\boldsymbol{\kappa}} = (\sin\,\theta,\, 0,\, \cos\,\theta)$$

We may now rewrite **(13C-5)** as

$$(n^2 - n_0^2)[n_0^2(n^2 - n_e^2)\sin^2\,\theta + n_e^2(n^2 - n_0^2)\cos^2\,\theta] = 0$$

Solutions of this equation are

$$n = n_0^2 \tag{13C-6}$$

$$\frac{1}{n^2} = \frac{\sin^2\,\theta}{n_e^2} + \frac{\cos^2\,\theta}{n_0^2} \tag{13C-7}$$

The surfaces described by **(13C-6)** and **(13C-7)** are a circle and an ellipse. The points of intersection of the circle and ellipse define a single optical axis for a uniaxial crystal.

REFRACTION IN CRYSTALS

We have derived Snell's law for light crossing a boundary from one isotropic medium to another by a number of techniques; here, we introduce yet another approach. The objective of this derivation is to derive what happens when light crosses the boundary between an isotropic and anisotropic material.

We assume that a plane wave is incident, from the isotropic medium, onto a planar boundary at an incident angle of θ_i with respect to the normal of the boundary. We will require, as the boundary condition, that the phase of the plane wave not change as the boundary is crossed. The incident and transmitted waves are

$$\mathbf{E}_i = \mathbf{E}_0 \exp\,[i(\omega t - \mathbf{k}\boldsymbol{\cdot}\mathbf{r})], \qquad \mathbf{E}_t = \mathbf{E}_0 \exp\,[i(\omega t - \mathbf{k}'\boldsymbol{\cdot}\mathbf{r})]$$

and the boundary condition is

$$\omega t - \mathbf{k}\boldsymbol{\cdot}\mathbf{r} = \omega t - \mathbf{k}'\boldsymbol{\cdot}\mathbf{r} \tag{13C-8}$$

A unit vector $\hat{\boldsymbol{\kappa}}$ is defined in the propagation direction

$$\mathbf{k} = \frac{n\omega}{c}\,\hat{\boldsymbol{\kappa}} \tag{13C-9}$$

where n is the index of the medium. To simplify the derivation, we will assume that the index of refraction of the first isotropic medium is 1, leading to the boundary condition

$$\mathbf{r}\boldsymbol{\cdot}(n\hat{\boldsymbol{\kappa}}' - \hat{\boldsymbol{\kappa}}) = 0 \tag{13C-10}$$

The surface of the boundary is positioned at $z = 0$, with its normal parallel to the z axis, and the vector $\mathbf{r}$ lies the x, y plane. From **(13C-10)** we find that for this geometry, the vector $(n\hat{\boldsymbol{\kappa}}' - \hat{\boldsymbol{\kappa}})$ must be normal to the surface.

A geometrical construction involving the index surfaces can be used to determine the allowed values for $\hat{\boldsymbol{\kappa}}'$ and $\hat{\boldsymbol{\kappa}}$; see Figure 13C-3. (We will

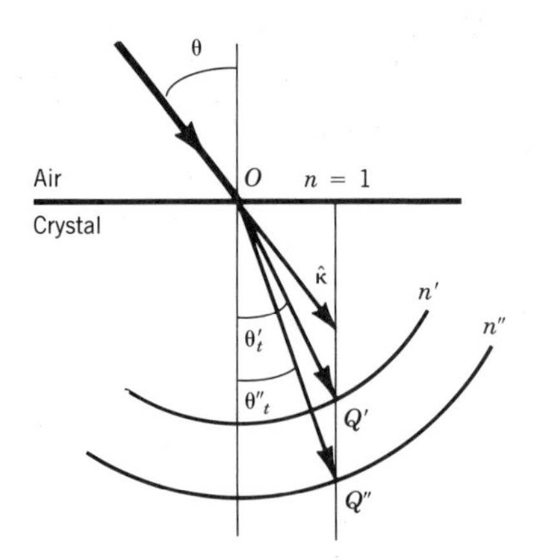

FIGURE 13C-3. Refraction at the boundary of an ansiotropic medium.

use the term "index surface" here to prevent confusion with the normal to a surface.) First, select a point O on the boundary and construct the two index surfaces centered on O. From O, construct the unit vector $\hat{\boldsymbol{\kappa}}$. The only vectors that meet the condition set by **(13C-10)** are those of length OQ' or OQ'' that terminate on the two index surfaces labeled n' and n''. Each incident wave will undergo double refraction and generate two refracted waves, $n'\hat{\boldsymbol{\kappa}}'$ and $n''\hat{\boldsymbol{\kappa}}''$.

Since we know that the frequency is unchanged across the boundary at $z = 0$, we can rewrite **(13C-8)** as

$$\mathbf{r}\boldsymbol{\cdot}\hat{\boldsymbol{\kappa}} = \mathbf{r}\boldsymbol{\cdot}n'\hat{\boldsymbol{\kappa}}' = \mathbf{r}''\boldsymbol{\cdot}n''\boldsymbol{\kappa}'' \tag{13C-11a}$$

$$x\kappa'_x + y\kappa'_y = n'(x\kappa'_x + y\kappa'_y) = n''(x\kappa''_x + y\kappa'_y) \tag{13C-11b}$$

This equation must be obeyed for all x and y, thus

$$\frac{\kappa'_x}{\kappa_x} = \frac{\kappa'_y}{\kappa_y} = \frac{\kappa''_x}{\kappa_x} = \frac{\kappa''_y}{\kappa_y}$$

From these relationships, we learn that the propagation vectors of the two transmitted waves must lie in the plane of incidence and that θ'_t and θ''_t, shown in Figure 13C-3, are given by

$$\frac{\sin\,\theta_i}{\sin\,\theta'_t} = n', \qquad \frac{\sin\,\theta_i}{\sin\,\theta''_t} = n'' \tag{13C-12}$$

The velocity of propagation is independent of direction for the spherical index surface (the ordinary wave) with index of refraction n'. For the orthogonal polarization, the index n'' is a function of direction, given by **(13C-7)** for a uniaxial crystal. Snell's law is no longer obeyed and the simple relationship must be replaced.

Appendix 13-D

RAY SURFACES

A useful construction that allows ray tracing in an anisotropic material is called the *ray surface*. It is based on the theoretical evaluation of wave propagation formulated around the orthogonal vectors **E** and **S**. In this appendix, we will see how the ray surfaces can be used along with Huygens' principle to follow the propagation of a ray in an anisotropic material.

In the same way that normal surfaces were generated in Appendix 13-C, we can generate ray surfaces. The procedure generates two surfaces by plotting in the direction of $\hat{\mathbf{s}}$ (the unit vector parallel to the Poynting vector **S**) the two ray velocities associated with that direction. Either of the surfaces, ray or normal, can be used to produce the other.

To generate the ray surfaces, we proceed as we did in Appendix 13-C for the normal surfaces. Starting with **(13B-3)**, we can generate a general equation for the ray surfaces

$$s_1^2 v_1^2 (v_r^2 - v_2^2)(v_r^2 - v_3^2) + s_2^2 v_2^2 (v_r^2 - v_1^2)(v_r^2 - v_3^2) + s_3^2 v_3^2 (v_r^2 - v_1^2)(v_r^2 - v_2^2) = 0 \quad \text{(13D-1)}$$

This equation has the same form as **(13C-1)** for the normal surfaces.

To determine the shape of the ray surfaces, a procedure, parallel to the one used in Appendix 13-C, is followed. The ray surfaces appear identical to the normal surfaces but they are not.

To demonstrate that the normal and ray surfaces are different, we will examine the cross section of the ray surfaces in the *xz* plane

$$\hat{\mathbf{s}} = (s_1, 0, s_3)$$

$$s_1^2 v_1^2 (v_r^2 - v_2^2)(v_r^2 - v_3^2) + s_3^2 v_3^2 (v_r^2 - v_1^2)(v_r^2 - v_2^2) = 0$$

One of the ray surfaces is described by the equation

$$v_r^2 = v_2^2 \quad \text{(13D-2)}$$

and the second surface is described by

$$\frac{s_1^2}{v_3^2} + \frac{s_3^2}{v_1^2} = \frac{1}{v_r^2} \quad \text{(13D-3)}$$

To display the ray surface, we assume that $v_1 > v_2 > v_3$. The ray surfaces just calculated for the *xz* plane are shown in Figure 13D-1. The points labeled *B* define the directions in which there is only one ray velocity. An optical axis is defined as the direction for which there is only one phase velocity, but the ray velocity is not normal to the phasefront and is thus not parallel to the phase velocity. Therefore, the directions determined by the points labeled *B* in Figure 13D-1 are not parallel to the optical axes.

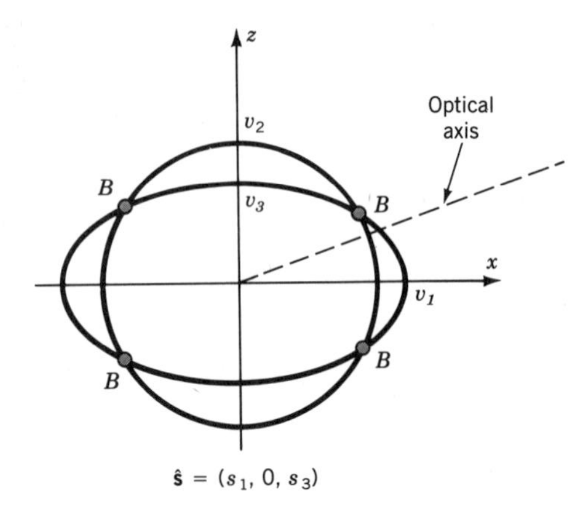

FIGURE 13D-1. Ray surface cross section for a uniaxial crystal. Note that the optical axis is not in the same direction as the direction for a single ray velocity.

The ray surface has the physical significance of being the locus of all points, reached in a unit time by a wave disturbance starting from a point source at the origin. Thus, the ray surface can be used with Huygens' principle to generate the ray path taken in an anisotropic medium. Figure 13D-2 displays how one would find the direction of the transmitted rays in a uniaxial crystal using the ray surfaces.

At time $t = 0$, the wavefront AA' reaches point O. At time t_1, point P' of the wavefront reaches point P. During the time t_1, two waves can be assumed to have radiated out from point O according to Huygens' principle. The ordinary ray surface would be a circle of radius PP'/n_0. The extraordinary ray surface would be an ellipse with major axis PP'/n_e. The refracted rays are the normals to planes from point P, tangent to the two ray surfaces. The direction of **D** is indicated on the two rays.

The ray geometry view of anistropy is useful for qualitative discussions of double refraction, but for quantitative work, the index geometry used in Appendix 13-C is more convenient. It is for this reason that we find an equivalent approach to the normal surfaces in other problems of wave propagation in anisotropic materials. For example, in solid state physics, the Fermi surface (an index surface) is used in the study of the propagation of electrons in solids.

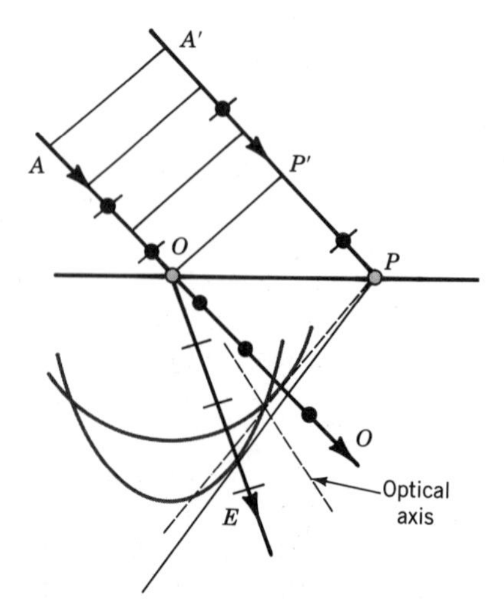

FIGURE 13D-2. Use of Huygens' principle in an anisotropic crystal.

Optical Modulation

Many applications incorporating optics require the capability of temporally modifying the amplitude, phase, direction, or frequency of the light wave. For example, in fiber optic communications, the amplitude of a light wave is encoded with a desired signal and the wave's frequency is adjusted to allow the multiplexing of a number of signals. In information storage, e.g., compact disks, a light wave's intensity is encoded with an audio signal that is then recorded by modifying the reflective properties of the disk. In printing and photocopying, a laser source's intensity is encoded with information while the beam is scanned over a photo-sensitive medium. Finally, for entertainment and displays, such as laser light shows, a number of laser beams are rapidly scanned over a display screen in artistic patterns.

When a light wave has one of its parameters modified with respect to time, it is said to be *modulated* and the device accomplishing this modification is called a *modulator*. Modulation may be accomplished by modifying the intensity of the light source, as is done, by varying the current in semiconductor injection lasers. The intensity, or the direction, of a wave can be modulated by mechanical means through the use of a simple shutter, a chopper (a disk with apertures cut in its circumference), or a moving optical component such as a mirror or hologram. Finally, an optical wave can be modulated by modifying the medium through which it propagates.

In this chapter, we will examine, in detail, the use of an electric, magnetic, or stress field to modify the optical properties of a medium. The application of one of these disturbances to the medium produces an anisotropy that requires a tensor description. The nonzero elements of the various tensors needed in this chapter are listed in Appendix 14-A.

The application of an electric field causes the electrons, ions, and permanent dipoles, if they exist, to reorient, inducing an electric polarization. The induced polarization produces an anisotropy that creates a birefringence. The field-dependent birefringence modifies the optical polarization of a light wave propagating through the medium. The theory describing the birefringence, induced by an electric field, will be developed by describing the changes in the optical indicatrix induced by the field. The theory will assume that the change in the index of refraction with applied field is small and can be described by a series expansion in powers of the applied electric field. The applied field contains both the optical field and an ac electric field. In this chapter, we will assume the optical field is small and neglect it. (In the next chapter, we will discuss the physical behavior of materials in the presence of large optical fields.)

In materials with a center of symmetry, only even powers of the ac electric field are found in the description of the index change. A very simple theory will be used to derive the quadratic dependence of the index of refraction. In materials without a center of symmetry, the quadratic field dependence of the index of refraction is overshadowed by a linear field dependence. A brief analysis will be made to demonstrate the modifications to the optical indicatrix that arise from this linear field dependence.

Both the linear and quadratic electric field effects induce a birefringence in the material. The induced birefringence results in an electric-field-dependent phase difference between two orthogonally polarized waves propagating in the material. If a polarized wave is incident on the material, the polarization of the wave exiting the material will be modified by an amount proportional to the electric field. The technique used to convert the modification of the polarization to an intensity modulation will be described and its implementation, in the design of an electrooptic modulator, is discussed.

To describe the modification of optical properties of a material by a magnetic field, we will add a static magnetic field to the molecular model introduced in Chapter 7. The magnetic field induces microscopic currents. The induced currents lead to an equation for the index of refraction that predicts different propagation velocities for right- and left-circularly polarized waves. The propagation velocities will be shown to depend on the "handedness" of the circularly polarized light, relative to the circulating currents.

Application of a stress to a material modifies the optical properties of the material by producing an anisotropy in the density of the material. The spatial density variation produces a birefringence in the material that can be treated in terms of a distortion of the optical indicatrix. The stress-induced birefringence modifies the polarization of a light wave as it propagates through the material. By observing the modification of polarized light by the material under stress, a measurement of the size and location of the stress can be obtained.

A time-varying stress can be treated as an acoustic wave. The theory used to explain light modulation by an acoustic wave is based on scattering theory. The scattering process modifies the direction and frequency of the light wave and the modification can be predicted through the use of the laws of conservation of energy and momentum. The use of an acoustooptic modulator as a beam deflector and a frequency modulator is discussed in light of the predictions of the scattering theory.

In Appendix 14-B, a phenomenological theory of the acoustooptic process will be given, based on the change in the index of refraction produced by the density modulation created by the sound wave. In Appendix 14-C, a figure of merit is defined that can be used to evaluate the performance of an acoustooptic device.

The first observation of externally modified optical properties was made by Michael Faraday in 1845. Faraday observed that the orientation of the plane of polarization of a light wave could be rotated by a nonactive material if a strong magnetic field was applied to the material, parallel to the light's propagation direction. This effect is now called the *Faraday* or *magnetooptic effect.*

In 1876 **John Kerr (1824–1907)** observed that optically isotropic materials exhibited uniaxial, birefringent behavior when placed in an electric field. This effect, called the *Kerr effect*, has a quadratic dependence on the applied electric field. A very simple theory will be presented that gives the proper wavelength dependence. The theory will be used to design an optical shutter.

Another electrooptic effect, first observed by **W.C. Roentgen** in quartz, is the linear electrooptic effect, called the *Pockels effect.* [Because the effect was studied extensively by **Carl Alwin Pockels (1865–1913)**, it bears his name.] This electrooptic effect varies linearly with the applied field.

In 1901 Kerr observed a second optical effect due to a magnetic field perpendicular to the optical propagation direction. The effect observed in fluids caused the material to exhibit a birefringence with a quadratic dependence on the magnetic field. This effect was studied, around 1905, by **Cotton** and **Mouton** and thus is associated with their names.

Brewster discovered in 1816 that an applied mechanical stress could modify the optical properties of a material. This behavior is known as *mechanical birefringence*, *photoelasticity*, or *stress birefringence*. The interaction's usefulness in optical modulation was suggested in 1921 by Brillouin, who developed a theory treating the compression waves of sound as a grating and describing the interaction of light and sound in terms of diffraction. Debye and Sears, and independently, Lucas and Biquard, observed the diffraction of light by sound waves in 1932.

ELECTROOPTIC EFFECT

In Chapter 7, a linear relationship between the applied electric field and material field **(7-16)** was adequate to describe the behavior of light propagating in a medium. The linear theory could be used because the only field under consideration was a small optical field. In this chapter, we will continue to assume that the optical field is quite small but will assume a second, larger, time-dependent electric field, the modulation field, is present. The modulation field increases the total applied electric field to a value sufficient to produce nonlinear behavior. The nonlinear behavior can be modeled by modifying the linear theory using a series expansion of the internal field, in terms of powers of the applied field.

The modulation field introduces a preferred direction, creating anisotropic behavior in the material under study. The dielectric constant generated by the model will therefore be a tensor, which will be shown to be symmetric if there are no losses in the material. An ellipsoid, similar to the one introduced in Chapter 13, can be used to describe the properties of the symmetric tensor. In this section, we will present the basic tensor equation for $1/n^2$ and simplify the equation by assuming that the applied light field is small. The resulting equation will be a function of only the modulation field.

We assume that two applied fields at frequencies ω_1, the modulation frequency, and ω_2, an optical frequency, are imposed on the optical material. The amplitude of the modulation field is assumed to be much larger than the light field $E_1 >> E_2$. The applied fields produce a small change in the index of refraction of the material so we may expand

$$\frac{1}{n^2} = \ni = \frac{\epsilon_0}{\epsilon}$$

the impermeability tensor in a power series of the total applied field.

$$\ni_{ik} = \frac{\epsilon_0}{\epsilon_{ik}} = \ni_{ik}^0 + \sum_{j=1}^{3} r_{ikj} E_j^{\omega} + \sum_{j=1}^{3} \sum_{m=1}^{3} \rho_{ikjm} E_j^{\omega} E_m^{\omega} + \cdots \tag{14-1}$$

where

$$\epsilon_{ik}^0 = \frac{\epsilon_0}{\ni_{ik}^0}$$

is the zero-field dielectric constant. The term linear in electric field describes the Pockels effect and the constants r_{ikj} are the *linear electrooptic coefficients* or the *Pockels coefficients*. The term displaying a quadratic dependence on the electric field describes the Kerr effect and the constants ρ_{ikjm} are the *quadratic electrooptic coefficients* or the *Kerr coefficients*. The definition of the Pockels and Kerr constants, in terms of the reciprocal dielectric tensor, allows the use of the index ellipsoid in the discussion of the electrooptical behavior of a crystal.

The applied field is made up of the optical and modulation fields

$$\mathbf{E}^{\omega} = \mathbf{E}^{\omega_1} \cos \omega_1 t + \mathbf{E}^{\omega_2} \cos \omega_2 t \tag{14-2}$$

where we have hidden any spatial dependence of the applied field in the amplitude. Using this applied field, we may rewrite **(14-1)** as

$$\begin{aligned}\ni_{ik} = \ni_{ik}^0 &+ \sum_{j=1}^{3} r_{ijk}(E_j^{\omega_1} \cos \omega_1 t + E_j^{\omega_2} \cos \omega_2 t) \\ &+ \sum_{j=1}^{3} \sum_{m=1}^{3} \frac{\rho_{ikjm}}{2} [(E_j^{\omega_1} E_m^{\omega_1} + E_j^{\omega_2} E_m^{\omega_2}) + E_j^{\omega_1} E_m^{\omega_1} \cos 2\omega_1 t \\ &\qquad + E_j^{\omega_2} E_m^{\omega_2} \cos 2\omega_2 t] \\ &+ \sum_{j=1}^{3} \sum_{m=1}^{3} \frac{\rho_{ikjm}}{2} (E_j^{\omega_1} E_m^{\omega_2} + E_j^{\omega_2} E_m^{\omega_1})[\cos(\omega_1 + \omega_2)t + \cos(\omega_1 - \omega_2)t]\end{aligned} \tag{14-3}$$

Each component of **(14-3)** is a function of the fundamental frequencies, their harmonics, or the sum and difference frequencies. In its most general form, we know from Chapter 13 that **(14-3)** must be a tensor. We will divide $\tilde{\ni}$ into a set of tensors associated with each of these frequencies

$$\tilde{\ni} = \tilde{\ni}^0 + \tilde{\ni}^{\omega_1} \cos \omega_1 t + \tilde{\ni}^{\omega_2} \cos \omega_2 t + \tilde{\ni}^{2\omega_1} \cos 2\omega_1 t + \cdots$$

Each tensor can be represented by an ellipsoid, as was done in Chapter 13, for naturally occurring birefringence. Each ellipsoid will have its own orientation, allowable planes of polarization, etc.

In this chapter, we will assume that the electric field associated with ω_1 is much larger than the optical field, $E^{\omega_1} >> E^{\omega_2}$, which will allow terms containing ω_2 to be ignored. By neglecting terms involving the optical field, we can write **(14-1)** as a function of only the modulation field $\omega_1 = \omega$

$$\mathfrak{z}_{ik} = \mathfrak{z}_{ik}^{0} + \sum_{j=1}^{3} r_{ikj} E_j \cos \omega t + \sum_{j=1}^{3} \sum_{m=1}^{3} \rho_{ikjm} E_j E_m (2 \cos^2 \omega t) + \cdots \quad (14\text{-}4)$$

where we have supressed the superscript ω_1. In the remainder of this chapter, the reader should remember that the applied field will only refer to the modulating field.

The coefficients of the two tensors $\tilde{r}$ and $\tilde{\boldsymbol{\rho}}$ are determined by the material's symmetry. Appendix 14-A contains tables showing the nonzero coefficients for the 32 crystal point groups that are optically distinguishable.

A *center of symmetry*, or *inversion center*, exists at a point if there is no change in the atomic arrangement when each atom, a vector $\mathbf{r}$ away from the point, is replaced by one $-\mathbf{r}$ away. If a material has a center of symmetry, reversing the field $\mathbf{E}$ should not change any physical property. Consider the one-dimensional case in which we assume that the application of a field $\mathbf{E}$ to a material produces a change in the index of refraction of

$$\Delta n_1 = rE$$

where r is a constant characterizing the linear electrooptic effect. If we reverse the applied field E, then

$$\Delta n_2 = -rE$$

However, if the material has inversion symmetry, $\Delta n_1 = \Delta n_2$ which implies $r = -r$. This can only be true for $r = 0$. Therefore, materials with inversion symmetry, for example, a liquid, will not exhibit a Pockels effect. The electrooptic effect displayed by materials with inversion symmetry is the Kerr effect. In materials with no inversion symmetry, the Kerr effect will be masked by the larger Pockels effect.

The derivation of the electrooptical properties of materials used in this chapter is based on an expansion of the impermeability tensor in terms of the electric field. An equivalent description, based on the material's dielectric polarization, can be made and is often encountered in the literature. The constants in the polarization-based expansion are called the *polarization-optic coefficients* and are simply related to the electrooptic coefficients that will be discussed here. The mathematical formalism associated with the dielectric polarization is identical to this presentation and the two descriptions can be interchanged. The similarity between the two descriptions make it unnecessary to discuss the polarization formalism.

ELECTROOPTIC INDICATRIX

In materials without a center of symmetry, the dominate interaction is the Pockels effect. The objective of the discussion in this section is to obtain an expression for the index of refraction change produced, through the Pockels effect, by the application of a modulation field to the material. In a later section, we will discuss how this index change can be used to produce amplitude modulation of a light wave.

The Pockels effect exhibits dispersion associated with a number of different physical processes. In the simple theory presented here, we will ignore dispersion. A brief discussion of the various physical processes leading to dispersion will explain why dispersion may be neglected.

All crystals that exhibit the linear electrooptic effect are also *piezoelectric*, that is, the application of an electric field will produce macroscopic deformations of the material called strains. In fact, the Pockels tensor and piezoelectric tensor have identical symmetries. The strains produced by the piezoelectric effect will modify the indices of refraction, producing a second interaction affecting the optical properties of the material, known as the *photoelastic effect* (or *strain-optic effect*). We can ignore the complications introduced by the feedback from the photoelastic effect, if we clamp the crystal or the modulation frequency is above the frequency of the fundamental mechanical resonance of the solid; typically mechanical resonances lie below 1 MHz.

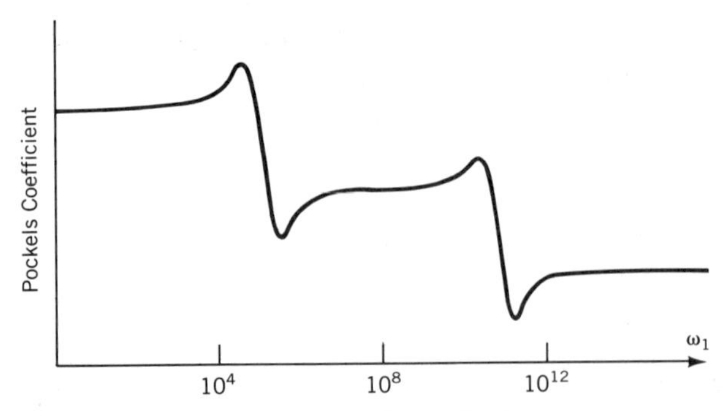

FIGURE 14-1. Dispersion curve for the Pockels coefficient for an idealized material. Only one mechanical resonance at about 1 MHz is shown. No electronic resonances are displayed.

Between 1 and 100 MHz, at harmonics of the mechanical resonance, the photoelastic effect can again become important but above 100 MHz, the crystal can no longer undergo macroscopic deformation. Above 1 THz, electron motion is the only contribution to the linear electooptical effect. Between 10 GHz and 1 THz, both atomic and electronic motion contribute to the field-induced polarization. In Figure 14-1, an idealized dispersion curve for the Pockels coefficient is displayed. We select an optical frequency ω_2 on the order of 10^{15} Hz and we restrict the range of modulation frequencies ω_1 to between 1 MHz and 10 GHz. By limiting the frequencies of the applied electric fields to these ranges, and by staying away from harmonics of the fundamental mechanical resonance, we eliminate the need to consider dispersion effects. Some values of Pockels coefficients for electrooptic materials used as optical modulators are shown in Table 14.1.

When an electric field is applied to a crystal, the general equation for the indicatrix can be obtained from **(13A-4)**

$$\sum_{i=1}^{3}\sum_{j=1}^{3}\sum_{k=1}^{3}\sum_{m=1}^{3}(\ni_{ij} + r_{ijk}E_k + \rho_{ijkm}E_kE_m)x_jx_i = 1 \qquad (14\text{-}5)$$

TABLE 14.1 Pockels Coefficients

Material	Crystal Class	n	r, 10^{-12} m/V
$BaTiO_2$	4mm	$n_1 = 2.480$ $n_3 = 2.426$	$r_{33} = 28$ $r_{13} = 8$ $r_{51} = 820$
$KNbO_3$	mm2	$n_1 = 2.279$ $n_2 = 2.329$ $n_3 = 2.167$	$r_{33} = 25$ $r_{13} = 10$
$LiNbO_3$	3m	$n_1 = 2.272$ $n_3 = 2.187$	$r_{33} = 30.8$ $r_{51} = 28$ $r_{13} = 8.6$ $r_{22} = 3.4$
ZnS	6mm	$n_0 = 2.135$	$r_{33} = 1.8$ $r_{13} = 0.9$
$Bi_{12}SiO_{20}$	23	$n_0 = 2.54$	$r_{41} = 5.0$
GaAs	$\bar{4}3m$	$n_0 = 3.6$	$r_{41} = -1.5$
Quartz	32	$n_0 = 1.544$ $n_e = 1.553$	$r_{41} = 1.4$ $r_{11} = 0.59$

where the time dependence of the electric field is not shown explicitly. The material is assumed to have no center of symmetry so that the linear electrooptic effect is dominate.

We define a coordinate system with the wave vector along x_3 and the applied field along x_1, as shown in Figure 14-2. The indicatrix cross section, taken through $x_3 = 0$, is described by the equation

$$\left(\frac{1}{n_1^2} + r_{11}E_1\right)x_1^2 + \left(\frac{1}{n_2^2} + r_{21}E_1\right)x_2^2 + 2r_{61}E_1x_1x_2 = 0 \tag{14-6}$$

We have used the definition $1/\ni = \epsilon/\epsilon_0 = n^2$ and are neglecting any simplification that may result due to crystal symmetry.

For the moment, we will ignore the term containing r_{61}. The effect of the term involving $r_{11}E_1$ on **(14-6)** is to modify the index of refraction along the x_1 axis

$$\frac{1}{(n_1 + \Delta n_1)^2} = \left(\frac{1}{n_1^2} + r_{11}E_1\right) \tag{14-7}$$

If we use the geometrical representation of the tensor, the dimension of the indicatrix along x_1 is changed by the applied field. Since $\Delta n_1 << n_1$ ($O[\Delta n] \approx 10^{-4}$ for fields below 20 kV/cm), we can use the binomial expansion to write

$$\frac{1}{(n_1 + \Delta n_1)^2} \approx \frac{1}{n_1^2} - \frac{2\Delta n_1}{n_1^3} = \left(\frac{1}{n_1^2} + r_{11}E_1\right)$$

The change in index due to the field is obtained by equating equivalent terms

$$\Delta n_1 \approx -\frac{n_1^3 r_{11}E_1}{2} \tag{14-8}$$

Using the same reasoning for the x_2 direction yields

$$\Delta n_2 \approx -\frac{n_2^3 r_{21}E_1}{2} \tag{14-9}$$

A polarized light wave propagating along the x_3 direction can be described by two linearly polarized waves: one with its electric field along the x_2 direction and the other with its electric field along the x_1 direction

$$E(x_2) = A_2 \exp[i(\omega t - k_0 n_2 x_3)], \qquad E(x_1) = A_1 \exp[i(\omega t - k_0 n_1 x_3)] \tag{14-10}$$

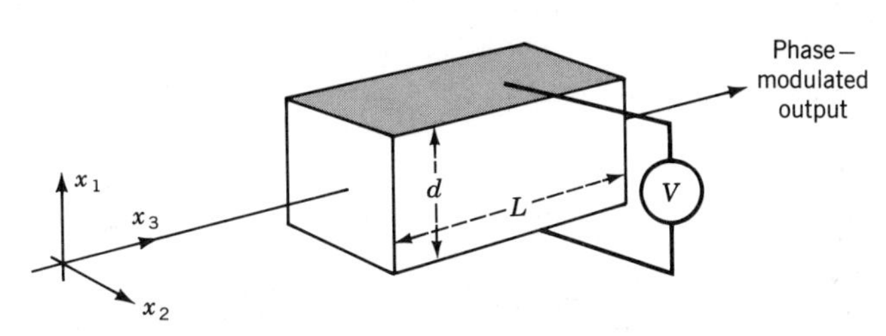

FIGURE 14-2. Geometry of electrooptic analysis.

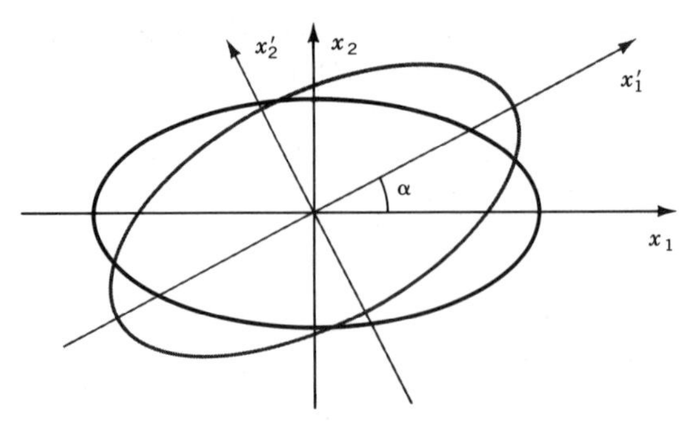

FIGURE 14-3. Rotation of indicatrix cross section by the applied electric field.

where k_0 is the magnitude of the wave vector in free space. The relative phase of the two orthogonal polarization components depends on n_i ($i =$ 1 or 2). After traveling a distance L in the x_3 direction in the crystal, the change in phase of the two polarization components in **(14-10)** is

$$-k_0 L[n_2 - (n_2 + \Delta n_2)] = -\frac{k_0 n_2^3 r_{21} E_1 L}{2}$$
$$-k_0 L[n_1 - (n_1 + \Delta n_1)] = -\frac{k_0 n_1^3 r_{11} E_1 L}{2} \tag{14-11}$$

If the applied field E_1 varies sinusoidally with time, then the relative phase between the two polarizations will also vary sinusoidally and phase modulation will occur. In a later section, we will describe how this phase modulation can be converted into amplitude modulation.

We now return to **(14-6)** to consider the previously ignored third term, $2r_{61}E_1x_1x_2$. The effect of this term on the elliptical cross section of the indicatrix, described by **(14-6)**, is to rotate the axes of the ellipse through an angle α (see Figure 14-3), where α is given by

$$\tan 2\alpha = \frac{2r_{61}E_1}{(1/n_1^2) - (1/n_2^2) + (r_{11} - r_{21})E_1} \tag{14-12}$$

The angle α is small if the natural birefringence of the crystal $(n_1 - n_2)$ is large. Any change in n_1 and n_2 due to the rotation is a second-order effect and can be negligible. However, we cannot simply ignore the effect of $r_{61}E_1$ because the index of refraction for a wave polarized along the direction 45° to x_1 and x_2 (denoted by n') will vary linearly with a change in E_1

$$\Delta n' = \frac{n'^3 r_{61} E_1}{2} \tag{14-13}$$

To eliminate the complication introduced by the third term in **(14-6)**, a coordinate transformation is found that eliminates all cross products x_ix_j. We will see how this step is accomplished a little later.

KERR EFFECT

When a material has inversion symmetry, the Pockels effect will be absent and the observable electrooptic effect will be the Kerr effect. The Kerr effect was the first electrooptic process to find application, due to the fact it was easy to obtain a modulator material with good optical properties. A liquid or glass can be used as a modulator material, and because natural birefringence is absent, the modulator constructed with these isotropic materials can handle a diverging, or converging, beam of light. The drawback of this effect is the large electric fields that must be used and the fact that the modulation is dependent on the square of the applied electric field. Today, because of the availability of large, strainfree, transparent crystals, the Pockels modulator has replaced the Kerr device in many applications.

In this section, we will derive a relationship for the index of refraction change due to the quadratic electrooptic effect. The theory presented here is a phenomenological theory that cannot be used for predictive purposes. However, it does provide some justification for the wavelength dependence of the Kerr constant observed in liquids.

One type of material with inversion symmetry is an isotropic material such as a fluid. The fieldfree dielectric constant of an isotropic material is a scalar so that **(14-4)** must be rewritten as

$$\backepsilon_{ik} = \backepsilon^0 + \sum_{j=1}^{3} \sum_{m=1}^{3} \rho_{ikjm} E_j E_m \tag{14-14}$$

From Table 14A.3 in Appendix 14-A, the nonzero coefficients of the Kerr tensor, for an isotropic medium, can be used to write **(14-14)** in matrix form

$$\begin{bmatrix} \backepsilon_{11} \\ \backepsilon_{12} \\ \backepsilon_{13} \\ \backepsilon_{14} \\ \backepsilon_{15} \\ \backepsilon_{16} \end{bmatrix} = \begin{bmatrix} \backepsilon^0 \\ \backepsilon^0 \\ \backepsilon^0 \\ 0 \\ 0 \\ 0 \end{bmatrix} + \begin{bmatrix} \rho_{11} & \rho_{12} & \rho_{12} & 0 & 0 & 0 \\ \rho_{12} & \rho_{11} & \rho_{12} & 0 & 0 & 0 \\ \rho_{12} & \rho_{12} & \rho_{11} & 0 & 0 & 0 \\ 0 & 0 & 0 & \rho_{11}-\rho_{12} & 0 & 0 \\ 0 & 0 & 0 & 0 & \rho_{11}-\rho_{12} & 0 \\ 0 & 0 & 0 & 0 & 0 & \rho_{11}-\rho_{12} \end{bmatrix}$$

$$\times \begin{bmatrix} E_1^2 \\ E_2^2 \\ E_3^2 \\ E_2E_3 \\ E_3E_1 \\ E_1E_2 \end{bmatrix} = \begin{bmatrix} \rho_{11}E_1^2 + \rho_{12}E_2^2 + \rho_{12}E_3^2 \\ \rho_{12}E_1^2 + \rho_{11}E_2^2 + \rho_{12}E_3^2 \\ \rho_{12}E_1^2 + \rho_{12}E_2^2 + \rho_{11}E_3^2 \\ (\rho_{11} - \rho_{12})E_2E_3 \\ (\rho_{11} - \rho_{12})E_3E_1 \\ (\rho_{11} - \rho_{12})E_1E_2 \end{bmatrix}$$

By orienting the coordinate system along the principal axes [see **(13A-5)**], we can make all terms in the tensor zero except for the principal elements, which, as we see in the above matrix, are all equal

$$\rho_{11} = \rho_{22} = \rho_{33} = \rho$$

We orient the applied electric field along the z axis ($E_1 = E_2 = 0,\ E_3 = E$), reducing **(14-14)** to three equations. The first of these equations yields the index of refraction parallel to the electric field

$$\backepsilon_{\parallel} = \backepsilon^0 + \rho E^2 = \frac{1}{n_e^2} \tag{14-15}$$

The second and third equations are identical and represent the index of refraction normal to the electric field

$$\backepsilon_{\perp} = \backepsilon^{0} = \frac{1}{n_0^2} \tag{14-16}$$

Thus, in the presence of an applied electric field, the isotropic material behaves like a uniaxial crystal with an optical axis parallel to the electric field. The change from the field free index of refraction n_0 to n_e is small, allowing the use of the binomial expansion to write

$$\begin{aligned} \frac{1}{n_e^2} &= \frac{1}{(n_0 + \Delta n_0)^2} \\ &= \frac{1}{n_0^2} - \frac{2\Delta n_0}{n_0^3} \\ &= \frac{1}{n_0^2} + \rho E^2 \end{aligned}$$

Experimentally, it has been determined that the index difference between the ordinary and extraordinary indices, due to the induced birefringence, is

$$\Delta n = \mathsf{K} E^2 \lambda$$

where K is called the *Kerr constant*. This index difference produces a phase difference between two orthogonally polarized waves of

$$\begin{aligned} \Gamma &= 2\pi \frac{\Delta n L}{\lambda} \\ &= 2\pi \mathsf{K} E^2 L \end{aligned}$$

From the above derivation, the relative phase difference between the ordinary and extraordinary waves produced by the Kerr effect is

$$\begin{aligned} \Gamma &= k_0(n_e - n_0)L \\ &= k_0(n_0 + \Delta n_0 - n_0)L \\ &= \frac{2\pi}{\lambda}\Delta n_0 L \\ &= \frac{\pi}{\lambda} n_0^3 \rho E^2 L \end{aligned} \tag{14-17}$$

Equating this result to the experimentally determined relationship yields a Kerr constant that is given by

$$\mathsf{K} = \frac{n_0^3 \rho}{2\lambda}$$

A molecular theory of the Kerr effect in fluids also predicts the λ^{-1} dependence.[59]

We can define a *half-wave voltage* as that voltage when applied to a Kerr cell with thickness d and length L that gives $\Gamma = \pi$

$$\mathsf{V}_{\lambda/2} = \frac{d}{\sqrt{2\mathsf{K}L}} \tag{14-18}$$

TABLE 14.2 Kerr Coefficient

Type	Material	K,[a] esu	$V_{\lambda/2}$, kV	
			3.2×5 cm	0.3×0.1 cm
Liquid	Nitrobenzene	33.0×10^{-6}	48.9	6.2
	o-Dichlorobenzene	4.6	120.2	16.5
	Acrylonitrile	7.8	97.9	12.5
	Benzonitrile	12.1	79.2	10.3
	Acetonitrile	4.7	128	16
Gas	CO_2 (STP)	2.5×10^{-11}	—	—
Solid	Glass (typical)	9.0×10^{-9}	—	—

[a]The units are those usually found in the literature. To convert to mks units, divide by 9×10^6. The Kerr coefficient then has units of m/V^2.

The Kerr effect parameters for some liquids used in Kerr modulators are listed in Table 14.2. The half-wave voltage for Kerr cells of two different dimensions are also listed.

AMPLITUDE MODULATION

We saw from **(14-11)** and **(14-17)** that the output wave of an electrooptic crystal was phase-modulated by the applied electric field. Intensity modulation of the light wave is of more general use; therefore, we wish to convert the phase modulation of the transmitted wave to an amplitude modulation. To do this, we use the experimental arrangement shown in Figure 14-4.

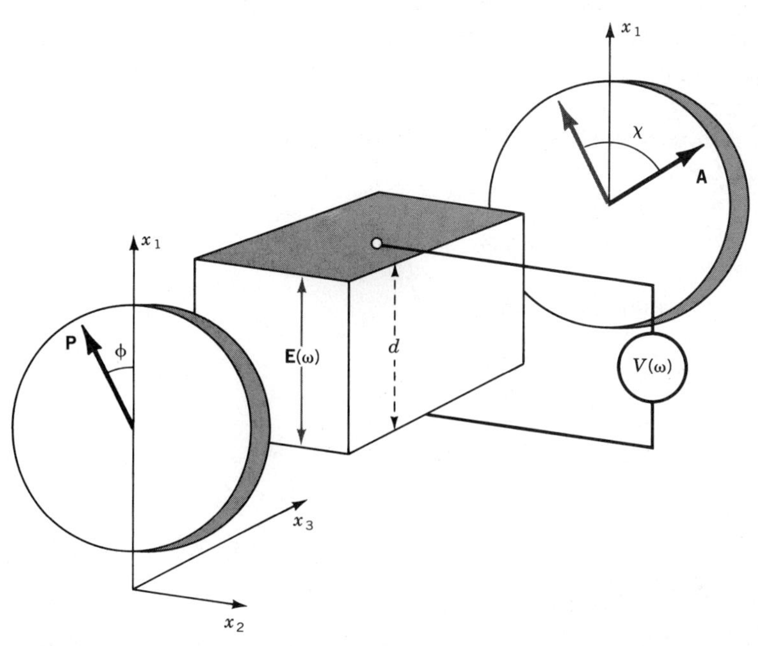

FIGURE 14-4. Use of a polarizer and analyzer to convert phase modulation to amplitude modulation. The polarizer is shown as a disk in the foreground of the picture and the analyzer is shown as a disk in the background. The parallelepiped represents the electrooptic modulator. The electric field is applied parallel to the x_1 axis and light propagates along the x_3 axis. The vectors, labeled **P** and **A**, denote the orientation of the electric field of a linearly polarized light wave when transmission through the polarizers is a maximum.

We first produce linearly polarized light by passing the incident light wave through a linear polarizer (see Figure 14-4). The polarizer is orientated so that its polarization axis makes an angle ϕ with the x_1 axis. (You will recall from the previous chapter that the polarization axis of a linear polarizer is defined as the direction, parallel to the electric field of a linearly polarized wave, when the transmission of that wave through the polarizer is a maximum.) The next step is to pass the polarized light through the modulator and at the same time apply an electric field to the modulator, parallel to the x_1 axis. Upon leaving the modulator, the wave passes through a second linear polarizer, called an *analyzer*, to differentiate the second polarizer from the first.

The modulator performs the role of a dynamic optical retarder and converts the linearly polarized wave into an elliptically polarized wave. The analyzer transmits one component of the elliptically polarized light. In this section, we will derive an equation for the intensity of the light transmitted by the analyzer as a function of the voltage applied to the modulator.

To analyze the total optical system, we will use the coordinate system shown in Figure 14-4 and redrawn in Figure 14-5. The applied electric field establishes a unique direction in the crystal along the x_1 axis. We will derive the phase difference between a wave polarized with its **D** vector parallel to the x_1 axis and one with its **D** vector polarized parallel to the x_2 axis, perpendicular to x_1. The incident wave is propagating along the x_3 axis. The initial polarizer is oriented such that the polarization axis makes an angle ϕ with respect to $\mathbf{D}_1$. The direction of the polarization axis is labeled **OP** in Figure 14-5. The analyzer, at the exit of the crystal, is oriented such that its polarization axis, **OA** in Figure 14-5, makes an angle χ with respect to **OP**. After passing through the polarizer, the light incident on the crystal has an amplitude component parallel to $\mathbf{D}_1$ given by $\mathcal{A} \cos \phi$ and in the $\mathbf{D}_2$ direction given by $\mathcal{A} \sin \phi$, where $\mathcal{A}$ is the wave amplitude after propagating through the polarizer.

After propagating through the crystal, the light is incident on the analyzer that only transmits those components of polarization parallel to **OA**. The component along **OA** due to $\mathbf{D}_1$ is

$$A \cos \phi \cos(\phi - \chi) \tag{14-19a}$$

and the component due to $\mathbf{D}_2$ is

$$A \sin \phi \sin(\phi - \chi) \tag{14-19b}$$

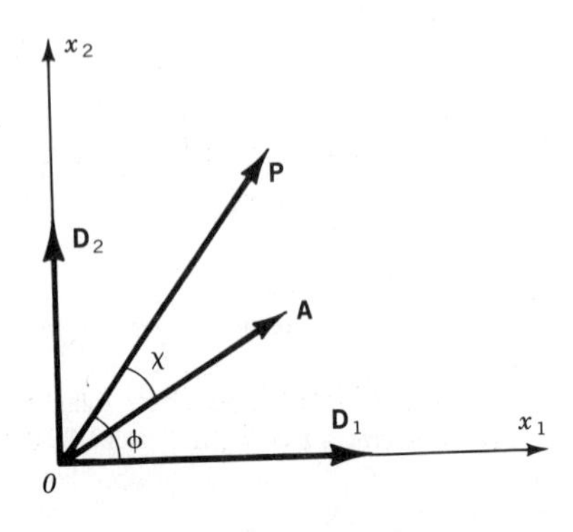

FIGURE 14-5. Geometry associated with an electrooptic modulator.

where A is the amplitude of the wave after propagating through the electrooptic crystal. We can write the phase difference between the components of $\mathbf{D}_1$ and $\mathbf{D}_2$, as we did in **(14-11)**

$$\begin{aligned}\Gamma &= k_0L[(n_1 + \Delta n_1) - (n_2 + \Delta n_2)] \\ &= k_0L[(n_1 - n_2) + (\Delta n_1 - \Delta n_2)]\end{aligned} \tag{14-20}$$

The components of $\mathbf{D}_1$ and $\mathbf{D}_2$ along the direction **OA** interfere coherently and yield, according to **(4-13)**, an intensity

$$I = A^2\{\cos^2\phi\cos^2(\phi - \chi) + \sin^2\phi\sin^2(\phi - \chi)$$

$$+2\cos\phi\sin\phi[\cos(\phi - \chi)\sin(\phi - \chi)\cos\Gamma]\}$$

We use the identity

$$\cos\Gamma = 1 - 2\sin^2\frac{\Gamma}{2}$$

to simplify the equation for the intensity

$$I = A^2\left[\cos^2\chi - \sin 2\phi\sin 2(\phi - \chi)\sin^2\frac{\Gamma}{2}\right] \tag{14-21}$$

Three limiting cases of **(14-21)** exist:

1. The electrooptic crystal is not present so that $\Gamma = 0$

$$I = A^2\cos^2\chi \tag{14-22}$$

   This equation is known as *Malus' law*. When $\chi = \pi/2$, the polarizers are said to be *crossed* and no light propagates through the polarizer/analyzer pair. When $\chi = 0$, the polarizers are said to be *parallel* and the maximum amount of light is transmitted by the pair.

2. The analyzer and polarizer are aligned so that their axes are parallel, $\chi = 0$ and $\Gamma \neq 0$. For the condition of parallel polarizers, the transmitted intensity is

$$I_\parallel = A^2\left(1 - \sin^2 2\phi\sin^2\frac{\Gamma}{2}\right) \tag{14-23}$$

3. The analyzer and polarizer are aligned perpendicular so that $\chi = \pi/2$ and $\Gamma \neq 0$. For the condition of crossed polarizers, the transmitted intensity is

$$I_\perp = A^2\sin^2 2\phi\sin^2\frac{\Gamma}{2} \tag{14-24}$$

The intensity $I_\parallel$ is the complement of $I_\perp$. The maximum of $I_\parallel$ (minimum of $I_\perp$) occurs whenever

$$\phi = 0, \frac{\pi}{2}, \pi, \ldots$$

The angle ϕ assumes these values whenever the polarizer is aligned parallel to $\mathbf{D}_1$ or $\mathbf{D}_2$. With the polarizer in these orientations, no light modulation takes place.

The minimum of $I_{\parallel}$ (maximum of $I_{\perp}$) occurs when

$$\sin 2\phi = \pm 1$$

$$\phi = \frac{\pi}{4}, \frac{3\pi}{4}, \frac{5\pi}{4}, \ldots$$

This orientation corresponds to the polarization axis aligned at 45° angle with respect to the electric field direction. When the polarizer is in this orientation,

$$I_{\parallel \min} = A^2\left(1 - \sin^2 \frac{\Gamma}{2}\right)$$

$$= A^2 \cos^2 \frac{\Gamma}{2} \tag{14-25a}$$

$$I_{\perp \max} = A^2 \sin^2 \frac{\Gamma}{2} \tag{14-25b}$$

These equations can be given a physical interpretation if we use **(14-11)** or **(14-17)** for the phase difference produced by the field-induced birefringence.

Kerr Modulation

If the parallelepiped in Figure 14-4 contains a Kerr medium, then the device shown in Figure 14-4 is called a Kerr cell. We can calculate the light intensity transmitted by the device as a function of the applied voltage by replacing the phase difference Γ in **(14-25)** with **(14-17)**

$$\Gamma = \frac{\pi}{\lambda} n_0^3 \rho E^2 L$$

$$= 2\pi E^2 KL$$

$$= 2\pi\left(\frac{V}{d}\right)^2 KL$$

Using **(14-18)**, we can write the phase difference as a function of the half-wave voltage of the Kerr cell

$$\Gamma = \pi\left(\frac{V}{V_{\lambda/2}}\right)^2$$

With this relationship for the phase difference, we can rewrite **(14-25b)** as

$$I_{\perp \max} = A^2 \sin^2 \frac{\pi}{2}\left(\frac{V}{V_{\lambda/2}}\right)^2 \tag{14-26}$$

The output intensity of the Kerr cell has the functional form shown in Figure 14-6. For comparison, the graph also displays the output intensity of a Pockels cell. The equation for the Pockels cell curve will be derived next.

Pockels Modulation

If the parallelepiped in Figure 14-4 contains a material that exhibits the Pockels effect, then we need to use **(14-20)** in **(14-25)**. By using the trigonometric identities

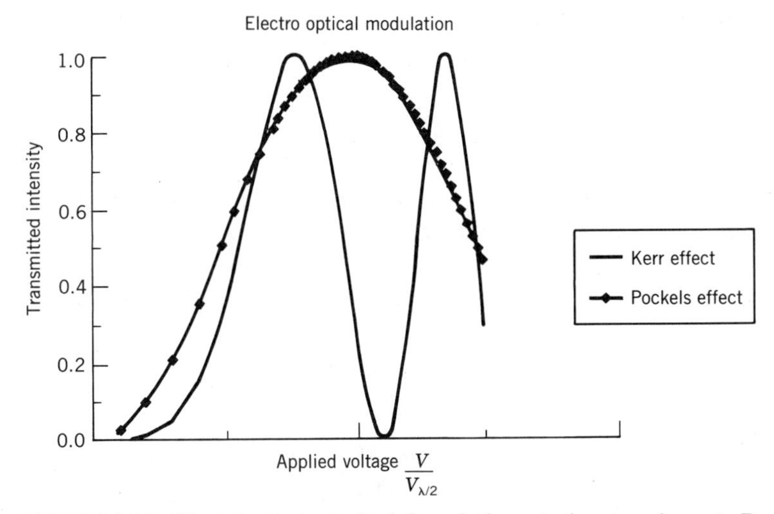

FIGURE 14-6. The intensity transmitted through the optical system shown in Figure 14-4. Curves for idealized Kerr and Pockels media are shown. The linearity of the two effects should be compared between zero and the first intensity maximum.

$$\cos(a + b) = \cos a \cos b - \sin a \sin b$$

$$\sin(a + b) = \sin a \cos b + \cos a \sin b$$

we can express the sinusoidal functions of the phase difference Γ in terms of the index modulation

$$\cos \frac{\Gamma}{2} = \cos \frac{k_0 L}{2}(n_1 - n_2) \cos \frac{k_0 L}{2}(\Delta n_1 - \Delta n_2) - \sin \frac{k_0 L}{2}(n_1 - n_2) \sin \frac{k_0 L}{2}(\Delta n_1 - \Delta n_2)$$

$$\sin \frac{\Gamma}{2} = \sin \frac{k_0 L}{2}(n_1 - n_2) \cos \frac{k_0 L}{2}(\Delta n_1 - \Delta n_2) + \cos \frac{k_0 L}{2}(n_1 - n_2) \sin \frac{k_0 L}{2}(\Delta n_1 - \Delta n_2)$$

This result for Pockels modulation is much more complicated than the result for Kerr modulation. Kerr modulation is simple because the fieldfree material is isotropic and the addition of an electric field creates an optically uniaxial material. The Pockels material may be birefringent in the fieldfree condition and the application of a field to a birefringent material will produce, in general, a biaxial medium. The added complexity, introduced by possible biaxial behavior, requires more effort to understand the physical significance of the Pockels modulation result.

The expressions for the index modulation Δn_1 and Δn_2 are found in **(14-8)** and **(14-9)**. The contribution from $(n_1 - n_2)$ may be used by the experimenter to control the functional form of the intensity output of the Pockels cell.

Several special cases will aid in interpretating the results we have just derived for the Pockels effect modulator. Assume initially that the applied electric field is such that $\Delta n_1 = \Delta n_2$. This could occur when a time-varying electric field passes through zero. Whenever $\Delta n_1 = \Delta n_2$, we have

$$\cos^2 \frac{\Gamma}{2} = \cos^2 \frac{k_0 L}{2}(n_1 - n_2), \qquad \sin^2 \frac{\Gamma}{2} = \sin^2 \frac{k_0 L}{2}(n_1 - n_2)$$

The fieldfree material produces a phase difference equivalent to that produced by a retarder plate **(13-21)**

$$\frac{k_0 L}{2}(n_1 - n_2) = \pi N$$

where N is the retardation defined in **(13-20)**. We will examine the modulation produced by the Pockels effect for three different retardations of the fieldfree material.

Full-Wave Retardation

$$N = 1, 2, \ldots$$

The intensity transmitted by the optical system, shown in Figure 14-4, is

$$I_{\parallel \min} = A^2 \cos^2\left[\frac{k_0 L}{4}(n_2^3 r_{21} - n_1^3 r_{11})E_1\right] \tag{14-27a}$$

$$I_{\perp \max} = A^2 \sin^2\left[\frac{k_0 L}{4}(n_2^3 r_{21} - n_1^3 r_{11})E_1\right] \tag{14-27b}$$

The geometry of Figure 14-4, allows the electric field to be written in terms of the applied voltage $E_1 = V/d$ so that **(14-27b)** becomes

$$I_{\perp \max} = A^2 \sin^2\left[\frac{k_0 L}{4d}(n_2^3 r_{21} - n_1^3 r_{11})V\right] \tag{14-28}$$

The intensity variation of **(14-28)** is the curve labeled Pockels effect in Figure 14-6. It describes the transmission of the optical system in Figure 14-4 when the polarizers are crossed.

Half-Wave Retardation

$$N = \frac{1}{2}, \frac{3}{2}, \ldots$$

The transmitted intensity is

$$I_{\parallel \min} = A^2 \sin^2\left[\frac{k_0 L}{4}(n_2^3 r_{21} - n_1^3 r_{11})E_1\right] \tag{14-29a}$$

$$I_{\perp \max} = A^2 \cos^2\left[\frac{k_0 L}{4}(n_2^3 r_{21} - n_1^3 r_{11})E_1\right] \tag{14-29b}$$

If the fieldfree retardation is a half-wave, the transmitted intensity is the complement of the full-wave retardation case. The transmitted intensity, when the polarizer and analyzer are parallel, for half-wave retardation is identical to **(14-28)**.

Quarter-Wave Retardation

$$N = \frac{1}{4}, \frac{3}{4}, \ldots$$

For this fieldfree condition, the amplitude transmission of the optical system, shown in Figure 14-4, is

$$\left.\begin{matrix}\sqrt{I_{\perp \max}} \\ \sqrt{I_{\parallel \min}}\end{matrix}\right\} =$$

$$\frac{E}{\sqrt{2}}\left\{\cos\left[\frac{k_0 L}{4}(n_2^3 r_{21} - n_1^3 r_{11})E_1\right] \pm \sin\left[\frac{k_0 L}{4}(n_2^3 r_{21} - n_1^3 r_{11})E_1\right]\right\}$$

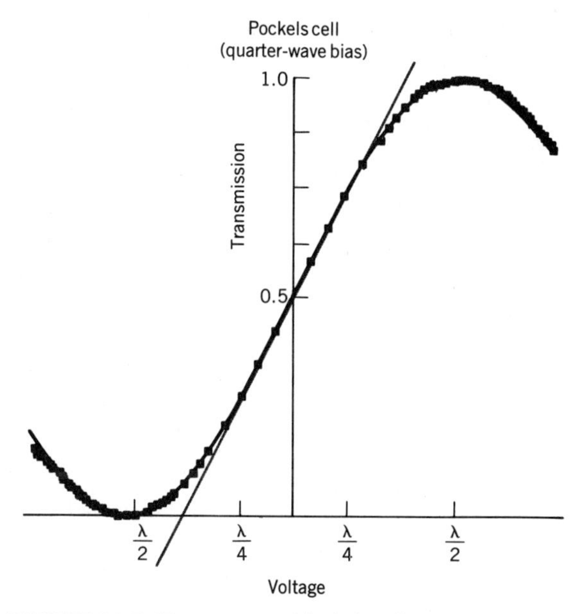

FIGURE 14-7. Transmission of Pockels cell as predicted by (14-30). A straight line has been drawn through the curve to demonstrate the ability to generate an intensity transmission that varies linearly with modulation voltage.

and the intensity transmitted by the system is given by

$$\left.\begin{array}{l} I_{\perp\max} \\ \\ I_{\|\min} \end{array}\right\} = \frac{A^2}{2}\left[1 \pm \sin\frac{k_0 L}{2}(n_2^3 r_{21} - n_1^3 r_{11})E_1\right] \tag{14-30}$$

The transmitted intensity predicted by **(14-30)** is shown in Figure 14-7. The quantity

$$\frac{k_0 L}{2}(n_1 - n_2)$$

is called the *modulation bias* of the modulation system. Any bias can be selected, even those we have not considered, but normally a bias of $\pi/4$ is the most useful. The straight line drawn on the plot of **(14-30)** in Figure 14-7 displays the reason for the selection of a quarter-wave bias. With this bias, the output intensity of the Pockels cell is proportional to the applied modulation voltage, if the applied voltage swing is not too large.

If the bias is positioned at the 50% transmission point in Figure 14-7, then linear modulation of the light intensity can be obtained. For example, if a sine wave modulation voltage creates a transmissivity variation between 12.5 and 87.5% of the transmission maximum, the frequency of the modulated light wave will contain less than 3% of higher harmonics of the modulation frequency.

The bias may be controlled by three methods. If the modulation crystal is birefringent in zero field, then the length of the modulation crystal can be selected to satisfy the equation

$$L = \frac{\lambda N}{n_1 - n_2} \tag{14-31}$$

A second method of obtaining the bias is through the application to the electrooptic crystal of a dc electric potential, along with the ac modulation field. Figure 14-8 displays the use of a dc bias voltage, in addition to the ac modulation voltage, to position the modulation on the most linear region of the transmission curve.

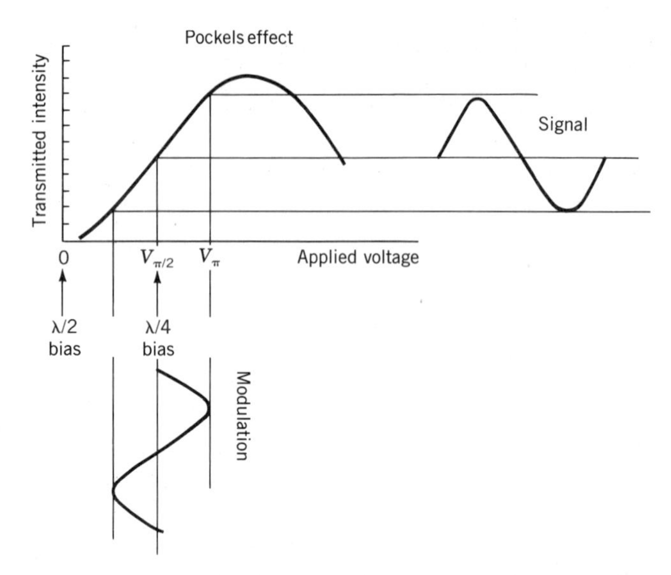

FIGURE 14-8. dc Field bias of an electrooptic modulator to produce an intensity modulation that is a linear representation of the modulation voltage.

The third method of producing a bias in an electrooptic modulator is the use of an optical retarder between the polarizer and analyzer (see Figure 14-9). This method of bias application converts the linearly polarized input wave into a circularly polarized wave, through the use of a quarter-wave plate, with its fast axis oriented at a 45° angle to the plane of polarization. The circularly polarized light is converted into elliptically polarized light by the modulator. The analyzer then transmits the component of the elliptically polarized wave parallel to its axis.

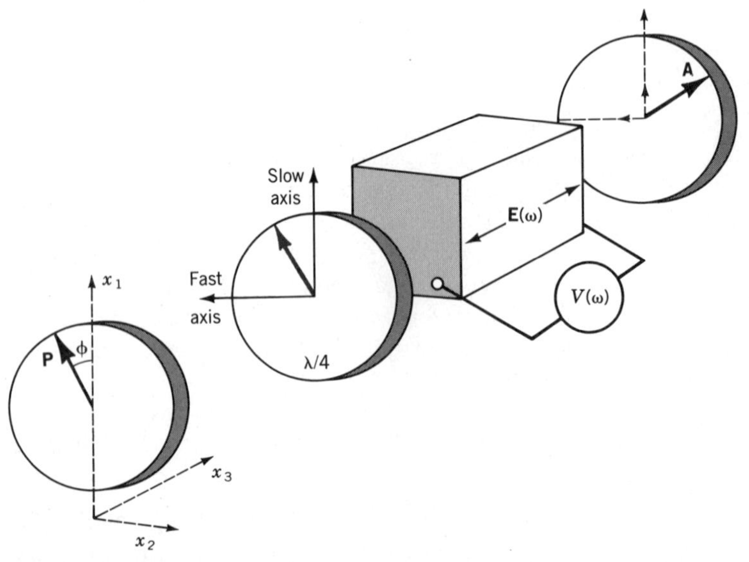

FIGURE 14-9. The use of a quarter-wave retarder to adjust the modulation bias of an electrooptic modulator.

MODULATOR DESIGN

In this section, and in the problems at the end of the chapter, we will discuss the design of two types of electrooptic modulators. The Kerr modulator will be designed using nitrobenzene as the Kerr liquid. The Pockels modulator will use one of the most popular crystal types: the $\bar{4}2m$ symmetry group. (A discussion of crystal structure, a definition of the 32 crystal point groups, and a list of their symmetries can be found in most introductory solid-state physics books.[28])

Jones calculus can be used to describe the electrooptic modulator in the same way as it can be used to describe static polarizing components. If δ is the retardation bias and Γ the phase shift introduced by the electrooptic modulator, then the Jones matrix is

$$\begin{pmatrix} e^{i(\delta/2)} & 0 \\ 0 & e^{-i(\delta+\Gamma)/2} \end{pmatrix} \qquad (14-32)$$

Kerr Modulator

A typical Kerr modulator contains the components shown in Figure 14-4. The polarizer is oriented with its axis at a 45° angle with respect to the applied electric field $\phi = 45°$. The analyzer and polarizer are crossed, $\chi = 90°$. The transmitted intensity for this geometry is given by **(14-26)**. The Kerr medium that fills the parallelepiped is nitrobenzene, selected because of its large Kerr coefficient; see Table 14.2. The dimensions of the parallelepiped are selected as $L = 3$ cm and $d = 2$ cm, which allow the calculation of a half-wave voltage using **(14-18)**

$$V_{\lambda/2} = \frac{2 \times 10^{-2}}{\sqrt{2\left(\dfrac{33 \times 10^{-6}}{9 \times 10^{6}}\right)(3 \times 10^{-2})}} = 55.2 \text{ Kv}$$

The transmissivity of this Kerr modulator is given by

$$\mathrm{T} = \frac{I_{\perp\max}}{A^2} = \sin^2\left(\frac{V^2}{1.73 \times 10^9}\right)$$

The transmitted light intensity can be switched from zero to a maximum value by the application of 52.2 Kv to the Kerr modulator. The transmitted intensity is not a linear function of the applied voltage; thus, the Kerr modulator is not an attractive device when complex signals are to be used. The device can operate as an effective shutter and has been used in this role to measure the speed of light and as a Q-spoiler in laser cavities.

Pockels Modulator

The popularity of the $\bar{4}2m$ symmetry group as a Pockels material is due, in part, to the fact that it is one of only two classes. The other is the $\bar{4}3m$ for which the electrooptic effect is not complicated by unwanted birefringence or optical activity. Several members of this symmetry group are listed in Table 14.3. The crystals listed in Table 14.3 are water-soluble and fragile, but can be handled without too much difficulty. In the absence of an applied

TABLE 14.3 Parameters of Pockels Modulator

	10^{-12} m/V		$\lambda = 550$ nm		
Type	r_{41}	r_{63}	n_0	n_e	T_C, °K
KDP (KH_2PO_4)	8.77	9.7	1.51	1.47	122
KD*P[a]	8.8	24.3	1.508	1.468	221
ADP($NH_4H_2PO_4$)	24.5	8.5	1.527	1.481	148

[a] The notation KD*P represents a KDP crystal with the hydrogen atoms replaced by deuterium atoms.

field, these crystals are uniaxial, as can be inferred from the pair of indices in the chart. The temperature T_C listed in Table 14.3 is the Curie temperature, below which KDP and KD*P become ferroelectric and ADP becomes antiferroelectric. If the temperature of the crystal is allowed to approach the Curie temperature, the electrooptic effect becomes very large

$$r \propto \frac{1}{T - T_C}$$

unfortunately, so does the dielectric loss.

The first step we must take in developing design equations for these crystals is to obtain, from Appendix 14-A, the nonzero elements of the Pockels tensor used to write the equation of the ellipsoid (see Problem 14-2). We will simplify this equation by assuming that the external electric field lies along the z axis, which is chosen to be the principal axis for the extraordinary index of refraction. (In crystallography, this axis is called the c axis.)

In the absence of an external electric field, the principal axes of the indicatrix are aligned along the crystallographic axes. Upon the application of the external field, the indicatrix will rotate, as is shown in Figure 14-3. The resulting indicatrix equation will have cross terms, as did **(14-6)**. To eliminate the cross terms, the tensor matrix must be rediagonalized, that is, we must find a new, rotated coordinate system that will eliminate the cross term involving x_1x_2 (see Problem 14-3). In the new coordinate system, the equation for the indicatrix is

$$\left(\frac{1}{n_0^2} + r_{63}E_3\right)x_1^2 + \left(\frac{1}{n_0^2} - r_{63}E_3\right)x_2^2 + \frac{x_3^2}{n_e^2} = 1 \tag{14-33}$$

Using the binomial expansion, as was done during the derivation of **(14-8)** and **(14-9)**, we obtain the relationships between the index change and applied electric field (see Problem 14-4). The equations for the index change, induced by the applied electric field, can be used to obtain the phase modulation, created by the electrooptic effect, once the propagation direction of the light wave is specified.

Longitudinal Modulator

If the incident light wave propagates in a direction parallel to the applied electric field, along x_3 in Figure 14-9, we have a *longitudinal electrooptic modulator.* To propagate a light wave along the electric field, the electrodes producing the field must be transparent or ring-shaped electrodes must be used. The phase change produced by this design is

$$\Gamma = k_0L(n_0^3r_{63}E_3) \tag{14-34}$$

This phase change may be substituted into the Jones matrix **(14-32)** to determine the polarization properties of the longitudinal modulator.

In this modulator design, the electric field is related to the applied voltage by $V = LE_3$. In terms of the applied voltage, the phase change produced by the Pockels cell is

$$\Gamma = \frac{2\pi V}{\lambda}(n_0^3r_{63}) \tag{14-35}$$

We see in **(14-35)** that the phase change produced by the longitudinal modulator is independent of the modulator dimensions.

We can define a half-wave voltage for the longitudinal modulator, as we did in **(14-8)** for the Kerr cell. This parameter, denoted by V_π to distinguish it from the Kerr half-wave voltage, equals the voltage that must be applied to the Pockels modulator to produce a phase change of one half-wave. The half-wave voltage of the longitudinal modulator is a characteristic of the Pockels material and is independent of crystal dimensions

$$V_\pi = \frac{\lambda}{2n_0^3 r_{63}} \tag{14-36}$$

Some typical values of Pockels half-wave voltage are given in Table 14.4.

The geometrical orientation of the Pockels material in the longitudinal modulator discussed here is such that the light wave propagates along the optical axis of the crystal. (A crystal cut in the geometry of this example is called a 0°, z-cut crystal.) In this orientation, $n_1 = n_2 = n_0$, so we cannot produce a bias by adjusting the length of the electrooptic material. We may produce a bias by the addition of a quarter-wave plate (see Figure 14-9) or the less desirable addition of a dc voltage to the modulation signal.

Electrical losses in the dielectric Pockels material introduce heating when a dc voltage bias is used. This heating can produce an unstable modulator because the electrooptic properties are a strong function of the temperature due to the ferroelectric behavior of the Pockels materials.

Transverse Modulator

When the light propagates transverse to the direction of the applied electric field, say, along x_2 in Figure 14-4, the modulator is called a *transverse electrooptic modulator*. This type of modulator is easier to construct because transparent or ring electrodes are not needed, as they are for a longitudinal modulator. We adjust the orientation of the modulator so that the direction of polarization of the incident wave, relative to the electric field, is $\phi = 45°$ and the electric field is along the x_3 axis. (The crystal cut to operate in this geometry is called a 45°, z-cut configuration.) The retardation produced by the modulation field is

$$\Gamma = \frac{2\pi}{\lambda}(n_0 - n_e)L - \frac{2\pi n_0^3 r_{63}}{\lambda}\left(\frac{VL}{d}\right) \tag{14-37}$$

Because the light wave is propagating normal to the optical axis of the Pockels material, this design allows the application of an optical bias through the proper choice of crystal dimensions. We can define a half-wave voltage for this design in terms of the half-wave voltage of **(14-36)**

$$V_\pi^t = \frac{2V_\pi}{L/d}$$

The retardation is not independent of crystal dimensions, as it was in the longitudinal modulator design, but this can be used to our advantage.

TABLE 14.4 Half-Wave Voltage for Longitudinal Modulator

Type	$\lambda = 546$ nm V_π, kV
KDP (KH_2PO_4)	7.5
KD*P	2.9
ADP($NH_4H_2PO_4$)	9.2
$LiNbO_3$	4.0
$LiTaO_3$	2.7

By selecting a large L/d ratio, the drive voltage requirements can be reduced below those required for a longitudinal modulator. The ordinary and extraordinary waves will travel different paths because of the natural birefringence (the result is Poynting vector walkoff; see Problem 13-9) and the use of a large L/d ratio will enhance the separation of the two waves, degrading the modulator's performance and preventing us from obtaining an arbitrarily large reduction in drive voltage.

MAGNETOOPTIC EFFECT

When linearly polarized light propagates through a medium, parallel to an externally applied magnetic field, the plane of polarization will be rotated, as if the material were optically active. This is called the *Faraday effect*. The behavior of an optically inactive material in a magnetic field resembles the optical properties of a naturally active, isotropic material with one important exception: Natural optical activity depends on the direction of propagation of the light, whereas the Faraday effect depends on the direction of the magnetic field but **not** the direction of propagation of the light.

Because of their similarity, we could analyze the magnetooptical effect, as we did natural optical activity, by introducing an antisymmetric tensor of rank two, or equivalently, the *gyration vector* **g**. This formalism would allow a relationship, equivalent to **(13-27)**, to be written for a component of the displacement

$$D_j = \sum_{k=1}^{3} \epsilon_{jk} E_k + i(\mathbf{E} \times \mathbf{g})_j \tag{14-38}$$

The medium described by **(14-38)** is said to be *gyrotropic*.[60]

We will use a different mathematical formalism, one that is based on the dispersion model of Chapter 7. To explain the magnetic field induced, optical activity, we will modify the classical dispersion model to include an externally applied magnetic field, parallel to the propagation direction. The incident, linearly polarized light wave is decomposed into a right- and left-circularly polarized wave. The rotating electric field associated with the light wave caused the electrons in the material to move in circular orbits, in the plane transverse to the magnetic field (the xy plane). The magnetic field exerts a radial force on the moving electrons that modifies the spring constant used in the dispersion model, increasing it for counterclockwise rotation and decreasing it for clockwise rotation about the **B** field. The modified spring constants lead to two resonant frequencies.

In a vapor, such as sodium, in the absence of a **B** field, there are certain resonant frequencies where absorption takes place (in the classical theory, developed in Chapter 7, there would be one frequency ω_0). **P. Zeeman (1865–1943)** in 1896 discovered that when a magnetic field is present, what had been a single absorption frequency splits into a number of absorption frequencies. The classical theory we will develop here predicts that an isotropic, three-dimensional oscillator's resonant frequency will split into three frequencies. Two new frequencies associated with motion normal to the magnetic field are observed; see Figure 14-10a. The absorption lines associated with these new frequencies absorb right- or left-circularly polarized light. The unmodified, zero-field frequency associated with motion parallel

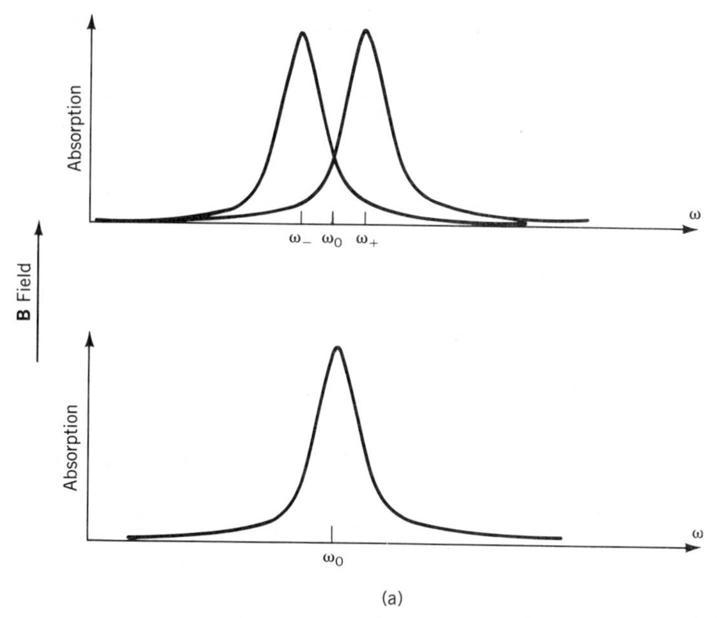

FIGURE 14-10*a*. The Zeeman effect. When the material with the resonant absorption shown in the lower drawing is placed in a magnetic field, then the absorption splits into two absorption lines. The displacement of the two new frequencies from the original frequency is a function of the applied magnetic field.

to the magnetic field is still present, but its absorption properties have been modified. It now absorbs linearly polarized light.

The classical theory explains the experimentally observed *normal Zeeman effect*. There are, however, some resonant absorptions that split into a number of new frequencies whose magnetic field behavior and polarization properties can only be explained by quantum theory.[61] This behavior is called the *anomalous Zeeman effect*. We will ignore the failure of classical theory to explain the anomalous Zeeman effect and use the theory to explain the normal Faraday effect. The classical theory will demonstrate the close relationship between the Faraday and Zeeman effects. The result will also provide a useful visualization of the physical processes related to the Faraday effect.

A vector relationship will be necessary because the addition of a magnetic field makes it impossible to treat the problem in one dimension. Starting with the equation of motion of bound charges in a noninteracting gas **(7-26)**, and adding to the electromagnetic field present a large magnetic field $\mathbf{B}$, we obtain the modified the equation of motion for the model

$$\frac{d^2\mathbf{r}}{dt^2} + \omega_0^2\mathbf{r} = -\frac{e}{m}\left(\mathbf{E} + \frac{d\mathbf{r}}{dt}\times\mathbf{B}\right) \qquad (14\text{-}39)$$

We assume that the magnetic field's frequency is much less than optical frequencies so we may ignore its time dependence. We will continue to ignore the magnetic field associated with the light wave; see **(2-33)**.

The coordinate system selected is oriented such that the magnetic field $\mathbf{B}$ lies in the positive z direction and the light wave propagates along this same direction. For this geometry, the equation of motion reduces to the following two equations:

$$\frac{d^2x}{dt^2} + \frac{e}{m} B_z \frac{dy}{dt} + \omega_0^2 x = -\frac{e}{m} E_x \tag{14-40}$$

$$\frac{d^2y}{dt^2} - \frac{e}{m} B_z \frac{dx}{dt} + \omega_0^2 y = -\frac{e}{m} E_y \tag{14-41}$$

If we multiply **(14-41)** by $\pm i$ and add **(14-40)** to **(14-41)**, we obtain an equation of circular motion

$$\frac{d^2\mathsf{R}_\pm}{dt^2} \mp \frac{e}{m} iB_z \frac{d\mathsf{R}_\pm}{dt} + \omega_0^2 \mathsf{R}_\pm = -\frac{e}{m} E_\pm \tag{14-42}$$

We now have a one-dimensional equation, with $\mathsf{R}_\pm$ equal to the coordinate for a charge moving in a circular orbit in the x, y plane and $E_\pm$ equal to the electric field for a circularly polarized wave

$$\mathsf{R}_\pm = x \pm iy, \qquad E_\pm = E_x \pm iE_y$$

The optical wave is assumed to be monochromatic, with a frequency ω.

At the surface of the magneto-optic material, which we select to be $z = 0$, the incident wave is linearly polarized

$$E_x = E_0 \cos \omega t, \qquad E_y = 0$$

In the material, we describe the wave using a circular polarization representation

$$E_\pm = E_0 \exp\left[i(\omega t - k_\pm z)\right]$$

The polarization induced by this electric field can be written with the help of **(7-29)**

$$P_\pm = P_x \pm iP_y = Ne(x \pm iy) = Ne\mathsf{R}_\pm \tag{14-43}$$

As we did in Chapter 7, we assume that the induced polarization is of the same form as the applied optical field

$$P_\pm = NeCE_\pm = NeC' \exp\left[i(\omega t - k_\pm z)\right]$$

so that the spatial coordinate can be written

$$\mathsf{R}_\pm = C' \exp\left[i(\omega t - k_\pm z)\right]$$

Substituting the assumed solution and its derivatives into **(14-42)** yields

$$-\omega^2 \mathsf{R}_\pm \pm \frac{e}{m} B_z \omega \mathsf{R}_\pm + \omega_0^2 \mathsf{R}_\pm = -\frac{e}{m} E_\pm$$

$$\mathsf{R}_\pm = \frac{\frac{e}{m} E_x}{(\omega_0^2 - \omega^2) \pm \frac{e}{m} B_z \omega} \tag{14-44}$$

Substituting **(14-44)** into **(14-43)** yields

$$P_\pm = \frac{N\frac{e^2}{m} E_\pm}{(\omega_0^2 - \omega^2) \pm \frac{e}{m} B_z \omega} \tag{14-45}$$

We define the *Larmor precession frequency* as

$$\omega_L = \frac{e}{2m}B_z \tag{14-46}$$

The Larmor frequency is much smaller than the applied field's frequency ω or the resonant frequency ω_0, both of which are optical frequencies on the order of 10^{14} Hz. For an electron in a magnetic field of 0.1 Weber/m^2,

$$\nu_L = \frac{\omega_L}{2\pi} = 1.4 \text{ GHz}$$

We can write

$$(\omega_0 \pm \omega_L)^2 = \omega_0^2\left[1 \pm 2\frac{\omega_L}{\omega_0} + \left(\frac{\omega_L}{\omega_0}\right)^2\right]$$

The small relative size of ω_L means that the term involving ω_L^2 can be neglected

$$(\omega_0 \pm \omega_L)^2 \approx \omega_0^2\left(1 \pm 2\frac{\omega_L}{\omega_0}\right)$$

The fact that $\omega_L << \omega$, ω_0 allows a replacement of $\omega\omega_L$ by $\omega_0\omega_L$. The denominator of **(14-45)** can now be rewritten as

$$\omega_0^2 - \omega^2 \pm 2\omega_L\omega \approx (\omega_0 \pm \omega_L)^2 - \omega^2$$

We now see that the magnetic field splits the resonant frequency into two new frequencies, located above and below the fieldfree resonance by a distance equal to the Larmor frequency.

Using the same procedure used to obtain **(7-30)**, we obtain the index of refraction of the magnetooptic material

$$n_\pm^2 = 1 + \frac{\omega_p^2}{(\omega_0 \pm \omega_L)^2 - \omega^2} \tag{14-47}$$

This equation describes the Faraday effect and demonstrates that the magnetic field creates unequal propagation velocities for left- and right-circularly polarized light.

When linearly polarized light passes through the Faraday medium, we can use this theory if we decompose the linear polarization into right- and left-circularly polarized components. Equation **(14-47)** states that the two circular polarizations travel at different velocities. After passing through a length of the material L, a linearly polarized wave will have the direction of its electric field rotated through an angle θ, from its orientation at the entrance to the material. The angle θ through which the electric field is rotated can be calculated using **(13-32)**

$$\theta = \frac{\pi L}{\lambda_0}(n_- - n_+) = \frac{\omega L}{c}(n_- - n_+)$$

Because ω_L is much smaller than ω_0, the index difference $(n_- - n_+)$, predicted by **(14-47)** is small and we can use the approximation

$$(n_-^2 - n_+^2) \approx 2n_0(n_- - n_+)$$

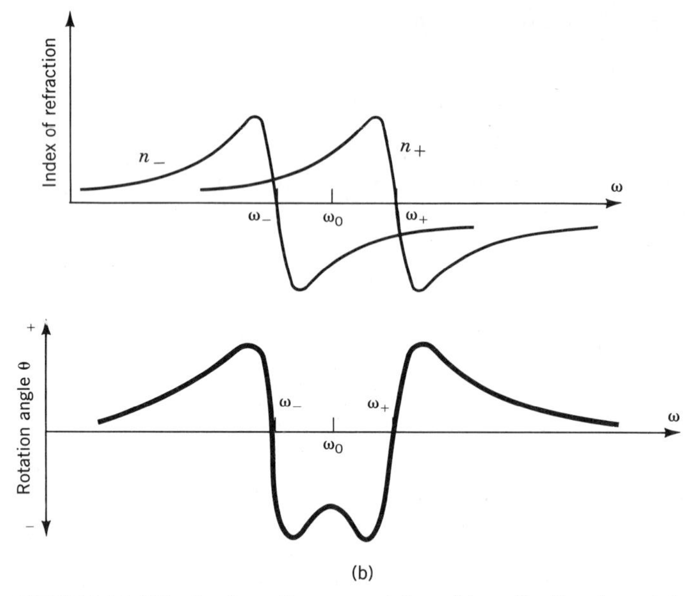

(b)

FIGURE 14-10*b*. A schematic representation of how the Faraday rotation behaves as a function of frequency. The absorption resonance of Figure 14-10*a* is the origin of the dispersion shown here.

to write

$$\theta = \frac{\omega\omega_p^2 L}{cn_0}\left[\frac{2\omega_0\omega_L}{({\omega_0}^2 + \omega_L^2 - \omega^2)^2 - 4\omega_0^2\omega_L^2}\right] \tag{14-48}$$

A schematic representation of the prediction of this classical model is shown in Figure 14-10*b*. The resonant absorption curve from Figure 14-10*a*, exhibits the normal Zeeman effect and splits into two resonances that absorb circularly polarized light. The dispersion curves associated with these two resonances, shown in Figure 14-10*b*, are used to calculate the Faraday rotation according to **(14-48)**.

The equation for the angle of rotation generated by the classical model agrees with the empirically determined relationship

$$\theta = VLH \tag{14-49}$$

The Verdet's constant V is positive when the direction of the electric field of a linearly polarized wave is rotated clockwise, measured when looking along the direction of the field *H*. (The polarization rotates in the same direction as the flow of the current producing the magnetic field.) The very simple theory presented here agrees with experiment which has demonstrated that, away from absorption bands, the Verdet's constant falls to zero with increasing wavelength. Some typical Verdet's constants are shown in Table 14.5.

A light modulator could be designed using the Faraday effect, but it would have a limited frequency range because it is very difficult to modulate a magnetic field at high frequencies. The major application of the Faraday effect is as an *isolator*, that is, a device that allows an electromagnetic wave to propagate in only one direction. It makes use of the fact that the angle

TABLE 14.5 Verdet's Constant

Substance	$\lambda = 589$ nm V, min. of arc/Oe-cm[a]
H_2O	0.0131
Crown glass	0.0161
Flint glass	0.0317
CS_2	0.0423
CCl_4	0.0160
NaCl	0.0359
KCl	0.02858
Quartz	0.0166
ZnS	0.225

[a] The units are those normally found in the literature. To convert Oe-cm to mks units, we multiply by $4\pi \times 10^{-1}$ to obtain (amp-turns/m)m. If we assume that $\mu = \mu_0$, then we can multiply by 10^6, allowing the use of units of Weber/m^2 for the magnetic field.

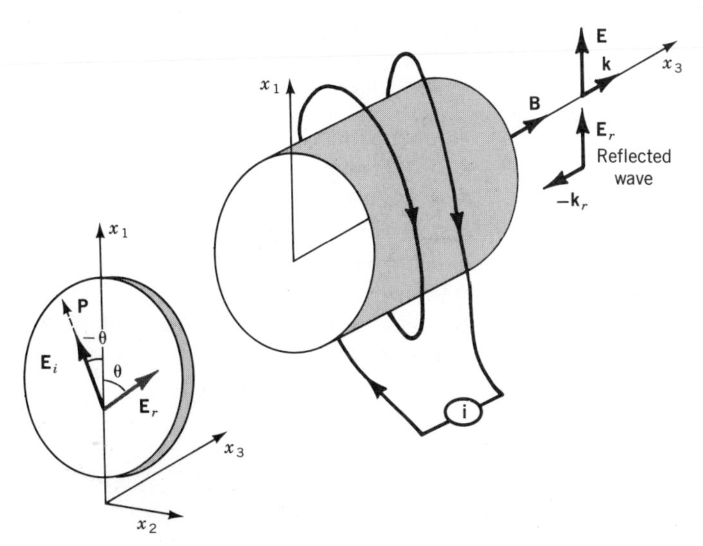

FIGURE 14-11. An isolator based on the Faraday effect. The wave initially encounters a polarizer that establishes the direction of the linearly polarized wave incident on the Faraday rotator. A magnetic field is applied to the rotator and the polarization is rotated through the angle θ so that it is now parallel to the x_1 axis. Any of the wave reflected back to the isolator is again rotated through the angle θ and arrives at the polarizer at an angle of 2θ with respect to the polarizer's axis. By proper choice of the **B** field, we can make $\theta = 45°$.

of rotation is the same for light propagating parallel and antiparallel to the applied magnetic field. A Faraday isolator is displayed in Figure 14-11.

The initial polarizer determines the polarization angle $-\theta$ of the incident light wave. It is normally adjusted so that it makes an angle of 45° with respect to the axis x_1. The **B** field of magnitude B_z, applied to the Faraday rotator, is such that the electric field of the wave is rotated through 45°. Any reflected wave that propagates back through the Faraday rotator is rotated through an additional 45°. This reflected wave arrives at the polarizer with its

TABLE 14.6 Cotton–Mouton Coefficient

Substance	$\lambda = 589$ nm C, $1/(\text{Oe}^2\cdot\text{cm})^a$
Acetone	37.6×10^{-13}
Benzene	7.5
Chloroform	−65.8
CS_2	−4.0

[a] The units are those normally found in the literature. To convert from Oe^2-cm to mks units, we multiply by $(4\pi)^2 \times 10^{-4}$ to obtain (amp-turns/m)^{2}m. When we can assume that $\mu = \mu_0$, we can multiply by 10^{10} to allow the use of Weber/m^2 for the magnetic field.

electric field oriented 90° to the polarizer's axis and is thus prevented from passing through.

Several other magnetic effects are observed in materials, but they have little technological importance. The *Cotton–Mouton effect* is the magnetic analog of the Kerr effect in that it depends on the square of the applied field (a magnetic field for the Cotton-Mouton effect). A birefringence is observed in the presence of a magnetic field if the light waves are propagating in a direction normal to the applied magnetic field. The birefringence is interpreted as due to the difference in the propagation velocities of two waves. One wave, with **D** parallel to the external magnetic field **H**, has an index of refraction represented by $n_{\|}$; the second wave, with **D** perpendicular to the magnetic field, has an index of refraction represented by $n_{\perp}$.

The phase shift that is produced when the two orthogonally polarized waves propagate, perpendicular to the applied magnetic field, through a distance d in the material, is given by

$$\delta = \frac{n_{\|} - n_{\perp}}{\lambda} d = CdH^2$$

Table 14.6 contains a few typical values for the Cotton–Mouton coefficient.

A similar magnetooptic effect observed in gases, near resonant absorptions, is called the *Voigt effect*. The Cotton–Mouton and Voigt effects are too small to have found any practical application (the Faraday effect is approximately 10^3 times greater than the Cotton–Mouton effect and the Voigt effect is about 10^3 times smaller than the Cotton-Mouton effect) and we will not discuss them further.

PHOTOELASTIC EFFECT

An isotropic material can become anisotropic when subject to an applied stress or induced strain, through an elastooptic interaction known as *stress birefringence*, or the *photoelastic effect*. The photoelastic effect was first observed by Brewster in 1815 and a microscopic theory of the effect was published by H. Mueller in 1935.[62]

The photoelastic effect reduces the performance of optical components and devices by introducing phase distortions. These distortions can arise from stresses produced by improper mounting, or by unequal thermal expansion of the mounts and components. Another source of distortion is strain that has been frozen into the optical component during manufacture. When possible, optical materials with a small photoelastic effect are used to reduce these unwanted distortions.

There are applications that utilize the elastooptic interaction and the materials used in these applications are selected for their large photoelastic effect. In this section, we will discuss very briefly one such application: the use of stress birefringence in the analysis of mechanical structures under load. The purpose of introducing this application is to demonstrate that the optical indicatrix can be used in the analysis of the photoelastic effect. Later, a completely different approach will be used to develop an important application of the photoelastic effect involving acoustic waves.

As was the case for the electrooptic effect, the change in the index of refraction can be visualized as a change in the size, shape, and orientation of the indicatrix. As was done for the electrooptic effect in **(14-1)**, the impermeability tensor in the presence of stress can be written as

$$\backepsilon_{ik} = \backepsilon_{ik}^{0} + \sum_{j=1}^{3}\sum_{m=1}^{3} q_{ijkm}\mathsf{T}_{jm} \tag{14-50}$$

$\tilde{\mathbf{T}}$ is a symmetric, second-rank tensor; its diagonal terms are the normal components and the off-diagonal terms are the shear components of stress. The typical sign convention treats compressive stress as negative. The q_{ikjm} are the coefficients of a fourth-rank tensor called the *stress-optical tensor*

$$O[q] \approx 10^{2}\frac{\text{mm}}{\text{N}}$$

There are 81 terms in this tensor, but because $\tilde{\mathbf{T}}$ and $1/\tilde{\epsilon}$ are symmetric, we have $q_{ikjm} = q_{ikmj}$ and $q_{ikjm} = q_{kijm}$. This reduces the number of independent terms to 36. Symmetry arguments reduce the number of required indices for q to two, as was the case for the electrooptic effect

$$\begin{bmatrix} \backepsilon_1 \\ \backepsilon_2 \\ \backepsilon_3 \\ \backepsilon_4 \\ \backepsilon_5 \\ \backepsilon_6 \end{bmatrix} = \begin{bmatrix} \backepsilon_1^0 \\ \backepsilon_2^0 \\ \backepsilon_3^0 \\ \backepsilon_4^0 \\ \backepsilon_5^0 \\ \backepsilon_6^0 \end{bmatrix} + \begin{bmatrix} q_{11} & q_{12} & q_{13} & q_{14} & q_{15} & q_{16} \\ q_{21} & q_{22} & q_{23} & q_{24} & q_{25} & q_{26} \\ q_{31} & q_{32} & q_{33} & q_{34} & q_{35} & q_{36} \\ q_{41} & q_{42} & q_{43} & q_{44} & q_{45} & q_{46} \\ q_{51} & q_{52} & q_{53} & q_{54} & q_{55} & q_{56} \\ q_{61} & q_{62} & q_{63} & q_{64} & q_{65} & q_{66} \end{bmatrix} \begin{bmatrix} \mathsf{T}_1 \\ \mathsf{T}_2 \\ \mathsf{T}_3 \\ \mathsf{T}_4 \\ \mathsf{T}_5 \\ \mathsf{T}_6 \end{bmatrix} \tag{14-51}$$

An equivalent formulation of the elastooptic interaction can be made by associating a change in the index of refraction with an internal strain

$$\backepsilon_{ik} = \backepsilon_{ik}^{0} + \sum_{j=1}^{3}\sum_{m=1}^{3} p_{ijkm}\mathsf{S}_{jm} \tag{14-52}$$

The S_{jm} in **(14-52)** are the components of the strain tensor. The diagonal components are tensile or stretching strains, and the off-diagonal components are shear. The *strain-optic* tensor components p_{ijkm} are dimensionless numbers; some typical values are listed in Table 14.7. The strain-optic com-

TABLE 14.7 Strain-optic Coefficients

Material								
	Isotropic							
	$p11$	$p12$						
Fused silica	0.121	0.27						
Water	0.31	0.31						
Polystyrene	0.3	0.31						
	Cubic							
	$p11$	$p12$	$p14$					
GaAs	−0.165	−0.14	−0.072					
Ge	0.27	0.235	0.125					
LiF	0.02	0.128	−0.064					
NaCl	0.11	0.153	−0.01					
	Trigonal							
	$p11$	$p12$	$p13$	$p14$	$p31$	$p33$	$p41$	$p44$
Ruby	−0.23	−0.03	0.02	0.00	−0.04	−0.2	0.01	−0.1
$LiNbO_3$	−0.02	0.08	0.13	−0.08	0.17	0.07	−0.15	0.12
	Tetragonal							
	$p11$	$p12$	$p13$	$p31$	$p33$	$p44$	$p66$	
KDP	0.251	0.249	0.246	0.225	0.221	—	0.058	
TeO_2	0.0074	0.187	0.34	0.09	0.24	−0.17	−0.046	

ponents p_{ikjm} and the stress-optic q's are linearly related through the elastic stiffness constants, $\mathbf{c}_{mnk\ell}$

$$p_{ijk\ell} = \sum_{m=1}^{3} \sum_{n=1}^{3} q_{ijmn} \mathbf{c}_{mnk\ell} \tag{14-53}$$

Reversing the applied stress or induced strain results in a change from tension to compression and, in general, produces a change in the index of refraction. This fact means that unlike the electrooptic effect, the photoelastic effect can be observed in a centrosymmetric crystal and in an isotropic material such as a glass or liquid.

From the strain formalism, the origin of the photoelastic effect is associated with a change in the density of the medium. As we demonstrated in Chapter 7 **(7-46)**, a change in the density of a material will result in a change in the index of refraction

$$\frac{n^2 - 1}{n^2 + 2} \propto \rho \tag{14-54}$$

However, this result cannot be simply applied to predict the elastooptic properties of a material. The index of refraction is also modified through a change in the polarizability, induced by the strain. The change in polarizability is not an important effect in liquids, but in solids the molecular polarizability normally decreases under compression, counteracting the modification produced by the density.

Another complication in predicting the elastooptic properties of a material occurs when the material is piezoelectric. There is a coupling between the electrooptic and photoelastic terms through the piezoelectric effect in noncentrosymmetric crystals. An applied strain produces an electric field via the piezoelectric effect; the electric field, generated by the strain, modifies the index of refraction through the electrooptic effect. The index change produced by the piezoelectric coupling can either add to or subtract from the photoelastic effect.

We will neglect these complications by examining the effects of an applied stress on an isotropic material. Without the stress, the optical indicatrix is a sphere and there is only one index of refraction. With a stress applied, the material will become birefringent. Appendix 14-A can be used to determine the nonzero elements of the $\tilde{\mathbf{q}}$ tensor because the Kerr matrix has the same nonzero terms as the q matrix. For an isotropic material, **(14-51)** becomes

$$\begin{bmatrix} \backepsilon_1 \\ \backepsilon_2 \\ \backepsilon_3 \\ \backepsilon_4 \\ \backepsilon_5 \\ \backepsilon_6 \end{bmatrix} = \begin{bmatrix} \backepsilon^0 \\ \backepsilon^0 \\ \backepsilon^0 \\ 0 \\ 0 \\ 0 \end{bmatrix} + \begin{bmatrix} q_{11} & q_{12} & q_{12} & 0 & 0 & 0 \\ q_{12} & q_{11} & q_{12} & 0 & 0 & 0 \\ q_{12} & q_{12} & q_{11} & 0 & 0 & 0 \\ 0 & 0 & 0 & q_{11} - q_{12} & 0 & 0 \\ 0 & 0 & 0 & 0 & q_{11} - q_{12} & 0 \\ 0 & 0 & 0 & 0 & 0 & q_{11} - q_{12} \end{bmatrix} \begin{bmatrix} T_1 \\ T_2 \\ T_3 \\ T_4 \\ T_5 \\ T_6 \end{bmatrix}$$

The proper choice of the coordinate system can simplify this equation. Assume that the elastooptic material is a parallelepiped of isotropic plastic; see Figure 14-12. A compression T is applied in the x_1 direction and light propagates through the plastic in the x_3 direction. From the tensor equation, the principal axes of the new ellipsoid are aligned with the applied stress and are equal to

$$\frac{1}{n_1^2} = \frac{1}{\backepsilon_1} = \frac{1}{n_0^2} + T_1 q_{11} \tag{14-55a}$$

$$\frac{1}{n_2^2} = \frac{1}{\backepsilon_2} = \frac{1}{\backepsilon^0} = \frac{1}{n_0^2}$$

$$\frac{1}{n_3^2} = \frac{1}{\backepsilon_3} = \frac{1}{\backepsilon^0} = \frac{1}{n_0^2} \tag{14-55b}$$

The formerly isotropic material now has the indicatrix of a positive uniaxial crystal. In Figure 14-12, the material is illuminated by linearly polarized light, with the electric field oriented at a 45° angle to the applied stress. This wave can be decomposed into two orthogonally polarized waves, polarized parallel to x_1 and x_2. The component of polarization along the x_1 axis and the component of polarization along the x_2 axis will travel at different velocities.

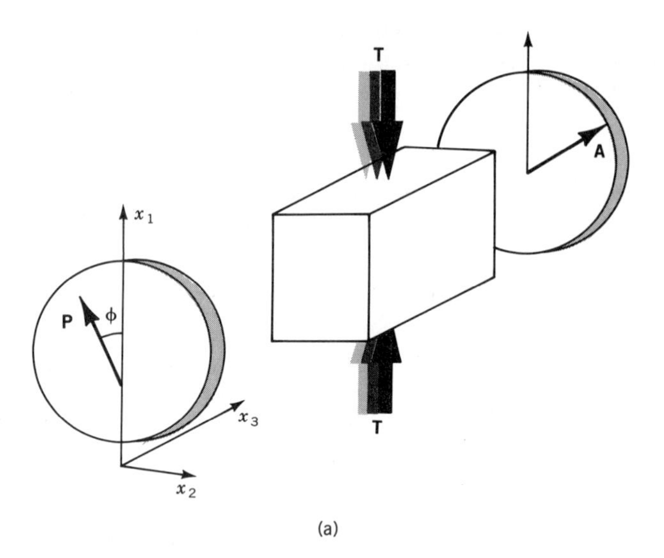

(a)

(b)

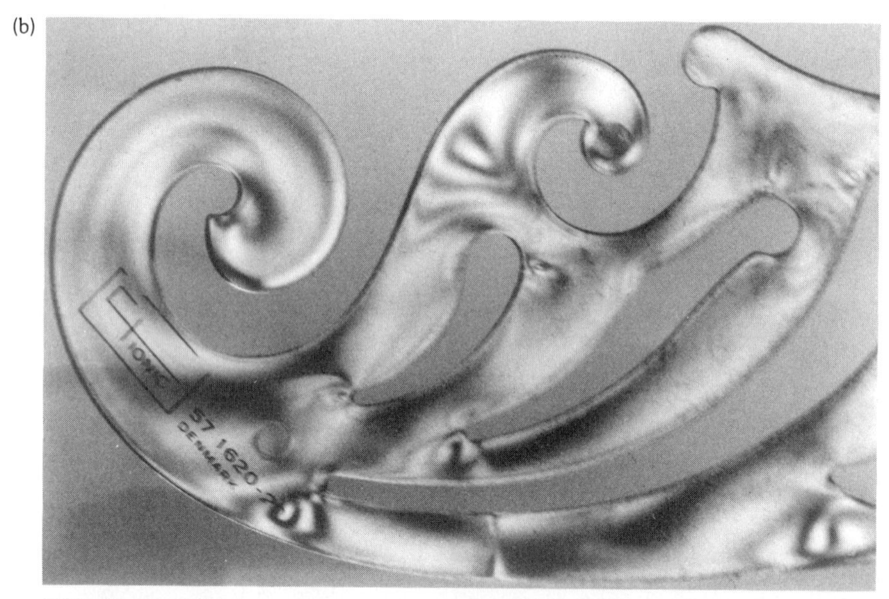

FIGURE 14-12. (a) A compressive stress is applied to an isotropic material located between two crossed polarizers. The stress produces a birefringence and the light transmitted by the system is a function of the applied stress. (b) In the manufacture of plastic French curves, components of stress are locked in the plastic. Here is shown a French curve, illuminated with polarized light and viewed through a polarizer oriented with its axis normal to the axis of polarization of the illumination. Sunlight Rayleigh scattered through 90°by the atmosphere is polarized and was used to illuminate the French curve. A color reproduction of this photo is found in the insert.

The difference in phase between the two waves is given by

$$\Gamma = k_0 L(n_1 - n_2)$$

We may use **(14-55)** to write

$$\frac{1}{n_1^2} - \frac{1}{n_2^2} = q_{11} \mathrm{T}_1 \tag{14-56}$$

The stress does not produce a large change in the index of refraction and we are allowed to make the following approximation:

$$\frac{1}{n_1^2} - \frac{1}{n_2^2} \approx \frac{2(n_2 - n_1)}{n^3}$$

With this approximation, the phase difference for the two polarizations propagating through the stressed plastic is

$$\Gamma = \frac{\pi L}{\lambda_0} n^3 q_{11} T_1 \tag{14-57}$$

The intensity transmitted by the crossed polarizers in Figure 14-12, after having passed through the stressed plastic, is given by **(14-25b)**

$$I = A^2 \sin^2\left(\frac{\pi L n^3}{2\lambda_0} q_{11} T_1\right)$$

In monochromatic light, the contours of equal intensity, observed through the crossed polarizers, represent contours of equal stress. We have treated the easy problem of a single stress, lying in a predetermined direction. Usually, the direction of the stress will not be known but can be determined by rotating the crossed polarizers. The size of the stress can be determined from the order of the fringe or, if white light is used, by the color of the fringes as seen in Figure 14-12*b* (the color photo is found in the insert). The strain-optic coefficients of a few materials are contained in Table 14.7.

ACOUSTOOPTICS

If the strain in a medium is time-dependent, then we can treat the strain as a pressure wave or, equivalently, an acoustic wave of wavelength Λ and frequency F, traveling at a velocity v

$$\frac{\Omega}{\kappa} = \frac{\Lambda}{2\pi} 2\pi F = v$$

Brillouin theoretically predicted the acoustooptic interaction in 1921. Concurrent verification of Brillouin's theory was made in 1932 by two pairs of investigators: **P. Debye** and **F. W. Sears** and **R. Lucas** and **P. Biquard**. A number of theoretical models have been developed for describing the acoustooptic interaction. (These theoretical models may also be used in describing holography, Chapter 12, and phase conjugation, Chapter 15.)

1. It is possible to treat acoustooptic diffraction as a parametric process, in which the optic and acoustic wave are mixed via the elastooptic effect. A *parametric process* occurs when an oscillator of frequency ω_0, described by the equation of motion **(1A-9)**, has one of its *parameters*, for example, the spring constant, modulated by a frequency ω_p, the *pump frequency*. A typical equation of motion for a parametric oscillator is

$$\frac{d^2y}{dt^2} + \gamma\frac{dy}{dt} + (\omega_0^2 + \alpha \sin \omega_P t)\, y = 0$$

When ω_P meets the proper conditions, the oscillations at ω_0 grow, using energy from the source of the pump frequency. A simple example of a parametric oscillator is a child's swing. The amplitude of the swing's excursions grows, as the child "pumps" the swing, by modulating the length of the swing, twice each cycle. In the acoustooptic interaction, the pump wave is the incident light wave. The *signal wave*, which grows at the expense of the pump wave, has a frequency of $\omega_P - \Omega$, where Ω is the acoustic wave's frequency. The acoustic wave is called the *idler wave* in the terminology of parametric processes. This model does not provide any unique insight into the acoustooptic effect so we will not develop its formalism.

2. A second model is based on a phenomenological theory that assumes the induced polarization is proportional to the product of the electric field and the acoustic strain. This is equivalent to assuming that the dielectric constant has the same functional form as the density modulation produced by the acoustic wave. This formalism, known as the coupled wave theory, will be discussed in Appendix 14-B.
3. The third approach uses the formalism of scattering theory: treating light waves as photons and sound waves as particles, called *phonons*. We will use this approach to discuss two limiting descriptions of acoustooptic scattering: Bragg scattering and Raman–Nath scattering.

We will conclude this section with a discussion of some applications that are based on the Bragg scattering condition, the only type of acoustooptic scattering that has found extensive technological application.

The scattering model of the acoustooptic effect is based on a quantum mechanical view of the interaction. Light of frequency ω and wavelength λ is treated as a collection of particles, called *photons*, with energy

$$U_L = \hbar\omega$$

and momentum

$$p_L = \hbar k_L = \frac{h}{\lambda}$$

The acoustic signal of frequency Ω and wavelength Λ is also treated as a collection of particles, called *phonons*, with energy

$$U_S = \hbar\Omega$$

and momentum

$$p_S = \hbar\kappa = \frac{h}{\Lambda}$$

Two types of inelastic scattering processes can take place:

1. *Absorption*. An incident photon of energy $\hbar\omega_i$ can collide with a phonon of energy $\hbar\Omega$. The phonon is destroyed and a photon of energy $\hbar\omega_d$ leaves the scattering site.
2. *Emission*. A photon of energy $\hbar\omega_i$ enters the scattering site and a phonon of energy $\hbar\Omega$ and a photon of energy $\hbar\omega_d$ leave the scattering site.

The solution of a scattering problem requires the application of conservation of momentum and energy. Both types of three-particle, inelastic, scattering processes can use the same momentum conservation diagram by changing the direction of the phonon momentum $\boldsymbol{\kappa}$; see Figure 14-13.

The energy conservation equations for emission and absorption are

$$\omega_i = \Omega + \omega_d, \qquad \omega_i + \Omega = \omega_d \tag{14-58}$$

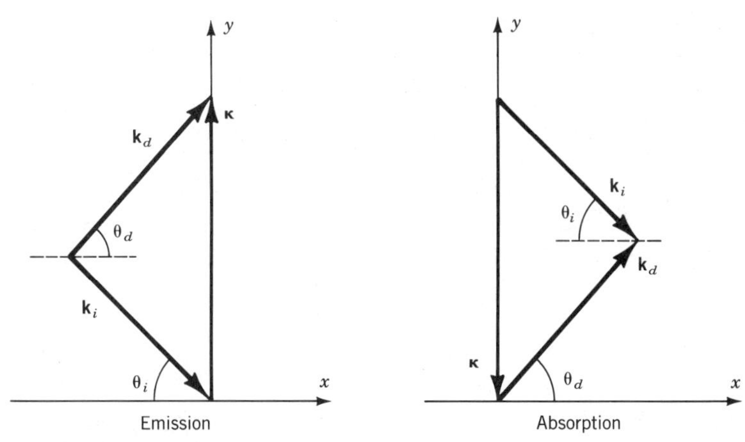

FIGURE 14-13. Momentum conservation diagram for the scattering of light from acoustic waves. The subscript *i* denotes the incident wave parameters and the subscript *d* the diffracted wave parameters.

The frequency of the scattered radiation is up- or down-shifted by the acoustic frequency. (Classically, the frequency shift associated with acoustooptic scattering can be viewed as a Doppler shift [see **(8-10)**] of the light wave reflected from the high-density region of the acoustic wave, traveling toward or receding from the incident light wave.)

Quantum mechanics states that the probability of the absorption of a phonon is $\sqrt{N}$ and the probability of emission of a phonon is proportional to $\sqrt{N+1}$, where N is the number of phonons in a given mode. If these are thermal phonons, i.e., phonons that exist because of lattice vibrations of the material at a temperature T, then the number of phonons with an energy $\hbar\Omega$ is

$$N = \exp\left[-\frac{1}{1-(\hbar\Omega/kT)}\right] \tag{14-59}$$

If T is room temperature (300°K) and Ω is a microwave frequency (10^9 Hz), then the approximation

$$N \approx \frac{kT}{\hbar\Omega} >> 1$$

is an accurate one and the emission and absorption processes can be treated as equally probable.

From Figure 14-13, the x components of the momentum vectors are

$$k_i \cos\theta_i = k_d \cos\theta_d \tag{14-60}$$

for both emission and absorption processes. The y components for a phonon absorption are

$$-k_i \sin\theta_i + \kappa = k_d \sin\theta_d$$

and for a phonon emission they are

$$-k_i \sin\theta_i = -\kappa + k_d \sin\theta_d \tag{14-61}$$

In this simple theory, we assume that the material is isotropic so that the velocity of the scattered light wave is identical to the incident wave.

Bragg Scattering

Since $\Omega << \omega$, we can to first order neglect the frequency shift predicted by **(14-58)** and assume that $\omega_i \approx \omega_d$. [This assumption, which treats the sound wave as stationary, is a reasonable approximation because the sound wave moves an infinitesimal amount during the passage of a light wave (see Problem 14-13).] The magnitudes of the momentum of the incident and scattered waves have identical values, $|k_i| = |k_d|$ because we neglect the frequency shift of the light wave caused by the inelastic scattering event. The solution of **(14-61)** can now be written as

In the discussion of holography **(12-13)**, a more intuitive analysis of diffraction from a grating was used to obtain the Bragg equation. That analysis could equally apply to sound waves if the holographic fringe recording is replaced by a periodic density variation. The key point of that derivation was that waves, diffracted from many planes of a thick grating, interfere constructively in only one direction, given by the Bragg angle.

$$\sin\,\theta_B = \frac{\kappa}{2k_i} \tag{14-62}$$

This is Bragg's law and describes diffraction of the light waves by the grating formed from the wavefronts of the acoustic wave. The incident wave vector in the scattering calculation is defined in the scattering medium, but normally, the wave vector is described in terms of its vacuum properties. If the material is isotropic and the sides of the material are parallel to the direction of sound propagation, we can demonstrate (Problem 14-11) that it is not necessary to modify **(14-62)** because, in this case, the equation is independent of the index of refraction of the material.

A possible physical realization of Bragg diffraction is shown in Figure 14-14. The lines forming the grating in Figure 14-14 represent the regions of high density produced by the sound wave. For an acoustic frequency of 500 MHz and a sound speed of 6×10^3 m/sec (the value for fused silica), the acoustic wavelength is

$$\Lambda = \frac{6 \times 10^3}{5 \times 10^8} = 1.2 \times 10^{-5} \text{ m}$$

If the illuminating wavelength is $\lambda = 632.8$ nm, then **(14-62)** predicts a Bragg angle of

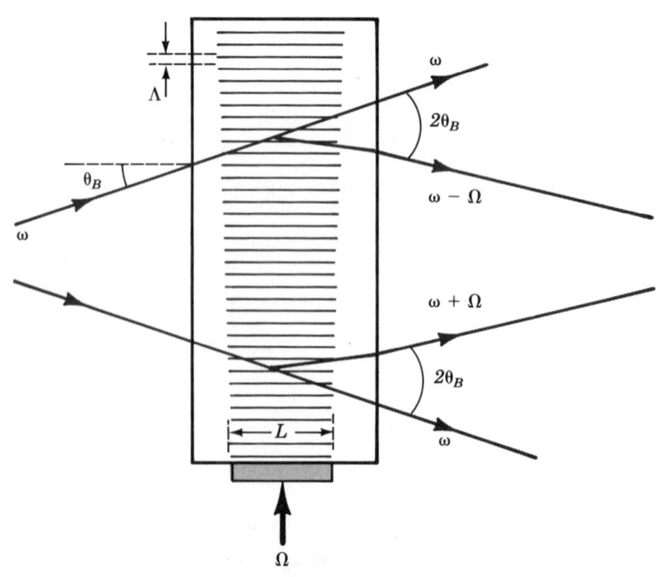

FIGURE 14-14. Bragg scattering of an optical wave from an acoustic wave.

$$\theta_B = \sin^{-1}\left[\frac{6.238 \times 10^{-7}}{2(1.46)(1.2 \times 10^{-5})}\right] = 1.02°$$

As will be shown in Appendix 14-B **(14B-21)**, with sufficient acoustic power all of the incident light can be diffracted into the Bragg angle.

Raman–Nath Scattering

A second limiting condition is approached when the width of the acoustic wavefront is decreased. [The first analysis of this type of scattering was carried out by **C. V. Raman** and **N. S. Nagendra Nath.**] It is through the analysis of this limiting condition that the scattering model provides a unique insight into the acoustooptic process. If the width of the sound wave L is decreased, diffraction will cause the sound wave to spread; see **(10-17)**. Additional plane waves of sound are needed in the Fourier transform description of the more rapidly spreading wave. In the analysis, we will assume that a finite Fourier series is adequate to describe the sound beam. As the width L of the beam is decreased, we will assume that we need to only add pairs of terms to the Fourier series, to continue to describe the beam.

The first effect of decreasing L is that the pair of propagation vectors κ_1 and κ_{-1} must be added to the fundamental κ_0 to describe the sound wave. When a photon is scattered from this acoustic disturbance, the propagation vectors $(\mathbf{k}_d)_1$ and $(\mathbf{k}_d)_{-1}$ both satisfy the momentum conservation requirement. A further decrease in L results in κ_2 and κ_{-2} appearing at reasonable amplitudes in the Fourier series. This increase in the number of propagation vectors $\kappa_{\pm j}$ continues as L decreases. A schematic portrayal of the acoustic wave and its propagation vectors is shown on the left of Figure 14-15.

For each additional sound propagation vector needed to describe the sound beam, another light propagation vector is able to satisfy the momentum conservation condition. Reducing the width of the acoustic beam results in an increase in the number of diffracted waves produced by the acoustoop-

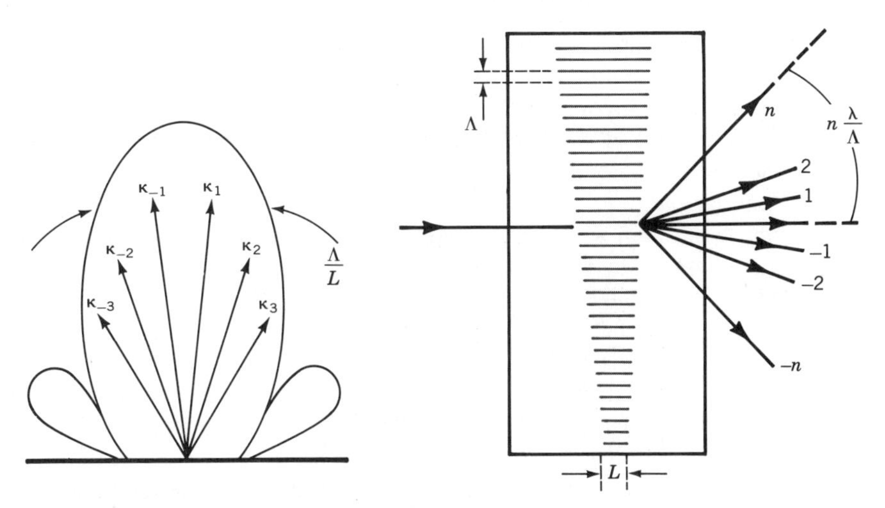

FIGURE 14-15. Raman–Nath scattering. The curve on the left shows the increase in number of propagation vectors due to diffraction of the acoustic wave. On the right, the increase in the number of diffracted orders of light.

tic interaction; see the drawing on the right of Figure 14-15. The scattering process resulting in a large number of diffracted orders is called *Raman–Nath scattering* and occurs for any grating whose thickness can be ignored. To meet the energy conservation condition, the nth diffracted order is Doppler-shifted in frequency by $n\Omega$.

The extent to which we can expect to observe a large number of diffracted orders is determined by the ratio of angular width of the sound beam (Λ/L) to the separation between diffracted orders (λ/Λ). Historically, the quantity

$$Q = \frac{\kappa^2 L}{n k_{0i}} = \frac{2\pi}{n} \bullet \frac{\lambda_0}{\Lambda} \bullet \frac{L}{\Lambda} \tag{14-63}$$

which contains this ratio is used to define the Bragg ($Q > 1$) and Raman–Nath ($Q < 1$) regions. If the acoustic wavelength were 1.2×10^{-5} m and the illuminating wavelength 623.8 nm, then $Q = (1.86 \times 10^4)L$. An acoustic beam wider than 0.1 mm would operate in the Bragg region. [The same quantity is used in holography to define thick ($Q > 10$) and thin ($Q < 10$) holograms. When Q is applied to holography, Λ is redefined as the fringe spacing and L as the hologram's thickness.]

Bragg scattering results in a more efficient light modulator because all of the light energy is contained in one diffracted beam. For this reason, most acoustooptic applications utilize the Bragg effect. The acoustooptic modulator excels over an electrooptic modulator in its capabilities to modify the frequency and direction of light waves. We will look at several applications of acoustic Bragg scattering that utilize these advantages provided by the acoustooptic device.

Acoustooptic Modulator

Any device that modifies the frequency of a wave is called a *frequency modulator*, or *fm modulator*, and it is characterized by its bandwidth, i.e., the frequency range over which it can operate. An acoustooptic modulator shifts the frequency of the diffracted light by an amount equal to the acoustic frequency and therefore can be used as an fm modulator. The bandwidth of an acoustooptic light modulator is determined by the angular spread over which one can satisfy the Bragg condition **(14-62)**. Usually, a large frequency bandwidth is needed to maximize the amount of information carried by the light wave; however, sometimes only a single frequency shift is required, i.e., the generation of the local oscillator frequency for a laser radar. We will determine the parameters affecting the operating bandwidth of an acoustooptic modulator by determining the angular bandwidth allowed by Bragg scattering. The analysis will demonstrate the technique for optimization of the acoustooptic cell as a frequency modulator.

The Bragg condition can be rewritten in terms of the acoustic frequency and velocity

$$\sin \theta_B = \frac{\lambda F_m}{2v}$$

where v is the acoustic velocity and F_m the modulation frequency. The derivative of this equation yields the angular spread created by a distribution of acoustic frequencies

$$\cos\,\theta\;\Delta\theta = \frac{\lambda}{2v}\Delta F_m$$

The maximum spread in acoustic frequencies defines the modulation bandwidth

$$\Delta F_m = \left(\frac{2v\,\cos\,\theta}{\lambda}\right)\Delta\theta \tag{14-64}$$

The angle of incidence must change by $\Delta\theta$ in order to satisfy the Bragg condition when the acoustic frequency changes by ΔF. An alternate interpretation of **(14-64)** is that a wide instantaneous bandwidth is associated with a diverging beam, that is, a large distribution of propagation vectors.

An increase in the bandwidth of the acoustooptic cell is accomplished by increasing the diffraction spreading of a finite dimensioned optical or acoustic wave. Assume that we have an optical beam of width w whose diffraction spread in the acoustooptics material is

$$\delta\theta = \frac{\lambda}{w}$$

and an acoustic beam of width L with a diffraction spread of

$$\delta\Phi = \frac{\Lambda}{L}$$

We will evaluate the effects of modifying the diffraction spread of the acoustic and/or light wave by the use of the momentum conservation diagrams in Figures 14-16 and 14-17. The diagrams are constructed through the use of a circle of radius k_i. Because we will neglect the frequency shift of the diffracted wave, the circle defines the length of both the incident and diffracted wave vectors. If we assume $\delta\Phi >> \delta\theta$, that is, that the acoustic beam is narrow and the optical beam wide, we obtain the conservation of momentum dia-

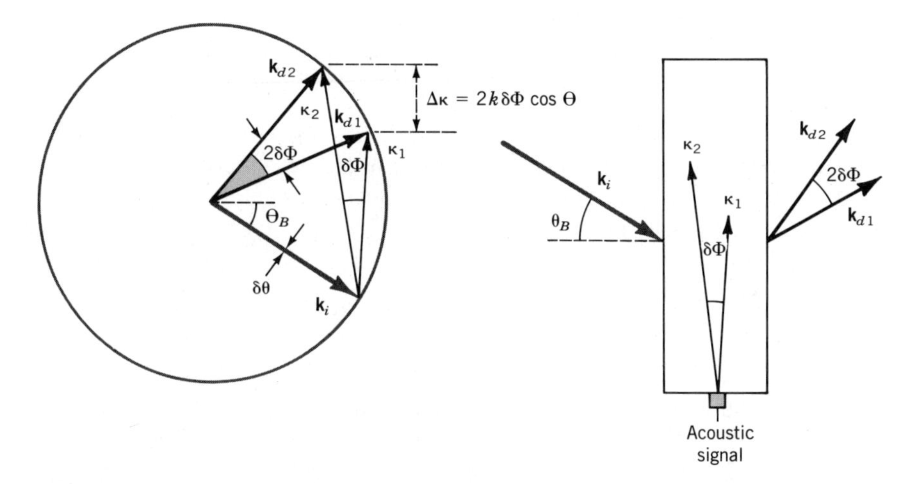

FIGURE 14-16. Momentum conservation diagram for an acoustooptic modulator in which the acoustic beam divergence dominates. A modulator with this property would be a good beam deflector.

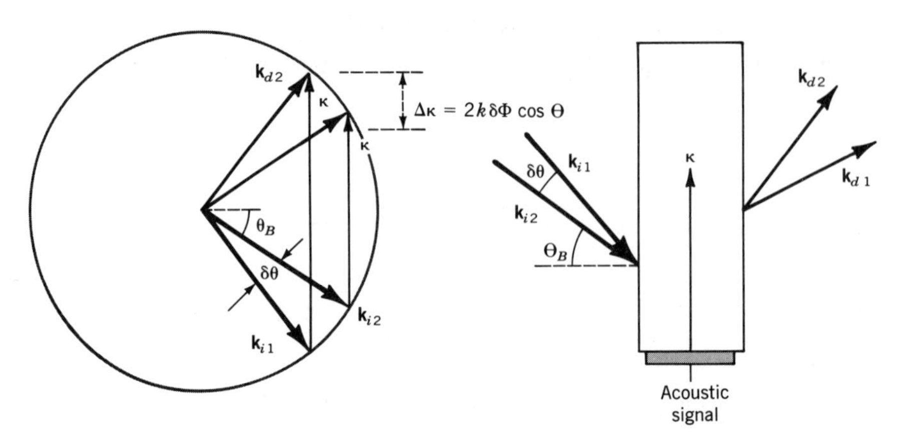

FIGURE 14-17. Momentum diagram for the design of a frequency modulator using acoustooptics. Here, we are assuming a very small spread in the acoustic beam so all of the κ's are parallel.

gram shown in Figure 14-16. There will be a range of values of **κ** falling within $\delta\Phi$ whose vectors fall on the circle of Figure 14-16 and thus satisfy the Bragg condition. The apparent frequency bandwidth of this device is determined by the acoustic spread $\Delta\theta = \delta\Phi$ and is given by

$$\Delta F = \frac{2v \cos \theta_B}{L}\left(\frac{\Lambda}{\lambda}\right) \tag{14-65}$$

All of this apparent bandwidth, however, is not available as a frequency-modulated signal. The acoustic waves diffract light into different directions, as is shown by $\mathbf{k}_{d1}$ and $\mathbf{k}_{d2}$ in Figure 14-16. Only light that is both frequency-shifted and collinear in direction can be used for fm modulation; thus, the frequency bandwidth is actually limited to the optical beam spread $\delta\theta$. If acoustic beam diffraction dominates, part of the acoustic energy is wasted.

If we assume that $\delta\theta >> \delta\Phi$, (Figure 14-17), a broad, well-collimated acoustic beam is used with a narrow, diverging or converging, optical beam. Separate propagation vectors of the optical beam fall on the circle of Figure 14-17 and satisfy the Bragg condition for different acoustic frequencies. The spread in the optical beam provides an apparent frequency bandwidth of

$$\Delta F = \frac{2v \cos \theta}{w} \tag{14-66}$$

Here also, the light is diffracted into a range of angles and the entire bandwidth is not usable. The available bandwidth is determined by the acoustic beam's divergence, and the light at larger diverging angles is wasted.

The most efficient design matches the light's divergence to the acoustic beam's divergence $\delta\theta = \delta\Phi$. The optimum design of a frequency modulator is thus given by

$$\frac{\lambda}{w} = \frac{\Lambda}{L}$$

and the modulation bandwidth can be calculated using either **(14-65)** or **(14-66)**. With this design, no optical or acoustic energy is wasted.

From **(14-66)**, we learn that a large bandwidth can be obtained by decreasing the width of the light beam; however, there is a minimum modulation frequency $F_{\min}$ that is determined by the beamwidth of the light wave.

TABLE 14.8 Acoustooptic Material Properties

Material	ρ, g/cm^3	v_s, km/sec	n	Attenuation (500 MHz), dB/μsec
Water	1.0	1.5	1.33	0.6
Fused silica	2.2	5.97	1.46	1.8
$LiNbO_3$	4.7	6.6	2.25	0.03
$PbMO_4$	6.95	3.66	2.3	1.2
TeO_2	5.99	4.26 (long)	2.27	1.0
		0.617(shear)[a]	3.0	

[a] An acoustic shear wave is a transverse acoustic wave.

The minimum modulation frequency occurs when one wavelength of sound Λ_{min} is contained in the light beam

$$\Lambda_{min} = \frac{2\pi v}{\Omega_{min}} = w \cos \theta$$

(This limit is a bit too liberal because it does not meet the sampling theorem requirement. A more useful minimum frequency is established by requiring three sound waves to be contained within the light beam.) Lower modulation frequencies cannot be resolved by the light beam because only a portion of the density variation associated with the sound wave will be contained in the beam.

The material selected for an acoustooptic frequency modulator should have a large speed of sound if a large bandwidth is desired. Accoustooptic properties of a number of important materials are found in Table 14.8. For example, the velocity of propagation for a longitudinal wave in TeO_2 is 4.26 km/sec; by using a beam focused to a width of 50 μm, the bandwidth of the modulator would be 170 MHz. In $LiNbO_3$ with the same optical parameters but a sound velocity of 6.6 km/sec, the bandwidth is 264 MHz. (A third method of expanding the bandwidth of an acoustooptic modulator uses an array of acoustic transducers. By adjusting the relative phase of the elements of the array, the acoustic beam direction is adjusted to insure that the Bragg condition is met.)

Spectrum Analyzer

The Bragg equation **(14-42)** states that each frequency in an acoustic wave will be diffracted through the elastooptic interaction into a unique angle. This fact reduces the effectiveness of an acoustooptic device used as a frequency modulator. However, what was a handicap is an advantage in a spectrum analyzer in which the objective is to measure the spectral content of the acoustic signal. The design goal of an fm modulator is to reduce the angular spread of the diffracted light, but the design goal of a spectrum analyzer is to maximize the angular spread to increase the accuracy of the analyzer. Figure 14-18 shows the optical system that performs the role of a spectrum analyzer.

The optical spectrum analyzer shown in Figure 14-18 consists of a plane wave of light, incident on an acoustooptic modulator. The acoustic wave is generated by applying an electrical signal $f(t)$ to a piezoelectric transducer. The acoustic wave produced diffracts the light wave via the elastooptic interaction. The Fraunhofer diffraction pattern produced by the

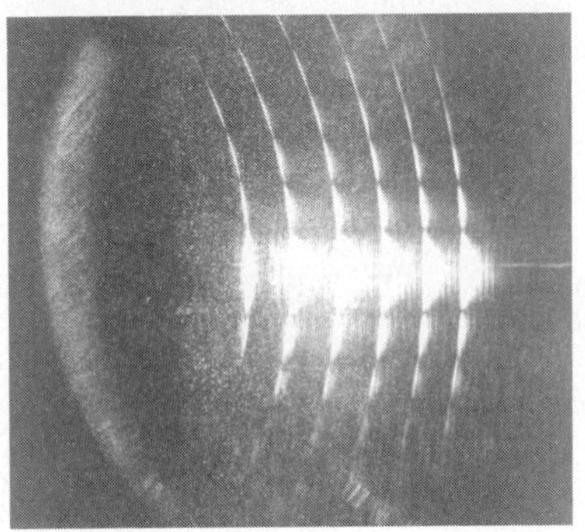

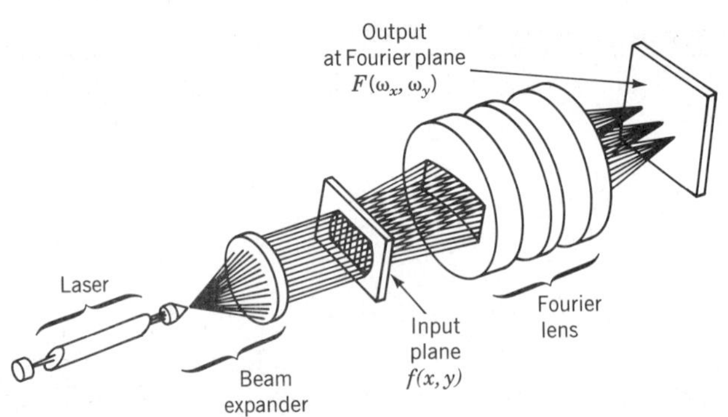

FIGURE 14-18. A spectrum analyzer. An electric signal is applied to a piezoelectric transducer that converts the electrical signal to an acoustic signal at the Input plane. A parallel beam of light is diffracted by the acoustic signal and the diffraction pattern produced appears in the back focal plane of the Fourier lens. The figures on the left show the spectrum produced when $f(t)$ contains one and five pure tones. The photos were intentionally overexposed to demonstrate that our theory has neglected other types of scattering processes. The engineer who wishes to use the acoustooptic effect in a spectrum analyzer must take all scattering processes into account because the undesired scattered light will limit the dynamic range of the analyzer. (Courtesy A. VanderLugt, North Carolina State University.)

acoustooptic interaction appears at the back focal plane of the lens labeled "Fourier lens."

On the left of Figure 14-18 are two typical diffraction patterns. The upper pattern is produced when $f(t)$ consists of a simple sine wave at a frequency of 435 MHz. The lower pattern is produced when $f(t)$ equals the sum of six sinusoidal waves at six distinct frequencies. If a linear detector array is positioned in the plane where these diffraction patterns were recorded, it will measure the frequency content of $f(t)$.

Acoustooptic Beam Deflector

A second application that seeks to maximize the deflection angle change for a given frequency change is the acoustooptic beam deflector. The acoustooptic modulator can quite naturally be used as a deflection or scanning device, and a number of applications are based on the ability to change the direction of the beam without modifying its intensity. If we rewrite **(14-64)**, we obtain an equation for the change in a light beam's direction as a function of the change in the sound frequency

$$\Delta\theta = \left(\frac{\lambda}{2v \cos \theta_B}\right) \Delta F$$

By changing the frequency of the acoustic wave, the diffracted light beam can be redirected. Two independent diffraction angles in space can be addressed by a deflected light beam if they are separated by more than the diffraction spread of the beam. Stated another way, the positions addressed by the diffracted light beam are unique if Rayleigh's criterion for resolution is met (Appendix 10-C). In the optimum design of a beam deflector, it is therefore necessary for the diffraction spread of the acoustic beam $\delta\Phi$ to be much larger than the diffraction spread of the optical beam, Figure 14-16. In this way, the positions addressed by the scanned light beam are determined by the diffraction spread of the light beam $\delta\theta$ which is determined by the size of the incident light wave. If we divide the maximum deflection angle $2\Delta\theta$ by the diffraction spread, we obtain the total number of resolvable positions that can be addressed by a beam deflector

$$N = \frac{2\Delta\theta}{\delta\theta} = \frac{w}{v \cos \theta_B}\Delta F \tag{14-67}$$

(This number also represents the total number of unique frequencies that can be measured by a spectrum analyzer.) To obtain a large value for N, the acoustooptic material should have a small acoustic propagation velocity. The modulator should also be very long so that the optical beam size w can be large.

A second measure of a deflector's performance is the speed with which each resolvable position can be addressed. The maximum rate that the frequency can be changed is limited by the time required for a sound wave to move across the light beam

$$\tau = \frac{w}{v \cos \theta_B}$$

(The acoustic wave at the old frequency must clear from the modulator before we are assured that the diffracted light beam is pointing in a new direction.) We are now faced with a trade-off that must be made. For the deflector to address a large number of spots, we wish w to be large, but to address the spots rapidly, we wish w to be small.

In recognition of the trade-off that must be made, a figure of merit for a deflector is defined to equal the number of separate positions that can be addressed in a unit time

$$\frac{N}{\tau} = \left(\frac{w}{v \cos \theta_B}\Delta F\right)\left(\frac{v \cos \theta_B}{w}\right) = \Delta F \tag{14-68}$$

The quantity

$$N = \tau\Delta F$$

called the *time-bandwidth product* is a fundamental measure of the information handling capability of any system.

In an imaging system, the equivalent measure of information throughput is called the *space-bandwidth* product. It is equal to the total number of pixals in an image, where a *pixal* is the smallest resolvable point in the image. In one dimension, a pixal would equal Δx the resolution of the optical system. If the maximum image size were x_{max}, then the space-bandwidth product would be

$$N = \frac{x_{max}}{\Delta x}$$

The resolution of an imaging system is given by **(10B-7)**

$$\Delta x = \frac{\lambda f}{D} = \frac{\lambda f}{2a}$$

where f is the focal length of the optical system and D its aperture diameter. From the geometrical optics' definition of a ray angle γ (see Figure 5-9), we can rewrite this relationship as

$$\Delta x = \frac{\lambda}{2\gamma}$$

so that the space-bandwidth product can be written

$$N = \frac{2x_{max}\gamma}{\lambda}$$

As the final step, recall the definition of the optical invariant **(5A-5)**

$$\mathcal{H} = nx\gamma$$

We see immediately that the space-bandwidth product of an imaging system is proportional to the optical invariant

$$N = \frac{2}{n\lambda}\mathcal{H}$$

We find that classical geometrical optics supports our modern view of an optical system as a linear system. The total information throughput is a constant proportional to the optical invariant.

One application of a beam deflector is the scanning of an image for electronic transmission, or of a photo-sensitive material to produce an image in reception. In such an application, the time bandwidth of the scanner must match the space bandwidth of the image and thus requires a large value for N. As an example, consider an acoustooptic modulator made of TeO_2 using a slow shear wave; the propagation velocity is 0.617 km/sec. If the optical beam width is 25 mm and the bandwidth 50 MHz, the number of addressable positions is then 2026. The beam direction cannot be changed in a time shorter than 32 μsec. This performance would be adequate for facsimile transmission applications or a high-resolution spectrum analyzer.

Another application of a beam deflector is the rapid diffraction of a light beam out of a laser cavity, as is shown in Figure 14-19. In such an application, the value of N is not important but τ should be very small. An acoustooptic beam deflector used for rapidly switching a light beam out of a laser cavity fulfills the role of a Q-switch, cavity dumper, or mode locker. The details of the design will be ignored here and only a general description of the role of the acoustooptic device will be discussed.

In the design shown in Figure 14-19, the laser operates normally without an acoustic signal present. The laser oscillations are established between the output mirror and the highly reflective back mirror; the prism performs the role of a frequency selector by refracting only one wavelength onto the back mirror. When an acoustic wave is launched in the prism, part of the light energy, labeled "dump" in Figure 14-19, is diffracted out of the cavity. This increases the loss in the laser cavity and the laser turns off. By interrupting the laser action at the proper times, high-power narrow pulses can be generated.

As an example of an acoustooptic mode locker, an acoustooptic beam deflector can be constructed of fused silica. The sound velocity is 6 km/sec, the bandwidth 40 MHz, and the width of the optical beam 4 mm. The total number of addressable positions is 27 but now $\tau = 0.7$ μsec.

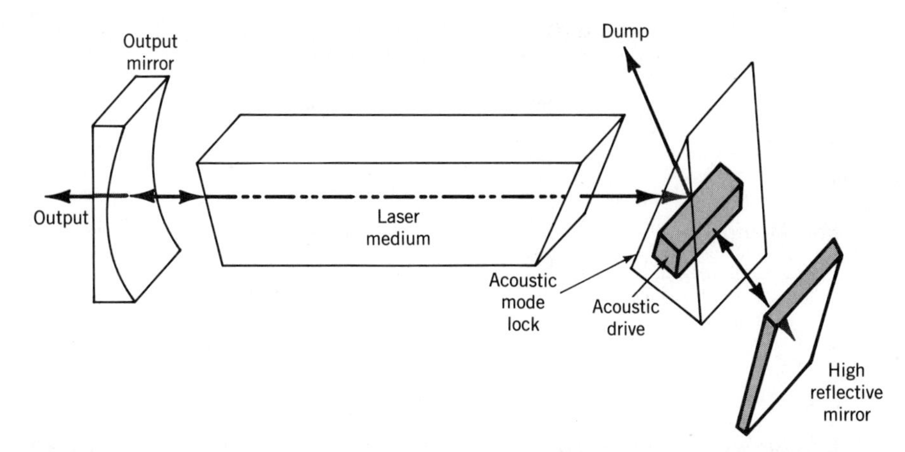

FIGURE 14-19. The use of an acoustooptic transducer in a laser cavity. In the mode locking application shown here, the acoustic wave periodically turns off the laser by diffracting part of the energy out of the laser cavity. This loss is removed when the acoustic signal passes through zero.

SUMMARY

In this chapter, the linear and quadratic electrooptic effects were found to modify the index of refraction in a material and produce birefringent behavior. This behavior creates an electrically controllable retardation that can be used to produce intensity-modulated light. If the electrooptic material is placed between crossed polarizers, the light transmitted through the polarizers is given by

$$\frac{I}{I_0} = \sin^2 \frac{\Gamma}{2}$$

where the phase change Γ is a function of the applied electric field.

For the quadratic electrooptic effect,

$$\Gamma = 2\pi \mathrm{K} E^2 L$$

$$= \frac{\pi}{\lambda} n_0^3 \rho E^2 L$$

$$= \frac{1}{2}\left(\frac{V}{V_{\lambda/2}}\right)^2$$

where

$$V_{\lambda/2} = \frac{d}{\sqrt{2\mathrm{K}L}}$$

is the voltage that must be applied across the electrooptic material of thickness d and length L to create a half-wave retardation. The parameter K is called the Kerr constant.

For the linear electrooptic effect, two possible modulator designs have been analyzed

1. The longitudinal modulator produces a phase change given by

$$\Gamma = \frac{2\pi V}{\lambda} n^3 r$$

$$= \pi \frac{V}{V_\pi}$$

where the voltage needed to create a half-wave retardation

TABLE 14.9 Order-of-Magnitude Estimate of Electro- and Magnetooptic Effects

Effect	Field	Strength Δn
Pockel	E	10^{-12} E
Kerr	E^2	10^{-19} E^2
Faraday	H	10^{-14} H
Cotton–Mouton	H^2	10^{-25} H^2
Voigt	H^2	10^{-28} H^2

$$V_\pi = \frac{\lambda}{2n^3 r}$$

is independent of the dimensions of the modulator.

2. The transverse modulator produces a phase change given by

$$\Gamma = \frac{\pi n^3 r}{\lambda}\left(\frac{VL}{d}\right)$$

The half-wave retardation voltage for the transverse modulator is

$$V_\pi^t = \frac{2V_\pi}{L/d}$$

A material placed in a magnetic field will exhibit a magnetooptical effect called the Faraday effect. This interaction causes the material to become optically active. A linearly polarized wave is rotated through an angle given by

$$\theta = \mathsf{V}LH$$

where H is the applied magnetic field and V is called the Verdet's constant.

Several magneto- and electrooptic effects were discussed in this chapter. A comparison of the size of each of the interactions is made in Table 14.9.

A strain field will also create birefringent behavior in a normally isotropic material. A time-dependent strain field is called an acoustic field, even when the frequency exceeds the range of human hearing. The acoustic wave will diffract a light wave into an angle given by the Bragg equation

$$\sin\,\theta_B = \frac{\lambda}{2\Lambda}$$

The frequency of the diffracted light is shifted by the acoustic frequency from the incident wave's frequency. The Bragg scattering condition only applies to an acoustic wave whose wavefront is wide relative to the acoustic wavelength. For waves that are narrow, a more complicated scattering condition called Raman–Nath scattering occurs. For Raman–Nath scattering, a number of diffracted orders are visible.

PROBLEMS

14-1. Derive the equation giving the index of refraction change produced by the linear electrooptic effect **(14-13)**.

14-2. Write the equation of the electrooptic indicatrix for a crystal belonging to the $\bar{4}2$m symmetry group.

14-3. Assume that the index of refractions for a uniaxial crystal, belonging to point group $\bar{4}2$m, are $n_1 = n_2 = n_0$ and $n_3 = n_e$. Also assume that an external electric field is applied along the z axis (label it E_3). Use these assumptions to modify the equation produced in Problem 14-2 and then find the coordinate transformation needed to put this equation into the form **(14-33)**.

14-4. Using **(14-33)**, derive the index change produced by the electric field applied along the z axis.

14-5. Derive the equation for the phase change produced by a longitudinal electrooptic modulator using the Pockels effect **(14-34)**.

14-6. Derive the equation for the phase change produced by a transverse electrooptic modulator using the Pockels effect **(14-37)**.

14-7. For a Pockels modulator, calculate the half-wave voltage for KDP and ADP for $\lambda = 500$ nm.

14-8. If we select the transverse electrooptic modulator design, what is the applied voltage needed to obtain a phase shift of π in KDP and ADP, assuming a crystal thickness of $d = 1$ mm? What is the length needed to give a $\pi/4$ bias?

14-9. Calculate the value of the Larmour precession frequency for a magnetic field of 1 Weber/m^2.

14-10. What colors would signify large values of stress if we used white light to observe stress in a plastic model? (The use of photoelastic models is an important tool in architectural research.)

14-11. Show that Bragg's equation **(14B-19)** is independent of the index of refraction of the material if the material is isotropic and the sides of the material are parallel to the direction of sound propagation.

14-12. Assume that the acoustical system has a center frequency of 200 MHz and a bandwidth of 100 MHz. What should the diameter of the light and the acoustic beams be to address 25 positions?

14-13. The sound velocity in glass is 3 km/s. If the sound wave is 1 cm wide, how far does the sound wave move during the passage of a light wave across the width of the sound wave? The index of refraction of the glass is $n = 1.5$.

14-14. Lithium niobate, a member of the 3m point group, is a popular material for electrooptic modulators. Calculate the indicatrix for this material with the electric field along the z direction.

14-15. Using the indicatrix produced in Problem 14-14, find the index change as a function of applied field in lithium niobate. Assume the light wave is propagating in the x direction and find the phase shift as a function of applied voltage.

14-16. GaAs is a cubic material belonging to the $\bar{4}3$m point group that is used as an electrooptic modulator in the infrared. Calculate the indicatrix for this material with the electric field along the z direction.

14-17. Using the indicatrix produced in Problem 14-16, find the index change as a function of applied field in GaAs. Assume the light wave is propagating in the z direction and find the phase shift as a function of applied voltage. What is the phase shift if the light is propagating in the x direction?

14-18. If 1 W of acoustic energy is sent into a TeO_2 crystal as an acoustic shear wave, what index of refraction change is produced? Assume that the sound beam is round with a diameter of 1 cm.

14-19. What is the size of the magnetic field required to rotate the polarization of light by $\pi/2$ in zinc sulfide?

Appendix 14-A

In Chapter 14, we demonstrated that for a lossless crystal

$$\backepsilon_{ik}^{0} = \backepsilon_{ki}^{0}$$

We can use a similar argument to demonstrate that

$$r_{ikj} = r_{kij}$$

$$\rho_{ikjm} = \rho_{kijm} = \rho_{ikmj}$$

This reduces the number of independent coefficients for the Pockels tensor to 18 and for the Kerr tensor to 36. Because of this symmetry requirement, we can reduce the number of subscripts by replacing *ik* by a single subscript and *jm* by a single subscript in the following manner:

$$\begin{array}{ll} 11 \Rightarrow 1 & 23 \Rightarrow 4 \\ 22 \Rightarrow 2 & 31 \Rightarrow 5 \\ 33 \Rightarrow 3 & 12 \Rightarrow 6 \end{array}$$

With this new notation, **(14-1)** can be rewritten in matrix notation as

$$\begin{bmatrix} \backepsilon_{11} - \backepsilon_1^0 \\ \backepsilon_{22} - \backepsilon_2^0 \\ \backepsilon_{33} - \backepsilon_3^0 \\ \backepsilon_{23} \\ \backepsilon_{31} \\ \backepsilon_{12} \end{bmatrix} = \begin{bmatrix} r_{11} & r_{12} & r_{13} \\ r_{21} & r_{22} & r_{23} \\ r_{31} & r_{32} & r_{33} \\ r_{41} & r_{42} & r_{43} \\ r_{51} & r_{52} & r_{53} \\ r_{61} & r_{62} & r_{63} \end{bmatrix} \begin{bmatrix} E_1 \\ E_2 \\ E_3 \end{bmatrix}$$

$$+ \begin{bmatrix} \rho_{11} & \rho_{12} & \rho_{13} & \rho_{14} & \rho_{15} & \rho_{16} \\ \rho_{21} & \rho_{22} & \rho_{23} & \rho_{24} & \rho_{25} & \rho_{26} \\ \rho_{31} & \rho_{32} & \rho_{33} & \rho_{34} & \rho_{35} & \rho_{36} \\ \rho_{41} & \rho_{42} & \rho_{43} & \rho_{44} & \rho_{45} & \rho_{46} \\ \rho_{51} & \rho_{52} & \rho_{53} & \rho_{54} & \rho_{55} & \rho_{56} \\ \rho_{61} & \rho_{62} & \rho_{63} & \rho_{64} & \rho_{65} & \rho_{66} \end{bmatrix} \begin{bmatrix} E_1^2 \\ E_2^2 \\ E_3^2 \\ E_2E_3 \\ E_3E_1 \\ E_1E_2 \end{bmatrix} \qquad (14A\text{-}1)$$

The material's symmetry reduces the number of nonzero coefficients in **(14A-1)**. Table 14A.1 lists the point groups[28] that do not exhibit a Pockels

TABLE 14A.1 Crystal Groups with No Pockels Effect

System	Group		
Triclinic		$\bar{1}$	
Monoclinic		2/m	
Orthorhombic		mmm	
Tetragonal	4/m	4/mmm	
Trigonal	$\bar{3}$	$\bar{3}$m	
Hexagonal	6/m	6/mmm	
Cubic	m3	432	m3m
Isotropic			

effect. Table 14A.2 displays the nonzero terms for each of the point groups without inversion symmetry.

Table 14A.2 applies to the Pockels tensor, the second-order susceptibility tensor d, which is discussed in Chapter 15, and the piezoelectric tensor that we will not discuss. Table 14A.3 displays the nonzero tensor elements for the Kerr and photoelastic tensors. Three of the fourth-rank tensors appear slightly more complicated than they actually are. For the trigonal and hexagonal systems,

$$\rho_{66} = \rho_{11} - \rho_{12}$$

For the trigonal system, the Kerr tensor has

$$\rho_{16} = 2\rho_{62}, \qquad \rho_{46} = 2\rho_{52}, \qquad \rho_{56} = 2\rho_{41}$$

For the hexagonal system, the Kerr tensor has

$$\rho_{61} = 2\rho_{26}$$

The photoelastic tensors for these systems differs from the electrooptic tensors in that the elements just identified differ by a factor of 2. Thus, for the trigonal and hexagonal systems

$$p_{66} = \tfrac{1}{2}(p_{11} - p_{12})$$

For the trigonal system,

$$p_{16} = p_{62}, \qquad p_{46} = p_{52}, \qquad p_{56} = p_{41}$$

and for the hexagonal system,

$$p_{61} = p_{26}$$

TABLE 14A.2 Pockeks and Second-order Susceptibility Tensors

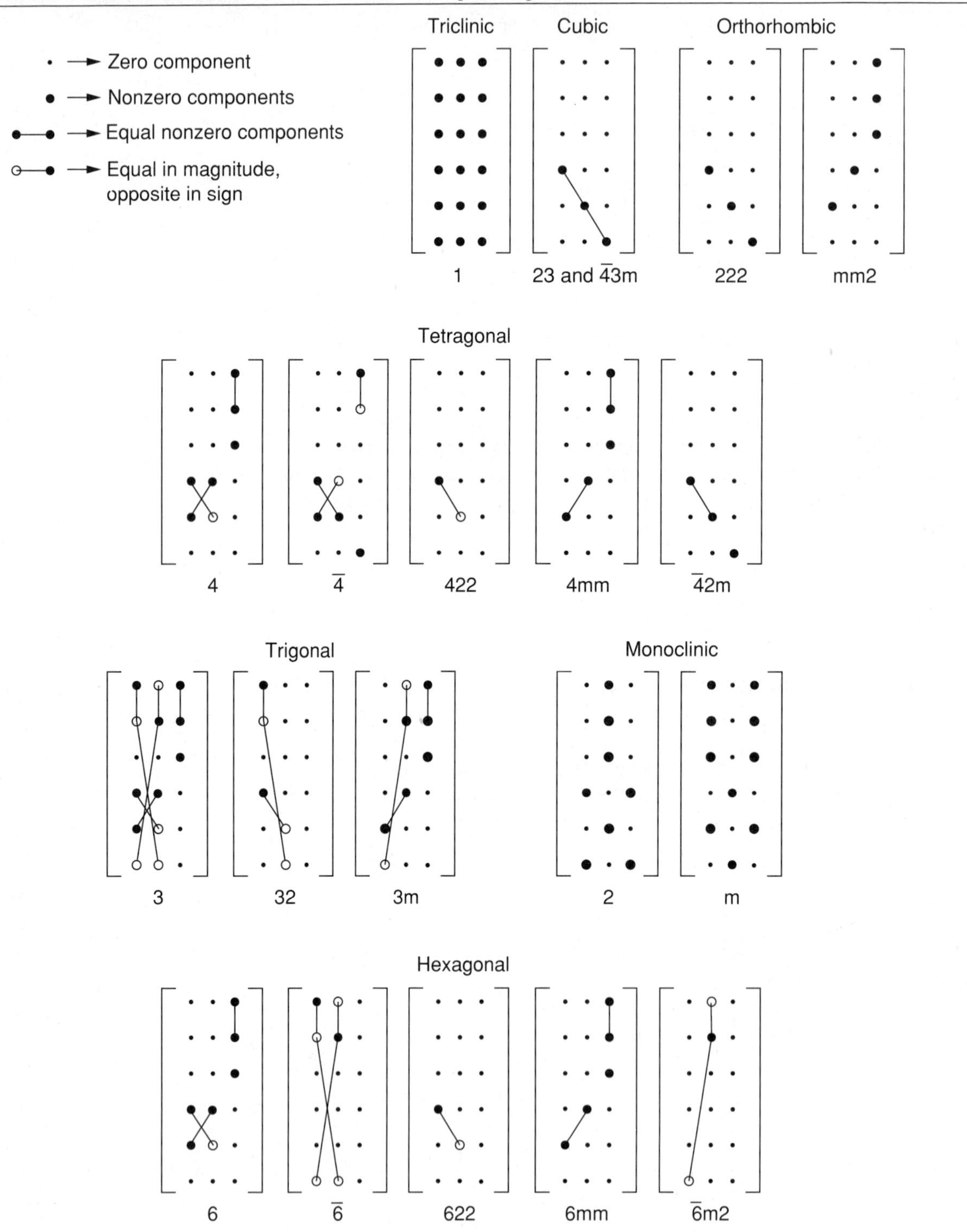

TABLE 14A.3 Kerr and Photoelastic Tensors

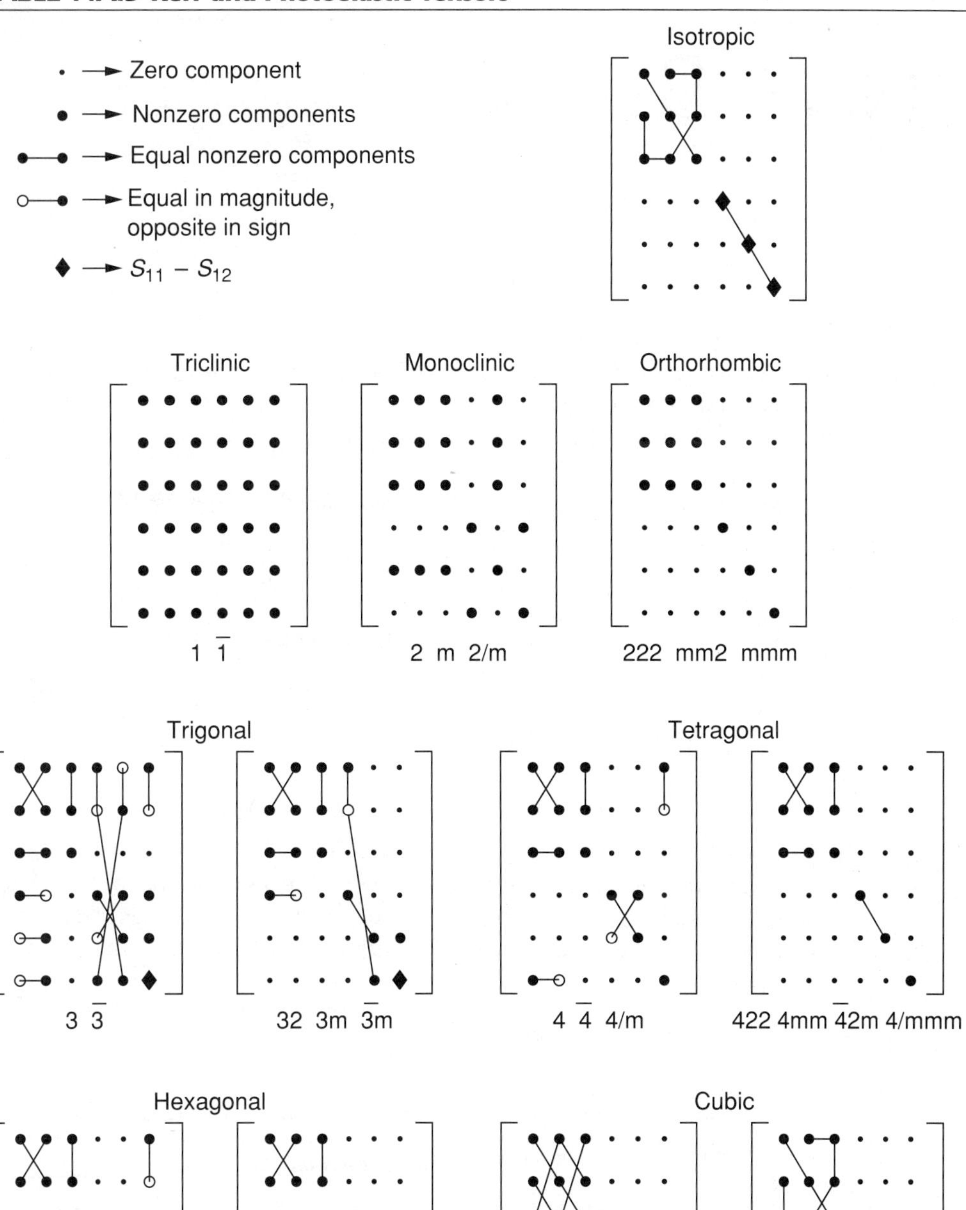

Appendix 14-B

PHENOMENOLOGICAL ACOUSTOOPTIC THEORY

An acoustic wave is a pressure wave that modulates the number density of the atoms of the medium

$$N(\mathbf{r}, t) = N_0[1 + \Delta \cos(\Omega t - \boldsymbol{\kappa} \cdot \mathbf{r})] \tag{14B-1}$$

where the acoustic wave variables are, as defined in Chapter 14,

$$\frac{\Omega}{\kappa} = \frac{\Lambda}{2\pi} 2\pi F = v \tag{14B-2}$$

We assume the dielectric constant has the same form as **(14B-1)**

$$\epsilon = \epsilon' + \epsilon_1 \cos(\Omega t - \boldsymbol{\kappa} \cdot \mathbf{r}) \tag{14B-3}$$

or in terms of the index of refraction,

$$\frac{1}{n^2} = \frac{1}{n_0^2} + \frac{1}{n_1^2} \cos(\Omega t - \boldsymbol{\kappa} \cdot \mathbf{r})$$

Since the dielectric constant is now a function of space and time, we must rederive the propagation properties of a wave from Maxwell's equations. Taking the curl of **(2-3)** and using **(2A-12)**, as was done in Chapter 2, produce

$$-\nabla^2 E + \nabla(\nabla \cdot \mathbf{E}) = -\mu \frac{\partial^2 \mathbf{D}}{\partial t^2}$$

We can find the divergence of **E** by first writing the divergence of **D**

$$\nabla \cdot \mathbf{D} = \epsilon \nabla \cdot \mathbf{E} + \mathbf{E} \cdot \nabla \epsilon = 0$$

where ϵ is no longer independent of the spatial coordinates. Solving for the divergence of **E** yields

$$\begin{aligned} \nabla \cdot \mathbf{E} &= -\frac{1}{\epsilon} \mathbf{E} \cdot \nabla \epsilon \\ &= -\mathbf{E} \cdot \nabla \ln \epsilon \end{aligned} \tag{14B-4}$$

The wave equation, which describes the propagation of light in a medium modulated by an acoustic wave, becomes

$$\nabla^2 \mathbf{E} + \nabla(\mathbf{E} \cdot \nabla \ln \epsilon) = \mu \frac{\partial^2}{\partial t^2} (\epsilon \mathbf{E}) \tag{14B-5}$$

If we assume **E** is a superposition of plane waves of wavelength λ/n in the medium and use the time- and space-dependent equation for ϵ, we find that $\nabla(\mathbf{E} \cdot \nabla \ln \epsilon)$ is of the order

$$O\,[\nabla(\mathbf{E} \cdot \nabla \ln \epsilon)] = \epsilon_1 \frac{\lambda}{\Lambda} \nabla^2 \mathbf{E}$$

Since

$$O\left[\frac{\lambda}{\Lambda}\right] = 10^{-2}$$

we may normally neglect that term and write the wave equation as

$$\nabla^2 \mathbf{E} = \mu \frac{\partial^2}{\partial t^2}(\epsilon \mathbf{E}) \tag{14B-6}$$

We will use the acoustooptic experiment shown in Figure 14B-1. The sound is traveling in the x direction. It occupies the region in the z direction of $-L/2 \le z \le L/2$. Light is incident on the sound column at an angle θ with respect to the z axis and is represented by

$$\mathcal{E} = \mathbf{E}_0 \exp\,[i(\omega t + kx \sin\,\theta - kz \cos\,\theta)] \tag{14B-7}$$

After passing through the sound field, the electric field is modulated by the periodicity of the sound field. To describe the periodic disturbance of the light's wavefront, we expand the wave in a Fourier series

$$\mathcal{E} = \sum_{\ell=-\infty}^{\infty} V_\ell(z) \exp\{i\,[(\omega + \Omega\ell)t + (k \sin\,\theta - \ell\kappa)x - kz\cos\theta]\} \tag{14B-8}$$

Equation **(14B-8)** represents an expansion of E in terms of plane waves. The index ℓ labels the ℓth Fraunhofer diffraction order or the ℓth spatial frequency needed to describe the wavefront.

The frequency of the ℓth order in **(14B-8)** is $\omega + \ell\Omega$. This frequency shift is the Doppler shift, produced by the reflection of the light wave from a moving acoustic wavefront.

The quantum mechanical formalism, used in the scattering model of acoustooptics, views the ℓth term of the expansion as a plane wave originating from the destruction or creation of ℓ phonons by the incident light beam.

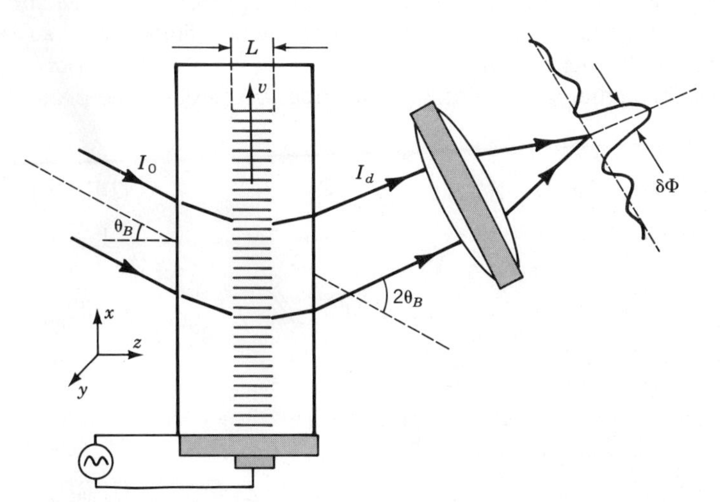

FIGURE 14B-1. Geometry for an acoustooptic experiment. We assume that the device is infinitely wide in the y direction.

The x component of the wave vector for the ℓth order is

$$k_{\ell x} = k \sin \theta - \ell\kappa$$
$$= -k_\ell \sin \phi_\ell$$

(In Figure 14B-1, the special case of Bragg scattering where $\phi_\ell = 2\theta_B$ is shown.) The ℓth wave vector has a magnitude given by

$$k_\ell = \frac{\omega_\ell}{c}$$
$$= \frac{\omega + \ell\Omega}{c}$$

and the angular deflection ϕ_ℓ of the ℓth wave vector is

$$\sin \phi_\ell = \frac{c(\ell\kappa - k \sin \theta)}{\omega + \ell\Omega}$$
$$= \frac{\ell\dfrac{\kappa}{k} - \sin \theta}{1 + \ell\dfrac{\Omega}{\omega}} \tag{14B-9}$$
$$\approx \ell\frac{\lambda}{\Lambda} - \sin \theta$$

For a given λ, the angular deflection ϕ_ℓ of the ℓth order decreases with increasing Λ. If Λ is sufficiently large, diffraction of the light by the sound waves will not be observed. In Chapter 14, the lower limit on observation of acoustooptic diffraction was established by requiring an acoustic wavelength no longer than the width of the optical beam. Here, the limit is established by the requirement that the diffracted and undiffracted beams not overlap.

We substitute **(14B-8)** into **(14B-6)**, neglecting terms in

$$\frac{d^2V_\ell}{dz^2}$$

The second derivative is neglected because, in this approximation, each diffracted plane wave changes slowly with z. Once the substitution $\omega = kc$ and $\Omega = \kappa v$ is made, the fact that $\Omega/\omega << 1$ and $v/c << 1$ is used to produce an equation describing the behavior of the amplitudes of the diffracted plane waves

$$\frac{\partial V_\ell(z)}{\partial z} + i\beta_\ell V_\ell(z) + \xi[V_{\ell+1}(z) - V_{\ell-1}(z)] = 0 \tag{14B-10}$$

where

$$\beta_\ell = -\frac{\kappa\ell}{\cos \theta}\left(\sin \theta - \frac{\ell\kappa}{2k}\right) \tag{14B-11}$$

and

$$\xi = \frac{1}{4}\left(\frac{\epsilon_1}{\epsilon'}\right)\frac{k}{\cos \theta}$$
$$= \frac{1}{4}\left(\frac{n_0}{n_1}\right)^2\frac{k}{\cos \theta} \tag{14B-12}$$

The physical significance of the two parameters β_ℓ and ξ will become apparent later in the analysis.

The general solution of the set of coupled equations, **(14B-10)** is very difficult. Normally, the problem is considered in two limits. In the first limit, we assume that $V_{\ell+1} << V_\ell$. This approximation results in a theory that describes Bragg scattering. In the second limit, we assume

$$\frac{\beta_\ell V_\ell(z)}{\xi} = 0 \tag{14B-13}$$

which occurs when $\kappa \to 0$, as is apparent by examining **(14B-11)**. This approximation describes Raman–Nath scattering.

Bragg Region

The Bragg scattering equation is obtained by solving **(14B-10)** using the assumption that $V_{\ell+1} << V_\ell$. In the derivation, we will only consider terms of the Fourier series for which $\ell \geq 0$. (This assumption limits the theory to a description of the light, upshifted in frequency by acoustooptic scattering. The resulting theory can be used for light downshifted in frequency by replacing ℓ by $-\ell$ in the theory.) The initial condition selected for this discussion assumes that, before encountering the sound wave, only the fundamental frequency of the Fourier series is present, that is, for $z < -L/2$, only $V_0 \neq 0$. These assumptions modify **(14B-10)**, reducing it to a simple differential equation

$$\frac{\partial V_\ell}{\partial z} + i\beta_\ell V_\ell = \xi V_{\ell-1} \tag{14B-14}$$

The solution of this equation is

$$V_\ell = e^{-i\beta_\ell z} \int_{-\infty}^{\infty} \xi V_{\ell-1} e^{i\beta_\ell z'}\, dz' \tag{14B-15}$$

The physical significance of the parameter ξ can be deciphered by first obtaining an expression for the index of refraction change produced by the acoustic wave. From **(14B-3)**, we can write the difference between the maximum and minimum dielectric constant produced by the acoustic wave

$$\epsilon_{max} - \epsilon_{min} = \left.\frac{1}{n^2}\right|_{max} - \left.\frac{1}{n^2}\right|_{min} = \frac{2}{n_1^2}$$

We define the difference between the index of refraction at its maximum and minimum values as

$$\Delta n = n_{max} - n_{min}$$

where

$$\left.\frac{1}{n^2}\right|_{max} = \frac{1}{n_{min}^2} \qquad \left.\frac{1}{n^2}\right|_{min} = \frac{1}{n_{max}^2}$$

The peak-to-peak dielectric constant variation produced by the acoustic wave is approximately equal to

$$\epsilon_{max} - \epsilon_{min} \approx \frac{2\Delta n}{n_0^3}$$

The index change produced by the acoustic wave is therefore

$$\Delta n = \frac{n_0^3}{n_1^2}$$

The parameter ξ, as defined in **(14B-12)**, can now be seen to be proportional to the index change produced by the acoustic wave. Rewriting **(14B-12)** in terms of the index change yields

$$\xi = \frac{1}{4}\left(\frac{\Delta n}{n_0}\right)\frac{k}{\cos\theta} \tag{14B-16}$$

If we treat the sound wave as a plane wave with a uniform amplitude across its wavefront, then over the acoustic beamwidth of $-(L/2) \leq z \leq (L/2)$, Δn must be a constant. Since ξ is proportional to the index change, it must be a nonzero constant over the acoustic beamwidth and zero outside of the beam.

If we limit our consideration to the case when $\ell = 1$, **(14B-15)** can be greatly simplified. The initial assumption of this model that $V_{\ell+1} << V_\ell$ now becomes $V_1 << V_0$. This assumption is equivalent to assuming that very little energy will be diffracted out of the incident light wave by the sound wave. This small scattering assumption allows V_0 to be treated as a constant in **(14B-15)**.

The fact that very little energy is diffracted by the acoustic wave means that Δn is small and any departure of the sound wave from a plane wave will be negligible. This fact allows us to treat ξ as a constant and remove it from the integral in **(14B-15)**. We can now rewrite **(14B-15)** as

$$V_1 = e^{-i\beta_1 z}\xi V_0 \int_{-L/2}^{L/2} e^{i\beta_1 z'}\,dz'$$

This integral can be recognized as the Fourier transform of the rect(x) function **(6-17)** and is equal to a sinc function

$$V_1(z) = \xi V_0 L \frac{\sin\left(\beta_1 \frac{L}{2}\right)}{\beta_1 \frac{L}{2}} e^{-i\beta_1 z} \tag{14B-17}$$

The sinc function is large only when its argument is approximately zero so we may write the approximate equation to find the condition for $(V_1)_{max}$

$$\beta_1 = -\frac{\kappa}{\cos\theta}\left(\sin\theta - \frac{\kappa}{2k}\right) \approx 0 \tag{14B-18}$$

The requirement that **(14B-18)** be zero can be rewritten as

$$\sin\theta_B = \frac{\kappa}{2k} \tag{14B-19}$$

This condition is equivalent to the Bragg equation, where θ is denoted as θ_B to emphasize its unique character [see **(13-13)** and **(14-62)**].

The fraction of light diffracted at the Bragg angle θ_B is a measure of the diffraction efficiency **(12-14)** of the Bragg process. The fraction of energy found in the diffracted beam is

$$\frac{I_1}{I_0} = \frac{V_1 V_1^*}{V_0 V_0^*}$$

$$= \xi^2 L^2 \frac{\sin^2\left(\beta_1 \frac{L}{2}\right)}{\left(\beta_1 \frac{L}{2}\right)^2} \qquad \text{(14B-20)}$$

The diffracted intensity distribution predicted by **(14B-20)** is sketched in Figure 14B-1. The maximum intensity occurs at the Bragg angle for which $\beta_1 = 0$

$$\left(\frac{I_1}{I_0}\right)_{max} = (\xi L)^2$$

$$= \left(\frac{\Delta n L}{4 \cos\theta} \frac{k}{n_0}\right)^2$$

The diffraction efficiency η is defined as this maximum value

$$\eta = \left.\frac{I_1}{I_0}\right|_{max} = \xi^2 L^2$$

The index change produced by an acoustic wave can be expressed in terms of the acoustic wave's parameters **(14C-5)**. The diffraction efficiency can be written in terms of the acoustic parameters,

$$\eta = \frac{\pi}{(\lambda n_0)^2 \cos^2\theta} \bullet \frac{P_{ac} p^2 n_0^6}{\rho v^3} \qquad \text{(14B-21)}$$

where we have assumed that the area of the acoustic wave is $\pi L^2/4$. For example, assume 1 W of acoustic power is applied to a TeO_2 crystal as a slow shear wave. The acoustic parameters are

$$P_{ac} = 1 \text{ W}, \qquad p = 0.09$$

$$v_s = 0.617 \text{ km/sec}, \qquad \rho = 5.99 \text{ g/cm}^3$$

and the optical parameters are

$$n_0 = 2.27, \qquad \lambda = 632.8 \text{ nm}$$

From **(14B-21)**, the diffraction efficiency is 12%. (The value for the photoelastic constant was obtained using M_2 from Table 14C.1.)

In the Bragg approximation, the diffracted wave amplitudes V_ℓ decrease so rapidly that only V_1 is important. The theory just developed describes the scattering of light from sound wavefronts moving toward the light beam. The diffracted beam, corresponding to the $\ell = 1$ term in **(14B-10)**, is Doppler-shifted up in frequency and is called the +1 order. As was mentioned earlier, to treat the case of scattering from a sound wave moving away from the incident light, the diffracted beam is represented by the negative index, $\ell = -1$ in **(14B-10)**, and is called the −1 order. This order is Doppler-shifted down in frequency by the sound frequency Ω.

The diffraction efficiency obtained by assuming a small index modulation is only an approximate relationship for η. It is based on the assumption that only a small amount of the incident energy is diffracted out of the incident beam. In a more complete theory,

$$\eta = \sin^2(\xi L)$$

In summary,

1. The Bragg angle can be expressed as the reciprocal of the sound wavelength because the Bragg angle is quite small. From **(14B-19)**,

$$\theta \propto \frac{2\pi}{\Lambda} = \frac{\Omega}{v} \tag{14B-22}$$

   A reasonable estimate of the Bragg angle is obtained, therefore, from the ratio of the elastic wave frequency to its velocity. This relationship demonstrates explicitly that the diffraction angle can be changed by changing the acoustic frequency, as discussed in Chapter 14. An additional application of **(14B-22)** is the use of Bragg scattering to measure the elastic properties of a material.

2. The intensity of the diffracted light is proportional to the square of the index change produced by the acoustic wave. In Appendix 14-C, the square of the index change is shown to be proportional to the acoustic intensity

$$I_1 \propto \frac{P_{ac}}{A}$$

   The relationship between diffracted energy and acoustic energy permits the modulation of the diffracted beam by amplitude-modulating the sound wave.

3. The frequency shift of the diffracted light, appearing in **(14B-8)**, can be used to produce frequency-modulated light.

Raman–Nath Region

Another limiting condition can be used to solve **(14B-10)**. Rather than assume that the amplitudes of the Fourier series rapidly decrease,

$$V_\ell << V_{\ell-1}$$

we define a new parameter

$$\chi = 2\xi z$$

a simple rescaling of the z coordinate, and rewrite **(14B-10)** in terms of this new parameter

$$2\frac{\partial V_\ell(\chi)}{\partial \chi} + V_{\ell+1}(\chi) - V_{\ell-1}(\chi) = \left(-\frac{\ell \beta_\ell}{\xi}\right) V_\ell(\chi) \tag{14B-23}$$

Since

$$O\left(\frac{\kappa}{k}\right) = O\left(\frac{\lambda}{\Lambda}\right) = 10^{-2}$$

we can assume that

$$\frac{\beta_\ell}{\xi} = \frac{4\kappa\ell}{k}\left(\frac{\epsilon'}{\epsilon_1}\right)\left(\sin\theta - \frac{\ell\kappa}{2k}\right) << 1 \tag{14B-24}$$

This assumption is correct in the limit as $\kappa \to 0$ for a given value of (ϵ'/ϵ_1). If the elastic wave amplitude increases, then (ϵ'/ϵ_1) decreases, if we assume κ

is fixed. Thus, this approximation is expected to be applicable when light is diffracted by large-amplitude (large ϵ_1), low-frequency (small $\Omega = \upsilon\kappa$) elastic waves. Using the assumption stated in **(14B-24)**, we set

$$\frac{\ell \beta_\ell V_\ell(\chi)}{\xi} = 0 \tag{14B-25}$$

allowing us to rewrite **(14B-23)** as a recursion relation

$$2\frac{\partial V_\ell}{\partial \chi} + V_{\ell+1}(\chi) - V_{\ell-1}(\chi) = 0 \tag{14B-26}$$

This recursion relation is satisfied by the Bessel functions of integral order.

We assume, as an initial condition, that only the fundamental component of the Fourier series is present before interaction of the light wave with the acoustic wave, i.e., only $V_0 \neq 0$ when $z \leq -L/2$. The acoustic wave is assumed to be a plane wave so that the parameter ξ is a nonzero constant over the region $-L/2 \leq z \leq L/2$ and zero outside this region. The intensity distribution for the first-order diffracted wave is given by

$$\frac{I_1}{I_0} = \mathbf{J}_1^2\left(\frac{\Delta nkL}{n_0 2 \cos\theta}\right) \tag{14B-27}$$

The intensity distribution of the ℓth order is described by the ℓth Bessel function J_ℓ. The total diffracted field is described by a series of Bessel functions, mathematically identical to the series obtained in Appendix 12-A, for a thin-phase hologram. The intensity distribution predicted by **(14B-27)**, for the first five Bessel functions is shown in Figure 14B-2. This regime is sometimes called the *Lucas–Bequard* or *Debye–Sears* region.

Because each diffracted order is described by a Bessel function of that order, the maximum intensity of the first-order diffraction peak occurs when J_1 is a maximum, i.e., when the argument of J_1 is equal to 1.8. All of the light is diffracted out of the incident beam when the argument is equal to the first zero of the zero-order Bessel function, that is,

$$\frac{\Delta nkL}{n_0 2 \cos\theta} = 2.4 \tag{14B-28}$$

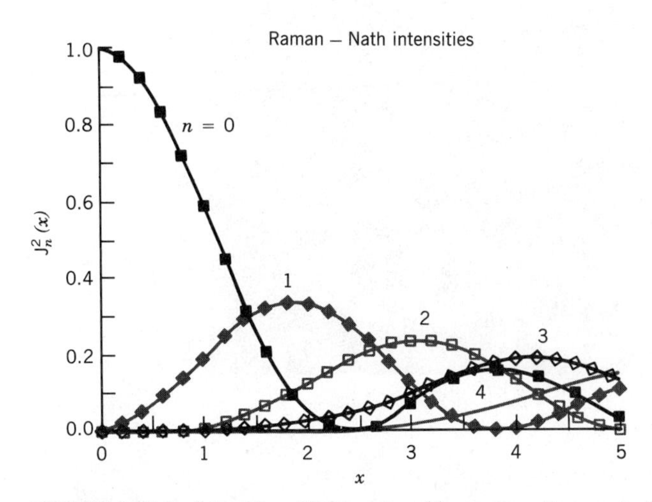

FIGURE 14B-2. Intensities of light scattered by an acoustic wave in the Raman–Nath regime.

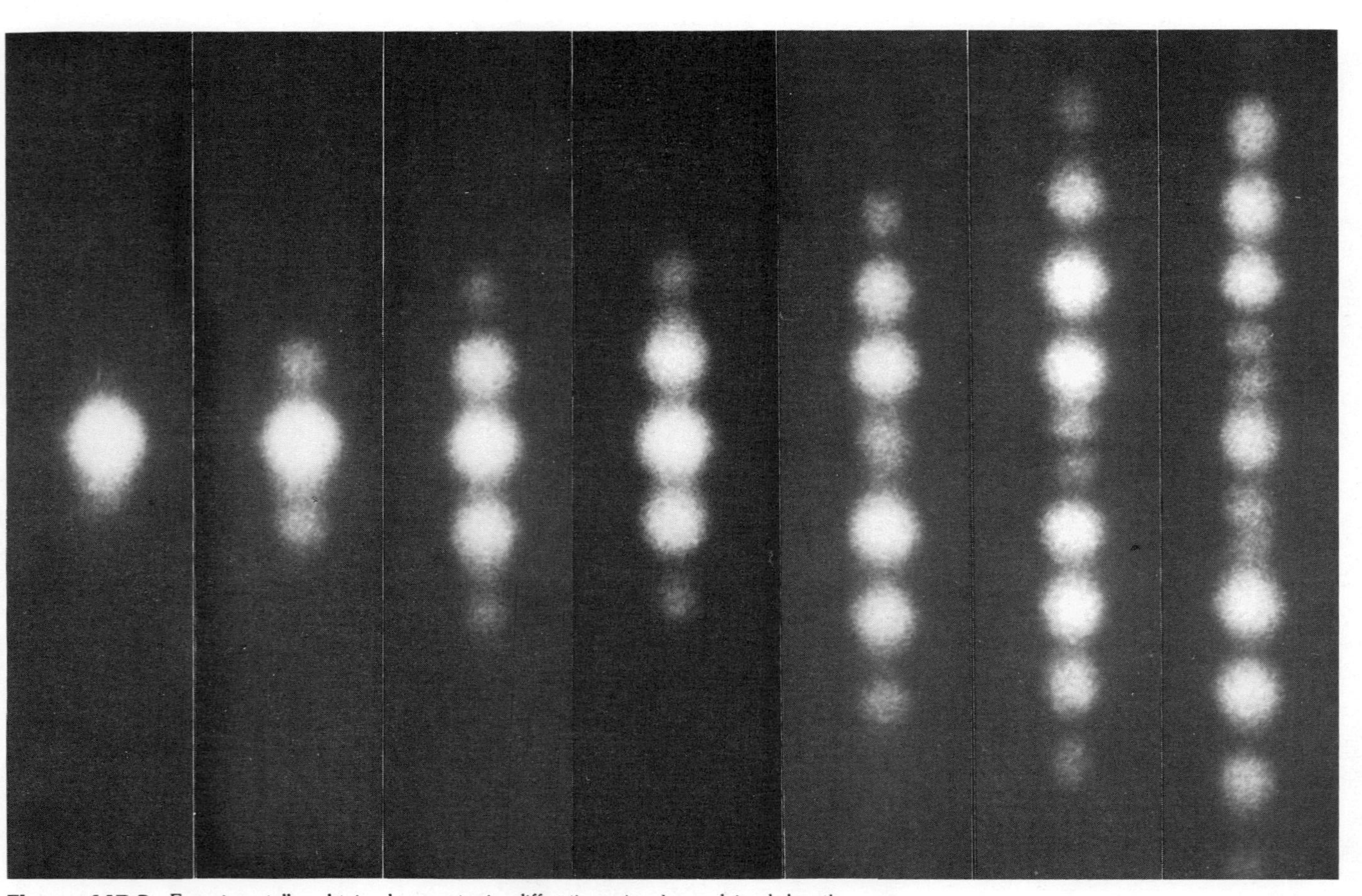

Figure 14B-3. Experimentally obtained acoustopic diffraction signals explained by the Raman-Nath theory. [Courtesy Eric Mazur, Harvard University].

In Figure 14B-3, the intensity pattern of light scattered by an acoustic beam under conditions for Raman–Nath scattering is shown. A schematic of the experiment used to produce the intensity patterns is shown in Figure 14B-4. The acoustic energy and therefore Δn was increased to produce the pictures that extend from left to right. It is easy to see that the energy in the incident beam is diffracted into the spatial sidebands as the acoustic energy is increased. At an acoustic energy that corresponds to the index modulation given by **(14B-28),** the intensity in the undiffracted beam goes to zero. The acoustic energy was increased further until, at higher acoustic modulation, the first-order diffracted intensity goes to zero.

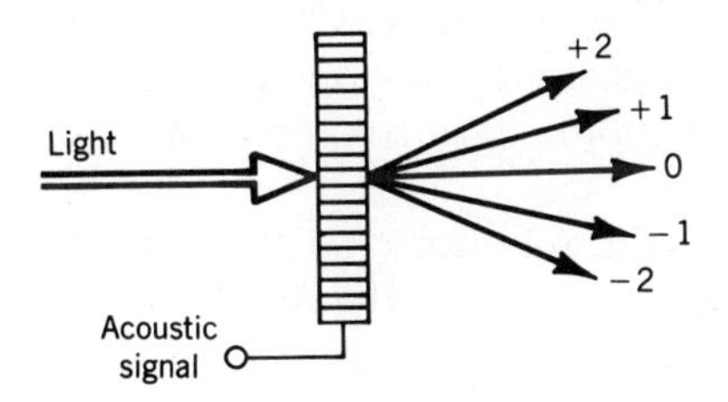

FIGURE 14B-4. A schematic representation of the experiment used to produce the Raman–Nath scattering of Figure 14B-3. Each diffraction order is labeled by an integer.

Appendix 14-C

ACOUSTIC FIGURE OF MERIT

In the acoustooptic literature, a number of figures of merit are used to characterize acoustooptic materials. To understand the reasoning that leads to these figures of merit, we will derive one of the most common figures of merit.

We begin by assuming that the medium is isotropic so that we can use a scalar strain and an average photoelastic constant. The result of these assumptions is that the change in index due to strain is given by

$$\Delta\left(\frac{1}{n^2}\right) = p|\mathbf{S}| \tag{14C-1}$$

or by using the approximation used to obtain **(14-57)**, we can write

$$\Delta n = \frac{pn^3}{2}|\mathbf{S}| \tag{14C-2}$$

The energy density in a strained medium is

$$\frac{1}{2}\mathbf{S}\mathbf{S}^*\rho v^2 = \frac{1}{2}|\mathbf{S}|^2\rho v^2 \tag{14C-3}$$

where v is the velocity of the pressure (sound) wave and ρ the material's density. The power in an acoustic wave is the energy density **(14C-3)** multiplied by the volume occupied by the wave

$$P_{ac} = Av\left(\frac{1}{2}|\mathbf{S}|^2\rho v^2\right)$$

where A is the area of the sound wavefront, v the velocity, and Av the volume containing the sound wave. Solving this equation for the strain yields

$$|\mathbf{S}| = \sqrt{\frac{2P_{ac}}{A\rho v^3}} \tag{14C-4}$$

This equation can be used in **(14C-2)** to calculate the index change produced by the sound wave

$$\Delta n = \sqrt{\frac{P_{ac}p^2n^6}{A\rho v^3}} \tag{14C-5}$$

The optical phase retardation of a light wave, propagating across the sound wave, is

$$\Gamma = \frac{2\pi L}{\lambda}\Delta n$$
$$= \frac{2\pi L}{\lambda}\sqrt{\frac{P_{ac}p^2n^6}{A\rho v^3}} \tag{14C-6}$$

If the sound beam is round with diameter L, then the beam's cross-sectional area is $A = 4\pi L^2/4$, giving a phase retardation of

$$\Gamma = 4\sqrt{P_{ac}\frac{\pi p^2 n^6}{\lambda^2 \rho v^3}}$$

The best material for an acoustooptic modulator is the one that produces the largest phase change Γ of a light wave, with the smallest sound power. The figure of merit that indicates a material's ability to perform this modulation is given by

$$M_2 = \frac{n^6p^2}{\rho v^3} \tag{14C-7}$$

Some typical values for this figure of merit are shown in Table 14C.1, along with two other figures of merit M_1 and M_3. These figures of merit are referenced to fused silica.

The other two figures of merit listed in Table 14C.1 have been developed to discriminate between materials in applications where device geometries introduce additional constraints. The definitions of these additional figures of merit are

$$M_1 = \frac{n^7p^2}{\rho v}, \qquad M_3 = \frac{n^7p^2}{\rho v^2}$$

All of these figures of merit neglect two important physical properties: acoustic and optical absorption. The two absorption coefficients reduce the efficiency of any modulator.

TABLE 14C.1 Acoustooptic Figures of Merit

Material	M_1	M_2, 10^{-15} sec^3/kg	M_3
Fused silica	1.0	1.51	1.0
Water	6.1	160	24
$LiNbO_3$	8.3	6.95	7.5
$PbMO_4$	15.3	35.8	24.9
TeO_2(long)	18.5	34.4	25.6
TeO_2(shear)	8.8	793	85

Nonlinear Optics

Nonlinear processes are involved in the detection of light and in all the light modulator devices discussed in Chapter 14. Before 1960 these nonlinear processes were the exception and conventional optical systems were analyzed using a linear theory. The neglect of nonlinear effects was due to the fact that large optical fields could not be produced. With the experimental demonstration of the ruby laser by **Theodore Harold Maiman** in July, 1960, a light source became available that would produce nonlinear optical behavior (a Q-switched ruby laser is 10^{12} times as bright as a high-pressure mercury arc lamp). The first nonlinear optics experiment was conducted by **Peter A. Franken**, along with several other investigators, at the University of Michigan. The Michigan group initially demonstrated second harmonic generation and later, optical rectification. The field of nonlinear optics quickly developed and today it is a large and active area of research.

The theory we have developed for optics has assumed a linear relationship between the applied electric field and the material's response, i.e., the polarization

$$\mathbf{P} = \epsilon_0 \chi \mathbf{E}$$

Using a typical laser, Nd:YAG, we can generate 1 Mw, 1 nsec pulses at a 1 μm wavelength. A typical beam diameter of 1 mm^2 results in a peak electric field strength of 10^7 V/m. If we focus this beam to an area of approximately λ^2, the electric field strength can reach 10^{10} V/m. This field strength is on the order of atomic field strengths and is too large to allow the relationship between **P** and **E** to be treated as linear. In this chapter, we will treat nonlinear effects by assuming that the relationship between **P** and **E** is described by a power series

$$\begin{aligned} P_i^{\omega} = P_i^0 &+ \sum_j \chi_{ij} E_j^{\omega} + \sum_{j,\ell} \chi_{ij\ell} \nabla_\ell E_j^{\omega} \\ &+ \sum_{j,\ell} \chi_{ij\ell} E_j^{\omega_1} E_\ell^{\omega_2} + \sum_{j,\ell,m} \chi_{ij\ell m} E_j^{\omega_1} E_\ell^{\omega_2} E_m^{\omega_3} + \sum_{j,\ell} \chi_{ij\ell} E_j^{\omega_1} B_\ell^{\omega_2} + \cdots \end{aligned}$$

This means that we are assuming that the nonlinear effects are a small perturbation on the linear behavior we have described. Each of the terms in this expansion can be identified with observable phenomena in optics.[63]

1. The first term in the expansion is associated with the spontaneous dc ($\omega = 0$) polarization of a material. The connection of this term to optics is not immediately apparent, but there are several important optical applications utilizing this polarization.

Pyroelectrics are materials that exhibit an electric polarization when the temperature of the crystal is changed. The materials exhibiting this effect have polar unit cells, i.e., the unit cell of the crystal has a net dipole moment. Rochelle salt, Wurtizite, and cane sugar are examples of pyroelectric materials. The temperature-induced polarization does not persist, but rather is neutralized by migration of charge to the crystal's surface. Thus, only temperature changes are observable by monitoring the polarization.

A detector constructed from a pyroelectric material is only sensitive to dT/dt and can only respond to modulated, chopped, or pulsed radiation. To construct a detector, electrodes are attached to the pyroelectric crystal, with an orientation perpendicular to the axis of polarization. Radiation incident on the detector surface causes a temperature change of the crystal that, through the pyroelectric effect, changes the spontaneous electric polarization. A current is generated to neutralize the thermally induced polarization that is proportional to the change in the crystal's temperature. Pyroelectrics are used in optics as broadband radiation detectors and are found commercially in thermally activated light switches. A popular detector material is $LiTaO_3$.

Ferroelectrics are pyroelectric materials that exhibit the long-range ordering of the dipole moments, creating a spontaneous polarization. The direction of spontaneous polarization can be reversed by an applied electric field. The polarization exhibits a hysteresis loop, when plotted as a function of electric field, due to domains in the crystal, similar to the domains found in magnetic materials. The ferroelectric is therefore an electric equivalent of a ferromagnet. The three major groups of ferroelectrics contain as their proto-typical members

Typical Members of Ferroelectric Classes

Name	Formula	Crystal Group
Barium titanate	$BaTiO_3$	m3m
KDP	KH_2PO_4	$\bar{4}2m$
Rochelle salt	$NaK(C_4H_4O_6) \cdot 4H_2O$	222

Through the rotation of the polarization of the individual domains, a spatial variation of the birefringence of a ferroelectric has been used to provide optical displays with nonvolatile storage.

---

2. The second term in the expansion is the normal linear optical response of a material.
3. The third term describes optically active materials.
4. The fourth term corresponds to the second-order processes listed in Table 15.1.

TABLE 15.1 Second-order Nonlinear Processes

Frequency of Incident Fields	Frequency of Polarization	Process
$\omega_1,\ \omega_2$	$\omega_1 = \omega_1 + \omega_2$	Sum frequency mixing
ω_1	$\omega\ = 2\omega_1$	Second harmonic generation
$\omega_1,\ 0$	$\omega\ = \omega_1$	Pockels effect
$\omega_1,\ \omega_2$	$\omega\ = \omega_1 - \omega_2$	Difference frequency mixing
ω_1	$\omega_1 = 0$	Inverse electrooptic effect, dc rectification

5. The fifth term corresponds to third-order processes listed in Table 15.2.
6. The sixth term corresponds to various magnetooptic effects. When $\omega_2 = 0$, the term describes the Faraday effect.

In this chapter, we will briefly examine some nonlinear processes and their applications. For more details, consult one of the several books on nonlinear optics.[64–66] The theoretical development of nonlinear optics presented in this chapter is based on perturbation theory, introduced in Appendix 15-A.

The polarization due to a nonlinear response of a material to an electromagnetic wave will be obtained from perturbation theory. The nonlinear behavior of the material will be characterized through a parameter, called the nonlinear dielectric susceptibility. A related parameter, called the nonlinear optical coefficient, is found to be more useful to experimentalists so a relationship between the two parameters will be derived. The calculation of an effective nonlinear optical coefficient for one crystal class will be presented in Appendix 15-C, whereas the coefficients for the other crystal classes will be simply listed in tabular form. The value of the nonlinear coefficient of a material can be estimated by using a rule to be discussed in Appendix 15-B.

Maxwell's equations will be used to obtain the equation of motion for a wave, propagating in a nonlinear medium. In its most general form, this equation involves waves of three frequencies. The conservation of energy

TABLE 15.2 Third-order Nonlinear Processes

Frequency of Incident Fields	Frequency of Polarization	Process
ω_1	$3\omega_1$	Third harmonic generation
$\omega_1,\ \omega_2$	$\omega\ = \omega_1 + \omega_2 - \omega_2$	Raman effect
ω_1	$\omega\ = \omega_1 + \omega_1 - \omega_1$	Optical Kerr effect
$\omega_1,\ \omega_2,\ \omega_3$	$\omega\ = \omega_1 + \omega_2 + \omega_3$	Three frequency mixing
$\omega_1,\ \omega_2,\ \omega_3$	$\omega_3 + \omega_1 = \omega_1 + \omega_2$	Four-wave difference frequency mixing
ω_1	$\omega\ = \omega_1$	Intensity-dependent refractive index (phase conjugation)
$\omega_1,\ 0$	$\omega\ = \omega_1$	Kerr effect

and conservation of momentum restrict the values of the frequencies and propagation vectors of the three waves.

We will restrict our attention to the situation for which two of the frequencies are identical and equal to ω. The conservation of energy requires that the frequency of the third wave be 2ω. The nonlinear process involving these waves is called second harmonic generation. We will discuss a number of methods for satisfying conservation of momentum for second harmonic generation.

Third-order nonlinearities can be observed whenever symmetry conditions prevent second-order nonlinearities. The discussion of third-order nonlinearities will be limited to the process called phase conjugation. This effect will be discussed in terms of holography and the effect will be interpreted as a dynamic holographic recording.

NONLINEAR POLARIZATION

The macroscopic polarization was defined in **(8-19)** as

$$P = Nex$$

We will use the first-order perturbation calculation of the oscillator displacement x for a forced oscillator with a second-order nonlinearity derived in Appendix 15-A **(15A-16)**. We can write the resulting polarization as

$$\begin{aligned} P &= Nex^{(1)} \\ &= P(\omega_1) + P(\omega_2) + P(\omega_1 + \omega_2) + P(\omega_1 - \omega_2) + P(2\omega_1) \\ &\quad + P(2\omega_2) + P(0) \end{aligned} \tag{15-1}$$

We will later learn that the individual terms of **(15-1)** occur separately, due to conservation of momentum (which in nonlinear optics is called *phase matching*), allowing the effects of each polarization term to be treated individually.

We introduce a new parameter called the *dielectric susceptibility*

$$\begin{aligned} \chi &= \frac{\epsilon}{\epsilon_0} - 1 \\ &= n^2 - 1 \end{aligned} \tag{15-2}$$

which is the proportionality constant between the applied electric field and induced polarization

$$P(\omega_i) = \chi(\omega_i)\epsilon_0 E_i \tag{15-3}$$

In general, χ is a complex function, but if the frequencies involved are far from resonance, as we will assume, it can be treated as a real function. To find the induced polarization $P(\omega)$ for a second-order nonlinearity, we will use the first-order perturbation results of Appendix 15-A, that is, the terms of $x^{(1)}$ **(15A-16)** to write susceptibilities for each of the frequency components.

The fundamental frequency term of the susceptibility is

$$\chi(\omega_i) = \frac{Ne}{\epsilon_0 E_i} X_i \tag{15-4}$$

We use **(15A-17)** to write

$$\chi^{(1)}(\omega_i) = \frac{Ne^2}{\epsilon_0 m} \frac{1}{\sqrt{(\omega_i^2 - \omega_0^2)^2 + \gamma^2 \omega_i^2}} \tag{15-5}$$

This is the *zero-order solution* that is called the *linear susceptibility* $\chi^{(1)}(\omega)$. It is a dimensionless number, independent of the system of units of **P** and **E**.

The second-order, nonlinear components of the susceptibility are defined by the equation

$$P(\omega_i \pm \omega_j) = P_{NL}(\omega_3)$$
$$= \chi^{(2)}(\omega_i \pm \omega_j)\epsilon_0 E_i E_j$$

Here, we denote the terms resulting from the first-order perturbation model as the second-order nonlinear susceptibility $\chi^{(2)}$. We can use **(15A-18)** or **(15A-19)** to write the *sum* or *difference frequency susceptibility*

$$\chi^{(2)}(\omega_i \pm \omega_j) = \frac{Ne}{\epsilon_0 E_i E_j} X_{i+j} =$$

$$\frac{Ne^3 a_2/\epsilon_0 m^2}{\sqrt{(\omega_i^2-\omega_0^2)^2+\gamma^2\omega_i^2}\sqrt{(\omega_j^2-\omega_0^2)^2+\gamma^2\omega_j^2}\sqrt{[(\omega_i \pm \omega_j)^2-\omega_0^2]^2+\gamma^2(\omega_i \pm \omega_j)^2}} \tag{15-6}$$

This term describe a *parametric interaction*. (Parametric processes were defined in Chapter 14 during the discussion of the acoustooptic effect.) Whenever $\omega_1 = \omega_2$, the process is said to be *degenerate*. The physical processes described by this susceptibility are grouped under five separate categories

1. The term *parametric amplification* is used to indicate that the transfer of power is from a pump wave at

$$\omega_3 = \omega_1 + \omega_2$$

to the signal and idler waves at ω_1 and ω_2. If we place the material producing parametric amplification in an optical resonator (e.g., a Fabry–Perot cavity) tuned to the signal, idler or both, then we will observe *parametric oscillation*.

2. If the energy flows into the harmonic frequency ω_3 from the signal and idler waves, the process is called *sum frequency* or *difference frequency generation*.

3. A high-intensity idler frequency ω_2 can be used to convert a low-level signal frequency ω_1 to the frequency ω_3 via *frequency up-conversion*. This process is identical to sum frequency generation except for the relative intensities of the three waves.

4. For degenerate sum frequency generation, where $\omega_1 = \omega_2$, the name *second harmonic generation* is used to describe the process. We can derive the susceptibility for second harmonic generation by setting $\omega_2 = 0$ and using the first term of **(15A-20)** in an equation similar to **(15-4)**. We obtain for the second harmonic component of the susceptibility

$$\frac{Ne^3 a_2/2\epsilon_0 m^2}{[(\omega_1^2-\omega_0^2)^2+\gamma^2\omega_1^2]\sqrt{(4\omega_1^2-\omega_0^2)^2+\gamma^2 4\omega_1^2}} \tag{15-7}$$

If we define $\chi^{(2)}(2\omega)$ as the degenerate susceptibility calculated using **(15-6)**, then **(15-7)** is equal to

$$\frac{\chi^{(2)}(2\omega)}{2}$$

The factor of 2 occurs because in **(15-7)**, only one fundamental wave is present, whereas in **(15-6)**, there are two fundamental waves. $\chi^{(2)}(2\omega)$ has dimensions of inverse electric field intensity (in mks units, this is m/V).

TABLE 15.3 Second Harmonic Nonlinear Susceptibility

Material	Class	Element	$\chi^{(2)}$, 10^{-12} m/V	λ, μm
Te	32	111	1.0×10^4	10.6
$LiNbO_3$	3m	222	6.14	1.06
		322	−11.6	
$BaTiO_3$	4mm	131	−34.4	1.06
		311	−36	
		333	−13.2	
KDP	$\bar{4}2$m	123	0.98	1.06
		312	0.94	
$LiIO_3$	6	331	−11.2	1.06
		333	−8.4	
CdSe	6mm	131	62	10.6
		311	57	
		333	109	
GaAs	$\bar{4}3$m	123	377	10.6

5. *Optical rectification* involves the zero frequency component of the susceptibility, obtained by using **(15A-21)**. *Optical rectification* can also be viewed as a special case of difference frequency generation, in which the difference frequency is zero. It is related to the Pockels effect and may be thought of as an inverse Pockels effect.

The nonlinear susceptibility can be enhanced by operating near a resonance so that $\omega_i^2 - \omega_0^2 \approx 0$. It should be noted that χ in **(15-6)** is not only a function of the applied frequencies but is also a function of the generated frequency. This means that $\chi(\omega_1, \omega_2, \omega_3)$ can be enhanced not only when ω_1 or ω_2 are near an absorption resonance, but also when ω_3 is near an absorption frequency.

The units usually used in the literature are cgs (Gaussian) units. To convert from cgs to mks units, we use the fact that the polarization is given by

$$P(\text{cgs}) = Nex \Leftrightarrow \frac{1}{\text{cm}^3}\bullet\text{statC}\bullet\text{cm} = \frac{\text{m}^3}{10^6\text{cm}^3}\bullet\frac{3\times 10^9\text{statC}}{\text{C}}\bullet\frac{10^2\text{cm}}{\text{m}} \Leftrightarrow P(\text{mks})$$

$$P(\text{cgs}) = (3\times 10^5)P(\text{mks}) \tag{15-8}$$

From Appendix 2-B, we obtain the relationship between the electric field in cgs and mks units, allowing us to write

$$\chi^{(n)}(\text{mks}) = \frac{\chi^{(n)}(\text{cgs})}{\epsilon_0(3\times 10^4)^n(3\times 10^5)} \approx \frac{4\pi}{(3\times 10^4)^{n-1}}\chi^{(n)}(\text{cgs}) \tag{15-9}$$

Some second harmonic nonlinear susceptibilities are listed in Table 15.3.

NONLINEAR OPTICAL COEFFICIENT

χ is the quantity calculated by the theorist, but the experimentalist measures the *nonlinear optical coefficient* d. The definition of the nonlinear optical coefficient is

TABLE 15.4 Relative Nonlinear Coefficient

Material	λ, μm	n_0	n_e	d, ref. KDP	M, $\times 10^{-46}$
Te	5.3	4.86	6.3		2835
	10.6	4.8	6.25	$d_{11} = 1298$	
ADP	0.265	1.59	1.54		0.08
	0.53	1.53	1.48		1.14
	1.06	1.51	1.47	$d_{36} = 1.21$	
$LiNbO_3$	0.53	2.33	2.23		3.03
	1.06	2.24	2.16	$d_{15} = 12.5$	
$AgGaSe_2$	5.3	2.61	2.58		8.58
	10.6	2.59	2.56	$d_{36} = 76$	
$ZnGeP_2$	5.3	3.11	3.15		146
	10.6	3.07	3.11	$d_{36} = 173$	
$CdGeAs_2$	5.3	3.53	3.62		810
	10.6	3.50	3.59	$d_{36} = 472$	

$$P(2\omega) = d(2\omega)E^2 \cos(2\omega t - 2kx) \tag{15-10}$$

Equating **(15-7)** and **(15-10)** will generate a relationship for the nonlinear optical coefficient in terms of the susceptibility

$$d(2\omega) = \frac{\dfrac{Ne^3 a_2}{2m^2}}{[(\omega^2 - \omega_0^2)^2 + \gamma^2\omega^2]\sqrt{(4\omega^2 - \omega_0^2)^2 + \gamma^2 4\omega^2}} \tag{15-11}$$

or

$$2d(2\omega) = \chi^{(2)}(2\omega)$$

We can rewrite d in terms of the linear susceptibility at the fundamental and the second harmonic of the light wave, given by **(15-5)**

$$d(2\omega) = \frac{a_2 m\epsilon_0^3}{2N^2e^3}\left[\chi^{(1)}(\omega)\right]^2\chi^{(1)}(2\omega) \tag{15-12}$$

Using the definition of χ given by **(15-2)**, we find that **(15-12)** states that materials with a high refractive index will also have a high nonlinear coefficient (see Table 15.4). It has been found empirically that the ratio of d to the two susceptibilities is approximately a constant for a wide range of materials. This fact, called *Miller's rule*, will be discussed in Appendix 15-B.

The relationship between the nonlinear optical coefficient in cgs and mks units is obtained from **(15-9)**

$$d_{\text{mks}}(2\omega) = (3.7 \times 10^{-15})d_{\text{cgs}}(2\omega) \tag{15-13}$$

We will discuss the nonlinear optical coefficient in terms of second harmonic polarization, but it can apply to any frequency component of the susceptibility. A more general expression for the nonlinear optical coefficient can be defined

$$d(\omega_3) = \frac{ma_2\epsilon_0^3\chi^{(1)}(\omega_1)\chi^{(1)}(\omega_2)\chi^{(1)}(\omega_3)}{N^2e^3}$$

which is expressed in terms of the three frequencies, associated with the pump, signal, and idler waves.

SYMMETRY PROPERTIES

We have treated the parameters χ and d as scalars, but in reality they are third-rank tensors, connecting two vectors to a third. For example, the polarization along the x direction, associated with a wave of frequency ω_3, is related to the electric field at ω_1 and ω_2 by the equation

$$
\begin{aligned}
P_x(\omega_3) = {} & d_{xxx}(\omega_3)E_x(\omega_1)E_x(\omega_2) + d_{xyy}(\omega_3)E_y(\omega_1)E_y(\omega_2) \\
& + d_{xzz}(\omega_3)E_z(\omega_1)E_z(\omega_2) + d_{xzy}(\omega_3)E_z(\omega_1)E_y(\omega_2) \\
& + d_{xyz}(\omega_3)E_y(\omega_1)E_z(\omega_2) + d_{xzx}(\omega_3)E_z(\omega_1)E_x(\omega_2) \\
& + d_{xxz}(\omega_3)E_x(\omega_1)E_z(\omega_2) + d_{xxy}(\omega_3)E_x(\omega_1)E_y(\omega_2) \\
& + d_{xyx}(\omega_3)E_y(\omega_1)E_x(\omega_2)
\end{aligned}
$$

Some authors define d without ϵ_0. With that definition the constant used in **(15-9)** will also relate d between the two systems of units. We have included ϵ_0 in the listed values of d.

In the nonlinear optics literature, the value of $d(2\omega)$ is often stated with reference to $d_{36}(2\omega)$ for KDP, which is equal to

$$d_{36}(2\omega) = 4.42 \times 10^{-24} \text{m/V}$$

Some nonlinear optical coefficients referenced to KDP are shown in Table 15.4.

In general, a tensor of rank 3 has 27 components, but a number of symmetry arguments can be invoked to reduce the number of nonzero terms in the tensor associated with the optical nonlinear interaction.

Loss-Free Medium In a lossless medium, we can use an argument similar to the one used in Chapter 13 to demonstrate that the order of writing E_j and E_k is immaterial and

$$d_{ijk} = d_{ikj}$$

for all combinations of i, j, and k; the frequencies must be permuted with the coordinates for this equality to hold. The equality allows the replacement of the last two indices of d and χ with one, as was done for the electrooptic coefficient [see Appendix 14-A **(14A-1)** to obtain the rule used in the replacement]. In matrix form, the polarization is given by

$$
\begin{bmatrix} P_x(\omega_3) \\ P_y(\omega_3) \\ P_z(\omega_3) \end{bmatrix} =
\begin{bmatrix} d_{11} & d_{12} & d_{13} & d_{14} & d_{15} & d_{16} \\ d_{21} & d_{22} & d_{23} & d_{24} & d_{25} & d_{26} \\ d_{31} & d_{32} & d_{33} & d_{34} & d_{35} & d_{36} \end{bmatrix}
\begin{bmatrix} E_x(\omega_1)E_x(\omega_2) \\ E_y(\omega_1)E_y(\omega_2) \\ E_z(\omega_1)E_z(\omega_2) \\ E_y(\omega_1)E_z(\omega_2) + E_z(\omega_1)E_y(\omega_2) \\ E_x(\omega_1)E_z(\omega_2) + E_z(\omega_1)E_x(\omega_2) \\ E_x(\omega_1)E_y(\omega_2) + E_y(\omega_1)E_x(\omega_2) \end{bmatrix}
\qquad (15\text{-}14)
$$

Crystal Symmetry The crystal symmetry determines the nonzero terms d_{ijk} of the tensor; for example, if the crystal has inversion symmetry, all the d_{ijk} are zero, reducing the total number of point groups that must be considered from 32 to 20. The application of symmetry arguments yields a tensor with the same nonzero terms as the piezoelectric and the Pockels effect tensors (see Appendix 14-A).

The actual values of the elements of the following three tensors, the piezoelectric, the Pockels effect, and the optical susceptibility tensors, differ because the physical origin of the processes they describe are different. The piezoelectric tensor describes the electrical behavior of a crystal when the ions are physically displaced. The optical susceptibility is due to the electronic polarizability, as is the Pockels effect, but dispersion causes the terms of these two matrices to differ.

Dispersion-Free Medium The nonlinear optical coefficient tensor has additional symmetry and is further simplified if dispersion in the material under study can be neglected. If the nonlinear polarization is due only to electronic motion and there are no absorption resonances in the spectral region containing ω_1, ω_2, and ω_3, the dispersion of the nonlinearity is negligible and all frequencies act in an equivalent manner. When the conditions for negligible dispersion are met, all elements d_{ijk} of the nonlinear optical coefficient tensor formed by permuting i, j, and k, are equal. This is called *Kleinman's symmetry condition.*[67]

TABLE 15.5 Effective Nonlinear Optical Coefficient

Class	2 e-rays, 1 o-ray	2 o-rays, 1 e-ray
6	0	$d_{15} \sin\theta$
4	0	$d_{15} \sin\theta$
6mm	0	$d_{15} \sin\theta$
4mm	0	$d_{15} \sin\theta$
622	0	0
644	0	0
$\bar{6}$m2	$d_{22} \cos^2\theta \cos 3\phi$	$-d_{22} \cos\theta \sin 3\phi$
3m	$d_{22} \cos^2\theta \cos 3\phi$	$d_{15} \sin\theta - d_{22} \cos\theta \sin 3\phi$
$\bar{6}$	$(d_{11} \sin 3\phi + d_{22} \cos 3\phi) \cos^2\theta$	$(d_{11} \cos 3\phi - d_{22} \sin 3\phi) \cos\theta$
3	$(d_{11} \sin 3\phi + d_{22} \cos 3\phi) \cos^2\theta$	$d_{15} \sin\theta + (d_{11} \cos 3\phi - d_{22} \sin 3\phi) \cos\theta$
32	$d_{11} \cos^2\theta \sin 3\phi$	$d_{11} \cos\theta \cos 3\phi$
$\bar{4}$	$(d_{14} \cos 2\phi - d_{15} \sin 2\phi) \sin 2\theta$	$-(d_{14} \sin 2\phi + d_{15} \cos 2\phi) \sin\theta$
$\bar{4}$2m	$d_{14} \sin 2\theta \cos 2\phi$	$-d_{14} \sin\theta \sin 2\phi$

We will simplify all further discussions by replacing the nonlinear optical coefficient tensor by the scalar parameter called the *effective* d, d_{eff}. The equations we derive using the effective d can be restored to a three-dimensional result by inserting the appropriate equation for d_{eff}. The calculation of d_{eff} assumes two linearly polarized waves incident on a birefringent crystal. The equation for d_{eff} depends on the crystal class and the waves' polarization and propagation vector orientation with respect to the optical axis. An example of the technique used to calculate d_{eff} is found in Appendix 15-C, along with the geometrical identification of the angles θ and ϕ. Table 15.5 shows d_{eff} for materials where Kleinman's symmetry is valid; a more complete listing can be found in the book by Zernike and Midwinter.[64]

WAVE PROPAGATION IN A NONLINEAR MEDIUM

The effective nonlinear optical coefficient results in an effective nonlinear polarization; see Appendix 14-C. This effective nonlinear polarization will be used to derive the equations for the electromagnetic radiation produced by the nonlinear interaction. The starting point of this calculation is Maxwell's equations. The constitutive relations from Chapter 2 are modified to allow for a nonlinear polarization

$$\mathbf{B} = \mu\mathbf{H}, \qquad \mathbf{J} = \sigma\mathbf{E}$$

$$\mathbf{D} = \epsilon_0\mathbf{E} + \mathbf{P} \tag{15-15}$$

$$= \epsilon_0\mathbf{E} + \epsilon_0\chi\mathbf{E} + \mathbf{P}_{NL}$$

We will assume the medium is homogeneous and nonmagnetic, allowing the parameters χ, σ, and μ to be treated as constants with respect to the ∇ operator. We can now use Maxwell's equations to develop the wave equation in much the same way as we did in Chapter 2.

The time derivatives of **(2-3)** and **(2-4)** are

$$\nabla\times\frac{\partial\mathbf{E}}{\partial t} = -\mu\frac{\partial^2\mathbf{H}}{\partial t^2}$$

$$\nabla\times\frac{\partial\mathbf{H}}{\partial t} = \epsilon_0(1+\chi)\frac{\partial^2\mathbf{E}}{\partial t^2} + \sigma\frac{\partial\mathbf{E}}{\partial t} + \frac{\partial^2\mathbf{P}_{NL}}{\partial t^2} \tag{15-16}$$

and the curl of these same equations are

$$\nabla\times\nabla\times\mathbf{E} = -\mu\nabla\times\frac{\partial\mathbf{H}}{\partial t}$$
$$\nabla\times\nabla\times\mathbf{H} = \epsilon_0(1+\chi)\nabla\times\frac{\partial\mathbf{E}}{\partial t} + \sigma\nabla\times\mathbf{E} \tag{15-17}$$

Combining **(15-16)** and **(15-17)**, we obtain

$$\nabla\times\nabla\times\mathbf{E} + \mu\epsilon_0(1+\chi)\frac{\partial^2\mathbf{E}}{\partial t^2} + \mu\sigma\frac{\partial\mathbf{E}}{\partial t} + \mu\frac{\partial^2\mathbf{P}_{NL}}{\partial t^2} = 0 \tag{15-18}$$

$$\nabla\times\nabla\times\mathbf{H} + \mu\epsilon_0(1+\chi)\frac{\partial^2\mathbf{H}}{\partial t^2} + \mu\sigma\frac{\partial\mathbf{H}}{\partial t} = 0$$

We are not interested in the equation involving the magnetic field **H** so we will only simplify **(15-18)**, using the vector identity **(2A-12)**. The result of this manipulation of Maxwell's equations is the wave equation for a nonlinear medium

$$\nabla^2\mathbf{E} = \mu\sigma\frac{\partial\mathbf{E}}{\partial t} + \mu\epsilon\frac{\partial^2\mathbf{E}}{\partial t^2} + \mu\frac{\partial^2\mathbf{P}_{NL}}{\partial t^2} \tag{15-19}$$

To interpret the physical significance of this nonlinear wave equation, we will evaluate the one-dimensional problem of three plane waves of frequency ω_1, ω_2, and ω_3 propagating in the positive z direction. The electric field **E** is composed of three plane waves of the form

$$\mathbf{E}(\omega_i) = E_i = \tfrac{1}{2}\left[E_i(z)e^{i(\omega_i t - k_i z)} + E_i^*(z)e^{-i(\omega_i t - k_i z)}\right] \tag{15-20}$$

(The subscript used in this equation denotes the frequency associated with the wave and not one of the Cartesian coordinates.)

By substituting these three plane waves into the wave equation, we will discover the properties of the waves when they are solutions to the nonlinear wave equation. The left side of **(15-19)** is evaluated using the assumption that the variation with z of the complex amplitude of the light waves is small, i.e.,

$$k_i\frac{\partial E_i}{\partial z} >> \frac{\partial^2 E_i}{\partial z^2}$$

The left side of **(15-19)** then becomes

$$\nabla^2 E_i = -\tfrac{1}{2}\left(k_i^2 E_i + 2ik_i\frac{\partial E_i}{\partial z}\right)\exp\left[i(\omega_i t - k_i z)\right] + \text{c.c.}$$

[We are using the notation introduced by **(1B-6)** and **c.c.** denotes the complex conjugate

$$\text{c.c.} \Rightarrow -\tfrac{1}{2}\left(k_i^2 E_i - 2ik_i\frac{\partial E_i}{\partial z}\right)\exp\left[-i(\omega_i t - k_i z)\right]$$

of the first term.]

The time derivatives of the electric field can be written by using **(1-22)**. The only term of **(15-19)**, that has not been evaluated previously involves P_{NL}. The definition of the nonlinear optical coefficient **(15-10)** can be generalized to the following expression for the nonlinear polarization due to the three waves:

$$\begin{aligned} P_{NL} &= P_{NL}(\omega_1) + P_{NL}(\omega_2) + P_{NL}(\omega_3) \\ &= \mathrm{d}[E^*(\omega_2)E(\omega_3) + E^*(\omega_1)E(\omega_3) + E^*(\omega_1)E(\omega_2)] \end{aligned}$$

If we use $P_{NL}(\omega_1)$ as an example, the second derivative with respect to time is

$$\frac{\partial^2 P_{NL}(\omega_1)}{\partial t^2} = -(\omega_3 - \omega_2)^2 \mathrm{d} E_2^* E_3 \exp\{-i[(\omega_3 - \omega_2)t - (k_3 - k_2)z]\} + \mathrm{c.c.} \tag{15-21}$$

The solution of the wave equation for the three plane waves would be quite easy if we could treat each frequency component separately, but P_{NL} involves the frequency combinations $\omega_1 + \omega_2$, $\omega_3 - \omega_2$, and $\omega_3 - \omega_1$. This prevents the simplification of the equation unless

$$\omega_3 = \omega_1 + \omega_2$$

When this equality holds, we can separate the wave equation into three coupled amplitude equations.

The coupled equation for $\mathbf{E}_1$ is

$$\begin{aligned} -\tfrac{1}{2}\left(k_1^2 E_1 + 2ik_1 \frac{\partial E_1}{\partial z}\right) e^{i(\omega_1 t - k_1 z)} + \mathrm{c.c.} = \\ (i\omega_1 \mu \sigma_1 - \omega_1^2 \mu \epsilon_1)\frac{E_1}{2} e^{i(\omega_1 t - k_1 z)} + \mathrm{c.c.} \\ -\left\{\frac{\omega_1^2 \mu \mathrm{d}}{2} E_3 E_2^* \exp\{i[\omega_1 t - (k_3 - k_2)z]\} + \mathrm{c.c.}\right\} \end{aligned} \tag{15-22}$$

where we have labeled ϵ and σ with the frequency label in case dispersion is important. If we multiply both sides of **(15-22)** by the plane wave

$$\frac{\exp\left[-i(\omega_1 t - k_1 z - \frac{\pi}{2})\right]}{k} = \frac{i}{k}\exp\left[-i(\omega_1 t - k_1 z)\right]$$

and use the identity

$$k_1^2 = \omega_1^2 \mu \epsilon_1$$

we can simplify the equation. Carrying out identical operations on all three equations yields

$$\frac{\partial E_1}{\partial z} = -\frac{\sigma_1}{2}\sqrt{\frac{\mu}{\epsilon_1}} E_1 - \frac{i\omega_1}{2}\sqrt{\frac{\mu}{\epsilon_1}} \mathrm{d} E_3 E_2^* \exp\left[-i(k_3 - k_2 - k_1)z\right] \tag{15-23}$$

$$\frac{\partial E_2}{\partial z} = -\frac{\sigma_2}{2}\sqrt{\frac{\mu}{\epsilon_2}} E_2 - \frac{i\omega_2}{2}\sqrt{\frac{\mu}{\epsilon_2}} \mathrm{d} E_3 E_1^* \exp\left[-i(k_3 - k_2 - k_1)z\right] \tag{15-24}$$

$$\frac{\partial E_3}{\partial z} = -\frac{\sigma_3}{2}\sqrt{\frac{\mu}{\epsilon_3}} E_3 - \frac{i\omega_3}{2}\sqrt{\frac{\mu}{\epsilon_3}} \mathrm{d} E_1 E_2 \exp\left[-i(k_1 + k_2 - k_3)z\right] \tag{15-25}$$

Equations **(15-23)** through **(15-25)** give the rate of change with distance of the wave amplitude at one frequency as a function of the wave amplitudes at two other frequencies. The equations, which are coupled through the nonlinear optical constant d, can be solved but the solutions involve elliptic integrals.[68] A great deal about the nonlinear paramagnetic interaction can be learned without finding the complete solution to these equations.

As was mentioned in Chapter 2, following **(2-20)**, the ratio

$$\sqrt{\frac{\mu_0}{\epsilon_0}} = 377\ \Omega$$

is defined as the impedance of free space. The ratios appearing in the coupled equations **(15-23)** through **(15-25)** can be simplified by using this definition

$$\sqrt{\frac{\mu}{\epsilon}} \approx \frac{1}{n}\sqrt{\frac{\mu_0}{\epsilon_0}} = \frac{377}{n}$$

The allowable frequencies are specified by the requirement that

$$\omega_3 = \omega_1 + \omega_2$$

There is a phase difference between the wave at ω_3 and the waves at ω_1 and ω_2 given by

$$\Delta k = k_3 - k_2 - k_1$$

This phase difference causes the amplitude of each wave to vary in a different sinusoidal fashion as the waves propagate along the z direction. The term involving σ, the medium's conductivity, is equal to the absorption loss that may take place as the waves propagate in the z direction.

The nonlinear interaction can be interpreted, in light of the discussion of acoustooptics in Chapter 14, as a scattering process. The requirement that

$$\omega_3 = \omega_1 + \omega_2$$

is a statement of conservation of energy. The phase difference Δk indicates the momentum change in the scattering event. If the scattering event is elastic, then $\Delta k = 0$, and momentum is conserved. If the scattering event is inelastic, then $\Delta k \neq 0$. In the next two sections, we will examine the impact of the two conservation laws on nonlinear optical processes.[69]

CONSERVATION OF ENERGY

In this section, we will assume that $\Delta k = 0$ and that the material is a perfect insulator so that $\sigma_1 = \sigma_2 = \sigma_3 = 0$. Using these assumptions and multiplying **(15-23)** through **(15-25)** by E_i^*, we obtain

$$E_1^* \frac{\partial E_1}{\partial z} = -\frac{i\mathrm{d}}{2} E_1^* E_2^* E_3\, \omega_1 \sqrt{\frac{\mu}{\epsilon_1}} \tag{15-26}$$

$$E_2^* \frac{\partial E_2}{\partial z} = -\frac{i\mathrm{d}}{2} E_1^* E_2^* E_3\, \omega_2 \sqrt{\frac{\mu}{\epsilon_2}} \tag{15-27}$$

$$E_3^* \frac{\partial E_3}{\partial z} = -\frac{i\mathrm{d}}{2} E_1 E_2 E_3^*\, \omega_3 \sqrt{\frac{\mu}{\epsilon_3}} \tag{15-28}$$

If we use the fact that

$$\frac{\partial}{\partial z}(E_i E_i^*) = E_i \frac{\partial E_i^*}{\partial z} + E_i^* \frac{\partial E_i}{\partial z}$$

we find that **(15-26)** through **(15-28)** lead to the identity

$$\frac{1}{\omega_1}\sqrt{\frac{\epsilon_1}{\mu}}\frac{\partial}{\partial z}(E_1 E_1^*) = \frac{1}{\omega_2}\sqrt{\frac{\epsilon_2}{\mu}}\frac{\partial}{\partial z}(E_2 E_2^*) = -\frac{1}{\omega_3}\sqrt{\frac{\epsilon_3}{\mu}}\frac{\partial}{\partial z}(E_3 E_3^*) \tag{15-29}$$

The definition of the average Poynting vector **(2-26)** can be used to rewrite **(15-29)** as an energy conservation law. We discover the requirement that

$\omega_3 = \omega_1 + \omega_2$ leads to an energy conservation law without the need to specify the type of interaction.

A quantum view of the Manley–Rowe relation can be obtained by noticing that EE^*/ω is a measure of the photon density

$$\frac{EE^*}{\omega} = \frac{I}{\omega} = \frac{N\hbar\omega}{\omega} = N\hbar$$

Two photons with energies of $\hbar\omega_1$ and $\hbar\omega_2$ combine to produce a single photon with energy $\hbar\omega_3$, the number of photons is not conserved but the energy is.

Manley–Rowe Relation

The conservation of energy law **(15-29)**

$$\frac{(\text{change in power at } \omega_1)}{\omega_1} = \frac{(\text{change in power at } \omega_2)}{\omega_2} = \frac{(\text{change in power at } \omega_3)}{\omega_3}$$

is known as the *Manley–Rowe relation* and is valid for all sum- or difference-frequency generation (mixing) processes. In words, it states that if waves at frequencies ω_1 and ω_2 are used to generate a wave at frequency ω_3, then the waves at frequencies ω_1 and ω_2 will lose the energy gained by the wave at frequency ω_3.

CONSERVATION OF MOMENTUM

We now will return to **(15-25)** and find an approximate form for the variation of the light wave E_3 at frequency ω_3 by assuming

1. $\omega_3 = \omega_1 + \omega_2$.
2. $\sigma = 0$, that is, the nonlinear medium is a lossless medium.
3. The nonlinear material occupies the positive half-plane to the right of $z = 0$, the boundary of the nonlinear material.
4. $E_3 = 0$ at $z \leq 0$, i.e., outside of the nonlinear material, only the waves E_1 and E_2 exist.
5. E_1 and E_2 are constant along the z direction, that is, very little energy is transferred into the amplitude E_3 from E_1 and E_2.

With these assumptions, we may integrate **(15-25)** to find the field's amplitude at a distance $z = L$ into the nonlinear material

$$\begin{aligned} E_3 &= -\frac{i\omega_3}{2}\sqrt{\frac{\mu}{\epsilon_3}}\,\mathsf{d}\int_0^L E_1E_2e^{i\Delta kz}\,dz \\ &= -\frac{\omega_3}{2}\sqrt{\frac{\mu}{\epsilon_3}}\,\frac{\mathsf{d}E_1E_2}{\Delta k}(e^{i\Delta kL} - 1) \end{aligned} \tag{15-30}$$

We will use the result of this integration to calculate the Poynting vector for the wave at frequency ω_3. The Poynting vector will allow the significance of the phase difference Δk to be given a physical interpretation in terms of its effect on energy flow. In addition, a figure of merit for evaluating nonlinear materials will be generated.

Poynting Vector

If we use the value of E_3 from **(15-30)** to calculate the Poynting vector of the wave at frequency ω_3, we will obtain the energy, per unit area, flowing across the plane at $z = L$ in the nonlinear medium

$$S_3 = \frac{n_3}{2c\mu}\left(\frac{\omega_3^2 \mu}{4\epsilon_3}\frac{d^2}{\Delta k^2}E_1^2 E_2^2\right)[(e^{i\Delta kL} - 1)(e^{-i\Delta kL} - 1)] \qquad (15\text{-}31)$$

We can use the three identities

$$\frac{1}{\epsilon_3}\frac{\epsilon_0}{\epsilon_0} = \frac{1}{n_3^2 \epsilon_0}, \qquad E_i^2 = \frac{2\mu c}{n_i} S_i, \qquad \frac{\mu^2 c}{\epsilon_0} \approx \sqrt{\left(\frac{\mu}{\epsilon_0}\right)^3}$$

to simplify **(15-31)**

$$S_3 = \frac{\omega_3^2}{2n_1 n_2 n_3}\sqrt{\left(\frac{\mu}{\epsilon_0}\right)^3}\frac{d^2 L^2 S_1 S_2}{(\Delta kL)^2}([2 - e^{-i\Delta kL} - e^{i\Delta kL}])$$

The term in parentheses can be replaced by

$$4\sin^2(\Delta kL/2)$$

yielding the final result for the energy in the nonlinearly generated wave at frequency ω_3

$$S_3 = \frac{\omega_3^2}{2n_1 n_2 n_3}\sqrt{\left(\frac{\mu}{\epsilon_0}\right)^3} d^2 L^2 S_1 S_2 \left[\frac{\sin^2(\Delta kL/2)}{(\Delta kL/2)^2}\right] \qquad (15\text{-}32)$$

The instantaneous value of **(15-32)** has units of J/m^2 and the average value has units of W/m^2.

The physical significance of the difference in propagation vectors Δk of the three waves can be discovered by examining **(15-32)**.

The factor d used in **(15-32)** is the effective nonlinear optical constant and must be replaced, in any calculation, with an expression derived using the method outlined in Appendix 15-C. Thus, implicit in d is information about the direction of the polarization of the output wave as a function of the input waves' polarization and propagation direction.

1. When $\Delta k = 0$, the output power contained in the wave of frequency ω_3 is proportional to the product of the input powers of the waves at ω_1 and ω_2. The output power is also proportional to the square of the propagation path length in the nonlinear medium.
2. *Figure of merit.* The energy in the nonlinearly generated wave is proportional to

$$S_3 \propto \frac{d^2}{n_1 n_2 n_3}$$

We can use this ratio as a figure of merit for nonlinear materials (see Table 15.4)

$$M = \frac{d^2}{n^3}$$

Miller's rule, discussed in Appendix 15-B, also suggests that the ratio of (d/n) is an appropriate figure of merit. This is because a material with a large nonlinear coefficient will also have a large refractive index according to Miller's rule. By dividing d by n, we remove the weighting given to the nonlinear optical coefficient by the index of refraction.

3. *Coherence length.* When $\Delta k \neq 0$, the output power varies as

$$\text{sinc}^2(\Delta kL/2)$$

The energy in the nonlinearly generated wave S_3 is a maximum at

$$L = \frac{\pi}{2\Delta k}$$

found by taking the derivative of **(15-32)** with respect to L and setting it equal to zero. We define this distance as *coherence length*

$$\ell_c = \frac{\pi}{2\Delta k} \tag{15-33}$$

In second harmonic generation, it equals the length of a nonlinear material effective in generating second harmonic radiation. Its size can be estimated by using the parameters associated with second harmonic generation

$$\omega_1 = \omega_2, \qquad \omega_3 = 2\omega_1$$

$$k_1 = k_2 = \frac{n_1\omega_1}{c}, \qquad k_3 = \frac{n_3\omega_3}{c} = 2\frac{n_3\omega_1}{c}$$

It is important to note that $k_3 \neq 2k_1$ because of dispersion

$$\begin{aligned} \Delta k &= k_3 - k_2 - k_1 \\ &= \frac{2\omega_1}{c}(n_3 - n_1) \end{aligned}$$

The coherence length for second harmonic generation is therefore

$$\ell_c = \frac{\lambda_1}{4(n_3 - n_1)} \tag{15-34}$$

where λ_1 is the vacuum wavelength of the fundamental frequency.

We can now use some real parameters to find a typical coherence length. We will double an infrared wavelength of 1 μm into the visible at 500 nm and will use KDP as the nonlinear material. For KDP (we will use the values for the ordinary ray),

$$\lambda_1 = 10^{-6} \text{ m}, \qquad n_1 = 1.50873$$

$$\lambda_3 = 5 \times 10^{-7} \text{ m}, \qquad n_3 = 1.529833$$

The coherence length is 1.18×10^{-5} m. This value is equal to the maximum useful crystal length for producing second harmonic power when $\Delta k \neq 0$.

4. The coherence lengths for second harmonic generation, shown in Table 15.6, are quite small. For second harmonic generation to be useful, it is necessary to obtain larger coherence lengths. This leads to the question of how close a match must be obtained between n_1 and n_3 to generate a useful coherence length. To produce a coherence length of 1 mm for

TABLE 15.6 Second Harmonic Coherence Length

Crystal	λ	ℓ_c, exp.	ℓ_c, calc.
KDP	0.694 μm	18.5 μm	18.8 μm
KD*P	0.694	20.6	
ADP	0.694	17.7	18.2
KDP	1.06	22.0	22.0
KD*P	1.06	21.2	
ADP	1.06	21.	20.6
$BaTiO_3$	1.06	5.8	

second harmonic generation using 1 μm, **(15-34)** predicts that the index for the fundamental wavelength must equal

$$n_1 = n_3 \pm 0.00025$$

The short coherence lengths listed in Table 15.6 demonstrate that it is important to keep $\Delta k \approx 0$ if the nonlinear properties of materials are to be used to generate new wavelengths. In nonlinear optics, the process of reducing the phase difference Δk is called *phase matching*. If we view harmonic generation as a scattering process, phase matching is equivalent to insuring that momentum is conserved.

Phase Matching

We will try to visualize the problem of phase mismatch by following two waves along a linear array of eight atoms. We will assume that the index of refraction for the material, at the fundamental frequency ω_1, is n_1 and the index at the harmonic wave ω_3 is $n_3 = (16/9)n_1$. To make the visualization easy, we assume that λ_1 is exactly equal to the atomic spacing. This requires that λ_3 be 8/9 of the atomic spacing. The phase mismatch between these two waves, as they propagate along the row of atoms, is

$$k_3 - k_1 = \frac{\pi n_1}{4\lambda_1}$$

The fundamental wave at a frequency of ω_1 induces a nonlinear polarization that lags behind the fundamental wave by $\pi/4$. This polarization generates a harmonic wave at ω_3. As we move from atom to atom, we will add the harmonic wave, generated by the polarization, at frequency ω_3 to the fundamental wave at frequency ω_1 using the vector approach introduced in Chapter 4; see Figure 4-1. The sum of these two waves will equal the local field experienced by the atom and this will, in turn, determine the induced polarization.

At atom 1, the only wave present is the fundamental wave; we label its electric field **F** in Figure 15-1*a*. At atom 2, the fundamental wave **F** is joined by the harmonic wave **H** produced by the polarization induced in atom 1. The harmonic wave lags the fundamental wave **F** by 45°, as shown in Figure 15-1*a*. The resultant of the two waves **H** and **F** is equal to a field of amplitude **R**. This resultant field induces a polarization in atom 2 that produces a harmonic wave at atom **3**, again lagging the inducing polarization by $\pi/4$; see Figure 15-1*b*. As the waves move on through the crystal, we reach atom **4**, where the resultant of **H** and **F** reaches its maximum value, (Figure 15-1*c*). The polarization induced by this wave is a maximum at atom 5. When we finally reach atom **8**, the induced polarization due to the

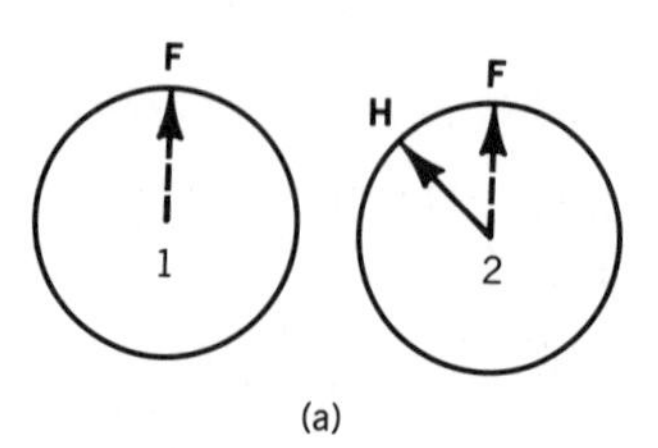

FIGURE 15-1*a*. The first two atoms in a nonlinear material are located at position 1 and 2. The electric field of the fundamental wave is denoted by **F**. It has the same phase at each atom. The polarization, induced by **F** in atom 1, generates a harmonic wave **H** at atom 2.

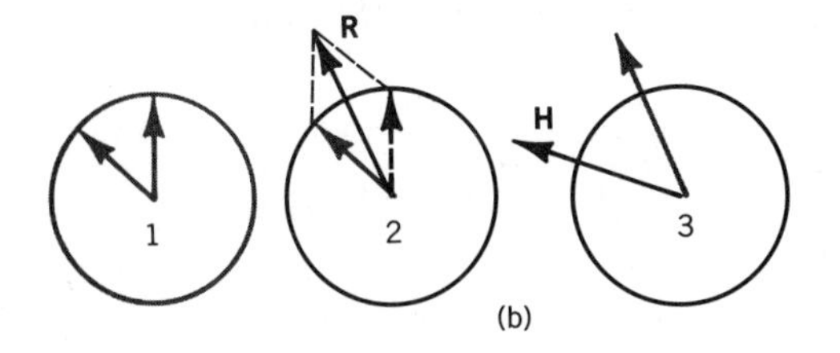

FIGURE 15-1*b*. Propagation of the harmonic wave along the chain of atoms.

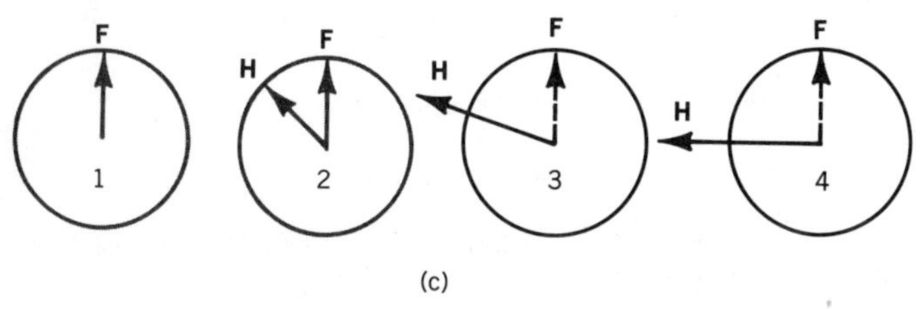

FIGURE 15-1*c*. Four atoms in a nonlinear material. The fundamental and harmonic waves have reached atoms 3 and 4. Atoms 1 and 2 from Figure 15-1*a* are shown for reference.

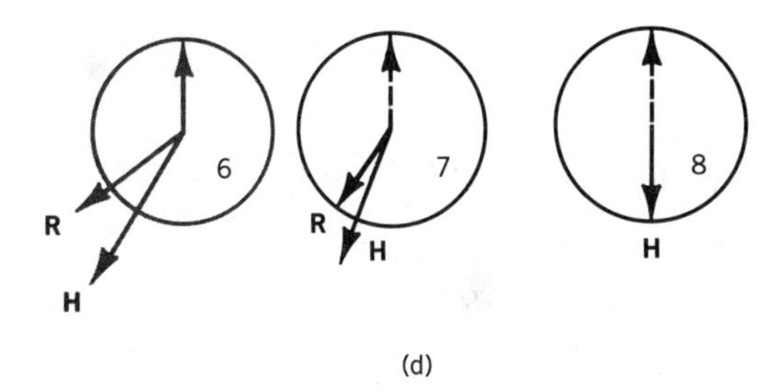

FIGURE 15-1*d*. The fundamental and harmonic waves in the nonlinear crystal have reached atom 8. The harmonic wave is completely out of phase with the fundamental so that the resultant field is zero. The polarization induced in atom 8 is therefore zero.

resultant, formed by adding the fundamental **F** and the harmonic **H** is zero. Atom 8 sees no field and therefore produces no harmonic wave.

This geometrical analysis agrees with the calculation of the coherence length using **(15-33)** and demonstrates that the coherence length is a measure of the distance over which the fundamental and harmonic waves propagate in phase through the nonlinear material.

SECOND HARMONIC GENERATION

In the derivation of **(15-32)**, we assumed low harmonic conversion efficiency, i.e., E_1 and E_2 remain near their original value and E_3 is small relative to them. This assumption prevents the calculation of how energy flows from the fundamental to the harmonic. A simple example, involving the second harmonic, will allow an estimate of the energy flow between the fundamental and harmonic waves. We will assume that

1. Second harmonic generation, $\omega_3 = 2\omega$, where $\omega_1 = \omega_2 = \omega$.
2. The material is a perfect dielectric, $\sigma = 0$. We need not worry about loss in the medium.
3. The nonlinear material occupies the positive half-plane, $z \geq 0$.
4. $E_1 = E_2 = E(\omega)e^{i\phi(\omega)}$ and $E_3 = E(2\omega)e^{i\phi(2\omega)}$.

5. There is no dispersion so that $\epsilon_1 = \epsilon_2 = \epsilon_3 = \epsilon$. This assumption allows perfect phase matching, $\Delta k = 0$.
6. $E_3 = 0$ and $E_1 = E_0 e^{i\phi(\omega)}$ at $z = 0$, the boundary of the nonlinear material.

These assumptions allow us to rewrite **(15-26)** and **(15-27)** as

$$\frac{\partial E(\omega)}{\partial z} = -i\sqrt{\frac{\mu}{\epsilon}}\frac{d\omega}{2}E(\omega)E(2\omega)e^{i\phi(2\omega)} \tag{15-35}$$

$$\frac{\partial E(2\omega)}{\partial z} = -i\sqrt{\frac{\mu}{\epsilon}}\, d\omega E^2(\omega)\exp\{i[2\phi(\omega) - \phi(2\omega)]\} \tag{15-36}$$

For growth of the second harmonic wave, we require

$$2\varphi(\omega) - \varphi(2\omega) = \frac{\pi}{2}$$

i.e., the phase of the generated wave must lag the polarization creating it by 90°. When this occurs, **(15-36)** is a real positive function.

We also use the energy conservation law, to be proven in Problem 15-3. This conservation law states that the sum of the power densities of the fundamental and second harmonic waves must be a constant over the propagation paths and allows us to rewrite **(15-36)** as

$$\frac{\partial E(2\omega)}{\partial z} = d\omega\sqrt{\frac{\mu}{\epsilon}}\,[E_0^2(\omega) - E^2(2\omega)] \tag{15-37}$$

By integrating **(15-37)** from $z = 0$ to $z = L$, we obtain an equation for the second harmonic amplitude at a distance L into the nonlinear material.

$$E(2\omega) = E_0(\omega)\tanh\left[E_0(\omega)Ld\omega\sqrt{\mu/\epsilon}\right] \tag{15-38}$$

The conservation law of Problem 15-3 permits the use of **(15-38)** to obtain the fundamental field's amplitude

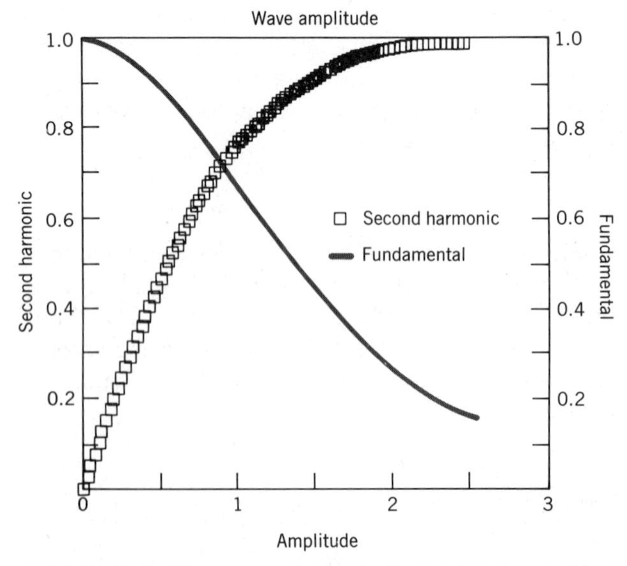

FIGURE 15-2. Wave energy for the fundamental and second harmonic waves as they travel into the nonlinear material.

$$E(\omega) = E_0(\omega) \operatorname{sech}\left[E_0(\omega) L d\omega \sqrt{\mu/\epsilon}\right] \qquad (15\text{-}39)$$

(15-38) and **(15-39)** are displayed in Figure 15-2. The initial slope of the second harmonic curve is given by **(15-32)** with $\Delta k = 0$.

The simple theory predicts the possibility of obtaining 100% conversion of the fundamental wave into the second harmonic wave. The quantum theory predicts that when the fundamental wave shrinks to the level of quantum noise, the energy flow will reverse and the fundamental wave will grow at the expense of the second harmonic. (These results have been derived for plane wave interactions. In practice, to increase the light fields, Gaussian beams are used.[70])

METHODS OF PHASE MATCHING

A number of different ways of obtaining phase matching have been proposed. We will discuss a large number of these because the problem of phase matching is key to the utilization of the second-order nonlinearity in harmonic generation.

Periodic Variation of the Nonlinear Coefficient

If the phase difference between the harmonic and the fundamental waves is changed by 180°, after the waves have propagated one coherence length, we can prevent the energy from draining, from the harmonic, back into the fundamental wave. One way to accomplish this would be to stack a series of crystal plates or thin-film layers, each one a coherence length thick (see Figure 15-3). In this design, the index of refraction is assumed to be the same in each layer so that there are no reflections at the interfaces.

To understand this method, we initially assume there is no nonlinear interaction in the shaded layers of Figure 15-3, $d_2 = 0$. After propagating a coherence length in material one, with nonlinear optical coefficient d_1, the fundamental and harmonic are out of phase by π. If we did nothing, energy would begin to flow back from the harmonic to the fundamental. At this point, the two waves enter the second material. Because we assumed $d_2 = 0$, there is no nonlinear interaction in the second material; the two waves are not coupled and do not interact. After traveling a coherence length in the linear medium, the waves are again in phase. The in-phase waves again enter the nonlinear region with the nonlinear coefficient d_1, and additional energy is transferred from the fundamental to harmonic wave.

The penalty paid for obtaining phase matching in this way is a reduction in the effective nonlinear coefficient to d_1/π. If the linear material layers were replaced by layers of a nonlinear material with a nonlinear coefficient $d_2 = -d_1$, then the effective nonlinear coefficient would be doubled.

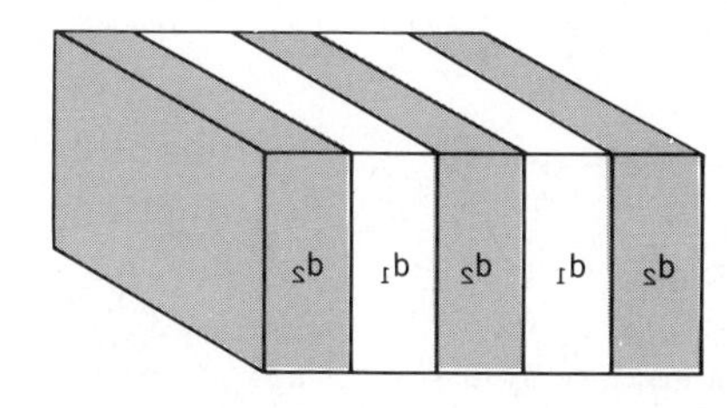

FIGURE 15-3. Multilayers of nonlinear material. The sign of the nonlinear coefficient is periodically modulated with the period of modulation being equal to the coherence length.

It is quite straightforward to show[71] that phase matching for difference frequency mixing is obtained when

$$\Delta k = k_3 - k_1 + k_2 = \frac{(2m + 1)\pi}{\ell}$$

where ℓ is the thickness of each plate and m is an integer constant. When $m = 0$, then ℓ is equal to the natural coherence length. The output intensity using this technique is, at best (when $m = 0$), only 40% of the value obtained if perfect phase matching existed. This type of phase matching has been demonstrated in GaAs.[72]

Total Reflection

We can use the fact that a wave undergoes a phase change during total reflection (see Figure 3-13) to obtain phase matching. Consider second harmonic generation; the relative phase change on reflection is given by

$$\varphi = 2\delta(\omega) - \delta(2\omega)$$

where δ is obtained from either **(3-54)** or **(3-55)**. The relative phase change between the fundamental and harmonic due to dispersion is

$$2\psi = [n(2\omega) - n(\omega)]\frac{2\omega L}{c}$$

where we have assumed a propagation distance of L. This phase change is compensated for by proper adjustment of θ, the incident angle. Phase matching by total reflection has been observed, using a CO_2 laser at $\lambda = 10.5915\ \mu m$, with the geometry shown in Figure 15-4.[73]

Dielectric Waveguide

The use of a waveguide for phase matching is a natural extension of the use of total reflection. Associated with each of the waveguide modes excited is a different effective index of refraction (see Chapter 5)

$$N = \frac{\beta}{k} \tag{15-40}$$

The effective index can range between the value for the substrate and

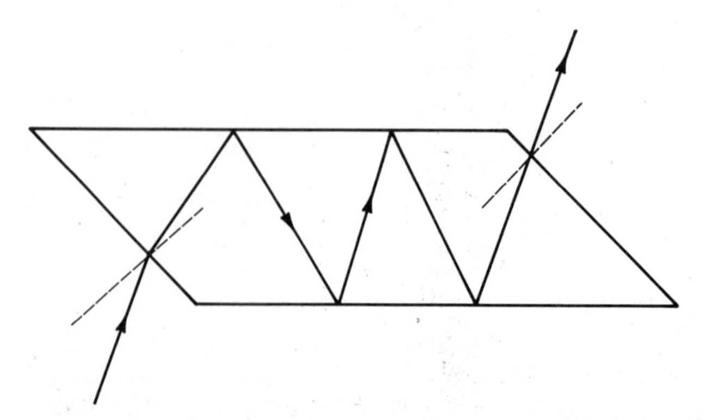

FIGURE 15-4. Configuration used to obtain phase matching by total reflection.

the value for the guide material in the bulk. A mathematical discussion of this type of phase matching can be found in a paper suggesting its use for operation in the submillimeter wavelength region of the spectrum.[74]

Noncollinear Phase Matching

The previous techniques are called *collinear techniques* because the waves are propagating in the same direction. It is possible to perform phase matching by *noncollinear techniques* when the waves propagate at an angle with respect to one another. The concept of noncollinear phase matching is quite straightforward, as can be seen in Figure 15-5. Instead of requiring the scalar momentum conservation equation to be satisfied, a vector equation must be satisfied.

$$\Delta \mathbf{k} = \mathbf{k}_3 - \mathbf{k}_1 - \mathbf{k}_2 = 0 \tag{15-41}$$

The restrictions that phase matching places on the angle α, between $\mathbf{k}_1$ and $\mathbf{k}_2$, and the angle β, between $\mathbf{k}_1$ and $\mathbf{k}_3$, can easily be derived by applying the law of cosines to Figure 15-5

$$\begin{aligned} k_3^2 &= k_1^2 + k_2^2 - 2k_1k_2 \cos \alpha \\ \cos \alpha &= 1 - 2 \sin^2 \frac{\alpha}{2} \\ &= \frac{n_1^2\omega_1^2 + n_2^2\omega_2^2 - n_3^2\omega_3^2}{2n_1\omega_1 n_2\omega_2} \end{aligned} \tag{15-42}$$

To evaluate the physical implications of this result, we will treat the case of difference frequency generation, $\omega_1 - \omega_2 = \omega_3$. For simplicity, we make the physically realistic assumption that $n_1 \approx n_2 = n$, $n_3 = n + \Delta n$, and $\Delta n^2 \approx 0$ (see Table 15.4 where index values for the case of second harmonic generation can be compared). With these assumptions about the index of refraction, the angle α is given by

$$\sin^2 \frac{\alpha}{2} \approx \frac{\Delta n \omega_3^2}{2n\omega_1\omega_2} \tag{15-43}$$

For small values of α, that is, nearly collinear waves,

$$\alpha \approx \frac{\omega_3}{\sqrt{\omega_1\omega_2}} \sqrt{\frac{2\Delta n}{n}}$$

Since α is real only if $\Delta n > 0$, noncollinear phase matching for difference frequency generation can be produced only in regions of anomalous dispersion.

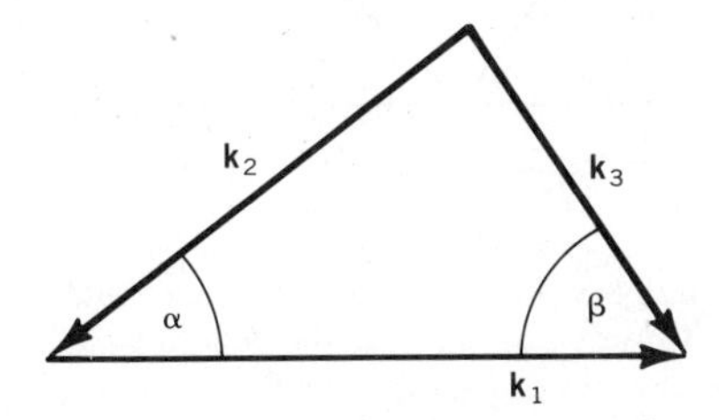

FIGURE 15-5. Noncollinear phase matching. The vector sum of the propagation vectors is set equal to zero rather than the scalar sum.

As we did for α, we can use the law of cosines to evaluate β

$$\sin^2 \frac{\beta}{2} \approx \frac{2n\Delta n\omega_1\omega_3 - 2n\Delta n\omega_3^2}{4n(n + \Delta n)\omega_1\omega_3}$$

$$= \frac{\Delta n}{2(n + \Delta n)} - \frac{\Delta n\omega_3}{2(n + \Delta n)\omega_1}$$

From the geometry used in Figure 15-5, $\omega_3 < \omega_1$, the second term is smaller than the first, and we may ignore it. If we also assume, as we did for α, that the waves are nearly collinear, $\sin(\beta/2) \approx \beta/2$, and the angle β is

$$\beta = \sqrt{\frac{2\Delta n}{n}} \tag{15-44}$$

The angle β is within the approximations used here independent of frequency. It is the broadband behavior implied by **(15-44)** that is the major advantage of the noncollinear phase-matching technique (see Problem 15-7).

Examples of geometries that have been used for noncollinear mixing are shown in Figure 15-6.

A useful application of noncollinear phase matching is in the measurement of picosecond optical pulses using second harmonic generation autocorrelation (see Figure 15-7 and Problem 15-7). Two samples of a picosecond pulse are obtained with a beam splitter. One of the samples is delayed in time by propagating it over a longer optical path. The two samples

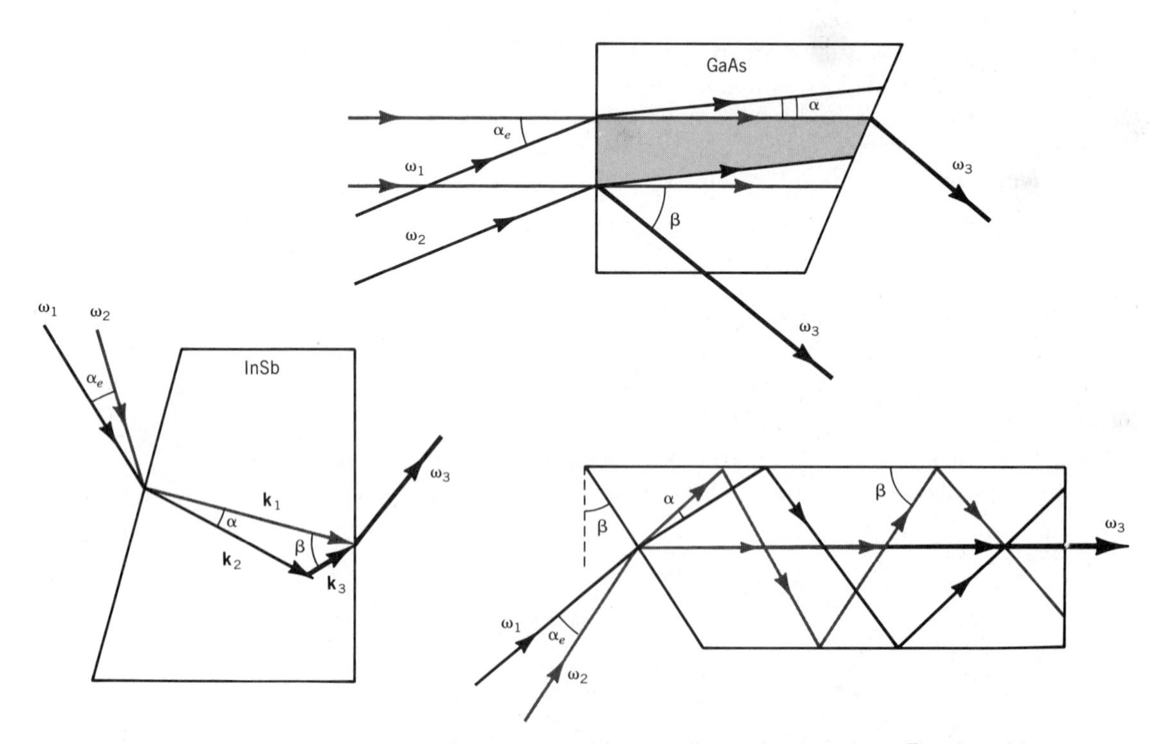

FIGURE 15-6. Geometries used for noncollinear phase matching. The slanted faces are at Brewster's angle, to allow efficient coupling. The guided-wave geometry in the bottom figure is to increase the effective interaction length.

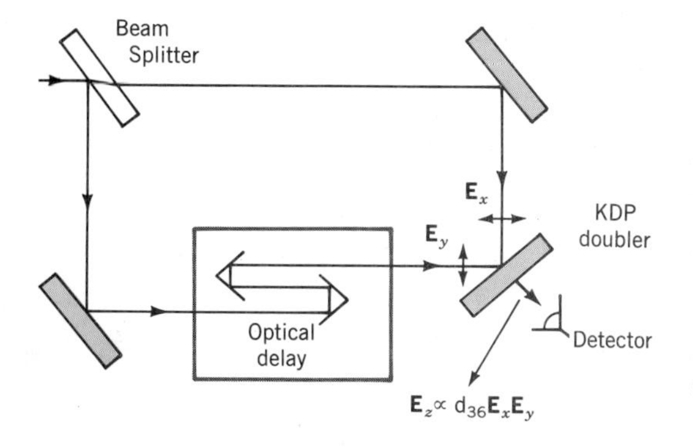

FIGURE 15-7a. Psec pulse detection using the autocorrelation of two samples of a pulse. One sample is delayed by a time τ with respect to the other sample. The two pulses are mixed in a nonlinear crystal and the sum wavelength is integrated on a detector to give the autocorrelation function.

are added together in a nonlinear crystal. The amount of second harmonic radiation produced is determined by the temporal overlap of the two pulses, Figure 15-7*b*. By varying the time delay, an autocorrelation of the picosecond pulse is produced. The temporal resolution is obtained by varying an optical path length to vary τ rather than through the temporal resolution of the detector.

Accidental noncollinear phase matching between the fundamental on-axis beam and energy, scattered from the second wave, can be observed in crystals where collinear phase matching occurs. In second harmonic generation, this leads to interesting ring-shaped patterns.[75]

Birefringent Phase Matching

The most popular method of phase matching utilizes the fact that, in an anisotropic crystal, the refractive index depends on both the wavelength and the polarization of a light wave. We derived an expression for the index of refraction, as a function of propagation direction, for a uniaxial crystal in Appendix 13-C. A plot of the extraordinary index of refraction as a function of the angle between the propagation direction and optical axis is presented in Figure 15-8. The curve was generated with **(13C-7)**. Figure 15-8 demonstrates that the index of refraction can be adjusted to any value between n_e and n_0 by varying the propagation direction. If the variation in the index of refraction due to birefringence is larger than the dispersive change of the index, then phase matching can be obtained. Phase matching is accomplished by using linearly polarized waves, propagating at the proper angle, to insure that

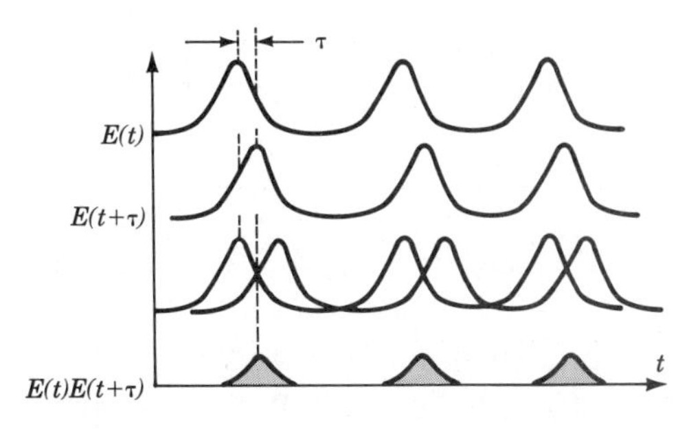

FIGURE 15-7b. The two pulses are overlapped in space, the second curve from bottom, by adjusting τ. The overlapping pulses produce the second harmonic emission shown as the bottom curve. Only the area under the pulse at the second harmonic is detected. Temporal resolution is obtained by varying τ.

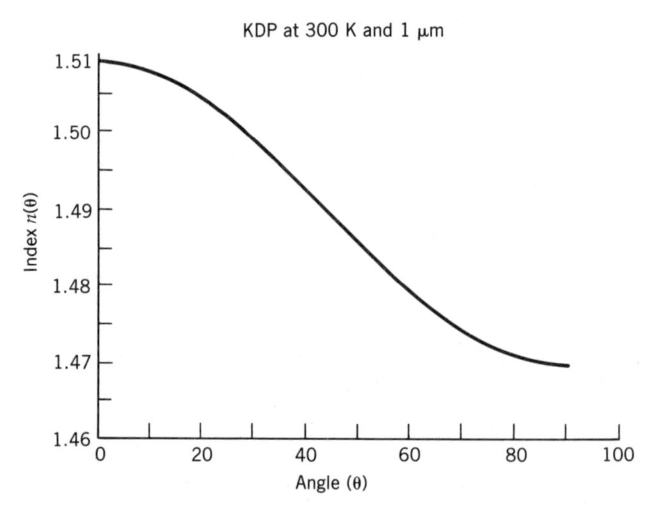

FIGURE 15-8. The variation in the value of the extraordinary index of refraction with propagation angle in KDP at room temperature. The index varies from n_0 at $\theta = 0^\circ$ to n_e at $\theta = 90^\circ$.

$$n(2\omega) = n(\omega)$$

There are two methods of achieving the phase-matched condition using birefringence

Type I. The waves at frequencies ω_1 and ω_2 both have the same polarization, i.e., both are ordinary or both are extraordinary waves. The third wave at frequency ω_3 is polarized in an orthogonal direction.

Type II. The waves at frequencies ω_1 and ω_2 are orthogonally polarized. The third wave at frequency ω_3 can have its polarization direction parallel to either of the other waves.

To demonstrate phase matching using birefringence, the phase-matching condition for three uniaxial crystals will be determined. For collinearly propagating waves, the phase-matching condition **(15-42)** reduces to

$$n_3\omega_3 = n_1\omega_1 \pm n_2\omega_2 \tag{15-45}$$

The indices of refraction depend on the type of phase matching and the type of birefringent crystal, biaxial or positive or negative uniaxial. For uniaxial crystals, the phase-matching equation becomes

$$n_e < n_0, \qquad\qquad n_e > n_0$$

Type I

$$n_e(\omega_3) = \frac{\omega_1}{\omega_3}n_0(\omega_1) + \frac{\omega_2}{\omega_3}n_0(\omega_2), \qquad n_0(\omega_3) = \frac{\omega_1}{\omega_3}n_e(\omega_1) + \frac{\omega_2}{\omega_3}n_e(\omega_2)$$

Type II

$$n_e(\omega_3) = \frac{\omega_1}{\omega_3}n_e(\omega_1) + \frac{\omega_2}{\omega_3}n_0(\omega_2), \qquad n_0(\omega_3) = \frac{\omega_1}{\omega_3}n_0(\omega_1) + \frac{\omega_2}{\omega_3}n_e(\omega_2)$$

The next three curves display the ordinary and extraordinary indices of refraction, as a function of wavelength, for three common birefringent crystals: Quartz, $LiNbO_3$, and KDP. In addition to plotting the dispersion

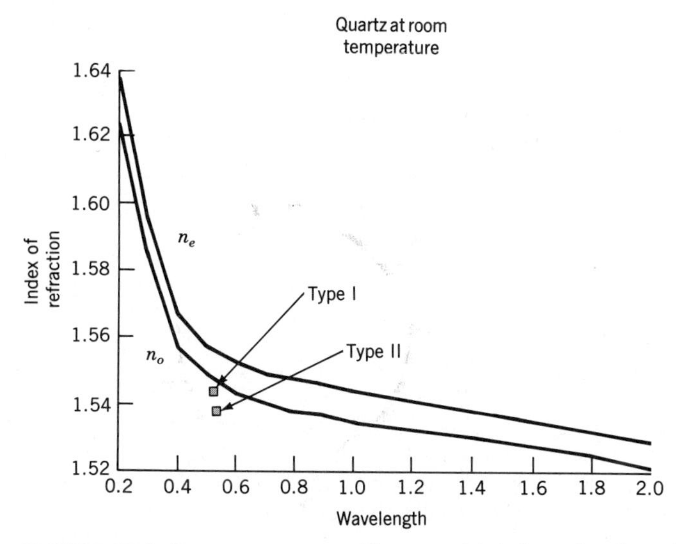

FIGURE 15-9. Dispersion in quartz. The points labeled type I and type II are the indices that would have to exist if we could have phase matching in quartz for doubling of 1.06 μm.

curves for these crystals, points are plotted on the graphs indicating the phase-matching conditions required to convert radiation of wavelength 1.06 μm to the wavelength of 530 nm.

The first curve, in Figure 15-9, shows the dispersion of the ordinary and extraordinary indices of quartz. These curves bound the indices of refraction that are accessible, at each wavelength, by varying the propagation direction. The points labeled "Type I" and "Type II" are the indices of refraction that must exist if phase matching is to occur for second harmonic generation with a fundamental wavelength of 1.06 μm. The indices that 530 nm light waves must have are smaller than any index available in quartz; therefore, phase-matching is not possible.

The next dispersion curve, Figure 15-10, is for $LiNbO_3$. From the position of the required second harmonic index for type-II phase matching, we see that only type-I phase matching can occur in $LiNbO_3$ for the second harmonic generation of 1.06 μm. In Figure 15-10, it is apparent that type-

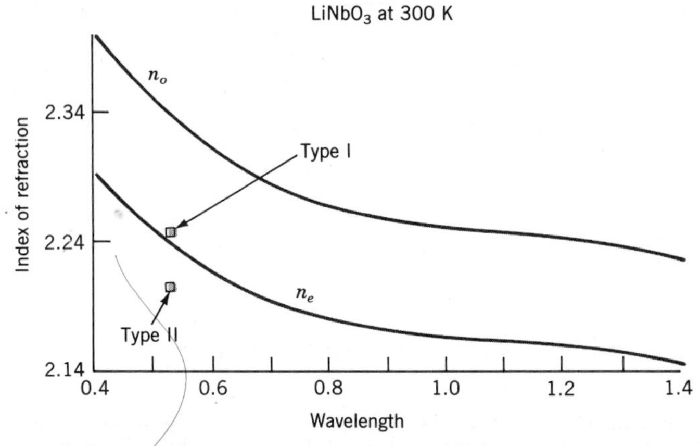

FIGURE 15-10. Dispersion in lithium niobate. As can be seen for the doubling of 1.06 μm, only type-I phase matching is possible.

I phase matching is approximately obtained when $\theta = 90°$. An exact phase matching at 90° can be obtained by adjusting the temperature of the crystal. Phase matching at 90° is called *noncritical phase matching. Critical phase matching* occurs in all other propagation directions.

Noncritical phase matching offers an advantage in second harmonic generation. When the light wave propagates normal to the optic axis of a birefringent crystal, the extraordinary and ordinary waves propagate in a collinear direction. At all other angles, except 0°, birefringence is observed and the two waves propagate at an angle with respect to each other. This is called Poynting vector walk-off [see Problem 13-9 and **(13-36)**]. For a typical birefringent crystal, when the propagation direction makes an angle of $\theta = 45°$ with the optical axis, the Poynting walk-off angle is about 2°. It is often necessary to focus the incident light wave to increase the nonlinear interaction. Poynting vector walk-off causes the ordinary and extraordinary waves associated with the focused beam to separate after a distance called the *aperture length*

$$\ell_a = \frac{\sqrt{\pi}\, w_0}{\phi}$$

where ϕ is the Poynting vector walk-off **(13-A)**, and w_0 is the beam waist. By obtaining phase matching at 90°, we can eliminate Poynting vector walk-off and increase the aperture length for second harmonic generation. Noncritical phase matching also offers the advantage of increasing the acceptance angle, that is, the angular spread of the fundamental beam, by as much as an order of magnitude over the acceptance angle for critical phase matching.

KDP is a material that can have either type-I or type-II phase matching, as we can see in Figure 15-11. For type-I phase matching, the phase matching angle is given by

$$\sin^2 \theta_m = \frac{\dfrac{1}{n_0^2(\omega)} - \dfrac{1}{n_0^2(2\omega)}}{\dfrac{1}{n_e^2(2\omega)} - \dfrac{1}{n_0^2(2\omega)}} \tag{15-46}$$

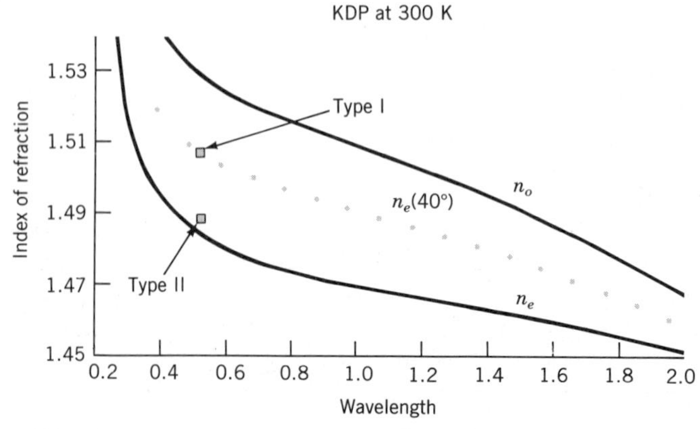

FIGURE 15-11. Dispersion in KDP. The birefringence is large enough in KDP to allow both type-I and type-II phase matching for the doubling of 1.0 μm. Note the angle for type-I phase matching is 42.4°.

and for type-II phase matching,

$$\sin^2\theta_m = \frac{\left[\dfrac{n_0(\omega)}{2n_0(2\omega) - n_0(\omega)}\right]^2 - 1}{\left[\dfrac{n_0(\omega)}{n_e(\omega)}\right]^2 - 1} \tag{15-47}$$

In those materials that allow both types of phase matching, how do we decide which type to use? One reason for selecting the type of phase matching is the ability to operate with the waves propagating in a direction 90° with respect to the optical axis. A second reason for selecting one type of phase matching over another can be found in Table 15.5. The magnitude of d_{eff} is a function of the phase-matching angle. For type-I phase matching in a member of the $\overline{4}2m$ group, we find from Table 15.5 that

$$d_{eff} = -d_{14} \sin\theta \sin 2\phi$$

Once θ is fixed by the phase-matching condition, we can maximize d_{eff} by selecting the direction of propagation so that it makes an angle of $\phi = 45°$ with respect to the x axis of the crystal. In Table 15.5, we find for a $\overline{4}2m$ member that the selection of $\theta = 90°$ results in a d_{eff} equal to zero for type-II phase matching, but results in the maximum value for d_{eff} for type-I phase matching. To maximize d_{eff} for type-II phase matching in a $\overline{4}2m$ member, we would select $\phi = 0°$ and select the phase-matching angle to be $\theta \approx 45°$. We are assuming that phase matching can occur at these angles.

PHASE CONJUGATION

A theoretical description of the third-order nonlinear properties of materials can be developed with the same procedures used to develop the second-order theory. In the second-order nonlinear theory, $\chi^{(2)}$ was found to exhibit both resonant and loss properties, similar to $\chi^{(1)}$. $\chi^{(3)}$ exhibits the same resonant behavior. The imaginary part of $\chi^{(3)}$, corresponding to either energy gain or loss, describes Raman, Brillouin, and Rayleigh scattering as well as two photon processes. The real part of $\chi^{(3)}$ describes the process of harmonic wave generation, the optical Kerr effect, and intensity-dependent changes in the index of refraction.

The nonlinear polarization is given by a number of terms of the form

$$\left[P_{NL}^{(3)}(\omega_4)\right]_i = D\left(\sum_{jk\ell} \chi_{ijk\ell}^{(3)}(\omega_1, \omega_2, \omega_3, \omega_4)E_j(\omega_1)E_k(\omega_2)E_\ell(\omega_3) + \text{c.c.}\right) \tag{15-48}$$

where D is a numerical constant that depends on the values of ω_1, ω_2, ω_3, and ω_4; see Table 15.7. $\chi^{(3)}$ is a fourth-rank tensor, relating one vector to three others.

Symmetry arguments can be applied to $\chi^{(3)}$, as they were to $\chi^{(2)}$, resulting in simplifications. [76] One important result of the symmetry arguments is that $\chi^{(3)}$ is nonzero for centrosymmetric systems, i.e., systems with inversion symmetry. The nonzero elements of the tensor are identified in Appendix 14-A. Far from resonant conditions,

$$O\{\chi^{(3)}\} \approx \begin{cases} 10^{-33}, & \text{cgs} \\ 10^{-58}, & \text{mks} \end{cases}$$

TABLE 15.7 Third-order Nonlinear Processes

D	ω_1	ω_2	ω_3	ω_4
$\frac{1}{4}$	ω	ω	ω	3ω
$\frac{3}{4}$	ω_1	ω_2	ω_1	$2\omega_1 + \omega_2$
$\frac{3}{2}$	ω_1	ω_2	ω_3	$\omega_1 + \omega_2 + \omega_3$
$\frac{3}{4}$	ω_1	ω_2	ω_1	$2\omega_1 - \omega_2$
$\frac{3}{2}$	ω_1	ω_2	ω_2	$\omega_1 + \omega_2 - \omega_2$
$\frac{3}{4}$	ω	ω	ω	$\omega + \omega - \omega$
3	ω	0	0	ω

TABLE 15.8 Third-order Nonlinear Susceptibility

Material	$\chi_{ijm\ell}$, $\times 10^{-13}$ esu[a]	$ijk\ell$
CS_2	2.7	1221
	3.9	1111
2-Methyl-4-nitroaniline	119	1111
Polydiacetylene	400	1111

[a]The cgs units shown here are the units normally found in the literature. To convert to mks units for χ, use (15-9).

$\boldsymbol{\chi}^{(3)}_{ijk\ell}(\omega, 0, 0, \omega)$ is the susceptibility associated with the dc Kerr effect, discussed in Chapter 14. We will only examine one additional term of the third order nonlinear polarization

$$\chi^{(3)}_{ijk\ell}(\omega, \omega, \omega, \omega)$$

which is associated with the optical Kerr effect. We will discover that the theory allows the effect to be interpreted as a holographic process. Some typical values of $\boldsymbol{\chi}^{(3)}$ are found in Table 15.8.

One of the processes associated with $\boldsymbol{\chi}^{(3)}_{ijk\ell}(\omega, \omega, \omega, \omega)$ is called *degenerate four-wave mixing* or *phase conjugation*. The mathematical details of the theory[19] are similar to the analysis of the second-order nonlinearity described earlier. The experimental geometry normally used in the development of the theory is shown in Figure 15-12.

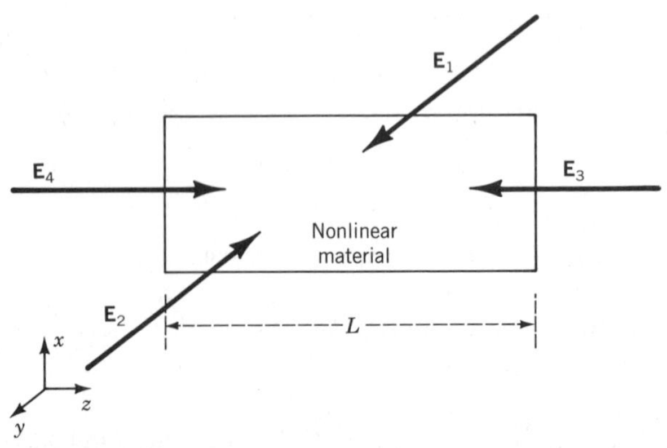

FIGURE 15-12. Geometry for phase conjugation.

The geometry of the experiment makes it possible to treat each wave separately, reducing the problem to the solution of a wave equation involving $P^{(3)}$. The fact that $\mathbf{k}_1 + \mathbf{k}_2 = 0$ in Figure 15-12 allows the problem to be reduced to one dimension. The orientation of the $\mathbf{k}$ vectors insures phase matching at all times.

The wave equation to be solved is **(15-19)**, with an applied electric field consisting of three waves, each described by **(15-20)**. The nonlinear polarization will involve terms of the form

$$\chi^{(3)}_{ijk\ell} E_{1j} E_{2k} E^*_{3\ell} e^{i(\omega t - kz)} \tag{15-49}$$

as well as terms involving 2ω. We will neglect the 2ω terms and give a justification for ignoring these higher-frequency terms later.

All of the waves are assumed to have the same polarization so that only one of the tensor components for $\boldsymbol{\chi}^{(3)}$ enters into the calculation. This assumption allows the replacement of the tensor with a scalar

$$\boldsymbol{\chi}^{(3)}_{1111} \rightarrow \boldsymbol{\chi}^{(3)}$$

(This assumption is more restrictive than we need. We could relax the assumption and require that wave pair 1 and 2 and pair 3 and 4 each have the same polarizations. The susceptibility would still only contain one tensor component

$$\boldsymbol{\chi}^{(3)}_{1221} \rightarrow \boldsymbol{\chi}^{(3)}$$

and the susceptibility could continue to be treated as a scalar.) A further simplification can be obtained by assuming that

$$|E_3|, |E_4| << |E_1|, |E_2|$$

Using an approach similar to the one used to derive **(15-23)** through **(15-25)**, the wave equation for the complex amplitude of the wave generated by the nonlinear interaction can be written as

$$\frac{dE_4}{dz} = -i\frac{\omega\chi^{(3)}}{2}\sqrt{\frac{\mu}{\epsilon}}\left[|E_1|^2 + |E_2|^2\right]E_4 - i\frac{\omega\chi^{(3)}}{2}\sqrt{\frac{\mu}{\epsilon}}E_1E_2E_3^* \tag{15-50}$$

The first term in **(15-50)** describes the optical Kerr effect. This term can be incorporated into the theory by modifying the propagation vector for wave 4 from k to

$$\mathscr{k} = k + \frac{\omega\chi^{(3)}}{2}\sqrt{\frac{\mu}{\epsilon}}\left[|E_1|^2 + |E_2|^2\right] \tag{15-51}$$

This new propagation constant can be included in the definitions of the complex amplitudes for the four waves by introducing a new complex amplitude of the form

$$A = Ee^{-i\mathscr{k}z} \tag{15-52}$$

With this new definition, **(15-50)** becomes for waves 3 and 4

$$\frac{dA_4}{dz} = -i\frac{\omega\chi^{(3)}}{2}\sqrt{\frac{\mu}{\epsilon}}A_1A_2A_3^* \tag{15-53}$$

$$\frac{dA_3^*}{dz} = -i\frac{\omega\chi^{(3)*}}{2}\sqrt{\frac{\mu}{\epsilon}}A_1^*A_2^*A_4 \tag{15-54}$$

We will not solve **(15-50)** for waves 1 and 2, but will instead treat them as "medium conditioners" and include their effects in a material constant defined by the equation

$$\boldsymbol{\kappa} = \frac{\omega\chi^{(3)*}}{2}\sqrt{\frac{\mu}{\epsilon}}\,A_1^*A_2^* \tag{15-55}$$

This material constant is a spatially dependent index of refraction that will affect the propagation of the other waves. It performs the role of a *coupling coefficient* between the various waves.

The solution of **(15-53)** and **(15-54)** can easily be obtained. The non-linear medium is assumed to occupy the positive half-plane with one edge at $z = 0$ and the other at $z = L$. As boundary conditions, we will assume that wave 4 is incident on the nonlinear medium from the left with a complex amplitude $A_4(0)$ at the boundary. We assume that the wave with complex amplitude A_3 is produced in the non-linear medium; therefore, we must have $A_3(L) = 0$ at the right boundary of the nonlinear medium. The solution of **(15-54)** for wave 3 is

$$A_3(z) = \frac{\cos|\boldsymbol{\kappa}\, z|}{\cos|\boldsymbol{\kappa}|L}A_3(L) + i\boldsymbol{\kappa}^*\frac{\sin|\kappa|(z-L)}{|\boldsymbol{\kappa}|\cos|\boldsymbol{\kappa}|L}A_4^*(0)$$

The solution of **(15-53)** is left as a problem; see Problem 15-5.

At the front surface of the nonlinear medium, we have

$$A_3(0) = -i\frac{\boldsymbol{\kappa}^*}{|\boldsymbol{\kappa}|}A_4^*(0)\tan|\boldsymbol{\kappa}|L \tag{15-56}$$

Equation **(15-56)** predicts that the third order nonlinear medium will act as a mirror with the very unique property of reflecting the conjugate of the incident wave. (See Chapter 12 for a discussion of conjugate waves.) The reflectivity of this strange mirror can be greater than 1 whenever

$$\frac{\pi}{4} \le |\boldsymbol{\kappa}|L \le \frac{3\pi}{4}$$

One of the interesting aspects of the phase conjugate system is the analogy with holography. Assume for the moment that $\boldsymbol{\kappa}L$ is small so that $\tan \boldsymbol{\kappa}L$ may be replaced by $\boldsymbol{\kappa}L$. Equation **(15-56)** describes the complex amplitude of the wave traveling to the left in Figure 15-12, as it leaves the nonlinear medium. For small $\boldsymbol{\kappa}L$, **(15-56)** becomes

$$A_3(0) = -i\frac{\omega\chi^{(3)}}{2}\sqrt{\frac{\mu}{\epsilon}}\,A_1A_2A_4^*(0) \tag{15-57}$$

Because of the similarity with terms in **(12-4)**, we treat the formation of A_3 as the diffraction of A_1 by a fringe system produced by interference between A_2 and A_4^*, as shown in Figure 15-13*a*. A second interpretation concerning the formation of A_3 views it as a wave originating from the diffraction of wave 2 by a fringe system created through the interference of A_1 and A_4^* (Figure 15-13*b*).

In this picture of phase conjugation, the terms involving 2ω that we ignored earlier describe a set of moving fringes. If the nonlinear medium has a temporal response that is too slow to respond to the motion of the fringes,

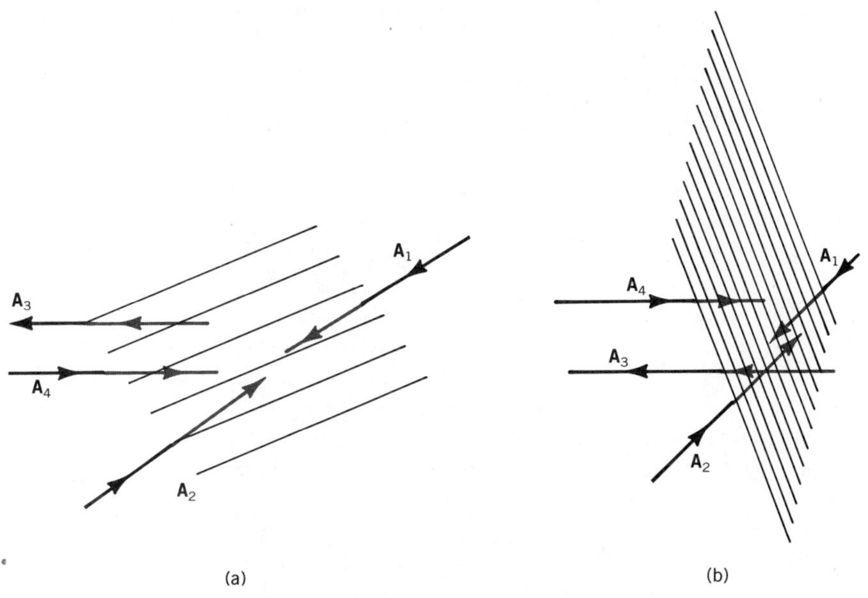

FIGURE 15-13. Phase conjugation from a holographic point of view. (a) Beams 1 and 4 conjugate interfere to produce a set of fringes and beam 2 diffracts off the fringes to produce beam 3. (b) Beams 2 and 4 conjugate produce a set of fringes and beam 1 diffracts off the fringes to produce beam 3.

due to waves of frequency 2ω (a most probable condition), then it is proper to ignore the nonlinear term involving 2ω.

If the conjugate wave $A_3 = A_4^*$ is made to propagate back along the same path taken by A_4 in reaching the conjugate medium, then A_3 upon arriving at the source of A_4 will have the same form as A_4. This occurs no matter how much the wave is distorted by the medium along the path, as long as the distortion does not change. The first demonstration of distortion removal by phase conjugation was made by Herwig Kogelnik in 1965. The demonstration was made by producing the interference fringes of Figure 15-13 in a photographic plate. Although this was an interesting demonstration of a novel application of holography, the fact that it did not occur in real time reduced its usefulness.

The ability to perform the same distortion removal but in real time makes the four-wave mixing we just described very appealing. Figure 15-14 is a demonstration of phase conjugation generated by Jack Feinberg of the University of Southern California. Figure 15-14*a* is the original undistorted image, a picture of a cat. The cat image was distorted by passing it through a microscropic slide smeared with glue. The distorted image, the wave we will use as A_4, is shown in Figure 15-14*b*. Through the use of phase conjugation, the distorted image was returned to nearly its original form by generating the wave A_3. The wave A_3 from the phase conjugation process is shown in Figure 15-14*c*.

Because the process shown in Figure 15-14 can be accomplished in real time it appears to be quite useful, but the need for two coherent light beams A_1 and A_2 make its implementation messy. Professor Feinberg discovered that by proper choice of materials and experimental conditions, only A_4 is needed to perform phase conjugation. In fact, the images shown in Figure

(a)

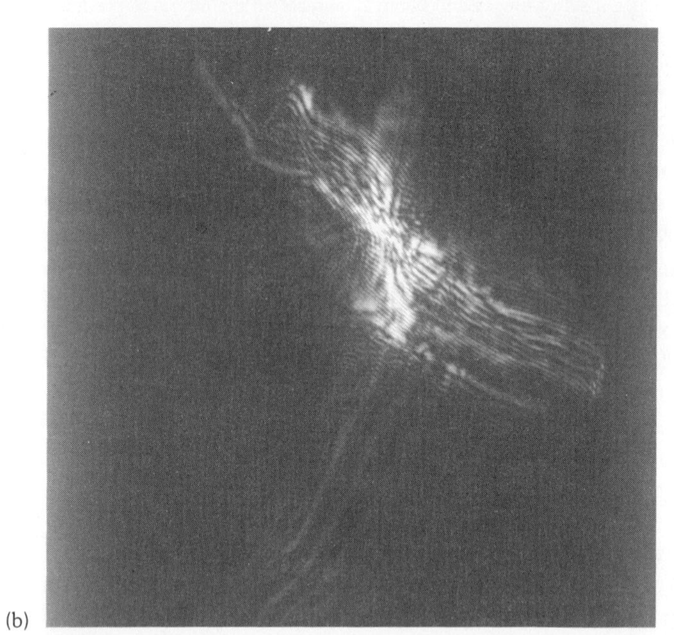

(b)

FIGURE 15-14. Image restoration using phase conjugation. (a) The original image of a cat. (b) The distorted image of a cat. The distortion was produced by smearing glue on a microscope slide. (c) The distortion is removed and the image is restored by the use of phase conjugation. A complete restoration has not been accomplished because some of the distorted signal missed the phase conjugator. (Courtesy **Jack Feinberg, University of Southern California**.)

(c)

15-14 were obtained using only A_4. This process is called *self-pumped phase conjugation.*[77]

To explain how and why self-pumped phase conjugation occurs requires the introduction of a new physical process and some modification to the theory associated with the coupling coefficient. There is no time left to explore this subject. The student should consult a very readable introduction to the subject by Professor Feinberg.[78]

SUMMARY

In this chapter, the effects of a large optical field are explored. The polarization induced by the optical field can no longer be treated as a linear function of the optical field, but now is described by a power series

$$P_i^{\omega} = P_i^0 + \sum_j \chi_{ij} E_j^{\omega} + \sum_{j,\ell} \chi_{ij\ell} \nabla_{\ell} E_j^{\omega}$$

$$+ \sum_{j,\ell} \chi_{ij\ell} E_j^{\omega_1} E_{\ell^{\omega_2}} + \sum_{j,\ell,m} \chi_{ij\ell m} E_j^{\omega_1} E_{\ell}^{\omega_2} E_m^{\omega_3} + \sum_{j,\ell} \chi_{ij\ell} E_j^{\omega_1} B_{\ell}^{\omega_2} + \cdots$$

We use the *dielectric susceptibility*

$$\chi = \frac{\epsilon}{\epsilon_0} - 1 = n^2 - 1$$

as the expansion parameter.

The *linear susceptibility*, defined by

$$P(\omega_i) = \chi^{(1)}(\omega_i)\epsilon_0 E_i$$

is obtained by neglecting nonlinear behavior

$$\chi^{(1)}(\omega_i) = \frac{Ne^2}{\epsilon_0 m} \frac{1}{\sqrt{(\omega_i^2 - \omega_0^2)^2 + \gamma^2 \omega_i^2}}$$

The second-order nonlinear polarization is defined as

$$P(\omega_i \pm \omega_j) = P_{NL}(\omega_3) = \chi^{(2)}(\omega_i \pm \omega_j)\epsilon_0 E_i E_j$$

The second-order susceptibility is found by applying first-order perturbation theory. If each of the applied fields is at a different frequency, then the theory yields the *sum* or *difference frequency susceptibility*.

$$\chi^{(2)}(\omega_i \pm \omega_j) = \frac{Ne}{\epsilon_0 E_i E_j} X_{i+j}$$

$$= \frac{Ne^3 a_2/\epsilon_0 m^2}{\sqrt{(\omega_i^2 - \omega_0^2)^2 + \gamma^2\omega_i^2}\sqrt{(\omega_j^2 - \omega_0^2)^2 + \gamma^2\omega_j^2}\sqrt{\left[(\omega_i \pm \omega_j)^2 - \omega_0^2\right]^2 + \gamma^2(\omega_i \pm \omega_j)^2}} \tag{15-6}$$

Experimentally, a parameter called the *nonlinear optical coefficient* is measured. It is defined by the equation

$$P(2\omega) = \mathrm{d}(2\omega)E^2 \cos(2\omega t - 2kz)$$

For second harmonic generation, we can write d in terms of the linear susceptibility at the fundamental and second harmonic of the light wave

$$\mathrm{d}(2\omega) = \frac{a_2 m \epsilon_0^3}{2N^2 e^3}\left[\chi^{(1)}(\omega)\right]^2 \chi^{(1)}(2\omega)$$

The wave equation for a nonlinear medium

$$\nabla^2 \mathbf{E} = \mu\sigma\frac{\partial \mathbf{E}}{\partial t} + \mu\epsilon\frac{\partial^2 \mathbf{E}}{\partial t^2} + \mu\frac{\partial^2 \mathbf{P}_{NL}}{\partial t^2}$$

was derived in this chapter, but its general solution was not obtained. The one-dimensional form of the wave equation was examined for the case when $\Delta k = k_3 - k_2 - k_1 = 0$. The result was an energy conservation law called the *Manley–Rowe relation*,

$$\frac{1}{\omega_1}\sqrt{\frac{\epsilon_1}{\mu}}\frac{\partial}{\partial z}(E_1 E_1^*) = \frac{1}{\omega_2}\sqrt{\frac{\epsilon_2}{\mu}}\frac{\partial}{\partial z}(E_2 E_2^*) = -\frac{1}{\omega_3}\sqrt{\frac{\epsilon_3}{\mu}}\frac{\partial}{\partial z}(E_3 E_3^*)$$

By assuming that the nonlinear interaction is weak, so that the incident two waves can be treated as constants, and by using the conservation of energy, the flux density of the nonlinearly generated wave is found to equal

$$S_3 = \frac{\omega_3^2}{2n_1 n_2 n_3}\sqrt{\left(\frac{\mu}{\epsilon_0}\right)^3}\,\mathrm{d}^2 L^2 S_1 S_2\left[\frac{\sin^2(\Delta kL/2)}{(\Delta kL/2)^2}\right]$$

Because the Poynting vector is proportional to d^2 and $1/n^3$, we define a *figure of merit* for nonlinear materials

$$M = \frac{\mathrm{d}^2}{n^3}$$

This figure of merit allows the performance of nonlinear materials to be compared.

If $\Delta k \neq 0$, then the fundamental waves at frequencies ω_1 and ω_2 travel at velocities that differ from the nonlinearly generated harmonic wave at frequency ω_3. The waves get out of phase in a distance called the *coherence length*. For second harmonic generation, the coherence length is equal to

$$\ell_c = \frac{\lambda_1}{4(n_3 - n_1)}$$

Over a distance equal to the coherence length, energy is transferred from the fundamental waves to the harmonic wave. For distances between ℓ_c and $2\ell_c$, energy is transferred back from the harmonic wave to the fundamental waves. This cyclic transfer of energy continues as long as the waves remain in the nonlinear material. Whenever $\Delta k = 0$, the waves are phase-matched and all of the energy is transferred from the fundamental waves to the harmonic wave. For example, the amplitude of the second harmonic wave generated under phase-matched conditions at a position L into the nonlinear medium is

$$E(2\omega) = E_0(\omega) \tanh\left[E_0(\omega) L d\omega \sqrt{\frac{\mu}{\epsilon}}\right]$$

Phase matching can be produced by

1. Periodic variation of the nonlinear coefficient.
2. The use of the phase changes produced by total reflection.
3. The use of the effective index of refraction for different modes in a dielectric waveguide.
4. The use of nonparallel propagation vectors to create $\Delta\mathbf{k} = 0$.
5. The use of the fact that the extraordinary index of refraction in a birefringent medium can be varied from n_e to n_0 by modifying the propagation direction.

Phase matching using birefringence is produced by satisfying one of the following four equations:

$$n_e < n_0, \qquad\qquad n_e > n_0$$

Type I

$$n_e(\omega_3) = \frac{\omega_1}{\omega_3} n_0(\omega_1) + \frac{\omega_2}{\omega_3} n_0(\omega_2), \qquad n_0(\omega_3) = \frac{\omega_1}{\omega_3} n_e(\omega_1) + \frac{\omega_2}{\omega_3} n_e(\omega_2)$$

Type II

$$n_e(\omega_3) = \frac{\omega_1}{\omega_3} n_e(\omega_1) + \frac{\omega_2}{\omega_3} n_0(\omega_2), \qquad n_0(\omega_3) = \frac{\omega_1}{\omega_3} n_0(\omega_1) + \frac{\omega_2}{\omega_3} n_e(\omega_2)$$

Only one third-order nonlinearity was discussed, the nonlinearity associated with phase conjugation. Third-order nonlinearities involve four waves; the complex amplitudes of wave 3 and 4 are given by the equations

$$\frac{dA_4}{dz} = -i\frac{\omega\chi^{(3)}}{2}\sqrt{\frac{\mu}{\epsilon}} A_1 A_2 A_3^*$$

$$\frac{dA_3^*}{dz} = -i\frac{\omega\chi^{(3)*}}{2}\sqrt{\frac{\mu}{\epsilon}} A_1^* A_2^* A_4$$

The effect of wave 1 and 2 is incorporated into the theory by the use of the coupling coefficient

$$\kappa = \frac{\omega\chi^{(3)*}}{2}\sqrt{\frac{\mu}{\epsilon}} A_1^* A_2^*$$

The complex amplitude of wave 3

$$A_3(z) = \frac{\cos|\kappa z|}{\cos|\kappa|L} A_3(L) + i\kappa^* \frac{\sin|\kappa|(z-L)}{|\kappa|\cos|\kappa|L} A_4^*(0)$$

is proportional to the conjugate of wave four at $z = 0$.

PROBLEMS

15-1. Show the steps leading to the more general expression for $d(\omega_3)$ given in Chapter 15 that involves $\chi^{(1)}(\omega_1)$, $\chi^{(1)}(\omega_2)$, and $\chi^{(1)}(\omega_3)$. Equation **(15-12)** is a special case of this more general expression.

15-2. Find d_{eff} for the crystal class $\bar{4}3m$. Assume the medium has a negative birefringence. The result will apply to a large number of the II-VI and the III-V semiconductors such as GaAs.

15-3. Using **(15-23)** and **(15-25)** for second harmonic generation, i.e., when $E_1 = E_2$ and $\omega_3 = 2\omega_1 = 2\omega_2$, show that the sum of the power densities of the fundamental waves and the second harmonic wave is a constant, with respect to the propagation path. Assume that the conductivities σ are zero and we have phase matching.

15-4. Franken demonstrated second harmonic generation by using quartz and a ruby laser operating at a wavelength of 694.3 nm. Was phasematching achieved? What is the coherence length for second harmonic phase matching in quartz?

15-5. Find the complex amplitude $A_4(z)$ using **(15-53)**. Find the relationship between $A_4(L)$ and $A_4(0)$ using the boundary conditions used in Chapter 15 to find the relationship for A_3.

15-6. Discuss the physical significance of the condition

$$|\kappa|L = \frac{\pi}{2}$$

from the theory of phase conjugation.

15-7. Examine noncollinear phase matching for second harmonic generation. Derive expressions for the angles α and β defined in Figure 15-5.

15-8. In the autocorrelator shown in Figure 15-7*a*, the length of one arm of the correlator must be varied to scan over a light pulse. Over what distance must the arm be varied to scan over a 1 psec pulse? How might you create this variation?

15-9. Beta Barium Borate (β–BaB_2O_4) is a popular nonlinear material. It belongs to the point group 3. What does its second-order susceptibility tensor look like if Kleinman's symmetry condition holds?

15-10. Find an expression for d_{eff} for the negative uniaxial crystal β–BaB_2O_4 for both type-I and type-II phase matching.

15-11. If we use 10.6 μm radiation to generate, through difference frequency mixing, 1.12 mm radiation, what is the maximum efficiency for the conversion based on the Manley–Rowe limit?

15-12. Calculate the phase-matching angle for type-I phase matching in KDP when the objective is to double 694 nm radiation. The dispersion data for KDP are given in Table 15.9.

TABLE 15.9 Dispersion Data for KDP

Wavelength	n_0	n_e
200	1.6226	1.5639
300	1.5456	1.4982
400	1.5245	1.4802
500	1.5149	1.4725
600	1.5093	1.4683
700	1.5052	1.4656
800	1.5019	1.4637

15-13. In Problem 15-12, what is the Poynting walk-off angle?

15-14. The dispersion relationships for the ordinary and extraordinary waves in β–BaB_2O_4 are given by Sellmeier relations

$$n_0^2 = 1.3416 + \frac{1.3812}{1 - \dfrac{0.015443}{\lambda^2}} + 0.00268\,\lambda^2$$

$$n_e^2 = -0.1552 + \frac{2.5077}{1 - \dfrac{0.006745}{\lambda^2}} + 0.01438\lambda^2$$

Calculate the phase matching angle for doubling 1 μm radiation using type-I phase matching.

15-15. Using the dispersion relations in Problem 15-14, neglecting the λ^2 dependence, find the wavelength where noncritical phase matching can be obtained.

15-16. We wish to frequency double a 1 psec pulse. If the fundamental and harmonic travel at different velocities, we have a temporal walk-off occurring as the wave propagates through the crystal. Calculate the temporal walkoff time per unit length for the doubling of a 1 psec pulse at a wavelength of 1 μm in β–BaB_2O_4.

Appendix 15-A

To describe dispersion in Chapter 7, the optical medium was modeled by a classical linear oscillator with a dissipative term **(7-26)**. The solution of the model's equation of motion was a damped sinusoidal wave **(7-27)**. The behavior of the oscillator, when it was driven by a simple periodic force, was determined by assuming a linear response to the driving force **(7-28)**. In this section, we will develop a model of a nonlinear optical medium by modifying the linear model used in Chapter 7. The linear model is first generalized to describe the effects of an arbitrary light wave as the driving force. The linear theory is then expanded to include a nonlinear response to the driving force by the use of a generalized potential energy function.

The solution of the nonlinear equation will be obtained by the use of perturbation theory. A brief description of the steps used in the solution will demonstrate that the second- and third-order nonlinearities can be treated independently. This result will be used to obtain the equation for the displacement of a nonlinear harmonic oscillator driven by a force containing two frequencies.

GENERALIZED LINEAR THEORY

To describe the behavior of materials in response to an arbitrary driving force, the simple harmonic function used in **(7-28)** is replaced by a generalized periodic function written as a Fourier series. This force is due to an applied electric field. The real electric field is written, using complex functions, in the notation of **(1B-6)**

$$E = \frac{1}{2}\Big[\mathrm{E}_1 \exp[i(\omega_1 t - k_1 x)] + \mathrm{E}_1^* \exp[-i(\omega_1 t - k_1 x)] + \cdots + \mathrm{E}_n \exp[i(\omega_n t - k_n x)] + \mathrm{E}_n^* \exp[-i(\omega_n t - k_n x)]\Big] \quad (15A\text{-}1)$$

We will assume a one-dimensional problem to reduce the complexity of the derivation. The notation may be simplified by using the identity

$$E(\omega_{-n}) \equiv E(-\omega_n) = E^*(\omega_n) = \tfrac{1}{2}\mathcal{E}_n^* e^{ik_n x}$$

The generalized harmonic driving force of **(15A-1)** can now be written in a very compact form

$$F(t) = \frac{e}{m} \sum_{n=-\infty}^{\infty} E(\omega_n) e^{i\omega_n t}$$

With the generalized driving force, the steady-state solution of the equation of motion describing the oscillator's displacement is a Fourier series, each term of which is identical to the solution obtained for a simple periodic driving force

$$x = \sum_{n=-\infty}^{\infty} X_n e^{i\omega_n t} = \sum_{n=-\infty}^{\infty} \frac{(e/m)E(\omega_n)}{\sqrt{(\omega_n^2 - \omega_0^2)^2 + \gamma^2\omega_n^2}} e^{i\omega_n t} \tag{15A-2}$$

There is a phase lag between the driving force and the nth component of the Fourier series describing the displacement, given by

$$\phi_n = \tan^{-1} \frac{\gamma\omega_n}{\omega_n^2 - \omega_0^2} \tag{15A-3}$$

NONLINEAR EQUATION OF MOTION

The model for a linear system can easily be modified to explain nonlinear behavior by replacing the harmonic oscillator potential used in the linear theory with a power series discussed in Appendix 1-A

$$V(x) = V_0 + \frac{1}{2}\left(\frac{d^2V}{dx^2}\right)x^2 + \frac{1}{6}\left(\frac{d^3V}{dx^3}\right)x^3 + \frac{1}{24}\left(\frac{d^4V}{dx^4}\right)x^4 + \cdots \tag{15A-4}$$

Figure 15A-1 shows a nonlinear potential constructed using an equation of the form

$$V(x) = \frac{m\omega_0^2}{2}x^2 + \frac{ma_2}{3}x^3 + \frac{ma_3}{4}x^4$$

In Figure 15A-1, the terms of this approximate potential were chosen to be

$$m\omega_0^2 = 1, \qquad ma_2 = \frac{1}{2}, \qquad ma_3 = \frac{1}{6}$$

The powers of x in the expansion that are needed to describe the potential are determined by the symmetry in the material. For example, if the material

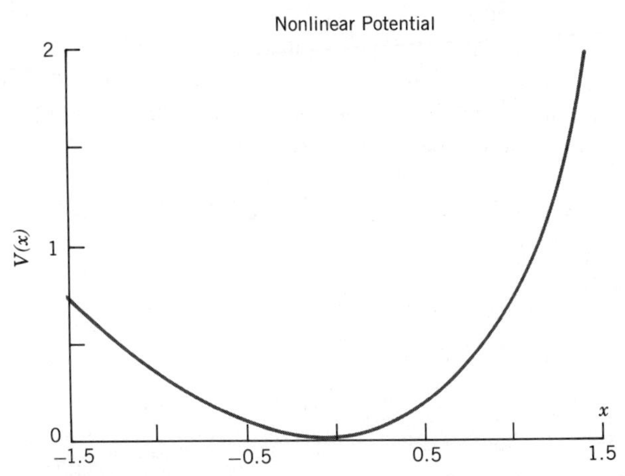

FIGURE 15A-1. A possible nonlinear potential based on (15A-4.)

has inversion symmetry, $V(x)$ is an even function and only even powers of x are found in the power series.

The nonlinear processes involved in a solid do not all have the same response speed. Electrons have the fastest response, whereas molecular motion is much slower. To handle both types of oscillators, we must replace the single coordinate x by coordinates for the electronic motion q_e and the ion motion q_i. For example, the nonlinear term ma_2x^3 would be replaced by four terms

$$Aq_i^3 + Bq_i^2q_e + Cq_iq_e^2 + Dq_e^3$$

In the simplified discussion presented here, we will ignore these complications. The nonlinear equation of motion that must be solved is therefore of the form

$$\frac{d^2x}{dt^2} + \gamma\frac{dx}{dt} + \omega_0^2x + a_2x^2 + a_3x^3 = 0$$

PERTURBATION TECHNIQUE

The application of the perturbation technique to the solution of the equation of motion for a nonlinear oscillator is based on the assumption that each of the additional terms in **(15A-4)** is much smaller than those preceding. The solution obtained by neglecting the additional term will therefore only need a small correction to satisfy the more complex potential.

In the description of the perturbation technique, a solution of the equation of motion of a nonlinear oscillator, with no dissipation $\gamma = 0$, will be obtained

$$\frac{d^2x}{dt^2} + \omega_0^2x + a_2x^2 + a_3x^3 = 0 \qquad \text{(15A-5)}$$

The first step, in the application of the perturbation technique, involves neglecting the nonlinear terms in the potential energy function by setting their coefficients equal to zero, $a_2 = a_3 = 0$. The resulting equation of motion describes a simple harmonic oscillator with a solution given by

$$x^{(0)} = X_1 \cos \omega_0 t \qquad \text{(15A-6)}$$

This solution is called the *zero-order approximate solution*.

The next step, in the application of the perturbation technique, requires the substitution of the zero-order solution **(15A-6)** into **(15A-5)** to produce a first-order approximation to the nonlinear equation. It is necessary to replace ω_0 by ω_1 in **(15A-6)** because the nonlinear oscillator will have a resonant frequency that differs from the linear oscillator's frequency

$$\frac{d^2x}{dt^2} + \omega_0^2x = -a_2X_1^2 \cos^2 \omega_1 t - a_3X_1^3 \cos^3 \omega_1 t$$

The use of the trig identities

$$\cos^2 \theta = \frac{1}{2}(1 + \cos 2\theta)$$

$$\cos^3 \theta = \frac{1}{4}(\cos 3\theta + 3 \cos \theta)$$

will make the results easier to interpret. Substituting the trig identities into the equation of motion produces an equation involving harmonics of ω_1.

$$\frac{d^2x}{dt^2} + \omega_0^2 x = -\frac{a_2 X_1^2}{2} - \frac{3a_3 X_1^3}{4} \cos \omega_1 t$$
$$-\frac{a_2 X_1^2}{2} \cos 2\omega_1 t - \frac{a_3 X_1^3}{4} \cos 3\omega_1 t \quad (15A\text{-}7)$$

The solution of this equation, called the *first-order approximate solution*, is assumed to be a finite harmonic series, containing the same frequencies as found in the driving force on the right side of **(15A-7)**

$$x^{(1)} = X_0 + X_1 \cos \omega_1 t + X_2 \cos 2\omega_1 t + X_3 \cos 3\,\omega_1 t \quad (15A\text{-}8)$$

To determine the values of the coefficients of the series, when **(15A-8)** is a solution of **(15A-7)**, we substitute **(15A-8)** and its derivatives into **(15A-7)**

$$(\omega_0^2 - \omega_1^2)X_1 \cos \omega_1 t - (\omega_0^2 - 4\omega_1^2)X_2 \cos 2\omega_1 t$$
$$-(\omega_0^2 - 9\omega_1^2)X_3 \cos 3\omega_1 t + \omega_0^2 X_0 = -\frac{a_2 X_1^2}{2} - \frac{3a_3 X_1^3}{4} \cos \omega_1 t$$
$$-\frac{a_2 X_1^2}{2} \cos 2\omega_1 t - \frac{a_3 X_1^3}{4} \cos 3\omega_1 t$$

Equating the coefficients of the cosine functions produces a set of equations, specifying the form of the coefficients X_0, X_2, and X_3 in terms of X_1 when **(15A-8)** is a solution.

The coefficient of the term X_0 is found by equating the terms, without time dependence, in the above equation. The time-independent, or dc, term in the first-order approximate solution must obey the equation

$$X_0 = -\frac{a_2 X_1^2}{2\omega_0^2} \quad (15A\text{-}9)$$

The physical interpretation given to this result is that the second-order nonlinear coefficient a_2 of the potential function causes a shift of the equilibrium position of the oscillator. The shift is a function of the amplitude of the fundamental frequency of the oscillator X_1.

The shift of the equilibrium position can be verified by constructing line segments parallel to the x axis, in Figure 15A-1, connecting points on the potential curve, on either side of $x = 0$, at equal values of $V(x)$. The centers of the line segments move to the left, as $V(x)$ increases. This is the behavior predicted by **(15A-9)**.

By equating terms involving $\cos \omega_1 t$, we discover that the frequency of the nonlinear oscillator is not equal to the frequency of the linear oscillator. The change in the oscillator frequency is proportional to the third-order nonlinear coefficient

$$\omega_1^2 = \omega_0^2 + \frac{3a_3 X_1^2}{4} \quad (15A\text{-}10a)$$

The oscillator frequency is also modified by the second-order nonlinearity. By equating terms involving $\cos 2\omega_1 t$, we discover that a second change in oscillator frequency occurs, proportional to the second-order nonlinear coefficient of the potential

$$\omega_1^2 = \frac{\omega_0^2}{4} - \frac{a_2 X_1^2}{8X_2} \tag{15A-10b}$$

It is difficult to keep all of the approximations separated. It is hoped that the discussion has made clear the difference between the order of the solution of **(15A-5)** and the order of the approximate representation of the nonlinear potential **(15A-4)**. Equation **(15A-11)** is the first-order perturbation solution of an equation involving a third-order expression of a nonlinear potential.

Both frequency changes are proportional to the square of the amplitude of the fundamental frequency of the oscillator. We would expect the frequency change associated with the second-order nonlinearity to dominate because the second-order nonlinearity should be larger than the third order.

Equation **(15A-10b)** can be solved for the functional form of the second harmonic coefficient X_2

$$X_2 = \frac{a_2 X_1^2}{2(\omega_0^2 - 4\omega_1^2)}$$

Finally, equating terms involving cos $3\omega_1 t$ yields the functional form of the coefficient of the third harmonic X_3

$$X_3 = \frac{a_3 X_1^3}{4(\omega_0^2 - 9\omega_1^2)}$$

We can now write the first-order solution of the nonlinear oscillator

$$\begin{aligned} x^{(1)} &= -(a_2/2\omega_0^2)\, X_1^2 + X_1 \cos \omega_1 t + (a_2/6\omega_0^2)\, X_1^2 \cos 2\omega_1 t \\ &\quad + (a_3/32\omega_0^2)\, X_1^3 \cos 3\omega_1 t \end{aligned} \tag{15A-11}$$

The displacement predicted by **(15A-11)** is a function, not only of the fundamental frequency of the oscillator, but also the second and third harmonics. In examining $x^{(1)}$, we see that the dc and second harmonic components are functions of the second-order nonlinear coefficient a_2, but not of the third-order coefficient a_3. The third harmonic of the oscillator frequency is only a function of the third-order nonlinear coefficient of the potential a_3. This means that, to the accuracy of this first-order theory, we can treat each nonlinear term separately.

A second-order perturbation calculation can be obtained by substituting the first-order solution **(15A-11)** into **(15A-5)** to form a second-order approximation to the differential equation. We will not carry out the second-order calculation.

SECOND-ORDER NONLINEARITY

We have demonstrated that the second- and third-order nonlinear contributions to the potential can be treated separately. We will examine the second-order nonlinearity using first-order perturbation theory. The driving field **E** to be used in the calculation is given by the Fourier series **(15A-1)**.

The zero-order solution of the differential equation

$$\frac{d^2x}{dt^2} + \gamma\frac{dx}{dt} + \omega_0^2 x + a_2 x^2 = -\frac{e}{m}E$$

is $x^{(0)}$ given by **(15A-2)**. The zero-order solution $x^{(0)}$ is used to rewrite the differential equation. The result will be the first-order approximation of the equation of motion of the nonlinear system

$$\frac{d^2x}{dt^2} + \gamma \frac{dx}{dt} + \omega_0^2 x = -\frac{e}{m}\sum_n E(\omega_n)e^{i\omega_n t}$$

$$-\frac{a_2 e^2}{m^2}\sum_n \sum_m \frac{E(\omega_n)E(\omega_m)\exp[i(\omega_n + \omega_m)t]}{(\omega_n^2 - \omega_0^2 + i\gamma\omega_n)(\omega_m^2 - \omega_0^2 + i\gamma\omega_m)} \tag{15A-12}$$

To solve this equation, the solution is assumed to have the same form as **(15A-8)**

$$x^{(1)} = \sum_k X_k e^{i\omega_k t} \tag{15A-13}$$

Substituting **(15A-13)** and its derivatives into **(15A-12)** generates the values of the coefficients X_n when **(15A-13)** is a solution. Whenever $\omega_k = \omega_n$, the coefficient of the nth harmonic must be

$$X_n = \frac{\frac{e}{m}E(\omega_n)}{\omega_n^2 - \omega_0^2 + i\gamma\omega_n} \tag{15A-14}$$

When $\omega_k = \omega_n + \omega_m$, the coefficient of the $n + m$ harmonic must be $X_{n+m} =$

$$\frac{\frac{e^2 a_2}{m^2}E(\omega_n)E(\omega_m)}{(\omega_n^2 - \omega_0^2 + i\gamma\omega_n)(\omega_m^2 - \omega_0^2 + i\gamma\omega_m)\left[(\omega_n + \omega_m)^2 - \omega_0^2 + i\gamma(\omega_n + \omega_m)\right]} \tag{15A-15}$$

The solution involves three frequencies ω_n, ω_m, and $\omega_n + \omega_m$ and the coefficients given above. It describes the behavior of the nonlinear oscillator for any pair of Fourier components of an arbitrary waveform. This result is too general to understand so we will make some simplifications.

To make it easier to obtain a physical interpretation of **(15A-15)**, we will limit the discussion to a nonlinear oscillator driven by two harmonic waves with frequencies ω_1 and ω_2. Making use of **(2B-6)** and the relationship between various coefficients

$$(X_{-n})^* = X_n, \qquad (X_{-n-m})^* = X_{nm}$$

(see Problem 15-3), we can write the solution of the nonlinear oscillator equation when driven by two frequencies as

$$\begin{aligned} x^{(1)} = 2\big[&|X_1|\cos\omega_1 t + |X_2|\cos\omega_2 t + 2|X_{1+2}|\cos(\omega_1 + \omega_2)t \\ &+ 2|X_{1-2}|\cos(\omega_1 - \omega_2)t + |X_{1+1}|\cos 2\omega_1 t + |X_{2+2}|\cos 2\omega_2 t \\ &+ |X_{1-1}| + |X_{2-2}|\big] \end{aligned} \tag{15A-16}$$

where we have assumed that ϕ_n, defined by **(15A-3)**, may be neglected.

Each term of **(15A-16)** is associated with a different frequency. The coefficient associated with each frequency is written explicitly below

1. The fundamental driving frequencies

$$\frac{\frac{e}{m}E_1\cos(\omega_1 t - k_1 x)}{\sqrt{(\omega_1^2 - \omega_0^2)^2 + \gamma^2\omega_1^2}} + \frac{\frac{e}{m}E_2\cos(\omega_2 t - k_2 x)}{\sqrt{(\omega_2^2 - \omega_0^2)^2 + \gamma^2\omega_2^2}} \tag{15A-17}$$

2. The sum frequency

$$\frac{\frac{e^2 a_2}{m^2} E_1 E_2 \cos\left[(\omega_1 + \omega_2)t - (k_1 + k_2)x\right]}{\sqrt{(\omega_1^2 - \omega_0^2)^2 + \gamma^2\omega_1^2}\sqrt{(\omega_2^2 - \omega_0^2)^2 + \gamma^2\omega_2^2}\sqrt{\left[(\omega_1 + \omega_2)^2 - \omega_0^2\right]^2 + \gamma^2(\omega_1 + \omega_2)^2}} \tag{15A-18}$$

3. The difference frequency

$$\frac{\frac{e^2 a_2}{m^2} E_1 E_2 \cos\left[(\omega_1 - \omega_2)t - (k_1 - k_2)x\right]}{\sqrt{(\omega_1^2 - \omega_0^2)^2 + \gamma^2\omega_1^2}\sqrt{(\omega_2^2 - \omega_0^2)^2 + \gamma^2\omega_2^2}\sqrt{\left[(\omega_1 - \omega_2)^2 - \omega_0^2\right]^2 + \gamma^2(\omega_1 - \omega_2)^2}} \tag{15A-19}$$

4. The second harmonics of the driving frequencies

$$\frac{\frac{e^2 a_2}{2m^2} E_1^2 \cos(2\omega_1 t - 2k_1 x)}{\left[(\omega_1^2 - \omega_0^2)^2 + \gamma^2\omega_1^2\right]\sqrt{(4\omega_1^2 - \omega_0^2)^2 + \gamma^2 4\omega_1^2}} + \frac{\frac{e^2 a_2}{2m^2} E_2^2 \cos(2\omega_2 t - 2k_2 x)}{\left[(\omega_2^2 - \omega_0^2)^2 + \gamma^2\omega_2^2\right]\sqrt{(4\omega_2^2 - \omega_0^2)^2 + \gamma^2 4\omega_2^2}} \tag{15A-20}$$

5. The dc terms

$$\frac{\frac{e^2 a_2}{m^2} E_1^2}{\omega_0^2\sqrt{(\omega_1^2 - \omega_0^2)^2 + \gamma^2\omega_1^2}} + \frac{\frac{e^2 a_2}{m^2} E_2^2}{\omega_0^2\sqrt{(\omega_2^2 - \omega_0^2)^2 + \gamma^2\omega_2^2}} \tag{15A-21}$$

Each of the terms discussed here can be treated separately because of the need to require conservation of momentum and conservation of energy. The conservation laws restrict solutions through the equations

$$\omega_3 = \omega_1 \pm \omega_2, \qquad \mathbf{k}_3 = \mathbf{k}_1 + \mathbf{k}_2$$

The results of the perturbation solution will be applied to the physical problem of calculating the macroscopic polarization of a nonlinear material in Chapter 15.

Appendix 15-B

MILLER'S RULE

R. C. Miller[79] observed that for second harmonic generation the ratio

$$\boldsymbol{\Delta} = \frac{d(2\omega)}{\left[\chi^{(1)}(\omega)\right]^2 \chi^{(1)}(2\omega)\epsilon_0^2} \tag{15B-1}$$

was a constant from material to material. Table 15B.1 displays some materials and their values of *Miller's coefficient* Δ. Examining the table, we see that for a large number of materials, Δ varies by a factor of 5, whereas the nonlinear coefficient varies by a factor of 10^3.

The classical theory derived in Chapter 15 yields an equation for $\boldsymbol{\Delta}$ in terms of the nonlinear coefficient

$$\boldsymbol{\Delta} = \frac{ma_2}{2N^2e^3} \tag{15B-2}$$

Treating $\boldsymbol{\Delta}$ as a constant implies that a_2/N_2 is also a constant. To obtain an estimate of the second-order nonlinear coefficient a_2, we assume that the second-order nonlinear term and harmonic term in the potential energy function are equal

$$\omega_0^2 x = a_2 x^2$$

If we assume that x is on the order of the lattice spacing

$$O[x] \approx \frac{1}{\sqrt[3]{N}}$$

then we can write a_2 as

$$a_2 = \frac{\omega_0^2}{x} \approx \omega_0^2 \sqrt[3]{N}$$

TABLE 15B.1 Miller's Coefficient

Material	d_{ij}, mks ($\times 10^{-22}$)	λ, μm	Δ_{ij}, mks ($\times 10^9$)	Crystal Class
KDP	0.1	1.06	9.7	$\bar{4}2m$
ZnS	5.66	1.06	9.5	$\bar{4}3m$
Te	940	5.0	11.6	32
CdTe	29.6	10.6	18.9	$\bar{4}3m$
GaAs	65	10.6	10	$\bar{4}3m$
CdSe	9.6	10.6	13	6mm
$LiNbO_3$	9.25	1.06	24.3	3m

This rather crude estimate of a_2 can be used to calculate an order-of-magnitude value for $\mathbf{\Delta}$. A more exact estimate of a_2 can be made by introducing an analytic potential for the anharmonic oscillator [80], but the more complicated potential provides the same order-of-magnitude estimate for $\mathbf{\Delta}$.

$$\mathbf{\Delta} \approx \frac{m\omega_0^2 \sqrt[3]{N}}{2N^2e^3} = \frac{m\omega_0^2}{2e^3 \sqrt[3]{N^5}} \tag{15B-3}$$

If

$$m = 10^{-30}\ \text{kg}, \qquad \omega_0 = 2\pi \times 10^{15}, \qquad N = \frac{10^{29}}{\text{m}^3}$$

then

$$\mathbf{\Delta} = 2.3 \times 10^9 \text{m/V}$$

or in cgs units $\mathbf{\Delta} \approx 10^{-6}$. A quantum mechanical model of the nonlinear oscillator, constructed using a two-level system, leads to similar results.

Appendix 15-C

NONLINEAR POLARIZATION IN THE 32 POINT GROUP

To demonstrate the procedure used to calculate the effective nonlinear coefficient d_{eff}, we will examine a crystal belonging to the crystal class 32. Examples of crystals in this class are Telurium and quartz.

We will assume that two linearly polarized plane waves, of the same frequency ω and with amplitudes E_0 and E_e, are copropagating in a crystal belonging to the point group 32. The propagation vectors of the waves make an angle θ with respect to the optical axis, which is parallel to the z axis (in crystallography, this is the c axis). We will assume further that the plane containing the propagation vectors and the optical axis makes an angle ϕ with respect to the x axis; see Figure 15C-1. The polarization of E_e (the extraordinary wave) lies in this plane at an angle θ to the xy plane and the polarization of E_0 is perpendicular to the plane (ordinary wave) lying in the xy plane; see Figure 15C-2.

The nonlinear susceptibility matrix for a crystal in the 32 symmetry group can be found in Appendix 14-A. The three components of the nonlinear polarization associated with the second harmonic are

$$\begin{aligned} P_x(2\omega) &= d_{11}E_{ex}E_{0x} - d_{11}E_{ey}E_{0y} + d_{14}E_{ey}E_{0z} + d_{14}E_{ez}E_{0y} \\ P_x(2\omega) &= -d_{14}E_{ez}E_{0z} - d_{14}E_{ez}E_{0x} - d_{11}E_{ex}E_{0y} - d_{11}E_{ey}E_{0x} \qquad (15C\text{-}1) \\ P_z(2\omega) &= 0 \end{aligned}$$

The geometric construction in Figure 15C-2 is used to obtain the field components.

$$\begin{aligned} E_{ex} &= E_e \cos\theta \cos\phi, & E_{0x} &= -E_0 \sin\phi \\ E_{ey} &= E_e \cos\theta \sin\phi, & E_{0y} &= E_0 \cos\phi \qquad (15C\text{-}2) \\ E_{ez} &= -E_e \sin\theta, & E_{0z} &= 0 \end{aligned}$$

By substituting **(15C-2)** into **(15C-1)**, we obtain the two components of P. These components are then used to calculate the polarization normal to the plane and parallel to E_0

$$\begin{aligned} P_\perp(2\omega) &= P_x \sin\phi - P_y \cos\phi \\ P_\perp(2\omega) &= E_e E_0 (d_{11} \cos\theta \cos 3\phi) \end{aligned}$$

The effective d for this polarization is

$$d_{eff} = d_{11} \cos\theta \cos 3\phi \qquad (15C\text{-}3)$$

The component of the polarization in the plane and parallel to $\mathbf{E}_e$ is

$$\begin{aligned} P_\parallel(2\omega) &= (P_x \cos\phi + P_y \sin\phi)\cos\theta + P_z \sin\theta \\ &= -E_e E_0 (d_{11} \cos^2\theta \sin 3\phi + d_{14} \sin\theta \cos\theta) \end{aligned}$$

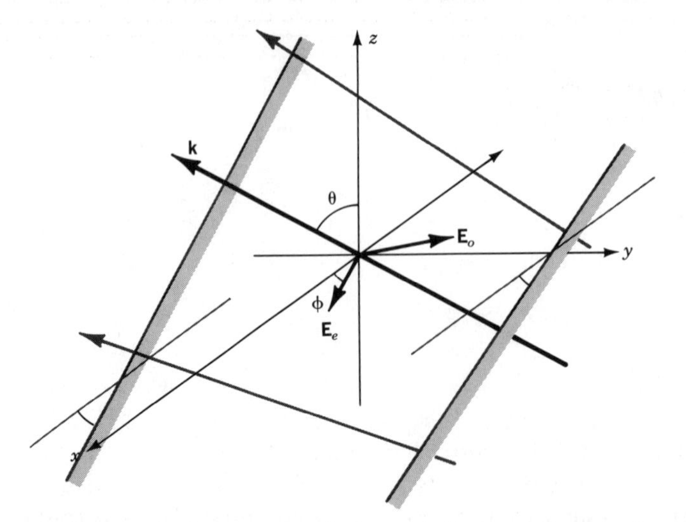

FIGURE 15C-1. Geometry for calculation of effective nonlinear coefficient.

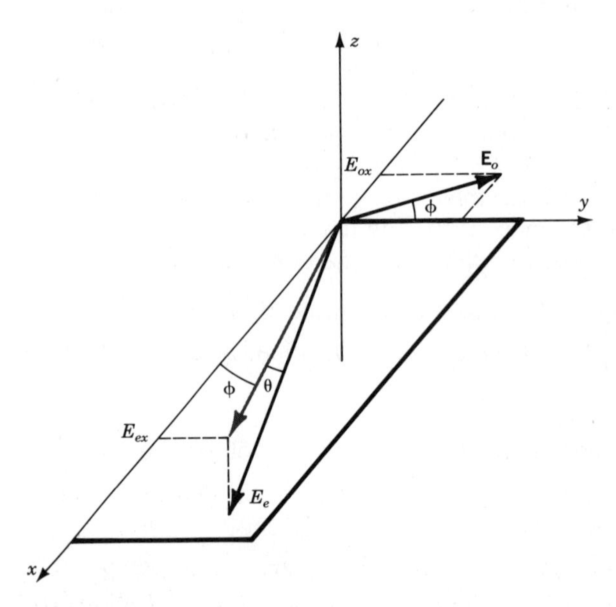

FIGURE 15C-2. An enlarged view of Figure 15C-1, showing the projection of the electric fields for two orthogonally polarized waves onto the coordinate axis.

The effective d for this polarization is

$$d_{eff} = d_{11} \cos^2 \theta \sin 3\phi + d_{14} \sin \theta \cos \theta \qquad \text{(15C-4)}$$

The 3ϕ has its physical basis in the fact that members of this group have a z axis with three-fold symmetry.

References

1. J. Clerk Maxwell, "A Dynamical Theory of the Electromagnetic Field," *Phil. Trans. Roy. Soc. London,* **155** (1865), 459–512.

2. Richard P. Feynman, *QED.* Princeton, N. J.: Princeton University Press, 1985.

3. Roald K. Wangsness, *Electromagnetic Fields*, Second Edition. New York: Wiley, 1986. J. D. Jackson, *Classical Electrodynamics*, Second Edition. New York: Wiley, 1975.

4. William H. McMaster, "Polarization and the Stokes Parameters," *Am. J. Phys.*, **22**, 351–362 (1954).

5. Miles V. Klein, *Optics.* New York: Wiley, 1970.

6. M. Born, and E. Wolf, *Principles of Optics.* New York: Pergamon Press, 1959.

7. George Stokes, *Trans. Cambridge Phil. Soc.*, **9**, 399 (1852).

8. H. Mueller, *J. Opt. Soc. Am.*, **38**, 661 (1948).

9. H. Poincaré, *Théorie Mathématique de la Lumiére* (Paris, Georges Carré), **2** (1892).

10. W. T. Doyle, "Scattering approach to Fresnel's equations and Brewster's law," *Am. J. Phys.*, **53**, 463–468 (1985).

11. C. L. Giles, and W. J. Wild, "Brewster Angles for Magnetic Media," *International J. Infra, and Millimeter Waves*, **6**, 187–197 (1985).

12. S. Zhu, A. W. Yu, D. Hawley, and R. Roy, "Frustrated Total Internal Reflection: A Demonstration and Review," *Am. J. Phys*, **54**, 601–607 (1986).

13. P. K. Tien, and R. Ulrich, "Theory of Prism-Film Coupler and Thin-Film Guides," *J. Opt. Soc. Am.*, **60**, 1325–1337 (1970).

14. Warren J. Smith, "How Flat is Flat," *Optical Spectra* (April–May 1974 and August 1978).

15. E. Delano and R. J. Pegis, "Methods of Synthesis for Dielectric Multilayer Filters," *Progress in Optics*, **7**, North-Holland, Amsterdam (1969).

16. O. S. Heavens, *Optical Properties of Thin Solid Films.* New York: Dover, 1965.

17. R. P. Feynman, R. B. Leighton, and M. Sands, *The Feynman Lectures on Physics.* Reading, Mass.: Addison–Wesley, 1964.

18. (a) A. Nussbaum, and R. A. Phillips, *Contemporary Optics for Scientists and Engineers*. Englewood Cliffs, N. J.: Prentice Hall, 1976. (b) D. C. O'Shea, *Elements of Modern Optical Design*. New York: Wiley, 1985.

19. A. Yariv, *Optical Electronics,* Third Edition. New York: Holt, Rinehart & Winston, 1985.

20. D. Bohm, *Quantum Theory*. New York: Prentice-Hall, 1951.

21. Allen H. Cherin, *An Introduction to Optical Fibers*. New York: McGraw-Hill, 1983.

22. P. M. Duffieux, *The Fourier Integral and Its Applications in Optics,* Second Edition, Masson and Cie, Paris (1970). This is an English translation of a book originally published in French as *L'Intégrale de Fourier et ses Applications à L'Optique*, Faculté des Sciences, Besancon, 1946.

23. R. K. Luneberg, *The Mathematical Theory of Optics*. Berkeley: University of California Press, 1964.

24. W. T. Cochran, J. W. Cooley, D. L. Favin, H. D. Helms, R. A. Kaenel, W. W. Lang, G. C. Maling, D. E. Nelson, C. M. Rader, and P. D. Welch, "What Is the Fast Fourier Transform?" *Proc. IEEE*, **55**, 1664–1674 (1967).

25. M. J. Lighthill, *Introduction to Fourier Analysis and Generalized Functions*. Cambridge: Cambridge University Press, 1964.

26. A. Papoulis, *The Fourier Integral and Its Applications*. New York: McGraw-Hill, 1962.

27. C. E. Shannon and W. Weaver, *The Mathematical Theory of Communications*. Champaign, Ill.: University of Illinois Press, 1949.

28. C. Kittel, *Introduction to Solid State Physics*, Third Edition. New York: Wiley, 1966.

29. J. I. Pankove, *Optical Processes in Semiconductors*. New York: Dover, 1971.

30. A. Sommerfeld, *Ann. Phys.*, **44**, 177 (1914).

31. L. Brillouin, *Ann. Phys.*, **44**, 203 (1914).

32. L. Brillouin, *Wave Propagation and Group Velocity*. New York: Academic Press, 1960.

33. H. Paul, "Interference Between Independent Photons," *Rev. Mod. Phys.*, **58**, 209–231 (1986).

34. F. Zernike, "The Concept of Degree of Coherence and Its Application to Optical Problems," *Physica*, **5**, 785 (1938).

35. H. H. Hopkins, *Proc. R. Soc.*, **A217**, 408 (1953).

36. L. Mandel, and E. Wolf, "Coherence Properties of Optical Fields," *Rev. Mod. Phys.*, **37**, 231 (1965).

37. Arvind S. Marathay, *Elements of Optical Coherence Theory*. New York: Wiley, 1982.

38. L. Mandel, and E. Wolf, *Proc. Phys. Soc. (London)*, **80**, 894 (1962).

39. John R. Klauder, and E. C. G. Sudarshan, *Fundamentals of Quantum Optics*. New York: Benjamin, 1968.

40. R. Hanbury Brown and R. Q. Twiss, "A New Type of Interferometer for Use in Radio Astronomy," *Phil Mag*, **45**, 663 (1954).

41. R. Hanbury Brown, and R. Q. Twiss, *Proc. R. Soc.*, **A243**, 291 (1957).

42. A. Rubinowicz, "The Miyamoto-Wolf Diffraction Wave," *Progress in Optics*, E. Wolf, ed., **IV** (1965).

43. Hermann A. Hauss, *Waves and Fields in Optoelectronics*, Englewood Cliffs, N. J.: Prentice-Hall, 1984.

44. A. Sommerfeld, *Optics*. New York: Academic Press, 1964.

45. Edward L. O'Neill, *Introduction to Statistical Optics*. Reading, Mass.: Addison–Wesley, 1963. Joseph Goodman, *Introduction to Fourier Optics*. New York: McGraw-Hill, 1968.

46. George B. Parrent, Jr., and Brian J. Thompson, "Physical Optics Notebook," Society of Photo-optical Instrumentation Engineers, 1969. Reprinted by the Institute of Optics, University of Rochester, 1971.

47. L. J. Cutrona, E. N. Leith, L. J. Porcello, and W. E. Vivian, "On the Application of Coherent Optical Processing Techniques to Synthetic-Aperture Radar," *Proc. IEEE*, **54**, 1026–1032 (1966).

48. M. Abramowitz, and I. A. Stegun, *Handbook of Mathematical Functions*, National Bureau of Standards Applied Mathematics Series-55, June 1964.

49. D. J. Stegliani, R. Mittra, and R. G. Semonin, "Resolving Power of a Zone Plate," *J. Opt. Soc. Am.*, **57**, 610–613 (1967).

50. Iazuo Sayanagi, "Pinhole Imagery," *J. Opt. Soc. Am.*, **57**, 1091–1099 (1967).

51. B. Rossi, *Optics*. Reading, Mass.: Addison–Wesley, 1957.

52. Dennis Gabor, "Holography, 1948–1971," *Proc. IEEE*, **60**, 655–668 (1972).

53. E. N. Leith, and J. Upatnieks, "Wavefront Reconstruction with Continuous-Tone Objects," *J. Opt. Soc. Am.*, **53**, 1377–1381 (1963). Also see the two articles by the same author in *J. Opt. Soc. Am.*, **52**, 1123 (1962) and *J. Opt. Soc. Am.*, **53**, 522 (1963).

54. P. Hariharan, *Optical Holography*. Cambridge: Cambridge University Press, 1984.

55. R. J. Collier, C. B. Burckhardt, and L. H. Lin, *Optical Holography*. New York: Academic Press, 1971.

56. R. Clark Jones, "A New Calculus for the Treatment of Optical Systems," *J. Opt. Soc. Am.*, **31**, 488 (1941).

57. W. Kauzmann, *Quantum Chemistry*. New York: Academic Press, 1957.

58. R. S. Longhurst, *Geometrical and Physical Optics,* Second Edition. New York: Wiley, 1967.

59. E. U. Condon, and H. Odishaw, *Handbook of Physics*. New York: McGraw-Hill, 1958.

60. L. D. Landau, and E. M. Lifshitz, *Electrodynamics of Continuous Media*. Reading, Mass.: Addison–Wesley, 1960.

61. Grant R. Fowles, *Introduction to Modern Optics*. New York: Holt, Rinehart & Winston, 1968.

62. D. Brewster, *Phil. Trans.*, 60 (1815) and H. Mueller, *Phys. Rev.*, **47**, 947 (1935).

63. P. A. Franken, and J. F. Ward, "Optical Harmonics and Nonlinear Phenomena," *Rev. Mod. Phys.,* **35**, 23–39 (1963).

64. F. Zernike, and K. E. Midwinter, *Applied Nonlinear Optics*. New York: Wiley, 1973.

65. N. Bloembergen, *Nonlinear Optics*. New York: Benjamin, 1965.

66. Y. R. Shen, *The Principles of Nonlinear Optics*. New York: Wiley, 1984.

67. D. A. Kleinman, *Phys. Rev.*, **126**, 1977 (1962).

68. J. A. Armstrong, N. Bloembergen, J. Ducuing, and P. S. Pershan, "Interactions Between Light Waves in a Nonlinear Dielectric," *Phys. Rev.*, **127**, 1918–1939 (1962).

69. N. Bloembergen, "Conservation Laws in Nonlinear Optics," *J. Opt. Soc. Am.*, **70**, 1429–1436 (1980).

70. G. D. Boyd, and D. A. Kleinman, "Parametric Interaction of Focused Gaussian Light Beams," *J. Appl. Phys.*, **39**, 3597–3639 (1968).

71. Y. Yacoby, R. L. Aggarwal, and B. Lax, "Phase Matching by Periodic Variation of Nonlinear Coefficients," *J. Appl. Phys.*, **44**, 3180–3181 (1973).

72. J. P. van der Ziel, and M. Ilegems, "Optical Second Harmonic Generation in Periodic Multilayer GaAs-AlGaAs Structures," *Appl. Phys. Letters*, **28**, 437–439 (1967).

73. G. D. Boyd, and C. K. N. Patel, "Enhancement of Optical Second-Harmonic Generation by Reflection Phase Matching in ZnS and GaAS," *Appl. Phys. Letters*, **8**, 313–315 (1966).

74. D. E. Thompson and P. D. Coleman, "Step-Tumble Far Infrared Radiation by Phase Matched Mixing in Planar Dielectric Waveguides," *IEEE Trans.*, **MTT-22**, 955–1000 (1974).

75. R. Trebino, "Second Harmonic Generation Rings and Refractive Index Measurement in Uniaxial Crystals," *Appl. Optics*, **20**, 2090–2096 (1981).

76. R. R. Briss, "Property of Tensors in Magnetic Crystal Classes," *Proc. Phys. Soc. (London)*, **79**, 946–953 (1962).

77. Jack Feinberg, "Self-pumped, Continuous-Wave Phase Conjugator Using Internal Reflection," *Optics Letters*, **7**, 486–487 (1982).

78. Jack Feinberg, "Photorefractive Nonlinear Optics," *Physics Today*, 6–12 (Oct. 1988).

79. R. C. Miller, "Optical Second Harmonic Generation in Piezoelectric Crystals," *Appl. Phys. Lett.*, **5**, 17 (1964).

80. C. G. B. Garret, and F. N. H. Robinson, "Miller's Phenomenological Rule for Computing Nonlinear Susceptibilities," *IEEE J. Quant. Elect.*, **QE-2**, 328 (1966).

Index

A

Abbe, Ernst, 193, 363, 397
Abbe's:
  number (value), 287–290
  Sine condition, 204
  theory, 397–407, 422
ABCD:
  fibers, 174–175
  law for gaussian waves, 343–350
  matrix, 144, 182–189
  principal planes, 187
  thin lens, 182, 345–346
Aberration, 140, 191, 212, 454, 457–458
  astigmatism, 206–207
  chromatic, 287–290
  coma, 203–206, 211
  distortion, 208–209
  field curvature, 208
  offense against sine, 204
  spherical, 198–202, 210
Absorption coefficient, 51
Acceptance angle:
  fiber, 152–154
  polarizer, 533
  2nd harmonic generation, 658
Achromatic lens, 288–290
Acoustic wave, 601, 620
Acousto-optic, 601–609
  beam deflector, 610–612
  effect, 601–603, 620–629
  mode locker, 612–613
  modification of index of refraction, 620, 623–624, 630
  modulator *see* Modulator
  scattering, 602–606, 621
Airy, Sir George Biddell, 373
Airy:
  disk, 202, 374, 398
  Formula, 373
  Function, 109. *See also* Fabry–Perot interferometer
  pattern, 196–197, 373
Alhazen, 62
Ampére's Law, 27
Analyzer, 579–580. *See also* Polarizer
Angular magnification, 184
Anisotropic:
  index of refraction, 530, 539
  material, 521
Aperture:
  function, 365, 391
  length, 658
  stop, *see* Stop
Apodization, 411–413
Arago, Dominique Francois, 325, 545
Argand diagram, 20
Array:
  function, 376. *See also* Grating, factor
  theorem, 375–376
Astigmatism, *see* Aberration
Atomic refractivity, 278. *See also* Lorenz–Lorentz, law of
Attenuation constant, 13
Autocorrelation, 235–236, 303–305
Axial ray, *see* Ray

B

Babinet, Jacques, 463
Babinet's:
  compensator, 542
  principle, 463–466
Bandwidth, 35, 114, 261, 606
  space, 485, 495–496, 519, 606
  time, 606–608
Barrel distortion, 209. *See also* Aberration; Distortion
Bartholinus, Erasmus, 521
Beam:
  divergence, *see* Gaussian wave parameters (Guassian wave), divergence angle
  waist, *also* width; or spot size, *see* Gaussian beam parameters (Guassian wave), waist
Bending, 209–210
Bessel function, *see* Functions, Bessel
Biaxial birefringence, 563–564
Biot, Jean Baptiste, 524, 545–546
Biot and Savart, law of, 27
Biquard, P., 571, 601
Biradial, 562
Birefringence, 521. *See also* Double refraction
Blazed:
  angle, 384–386
  grating, 379, 383–386
Boundary conditions:
  four wave mixing, 662
  Kirchhoff, 355, 358
  Maxwell's equation, 68–70, 124–125, 163
  nonlinear wave, 645, 649
  string:
    standing wave, 106
    traveling wave, 63–65
    three dimensional wave, 66, 565
Boyle, Robert, 87
Bragg, William Lawrence, 487
Bragg:
  angle, 626
  Equation, 381, 387, 606, 609, 614, 624
  law, 487–488, 604
  scattering, 604, 606, 622–626
Brewster, Sir David, 77, 571, 596
Brewster's:
  angle, 61, 77–78, 552, 604–605
  law, 526–528
Brillouin, L., 280, 571
Brillouin precurrsor, 280–281

C

Cardinal points, 185–188. *See also* Nodal, points; Focal point; Principal, points
Carrier frequency, 226–227, 256–257
Cauchy, Augustin Louis, 252
Cauchy's equation, 252–253
Chief ray, 190
Chiral, 546. *See also* Optical activity
Chirped grating, 493
Chromatic aberration, *see* Aberration
Chromatic resolving power, 111. *See also* Resolving power, chromatic
Circular dichroism, 548, 552
Circular polarization, 26–48
Classical turning points, 172–173, 176
Clausius–Mossotti relation, 278
Cleomedas, 61
Clerk Maxwell, James, *see* Maxwell, James Clerk
Coddington, Henry, 209
Coddington's:
  position factor, 210–211
  shape factor, 209–211
Coherence, 93, 104, 291, 362, 421
  degree, 296, 298, 307–308
    spatial, 305, 504–506
    temporal, 297–305, 401, 403–404, 503, 506–507
  longitudinal, 303, 503–504, 507
  mutual, 307, 421
  reduction property, 314
  transverse, 311–312, 506
Coherence length, 303, 313, 646–648, 666
Coherence time, 301–303
Coherency matrix, 45
Coherent source size, 312
Coma, *see* Aberration
Comb function, *see* Functions
Compensator, *see* Retarder
Complex index of refraction, 51–52, 83–84, 264, 268–276. *See also* Index of refraction
Conductivity, 28, 49–53, 60, 251–252
Confocal parameter, *see* Gaussian beam parameters (Gaussian wave), Rayleigh range
Conjugate:
  planes, 184, 188, 366, 396
  points, 186, 491
Conservation laws, *see also* Optical invariant
  energy, 94, 602, 606, 644–645, 676

information, 612
momentum, 603, 605, 607, 644–649, 651–659, 676
Constitutive relations, 28–29, 49
Constringence, *see* Abbe's, number (value)
Constructive interference, 87, 94
Contrast, 482, 490, 399, 405–406, 426–427
Convolution:
integral, 237–239, 241, 249, 375, 415, 518
theorem, 248, 375
Cornu, Marie Alfred, 440
Cornu spiral, 434, 439–442, 459–60, 468
Correlation, 235–238, 415, 421, 517–518
auto, 303–305, 318, 425
cross, 249, 304, 319
Cosine transform, 223, 226, 298, 310
Cotton–Mouton effect, 571, 596. *See also* Magneto-optic effect
Coulomb's Law, 26
Coupling coefficient, 662, 667
Critical angle, 78–80
Critical phase matching, 658
Crown glass, 288
Curie temperature, 588
Cut off condition, 158

D

Damped oscillator, 20
Damping force, 13, 20
Dark-ground illumination, 413
DaVince, Leonardo, 325
Dc stop, 413. *See also* Filter, spatial; Apodization
Debye, Peter, 149, 571, 601
Debye–Sears effect, 571
Decay constant, 81, 159–163, 571
Degree of coherence, *see* Coherence
Degree of polarization, 47–48
Delta function, *see* Functions
Density:
charge, 27
current, 27–28
energy, 33, 536, 630
flux, 34–36, 46–47, 666. *See also* Intensity
momentum, 37
number, 263
optical, 475
Density matrix, 45, 49
Depolarizer, 545
Descartes, René, 62
Destructive interference, 87, 94
Dextrorotatory, 546, 550
Dichroic polarizer, 523–526
Dichroism, 522, 524, 552
Dielectric constant, 28, 30, 60, 252, 263, 267–268, 273, 276–278, 620. *See also* Index of refraction
Dielectric polarization, 28, 262–263, 267, 273, 573, 592, 634, 638
2nd order nonlinear, 633, 635–640, 666
3rd order nonlinear, 633, 635, 659
Dielectric susceptibility, 267, 636–638, 665
Difference-frequency generation, 637
Diffraction, 310, 323, 362
angle, 341–342, 607–608, 610. *See also* Gaussian beam parameters (Gaussian wave), divergence angle
efficiency, 489–490, 508, 514, 625
grating, 377, 379–386, 401, 472
limit, *see* Rayleigh, limit; Strehl ratio
patterns, 370–374, 377–381, 386–387, 399–403, 407, 610, 442–443, 453–454, 464–466, 482–485
Diopter, 141–143
Dirac Delta function, 229–232. *See also* Functions, delta
Dispersion, 5, 251–290, 573–574, 590–593, 640
anomalous, 252, 271, 275
material, 262–279, 657–658
mode, 260–261
partial, 288, 290
profile, 260
Dispersion equation, 13, 157, 159–165, 253–254
Displacement current, 27
Displacement (electric), 27, 262–263, 522, 529, 534–541, 556, 560, 590. *See also* Dielectric polarization
Distortion, 596–597, 664. *See also* Aberration
by a hologram, 499, 501
in a pinhole camera, 457–458
Divergence angle, 341–342. *See also* Diffraction, angle
Dollond, John, 193, 289
Doppler, Christian, 299

Doppler:
- broadening, 299–300
- effect, 299, 603, 621

Double refraction, 521. *See also* Birefringence
Driffield, V. C., 475
Drude, Paul Karl Ludwig, 253, 547
Drude's equation, 267
Duffieux, Pierre-Michel, 214, 363

E

Edge enhancement, 409, 413. *See also* Filter, low-pass
Effective focal length, 187
Effective guide index, 159, 652
Effective nonlinear optical coefficient, 641, 646
Eikonal, 132
Eikonal equation, 131–133
Elastic stiffness constant, 598
Electro-optic:
- effect on index of refraction, 571–579, 614
- modulator, 582–590. *See also specific type*

Elliptical polarization, 43–45, 47–48
Ellipticity, 41, 43
End fired coupling, 151–155
Energy density, *see* Density
Entrance pupil, 188–190
Epoch angle, 18
Etalon, 117
Euclid, 61
Euler, Leonard, 1
Euler formula, 22, 216
Evanescent wave, 81–82, 161, 163, 255
Exit pupil, 188–190
Extinction coefficient, 51
Extinction ratio (of polarizers), 526
Extraordinary wave, 521, 530

F

f/#, 189. *See also* f-number
Fabry, Marie Paul Auguste Charles, 110
Fabry–Perot interferometer, *also* resonator or cavity
- contrast, 109
- finesse, 114–115
- free spectral range, 113–114
- spectral resolution, 111–115
- stability, 144–148, 346–347

False detail, 406–407
Faraday, Michael, 25, 571
Faraday:
- effect, 590, 594, 614. *See also* Magneto-optic effect
- isolator, 595
- law, 25

Far-field, 365, 386, 464. *See also* Fraunhofer, diffraction
Fast axis, 532
Fermat, Pierre de, 62, 130
Fermat's principle, 130, 133–138, 458–459
Ferroelectric, 634
Feussner, *see* Polarization prism
Fiber:
- cladding, 152
- core, 152
- graded index, 171
- step index, 151

Field flattener, 208
Figure, optical, 102, 114–115
Figure-of-merit, *see also* Bandwidth
- acoustic, 630–631
- nonlinear, 646, 666

Filter, 238–239
- bandpass, 161
- matched (phase and amplitude), 515–519
- spatial, 408–411, 415
  - high-pass, 408–409
  - low-pass, 408
- wavelength, 106, 120

Finesse, 114–115
Finite ray trace, 193
Fizeau, Armand H. L., 95
Fizeau fringes, *see* Fringes
Flint glass, 288
Flux density, 34. *See also* Density
Focal length of:
- hologram, 492
- lens, 184–185
- mirror, 146
- pinhole camera, 458
- zone plate, 455–456

Focal point, 184. *See also* Cardinal points
Forrester, A. T., 292, 294
Foucault, Jean Bernard Léon, 1
Foucault knife-edge, 413
Fourier, Jean Baptiste Joseph Baron de, 213
Fourier:
- integral, 221–230

series, 214–221, 670–671
transform, 221–230, 232–234, 237, 239, 241–244, 248–249, 305, 310, 366, 448, 492
Fourier transform spectroscopy, 297–300
Four-wave mixing, *see* Phase conjugation
Frankin, Peter, 633
Franklin, Benjamin, 1
Fraunhofer, Joseph, 120, 287, 362, 379
Fraunhofer:
approximation, 363–366
diffraction, 362, 386, 451, 459, 483, 609–610, 621
lines, 288
Free spectral range, 113–114, 382
Frequency plane, 399
Fresnel, Augustin Jean, 1, 325, 522, 546
Fresnel:
approximation, 434–437
diffraction, 433–447, 453–454, 464–466
ellipsoid, 561
equation, 537–541, 561
formula, 67–72
integral, 331, 438, 441, 467
mirror, 94
zone, 444, 448–494
zone plate, 454–457, 491–494
Fresnel Kirchhoff diffraction formula, 356
Fringe, 87, 96, 101–105, 109–112
Fizeau, 95–99, 101, 105, 116
Haidinger's, 99, 116
Fringe contrast in multiple reflections, 109
Fringe visibility, 109, 301, 490
Frustrated total reflection, 78–82, 85
Functions:
Bessel, 163, 242–244, 373–374, 513, 627
Airy, 109. *See also* Fabry–Perot interferometer
Comb, 230, 232, 234, 376
delta, 229–232, 238–341. *See also* Dirac Delta function
gaussian, 228–229, 300, 338–339
sinc, 224–225, 370–371, 374
square, 218–219, 224, 226, 235–236. *Also* rectangular
f-number, 425–426. *See also* f/#

G

Galilei, Galileo, 129
Gauss's law:
for the electric field (Coulomb's), 26
for the magnetic field, 27
Gaussian beam parameters (Gaussian wave):
complex curvature, 345
complex size, 339, 344
divergence angle, 341–342. *See also* Diffraction, angle
radius of curvature, 341, 349
Rayleigh range, 342, 349, 461
waist, 339, 341–342, 346, 349–350, 374, 404, 423
Gaussian constants, 144
Gaussian function, *see* Functions
Gaussian spectral distribution, 299–300
Geometrical optics:
and birefringence, 567–568
and holography, 469–502
Gladstone and Dale, law of, *see* Lorenz–Lorentz Law
Glan–Foucault prism, *see* Polarization prism
Glan–Thompson prism, *see* Polarization prism
Glass, dispersion in, 274, 287–290
Glauber, R. J., 292
Gradient index of refraction fiber, 169–176
Grating, 377, 379–386
equation, 387, 486
factor, 376, 378
formula, 381
Green's function, 240, 352–355, 357–358. *See also* Impulse response; Point spread function
Green's theorem, 359–360
Grimaldi, Francesco Maria, 94, 252, 323, 325
Group, 256, 301–302
Group velocity, 256. *See also* Velocity
Guided wave index of refraction, 149–151, 164–166
Guided wave mode, 155–158, 160, 162
Gyration vector, 590

H

Haidinger's fringes, *see* Fringe
Half-wave plate, 542. *See also* Retarder
Half-wave voltage, 578, 589, 613

Hamilton, William Rowan, 136
Hamilton's principle, 166–168
Hanbury Brown, R., 292, 314
Hanbury Brown and Twiss interferometry, *see* Interferometer, Intensity
Harmonic motion, 17, 19
Harmonic oscillator, *see* Oscillator
Helicity, 43
Helmholtz, Hermann Ludwig Ferdinand von, 9
Helmholtz equation, 9–10, 50, 131, 336
Hero of Alexandria, 133
Hertz, Heinrich Rudolf, 26
Heterodyne, 295
High-pass filter, *see* Filter, spatial
Hologram:
- absorption, 474, 486
- Fourier transform, 501–502, 515
- phase, 487, 475, 512–514
- reflection, 488, *also* white light
- thick, 487
- thin, 486–487

Holographic image, 482–483, 493–494, 498–502
Holography, 662–665
- bandwidth, 494–496
- coherency requirements, 503–507
- nonlinear effects, 485, 512–513

Homodyne, 295
Homogeneous wave, 9
Hooke, Robert, 87
Hopkins, H. H., 292
Hurter, F., 475
Hurter–Driffield (HD) curve, 475–476
Huygens, Christian, 1, 521
Huygens:
- principal, 323, 326–329, 567–568
- wavelet, 323

Huygens–Fresnel integral, 329–336, 435–437

I

Idler wave, 602, 637
Illuminance, 34. *See also* Intensity
Image:
- blur, 505–507
- contrast, 426–427

Impedance, 33, 73, 77, 124, 644
Impermeability tensor, 536, 597
Impulse response, 206, 240–241, 335, 349, 395–396, 420–421, 427, 505
Incoherent image, 420–428
Index of refraction, 30, 123. *See also* Complex index of refraction
- effective guide, 159
- frequency dependence, 266–276, 593. *See also* Dispersion
- gradient, 171, 261
- and the nonlinear coefficient, 575–576, 578
- normalized, 163–166
- principal, 539
- step, 151

Index surface, 523, 548–549, 563–565
Indicatrix, *see* Optical indicatrix
Inhomogeneous wave, 9
Intensity:
- from acoustic scattering, 625, 627
- due to interference, 91–92, 103–104, 108–110, 301, 318
- of light wave, 34. *See also* Flux density
- modulator, 579–582, 584–585
- photoelectric mixing, 293–295
- quantum, 35
- of wave, 9, 14

Interface matrix, 126
Interference, 235–237, 256
- and coherence, 295–297, 305–307
- division of amplitude, 96
- division of wavefront, 97
- of electromagnetic wave, 91–92
- fringe, *see* Fringe
- order, 103, 381

Interferometer:
- Fabry–Perot, 111–115
- Intensity, *also* Hanbury Brown and Twist, 313–317
- Mach–Zender, 105
- Michelson, 102–106, 295–305
- Michelson's stellar, 312–313, 315, 377
- Twyman–Green, 104–105

Irradiance, 34. *See also* Intensity,
Isolator, *see* Faraday, isolator

J

Jones, R. Clark, 49
Jones:
- calculus, 544–545, 549–550, 587
- vector, 48–49, 544, 549

K

Kepler, Johannes, 62
Kerr, John, 571

Kerr:
coefficients, 572, 599, 619
constants, 578, 613
effect, 572, 576–579, 660, *also* the quadratic electro-optic coefficient
modulator, 582, 587
Kirchhoff, Gustav, 324, 352
Kirchhoff integral, 352–354
Kleinman's symmetry, 640–641
Kogelnik, Herwig, 663
K-ratio, 490

L

Lagrange, Joseph Louis, 166
Lagrange:
equation, 166
integral invariant, 133
invariant, *see* Optical invariant
Lagrangian, optical, 167
Land, Edwin Herbert, 524
Larmor frequency, 593
Laterial magnification, 182
Law of:
rectilinear propagation, 327
reflection, 65–67, 327
refraction, 65–67, 140–141, 328
Left-handed, *see* Levorotatory
Leith, Emmett N., 469
Lens, transform properties, 367–369, 391–396
Lens equation:
Newtonian form, 185–186
Gaussian form, 185
Lensmaker equation, 184–185
Lens matrix, *see* ABCD, matrix
Lens splitting, 211–212
LeRous, P., 253
Levorotatory, 546, 550
Linear electro-optic coefficient, *see* Pockels, effect
Linear polarization, 41–42, 47–48
Linear system theory, 239–241, 396, 420
Lippershey, Hans, 129
Lissajous' figure, 39–40, 42–44
Littrow condition, 386
Lloyd's mirror, 94
Longitudinal coherence, *see* Coherence
Longitudinal electro-optic modulator, 613
Longitudinal magnification, 499
Longitudinal spherical aberration, *see* Aberration, spherical
Lord Rayleigh, *see* Rayleigh, Lord (John William Strutt)
Lorentz, Hendrik Antoon, 253
Lorentz force, 36–37
Lorenz–Lorentz Law, 278–279
Low-pass filter, *see* Filter, spatial
Lucas, R., 571, 601
Luneberg, R.K., 214

M

Magneto-optic effect, 590–596, 614. *See also* Cotton–Mouton effect; Faraday, effect; Voight effect
Magnification:
angular, 184, 188
lateral, 182, 287, 498–500, 509
longitudinal, 499, 509
Maiman, Theodore Harold, 633
Malus, Ètienne Louis, 522
Malus' Law, 581
Manley–Rowe relation, 645, 666
Marginal ray, see ray
Matched filter, *see* Filter, matched (phase and amplitude)
Maxwell, James Clerk , 25
Maxwell's equations, 26–29, 49–50, 537–538, 641–642
Meridional ray, *see* Ray
Michelson, Albert Abraham, 102, 295
Miller, R.C., 677
Miller's:
coefficient, 677
rule, 639, 646, 677–678
Mode:
Gaussian, 346–347
propagation, 150–151, 342
radiation (air), 149–150
substrate, 151
waveguide, 151, 160, 162
Modulation, 569. *See also* Contrast
amplitude, 226, 228, 256–259, 293, 399, 426, 579–586
bias, 469, 585
frequency, 606–609
phase, 576, 578
Modulation Transfer Function (MTF), 427–428
Modulator, 569
acousto-optic, 606–609
linear electro-optic, *see* Pockels
longitudinal, 588–589
magneto-optic (Faraday), 594
quadratic electro-optic (Kerr), 582, 587

Modulation (*Continued*)
- transverse, 589

Molar:
- polarization, 278
- refractivity, 278

Momentum, wave, 8
Momentum density, *see* Density, momentum
Monochromatic aberrations, *see* Aberration
Moor–Hall, Chester, 193, 289
Mueller, H., 48, 596
Mueller:
- calculus, 543–544
- matrix, 544

N

Nath, N.S. Nagendra, 605
Natural rotation, *see* Optical activity
Negative uniaxial crystal, 531
Newton, Isaac, 1, 130, 521
Newton's rings, 101–102
Nicol, William, 533
Nicol prism, *see* Polarization prism
Nodal:
- planes, 187
- points, 187–188. *See also* Cardinal points

Noncollinear phase matching, 653–655
Noncritical phase matching, 658
Nondispersive, 5
Nonlinear:
- optical coefficient, 638–641, 666, 677, 679–680
- polarization, 633, 636, 641–643, 661, 679
- susceptibility, 637–638

Normalized:
- aperture, 153–154, 189
- film thickness 157, *also* normalized frequency or V-number
- index, 163

Normal surface, 523, 563–565

O

Object-image matrix, 183, 185
Obliquity factor, 331–335, 356, 358, 444–447
Offense against sine, *see* Aberration
Optical activity, 545–552. *See also* Chiral
Optical axis, 139, 529, 537
Optical indicatrix, 535–537, 573–578, 597–599.
Optical invariant, 183–184, 208, 612
Optical path, 93
- dependence on index of refraction, 133–134
- length, 92–93, 133–134

Optical rectification, 638
Optical Transfer Function (OTF), 424, 427, 430
Optical sine theorem, 204
Optic ray axes, 562
Ordinary wave, 521, 530
Oscillator:
- damped, 20–21, 267
- harmonic, 17–20, 263, 266
- nonlinear, 670–676
- strength, 272

P

Parametric process, 601, 637
Paraxial approximation, 139, 364
Paraxial wave equation, 336
Partial dispersion, *see* Dispersion, partial
Pasteur, Louis, 546
Penetration depth, 81
Period:
- spatial, 3, 401, 405
- temporal, 19

Permeability, 28
Perot, Jean Baptiste Gaspard Gustave Alfred, 110
Petzval, Josef Max, 193
Petzval curvature, *see* Aberration, field curvature
Phase conjugation, 660–665
Phase contrast microscope, 413–415
Phasefront, *see* Wavefront, 131–132
Phase matching, 636, 648–649, 651–659
- birefringent, 655–659
- critical, 658
- noncollinear, 653–655
- noncritical, 658
- Type I, 656, 658
- Type II, 656, 659

Phase plate, 415
Phase velocity, 4. *See also* Velocity
Phonon, 602
Photoelastic:
- effect, 573, 596–601
- tensor, 619

Photometry, 34
Photon, 35, 114, 315–317, 602

Piezoelectric effect, 573, 598–599
Pin cushion distortion, 209
Pixal, 611
Plane:
  of incidence, 67–68, 70–71, 77, 522
  of polarization, 38, 522
  of vibration, 522
Plane wave, 10–11, 621
Plasma frequency, 263–266, 268–271, 280, 283
Pleochroism, 524, 529
Pockels, Friedrich Carl Alvin, 57
Pockels:
  coefficients, 572–574, 616–618
  effect, 572–576, 638
  modulator, 582–587
Poincaré, H., 47
Poincaré sphere, 47
Point spread function, 240, 505
Poisson, Simeon Denis, 325
Poisson's spot, 325, 453–454, 464–466
Polarization, 38–49, 67–74, 77–78, 81–85, 522. *See also specific type*
  axis, 526, 580
  degree, 47–48
  effect on interference, 91–92
  ellipse, 38–41, 46–47
Polarization-optic coefficients, 573
Polarization prism:
  Bertrand, 534
  Feussner, 534
  Glan-Foucault, 533
  Glan–Thompson, 534
  Nicol, 533
  Rochon, 532–533
  Wollaston, 532–533
Polarizer, 46, 579. *See also* Analyzer
  crossed, 581
  parallel, 581
  polaroid:
    h sheet, 524–526
    j sheet, 525–526
Positive uniaxial crystal, 532
Potential energy function, 17–18, 671–672
Power, 141–143, 287–290
Power spectrum, 298
Poynting, John Henry, 33
Poynting:
  theorem, 33
  vector, 33–36, 73, 80–81, 133, 535–536, 560, 645–646
  walk-off, 554, 590, 658
Principal:
  points, 186. *See also* Cardinal points
  planes, 186–187
Principal dielectric axes, 536
Principal ray, *see* Ray
Principle of reciprocity, 136
Principal refractive index, 539
Principal velocity, 561
Prism coupling, 82
Profile dispersion, 260
Propagation constant, 3, 254–255, 661
Propagation mode, 150, 342
Ptolemy, Claudius, 62
Pump:
  frequency, 601
  wave, 637
Pupil, *see* Entrance pupil; Exit pupil
Pyroelectric, 634

Q

Q-factor, 606
Quadratic electro-optic:
  coefficients, 572
  effect, 613. *See also* Kerr, effect
Quarter-wave plate, *see* Retarder

R

Radiant flux density, 34
Radiation pressure, 37
Radiometry, 34
Raman, C.V., 605
Raman–Nath scattering, 487, 605–606, 626–629
Ray, 130–131, 560
  axial, 140
  chief, *see* Principal
  equation, 561
  index of refraction, 560
  marginal, 188–189
  meridional, 149, 153
  principal, 188, 190
  skew, 149, 153–154, 203
  surface, 523, 567–568
  velocity, 560
Ray ellipsoid, *see* Fresnel, ellipsoid
Rayleigh, Lord (John William Strutt), 457, 397
Rayleigh:
  Criterion, 404, 422–424, 429, 610
  limit, 196
  range, *see* Gaussian beam parameters (Gaussian wave)

Rayleigh Sommerfeld formula, 357–358
Real image, 479, 498
Reciprocal dispersive power, *see* Abbe's number (value)
Rectangular function, *see* Functions, square
Rectilinear propagation, 327
Reflectance, 73–74. *See also* Reflectivity
Reflection coefficient, 64, 71–73, 75–76, 81–82, 120–121
  law of, 65–67, 327
  for a multilayer, 126–128
  total, 65, 254, 652
Reflectivity, 73–74, 76–77, 80, 83–84
  dependence on index of refraction, 75–76, 128
refraction, 61, 521, 565–566
  law of, 65–67, 140–141, 328. *See also* Snell's law
  matrix, 142–143
Refractivity:
  atomic, *see* Lorenz–Lorentz law
  molar, 278
Refringence, 521. *See also* Refraction
Relative partial dispersion, 290
Renormalization, 273
Resolution, 401. *See aslo* Rayleigh; Sparrow's Criterion
Resolving power:
  chromatic, 111–115, 297, 381–382, 388
  spatial, 332–333, 401, 404, 426, 429–430, 611
Retardation, 541, 583–584, 587, 589, 631
Retardation time, 104
Retarder, 541–543, 580
  compensator, Soleit-Babinet, 542
  half-wave, 542
  quarter-wave, 541
  rhomb-type, 543
Right-handed, *see* Dextrorotatory
Rittenhouse, David, 379
Ronchi ruling, 402
Röntgen, Wilhelm Conrad, 571
Rotational strength, 547
Rotatory:
  dispersion, 551
  power, 550. *See also* Specific rotation
Rowland, H.A., 380
Royen, Willebrord Snell van, *see* Snell (Willebrord Snell van Royen)

S

Saggital:
  approximation, 392
  field, 207
  focus, 206
  plane, 194
  ray, 203–207
Sag (sagitta), 207–208, 392
St. Venaut's Hypothesis, *see* Boundary conditions, Kirchhoff
Sampling theorem, 234, 426
Schlieren method, 213
Sears, F.W., 571, 601
Secondary spectrum, 290
Second harmonic generation, 637, 649–651
Seidel, Ludwig Philipp von, 193
Seidel coefficients, 197
Sellmeier, 253
Shape factor, *see* Coddington's, shape factor
Sign convention, 139–140
Sinc function, *see* Functions, sinc
Sine transform, 223
Skew ray, *see* Ray
Skin depth, 52–53
Slow axis, 531
Smith Chart, 124
Snell's Law, 67, 69, 84
Snell (Willebrord Snell van Royen), 62
Soleit-Babinet compensator, *see* Retarder
Sommerfeld, Arnold Johannes Wilhelm, 280, 324
Sommerfeld:
  integral, 358
  precursor, 280–281
  radiation condition, 355
Space bandwidth product, 611–612. *See also* Bandwidth
Sparrow, C., 423
Sparrow's Criterion, 423–424, 429
Spatial:
  coherence, *see* Coherence
  filter, *see* Filter, spatial; Apodization
  frequency, 258–259, 366, 369–370, 386, 405, 485, 495–496
Specific rotation, 550–552
Speckle noise, 376, 430
Spectral distribution function, 298
Spectrum analyzer, 609–610
Speed of light, 1, 29–30
Spherical aberration, *see* Aberration

Spherical wave, 12
Spring constant, 17
Stokes, George Gabriel, 46
Stokes:
  parameters, 46–49, 543
  vector 48, 543–544
Stop:
  aperture, 188–189
  dc, 413
  field, 190
  glare, 190
Strain-optic:
  effect, 573
  tensor, 597
Strehl ratio, 196
Stress birefringence, 596, 600–601
Stress-optical tensor, 597
Substrate mode, *see* Mode
Sum frequency generation, 637
Superposition principle, 16, 65
Susceptibility:
  dielectric, 636, 665
  linear, 636, 665
  2nd order tensor, 637–638, 666
  3rd order, 659

T

Tangential:
  field, 206–207
  focus, 207
  plane, 194
  ray, 203
Taylor, Dennis, 120
Telegraph equations, 50
Temporal coherence, *see* Coherence
TEM wave, 70
Tensor, 535–536, 547–548, 556–559
TE wave, 70
Time-bandwidth product, 611–612. *See also* Bandwidth
TM wave, 71
Total reflection, 78–83
Transfer function, 241
Transfer matrix (translation matrix), 143
Transmission coefficient, 64, 70–71, 75, 121, 478, 482, 484, 512
Transmission matrix, 126
Transmissivity, 587
Transmittance, 73–74
Transverse coherence length, 311
Transverse electro-optic modulator, 614
Twist, Richard Q., 292, 314
Twyman–Green, *see* Interferometer
Tyndall, 149

U

Uniaxial birefringence, 529, 564–565
Unit planes, 182, 186
Upatnieks, Juris, 469, 472
Up-conversion, 637

V

Van Cittert-Zernike theorem, 310
Vander Lugt, Anthony, 415
Vander Lugt filter, *see* Filter, matched
Velocity:
  group, 256, 258, 279
  of light, *see* Speed of light
  phase, 4–10, 52, 279, 281–282, 560, 567
Verdet's constant, 594, 614
Vertex, 139
Vignetting, 190
Virtual image, 479, 498
V-number (dispersion) *see* Abbe's, number (value)
V-number (waveguide) *see* Normalized, film thickness
Voigt, W., 253
Voigt effect, 596

W

Waist, *see* Gaussian beam parameters (Gaussian wave)
Wave:
  conjugate, 662
  electromagnetic, 30–35, 537–541, 620, 642
  extraordinary, 521, 530, 541
  Gaussian 336–343
  harmonic, 3, 9, 11, 14
  ordinary, 521, 530, 541
  plane, 10–11, 362, 365, 369–370, 449, 537–538, 642
  pump, 601
  signal, 602
  spherical, 10–12, 437, 444–445
  standing, 123–124
Wave equation, 5, 9
  in a conductor, 50
  with dispersion, 254, 263, 267
  from Maxwell's equation, 29
  modulated by acoustics, 620–621
  in nonlinear material, 641–642, 666
  paraxial, 336

Wavefront, 131–132, 326
  plane, 10–11
  spherical, 10–12
Waveguide mode, 151
Wavelength, 3
Wave number, *see* Propagation constant
Wave vector, 11, 536–537
Wollaston prism, *see* Prism coupling
Wood, Robert Williams, 380, 384, 457

Y

Young, Thomas, 1, 87, 325, 522
Young's two slit experiment, 94–97, 305–312, 368–369, 377–379, 405–406

Z

Zeeman, P., 590
Zeeman effect, 590–593
Zernike, Fritz, 292, 363, 413